MODELLING OF ENGINEERING AND TECHNOLOGICAL PROBLEMS

To learn more about AIP Conference Proceedings, including the Conference Proceedings Series, please visit the webpage **http://proceedings.aip.org/proceedings**

MODELLING OF ENGINEERING AND TECHNOLOGICAL PROBLEMS

International Conference on Modelling and
Engineering and Technological Problems (ICMETP)
and the 9th Biennial Conference of Indian Society of
Industrial and Applied Mathematics (ISIAM)

Agra, India   14 - 16 January 2009

EDITORS

A. H. Siddiqi
A. K. Gupta
BMAS Engineering College, Agra, India

M. Brokate
Munich Technical University, Garching, Germany

All papers have been peer-reviewed

SPONSORING ORGANIZATIONS
NBHM(National Board of Higher Mathematics, Govt. of India)
AICTE (All India Council of Technical Education, Govt. of India)
CSIR (Council of Scientific and Industrial Research, Govt. of India)
INSA (Indian National Science Academy, Govt. of India)
ICTP (International Center of Theoretical Physics, Joint Venture of UNESCO and Italian Govt.)
ICIAM (International Council of Industrial Applied Mathematics-Apex Body of the World for
Promotion of Industrial and Applied Mathematics)
ISIAM (Indian Society of Industrial and Applied Mathematics)

Melville, New York, 2009
AIP CONFERENCE PROCEEDINGS ■ VOLUME 1146

Editors:

A. H. Siddiqi
Dean of Academics
BMAS Engineering College
Agra Mathura Road (NH-2)
Keetham, Agra-282007
India

e-mail: siddiqi.abulhasan@gmail.com

A. K. Gupta
Principal
BMAS Engineering College
Agra Mathura Road (NH-2)
Keetham, Agra-282007
India

e-mail: ajit_46@yahoo.com

M. Brokate
Munich Technical University
Boltzmannstr. 3
D-85748 Garching
Germany

e-mail: brokate@ma.tum.de

L.C. Catalog Card No. 2009904925
ISBN 978-0-7354-0683-4
ISSN 0094-243X
Printed in the United States of America

CONTENTS

SECTION D: DYNAMICAL SYSTEMS

SECTION E: NUMERICAL METHODS

SECTION F: TYPICAL METHODS OF ENGINEERING AND TECHNOLOGICAL
PROBLEMS

PREFACE

Agra is world famous city for having Taj Mahal. Now it is becoming popular for excellent academic and professional institutions. Sharda Group of Institutions (SGI) is the largest educational group in northern India. The present international conference was organized at one of their institutions BMAS Engineering College, jointly with Indian society of Industrial and Applied Mathematics (ISIAM). 9th National Conference of ISIAM was also organized simultaneously during 14 -16 Jan 2009. The inaugural function was held on 14th January 09 at Ramanujan Auditorium. Dr. A.R. Kidwai, His Excellency, Governor of Haryana and a distinguished academician, was the Chief Guest. 21 invited speakers from different parts of the world and different regions of India besides 172 contributors of research papers were present during this conference. A souvenir was distributed containing the messages of the President of India Smt. Pratibha Devi Singh Patil; Honorable Shri Arjun Singh, Minister of Human Resource Development India; Honorable Sh. Kapil Sibal, Minister for Science and Technology and Earth Sciences, Government of India New Delhi; Dr. Bal Ram Jakhar, His Excellency Governor of Madhya Pradesh; Honorable Sh. M.A.A. Fatmi, Minister of State Human Resource Development, Government of India; Prof. K.R. Sreenivasan Professor and Director ICTP; Dr. S.Z. Qasim Former Member Planning Commission (Science); Prof. Prem Vrat, Vice Chancellor UP Technical University Mr. Pradeep Kumar Gupta, Chairman SGI, Mr. Y.K. Gupta, Vice Chairman SGI, Prof. M.S. Teotia Principal Advisor SGI; His Excellency Khurshid Alam Khan Ex-Governor of Karnataka and President Duty Society.

Prof. Roddam Narasimha FRS Head Mechanical Unit Jawahar Lal Nehru Centre for Advanced Scientific Research, Banglore and Prof. H.P. Dikshit, Director General School of Good Governance and Policy Analysis, were presented Dr. Zakir Husain Award 2007 & 2008 respectively. The Souvenir also contained biodata of distinguished invited speakers and persons associated with society along with 172 abstract of papers submitted for presentation to the conference. It is heartening to note that most prestigious organizations of our country such as National Board of Higher Mathematics (NBHM), Atomic Energy Commission, Government of India, Mumbai; All India Council for Technical Education (AICTE), Government of India; Council of Industrial and Scientific Research (CSIR) and Indian National Science Academy (INSA) have sanctioned funds to organize the conference. The two most important international agencies related to the theme of the conference namely, Abdus Salam International Centre of Theoretical Physics, Trieste Italy (ICTP-UNESCO Organization) and International Council of Industrial and Applied Mathematics (ICIAM) provided financial support for holding this conference. We are pleased to receive an overwhelming response from across India and abroad. Besides invited talks, several parallel sessions were arranged for contributory talks. The quality of both invited as well as contributory talks was appreciated by everyone who spoke in the valedictory function on 16th January 09 at 07:00 pm.

A special discussion was arranged on the theme of "Mathematics, Science, Computer and Society" on 15th January 09 at the Banquet Hall of Amar Hotel from 06:30 to 08:30 pm. Panelists were Prof. H.P. Dikshit, Prof. I. Sloan, Prof. Jeltsch, Prof. Arvind Gupta, Prof. Marie Farge, Prof. Dinesh Singh, Prof. Sundar Lal, Prof. H. Feitchtinger, Mrs. R. Mittelmann. The gist of the point of view of all speakers was the emphasis on the role of Mathematics in real world problems and to make Mathematics more popular in society. The need for immediate and appropriate steps was suggested. After this session and dinner Mr. Y.K. Gupta, Vice Chairman SGI, interacted with invited speakers and explored the possibility of long term cooperation and collaboration with invited speakers individually and their institutions as a whole.

The theme of the conference is an important ingredient of scientific computing also known as computational science and engineering. Out of 100 papers submitted for publication

36 have been included in this proceeding. On the basis of the subject matter of the selected papers, we have divided them in to six sections.

Section A is based on two Dr Zakir Hussain Award Lectures; one by Prof. R. Narasimha FRS on the theme IN WHAT SENSE IS A LOW-REYNOLDS MIXING LAYER STABLE? and other by Prof. H P Dikshit on Geometric Modeling and Shape Preserving Properties of Wachspress Functions.

Section B is named as Modelling of Real World Systems. There are nine papers in this section. First paper of this section is on Protein Designs in HP Models by Prof Arvind Gupta and his collaborators. The second paper of this section is on graph modeling for quadratic assignment problems associated with hypercube. Third paper by Prof. Aoyama and his coauthor T Koga deals with integrate system modeling for design and production planning of high quality products considering failure data. Paper of Prof. Achutan and Karuppath comments on mathematical modeling of quantum systems. Prof. B S Chaudhary and his co-workers discuss Visualization and Analysis of Wireless Sensor Network. Data for Smart Civil Structure Applications Based On Spatial Correlation Technique. Prof. Masaudi presents in his paper a stability result in a weakly damped nonlinear Timoshenko system. Dr. Mrs Sontakke, Dr. Nityanand Singh and have studied rain fall spatial variability and its impact on important environment process in India. Dr. Shelly Arora et al and Dr. Millie Pant et al present their models of real world problems.

Section C entitled wavelet analysis with applications has six papers. In this section Prof. Hans Feichtinger has explained in an excellent way that how abstract notions of mathematics could be very useful in practical situations. His paper entitled Banach Gelfand Triples for Applications in Physics and Engineering elaborates Gabor system. Dr. M Melek along his coworkers Dr A Tokgozlu and Prof. Aslan has studied variation of oil price an US dollor using Wavelet methods. In her paper Dr Mani Mehra presents a resume of her research on applications of wavelet methods to partial differential equations on flat region. Prof. Manchanda with her PhD student Ms Meenakshi presents a summary of her recent research work on Haar Vilenkin System and non-uniform multiresolution analysis. Dr. Kumar and Prof. Manchanda have studied role of Hurst Parameter for financial time series. Section C is concluded by a paper on reservoir flow simulation based on wavelet methods.

Section D on dynamical system comprise of five papers. First paper of this section by Prof. R Lozi and his co-worker Dr. Clarisses Fiol devoted to their valuable research work on Global Orbit Patterns for Dynamical Systems on Finite Sets. They have illustrated application of their results in engineering problems. Prof Brokate and Rasulova present their contribution on the Cauchy problem for BBGKY hierarchy of quantum kinetic equations with Coulmb potential in second paper of this section. Prof. P N Srikanth Dean, center of applicable mathematics of TFIR, Bangalore highlights how the topological information of the Mountain pass solutions can be used to obtain interesting break of symmetry results. In the next paper Prof R C Singh, Dean Academics of HIT, Greater Noida, India has given his valuable contribution on Intregral equation approach to orientational phase transitions in quadruplor gay-berne fluid using density functional theory. Contribution of Dr Manoj Kumar Singh on three Body Problem is the last paper of this section.

Section E devoted to Numerical Methods has five papers. First paper of this section by Dr. Prof. Siddiqi and Dr. S S Irfan is devoted to a generalized class of variational inequalities. For a large class of problems, variational inequalities provide foundation for numerical solution. In second paper of this section Dr. Khanday and Prof Saxena have studied finite element solution of a practical problem. Dr. V Kukreja et al have studied numerical solution of a boundary value problem using Hermite basis. Deepak Kumar et al and Dr Mudagi have studied certain real world problems in their papers listed in this section.

Last section F based on typical methods of engineering and technology contains nine papers. In the first paper, Prof. Kamran Asghar, Bokhari and Zama have developed algebraic computational packages which compute some tensors in geometry and general relativity. Prof. C T Bhunia and his coworkers have given nice mathematical models Max Packet generation process. Dr Reza Peygham has made a nice presentation on an intrior point method for semi definite programming based on new Kernel functions. Prof. B Ishwar and Dr Sanjay Kumar have discussed solution of generalized photo gravitational elliptic restricted three body problem. Prof V P Singh with his collaborators Dr Ms Millie Pant and Dr. Raja Thangaraj have used computational intelligence for certain real world problems. Effect of gravity modulation has been studied by Dr. T. Siva Kumar and Dr S Sarananan. In her paper Dr Anjana Solanki has studied transient behavior of batch arrival. Prof A K Wadhwani Dr. Neelam Sharma and Rakesh Saxena have studied fast arithmetic using signed digit number and ternary logic. Mr Pankaj Sharma and Vivek Sharma have studied self similar solution of self gravitating magneto gas dynamics in their paper.

We take this opportunity to thank all referees who have spared their valuable time to go through paper submitted for publication in this volume. We also thank to American Institute of Physics specially its Director Special Publications and Proceedings, Maya Flikop. We owe to Mr Mohd Kamaruddin EC department of BMAS Engineering College for extending technical support in compilation of this volume. Prof Mrs Pammy Manchanda joint secretary ISIAM deserves our compliment for providing her full support in organizing the conference and completion of this volume. In the end we express our gratitude to all external funding agencies and SGI group in extending whole hearted support for this conference.

A H Siddiqi
A K Gupta
M Brokate

SECTION A:
DR. ZAKIR HUSSAIN AWARD LECTURES

In What Sense is a Low-Reynolds Mixing Layer Stable?

BHATTACHARYA. P., GOVINDARAJAN. R., NARASIMHA. R.

Jawaharlal Nehru Centre for Advanced Scientific Research, Bangalore

1. Introduction

The Orr–Sommerfeld equation governs the evolution of linear disturbance modes in a strictly parallel flow. For the plane incompressible mixing layer this equation yields a critical Reynolds number equal to zero (Betchov & Szewczyk 1963). This result has been difficult to understand for a long time, for 'energy' arguments show that there exists a non-zero Reynolds number, however small, below which viscosity would damp out any disturbance. It has often been suggested (e.g. Drazin & Reid, 2004) that the resolution of this paradox is possible only by considering the non-parallelism of the mixing layer. As the rate of growth of the mixing layer and hence also is infinite at zero Reynolds, hence also at the critical Reynolds number predicted, and the degree of non-parallelism of the flow increases at lower Reynolds numbers, the Orr-Sommerfeld result is derived making an assumption that is grossly inadequate.

Recently, Bhattacharya et al. (2006) – henceforth referred to as I – used a non-parallel analysis to show that the critical Reynolds number Re_{cr} for the mixing layer is not zero. When expressed with the velocity difference Δ and vorticity thickness δ_ω as velocity and length scales respectively, they found that Re_{cr} is about 30. The analysis therein follows the minimal composite theory developed in Govindarajan & Narasimha (1997, 2001, 2005) chiefly for boundary layers. As is well-known, the marginal stability curve in non-parallel theory depends on a number of factors: the precise disturbance quantity that is being monitored, the choice of trajectory along which the quantity is monitored, etc. To avoid this complexity the quantity used in I to define Re_{cr} was an integral of the disturbance kinetic energy over a similarity coordinate y, i.e. a function only of the streamwise distance x.

In the present work, we propose to discuss the spatial evolution of the disturbance field. Results on three definitions of growth-rate were given by Gaster (1974, (a)–(c), p. 472 and Figures 2, 3) in his non-parallel stability analysis of the boundary-layer. Although the differences in the stability curves were not entirely negligible, there was a lower-bound in R_{cr} of about 500. In the present case, where the critical Reynolds number is much lower, we show that alternative definitions for the growth-rate give rise to *qualitatively* different stability loops of the mixing layer. Care must therefore be exercised when interpretations are offered regarding the physical processes underlying the instability of the mixing layer.

We investigate here the physical behaviour of the disturbance kinetic energy in order to gain insight into the character of the stability of the flow.

CP1146, *Modelling of Engineering and Technological Problems,* edited by A. H. Siddiqi, A. K. Gupta, and M. Brokate
© 2009 American Institute of Physics 978-0-7354-0683-4/09/$25.00

2.   Problem Statement

The mean flow in the incompressible mixing layer (see Figure 1 for notation) we consider possesses a similarity solution in terms of a similarity function $\phi(y)$, related to the dimensional streamfunction

$$\Phi_d(x_d y_d) = \frac{2\Lambda}{1+\Lambda} U_\infty \delta_\omega(x_d)\Phi(y), \quad \Phi'(y) = U/U_\infty.$$

where $\Lambda = (U_\infty - U_{-\infty}) / (U_\infty + U_{-\infty})$ is the velocity ratio parameter.

As a first step we simulate the flow as it might be observed in a wave-maker experiment. To this end a single disturbance mode, of dimensional frequency ω_d, is introduced into the flow at an appropriate station. Since the temporal variation of the dimensional kinetic energy of the disturbance motion k_d is purely sinusoidal, it is sufficient to analyse its average over one time period.

Following the minimal composite theory applied to the mixing layer as in I, the amplitude ϕ of the disturbance streamfunction is decomposed into a lowest-order solution ϕ_m and a higher-order correction ϕ_h, that is

$$\phi(x,y) = A(x)\phi_m(x,y) + \epsilon\phi_h(x,y), \qquad (1)$$

where

$$\hat{\phi}_d = (\delta_\omega \Delta)\phi(x,y) \exp\left[\mathrm{i}\left(\int \alpha \mathrm{d}x - \omega_d t_d \right)\right]$$

is the dimensional streamfunction corresponding to the mode ω_d. Further, $A = A(x)$ and $\alpha = \alpha(x) = \alpha_r + \mathrm{i}\alpha_i$ represent the amplitude function and the non-dimensional wavenumber of the disturbance respectively. The analysis in I shows that the higher-order correction $\epsilon\phi_h$ in (1) is $O(R^{-2/3})$ compared to $A\phi_m$. The non-dimensionalisation adopted here is identical to that in I, namely

$$\mathrm{d}x_d = \delta_\omega \mathrm{d}x, \quad y_d = \delta_\omega y \quad \text{and} \quad R \equiv \delta_\omega \Delta/\nu.$$

Note that y is (by definition) the similarity coordinate for the mean flow in the mixing layer.

Now the local kinetic energy density of the disturbance is

$$k_d \equiv \frac{1}{2}\left(u_d'^2 + v_d'^2\right)$$

where

$$u'_d = \left(\frac{\partial \hat{\phi}_d}{\partial y_d}\right)_r \quad \text{and} \quad v'_d = \left(\frac{\partial \hat{\phi}_d}{\partial x_d}\right)_r , \tag{2}$$

subscript r standing for 'real part of'.

Substituting for the velocity components using (1), we then write the spatial distribution of the disturbance energy in the form

$$\langle k_d \rangle (x_d, y_d) \equiv \frac{1}{4}\Delta^2 \left(k_0 + k_\epsilon + k_R\right) \exp\left(-2\int_{x_0}^{x} \alpha_i dx\right) + o(R^{-1}), \tag{3}$$

where

$$k_0 = |AD\phi_m|^2 + |\alpha A \phi_m|^2, \tag{4}$$

$$k_\epsilon = 2\epsilon \left(AD\phi_m D\phi_h^* + |\alpha|^2 A\phi_m \phi_h^*\right)_r. \tag{5}$$

$$k_R = 2|A|^2\left(i\alpha\phi_m\left[\frac{p}{R} + \frac{A'^*}{A^*} + \partial_x - \frac{1}{R}pyD\right]\phi_m^*\right)_r, \tag{6}$$

and the angular brackets $\langle \ \rangle$ denote the average over one time period $2\pi/\omega_d$. In these expressions the asterisk denotes the complex-conjugate, the prime on A denotes a derivative taken with respect to x, and $D = \partial/\partial y$. Note that k_0, k_ϵ and k_R are contributions to the disturbance kinetic energy at $O(1)$, $O(\epsilon)$ and $O(R^{-1})$ levels respectively. Fixing the amplitude level of the disturbance to be A_0 at the location where it is introduced, say at $x_d = x_{d0}$, we further define

$$\kappa \equiv \frac{\langle k_d \rangle}{|A_0|^2\Delta^2}, \tag{7}$$

as the non-dimensional disturbance kinetic energy. Two integral quantities are also defined as follows :

$$K(x) = \int_{-\infty}^{\infty} \kappa \, dy, \tag{8}$$

5

$$\overline{K}(x) = \int_{-\infty}^{\infty} \kappa\, d\left(\frac{y_d \Delta}{\nu}\right) = R(x)\, K(x). \tag{9}$$

Both these integrals are non-dimensional, but while K is an integral over the similarity variable y, $\overline{K}$ uses the length-scale (ν/Δ) which is independent of x, unlike δ_ω increases with the streamwise distance like $x^{1/2}{}_d$. Thus

$$d\delta_\omega / dx_d = p/R, \tag{10}$$

where p is a constant depending on the velocity ratio parameter $\Lambda = \Delta / (2U_\infty - \Delta)$ and is given by

$$p = \left(\frac{\Lambda}{1+\Lambda}\right)\left(\frac{|2\Lambda/(1+\Lambda)|}{\max|\Phi''|}\right)^2,$$

where $\Phi'(y) = U_d(y_d)/\Delta$ is the derivative in y of the similarity streamfunction $\Phi(y)$ used to specify the mean-flow, U_∞ is the free-stream velocity of the faster stream and the prime on Φ denotes a derivative with respect to y. Equation (10) implies that the rate of growth of $\overline{K}$ in the downstream direction includes the effect of the spreading of the mixing layer, or

$$\frac{1}{\overline{K}}\frac{d\overline{K}}{dx} = \frac{1}{K}\frac{dK}{dx} + \frac{p}{R}. \tag{11}$$

In I the growth-rate g corresponds to the quantity K above. Using (4)–(6) and (7)–(9) it is straightforward to show that the spatial growth-rates of κ and K are given by

$$\frac{1}{\kappa}\frac{\partial\kappa}{\partial x} = -2\alpha_i + 2\left(\frac{A'}{A}\right)_r$$
$$+ \frac{2\left(\partial_x D\phi_m D\phi_m^* + |\phi_m|^2\alpha^*\alpha' + |\alpha|^2\phi_m^*\partial_x\phi_m\right)_r}{|D\phi_m|^2 + |\alpha\phi_m|^2} \tag{12}$$
$$+ o(R^{-1}),$$

$$\frac{1}{K}\frac{dK}{dx} = -2\alpha_i + 2\left(\frac{A'}{A}\right)_r$$
$$+ \frac{\int_{-\infty}^{\infty} 2\left(\partial_x\phi_m D\phi_m^* + |\phi_m|^2\alpha^*\alpha' + |\alpha|^2\phi_m^*\partial_x\phi_m\right)_r dy}{\int_{-\infty}^{\infty}(|D\phi_m|^2 + |\alpha\phi_m|^2)dy}$$
$$+ o(R^{-1}); \tag{13}$$

6

and the growth-rate of $\overline{K}$ is obtained by using (13) in (11).

The determine the evolution of κ we start with the equation for the disturbance energy in two dimensional flow, obtained by taking the scalar product of the disturbance momentum equation with the disturbance velocity. We then obtain (e.g. from Schlichting & Gersten, 2004, pp. 503-4),

$$\rho\left(U\frac{\partial\langle k_d\rangle}{\partial x_d}+V\frac{\partial\langle k_d\rangle}{\partial y_d}\right)=-\left\langle u'_d\frac{\partial\hat{p}_d}{\partial x_d}+v'_d\frac{\partial\hat{p}_d}{\partial y_d}\right\rangle$$

$$\underbrace{\qquad\qquad}_{\text{advection}} \qquad \underbrace{\qquad\qquad\qquad\qquad}_{\substack{\text{pressure}\\\text{transport}}}$$

$$+\rho\nu\left[\frac{\partial^2}{\partial x_d^2}\left(\langle k_d\rangle+\langle u'^2_d\rangle\right)+2\frac{\partial^2}{\partial x_d\partial y_d}\langle u'_d\,v'_d\rangle+\frac{\partial^2}{\partial y_d^2}\left(\langle k_d\rangle+\langle v'^2_d\rangle\right)\right] \quad \begin{array}{l}\text{viscous}\\\text{diffusion}\end{array}$$

$$-\rho\underbrace{\left(\langle u'^2_d\rangle\frac{\partial U}{\partial x_d}+\langle u'_d\,v'_d\rangle\frac{\partial V}{\partial x_d}+\langle u'_d\,v'_d\rangle\frac{\partial U}{\partial y_d}+\langle v'^2_d\rangle\frac{\partial V}{\partial y_d}\right)}_{} \quad \begin{array}{l}\text{turbulent}\\\text{production}\end{array}$$

$$-\rho\nu\left[2\left\langle\left(\frac{\partial u'_d}{\partial x_d}\right)^2+2\left(\frac{\partial v'_d}{\partial y_d}\right)^2\right\rangle+\left\langle\left(\frac{\partial u'_d}{\partial y_d}\right)^2+2\left(\frac{\partial v'_d}{\partial x_d}\right)^2\right\rangle\right] \quad \text{dissipation.} \quad (14)$$

The differences between the above expressions and those in (16.14)-(16.19) in Schlichting & Gersten (2004), are as follows:

1. The equations above are two-dimensional, $\partial/\partial z$ W, w'_d are identically zero everywhere in the flow.
2. The contribution to turbulent diffusion from the advection of disturbance energy, by the disturbance field itself, is neglected because it is of a higher-order (in disturbance magnitude units) compared to the other terms in the equation. Also note that the disturbance pressure is of a lower order than the disturbance velocity field.

To derive the expression for the pressure transport term, we write the dimensional disturbance pressure field as a waveform

$$\hat{p}_d=\rho\Delta^2\Pi(x,y)\exp\left|\left(\int\alpha dx-\omega_d t_d\right)\right|.$$

The pressure gradient in both x and y directions can be computed using the momentum equations, written in vector form as

$$\overline{\nabla}_d \hat{p}_d = \frac{\partial \overline{u}_d}{\partial t_d} - (\overline{U}.\overline{\nabla}_d)\overline{u}_d - (\overline{u}_d.\overline{\nabla}_d)\overline{U} + \frac{1}{R}\overline{\nabla}_d^2\overline{u}_d \tag{15}$$

where $\overline{\nabla}_d = \{\partial/\partial x_d, \partial/\partial y_d\}, \overline{U} = \{U, V\}$ and $\overline{u}_d = \{u'_d, v'_d\}$. By substituting for the velocity components using (1) and (2), and similarly for the mean flow and disturbance pressure, followed by expanding and retaining terms only upto $O(R^{-1})$, we get the normal and streamwise directions respectively

$$D\Pi = \underbrace{-\alpha^2(\Phi'-c)A\phi_m}_{O(1)} \underbrace{-\alpha^2(\Phi'-c)\varepsilon\phi_h}_{O(\varepsilon)} + i\Phi'\partial_x(A\alpha\phi_m)$$

$$+ i\alpha A(\Phi'-c)\left[\frac{p}{R} + \frac{A'}{A} + \partial_x - \frac{py}{R}D\right]\phi_m$$

$$- \frac{1}{R}i\alpha Ap\Phi D\phi_m + \frac{1}{R}i\alpha Apy\Phi''\phi_m$$

$$- \frac{1}{R}i\alpha A(D^2 - \alpha^2)\phi_m,$$

$$(i\alpha + \partial_x - \frac{py}{R}D)\Pi = \underbrace{-i\alpha[(\Phi'-c)AD\phi_m - \Phi''A\phi_m]}_{O(1)}$$

$$\underbrace{-i\alpha[(\Phi'-c)\varepsilon D\phi_h - \Phi''\varepsilon\phi_h]}_{O(\varepsilon)}$$

$$+ \left[\begin{array}{l} \dfrac{1}{R}p\Phi AD^2 - A\Phi'\left(\partial_x + \dfrac{A'}{A}\right)D + \dfrac{1}{R}pA\Phi'' \\[2mm] + A\Phi''\left(\partial_x + \dfrac{A'}{A}\right) + \dfrac{1}{R}pAD^3 - \dfrac{1}{R}A\alpha^2 D \end{array}\right]\phi_m \tag{16}$$

where $c(= \omega/\alpha)$ is the non-dimensional phase speed of the wave. Using these in the expression for pressure-transport and averaging over one time-period we obtain the expression for T_p given in (18) below. While averaging it should be noted that $\partial_x\omega = \omega p/R$, since ω_d (not ω) is invariant in x. (In (16), and in the sequel, $O(1)$ terms appear in blue, $O(\varepsilon)$ terms appear in red and $O(R^{-1})$ terms appear in black.) Using these in (14), and again keeping terms upto $O(R^{-1})$, we can write

8

$$\Phi' \frac{\partial \kappa}{\partial x} = P - \varepsilon + D_p + T + D_\nu, \tag{17}$$

where

$$
\begin{aligned}
P &= -\frac{1}{2}\Phi''|A|^2 (D\phi_m i\alpha^* \phi_m^*)_r - \frac{1}{2}\epsilon \Phi''(i\alpha^* A D\phi_m \phi_h^* + i\alpha A\phi_m D\phi_h^*)_r \\
&\quad -\frac{1}{2}\Phi''\frac{py}{R}|A\alpha\phi_m|^2 + \frac{1}{2}\Phi''|A|^2 \left\{ D\phi_m^* \left(\partial_x + \frac{A'}{A} + \frac{p}{R} \right)\phi_m \right\}_r, \\
\varepsilon &= \frac{1}{2R}\left\{ 4|\alpha D\phi_m|^2 + |D^2\phi_m|^2 + |\alpha^2 \phi_m|^2 + 2\left(\alpha^{*2}\phi_m^* D^2\phi_m \right)_r \right\}, \\
T_p &= -\frac{1}{2}\alpha_i \Phi'\left(|AD\phi_m|^2 + |\alpha A\phi_m|^2 \right) - \frac{1}{2}\Phi''|A|^2 (i\alpha\phi_m D\phi_m^*)_r \\
&\quad -\frac{1}{2}\epsilon\left[2\alpha_i \Phi'|\alpha|^2 \{A\phi_m\phi_h^*\}_r + 2\alpha_i \Phi'\{AD\phi_m D\phi_h^*\}_r \right. \\
&\quad \left. + \Phi''(i\alpha A^* D\phi_m^* \phi_h)_r + \Phi''(i\alpha A\phi_m D\phi_h^*)_r \right] \\
&\quad -\frac{1}{2R}|A\alpha|^2 \left(\phi_m^* D^2\phi_m \right)_r - \frac{1}{2R}|A|^2 \left(D\phi_m^* D^3\phi_m \right)_r
\end{aligned}
$$

$$
\begin{aligned}
&\quad +\frac{1}{2R}\left(\alpha^2 \right)_r \left[|AD\phi_m|^2 + |\alpha A\phi_m|^2 \right] \\
&\quad -\frac{1}{2R}p\Phi|A|^2 \left(D\phi_m^* D^2\phi_m \right)_r - \frac{1}{2R}p\Phi|\alpha A|^2 (\phi_m^* D\phi_m)_r \\
&\quad +\frac{1}{2R}py\Phi''|\alpha A\phi_m|^2 - \frac{1}{2}\Phi''|A|^2 \left(D\phi_m^* \left\{ \frac{p}{R} + \partial_x + \frac{A'}{A} \right\}\phi_m \right)_r \\
&\quad +\Phi'\alpha_i^2 \frac{p}{R}|A\phi_m|^2 + \frac{1}{2}\Phi'|\alpha A|^2 (\phi_m^* \partial_x \phi_m)_r + \frac{1}{2}\Phi'|A\phi_m|^2 (\alpha^*\alpha')_r \\
&\quad +\frac{1}{2}\Phi'|A|^2 \left(D\phi_m^* \left\{ \partial_x + \frac{A'}{A} \right\} D\phi_m \right)_r + \frac{1}{2}\Phi'|\alpha A\phi_m|^2 \left(\frac{A'}{A} \right)_r \\
&\quad -\Phi'\alpha_i|A|^2 \left(i\alpha\phi_m \left\{ \partial_x - \frac{py}{R}D + \frac{A'^*}{A^*} \right\}\phi_m^* \right)_r, \\
T_V &= \frac{2p\Phi}{R}\left[(D\phi_m D^2\phi_m^*)_r + |\alpha|^2 (D\phi_m \phi_m^*)_r \right] \\
T_\upsilon &= \frac{1}{2R}\left\{ |D^2\phi_m|^2 + |\alpha^2\phi_m|^2 + 4|\alpha D\phi_m|^2 + 2\left(\alpha^{*2}\phi_m^* D^2\phi_m \right)_r \right. \\
&\quad -\left(\alpha^2 \right)_r |\alpha\phi_m|^2 - \left(\alpha^2 \right)_r |D\phi_m|^2 + |\alpha|^2 \left(\phi_m^* D^2\phi_m \right)_r \\
&\quad \left. + \left(D\phi_m D^3\phi_m^* \right)_r \right\}.
\end{aligned}
$$

(18)

The terms here denote respectively the gain in disturbance kinetic energy due to shear-production (P), loss due to dissipation (ε) and transport of disturbance kinetic energy rspectively by the disturbance pressure field (T_p), by the mean-flow in the normal direction (T_V) and by viscous diffusion (T_ν). The only difference between (14) and (18) in the advection term in the former, and that labeled as T_v in the latter. Following our choice of non-dimensionalization, the advection term in (14) can be expanded as

$$\rho \left(\frac{U}{\delta_\omega} \frac{\partial \langle k_d \rangle}{\partial x} - \frac{U}{\delta_\omega} \frac{py}{R\delta_\omega} D\langle k_d \rangle + V \frac{\partial \langle k_d \rangle}{\partial y} \right). \tag{19}$$

The last two terms above are moved to the right hand side of the equation and appear as transport of the disturbance energy by the mean flow in the normal direction. Equation (12) is now retrieved by dividing the right hand side of (17) by the produce ($\Phi'\kappa$) and retaining terms upto $O(R^{-1})$.

Note that the $O(\varepsilon)$ contributions to shear production P and pressure transport T_p depend on the higher order eigen-function $\varepsilon\phi_h$. This is obtained by using the amplitude evolution equation (I),

$$\frac{\epsilon}{A}\mathcal{M}\phi_h = -\mathcal{H}\phi_m - \frac{A'}{A}\mathcal{S}\phi_m. \tag{20}$$

where $\mathcal{M}$, $\mathcal{H}$ and S are respectively . . . The non-homogeneous linear equation (20) can be solved since φ_m is known from the solution of the lowest-order equation

$$\mathcal{M}\phi_m = 0, \tag{21}$$

where $\mathcal{N} = \mathcal{M} + \mathcal{H}$ determines ϕ correct to $O(R^{-1})$,

$$\mathcal{N}\phi = o(R^{-1}).$$

The numerical procedure for solving (20) yields a solution of the form ϕ_{hsol} ($= B\phi_m + \phi_h$), containing an arbitrary multiple B of ϕ_m, where B is a complex-valued constant. The multiple of ϕ_m can be removed using a Gram-Schmidt orthogonalization

$$\phi_h(y) = \phi_{hsol}(y) - \left(\frac{\int \phi_{hsol}\phi_m^* dy}{\int |\phi_m|^2 dy} \right) \phi_m(y),$$

the integrals being taken from $-\infty$ to ∞.

3.   Analysis

We now choose a value of the x-independent frequency parameter $\upsilon\omega_d\Delta^{-2}$ such that the corresponding mode is marginally growing near the critical Reynolds number reported in I. The eigenfunctions ϕ_m and ϕ_h, and the wavenumber α, are obtained at different streamwise locations. A rectangular domain $(x, y) \in [x_0, x_1] \times [-L, L]$ in non-dimensional space is considered. We use a forward-difference scheme to obtain $\kappa_{i+1, j} \equiv \kappa(x_{i+1}, y_i)$ from $\kappa_{i,\,j}$ from the relation

$$\kappa_{i+1,j} = \kappa_{i,j}\left\{1 + \left(\frac{1}{\kappa}\frac{d\kappa}{dx}\right)_{i,j}(x_{i+1} - x_i)\right\} + o(1)\,.$$

(22)

Similar formulations allow us to determine K and $\overline{K}$.

While evaluating integrals, the contributions from the infinite tails are carefully included after verifying that the integrand has attained the asymptotic exponential decay at the boundary of the computational domain. The finite domain is taken to be sufficiently large so that this assumption holds. In the results reported the contribution from the tail-region contributed less than 10^{-3} percent to the total integral.

The analysis is carried out for three illustrative values of the velocity ratio parameter $\Lambda = 49/50$, $1/3$ and $1/39$. In I we had noted that the critical Reynolds number expressed in the conventional scales U_∞ and $l_d\ (= \sqrt{x_d\nu/U_\infty})$, that is $R^* = U_\infty l_d/\nu$, is about 6 for $\Lambda = 1$ and ∞ for $\Lambda = 0$. Therefore the half-jet ($\Lambda = 1/39$) is close to the limit of shear-less flow, which is stable at very high R^*.

For the half-jet a numerical difficulty was encountered while obtaining the growth rate $(1/\kappa)(\partial\kappa/\partial x)$ for $y \ll -1$, as division by the streamwise mean velocity Φ' causes the growth-rate to diverge. [Theoretically κ itself is very small in this region, making the product $\kappa\Phi'$ finite. But computationally this was not obtained.] For this reason results are not reported in the region $y \ll -1$ for the half-jet case; instead we consider $\Lambda = 49/50$ which is deemed sufficiently close to be representative of that extremity.

4.   Results

The contours of κ in the half-jet flow $\Lambda = 49/50$ and nearly shearless flow $\Lambda = 1/39$ are shown in Figures 2 and 3 for near-critical disturbance modes. To obtain this flow picture of the spatial distribution of disturbance energy the growth-rate was computed using (17), taking care to include only terms correct to $O(R^{-1})$. The same quantity was also computed using the growth rate given by (12). The two computations were verified to be identical to six decimal places.

It is apparent that upstream (/ downstream) of the streamwise location corresponding to $R \approx 39$ there is hardly any streamline in the core of the flow along which energy decays (/ grows) in the streamwise direction. Thus the picture is intuitively faithful to the concept of stability (/ instability). Before we fix this idea, we consider Figure 2 in some more detail.

The edges separating differently coloured regions correspond to trajectories in (x_d– y_d) space along which the mode is marginally stable. Consider now the isotachs (corresponding to constant y values) that are superposed on this picture. The inflexion point is at $y = -0.11$. At every point in a neighbourhood of the inflexion point and upto a certain Reynolds number, the disturbance energy decays along an isotach; beyond the same Reynolds number it increases. The value of R where the change occurs depends in general on position. In some regions, the disturbance energy starts amplifying at an earlier streamwise location compared to others. It will be seen that over a considerable portion of the mixing layer the energy decreases to a minimum and then starts increasing again along the streamlines. It is in this sense that we propose that it is appropriate to speak of the stability of the mixing layer below a Reynolds number of the order of 30.

Figures 4 and 5 show the variation of the disturbance kinetic energy along selected streamlines. This variation confirms the conclusion drawn above. It is therefore interesting to see how the integrated effect, of the decaying nature of the disturbance in the core and the strong non-parallelism of the mixing layer, manifests itself. In I we had defined the quantity to be monitored as K, which owing to the integration over the similarity coordinate y, neglected the growing nature of the mixing layer. The thickness of the mixing layer is proportional to the vorticity-thickness and therefore its logarithmic rate of growth is also p/R. Hence, at low values of R this quantity becomes progressively larger. Below we consider the integrated quantity $\overline{K}$ as defined in (9), the growth rate of which is given by (11).

For the same frequency ω_d considered above, the variation of the two integrated kinetic energies are shown in Figures 4 (corresponding to $\Lambda = 49/50$) and 7 ($\Lambda = 1/39$) across the same range of R as above. We see that K decays upto $R \approx 30$ and then amplifies, but at relatively small rate. On the other hand $\overline{K}$ amplifies monotonically at all R. Note that a stretched scale is used to highlight the variation in K in comparison to that of $\overline{K}$. Secondly the rate of growth of $\overline{K}$ is indistinguishable from p/R.

Note that in computing κ, the error made is $o(1)$, which is due to the integration over an x-domain that scales as $O(R)$. Therefore, the accuracy in predicting the point of instability by observing K or $\overline{K}$ is also correct to that order. The Reynolds numbers at which the quantities κ and K start amplifying, in the contour and integral plots shown herein, must be interpreted with this in mind. In I, we had used the local growth rate g, which is obtained correct to $o(1/R)$, to identify marginal stability.

5.   Conclusion

The present study puts into perspective what one understands by the stability of highly non-parallel flows. It is seen that the definition of growth rate, already known to be dependent on the choice of monitoring parameters, might give results that lead to qualitatively different interpretations regarding the nature of the stability of the flow. The mixing layer provides a striking example to illustrate this point. Thus it was found in the foregoing analysis that the integral of the disturbance energy over the physical coordinate y_d grows continuously from very low Re, and hence points toward instability. On the other hand an integral taken over the similarity coordinate y, or the maximum κ_{max} in y at any station x, shows stability upto $R \approx 30$. It is in this sense that the mixing layer has a non-zero critical Reynolds number. There is no inherent contradiction between these results in a strongly non-parallel flow, for one simple reason. The maximum disturbance kinetic energy density κ_{max} may drop substantially over a given streamwise extent of the flow, as shown in Figure 2, but its integral in y_d may increase if the flow is rapidly thickening (see Figure 1 again). Thus the streamwise rate of change of the integral depends not only on the rate of change of the disturbance kinetic energy, but also on that of the local vorticity thickness. As $\overline{K} \propto RK$, it is possible that K decays, even as $\overline{K}$ increases. The result that $\overline{K}$ grows at all R seems to echo the parallel stability result of Betchov & Szewczyk (1963). (Results from present computations for $R < 10$ may not be accurate because the underlying minimal composite theory is based on large R asymptotics.) However, we believe that such a conclusion is misleading because of two reasons. Firstly, within substantial regions of xy space the disturbance energy decays continuously upto a significantly high Reynolds number. Secondly, it is the high degree of non-parallelism of the mean flow, given by the quantity $d\,\delta_\omega / d\,x_d = p/R$, that accounts for the growth of $\overline{K}$, which is explained by the spreading of a basically smaller disturbance over a rapidly growing y-domain.

We can state these results in slightly different terms. Define the L_p norm of κ by

$$\overline{K}_p = (\int_{-\infty}^{+\infty} |\kappa|^p \; d\,Y)^{1/p}, \quad Y \equiv y_d\,\Delta/\nu .$$

Then we have

$$\overline{K}_1 = \overline{K}$$

$$\overline{K}_\infty = \sup_Y \kappa .$$

The results presented here show that while $\overline{K}_\infty$ exhibits a non-zero critical Reynolds number, $\overline{K}_1$ is consistent with a value of zero for R_{cr}. We cannot assert that $\overline{K}_1$ grows monotonically from $R = 0$, because our calculations cannot be extended to extremely low values of R: the minimal composite theory we use is not valid for small R. Nevertheless, it is a fact that $\overline{K}_1$ exhibits no minimum anywhere at least for $R > 20$.

Thus, if $\overline{K} = \overline{K}_1$ is chosen as the norm, it points towards instability over the whole range of R considered. On the other hand the integral K over the similarity coordinate y decays upto a value of R in the range 28 to 39 depending on Λ, and hence possesses a critical R in the same range. It is in this sense that the mixing layer has a non-zero critical

Reynolds number. Thus, while the maximum disturbance kinetic energy density $\kappa_{\max} = \overline{K}_\infty$ drops substantially over a given streamwise extent of the flow (Figure 2), its integral in y_d can increase if the flow is rapidly thickening (see Figure 4), as it does at low R.

Stability limits thus depend on how the norm is defined.

To conclude, the interpretation of the stability of highly non-parallel flows is not necessarily straightforward, and so it is revealing to analyse the disturbance energy evolution in different ways. Nevertheless, we wish to suggest that, from the point of view of predicting transition, the simple $\kappa_{\max}$ is likely to be the physically relevant norm.

References

Betchov R and Szewczyk A. 1963 Stability of a shear layer between parallel streams. *Phys. Fluids* 6:1391–1396.

Bhattacharya P, Manoharan M P, Govindarajan R and Narasimha R 2006 The critical Reynolds number of a laminar incompressible mixing layer from minimal composite theory *J. Fluid Mech.* 565:105-114.

Govindarajan R and Narasimha R 1997 A low-order theory for stability of non-parallel boundary layer flows. Proc. R. Soc. Lond. A 453:2537–2549.

Govindarajan R, Narasimha R 2001 Estimating amplitude ratio in boundary layer stability theory: a comparison between two approaches. *J. Fluid Mech.* 439:403-412.

Govindarajan R, Narasimha R 2005 Accurate estimate of disturbance amplitude variation from solution of minimal composite stability theory. *Theor. Comput. Fluid Dyn.* 19:229-235.

Schlichting H and Gersten K 2004 *Boundary Layer Theory*, 8th Ed. Springer. McGraw-Hill, New York.

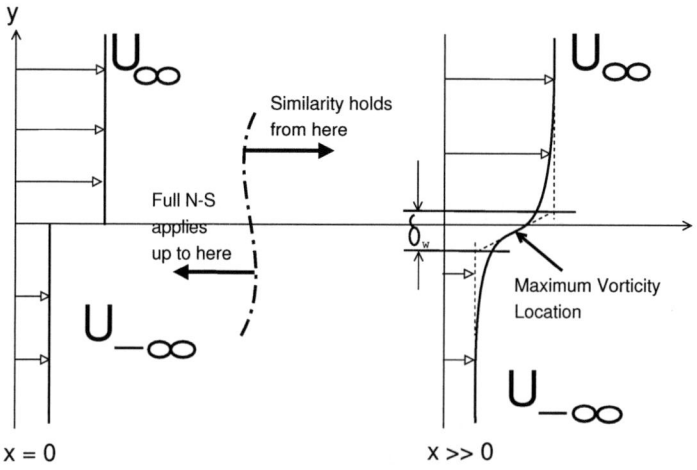

Figure 1: Geometry of the mean flow.

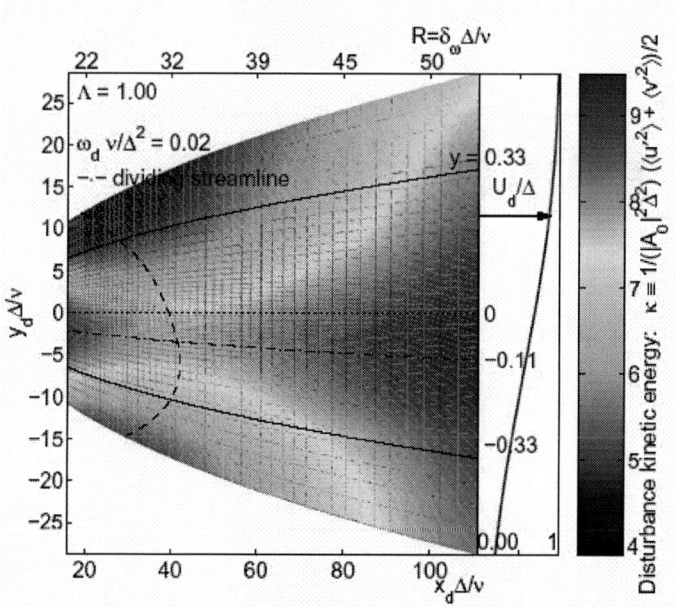

Figure 2: Spatial distribution of disturbance kinetic energy, half jet.

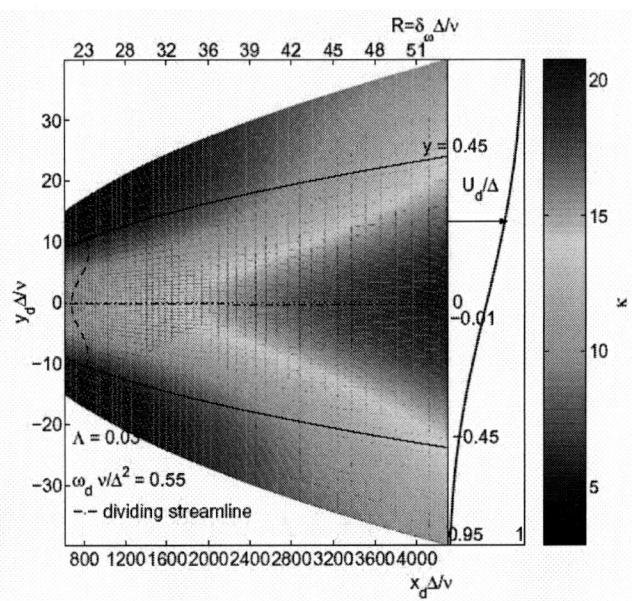

Figure 3: Spatial distribution of disturbance kinetic energy, nearly shear-less flow.

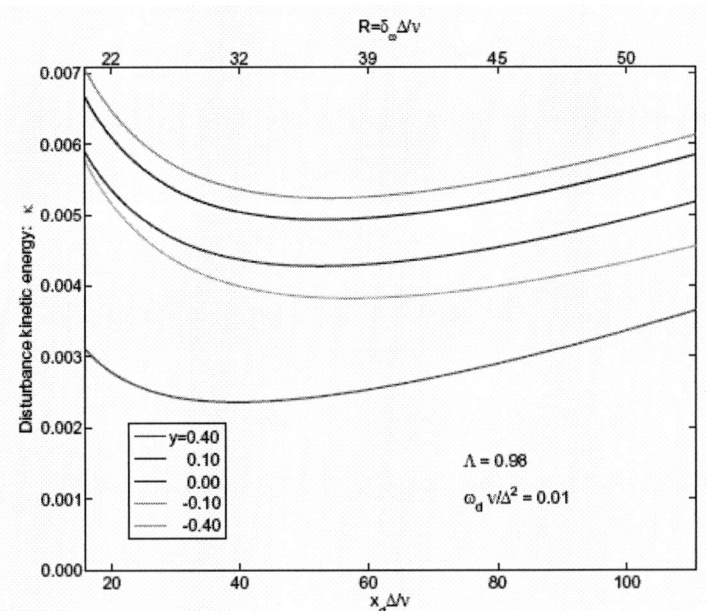

Figure 4: Kinetic energy density variation along isotachs in a near half-jet.

16

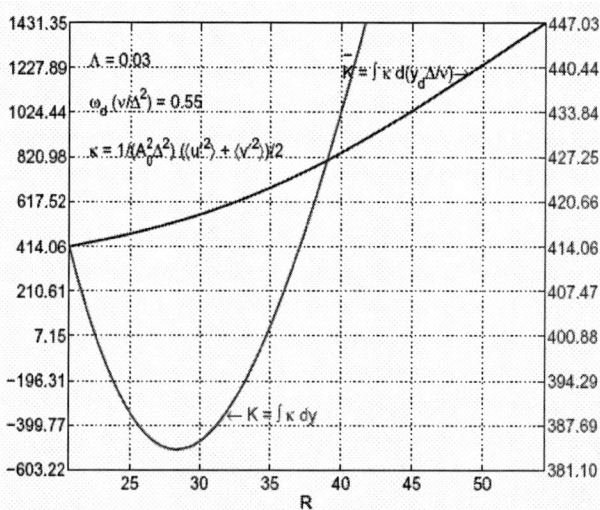

Figure 6: Comparison of the streamwise variation of $\overline{K}$ and K. near half-jet.

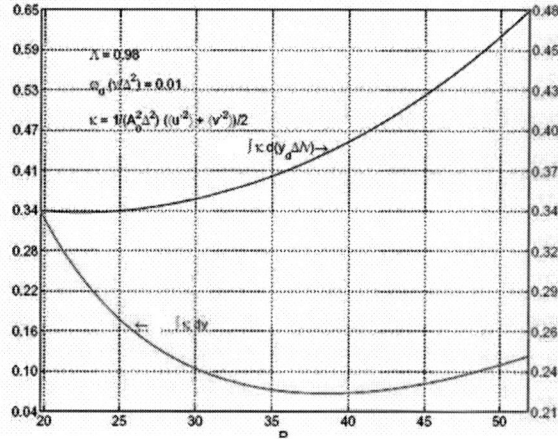

Figure 7: Comparison of the streamwise variation of $\overline{K}$ and K. nearly shear-less flow, legend on curves same as in Figure 6.

Geometric Modelling and Shape Preserving Properties of Wachspress Functions

H.P.Dikshit

School of Good Governance and Policy Analysis, C-403 Narmada Bhawan,59 Arera Hills, Bhopal 462011, India

Abstract. Wachspress rational basis functions have been studied with the following two objectives. We first introduce barycentric coordinates for polygons in terms of Wachspress rational functions and also give a more general definition in terms of volumes. We use these to develop quadrilateral and pentagonal patches for geometric modelling to construct surfaces of desired shape and smoothness. We also compare this method with the well known Bézier-Bernstein methods. Next we define rational complex planar splines using complex analogues of real Wachspress functions and study their shape preserving properties like quasiregularity and quaziconformality of the spline interpolants. Finally, we present a beautiful application of quintic splines in obtainig a sharp result on bounds for Lebesgue type constant for the Nörlund transformation of Fourier series.

Keywords: Wachspress basis functions; Barycentric coordinates; Quadrilateral patches; Pentagonal patches; Geometric modelling; Complex planar splines; Quasiregular and quasiconformal mappings; Bounds for Lebesgue type constant
PACS: D 87.15 Aa, D 87.16 Ac

1. INTRODUCTION

In this 2009 Zakir Hussain Award lecture, we would present some beautiful properties of Wachspress rational polynomial basis functions which were first introduced in [25]. For brevity we call these as Wachspress functions. In Section 2, we will define Wachspress functions and present some of their important properties useful for Computer Aided Geometric Design (CAGD) applications [17]. The well known Bézier-Bernstein technique which has been extensively studied with a variety of applications to geometric modelling depends on Bernstein polynomial basis functions defined on simplexes, in particular, on intervals on the real line and triangles in the plane. This is because of the availibilty of barycentric coordinates expressible in terms of the end points of intervals and vertices of triangles. Thus, Bézier curves on intervals and Bézier patches on triangles are respectively used to design composite curves and surfaces of desired shapes. Nice smoothness conditions for such curves and surfaces are known. The dependence of the shapes of these curves and surfaces on control points, control polygons or polytopes is another advantage from the point of view of geometric modelling applications. Although, rectangular patches could be defined in terms of bilinear Bernstein polynomials but such a construction on polygons other than rectangles or triangles was not possible.

Therefore as a first step for developing other polygonal patches, we require a barycentric coordinate system for the underlying polygon in terms of its vertices. In one of our pioneering efforts to develop quadrilateral and pentagonal patches in [5, 6] we used a system of barycentric coordinates for quadrilaterals and pentagons with some geometric

CP1146, *Modelling of Engineering and Technological Problems,* edited by A. H. Siddiqi, A. K. Gupta, and M. Brokate

constraints which is due to Wachspress [25].

It may be observed that the problem of defining barycentric coordinates for different types of polygons has recieved considerable recent attention (see [20, 27]). In Section 3, we would present barycentric coordinate system using Wachspress functions and obtain their connection with bilinear Bernstein polynomials. This is helpful in comparing Wachspress quadrilateral patches and tensor product rectangular patches defined in terms of bilinear Bernstein polynomials. In Section 3, we also give some more recent results concerning barycentric coordinates for polygons and compare them with the definition given in terms of Wachspress functions. In Section 4, we introduce Wachspress patches for geometric modelling applications and discuss smoothness of composite Wachspress surfaces. We also give subdivision formula for quadrilateral patches. In Section 5, we consider certain complex analogues in z and its conjugate $\bar{z}$ of real polynomials and Wachspress rational polynomials and study their nice shape preserving properties, like *quasiregularity* and *quasiconformality* . In Section 6, we present some results giving bounds for certain integrals which are in some sense related to Lebesgue constants. These bounds lead to necessary and sufficient conditions for convergence of transformed sequences of partial sums of Fourier series at a point under assumptions on the generating function.

2. WACHSPRESS RATIONAL BASIS FUNCTIONS

Wachspress initiated the study of rational polynomial basis functions for finite element construction over quadrilaterals and more general polygonal and curved elements. Later Apprato et al. [3] and Gout [18] studied the interpolatory and convergence properties of lower degree rational finite elements and their applications in solving second order boundary value problems. Some recent studies include applications of linear Wachspress functions to problems of solid mechanics and biomechanics [7, 21, 24] . An interesting application of these rational polygonal elements is in boundary color interpolation problems of computer graphics [7].

One of the important properties of Wachspress rational polynomial functions is that along the edges of the underlying polygonal partition the rational polynomials reduce to polynomials, which greatly facilitates the study of interpolation at vertices and smoothness properties across the edges. The foregoing important property of Wachspress rational functions is ensured by the following important consequence of the well known *Bezut's Theorem* for real polynomials.

Lemma 1. *Given positive integers s,k,m,n, such that $0 < s \le k \le m \le n$. We denote by p_r polynomial of dgree r. Then the condition:*

$$p_n(x_j) = p_m(x_j) = p_k(x_j); \quad j = 1,2,\ldots,s; \tag{1}$$

implies that the rational polynomial p_n/p_m along $p_r(x) = 0$, is some rational polynomial of reduced degree p_{n-s}/p_{m-s}. In particular, p_n/p_m reduces to a polynomial p_{n-m}, along $p_m(x) = 0$.

Considering a closed convex quadrilateral $Q \subset \mathbb{R}^2$, we first introduce Wachspress function over Q. Two sides of Q are said to be opposite, if they do not share a common

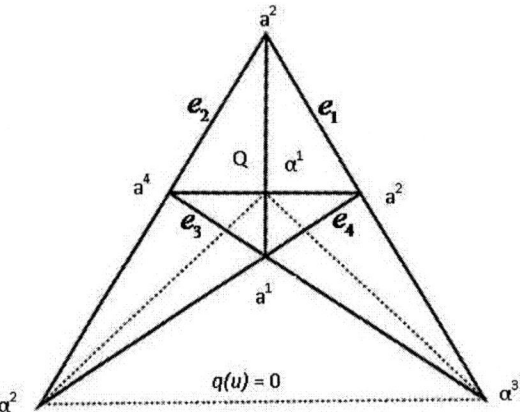

FIGURE 1. Admissible quadrilateral Q, exterial diagonal q, associate quadrangle and the reference triangle.

vertex. We will say Q is admissible, if its opposite sides intersect, that is, it is different than a trapezium or a parallelogram. We would consider only admissible quadrilaterals unless stated otherwise. The linear form which vanishes on the line passing through the points of intersections of opposite sides of Q will be denoted by q. This line $q(u) = 0$, will be called the exterior diagonal of Q. We label the vertices $\{a^i\}_{i=1}^4$ in counterclockwise direction in such a way that a^3 is the most distant vertex from the exterior diagonal q of Q. Let e_i be the linear form which vanishes on the line joining the points a^{i+1} and a^{i+2}, $i \in Z_4$. We associate with Q, a complete quadrangle, which is a collection of four distinct points joined in pairs by six distinct lines. The four vertices of Q are also called vertices of the quadrangle and the lines are called its sides. Any point of intersection of two opposite sides is called a diagonal point. Let α^1 denote the diagonal point of the quadrangle associated with Q, lying in the interior of Q, while α^2 (α^3) denote the diagonal point obtained by joining the lines e_2 and e_4 (e_1 and e_3) (see Fig. 1). The triangle formed by three diagonal points is called the diagonal triangle of the quadrangle. We will use this diagonal triangle as reference triangle for both projective and affine systems of coordinates. Linear Wachspress basis functions for interpolation of data specified at the vertices of Q are defined in the following.

$$W_i(u) = K_i \, e_i(u) \, e_{i+1}(u)/q(u), \;\; i \in Z_4; \;\; u \in Q,$$

where K_i is chosen so that $W_i(a^i) = 1$. These basis functions posses following interesting properties, which make them especially useful for geometric modelling applications [6].

$$\left.\begin{array}{l} W_i(a^j) = \delta_{i,j}; \;\; \text{where } \delta_{i,j} \text{ is 1 for } i = j \text{ and 0 otherwise;} \\ W_i \text{ is linear along each edge of } Q; \\ W_i(u) \geq 0, \; \sum_{i=1}^4 W_i(u) = 1, \; u \in Q. \end{array}\right\} \tag{2}$$

We give a simple proof of the following which enables us to define barycentric coordinates for Q. Any linear polynomial p defined on Q may be written as

$$p(u) = \sum_{i=1}^{4} p(a^i)\, W_i(u),\ u \in Q. \tag{3}$$

From equation (3), we conclude that linear polynomials are reproduced. This is of particular importance in developing barycentric coordinates using Wachspress functions. We give below an interesting proof of (3). Let $P(u)$ denote the polynomial which is the difference of the polynomials appearing on the two sides of the equality sign in (3). Now observing that $P(a^i) = 0, i \in Z_4$ and $P(u)$ is linear along the edges by virtue of (2), we coclude that $P(u)$ vanishes along the four edges of Q so that it is a polynomial of at least degree 4, whereas by construction its numerator is a polynomial of degree at most 2. Thus, $P(u) = 0$, and we prove (3). The above properties of the Wachspress functions share many properties of Bernstein polynomial basis functions which play an important role in the well known *Bézier-Bernstein* technique [17]. This motivated us to use the Wachspress functions to define blending functions for developing quadrilateral patches for geometric modelling applications in Section 4.

We next introduce Wachspress functions on closed convex pentagons $P \subset \mathbb{R}^2$, which have no two parallel sides. Such pentagons will be called admissible and we will always assume that the pentagons are admissible unless stated otherwise. Let $\{a^i\}_{i=1}^5$, be the vertices of P, labelled in counterclockwise direction. Any two sides which are not adjacent are called opposite sides. Let e_i dnote the linear form which vanishes on the edge of P joining a^i and a^{i+1}. Consider the five points of intersection $\{\alpha^i\}_{i=1}^5$, of opposite sides, called the exterior extension points (EIP) of P. Let $r(u)$ denote the quadratic form vanishing on the conic passing through the points $\{\alpha^i\}_{i=1}^5$. We are now set to introduce the following linear Wachspress basis (wedge) functions for interpolation of data specified at the vertices of pentagon P.

$$V_{1,i}(u) = M_i\, e_{i+1}(u)\, e_{i+2}(u)\, e_{i+3}(u)/r(u),\ i \in Z_5,\ u \in P, \tag{4}$$

where M_i is chosen so that $V_{1,i}(a^i) = 1$. As in the case of W_i's, the basis functions $V_{1,i}$'s are nonnegative, form a partition of unity and exhibit the property similar to (3), that is they reproduce linear polynomials. By an application of Lemma 1, it is also seen that $V_{1,i}$'s reduce to polynomials on the edges of P. These pentagonal Wachspress functions on pentagons have found interesting applications in web enabled color interpolation schemes useful in instrumentation imaging and internet graphics [7]. We will use these Wachspress functions to define blending functions for developing pentagonal patches for geometric modelling applications in Section 4.

3. BARYCENTRIC COORDINATES FOR POLYGONS

One of the pioneering methods for defining barycentric coordinates is by using Wachspress functions defined over convex polygons with certain restrictions about their geometry [25]. For developing quadrilateral patches and more recently pentagonal patches, for geometric modelling applications, we use the foregoing type of barycentric coordinates. For defining barycentric coordinates for quadrilaterals, we take $p(u) = u$ in (3), so that

$$u = \sum_{i=1}^{4} a^i \, W_i(u), \; u \in Q. \tag{5}$$

This shows that $(W_1(u), W_2(u), W_3(u), W_4(u))$ may be used as a system of barycentric coordinates for any point $u \in Q$. Therefore, it is natural to propose a de Casteljau type algorithm for generating higher degree Wachspress funstion by iterations. Thus, for convenience, we introduce the following notations.

$$W_1 = W_{0,0}, \; W_2 = W_{1,0}, \; W_3 = W_{1,1}, \; W_4 = W_{0,1}.$$

Using a projective map $A : Q \to [0,1]^2$, a relationship between Wachspress basis functions and tensor product bilinear Bernstein polynomials has been presented by us in [6], which in its turn is used to define higher degree Wachspress basis functions in an iterative way. In order to write the relationship, we begin by defining

$$A(u) = \frac{1}{2} \left(1 + \frac{\mu_1 \lambda_2}{\mu_2 \lambda_1}, 1 + \frac{\mu_1 \lambda_3}{\mu_3 \lambda_1} \right) = (x_1, x_2) = x, \tag{6}$$

say, where $(\lambda_1, \lambda_2, \lambda_3)$ are the barycentric coordinates of a point $u \in Q$ with respect to the diagonal reference triangle with vertices α^1, α^2, α^3 whereas (μ_1, μ_2, μ_3) are the barycentric coordinates of a^3 with respect to the foregoing triangle. It may be verified that

$$A(a^1) = (0,0), \; A(a^2) = (1,0), \; A(a^3) = (1,1), \; A(a^4) = (0,1).$$

The exterior diagonal is projected to line at infinity. It has been shown in [18], that the map is a $C^\infty-$ diffeomorphism from Q onto $[0,1]^2$. Using the above transformation, one may express the equations $e_i(u) = 0$, in the following barycentric form:

$$e_1(u) = (\mu_2 \lambda_1 - \mu_1 \lambda_2) = 0, \; e_2(u) = (\mu_3 \lambda_1 - \mu_1 \lambda_3) = 0,$$
$$e_3(u) = (\mu_2 \lambda_1 + \mu_1 \lambda_2) = 0, \; e_4(u) = (\mu_3 \lambda_1 + \mu_1 \lambda_3) = 0.$$

Therefore, the Wachspress basis functions may also be written in barycentric coordinates form using bilinear Bernstein basis as follows [6].

$$W_{0,0}(u) = s_{0,0} \, (1 - x_1) \, (1 - x_2)/s(x) = s_{0,0} \, B_{0,0}(x)/s(x),$$
$$W_{1,0}(u) = s_{1,0} \, x_1 \, (1 - x_2)/s(x) = s_{1,0} \, B_{1,0}(x)/s(x),$$
$$W_{1,1}(u) = x_1 \, x_2/s(x) = B_{1,1}(x)/s(x),$$
$$W_{0,1}(u) = s_{0,1} \, (1 - x_1) \, x_2/s(x) = s_{0,1} \, B_{0,1}(x)/s(x),$$

where $B_{i,j}(x) = (1 - x_1)^{1-i} x_1^i \, (1 - x_2)^{1-j} x_2^j$, and

$$s(x) = s(x_1, x_2) = \mu_1/\lambda_1 = \mu_1 + \mu_2(2x_1 - 1) + \mu_3(2x_2 - 1); s_{i,j} = s(i,j). \tag{7}$$

We introduce some recent methods of defining coordinates of a point u of a polygon P with vertices given by $\{a^1, a^2, ..., a^n\}$ (see for example [20] and [27]). For computer graphics and geometric modelling, we need to express $u \in P$, as an affine combination of the vertices. Therefore, a very general first requirement would be to look for weights $\lambda_i : R \to \mathbb{R}$, that genertalize equation (5).

$$\sum_{i=1}^{n} \lambda_i(u)(a^i - u) = 0,$$

and associated normalized barycentric coordinates

$$w_i(u) = \frac{\lambda_i(u)}{\sum_{j=1}^{n} \lambda_j(u)}, \text{ such that, } w_i(a^j) = \delta_{i,j}.$$

Thus, assuming P to be a convex polygon the following choice of λ_i, in particular, gives the barycentric coordinates expressed in terms of Wachspress rational functions.

$$\lambda_i(u) = \frac{A(a^{i-1}, a^i, a^{i+1})}{A(a^{i-1}, a^i, u) \, A(a^i, a^{i+1}, u)},$$

where $A(u_1, u_2, u_3)$ denotes the signed area of the triangle $[u_1, u_2, u_3]$. By multiplying by the product of all areas $A(a^i, a^{i+1}, u)$, we obtain the following equivalent choice of λ_i.

$$\lambda_i(u) = A(a^{i-1}, a^i, a^{i+1}) \prod_{j \neq i-1, i} A(a^i, a^{i+1}, u).$$

It is important to note that this definition does not require the conditions of admissibility on quadrilaterals and pentagons. For example, in the case of quadrilateral Q such that two of its opposite edges are parallel where as the other two intersect at a point u_0, then Q is not admissible. But from the above general definition it follows that in this case, the role of exterior diagonal is played by the linear form which lies on the line passing through u_0 and is parallel to the sides of Q, which are parallel [14]. With this geometric interpretation, we can apply Lemma 1 to conclude that the Wachspress elements on Q reduce to linear polynomials on its edges.

4. WACHSPRESS POLYGONAL PATCHES FOR GEOMETRIC MODELLING

In the previous section we have discussed nice properties of Wachspress basis functions defined on admissible closed convex quadrilaterals and pentagons. In this section we will define linear Wachspress patches over quadrilaterals and pentagons and through iteration we will define higher degree Wachspress surfaces. These have many nice properties common to *Bézier-Bernstein* methods.

4.1. Quadrilateral patches, C^1 smooth composite surfaces and subdivison formula

Considering a given sequence of scalars or vectors $\{c_{i,j}\}$, we define an operator E by requiring that $E_{m,n} c_{i,j} = c_{i+m,j+n}$; $m, n = 0$, or 1. Now in order to use de Casteljau type algorithm, we set

$$W_1(c,u) = (W.E)c_{0,0} = c_{0,0}W_{0,0} + c_{1,0}W_{1,0} + c_{0,1}W_{0,1} + c_{1,1}W_{1,1},$$

and iterate $(W.E)$ to obtain higher degree n, Wachspress blending function $W_{i,j}^n$ in terms of the Bernstein basis polynomials given by

$$B_{i,j}^n(x) = \binom{n}{i}\binom{n}{j}(1-x_1)^{n-i}x_1^i(1-x_2)^{n-j}x_2^j. \tag{8}$$

In [6], we obtained the following explicit representation:

$$W_{i,j}^n(u) = K_{i,j}^n B_{i,j}^n(x)/(s(x))^n, \tag{9}$$

where $s(x)$ is defined by (7) and

$$K_{i,j}^n = \binom{n}{i}^{-1} \sum_{p=(i+j-n)_+}^{min(i,j)} \binom{n-j}{i-p}\binom{j}{p} Q_{i,j}^n(p); \quad Q_{i,j}^n(p) = s_{0,0}^{n+p-i-j}s_{1,0}^{i-p}s_{0,1}^{j-p}.$$

Here, u and x are related by $A(u) = x$, as given in (6). For convenience let $W_{i,j}^r(u) = 0$ for $i, j < 0$ or $i, j > r$. We observe that the above Wachspress blending functions have the following properties including a nice recurrence relation, for useful geometric modelling applications [6].

$$\left. \begin{array}{l} \sum_{i=0}^n \sum_{j=0}^n W_{i,j}^n(u) = 1; u \in Q; \ W_{i,j}^n(u) \geq 0; \ 0 \leq i,j \leq n; \\ W_{i,j}^n \text{ reduces to a polynomial of degree } n \text{ along each edge of } Q; \\ W_{i,j}^n = W_{0,0}W_{i,j}^{n-1} + W_{1,0}W_{i-1,j}^{n-1} + W_{0,1}W_{i,j-1}^{n-1} + W_{1,1}W_{i-1,j-1}^{n-1}. \end{array} \right\} \tag{10}$$

We give in the next page two graphical representations of degree three Wachspress blending functions [6], which show that each one of them has a sigle maximum on the underlying quadrilateral. We tested this property in a large number of Wachspress blending functions, but we do not have an analytical proof for this. We now set for convenience,

$$v_n(x) = \sum_{i=0}^n \sum_{j=0}^n c_{i,j}K_{i,j}^n B_{i,j}^n(x); \quad r_n(x) = (s(x))^n; \tag{11}$$

and give an explicit representation of the Wachspress patch $W_n(c,u)$ of degree n over Q in the following [6].

Theorem 1. *For any sequence of control points* $c = \{c_{i,j}\}_{i,j=0}^n$,

$$W_n(c,u) = \sum_{i=0}^n \sum_{j=0}^n c_{i,j} W_{i,j}^n(u) = v_n(x)/r_n(x). \tag{12}$$

24

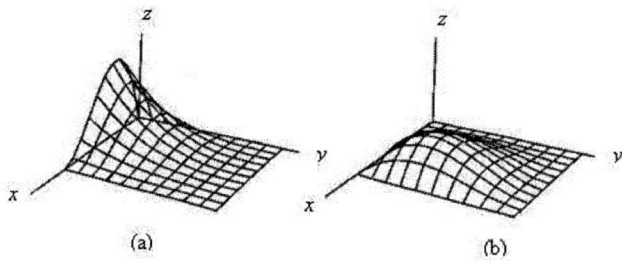

FIGURE 2. Wachspress basis functions (a) $W_{1,0}^3$ and (b) $W_{2,1}^3$ of degree three with single maximum.

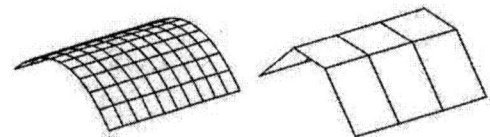

FIGURE 3. Wachspress surface lies in the convex hull of the control points and mimicks the control polytope.

In view of the property (10), the representation of $W_n(c,u)$ is a convex combination of $W_{i,j}^n(u)$. Thus the surface lies in the convex hull of the control points. This gives a good tool to design surfaces of desired shapes by choosing control points suitably (see Fig. 3). Again setting

$$c_{i,j}^{(1)} = \sum_{r=i-1}^{i} \sum_{t=j-1}^{j} \lambda_{r,t}^n c_{r,t}; \ 0 \le i, j \ \le n+1 \text{ with } \lambda_{r,t}^n = 0, \text{for } r,t < 0.$$

We also proved the following degree elevation property for Wachspress quadrilateral patches in [6].

Theorem 2. *Given a sequence of control points* $c = \{c_{i,j}\}_{i,j=0}^n$, *we have* $W_n(c,u) = W_{n+1}(c^{(1)},u)$. *Further, the control points in* $c^{(1)}$ *are convex combinations of control points in* c.

In order to develop C^1-smooth composite surfaces using Wachspress patches, we need to find directional cross boundary derivatives for such blending functions. For this let d be a given unit vector in a certain direction in $\mathbb{R}^2$. We denote the k-th order directional derivative of $W_n(c,\cdot)$ in the direction d by $D_d^k W_n(c,\cdot)$ with $D_d^1 W_n(c,\cdot) = D_d W_n(c,\cdot)$. Computing first $D_d^1 W_n(c,\cdot)$, and using mathematical induction, expression for k-th order mixed directional derivative of $W_n(c,\cdot)$ has been obtained in [12]. Using this result, the k-th order cross boundary directional derivative in the direction d is given by

$$D_d^k W_n(c,u) = \frac{1}{r_n(x)} \sum_{j=0}^k \binom{k}{j} \frac{n!}{(n-k+j)!} (\delta_1 s(x))^{k-j} D_d^j v_n(x), \tag{13}$$

25

where $\delta_1 = \frac{1}{\mu_1} D_d \lambda_1$. For studying C^1-continuity property of composite surface, we now assume that admissible quadrilateral Q shares a common edge e with another admissible quadrilateral $\tilde{Q}$. We further assume that the exterior diagonal q of Q, the line e and the exterior diagonal $\tilde{q}$ of $\tilde{Q}$ meet at a point. We also define Wachspress patch $\tilde{W}_n(\tilde{c}, \cdot)$ of degree n over $\tilde{Q}$, using (12) with the control sequence $c = \{c_{i,j}\}_{i,j}^n$ replaced by $\tilde{c} = \{\tilde{c}_{i,j}\}_{i,j}^n$. All the entities γ defined for Q will also be defined for $\tilde{Q}$ with γ replaced by $\tilde{\gamma}$. For example, the projective map $A : Q \to [0,1]^2$ will be replaced by $\tilde{A} : \tilde{Q} \to [0,1]^2$ with $\tilde{A}(u) = \tilde{x}$ and $\tilde{\lambda} = (\tilde{\lambda}_1, \tilde{\lambda}_2, \tilde{\lambda}_3)$. Note that u will remain unchanged as it is a global variable over $Q \cup \tilde{Q}$.

We further assume that the common edge for both the quadrilaterals Q and $\tilde{Q}$ corresponds to the edge $x_1 = \tilde{x}_1 = 1$ on the standard square. It may be observed that the assumption that the edge e, q and $\tilde{q}$ meet at a point leads to $\lambda_1/\tilde{\lambda}_1 = $ constant $= \mu_1/\tilde{\mu}_1$ on e and therefore, we have [6].

$$\frac{s(x)}{\tilde{s}(\tilde{x})} = \frac{\mu_1 \tilde{\lambda}_1}{\tilde{\mu}_1 \lambda_1} = 1 = \frac{r_n(x)}{\tilde{r}_n(\tilde{x})}, \quad (\mu_1 \lambda_3 + \mu_3 \lambda_1)/(\tilde{\mu}_1 \tilde{\lambda}_3 + \tilde{\mu}_3 \tilde{\lambda}_1) = \text{a constant}, \qquad (14)$$

along the common edge e. Further, computation shows that $x = \tilde{x}$ along e.

We now turn to discuss the conditions under which the composite patch defined over $Q \cup \tilde{Q}$ will be C^1-smooth. It is readily seen that C^0- continuity leads to

$$c_{n,j} = \tilde{c}_{n,j}, j = 0, 1, \cdots, n.$$

From this and (14), it follows that $v_n(x) = \tilde{v}_n(x)$ along e. We also notice that by (12), $W_n(c, u) = v_n(x)/r_n(x)$ and $\tilde{W}_n(\tilde{c}, u) = \tilde{v}_n(\tilde{x})/\tilde{r}_n(\tilde{x})$. Let d be a cross boundary unit vector. Then C^1- continuity requires $D_d W_n(c, u) = D_d \tilde{W}_n(\tilde{c}, u)$ along e. After some computation, we see that this leads to the condition:

$$D_d v_n(x) - D_d \tilde{v}_n(x) + n s(x) v_n(x)(\delta_1 - \tilde{\delta}_1) = 0. \qquad (15)$$

Thus, the set of condition (15) along with the C^0- continuity condition provides a characterization for C^1- continuity of composite Wachspress patch. Proceeding in a similar fashion and using (13), we observe that the condition for continuity of $k-th$ cross boundary derivative is given by $D_d^k W_n(c, u) - D_d^k \tilde{W}_n(\tilde{c}, u) = 0$, that is,

$$\sum_{j=0}^{k} \binom{k}{j} \frac{n!}{(n-k+j)!} (s(x))^{k-j} [\delta_1^{k-j} D_d^j v_n(x) - \tilde{\delta}_1^{k-j} D_d^j \tilde{v}_n(x)] = 0,$$

along the common edge e.

We now proceed to determine conditions for C^1-continuity using explicit forms of the patches $W_n(c, u)$ and $\tilde{W}_n(\tilde{c}, u)$. This will provide us conditions in terms of Wachspress control points. For this, let us denote the following by $L_{n,j}$.

$$(\alpha - \beta)b_{n-1,j}^n + \frac{j}{n}(\alpha - \gamma)b_{n,j-1}^n + \left(\beta + \frac{2j-n}{n}\gamma\right)b_{n,j}^n + \frac{n-j}{n}(\alpha + \gamma)b_{n,j+1}^n,$$

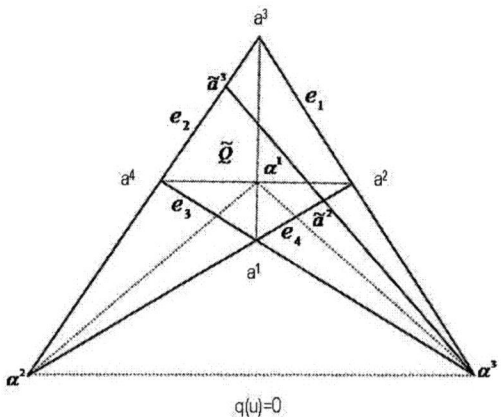

$q(u)=0$

FIGURE 4. Quadrilateral Q and its subquadrilateral $\tilde{Q}$.

where $b^n_{i,j} = c_{i,j} K^n_{i,j}$. Setting $\tilde{b}^n_{i,j} = \tilde{c}_{i,j} K^n_{i,j}$, we define $\tilde{L}_{n,j}$ by replacing α, β, γ and b in $L_{n,j}$ by $\tilde{\alpha}, \tilde{\beta}, \tilde{\gamma}$ and $\tilde{b}$, respectively. Now observing that $x_2 = \tilde{x}_2$ along the edge e, further computation shows that the C^1- continuity of composite surface imposes following constraints on the control points:

$$L_{n,j} = \tilde{L}_{n,j} \; j = 0, 1, \cdots, n.$$

Some examples of C^1 quadrilateral patches have been given in [12].

For studying the subdivision property, we refer to the quadrilateral Q already introduced in Sections 2 and 3 and consider a line joining α^3 and an interior point $\tilde{a}^3$ of the edge e_2, which divides the quadrilateral Q into two subquadrilaterals. Let $\tilde{a}^2$ be the point of intersection of the the edge e_4 and the line joining α^3 and $\tilde{a}^3$ (see Fig. 4) and the sub-quadrilateral formed by the vertices $a^1, \tilde{a}^2, \tilde{a}^3$ and a^4 be denoted by $\tilde{Q}$. For obtaining the subdivision formula, we proceed to find out control points of the restriction of $W_n(c,.)$ on $\tilde{Q}$, if we view it as a Wachspress patch defined on $\tilde{Q}$. The reference triangle of $\tilde{Q}$ has vertices $\tilde{\alpha}^1, \alpha^2, \alpha^3$. Let (μ_1, μ_2, μ_3) be the barycentric coordinates of a^3 with respect to $\langle \alpha^1, \alpha^2, \alpha^3 \rangle$ and $(\tilde{\mu}_1, \tilde{\mu}_2, \tilde{\mu}_3)$ be the barycentric coordinates of $\tilde{a}^3$ with respect to $\langle \tilde{\alpha}^1, \alpha^2, \alpha^3 \rangle$. In [6], we have shown that the barycentric coordinates of a^4 with respect to $\langle \alpha^1, \alpha^2, \alpha^3 \rangle$ are $(\mu_1, -\mu_2, \mu_3)/(1 - 2\mu_2)$.

Suppose that the interior point $\tilde{a}^3$ divides the edge from a^3 to a^4 in the ratio $\theta : 1 - \theta$ for some $0 < \theta < 1$ so that

$$\tilde{a}^3 = (1 - \theta)a^3 + \theta a^4; 0 < \theta < 1.$$

Then writing $m(\theta) = 1 - 2(1 - \theta)\mu_2$, we see that the barycentric coordinates of $\tilde{a}^3$ with respect to $\langle \alpha^1, \alpha^2, \alpha^3 \rangle$ are given by

$$\frac{1}{1 - 2\mu_2}\left(\mu_1 m(\theta), \mu_2(m(\theta) - 2\theta), \mu_3 m(\theta)\right).$$

Consider now the $C^{\infty}-$ projective map $A : Q \to [0,1]^2$ defined in (6). Then, we see that

$$A\left(\tilde{a}^3\right) = (\delta(\theta), 1), \delta(\theta) := \frac{(1-\theta)(1-2\mu_2)}{1-2(1-\theta)\mu_2}. \tag{16}$$

Thus, under the map $A : Q \to [0,1]^2$, the image of $\tilde{Q}$, that is, $A(\tilde{Q}) = [0, \delta(\theta)] \times [0, 1]$. We therefore introduce the restriction of A on $\tilde{Q}$ as follows:

$$A_{\tilde{Q}} : \tilde{Q} \to [0, \delta(\theta)] \times [0, 1] \; ; \; A_{\tilde{Q}}(u) = A(u), u \in \tilde{Q} \subset Q.$$

We shall also consider the map $\tilde{A} : \tilde{Q} \to [0,1]^2$ defined by

$$\tilde{A}(u) = \frac{1}{2}\left(1 + \frac{\tilde{\lambda}_2\tilde{\mu}_1}{\tilde{\lambda}_1\tilde{\mu}_2}, 1 + \frac{\tilde{\lambda}_3\tilde{\mu}_1}{\tilde{\lambda}_1\tilde{\mu}_3}\right) = \tilde{x} = \left(\tilde{x}_1, \tilde{x}_2\right), \; u \in \tilde{Q}, \; \tilde{x} \in [0,1]^2,$$

say, where as defined earlier $\left(\tilde{\mu}_1, \tilde{\mu}_2, \tilde{\mu}_3\right)$ and $\left(\tilde{\lambda}_1, \tilde{\lambda}_2, \tilde{\lambda}_3\right) = \left(\tilde{\lambda}_1(u), \tilde{\lambda}_2(u), \tilde{\lambda}_3(u)\right)$ are respectively the barycentric coordinates of $\tilde{a}^3$ and u with respect to the reference triangle: $\langle \tilde{\alpha}^1, \alpha^2, \alpha^3 \rangle$. Now introducing the following map $B : [0, \delta(\theta)] \times [0, 1] \to [0, 1]^2$ defined by

$$B(x) = \left(\frac{x_1}{\delta(\theta)}, x_2\right) \in [0, 1]^2, \; x = (x_1, x_2) \in [0, \delta(\theta)] \times [0, 1],$$

we see that $\tilde{A} = BA_{\tilde{Q}}$. Finally, after further computation we see that the following formula gives the control points $\tilde{c}_{i,j}$, of the subdivided part of the Wachspress patch on $\tilde{Q}$, in terms of the control points $c_{i,j}$ of the Wachspress patch on Q.

$$\tilde{c}_{i,j} = \frac{\sum_{k=0}^{i} c_{k,j} K_{k,j}^n B_k^i(\delta(\theta))}{(1-\theta)^{-n}(\delta(\theta))^n \tilde{K}_{i,j}^n},$$

where $\delta(\theta)$ is as given in equation (16) . This provides a subdivision formula for Wachspress quadrlateral patches.

Very recently, we have studied quadrilateral patches on convex quadrilaterals which are not admissible in [14].

4.2. Pentagonal patches and C^1-smooth composite surfaces

In view of the nice properties of Wachspress basis functions defined on admissible convex pentagon P, we are motivated to introduce quadratic Wachspress blending functions over underlying pentagon as follows.

$$W_{2,i}(u) = (W_{1,i}(u))^2; \; W_{2,i,j}(u) = 2 \; W_{1,i}(u) \; W_{1,j}(u)); \; i \neq j; i, j \in Z_5.$$

Since we will be using the foregoing definitions, it will be assumed throught this subsection that $i \neq j$. It can be easily checked that $W_{2,i}$, and $W_{2,i,j}$ are nonnegative on P and reduce to quadratic polynomials along its perimeter. In the following we obtained some nice properties of pentagonal blending functions. We first observe that the set G, of all the 15 blending functions is $\left\{W_{2,i}, W_{2,i,i+1}, W_{2,i,i+2}, i \in Z_5\right\}$. We shall consider the following subsets E and F of G, where $E = \left\{W_{2,i}, W_{2,i,i+1}, i \in Z_5\right\}$ consists of 10 blending functions, whereas $F = \left\{W_{2,i,i+2}, i \in Z_5\right\}$ consists of the remaining 5 blending functions. Then, we have the following [5].

Theorem 3. *The set G is linearly dependent. However, any subset of G, consisting of 10 blending functions of E together with any 4 blending functions of F, is linearly independent. Further, the blending functions in F, do not contribute to a non-zero value on the perimeter of the pentagon.*

We now proceed to examine interploatory properties of the Wachspress pentagonal patches, For any given real valued function f, $D_i(f)$ represents the directional derivative of f in the direction $a^{i+1} - a^i$ for $i \in Z_5$. Observing that the four blending functions of F do not contribute at the boundary of P, we need to consider only the ten blending functions of E. We therefore consider the following definition in terms of the coefficients: $c_i : i \in Z_5$ and $c_{i,i+1} : i \in Z_5$.

$$W = \sum_{i \in Z_5} c_i \, W_{2,i} + \sum_{i \in Z_5} c_{i,i+1} \, W_{2,i,i+1}. \tag{17}$$

It is directly seen that interpolation at the verticies, that is $W(a^i) = f(a^i)$, implies that $c_i = f(a^i)$. In [5], we have fully determined the function W, under the following interpolatory conditions.

$$W(a^i) = f(a^i), \; D_i W(a^i) = D_i f(a^i); \; i \in Z_5.$$

We will illustrate construction of C^1- smooth composite surface by using the foregoing conditions of interpolation. Given positive numbers h, λ such that $0 < h < 1$ and $1 < \lambda$, but $\lambda \neq 4$, then we consider the admissible convex pentagon P with vertices defined by $a^1 = (0,0), a^2 = (h,0), a^3 = ((2+\lambda)h/3, 2(\lambda-1)h/3), a^4 = (2(\lambda-1)h/3, (2+\lambda)h/3), a^5 = (0,h)$. The condition that $\lambda \neq 4$, is essential, for otherwise the vertices a^3 and a^4 coincide with common coordinates $(2h, 2h)$ and P is no more a pentagon. Let $\tilde{P}$ with vertices $\tilde{a}^i, i \in Z_5$ be the mirror image of P about the x-axis, so that the vertices $a^j = \tilde{a}^j, j = 1, 2$ and the edge joining a^1, a^2 or $\tilde{a}^1, \tilde{a}^2$ which lies on the x-axis, is the common edge between P and $\tilde{P}$. Analogues of $W_{2,i}$ and $W_{2,i,i+2}$ on $\tilde{P}$ are respectively denoted by $\tilde{W}_{2,i}$ and $\tilde{W}_{2,i,i+2}$. Thus the pentagonal patch $\tilde{W}$ on $\tilde{P}$ is given by the equation (17) with $W_{2,i}, W_{2,i,i+2}$ replaced by $\tilde{W}_{2,i}, \tilde{W}_{2,i,i+2}$ and $c_{i,j}$ replaced by the new control points $\tilde{c}_{i,j}$. For continuity condition, we match the restrictions of W and $\tilde{W}$, along the common edge, which are polynomials. This imposes the conditions:

$$c_0 = \tilde{c}_0; \; c_1 = \tilde{c}_1; \; c_{0,1} = \tilde{c}_{0,1}.$$

The condition for the continuity of the cross derivative on the common edge which is required for C^1−continuity for the composite surface on $P \cup \tilde{P}$ imposes the following set of additional conditions.

$$\left.\begin{array}{l} d_0(\lambda)(c_0 + \tilde{c}_0) + 20\, d_1(\lambda)(c_{4,0} + \tilde{c}_{4,0}) = 0, \\ d_0(\lambda)(c_1 + \tilde{c}_1) + 3\, d_2(\lambda)(c_{4,0} + \tilde{c}_{4,0}) = 0, \\ d_0(\lambda)(c_{0,1} + \tilde{c}_{0,1}) + 12\, d_3(\lambda)(c_{4,0} + \tilde{c}_{4,0}) = 0, \\ d_0(\lambda)(c_{1,2} + \tilde{c}_{1,2}) + 20\, d_4(\lambda)(c_{4,0} + \tilde{c}_{4,0}) = 0, \end{array}\right\} \tag{18}$$

where $d_j(\lambda), j = 1,2,3,4$, are polynomials in λ, with $\lambda > 1$ given by the following set of equations.

$$d_0(\lambda) = 2\lambda^3 - 9\lambda^2 + 18\lambda - 56;\; d_1(\lambda) = \lambda(\lambda - 1),$$
$$d_2(\lambda) = 2\lambda^2 - 11\lambda + 24,\; d_3(\lambda) = (\lambda - 1)(\lambda - 2),\; d_4(\lambda) = \lambda(\lambda - 1),$$

Since by definition $\lambda > 1$, we see that $d_1(\lambda), d_4(\lambda)$ are both positive. $d_3(\lambda)$ is zero at $\lambda = 2$, and is negative for $1 < \lambda < 2$ and positive for $\lambda > 2$. Further, $d_2(\lambda) > 0$, for all values of λ. We notice that the first derivative of $d_0(\lambda)$ is positive, for all values of λ, so that $d_0(\lambda)$ is increasing. We also observe that $d_0(\lambda)$ is zero for $\lambda = 4$, which is not an admissible value of λ. Thus, $d_0(\lambda) \neq 0$ for relevant values of λ.

We now set $g_j(\lambda) = d_j(\lambda)/d_0(\lambda), j = 1,2,3,4$. Thus, combining the conditions (18) with the continuity condition, we observe that

$$c_0 + 10\, g_1(\lambda)(c_{4,0} + \tilde{c}_{4,0}) = 0,$$
$$2c_1 + 3\, g_2(\lambda)(c_{4,0} + \tilde{c}_{4,0}) = 0,$$
$$c_{0,1} + 6\, g_3(\lambda)(c_{4,0} + \tilde{c}_{4,0}) = 0,$$
$$c_{1,2} + \tilde{c}_{1,2} + 20\, g_4(\lambda)(c_{4,0} + \tilde{c}_{4,0}) = 0.$$

As an example for a choice of control points ensuring C^1-smoothness we deduce the following from the above.

$$c_j = \tilde{c}_j = 0, j = 0, 1;\; c_{0,1} = \tilde{c}_{0,1} = 0;\; c_{1,2} = -\tilde{c}_{1,2};\; c_{4,0} = -\tilde{c}_{4,0}.$$

Very recently, we have studied pentagonal patches on convex pentagons which are not admissible in [14].

5. COMPLEX PLANAR SPLINES AND SHAPE PRESERVING PROPERTIES

The complex analogues of real splines were introduced by Ahlberg, Nilson and Walsh in [1] by considering smooth piecewise complex polynomial functions defined on a given partition of a rectifiable closed Jordan arc in subarcs. Such splines were extended to the interior of the region closed by the arc by Cauchy integral formula and were called analytic splines. Interpolatory, convergence and other approximating properties of these splines have been extensively studied and elegant representations of such functions in terms of logrithmic functions have been obtained. However, in the above process of extension of splines from its definition on the boundary to the interior of the region, the functions so constructed even lose the basic expectation of being a piecewise polynomial (pp)-function. Further, the construction is computationally more involved.

We would not be able to discuss further works on analytic splines here but would prefer to concentrate on a different but computationally convenient approach for defining complex splines. One possible natural approach for this purpose seems to be the use of complex polynomial pieces defined over different elements of a given partition of the complex domain. However, the difficulty encountered in employing such a definition is that even continuity requirement across the common edges of the partion will force the adjacent complex polynomial pieces, which are analytic, to be identical due to the well known Identity Theorem. We have to therefore choose nonanalytic functions for defining pp-functions over any given partition of the complex domain. We are thus lead to the reasonable choice of complex analogues of real polynomials as functions of z and its conjugate $\bar{z}$ which are nonanalytic and computationally convenient. Such complex splines were first introduced in [22] and are called *complex planar splines*. Complex planar splines defined by employing analogues of real polynomials and complex rational planar splines defined by employing analogues of real Wachspress rational functions will be discussed in the following subsections 5.1 and 5.2, respectively. Indeed both the complex planar splines and complex rational planar splines exhibit nice interpolating, convergence and shape preserving properties of *quasiregularity* and *quasiconformality*.

We now introduce some preliminaries for the next two subsections. Let $\bar{D}$ be any simply connected domain in the complex plane. It will be assumed that we always have a polygonally bounded compact subset $D \subset \bar{D}$ (usually close to $\bar{D}$). For a given positive integer $k \geq 3$, we say that a finite number of k−gons $\{\delta_j\}$ define a partition Δ of D, if $D = \underset{j}{\cup} \delta_j$ and for $i \neq j$, $\delta_i \cap \delta_j$ is either empty or consists of one common vertex or one common edge. Whenever we refer to any specific partition Δ of D, it will be further assumed that D accepts that partition.

We introduce some additional notations. Unless stated otherwise, throughout this section, h will denote the *diameter of any partition* Δ under consideration. While discussing, the convergence and *quasiregular* and *quasiconformal* properties of interpolants , we will always assume that the families of partitions Δ_h are regular, that is, the quotient of the radii of circumscribed and inscribed circles for any partition is uniformly bounded independent of h.

We need to give the following defintions [2, 11]. A non constant continuous function $f : R \to \mathbb{C}$ will be called *quasiregular*, if it satisfies the following two properties:

1. On each closed rectangle $\subset R$ with sides parallel to the coordinate axes, f is absolutely continuous on a.e. horizontal and vertical lines. In addition,

$$\int \int_D (|f_z|^2 + |f_{\bar{z}}|^2)\, dxdy < \infty.$$

2. f satisfies the *Beltrami equation* $f_{\bar{z}} = \mu_f\, f_z$ a.e. in R for some measurable function μ_f, with

$$\|\mu_f\| = \text{ess}\sup_R |\mu_f| < 1.$$

31

As a consequence of the foregoing definition $f_z(z) \neq 0$ for almost every $z \in R$. Thus the function $\mu_f = f_z/f_{\bar{z}}$, which is associated with f is called its *complex dilation*.

A *quasiconformal* mapping is a homeomorphic *quasiregular* mapping.

Throughout $\| \cdot \|$ will represent sup norm. For any complex function f defined on R and a given positive number ε, we write

$$\omega_\infty(f, D, \varepsilon) = \sup\{|f(z+t) - f(z)| : z, z+t \in D, |t| \leq \varepsilon\}.$$

5.1. Complex polynomial planar splines

The main objective of this section is to study the interpolatory, convergence and shape preserving properties of complax polynomial planar splines. A polynomial complex planar spline of degree $m > 0$ is a piecewise defined continuous function s such that its restriction on each element of the partition Δ is a polynomial in z and its conjugate $\bar{z}$ of the type:

$$\sum_{j,k=0}^{n} a_{j,k} z^j (\bar{z})^k, \ a_{j,k} \in \mathbb{C},$$

with degree $\leq m$. Here the degree of the polynomial is given by $max\{j+k : a_{j,k} \neq 0\}$.

In this subsection, we shall mainly consider those partitions Δ of D for which all δ_j's are either triangles or rectangles with sides parallel to the axes. Δ_1 and Δ_2 respectively denote partitions of D into triangles and rectangles with sides parallel to the axes. The refinement of Δ_2 obtained by adding one diagonal edge in every rectangle of Δ_2 will be denoted by Δ_3. Given a complex function f defined on R, we now recall the following problems of interpolation concerning complex planar splines on partitions $\Delta_j; j = 1, 2, 3$.

I. Continuous linear complex planar spline $s^1 f$ over the partition Δ_1 and with polynomial pieces of the type :

$$a_1 + a_2 z + a_3 \bar{z},$$

which interpolates f at all the vertices of Δ_1.

II. Continuous quadratic complex planar spline $\hat{s}^2 f$ over the partition Δ_2 and with polynomial pieces of the type:

$$a_1 + a_2 z + a_3 \bar{z} + a_4 (z^2 - \bar{z}^2), \tag{19}$$

which interpolates f at all the vertices of Δ_2.

III. Continuous quadratic complex planar spline $\tilde{s}^2 f$ over a triangulation Δ_3 and with polynomial pieces of the type (19) which interpolates f at all the vertices of Δ_3 as also at the mid points of edges not parallel to the axes.

IV. Continuous total degree 2 complex planar spline $s^2 f$ over the triangulation Δ_3 and with polynomial pieces of the type:

$$a_1 + a_2 z + a_3 \bar{z} + a_4 z^2 + a_5 z \bar{z} + a_6 \bar{z}^2, \tag{20}$$

which interpolates f at all the vertices and mid points of all the edges of the partition Δ_3.

The interpolation problems I and II were respectively studied in [22, 23] whereas the problems III and IV were studied by us in [11]. Comparing the problems II and III, it may be observed that an additional degree of freedom in construction of complex planar spline is achieved for every rectangle in the partition Δ_2. This amounts to substantial gain in the total additional degrees of freedom which equals the total number of elements in the partition Δ_2 or equivalently half of the total number of elements in the partition Δ_3. Existence, uniqueness and convergence properties of interpolants in problems I–IV as also their explicit representations have been obtained.

Introducing the error functions:

$$\hat{e}(\hat{s}^2)f = \hat{s}^2 f - f, \quad \tilde{e}(\tilde{s}^2 f) = \tilde{s}^2 f - f, \text{and } e(s^2 f) = s^2 f - f,$$

we observe that the following convergence properties for $\hat{s}f$, $\tilde{s}f$ and $s^2 f$ have been obtained in [11, 22, 23].

Theorem 4. *If f is continuous, then*

$$\| e(s^2 f) \| \leq 9\,\omega_\infty\,(f, D, h); \quad \| \hat{e}(\hat{s}^2)f \| \leq c_1 \omega_\infty\,(f, D, h),$$

where h is the diameter of the respective partitions and c_1 is an appropriate positive constant.

If the second partial derivatives of f with respect to z and $\bar{z}$ exist and are Lipschitz continuous, then we say that f is a *smooth function*. For smooth functions we have the following [11, 23].

Theorem 5. *If f is a smooth function, then we have*

$$\| \tilde{e}(\tilde{s}^2 f) \|, \| \hat{e}(\hat{s}^2 f) \| \leq c_2 h^2, \quad \| \tilde{e}(\tilde{s}^2 f)_z \|, \| \hat{e}(\hat{s}^2 f)_z \| \leq c_3 h,$$

$$\| \tilde{e}(\tilde{s}^2 f)_{\bar{z}} \|, \| \hat{e}(\hat{s}^2 f)_{\bar{z}} \| \leq c_4 h,$$

where h is the diameter of the respective partitions and c_2, c_3, c_4 are appropriate positive constants.

The complex planar spline interpolant $s^2 f$ of a given smooth function f exhibits the following improved convergence properties [11].

Theorem 6. *If f is a smooth function then*

$$\| e(s^2 f) \| \leq c_5 h^3, \quad \| e(s^2 f)_z \|, \| e(s^2 f)_{\bar{z}} \| \leq c_6 h^2,$$

where h is the diameter of the respective partitions and c_5, c_6 are appropriate positive constants.

Complex planar spline interpolants of a general degree on partitions consisting of k-gons have been studied in [26].

Quasiregular and quasiconformal properties of $\hat{s}^2 f$ were studied in [23]. In order to study these properties for $s^2 f$ where more general polynomials of the type (20) were used in place of (19), one requires to first obtain precise conditions for various components of the definition of quasiconformality of such a polynomial p defined over a compact set D. Thus in [11], a set of sufficient conditions were obtained such that p is an orientation preserving homeomorphism which is quasiconformal on the compact set D, if and only if, $D \subset H_p^+$ (i.e. the orientation preserving region of the polynomial function p). As observed in ([11], p.354), the conditions used in the foregoing result are in a way best possible. Thus following is proved in [11].

Theorem 7. *If f is a smooth function such that $f_z(z) \neq 0$ for $z \in D$, then for $z \in D, s^2 f(z) \neq 0$ and the complex dilations of f and $s^2 f$ satisfy the following inequality*

$$\mid \|\mu_f\| - \|\mu_{s^2 f}\| \mid \leq c_7 h^2,$$

for sufficiently small diameter h of the partitition.

Finally, in [11], we prove the following shape preserving property of the interpolant $s^2 f$.

Theorem 8. *If f is a smooth quasiconformal mapping such that $f_z(z) \neq 0$ for $z \in D$, then $s^2 f$ on D with partition Δ_3 is quasiconformal, open, and the maximum of $|s^2 f(z)|$ is taken on the boundary of D, provided the diameter h of the partition is sufficiently small.*

A similar result for concluding quasiregularity of $s^2 f$ from the quasiregularity of f is also proved in [11].

5.2. Complex Wachspress rational polynomial planar splines

Interesting properties of the real Wachspress rational finite elements motivated us to study the corresponding rational complex planar splines defined in [15]. Let the compact subset D of a simply connected domain R (usually close to R), accepts a partition into squares with sides of length $2h$ and parallel to the axes. Then consider a finite partition $\Delta(t,h) = \Delta = \left\{ \delta_j \right\}_{j=1}^{N}$, of D which consists of N quadrilaterals with the following properties (see Fig. 4).

(i) One vertex of δ_j is at the centre of a square S, with sides of length $2h$.

(ii)δ_j has two equal edges each of length $t \neq h$ along two adjacent edges of S, with $0 < t < 2h$.

The partition of squares adjacent to S and sharing one side with it is obtained by rotating Fig. 4 through $\pi/2$. One may note that $AB = AH = ED = EF = t$ and $CB = CD = GF = GH = 2h - t$. We shall say that the grid element δ_j is of type I or II according as to whether t is less or greater than h. Let the vertices A, C, G and E of the square S be given by $\theta, \theta + 2h, \theta + 2ih$ and $\theta + 2(1+i)h$. Setting $\alpha = t/h$, we observe that the affine function $g(z) = hz + \theta$ maps the quadrilateral $\delta_0 = \delta_0(\theta)$ with vertices at $0, \alpha, i\alpha, 1 + i$ onto the unique quadrilateral in the partition of S with a vertex at θ. To map δ_0 onto the quadrilateral with a vertex in $\theta + 2(1+i)h$, it suffices to translate $g(\delta_0)$ by $(1+i)h$ and then rotate by an angle of π. Similarly, setting $\alpha = 2 - t/h$, we obtain the other two quadrilaterals in the partition of S by suitable translations and rotations of $g(\delta_0)$.

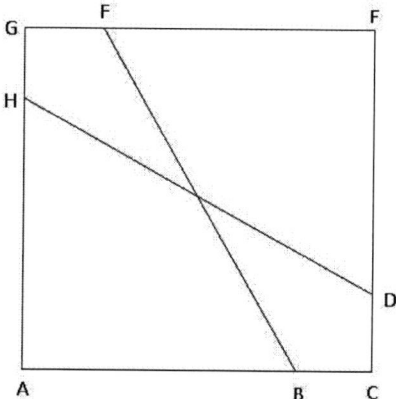

FIGURE 5. Partition of a square in to admissible quadrilaterals.

Thus, for constructing piecewise rational functions, we fix rational function on δ_0 and then recover the same for any δ_j through affine functions. The restriction of a piecewise rational function s defined on the partition Δ over δ_j will be denoted by s_j.

In [15], we have considered the following complex analogue of the Wachspress rational function on the quadrilateral δ_0

$$r_0(z) = (z^2 - \bar{z}^2)/q(z); \quad q(z) = (1-i)z + (1+i)\bar{z} + 2\beta,$$

where $\beta = \alpha/(1-\alpha)$. In this subsection, we will consider continuous complex rational planar spline functions s^1 and s^2 defined over the partition Δ such that their restrictions s_j^1, s_j^2 on δ_j, for $j = 1, 2, \cdots, N$, are certain affine transformations (as already described) of the following

$$s_0^1(z) = a + bz + c\bar{z} + dr_0(z);$$

$$s_0^2(z) = a + bz + c\bar{z} + (lz + m\bar{z} + n)r_0(z).$$

Given a function f. A rational complex planar spline s^1 interpolating f, at the vertices of all δ_js will be denoted by s_f^1. A rational complex planar spline s^2 interpolating f, at the vertices and mid points of sides not parallel to axes of all δ_js, will be denoted by s_f^2. These interpolants have ben determined in [15, 16]. [1]. Assuming Δ to be a regular partiotion, following convergence propertits have been established by us in the foregoing works.

$$\left\| s_f^1 - f \right\| \leq K_1 \omega(f, D, \delta); \quad \left\| s_f^2 - f \right\| \leq K_2 \omega(f, D, \delta),$$

[1] The work in [16] was carried out with the aid of MACSYMA, a large symbolic manipolation program developed at the MIT Laboratory for Computer Science and supported from 1975 to 1983 by the National Aeronautics and Space Administration under grant NSG 1323 by the Office of the Naval Research under grant N00014-77C-0641, by the U.S. Department of Energy under grant ET-78-C-02-4687, and by the U.S. Air Force under grant F49620-79-C-020, and since 1982 by Symbolics, Inc. of Burlington, MA.

35

where δ is the diameter of the partition and K_1, K_2 are positive constants. Further, if f is a smooth function as defined earlier, then the following have been proved in [15, 16].

$$\left\|s_f^1 - f\right\| \le K_3 \delta^2; \quad \left\|(s_f^1)_z - f_z\right\| \le K_4 \delta; \left\|(s_f^1)_{\bar{z}} - f_{\bar{z}}\right\| \le K_5 \delta;$$

$$\left\|s_f^2 - f\right\| \le K_6 \delta^2; \quad \left\|(s_f^2)_z - f_z\right\| \le K_7 \delta; \left\|(s_f^2)_{\bar{z}} - f_{\bar{z}}\right\| \le K_8 \delta;$$

where $K_3, K_4, K_5, K_6, K_7, K_8$ are positive constants.

In order to compare the regions for convergence in the cases: s_f^1 and s_f^2, for a given positive number ε, it follows from the properties discussed above that it would be enough to consider the restrictions of these rational complex planar splines on δ_0. For convenience, we denote these by s_f^j, $j = 1.2$. Let D_0^j denote the regions in which

$$\left\|s_f^j - f\right\| \le \varepsilon, \text{ for } j = 1, 2.$$ Then we show in [16] that D_0^2 is larger than D_0^1, in the cases when $f(z) = z^2$; $\sin z$; $\log(1+z)$; or $\exp z$. Results similar to Theorems 7-8 continue to hold for the rational complex planar splines. In particular the following shape preserving property holds.

Theorem 9. *If f is a smooth function such that $f_z \ne 0$, then s^j is quasiregular for $j = 1, 2$, if the diameter of the partition is sufficiently small.*

In fact a stronger version of the foregoing theorem can be obtained to ensure quasi-conformality by using a result in ([23], Lemma 3.2).

6. AN APPLICATION OF SPLINES FOR LEBESGUE-TYPE CONSTANTS

Consider a 2π periodic function $f \in L[0, 2\pi]$. E. Hille and J.D.Tamarkin [19] introduced the definition of *Fourier-effectiveness* of Nörlund Transformation (N, p) defined below. *Fourier-effectiveness* of (N, p) means that the (N, p) transformations of a class of series consisting of the Fourier series and other associated series were convergent under certain conditions involving f. They also gave a beautiful result for bounds of kernels associated with transformed Fourier series which in particular reduced to the Dirichlet kernel and thus provided bounds for Lebesgue-type constants. This sharp result was applied to obtain a set of necessary and sufficient condition for the *Fourier-effectiveness* of (N, p) transformation. Given a sequence of numbers $\{p_n\}$. The (N, p) transformed sequence $\{t_n\}$ of the series $\sum_j a_j$ or sequence of its partial sums $\{s_n\}$ is defined by

$$t_n = \sum_{j=0}^{n} p_{n-j} s_j / P_n, n = 0, 1, 2, ...; \text{ with } P_n = \sum_{j=0}^{n} p_j \ne 0; p_{n-j} = 0, \text{ for } j > n.$$

Let $\phi_0(t; x) = \phi_0(t) = f(x+t) - f(x-t); \quad \phi_1(t) = t^{-1} \int_0^t \phi_0(u) du.$ We also set

$$R_n = n p_n / P_n; \quad P_n S_n^r = \sum_1^n P_k / k^r, r = 1, 2; \quad P_n V_n^r = n^{r-1} \sum_{k=1}^{n} k |\Delta^r p_{k-r}|; r = 0, 1.$$

Fourier-effectiveness of (N,p) transformation included in particular convergence of (N,p) transform of the Fourier series of f at the point x under the assumption that $\phi_0(t;x) = o(1), t \to 0$. *Total-effectiveness* of (N,p) transformation which included in particular convergence of (N,p) transform of the Fourier series of f at the point x under the assumption that $\phi_1(t) = o(1), t \to 0$ has been studied in [4, 8, 9]. These results provide only sufficient conditions for *Total-effectiveness*. For corresponding necessary conditions we give below sharp bounds for Lebesgue-type constants associated with certain (N,p) transformation.

Denoting the (N,p) transformation of the series $\frac{1}{2} + \sum_k \cos kt$ by $N_n^p(t;1)$, Hille and Tamarkin have proved the following in [19].

Theorem 10. *Let (N,p) be a regular transformation such that $p_n > 0, \{R_n\} \in B$, and $\{V_n^1\} \in B$ then*

$$K_0 S_n^1 < \int_0^\pi |N_n^p(t;1)| dt < K_1 S_n^1; \tag{21}$$

where K_0 and K_1 are some positive constants.

As observed by us in [10], the conditions $\{R_n\} \in B$ and (N,p) is regular, used by Hille and Tamarkin in [19] for the proof of Theorem 10 follow from $\{V_n^1\} \in B$, since

$$np_n = -\sum_0^{n-1} \Delta(kp_k) = P_{n-1} - \sum_{k=1}^n k(\Delta p_{k-1}).$$

Denoting the (N,p) transformation of $\{k \cos kt\}$ by $N_n^p(t;2)$ we proved the following in [10].

Theorem 11. *Let $p_n > 0$, and $\{V_n^2\} \in B$ then*

$$K_0 S_n^2 < \int_0^\pi |N_n^p(t;2)| dt < K_1 S_n^2; \tag{22}$$

where K_0 and K_1 are some positive constants.

We call the integrals appearing in (21)-(22), as Lebesgue type constants associated with the (N,p) transformation, since in the special case when $p_0 = 1$, and all other p_j's being zero the integral in (21) reduces to the Lebesgue constant. Bounds for Lebesgue constants have been extensively investigated and we refer to a recent paper [28] in this direction.

Using the sharp bounds given in Theorem 11, we are able to prove the necessity part of the following, sufficiency part of which was proved by M. Astrachan in [4].

Theorem 12. *The (N,p) transform of Fourier series of f at the point x, is convergent under the condition $\phi_1(t) = o(1)$ as $t \to 0$, if and only if, $\{S_n^2\} \in B$.*

We briefly sketch the interesting analysis evolved in the proof given in [10]. For positive integers r, we consider disjoint intervals $I_r = ((2r+1/3)\pi/n, (2r+4/9)\pi/n) \subset (2/n,1)$, so that $t \in I_r$ implies that $\cos(4\pi/9) \leq \cos(n-k)t$, since $(n-k)t \in ((2r+1/3)\pi - 1, (2r+4/9)\pi)$. Writing $\tau = \pi/9n$, we observe that each interval I_r is of length

37

τ and any two consecutive intervals I_r, I_{r+1} are separated by a distance of 17τ. For determining the lower bound in Theorem 11, we move the intervals I_r to the left by applying the translation: $t - 17(r-1)\tau$ so that all the intervals I_r abut upon each other. We then integrate over the continuous interval obtained this way to obtain the lower bound of a certain expression which appears in the proof of Theorem 11. The final step in the proof of Theorem 11 involves an increasing sequence of positive integers $\{n(r)\}_{r=1}^{\infty}$ and a decreasing sequence of numbers $\{x_0(r)\}_{r=1}^{\infty}$. As an example of a crucial function meeting preassigned conditions which is required for the proof, we propose construction of a deficient quintic cardinal spline interpolant with the sequence of knots as $\{x_0(r)\}_{r=1}^{\infty}$.

REFERENCES

1. J. H. Ahlberg, E. N. Nilson, and J. L. Walsh, *Trans. Amer. Math. Soc.* **129**, 391–413 (1967).
2. L. V.Ahlfors, *Lectures on Quasiconformal Mappings*, Mathematical Studies **10**, Van Nostrand, 1966.
3. D. Apprato, R. Arcangeli, and J. L.Gout, *Numer. Math.* **32**, 247-270 (1979).
4. M. Astrachan, *Duke Math. J.* **2**, 543-568 (1936).
5. N. Choubey, H. P. Dikshit, and A. Ojha, *PAMM–Proc. Appl. Math. Mech.* **7**, 2020099–2020100 (2007).
6. W. Dahmen, H. P. Dikshit, and A. Ojha, *Computer Aided Geometric Design* **17**, 879–890 (2000).
7. G. Dasgupta, and A. E. Malsch, *Engineering Anal. with Boundary Element* **26**, 379-389 (2002).
8. H. P. Dikshit, *Math. Proc. Camb. Phil. Soc.* **65**, 495–505 (1969).
9. H. P. Dikshit, *Math. Annalen* **186**, 101–113 (1970).
10. H. P. Dikshit,and A. Kumar, *Pacific J. Math.* **97**, 339–347 (1981).
11. H. P. Dikshit,and A. Ojha, *Math. Proc. Camb. Phil. Soc.* **99**, 347–356 (1986).
12. H. P. Dikshit, and A. Ojha, *Computer Aided Geometric Design* **19**, 207–222 (2002).
13. H. P. Dikshit, and A. Ojha, *Computer Aided Geometric Design* **20**, 207–222 (2003).
14. H. P. Dikshit, and A. Ojha, *A note on Wachspress patches* , Preprint (2009).
15. H. P. Dikshit, A. Ojha, and A. Sharma, *Math. Proc. Camb. Phil. Soc.* **101**, 141–149 (1987).
16. H. P. Dikshit, A. Ojha, and R. A.Zalik, *Advances in Computational Math.* **2**, 235–249 (1994).
17. G. Farin, *Curves and Surfaces for Computer Aided Geometric Design: A Practical Guide*, Academic Press, 1993.
18. J. L.Gout, *Comp. Math. & Appl.* **5**, 337-347 (1979).
19. E. Hille, and J. D.Tamarkin, *Trans. Amer. Math. Soc.* **34**, 757-783 (1932).
20. K. Hormann, and M. S. Floater, *ACM Transactions on Graphics* **25**, 1424–1441 (2006).
21. E. A. Malsch, and G. Dasgupta, *Int. J. of Solids and Structures* **8**, 2165–2188 (2004).
22. G. Opfer, and M. L. Puri, *J. Approx. Theory* **31**, 235–249 (1994).
23. G. Opfer, and G. Schober, *Math.Z.* **180**, 383–402 (1981).
24. N. Sukumar, and E. A. Malsch, *Archives of Computer Methods in Engineering* **13**, 129–163 (2006).
25. E. L. Wachspress, *A Rational Finite Element Basis*, Academic Press, Bostan, 1975.
26. G. Walz, *Spline Funktionen im Komplexen*, Vorbereitung, 1993.
27. J. Warren, S. Schaefer, A. N. Hirani, and M, Desbrun, *Advances in Computational Mathematics* **27**, 319-338 (2007).
28. D. Zhao, *J. Math. Anal. Appl.*, appeared in (2008).

SECTION B:
MODELLING OF REAL WORLD SYSTEMS

Protein designs in HP models

Arvind Gupta*, Alireza Hadj Khodabakhshi*, Ján Maňuch*, Arash Rafiey* and
Ladislav Stacho†

*School of Computing Science, 8888 University Drive, Simon Fraser University, Burnaby BC, V5A 1S6, Canada
†Department of Mathematics, 8888 University Drive, Simon Fraser University, Burnaby BC, V5A 1S6, Canada

Abstract. The inverse protein folding problem is that of designing an amino acid sequence which folds into a prescribed shape. This problem arises in drug design where a particular structure is necessary to ensure proper protein-protein interactions and could have applications in nanotechnology. A major challenge in designing proteins with native folds that attain a specific shape is to avoid proteins that have multiple native folds (unstable proteins). In this technical note we present our results on protein designs in the variant of Hydrophobic-Polar (HP) model introduced by Dill [6] on 2D square lattice. The HP model distinguishes only polar and hydrophobic monomers and only counts the number of hydrophobic contacts in the energy function. To achieve better stability of our designs we use the Hydrophobic-Polar-Cysteine (HPC) model which distinguishes the third type of monomers called "cysteines" and incorporates also the disulfid bridges (SS-bridges) into the energy function. We present stable designs in 2D square lattice and 3D hexagonal prism lattice in the HPC model.

Keywords: inverse protein folding,lattice models,HP model

INTRODUCTION

It has long been known that protein interactions depend on their native three-dimensional fold and understanding the processes and determining these folds is a long standing problem in molecular biology. Naturally occurring proteins fold so as to minimize total free energy.

In many applications such as drug design and nanotechnology, we are interested in the complement problem to protein folding: *inverse protein folding* (IPF) or *protein design*. The inverse protein folding involves starting with a prescribed target fold or shape and designing an amino acid sequence whose native fold is the target (positive design). A major challenge in designing proteins that attain a specific native fold or shape is to avoid proteins that have multiple native folds (negative design). We say that a protein is *stable* if its native fold is unique.

Many forces act on the protein which contribute to changes in free energy including hydrogen bonding, van der Waals interactions, intrinsic propensities, ion pairing, disulfide bridges and hydrophobic interactions. Of these, the most significant is hydrophobic interaction [7]. This led Dill to introduce the *Hydrophobic-Polar (HP) model* [6]. Here the 20 amino acids from which proteins are formed are replaced by two types of monomers: hydrophobic (H or '1') or polar (P or '0') depending on their affinity to water. To simplify the problem, the protein is laid out on vertices of a lattice with each monomer occupying exactly one vertex and neighboring monomers occupy neighboring vertices. The free energy is minimized when the maximum number of hydrophobic monomers are adjacent in the lattice. Therefore, the "native" folds are those with the maximum number of such HH contacts.

Even though the HP model is the simplest model of the protein folding process, protein folding is NP-hard for both the 2D square [4] and the 3D cubic [2] lattices. The hardness of the inverse protein folding under the standard definition of the HP model is still unknown but it is conjectured to be an NP-hard problem. Several heuristic based algorithms have been described that attempt to solve IPF problem but neither of them guarantees that the designed sequences achieve their minimum energy when folded into the target conformations (positive design criteria) and that they have unique minimum energy conformations (negative design criteria). These heuristic methods can be classified into two categories. The methods in the first category use observations about the properties of proteins to justify algorithms that design sequences [16, 30]. The second category of heuristic methods are those in which an alternative formulation of IPF, a heuristic sequence design (HSD), is considered [5, 20, 27, 28, 12, 3].

Two HSD problems, in the *canonical* and the *grand canonical* models, were introduced in [27] and [28], respectively. The HSD problems look for protein sequence with the smallest energy when folded into the target conformation. In the canonical model the number of hydrophobic monomers that can be used in a protein sequence is limited by fixing the maximum ratio between hydrophobic and hydrophilic amino acids. This condition is needed because the confor-

CP1146, *Modelling of Engineering and Technological Problems*, edited by A. H. Siddiqi, A. K. Gupta, and M. Brokate
© 2009 American Institute of Physics 978-0-7354-0683-4/09/$25.00

mational energy can be minimized simply by using the sequence of all hydrophobic monomers, but this sequence is unlikely to achieve its lowest energy with the given target conformation. In the grand canonical model [28], the number of hydrophobic monomers is limited by adjusting the energy function instead. For instance, a simplified formulation of the grand canonical model used in [12] assumes that every hydrophobic contact contributes -2 to the total energy, every solvent accessible site of a hydrophobic amino acid contributes 1, and all other interactions do not contribute to the total energy. Since the hydrophobic monomers are penalized for their exposure to solvent, this contact potential implicitly limits the number of hydrophobic monomers in the sequence. It has been shown that the protein sequence design problem can be solved in polynomial time in the grand canonical model for both 2D and 3D square lattices, cf. [12], and in polynomial time for 2D lattices while the problem is NP-hard for 3D square lattice in the canonical model, cf. [3]. Note however, that the designed heuristic sequences under these two models are not guaranteed to satisfy the two criteria (positive and negative design) of the IPF problem.

Design of stable proteins of arbitrary lengths in the HP model was studied by Aichholzer *et al.* [1] (for 2D square lattice) and by Li *et al.* [21] (for 2D triangular lattice), motivated by a popular paper of Brian Hayes [13].

In natural proteins, sulfide bridges between two cysteine monomers play an important role in improving stability of the protein structure [14]. In [18], we refined the HP model by including a third type of monomers called cysteine (C or '2') in the protein sequences and incorporating the sulfur bridges (SS-bridges) between adjacent, but not consecutive pairs of cysteine monomers in the energy model. Every cysteine can be part of at most one SS-bridge. The total energy is calculated as c_1 times the number of contacts plus c_2 times the number of SS-bridges, where c_1, c_2 are some negative constants (our results are independent on values of these constants). This new *HPC model* increases the stability of designs by limiting the number of possibilities how the contacts can be formed.

In this technical paper we briefly review our recent results on designs in the HP and HPC models. In our work we study the IPF problem from a different perspective. Instead of designing a sequence directly for the target fold and relaxing conditions on the sequence we introduce a design method in 2D and 3D lattices under the HPC model that can approximate any target conformation. We call this problem *structure approximating inverse protein folding*. We show that our approximated structures, are native for designed proteins (positive design). The main challenge however, is to prove that the approximated structures are stable (negative design). We solve this for a robust subclass of our designing structures called *wave* structures in 2D square lattice under the HPC model. We also formally prove that an infinite subclass of our designing structures in 3D hexagonal lattice is stable under the HPC model. In the following sections we review these results in more details.

STRUCTURE APPROXIMATING IPF IN THE HP AND HPC MODELS ON 2D SQUARE LATTICE

In this section we review our designs in the original HP model of Dill [6] on the 2D square lattice. We first define the HP model formally.

Hydrophobic-Polar (HP) model

A critical step in the folding pathway of globular proteins is the formation of a tightly packed hydrophobic core. This core is formed due to hydrophobic interactions that draw the hydrophobic (water repelling) amino acids together and drive the hydrophilic (water attracting) amino acids to the surface of the protein. It is believed that hydrophobic interactions are the major driving forces in determining the native tertiary conformation of proteins. Based on this, Dill [6] introduced the simplified model of hydrophobic-polar (HP) for protein prediction and design problems. Two major simplifying assumptions are made in the HP model: first the 20 amino acids from which proteins are made are replaced by two types of monomers: hydrophobic (H, "1") and polar (P, "0"). Second the protein sequence is laid out on a spatial lattice with each monomer occupying exactly one vertex of the lattice and neighboring monomers occupy neighboring vertices such that the chain of monomers makes a self avoiding walk in the lattice.

The energy of a fold in HP model is defined as -1 times the number of pairs of H monomers that are adjacent in the lattice but are not consecutive in the sequence. Such pairs are called *HH contacts* or simply *contacts*. Figure 1 shows an example of a fold of a protein sequence in 2D square lattice and illustrates how the energy is of the fold is determined. Therefore, a fold with the maximum number of HH contacts is the native fold for a protein sequence.

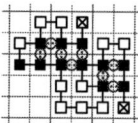

FIGURE 1. An example of a fold in 2D HP model. Hydrophobic (polar) monomers are depicted as black (white) squares. The ends of the protein sequence are marked with the crosses. The hydrophobic contacts are depicted with the circles. The energy of this fold is -8, which is the best possible for the protein sequence 01100110100001110100110.

There might be several native fold of a protein and in fact a random protein sequence will most likely have multiple native folds. A protein with a unique native fold is called *stable*.

Note that the HP model can be defined under various types of underlying 2D or 3D lattices such as square, triangular, cubic, hexagonal prism, FCC lattices and many more.

Constructible structures

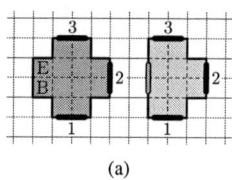   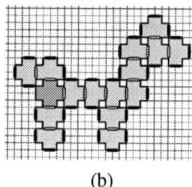

(a)   (b)

FIGURE 2. (a) The starting tile (left) and the regular tile (right) for constructible structures (b) An example of constructible structure: a tree build from basic tiles.

In [10], we introduced a wide class of structures on 2D square lattice called the *constructible structures*. These structures can be used to approximate any given shape. We showed that each constructible structure is a native fold for its protein sequence (positive design). The constructible structures are built from tiles. We use two types of tiles: a starting tile in the shape of "+" and a regular tile in the shape of "⊤", both depicted in Figure 2(a). Both tiles have three *ligands* depicted with black lines, two of which are side ligands marked with "*S*" and one forward ligand marked with "*F*". In addition, the regular tile has one *receptor*, depicted with a gray line.

A constructible structure is a partial tiling of the 2D square lattice *L* obtained by the following procedure:

1. Place the starting tile into *L*.
2. Place a regular tile into the square lattice so that its receptor is attached to a ligand of a tile already in the square lattice and it does not overlap with any other tile.
3. Continue with step 2., or end the procedure.

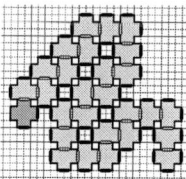   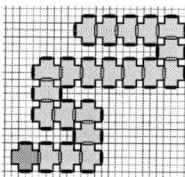

FIGURE 3. Two example of linear constructible structures.

43

An example of a constructible structure is shown in Figure 2(b). A constructible structure is called *linear* if it is constructed such that every regular tile is attached to the ligand of the last placed tile. Let $\mathscr{L}$ be the class of all linear constructible structures. Figure 3 depicts two examples of linear constructible structures. In a linear constructible structure the i-th tile is called *bending* if the $(i+1)$-th tile is attached to one of its side ligands (numbered as 1 or 3 in Figure 2(a)). We can classify the linear constructible structures $\mathscr{L}$ by the number of bending tiles they contain such that $\mathscr{L}_i$ represents the class of linear structures with exactly i bending tiles.

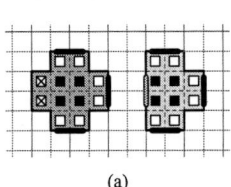

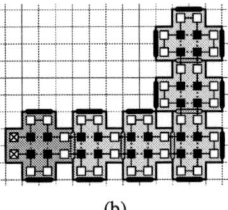

(a)   (b)

FIGURE 4. Constructing the protein sequence: (a) The pattern used to fill every tile. (b) Connecting the monomers into one protein sequence: each H monomer connects to neighboring P monomers and P monomers at unused ligands connect to each other.

Each constructible structure is assigned a unique protein sequence which defined as follow. The central part of the starting tile or each regular tile is filled with 4 hydrophobic monomers and remaining 8 (6) monomers are set to be polar, cf. Figure 4(a). Next, we need to connect monomers with peptide bonds. First, a neighboring hydrophobic and polar monomers are connected by a peptide bond, then for each unconnected ligand, the pair of two polar monomers adjacent to it are connected. An example of such sequence is shown at Figure 4(b). As we will see in the moment, the resulting protein sequence has a native fold that exactly fills the corresponding constructible structure.

The fold of constructible structures have a special property that implies that the fold is native and also helps in proving the stability of its protein (uniqueness of the fold):

- Every hydrophobic monomer in the fold of a constructible structure has two HH contacts.

Note that this is the maximum number of contacts a (non-terminal) hydrophobic monomer can make in any fold of the protein in 2D square lattice as it has only 4 neighbors and two of them are connected with peptide bonds. Therefore, the constructible structures has the minimum possible energy with respect to the number of hydrophobic monomers. Such folds are called *saturated*. Clearly, the saturated fold are native (positive design). In addition, any native fold must be also saturated, which helps in proving the stability of its protein sequence.

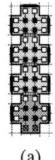

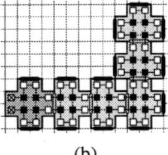

(a)   (b)

FIGURE 5. (a) An example of "I"-shaped constructible structure from class $\mathscr{L}_0$. (b) An example of "L"-shaped constructible structure from class $\mathscr{L}_1$.

We conjecture that the proteins of the constructible structures are stable. To support this conjecture in [10] we proved that the proteins of two simple but infinite subclasses of linear structures, namely $\mathscr{L}_0$ and $\mathscr{L}_1$ depicted in Figure 5, are stable. Furthermore, the stability of over 50,000 structures (including all structures with up to 9 tiles) was computationally tested as well. The stability proof for the $\mathscr{L}_0$- and $\mathscr{L}_1$-structures is based on the induction scanning diagonally through all hydrophobic monomers starting with the most outermost ones and showing that each such monomer is a part of HH-cycle created by 4 hydrophobic monomers surrounded by polar monomers (we refer to such configuration as a *core*).

The $\mathscr{L}_0$- and $\mathscr{L}_1$-structures are too simple to be able to approximate any given shape. At the same time the conjecture showing that all constructible structures are stable appears to be very hard to prove. Therefore, we have searched for a subclass of constructible structures which is rich enough to approximate (although more coarsely) any given shape and at the same time simple (repetitive) enough to restrict many cases in the case analysis required for the induction proof. We note that the proof of stability of $\mathcal{L}_0$-structures ($\mathcal{L}_1$-structures) required about 2 (6) pages. To overcome the problem with the analysis of huge number of cases, we use the following strategy:

- Consider a subclass of constructible structures with a large set of forbidden subsequences in their proteins (for instance, for constructible structures, subsequences 000, 11, 1010101, etc. are not occurring).
- Use the HPC model incorporating additional effective forces in the folding process: SS-bridges.
- Develop a program (2DHPSolver) for semi-automatic analysis of the cases arising in the proof process.

We formally define the HPC model in the next subsection.

Hydrophobic-polar-cysteine (HPC) model

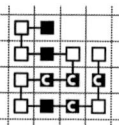

FIGURE 6. An example of energy calculation of a fold in the HPC model. There are 5 contacts between hydrophobic monomers, thus the contact energy is -5. There are three potential sulfide bridges sharing a common vertex, hence only one can be formed. Thus the sulfide bridge energy is -1 and the total energy is $-5 - 1 = -6$. The energy of this fold under the strong HPC model is -7 as it has one hypothetical non-cysteine bridge.

Cysteine is a hydrophobic amino acid [25] which contains a *thiol* group that can bind with the thiol group of another cysteine and form a *disulfide* bond or bridge. Disulfide bridges are another significant forces in the folding process of the proteins which play an important role in the stability of the protein structure [14, 15]. We extended the HP model by adding a third type of monomers, cysteines (C or '2'), to the designed protein sequences and incorporating disulfide bridges between two cysteines into the energy model. We call this model the hydrophobic-polar-cysteine (HPC) model. We represent a protein chain in HPC model as a sequence $p = p_1 p_2 \ldots p_{|p|}$ in $\{0, 1, 2\}^*$, where "0" represents a polar monomer, "1" a hydrophobic non-cysteine monomer and "2" a cysteine monomer. In the HP model only the HH contacts are considered in the energy model, with each contact contributing with $E_{HH} < 0$ to the total energy. In the HPC model cysteines act as hydrophobic monomers and can form the HH contacts, i.e., $1 \cdot 1$, $1 \cdot 2$ and $2 \cdot 2$ are all valid contacts. In addition, two adjacent non-consecutive cysteines can form a disulfide bridge contributing with another -1 to the total energy. However, each cysteine can be involved in at most one bridge. More formally, any two adjacent non-consecutive cysteines form a *potential* disulfide bridge and the disulfide-bridge energy is equal to $E_{SS} < 0$ times the number of pairs in the maximum matching in the graph of potential disulfide bridges. The total energy of the fold is calculated as E_{HH} times the number of contacts plus E_{SS} times the number of bridges. For example, the energy of the fold in Figure 6 is $5E_{HH} + E_{SS}$. All our results are independent on the values of E_{HH} and E_{SS} (as long as they are both negative), and therefore, for simplicity, we can assume that $E_{HH} = E_{SS} = -1$, and the energy of the fold in Figure 6 is -6.

Snake structures

In [18] we introduced a rich subclass of linear constructible structures called the *snake* structures. The snake structures are linear constructible structures in which every odd-numbered tile is a bending tile. In addition, since we consider the HPC model, the hydrophobic monomers of the bending tiles and the terminal tiles are set to be cysteines, and all other hydrophobic monomers are non-cysteines, cf. Figure 7.

Note that the snake structures can still approximate any given shape, although more coarsely than the linear structures. The idea of approximating a given shape with a linear structure is to draw a non-intersecting curve consisting of horizontal and vertical line segments. Each line segment is a linear chain of basic tiles depicted in Figure 2(a). At first glance, the snake structures seem more restricted than linear structures, as the line segments they

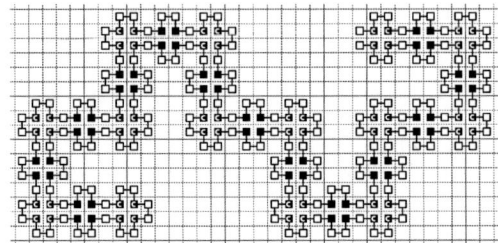

FIGURE 7. An example of a snake structure. Cysteines are marked with C.

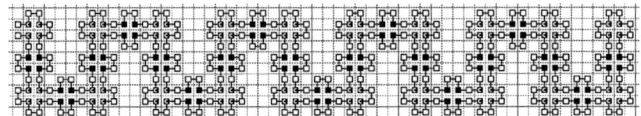

FIGURE 8. Simulation of a straight line segment with a snake structure.

use are very short and have the same size (3 tiles long). However, one can simulate arbitrary long line segments with snake structures forming a zig-zag pattern, cf. Figure 8.

The strong HPC model

We conjecture that the proteins of snake structures are stable in the HPC model, and furthermore, that it can be proved with techniques that we will present in the following sections. As an evidence to the correctness of this conjecture, we present the proof that the snake proteins are stable in an artificial variant of the HPC model called the *strong* HPC model. In this model, the energy function consists of three parts (first two are the same as in the HPC model): (i) the contact energy, (ii) the SS bridge energy and (iii) non-cysteine bridge energy. The last part is equal to -1 times the number of pairs in the maximum matching of the graph of potential non-cysteine bridges. There is a potential non-cysteine bridge between any two adjacent ordinary hydrophobic monomers. Thus, the fold in Figure 6 has energy $-5 - 1 - 1 = -7$ in the strong HPC model. This energy model can be interpreted as follows: we assume that we have two types of cysteine-like hydrophobic monomers each forming bridges, but no bridges are possible between cysteines of different types.

In the snake structures, approximately 40% of all monomers are hydrophobic and half of those are cysteines. Thus approximately 20% of all monomers are cysteines. Although, most of the naturally occurring proteins have much smaller frequency of cysteines, there are some with the same or even higher ratios: 1EZG (antifreeze protein from the beetle [22]) with 19.5% ratio of cysteines and the protein isolated from the chorion of the domesticated silkmoth [26] with 30% ratio.

Wave structures

Despite the fact that the snake structures are more restricted, the proof of their stability under the strong HPC model still required the analysis of huge number of cases and this number rapidly increases in the proper HPC model, and thus we were unable to finalize the proof. Therefore, in [19] we consider a subclass of the snake structures, called the *wave structures* and formally prove that they are stable under the proper HPC model.

The wave structures are instances of the snake structures that do not contain occurrence of the four forbidden motifs in Figure 9. The wave structures can also be constructed using a set of four *super-tiles* and their flipped

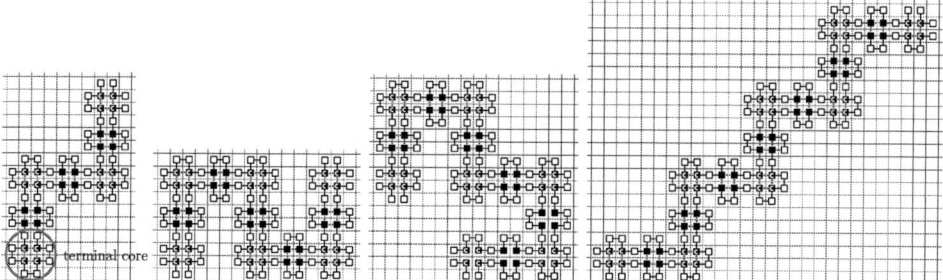

FIGURE 9. Forbidden motifs in wave structures.

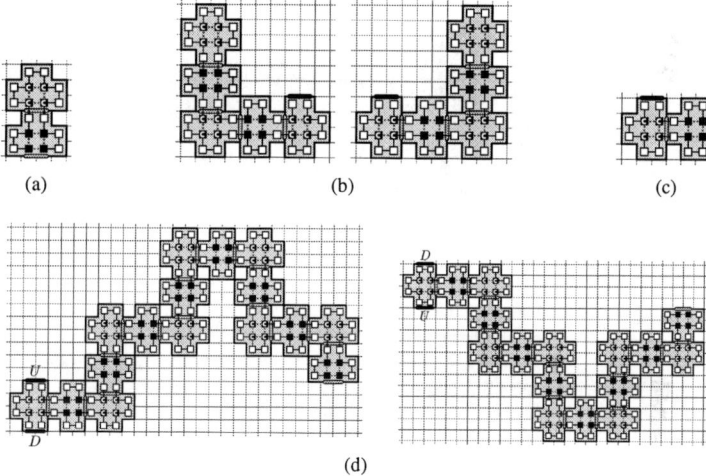

(a)   (b)   (c)

(d)

FIGURE 10. Super-tiles used to construct wave structures: (a) starting super-tile; (b) non-flipped and flipped versions of terminating super-tile; (c) bending super-tile; and (d) flipped and non-flipped versions of regular tile. The receptors are depicted with white ovals and ligands with black ovals.

versions depicted in Figure 10. Each super-tile has at most one position called "receptor", which connects to the next super-tile and at most two positions called "ligands", which can connect to the previous super-tile. The *starting* super-tile has one receptor and consists of two basic tiles (Figure 10(a)), the *terminating* super-tile has one ligand and consists of 5 basic tiles (Figure 10(b)), the *bending* super-tile has one ligand and one receptor and consists of two tiles (Figure 10(c)), and finally the *regular* super-tile has two ligands ("U" and "D") and one receptor and consists of 16 basic tiles (Figure 10(d)). The receptor of one super-tile can connect to the ligand of another one however, a regular super-tile must only connect through one of its ligands.

A wave structure is a partial tiling of the 2D square lattice obtained by the following procedure.

1. Place the starting super-tile into the square lattice and place a regular super-tile into the square lattice so that its "U" ligand is attached to the receptor of the staring gadget.

2. Let the last placed super-tile be a (flipped) regular super-tile *R*; either place a (flipped) regular super-tile so that its "U" ligand is attached to the receptor of *R* and continue with step 4 or place a (flipped) bending super-tile such

47

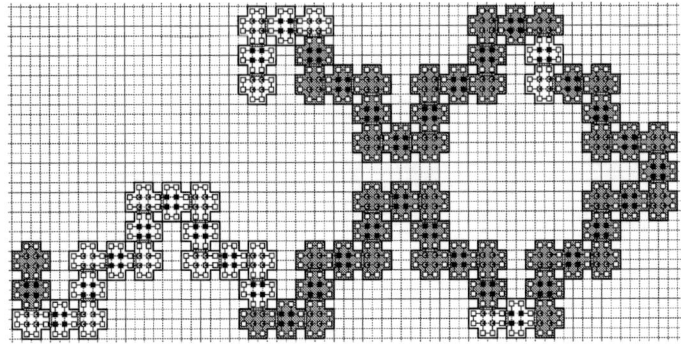

FIGURE 11. An example of a wave structure. It consists of 8 super-tiles. The borders between super-tiles are marked by the change of underlying color of the core tiles.

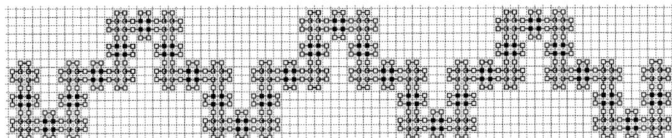

FIGURE 12. Simulation of a straight line segment with a wave structure.

that its ligand is attached to receptor of R and continue with step 3.

3. Let the last placed super-tile be a bending super-tile B. If B is a flipped tile attach a new regular super-tile otherwise attach a flipped regular super-tile to B. The new super-tile can be attached either with "U" or "D" ligand depending on intended direction of the bend.

4. Continue with step 2 or end the structure by attaching a (flipped) terminating super-tile to the last placed (flipped) regular super-tile.

In the above procedure the super-tiles are placed into the square lattice such that they do not overlap. An example of a wave structure is depicted in Figure 11. As snake structures, wave structures can approximate (although more coarsely) any given shape using line segments depicted in Figure 12.

Proof techniques

In this section we explain the concepts and techniques we use to prove the stability of snake and wave proteins. We also introduce 2DHPSolver, a semi-automated proving tool developed to analyze the huge number of cases required for the proofs. The protein sequences of snake and wave structures are saturated under the strong and proper HPC model. In saturated folds under the (strong) HPC model all parts of energy function produce minimum possible values. This means: (i) every hydrophobic monomer (cysteine or non-cysteine) has two contacts with other monomers; (ii) every cysteine is involved in disulfide bridge (iii) in addition, in strong HPC model every non-cysteine hydrophobic monomer is involved in a non-cysteine bridge. Clearly, a saturated fold of a protein must be native, and furthermore, if there is a saturated fold of a protein, then all native folds of this protein must be saturated. Notice that the snake and wave structures are saturated. This property greatly simplifies the stability proof of the snake and wave structures. Before we present the proof ideas we introduce some notations and definitions in the following paragraph that we will use in the proofs.

Let F be a fold of a snake or wave protein p in 2D square lattice $\mathcal{L}$. Define a path in F as a sequence of vertices such that no vertex appears twice and any pair of consecutive vertices in the path are connected by peptide bonds. A cycle is a path whose start and end vertices are connected by a peptide bond. Note that F has only one cycle which is the entire sequence p. For $i \in \{0,1,2\}$, an i-vertex in the fold F is a lattice vertex (square) containing a monomer i. For instance, a square containing a cysteine monomer in F is called a 2-vertex. An H-vertex is a vertex which is either 1-vertex or 2-vertex. Define an H-path in F to be a sequence of H-vertices such that each H-vertex appears once and any pair of consecutive hydrophobic (1 or 2) monomers form an HH contact. An H-cycle in F is an H-path whose first and last vertices form an HH contact. An H-cycle of length 4 is called a core in F. Clearly, every H-path in a saturated fold is part of an H-cycle.

The stability proof of the snake and wave structures consists of two main steps. First we show that the every H monmer in a saturated for of a snake or wave structure belong to a core and second we show that these cores are monochromatic and that they are connected uniquely to form the designed snake or wave structures. We use induction on the boundaries of diagonal rectangle surrounding the folds to prove the above mentioned properties of native folds. We have developed a tool 2DHPSolver to assist us in analyzing the huge number of cases arise in this induction process. In the following subsection we describe the 2DHPSolver and the induction techniques.

2DHPSolver: a semi-automatic prover

2DHPSolver is a semi-automatic tool used for proving the uniqueness of a protein design in 2D square lattice under the HP, HPC or strong HPC models. 2DHPSolver is not specifically designed to analyze the wave structures or even the constructible structures. It can be used to prove the stability of any design in 2D HP models. We use induction on the boundaries of diagonal rectangle surrounding the folds to first prove some properties of native folds and then use them to prove the uniqueness of the folds. We use 2DHPSolver in the following scenario; for any integer i, we define two *boundaries* SW_i and SE_i as the set of lattice vertices $\{[x,y]; x+y=i\}$ and $\{[x,y]; x-y=i\}$, respectively. Then we prove a property Π (for instance, every H-monomer belongs to a core) of all native conformations of designed proteins using induction on the boundary indices. To do this, we assume that the property Π holds for the part of the conformation that lies on boundary SW_k (SE_k) for any $k < i$ (the induction hypothesis) and prove that Π holds for the part of the conformation that lies on SW_i (SE_i). In what follows we explain how 2DHPSolver can be used to assist in the proof process. More details can be found in the help document in the software package (see below for download information).

2DHPSolver has three sets of inputs: the design rules, the initial configuration, and the run-time parameters. The design rules are the set of rules that specifies the properties of the designed sequence. For instance, the subsequences that cannot be part of any designed sequence are included in the design rules (these are called *forbidden* subsequences).

2DHPSolver maintains a list of current configurations. Initially this list contains only the initial configuration which is normally just a hydrophobic monomer on a SW (SE) boundary. Each configuration that 2DHPSolver stores and processes is in fact part of a potential saturated configuration of a designed protein. The proof process is completed when the list only contains configurations that satisfy the property Π. In each iteration, one of the current configurations is replaced by all possible extensions at one square in the configuration specified by user. Note that in displayed configurations red 1 represents a cysteine monomer, blue 1 a non-cysteine hydrophobic monomer, and uncolored 1 is hydrophobic monomer, but it is not known whether it is a cysteine or not. The following types of extensions are used in 2DHPSolver:

- extending a path;
- extending an H-path;
- coloring an uncolored H monomer.

Since a path and a 1-path can continue in 3 directions there are 6 ways to extend a path (with a 0 or 1 at each direction) and 3 ways to extend a 1-path. Furthermore, there are 2 ways to color and uncolored H monomer. For each of these possibilities, 2DHPSolver creates a new configuration which is then checked to see if it violates the rules of the design. Those which do not violate the design rules will replace the original configuration.

However, this approach will result in producing too many configurations, which makes it hard for the user to keep track of. Therefore, 2DHPSolver contains utilities for automatically finding extending sequences for each configuration which either leads to no valid configurations, in which case the configuration is automatically removed, or to only one valid configuration, in which case the configuration is replaced by the new more-completed configuration. This process is referred to as a *self-extension*. The time required for searching for such extending sequence depends on the depth of the search, which can be specified by user through two parameters "depth" and "max-extensions". Thus, leaving the whole process of proving to 2DHPSolver by setting the parameters to high values is not practical as it could take enormous amount of time. Note that the search space is infinite, and thus cannot be searched completely automatically. Instead, the user should set run-time parameters to moderate values and use intuition in choosing the next extension point when 2DHPSolver is unable to automatically find self-extending sequences. These parameters can be changed at any time during the use of the program by the user. Figure 13 depicts the interface menu of the 2DHPSolver.

```
Total # of fields: 6 Active field: 4
=== Field 4===
Comment:
ID: 161+
       ▮ ▮ ▮▮
   3 2 1 0 1 2

 1 . 0 ▮ 0 . . . 1
     | |  ▮▮
 0 1 1 1 1 0 . . 0
     | |    ▮▮
-1 . 0 0-1 ▮-0 -1
         |
-2 . . . . 0-1-▮ -2

-3 . , . 0-1-0 -3

-4 . . . . 1 . -4
       ▮ ▮ ▮▮▮
   3 2 1 0 1 2

p) Extend path  o) Extend one-path    c) Extend colors    s) Save        D) Delete
x) Self-Extend  u) Undo               d) set depth        m) set max-ext r) set mark    q
) Quit
Enter your choice: ▮
```

FIGURE 13.   A snapshot of 2DHPSolver interface.

2DHPSolver is written in C++ and its source code is freely available under the GNU Public License (GPL). For more information and to access the source codes please visit http://www.sfu.ca/~ahadjkho/2dhpsolver/.

The stability proof of the wave structures is a partial solution to our conjecture from [10], as the wave structures can be used to approximate any given shape (although more coarsely).

STRUCTURE APPROXIMATING IPF IN THE HP AND HPC MODELS ON 3D HEXAGONAL PRISM LATTICE.

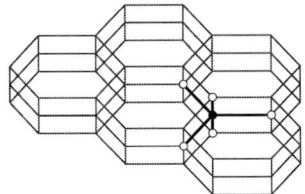

FIGURE 14.   A small portion of hexagonal prism lattice. The thicker lines depict 5 neighbors of one vertex, i.e., the degree of this lattice is 5.

The properties of the proteins in 2D HP model show remarkable similarity to the properties of real proteins [8]. However, to get one step closer to designing real proteins, we have to consider 3D models. Based on our previous research in [24], we have chosen the hexagonal prism lattice for our designs for two reasons. First, it has a relatively low degree (the number of neighbors of a vertex) of 5. The cubic lattice, for example, has a degree of 6. This low degree simplifies our designs. At the same time, relative to its degree it can well represent a large class of natural protein structures [24]. The hexagonal prism lattice is composed by stacking horizontal hexagonal grids ("honeycomb nets") on top of each other, cf. Figure 14. Achieving the stability on 3D models is more difficult, therefore, in what follows we relax the condition of stability: instead of require that the folds of all native folds follow the same path, we will just require that the placement of the amino acids in each native fold is identical. This concept is formalized in the next subsection.

Structural stability

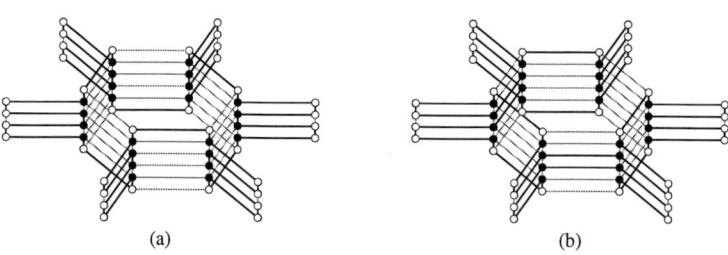

(a)   (b)

FIGURE 15. All native folds of the substring $t = (0100110010)^6$. These two folds are structurally similar, therefore the protein sequence is structurally stable. The hydrophobic (polar) amino acids are depicted are black (white) circles.

Every protein and every of its folds define a partial mapping from the lattice vertices to the set $\{0, 1, 2\}$. We say that two folds of the same protein are *structurally similar* if they define the same partial mapping. If all native folds of a given protein are pairwise structurally similar, then the protein is called *structurally stable*. Note that all native folds of a structurally stable protein have completely same shape (from outside their appear as a same fold). Since, we are trying to design a protein sequence which would fold into a fold having a prescribed shape, we do care that the protein sequence is not completely stable, as long as all its folds attain the prescribed shape. Therefore, this relaxed notion of stability is sufficient for our designs.

For instance, the protein sequence $t = (0100110010)^6$ is structurally stable, but not stable. Figure 15 depicts all two native folds of this sequence. It is easy to see that the mappings defined by t and its two folds are identical, i.e., the folds are structurally similar.

Tubular structures in the HP model

The HP model on the 3D hexagonal prism lattice is a straightforward extension of the HP model on the 2D square lattice defined in Section . A protein in the 3D HP model is represented as a string $p = p_1 p_2 \ldots p_{|p|}$ in $\{0, 1\}^*$, where "0" represents a polar monomer (depicted in figures as a white circle) and "1" a hydrophobic monomer (depicted as a black circle).

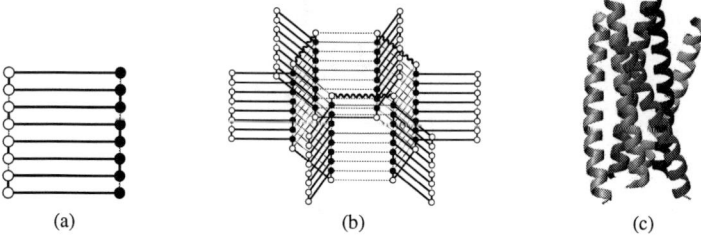

(a)   (b)   (c)

Source: Wikipedia (figure (c) only)

FIGURE 16. Illustration of a tube: (a) one of 6 zig-zag plates used to construct a tube; (b) a tube with the loops at top shown as waved lines and at the bottom as dashed lines; (c) a coiled coil structure of GP41 hexamer formed by 6 alpha-helices in protein 1AIK. This structure is somewhat similar to our tube (considering the restrictions imposed by the hexagonal prism lattice).

51

In [9] we have introduced a basic building block: a *tube*. A tube T_n is build from 6 zig-zag plates of size $2 \times 2n$ (which resemble alpha-helices fitted into the hexagonal prism lattice, cf. Figure 16(a)) connected at the top and bottom with 6 short loops, cf. Figure 16(b). The hydrophobic core is completely surrounded by polar monomers, i.e., the fold is saturated. The protein sequence of each tube is $p_n = (H00H)^n$ and the complete protein string for the tube is $t_n = (0p_n0)^6$. The height of (the hydrophobic core of) the tube T_n is $2n$.

Interestingly, a similar design exists in nature as a *coiled coil* structural motif in which 2–7 alpha-helices are coiled together, cf. Figure 16(c). Many coiled coil type proteins are involved in important biological functions such as the regulation of gene expression e.g. transcription factors [29, 23].

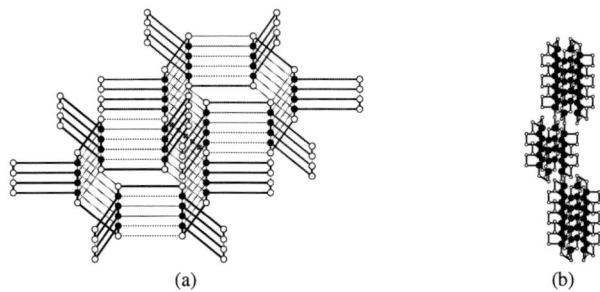

(a)   (b)

FIGURE 17. (a) Interconnection of tubes into one more complex structure. Two loops one from each tube are overlapped and disconnected (marked with (red) crosses in the figure). (b) An example of a tubular structure with 3 tubes.

In [9] we showed that a tube with any height is structurally stable. An example in Figure 15 shows that tubes are not stable when considering the strict definition of stability. In [9] we have also described a way how to connect two tubes together by overlapping a topmost loop of one tube with the bottommost loop of the other tube, cf. Figure 17(a). The structures built from tubes interconnected in this way are called *tubular structures*. Figure 17(b) shows an example of a tubular structure with 3 tubes. Since again the folds of tubular structures are saturated, it is easy to see that their protein sequences fold into them. Note that here we assume that proteins of tubular structures are closed chains of monomer, a similar assumption as used in [1], i.e., that the beginning and the end of the sequence are adjacent in the lattice. This can be easily achieved by using an additional force in the design not considered in the HP or HPC model. For instance, if the first monomer is positively charged polar monomer and the last is negatively charged, while all other polar monomers have neutral charge, the terminal monomers will stick together in any native fold.

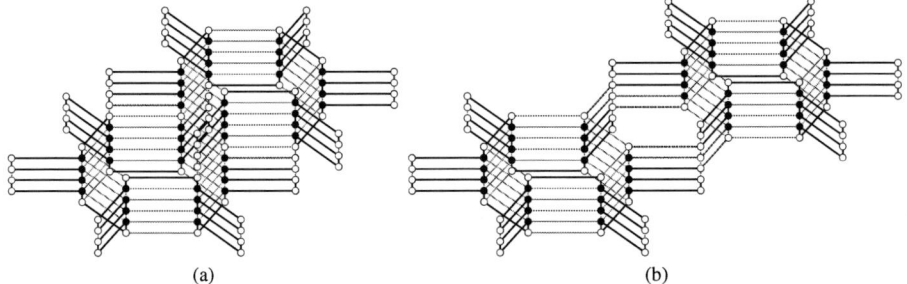

(a)   (b)

FIGURE 18. Two folds of the tubular structure with two tubes: (a) the intended fold in the form of a tubular structure; (b) alternative fold which is not structurally similar to the intended one. The (red) dashed lines shows the parts of the sequence which is folded differently.

However, these protein sequences might fold into other structures as well. For instance, the tubular structure can fold into 2 different conformations depicted in Figure 18. The alternative is similar to the intended fold, but it is not structurally similar as the relative position of tubes is shifted. Another example is the tubular structure with 4 tubes of the same height which could fold into at least two different folds which are even less similar to each other, cf. Figure 19. Hence, the protein sequences of these tubular structures are not structurally stable. We conjecture that these structures could become stable under the HPC model, however the choice of which hydrophobic monomers should be replaces with cysteines is not obvious. For instance, for the tubular structure with two tubes it is enough change one hydrophobic monomer belonging to the dashed sequence and one of its neighbors not belonging to the dashed sequence two cysteines and both folds in Figure 18 become stable. This works because two hydrophobic monomers in each dashed sequence are swapped in the alternative fold, and hence if one of them is cysteine, it will not find a pair in the other fold.

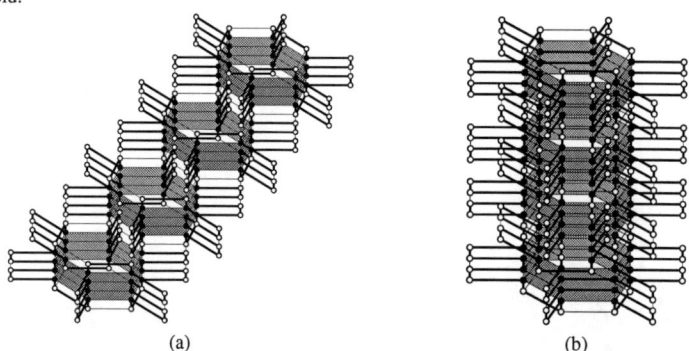

(a)   (b)

FIGURE 19. Two possible folds of a tubular structure with 4 tubes with heights 4: (a) the intended native fold; (b) alternative native fold — the four tubes are stack on top of each other. For clarity, the hydrophobic cores of the for tubes in the alternative native fold are highlighted.

Another drawback of tubular structures is the fact that a tube can only connect to one other tube at each end. Hence, the shapes of tubular structures are very limited (staircases of tubes). Thus, while tubular structures show that it is possible to construct arbitrary large structurally stable objects in the 3D HP model, they cannot be directly use for inverse protein folding. To overcome their shortcoming, we have generalized these structures as described in the next subsection.

Generalized tubular structures in the HPC model

The HPC model on the 3D hexagonal prism lattice is a straightforward extension of the HPC model on the 2D square lattice defined in Section . A protein in the 3D HPC model is represented as a string $p = p_1 p_2 \ldots p_{|p|}$ in $\{0,1,2\}^*$, where "0" represents a polar monomer (depicted in figures as a white circle), "1" a hydrophobic-none-cysteine monomer (depicted as a black square) and "2" a cysteine monomer (depicted as a black triangle). We use also H to represent a hydrophobic monomer which could be either 1 or 2 — either it was not yet decided or it does not matter which one it is (depicted in figures as a black circle).

A lattice vertex containing an $X \in \{0,1,2\}$ monomer is called an X-vertex. An H-vertex is either a 1-vertex or a 2-vertex. A neighbor of a vertex v which is an X-vertex is called X-neighbor.

The generalized tubular structures introduced in [11] are built from the tubes and connectors. The tubes are refined for the HPC model as follows: we change the first and the second hydrophobic monomers of one of the plates of each tube to cysteine monomers (2). The second building block used to build generalized tubular structures is a *connector*, cf. Figure 20(a). The hydrophobic core of the connector consists of 2 layers of two adjacent hexagons. Two tubes or a tube and a connector can be connected to one protein structure in two ways as follows. First, one top loop of the first tube is overlapped with a bottom loop of the second tube/connector, vice versa, and the peptide bonds between two polar monomers of each loop are disconnected, similarly as it was done when connecting two tubes in tubular structures. This way of connecting two components is called *vertical* connection. For instance, tubes in Figure 18(a) are vertically connected and tubes T_1 and T_2 in Figure 20(b) are vertically connected to the connector.

In the second way, called *horizontal* connection, the tubes or the tube and the connector are placed beside each other such that they have H-vertices in exactly one common plane H_i and exactly two H-vertices of the first component are connected to two H-vertices of the other component each through one 0-vertex. For instance, tubes in Figure 18(b) are horizontally connected and the tube T_3 in Figure 20(b) is horizontally attached to the connector. Repetitively,

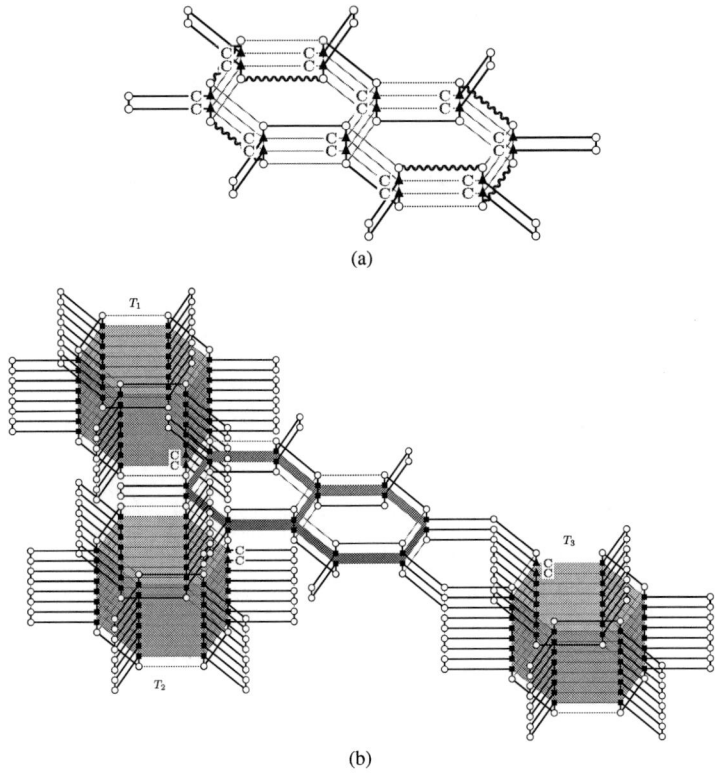

(a)

(b)

FIGURE 20. (a) A connector — the second building block for the generalized tubular structures. (b) An example of a generalized tubular structure showing the ability to branch. Polar, hydrophobic and cysteine monomers are depicted as empty circles, squares and triangles, respectively. Hydrophobic cores of 3 tubes and a connector are highlighted.

connecting tubes and connectors (such that no space violation occurs) we obtain the class of *tubular structures*. We choose to vertically or horizontally connect a tube to a component in a tubular structure such that no pair of H-vertices in the same plane and in middle layers of different tubes are at distant three of each other. Since, there is no substring 000 in the protein of any tubular structure, this condition ensures that the tubes in a tubular structures do not directly connect to each other through the H-vertices in their middle layers. This will greatly simplifies the stability proof of the structures.

An example of a generalized tubular structure with 3 tubes attached to the connector is shown in Figure 20(b). Since, the connector allows branching and changing the direction, the class of generalized tubular structures is sufficiently rich to roughly approximate any given shape. Furthermore, since again the generalized tubular structures are saturated, each such structure is one of the native folds of its protein. Proving structural stability of the structures is again very difficult task. In our recent manuscript [17] we were able to prove this formally for infinite subclass of the simple structures (consisting of one connector and three tubes, cf. Figure 20(b)) under the assumption that each of the three tubes is sufficiently long. We conjecture that the proteins of all generalized tubular structures are structurally stable.

Note that tubular structures from this subclass are not stable under the HP model, thus our results show that using disulfide bridges in protein designs improves their stability.

In the following subsections we first introduce some terminology and then we show the ideas of the stability proof of the generalized tubular structures with three tubes and one connector from [17].

Terminology

Consider a fold F. A path in F is a sequence of vertices $(x_1, x_2, \ldots, x_k)$ such that consecutive vertices are connected by peptide bonds. We say that F contains an occurrence of substring $w_1, w_2, \ldots, w_k$ if there is a path $(x_1, x_2, \ldots, x_k)$ in F such that x_i is a w_i-vertex. We number hexagonal grids of the lattice (also referred to as *planes*) with integer numbers, and denote the i-th grid by H_i. Consider vertex $x \in H_i$. We denote the vertical neighbor of x in H_{i+1} (above x) by x^1, and recursively, the vertical neighbor of x^j in H_{i+j+1} by x^{j+1}. Similarly, we denote the neighbor of x in H_{i-1} by x^{-1}, and the neighbor of x^{-j} in H_{i-j-1} by x^{-j-1}.

Let G_x be the graph of all H-vertices in H_i which are reachable from x by a path of H-vertices in H_i. Let G be a set of vertices in H_i. Then for $j \geq 1$, let G^j be the graph of all vertices in H_{i+j} which have a neighbor in G^{j-1}, and G^{-j} be the graph of all vertices in H_{i-j} which have a neighbor in G^{-j+1}, i.e., G^j and G^{-j}, $j \neq 0$, are vertical copies of the set G. Note that G_x is a planar graph (as H_i is as well). The degree of a vertex in G_x is called a *plane degree*. Let B_x be the boundary cycle of G_x, i.e., the set of vertices of G_x which lie on the outer face of G_x. A *component* in a fold F is a maximal set of H-vertices for which there is a path of H-vertices between any pair of them.

Let C be a component that lies in the planes H_{j+1} to H_{j+r}. Let layer C_i be a graph of all vertices of C in plane H_{j+i+1}. We say that layers C_i and C_k are the same if $C_i^{k-i} = C_k$, i.e, C_k is a copy of C_i. When we say that we are comparing layers C_i and C_j of component C, we mean comparing the sets C_i and C_j^{i-j}. For example, when we say C_i is identical to (respectively, a subset of) C_j we mean whether C_i is identical to (respectively, a subset of) C_j^{i-j}, and we write simply $C_i = C_j$ (respectively, $C_i \subseteq C_j$). The plane containing C_i will be denoted by $H(C_i)$.

Stability proof techniques

In this section we will present the ideas for proving that the protein of one basic generalized tubular structure: the structure built from one connector and three tubes depicted in Figure 20(b) is structurally stable. We will assume that three tubes $T_{k_1}, T_{k_2}, T_{k_3}$ used to construct this structure are sufficiently long. In particular, we will assume that $k_1, k_2, k_3 \geq 712$. We conjecture that this structure is structurally stable also for other values of k_1, k_2, k_3 and that all generalized tubular structures are structurally stable. Let q be the protein string of this structure and Q be its intended fold (as a generalized tubular structure).

Definition 1 (sparse protein) *We say that a protein is* sparse *if does not contain* HHH *as a substring and does not start or end with* H.

The proof techniques we used consist of the following three main steps:

- Analyzing possible component arising in a saturated fold of F.
- Proving that any saturated fold of F contains the same components as in the original fold.
- Proving that the components are connected the same way as the original fold and thus the protein of the structure is structurally stable.

We state the main lemmas that we used in performing each of the above states. However, before that we introduce some notations and definitions.

Types of H-*vertices*

Let F be a saturated fold of a sparse protein. Then each H-vertex has exactly three contacts, i.e., it has at least three H-neighbors and the remaining two neighbors are connected (via a peptide bond) and at most one of the two is an

H-vertex. We can classify every H-vertex x of S to one of the five types based on the position of its 0-neighbor(s), cf. Figure 21:

(a) vh-type: x has one vertical 0-neighbor (on top or below) and one horizontal 0-neighbor (in the same hexagonal grid);

(b) vv-type: x has two vertical 0-neighbors;

(c) hh-type: x has two horizontal 0-neighbors;

(d) h-type: x has one horizontal 0-neighbor;

(e) v-type: x has one vertical 0-neighbor.

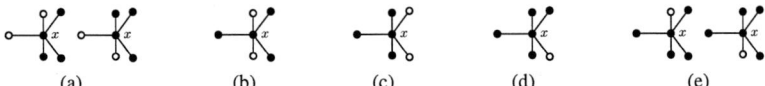

(a)   (b)   (c)   (d)   (e)

FIGURE 21. Five types of possible neighborhood of an H-vertex x: S-vertices: (a) vh, (b) vv, (c) hh; and D-vertices: (d) h and (e) v.

For every $X \in \{vv, hh, h, v\}$ an H-vertex of type X, will be called X-vertex. Furthermore, any H-vertex with two 0-neighbors is called a S-*vertex* and an H-vertex with one 0-neighbor is called a D-*vertex*.

Definition 2 (connections) *Let $u,v \in \{0,1,2,H,S,D,vv,hh,h,v\}$ and $s \in \{0,1,2,H\}^+$. We say that two vertices x and y are s-connected if there is a path $x, v_1, v_2, \ldots, v_k, y$ such that v_i is an s_i-vertex. If x is a u-vertex and y is a v-vertex, this path is called an usv-connection. If the end points x and y are H-vertices and belong to two different components, we say that these components are usv-connected. If $s = 00$ and $u,v \neq 0$, we will shorten this notation as $(u \backslash v)$-connection. In particular, we will be interested in H0H-connections and (S\h)-connections.*

A usv-connection with end points x and y is called internal, *if x and y are in the same component, and otherwise it is called* external. *We say that two usv-connections with end points at x,y and x',y', respectively, are* parallel *if x (y) is directly above/below x' (y'), i.e., $x' = x^i$ and $y' = y^j$, for some integers i, j, and all vertices between x and x' (y and y') are H-vertices. Note that it is also possible that x and y' are u-vertices and x' and y are v-vertices.*

We have the following observations:

Observation 1 *Let F be an arbitrary saturated fold of q. Then F contains 6 H0H-connections, 52 S-vertices, the number of D-vertices is 4 modulo 6 and it contains 36 (S\D)-connections. F does not contain HHH, 000, H0H0H and H0HH, but it does contain one occurrence of 20100101.*

Observation 2 *Let F be a saturated fold of a sparse protein. Then every H-vertex of F is either a vh-vertex, vv-vertex, hh-vertex, h-vertex or v-vertex. Furthermore, any neighboring 0-vertex and H-vertex are connected by a peptide bond.*

Using simple case analysis we can prove the following lemma.

Lemma 1 *Let F be a saturated fold of a sparse protein with no H0HH as a substring. Any occurrence of substring $(00HH)^k$ in F contains either only v-vertices or only h-vertices.*

Analyzing types of components

In this section we study all possible components that can arise in saturated folds of q. We first classify all components to three categories and then study which of these can appear in saturated folds of q.

Let F be a saturated fold of a sparse protein and C a component in F. Assume that C lies in the planes $H_s, \ldots, H_e$. Note that any H-vertex of plane degree one in the first or last layer of C is adjacent to at least three 0-vertices, a contradiction. Hence, we have the following observation.

Observation 3 *Let F be a saturated fold of a sparse protein and let C be a component in F. Then all vertices of the first or last layer of C have plane degree 2 or 3.*

The following definition defines several types of components.

Definition 3 (tube, simple tube, 2-layer component, wall, and complex component) *A* tube *is a component such that all its layers are identical and each layer contains only vertices of plane degree two (a cycle). A* simple tube *is a tube with only one hexagon in each layer. A* 2-layer *component is a component with two identical layers which have no vertex with plane degree 1 and at least one vertex with plane degree 3. A* wall *is a component such that all its layers are identical and each layer is a single path. Finally, a* complex *component is a component C such that there is i for which C_i and C_{i+1} are different.*

We have the following observations.

Observation 4 *Any component C in a saturated fold of a protein is one of the following three types: a tube, a 2-layer component or a complex component.*

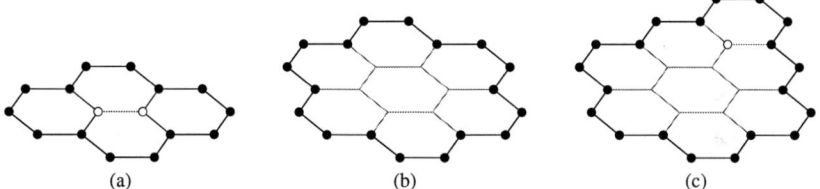

(a)   (b)   (c)

FIGURE 22. One layer of (a) the smallest non-simple tube; (b) the smallest non-simple tube without occurrences of HOH; and (c) the smallest non-simple tube with one occurrence of HOH per layer.

Observation 5 *Let F be a saturated fold of a sparse protein. If F contains a tube then the height (number of layers) of this tube is at least 2. One layer of the smallest non-simple tube is depicted in Figure 22(a). It contains two occurrences of HOH per layer, i.e., at least 4 such occurrences. One layer of the smallest non-simple tube with no occurrences of HOH is depicted in Figure 22(b). One layer of the smallest tube with one occurrence of HOH per layer is depicted in Figure 22(c).*

Different types of complex components

In what follows we further classify different types of complex components which can occur in saturated folds of sparse proteins with at most six occurrences of substring HOH.

Complex components with a vv-vertex

Lemma 2 *Let F be a saturated fold of a sparse protein with no occurrences of substrings HOHH and HOHOH and at most six occurrences of substring HOH. Consider a complex component C of F containing a vv-vertex. Then C has 6 vv-vertices forming a hexagon, lies in two layers which are almost identical, except for the six vv-vertices which are replaced with 0-vertices in the other layer, and neither layer contains a vertex of plane degree 1. We will call such a complex component, a* vv-component. *A* vv-*component contains 6 occurrences of HOH.*

Note that a vv-component is essentially a 2-layer component which is missing vertices of one hexagon in one of the two layers.

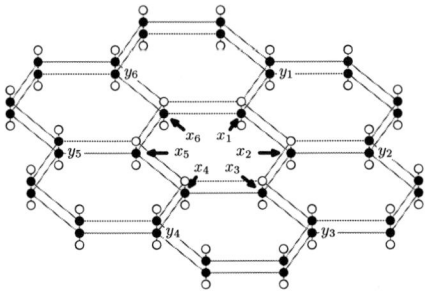

FIGURE 23. Part of a complex component with a vv-vertex. The arrows are pointing at six vv-vertices.

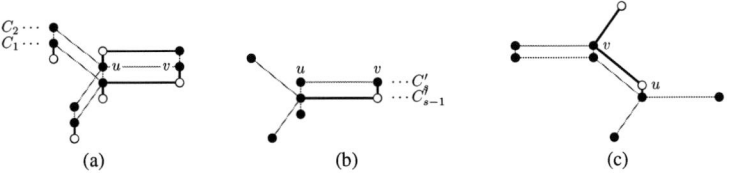

(a)   (b)   (c)

FIGURE 24. Analysis of a complex component without a vv-vertex: (a) the case in which $C_2' \neq C_1$; (b) the case in which C_i' is not a subset of C_1; (c) the case when C_s' is not a subset of $V^{2,2,2}$.

Complex components without a vv-vertex

Lemma 3 *Let F be a saturated fold of a sparse protein with no HOHH as a substring. Let C be a complex component of F without a vv-vertex and $C_1,\ldots,C_r$ its layers. Let $V^{2,2,2}$ be the set of all H-vertices in F with plane degree 2 such that both its horizontal H-neighbors have plane degree 2 as well.*

 (a) For $k \geq 1$, let C_k' be a subset of C_k consisting of components of C_k which are intersecting C_1. Let s be the smallest integer such that layer C_s' is different from C_1. Then $s > 2$ and C_s' is a collection of paths where each path is a subset of $C_1 \cap V^{2,2,2}$.

 (b) For $k \leq r$, let C_k'' be a subset of C_k consisting of components of C_k which are intersecting C_r. Let e be the largest integer such that layer C_e'' is different from C_r. Then $e < r - 1$ and C_e'' is a collection of paths where each path is a subset of $C_r \cap V^{2,2,2}$.

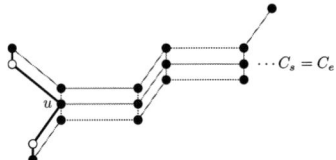

FIGURE 25. A complex component: the case when layers C_s and C_e are identical.

58

Definition 4 (basic complex component) *Let F be a saturated fold of a sparse protein with no* HOHH *as a substring. Let C be a complex component of F without a vv-vertex with layers $C_1,\dots,C_r$. Let s be the smallest integer such that C_s is different from C_1 and let e be the largest integer such that C_e is different from C_r. If C_s is a path and for any $i \in s+1,\dots,e$, C_i is identical to C_s then we call C a* basic complex component.

Note that a basic complex component consists of three parts stack vertically on each other: (1) a tube or 2-layer component; (2) a wall; and (3) a tube or 2-layer component.

Observation 6 *Let F be a saturated fold of a sparse protein with no* HOHH *as a substring. Any basic complex component of F contains at least 20 S-vertices (the lower and upper part at least 8 each and the wall at least 4) and at least 4 occurrences of substring* HOH.

In this subsection, we show that if a complex component C without vv-vertices is not basic, then its layers change exactly four times, i.e., it consists of five parts stacked on top of each other: (1) a 2-layer component or a tube; (2) a wall; (3) a pseudo 2-layer component with exactly one vertex with plane degree 1 in each of two layers; (4) another wall; and (5) a 2-layer component or a tube. The part in the middle (3) will be called an *appendix*, and such a complex component will be called an *appendix component*. An example of an appendix component is in Figure 26(a). Let us start with the formal definition of an appendix component.

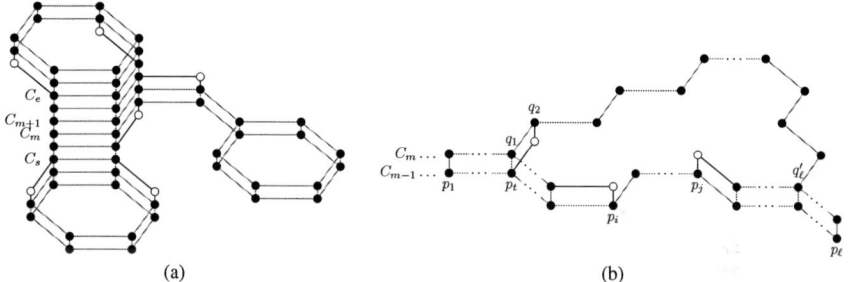

FIGURE 26. (a) An example of an appendix component and the six occurrences of HOH contained in it. (b) Illustration what happens if C_{m-1} is not a subset of C_m.

Definition 5 (appendix component) *Let F be a saturated fold of a sparse protein with no occurrence of the substring* HOHH. *Let C be a complex component of F without a vv-vertex with layers $C_1,\dots,C_r$. Let s be the smallest integer such that C_s is different from C_1 and let e be the largest integer such that C_e is different from C_r. Assume that both C_s and C_e contain only one path, and that there is an integer $s < m < e-1$ such that $C_s = C_{s+1} = \dots = C_{m-1}$, $C_m = C_{m+1}$, $C_{m+2} = C_{m+3} = \dots = C_s$, and either C_s is a subset of C_e or C_e is a subset of C_s and both of them are subsets of C_m. Furthermore, assume that C_m has exactly one vertex with plane degree 1 and this vertex is an end point of the paths in C_s and C_e. Such a complex component will be called an* appendix component *and the layers C_m and C_{m+1} we be called an* appendix. *Consider a path in C_m (C_{m+1}) starting at the vertex with plane degree 1 and ending before the first vertex with plane degree 3. These paths in C_m and C_{m+1} will be called the* arm *of the appendix.*

Note that an appendix without its arm is a proper 2-layer component.

Lemma 4 *Let F be a saturated fold of a sparse protein with no occurrence of the substring* HOHH, *and at most six occurrences of the substring* HOH. *Every non-basic complex component without a vv-vertex in F is an appendix component.*

In this subsection we show that only a limited number of configurations are possible for a complex component. This will greatly simplifies our analysis in the later sections. We used the following important lemma for proving the lemmas in this section.

Lemma 5 *Let F be a saturated fold of q. No v-vertex can be part of substring* $(00HH)^{356}$. *Consequently, there are at most 4 v-vertices in F.*

In the following arguments we say that a path has length k if it contains k vertices.

Lemma 6 *Let F be a saturated fold of q. Then F does not contain any vv-component.*

Lemma 7 *Let F be a saturated fold of q and let C be a complex component in F with layers* $C_1, C_2, \ldots, C_r$. *Layer* C_1 *and similarly* C_r *is either one hexagon or consists of two hexagons attached by one edge or connected by a path (cf. Figure 28).*

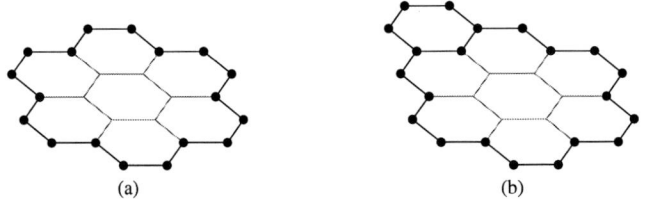

FIGURE 27. (a) The second smallest cycle without H0H occurrences. (b) The smallest possible layer C_1 of a complex component with the lower part being a 2-layer component containing a large cycle.

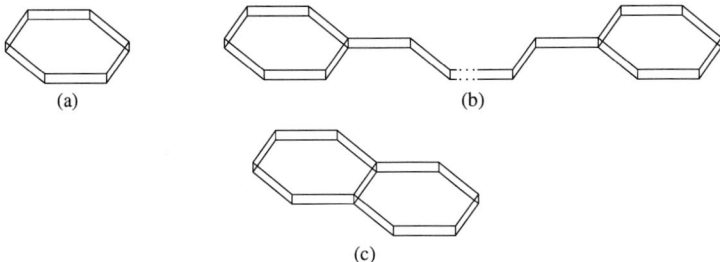

FIGURE 28. Possible configurations for the upper and lower part of a complex component.

Lemma 8 *Let F be a saturated fold of q and let C be a complex component in F. Then the lower and upper part of C are simple tubes.*

Lemma 9 *Let F be a saturated fold of q and let C be a complex component in F. The width of the wall in C is either 2 or 4.*

Lemma 10 *Let F be a saturated fold of q. Then F does not contain any appendix component.*

No other type of possible components can introduce 6 occurrences of H0H, hence, a saturated fold of F contains at least two components. On other hand, since any of possible components has at least 12 S-vertices, we have the following corollary.

Corollary 1 *Any saturated fold of q has at least 2 and at most 4 components.*

In what follows we will analyze all three possibilities. But first, let us have a closer look at tubes.

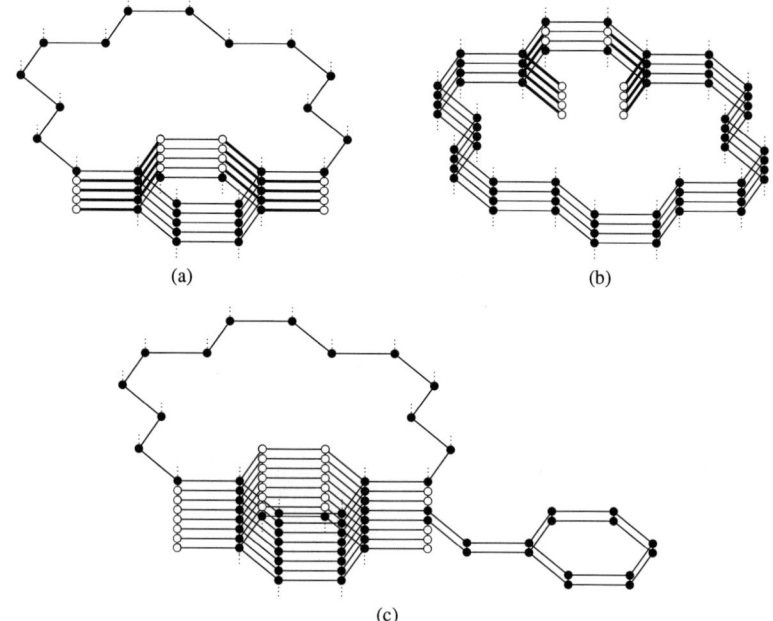

FIGURE 29. (a) A basic complex component with the second smallest tube as upper part and a simple tube as the lower part. (b) A basic complex component with the second smallest tube as upper and lower part. (c) An appendix component with the second smallest tube as upper part and a simple hexagon as lower part.

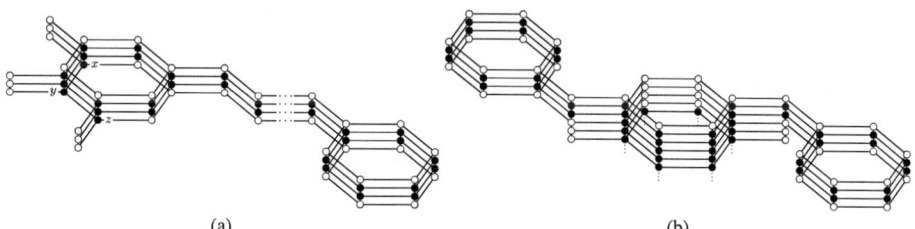

FIGURE 30. Examples of complex components with a 2-layer component consists of two hexagons connected by a path as the upper part: (a) wall is attached to one of the hexagons; (b) wall is attached to the path connecting hexagons.

Tubes

Lemma 11 *Let F be a saturated fold of q. Any tube in F has either 12 or at least 36 S-vertices.*

Lemma 12 *Let F be a saturated fold of q. Two HOH-connected tubes in F are both simple and furthermore, they make exactly two HOH-connections.*

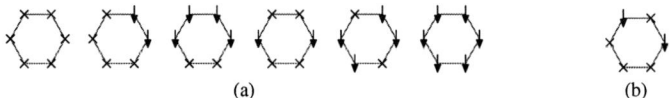

(a)   (b)

FIGURE 31. (a) All possible patterns (up to rotation) for vertical connections between two consecutive layers of a simple tube. The "x" means vertical connection is not present, arrow means it is present. (b) Pattern required to connect to the last layer of a simple tube which is connected to a path of length 3.

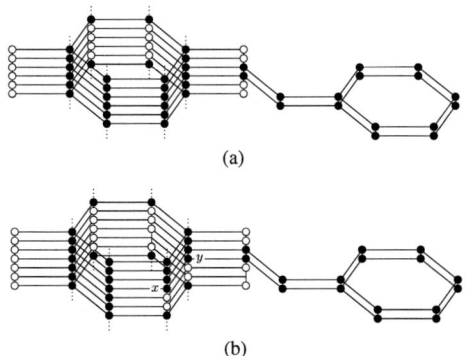

(a)

(b)

FIGURE 32. A part of (a) an appendix component with wall of width 4 all along; (b) an appendix component with wall of width 4 and 2 on different sides of appendix.

Connections of the components

So far we have proved that any saturated fold F of q must have exactly four components. In this section we prove that the fold F is similar to the designed fold, i.e., that q is structurally stable. First, we show that the components in F are the same as the components in the designed fold.

Lemma 13 *Let F be a saturated fold of q, then F has three simple tubes and a connector.*

Note that the above lemma is true even in the HP model when we do not use properties of cysteines. Next, we prove that in the HPC model the components in F must connect the same way as in the designed fold.

In Lemma 12, two tubes in F can connect with at most two HOH-connections. We will show the same for a tube and connector.

Claim 1 *Let F be a saturated fold of q. A tube and a connector in F can create at most two HOH-connections.*

Lemma 14 *Let F be a saturated fold of q. The tubes in F have more than 3 layers.*

Finally, we proving the following two lemmas and the final theorem.

Lemma 15 *Let F be a saturated fold of q. Any component in F must be HOH-connected to at least one other component.*

Lemma 16 *Let F be a saturated fold of q. All tubes in F must be HOH-connected to the connector.*

So far we have shown that all tubes must HOH-connect to C. We prove the final theorem.

Theorem 1 *The protein string q is structurally stable.*

CONCLUSIONS

Given a target conformation, the protein design process involves constructing protein sequences that fold into the target conformation (positive design) and has no native fold other than the target conformation (negative design). Since, this problem most likely NP-hard [3], we consider an approximation version of this problem: for a given shape find a stable protein sequence with a native fold approximating the shape. In this note, we reviewed our recent results on structure-approximating designs in HP models on the 2D square and 3D hexagonal prism lattices.

For 2D square lattice, we designed a system of waved structures which can, although quite coarsely, approximate any given 2D shape and the designed sequence are stable [19]. To obtain this result we have incorporated cysteines into the HP model (the HPC model introduced in [18]). This gives a partial affirmative answer to our conjecture posed in [10].

For 3D hexagonal prism lattice, we designed tubular and generalized tubular structures build from tubes and connectors. We showed that some of simplest (but arbitrary large) instances of these structures are stable [9, 11]. However, in [17], an alternative native fold was found for a tubular structure with 4 tubes. Thus, it is unlikely that more complex generalized tubular structures than the one studied in [11] would be stable under the simple HP model. The stability results for 2D HP model designs [18, 19] suggests the use of HPC model in which also disulfide bridges are used to calculate the energy of the structure. A completely different approach could be to rely on the "strong" stability of simple tubes (indicated by the shortness of the stability proof for single tube) and on attraction of complementary (negative and positive) chargers of polar amino acids on the surface of the tubes. In this design, a final structure would be a protein complex built from simple tubes with carefully designed chargers on the surfaces. This approach is especially promising since it allows to design protein complexes which are much more compact and dense than the generalized tubular structures in which large gaps between tubes are unavoidable.

REFERENCES

1. O. Aichholzer, D. Bremner, E.D. Demaine, H. Meijer, V. Sacristán, and M. Soss. Long proteins with unique optimal foldings in the H-P model. *Computational Geometry: Theory and Applications*, 25(1-2):139–159, 2003.
2. B. Berger and T. Leighton. Protein folding in the hydrophobic-hydrophilic (HP) model is NP-complete. *J. Comp. Biol.*, 5(1):27–40, 1998.
3. P. Berman, B. DasGupta, D. Mubayi, R. Sloan, G. Turán, and Y. Zhang. The inverse protein folding problem on 2D and 3D lattices. *Discr. Appl. Math.*, 155:719–732, 2007.
4. P. Crescenzi, D. Goldman, C. Papadimitriou, A. Piccolboni, and M. Yannakakis. On the complexity of protein folding. In *Proc. of STOC'98*, pages 597–603, 1998.
5. J.M. Deutsch and T. Kurosky. New algorithm for protein design. *Physical Review Letters*, 76:323–326, 1996.
6. K.A. Dill. Theory for the folding and stability of globular proteins. *Biochemistry*, 24(6):1501–1509, 1985.
7. K.A. Dill. Dominant forces in protein folding. *Biochemistry*, 29(31):7133–7155, 1990.
8. K.A. Dill, S. Bromberg, K. Yue, K.M. Fiebig, D.P. Yee, P.D. Thomas, and H.S. Chan. Principles of protein folding: A perspective from simple exact models. *Protein Science*, 4:561–602, 1995.
9. A. Gupta, M. Karimi, A. Khodabakhshi, J. Maňuch, and A. Rafiey. Design of artificial protein structures in 3D hexagonal prism lattice under HP model. In *Proc. of BIOCOMP 2007*, pages 362–369, 2007.
10. A. Gupta, J. Maňuch, and L. Stacho. Structure-approximating inverse protein folding problem in the 2D HP model. *Journal of Computational Biology*, 12(10):1328–1345, 2005.
11. A. Hadj Khodabakhshi, J. Maňuch, A. Rafiey, and A. Gupta. Inverse protein folding in 3D hexagonal prism lattice under HP model. In *Proc. of BIOCOMP (Las Vegas, 2008)*, 2008.
12. W.E. Hart. On the computational complexity of sequence design problems. In *Proc. of Comp. Molecular Biology*, pages 128–136, 1997.
13. B. Hayes. Prototeins. *American Scientist*, 86:216–221, 1998.
14. R. Jaenicke. Protein stability and molecular adaptation to extreme conditions. *Eur. J. Biochem.*, 202:715–728, 1991.
15. Hiroshi Kadokura. Oxidative protein folding: Many different ways to introduce disulfide bonds. *Antioxidants and Redox Signaling.*, 8:731–733, 2006.
16. S. Kamtekar, J.M. Schiffer, H. Xiong, J.M. Babik, and M.H. Hecht. Protein design by binary patterning of polar and nonpolar amino acids. *Science*, 262:1680–1685, 1993.
17. A. Hadj Khodabakhshi, J. Maňuch, A. Rafiey, and A. Gupta. Inverse protein folding in 3D hexagonal prism lattice under HPC model. (manuscript).
18. A. Hadj Khodabakhshi, J. Maňuch, A. Rafiey, and A. Gupta. Structure-approximating design of stable proteins in 2D HP model fortified by cysteine monomers. In *Proc. of APBC 2008*, pages 49–58, 2008.
19. A. Hadj Khodabakhshi, J. Maňuch, A. Rafiey, and A. Gupta. Stable structure-approximating inverse protein folding in 2D Hydrophobic-Polar-Cysteine (HPC) model. *J. Comp. Biol.*, 16, 2009. (to appear).

20. T. Kurosky and J.M. Deutsch. Design of copolymeric materials. *Physics A: Mathematical and General*, 28:387–393, 1995.
21. Z. Li, X. Zhang, and L. Chen. Unique optimal foldings of proteins on a triangular lattice. *Appl. Bioinformatics*, 4(2):105–16, 2005.
22. Y.C. Liou, A. Tocilj, P.L. Davies, and Z. Jia. Mimicry of ice structure by surface hydroxyls and water of a beta-helix antifreeze protein. *Nature*, 406:322–324, 2000.
23. J.M. Mason and K.M. Arndt. Coiled coil domains: Stability, specificity, and biological implications. *ChemBioChem*, 5(2):170–176, 2004.
24. C.R. Mead, J. Maňuch, X. Huang, B. Bhattacharyya, L. Stacho, and A. Gupta. Investigating lattice structure for inverse protein folding. *FEBS Journal*, 272((s1)):4739_1_380, 2005.
25. Nozomi Naganoa, Motonori Otaa, and Ken Nishikawa. Strong hydrophobic nature of cysteine residues in proteins. *FEBS Letters*, 458(8):69–71, 1999.
26. G.C. Rodakis and F.C. Kafatos. Origin of evolutionary novelty in proteins: How a high-cysteine chorion protein has evolved. *Proc. Natl. Acad. Sci. USA*, 79:3551–3555, 1982.
27. E.I. Shakhnovich and A.M. Gutin. Engineering of stable and fast-folding sequences of model proteins. *Proc. Natl. Acad. Sci.*, 90:7195–7199, 1993.
28. S. Sun, R. Brem, H.S. Chan, and K.A. Dill. Designing amino acid sequences to fold with good hydrophobic cores. *Protein Engineering*, 8(12):1205–1213, 1995.
29. Y.B. Yu. Coiled-coils: stability, specificity, and drug delivery potential. *Advanced Drug Delivery Reviews*, 54(8):1113–1129, 2002.
30. K. Yue and K.A. Dill. Inverse protein folding problem: Designing polymer sequences. *Proc. Natl. Acad. Sci. USA*, 89:4163–4167, 1992.

Graph Modeling for Quadratic Assignment Problems Associated with the Hypercube

Hans Mittelmann*, Jiming Peng† and Xiaolin Wu**

*Department of Mathematics, Arizona State University, Tempe, AZ 85287-1804, USA
†Department of Industrial and Enterprise System Engineering , University of Illinois at Urbana-Champaign. Urbana, IL, 61801
**Department of Electrical & Computer Engineering, McMaster University, Ontario, Canada

Abstract. In the paper we consider the quadratic assignment problem arising from channel coding in communications where one coefficient matrix is the adjacency matrix of a hypercube in a finite dimensional space. By using the geometric structure of the hypercube, we first show that there exist at least n different optimal solutions to the underlying QAPs. Moreover, the inherent symmetries in the associated hypercube allow us to obtain partial information regarding the optimal solutions and thus shrink the search space and improve all the existing QAP solvers for the underlying QAPs.

Secondly, we use graph modeling technique to derive a new integer linear program (ILP) models for the underlying QAPs. The new ILP model has $n(n-1)$ binary variables and $O(n^3 \log(n))$ linear constraints. This yields the smallest known number of binary variables for the ILP reformulation of QAPs. Various relaxations of the new ILP model are obtained based on the graphical characterization of the hypercube, and the lower bounds provided by the LP relaxations of the new model are analyzed and compared with what provided by several classical LP relaxations of QAPs in the literature.

Keywords: Quadratic Assignment Problem (QAP),Integer Linear Program (ILP), Graph Modeling, Relaxation, Lower Bound
PACS: 02.60.Pn

1. INTRODUCTION

The standard quadratic assignment problem takes the following form

$$\min_{X \in \Pi} \mathrm{Tr}\left(XAX^{\mathrm{T}}B\right) \tag{1}$$

where $A, B \in \Re^{n \times n}$, and Π is the set of permutation matrices. This problem was first introduced by Koopmans and Beckmann [27] for facility location. The model covers many scenarios arising from various applications such as in chip design [22], image processing [38], and keyboard design [9]. For more applications of QAPs, we refer to the survey paper [10] where many interesting QAPs from numerous fields are listed.

In this work we focus on a special class of QAPs where the matrix B is the adjacency matrix of a hypercube in space $\Re^d$ for a positive integer d. The problem arises from channel coding in communication where the purpose is to minimize the total channel distortion caused by the channel noise [6, 33]. Let us consider a given source alphabet (*binary codebook*) of fixed length (d)

$$\mathscr{C} = \{c_1, \cdots, c_n\}, n = z^d.$$

Since the the communication system is imperfect, there is a certain probability that the received message is different from the transmitted message. If we use the so-called memoryless binary-symmetric channel (BSC), the conditional probability of decoding c_j when transmitting c_i is

$$p(j/i) = q^{\delta(c_i, c_j)} (1-q)^{d-\delta(c_i, c_j)},$$

where q is the error bit rate, and $\delta(c_i, c_j)$ is the Hamming distance between c_i and c_j. Suppose that the codeword c_i occurs with a probability P_i and the assignment of a binary code to each codeword in $\mathscr{C}$ is obtained by a permutation $i \to \pi(i), i \in \{1, \cdots, n\}$. The problem of minimizing the channel distortion is defined by

$$\min_{\pi \in \Pi} \sum_{i=1}^{n} P_i \sum_{j=1}^{n} p(\pi(j)/\pi(i)) \delta(c_i, c_j). \tag{2}$$

CP1146, *Modelling of Engineering and Technological Problems*, edited by A. H. Siddiqi, A. K. Gupta, and M. Brokate

One can easily see that the above problem can be cast as a special case of problem (1) where the matrix B is the Hamming distance matrix, and A is the probability matrix. When the error bit rate is very small, the distortion corresponding to a large Hamming distance can be ignored. In such a case we can approximate problem (2) by using a simplified matrix B, derived from the adjacency matrix of a hypercube in $\Re^d$.

The problem (2) has been studied by many experts in the communication community. For example, in [33], Potter and Chiang showed that the problem is NP-hard in general. However, if the probability matrix A has also a very special structure such as the case of Harper code [23], then it can be solved in polynomial time. Numerous algorithms for general QAPs have been applied to solve the problem. For more engineering background and solving techniques for problem (2) we refer to [6, 33].

Since most existing solvers for problem (2) are based on algorithms for general QAPs, in what follows we give a brief review on these algorithms. It is known that QAPs are among the hardest discrete optimization problems. For example, a QAP with $n = 30$ is typically recognized as a great challenge from a computational perspective. The paper [5] was probably the first one where the optimal solutions to QAPs of size $n = 30$ were reported. A popular technique for finding the exact solution of the QAPs is branch and bound (B&B), which has been employed in most existing methods for the QAPs. In a typical B&B approach, we need to solve a relaxation of the original QAP whose solutions can further provide a lower bound for QAPs and help us in the process. There are several relaxations for QAPs. One relaxation is the well-known Gilmore-Lawler bound (GLB) which can be found via solving a linear programming problem [18]. Though the GLB bound can be obtained very fast, the quality of the GLB bound deteriorates fast as n increases. Other bounds include the bounds based on eigenvalues of the coefficient matrices [15], bounds based on convex quadratic programming [4] and semidefinite programming [41]. It has been observed that all these bounds are tighter than the GLB, but the computational expense to obtain these bounds is also much higher than that of the GLB. This prevents the B&B algorithm based on more complex relaxations from solving large scale QAPs with reasonable effort.

In this paper, we are particularly interested in methods based on ILP reformulations of the original QAP. There are several different ILP approaches for the QAP in the literature based on linearizing the quadratic form of the objective function. For example, Lawler [28] introduced n^4 additional binary variables and reformulated the QAP as an ILP with $n^4 + n^2$ binary variables and $n^4 + 2n + 1$ constraints. Kaufman and Broeckx [26] introduced n^2 new variables and reformulate the original QAP as an ILP with $O(n^2)$ constraints. Most ILP approaches for the QAPs are closely related to the following mixed ILP formulation

$$\min \quad \sum_{i=1}^{n} \sum_{j=1}^{n} \sum_{k=1}^{n} \sum_{l=1}^{n} a_{ik} b_{jl} y_{ijkl} \qquad (3)$$

$$
\begin{aligned}
s.t. \quad & \sum_{i=1}^{n} x_{ij} = 1, & & \forall j \in \{1, \cdots, n\}; \\
& \sum_{j=1}^{n} x_{ij} = 1, & & \forall i \in \{1, \cdots, n\}; \\
& x_{ij} \in \{0, 1\}, & & \forall i, j \in \{1, \cdots, n\}, \\
& \sum_{i=1}^{n} y_{ijkl} = x_{kl}, & & \forall j, k, l \in \{1, \cdots, n\}; \\
& \sum_{j=1}^{n} y_{ijkl} = x_{kl}, & & \forall i, k, l \in \{1, \cdots, n\}; \\
& y_{ijkl} = y_{klij}, y_{ijkl} \geq 0, & & \forall i, j, k, l \in \{1, \cdots, n\},
\end{aligned}
$$

which was proposed by Adams and Johnson [1] based on the linearization techniques introduced by Adams and Sherali [2, 3]. A similar model was proposed in [16] where the authors imposed the constraints $y_{ijij} = x_{ij}$. Noting the transformation $y_{ijkl} = x_{ij}x_{kl}$ where x is the original assignment matrix, Hahn and Grant [20] introduced extra constraints $y_{ijil} = 0, j \neq l$. In [21], the authors further lift model (3) to higher dimension, with n^6 variables based on the technique in [3] and proposed algorithms based on Lagrangian relaxation to attack the lifted mixed ILP in higher dimension. It has been reported that the lower bound provided by the level-2 reformulation model in [21] is the tightest [30].

Note that if we directly apply the above algorithms for general QAPs to the special case in this work, then we might not be able to explore the specific structure of the associate hypercube to improve the algorithm. We also note that the idea of exploring the structure of the problem has been employed in the study of QAPs. For example, it has been proved that several special classes of QAPs can be solved in polynomial time [7, 8]. In [25], Karisch and Rendl used the triangle decomposition for QAPs with rectangular grids to design an algorithm. In particular, the papers [17, 13] used the idea of graph modeling to derive lower bounds for QAPs.

The distinction between our ILP formulation and the existing ILP formulations of QAPs lies in the linearization process. While most existing ILP formulations of QAPs perform the linearization on the quadratic objective function $\mathrm{Tr}(AXBX^T)$, we try to characterize the set $X^T BX$ (or XAX^T) in a certain space, by using the structure of the matrices

A or B. As an initial step in our new approach, we also had a closer look at all the problems in the QAP library [10]. Not to our surprise, many problems in the QAP library have well-structured coefficient matrices. For example, one matrix in the problem series Chr# (introduced in [12]) is the adjacency matrix of a weighted tree. The matrix in problems Nug# [32], Scr# [36] and Sko# [37] is the Manhattan distance matrix of rectangular grids. It is reasonable to expect that if we use the special structure of the QAPs, more efficient models and algorithms could be developed.

One of the major contributions of this work is the existence of at least n different optimal solutions to the underlying QAPs. In particular, we show that for any fixed index pair (i, j), there exists a optimal permutation matrix X^* satisfying $x_{ij}^* = 1$. This observation can help us to restrict the search space and speed up all the existing QAP solvers when applied to the underlying problems in the present work. It should be pointed out that due to the discrete feature of constraint set involved in QAPs, the existence of multiple optimal solutions to general QAPs is not a big surprise. For example, in [31], Mautor and Roucairol observed that for several special classes of QAPs associated with some graphes, the symmetries in the graph can lead to the multiplicity of optimal solutions. Further, the authors of [31] used the symmetries in the associated graphes to improve the B&B approach.

The second major contribution is the introduction of a new ILP model for the index assignment problem based on the geometric feature of the hypercube. Several LP relaxations of the new model, derived from the graphical characterization of the hypercube are suggested. The lower bounds provided by these LP relaxations of the new model are analyzed and compared with what provided by several classical LP relaxations of QAPs.

The paper is organized as follows. In Section 2, we first describe the basic geometric structure of the hypercube. Then, we use the geometric structure of the hypercube to explore the symmetries in the solution set of the underlying QAP. In Section 3, we discuss how to characterize a hypercube via its adjacency matrix. In Section 4, we reformulate the original QAP as an equivalent ILP. Various LP relaxations of the new ILP model based on the graphical features of the hypercube are derived. In particular, we construct a simple LP with n^2 variables, which can be viewed as a relaxation of the original problem based, can provide a stronger bound than the LP relaxation of model (3). Based on such a observation and the graphical characterization of the hypercube, we then lift the simple LP model to higher dimension. In Section 4 we report some numerical experiments that show that our new model can provide a tighter lower bound than most existing lower bounds in the literature. In Section 5 we conclude our paper with some remarks.

2. NON-UNIQUENESS OF OPTIMAL SOLUTIONS TO QAPS ASSOCIATED WITH HYPERCUBE

In this section, we first describe several basic geometric properties of a hypercube in space $\Re^d$. Then we use these geometric properties of the hypercube to explore the symmetries in the solution set of the associated QAPs.

First we note that corresponding to every vertex v is a binary code c_v of length d. By the definition of the hypercube, the *degree* of every vertex in the hypercube is d, i.e., $deg(v_i) = d$. Let us also define the distance between two vertices v_i and v_j to be the Hamming distance between two binary codes corresponding to the two vertices, i.e.,

$$\delta(v_i, v_j) = \sum_{k=1}^{d} |c_{v_i}^k - c_{v_j}^k|.$$

We next present several technical results regarding the distance functions that will be used later on to reformulate the QAP associated with a hypercube as an ILP. The following result is an immediate consequence of the above definition of the distance function.

Lemma 2.1 *Suppose that v_1, v_2, v_3 are three different vertices of a hypercube. There exists a unique index $l \in \{0, 1, \cdots, \lfloor |\delta(v_1, v_2) - \delta(v_2, v_3)|/2 \rfloor \}$ such that*

$$\delta(v_1, v_3) = |\delta(v_1, v_2) - \delta(v_2, v_3)| + 2l.$$

In particular, if $\delta(v_1, v_2) = \delta(v_2, v_3) = 1$ or $\delta(v_1, v_2) = \delta(v_2, v_3) = d - 1$, then

$$\delta(v_1, v_3) = 2.$$

We next consider a vertex pair whose distance is k. We have

Lemma 2.2 *Suppose that v_i, v_j are two vertices of the hypercube satisfying $\delta(v_i, v_j) = k$ with $1 < k \leq d$, then there exist exactly k different vertices $v_{i_1}, v_{i_2}, \cdots, v_{i_k}$ such that*

$$\delta(v_i, v_{i_1}) = \delta(v_i, v_{i_2}) = \cdots = \delta(v_i, v_{i_k}) = k - 1,$$
$$\delta(v_j, v_{i_1}) = \delta(v_j, v_{i_2}) = \cdots = \delta(v_j, v_{i_k}) = 1.$$

Moreover, for every vertex v of the hypercube, we have

$$|S(v,l)| = C(d,l), \forall l = 1, \cdots, d,$$

where

$$S(v,l) = \{v' \in V : \delta(v,v') = l\},$$

$|S(.,.)|$ *denotes the cardinality of the set, and $C(d,l)$ is the combinatorial function.*

Proof: Without loss of generality, we can assume that v_i is the origin of the space $\mathfrak{R}^d$, while v_j is the vertex whose first k elements are 1 and the others are 0. Let us choose $v_{k_i} = (1, \cdots, 1, 0, 1, \cdots, 1, 0, \cdots, 0)^T$ to be the vector whose first k elements are 1 except its i-th element, which has value 0. One can easily verify that our choice satisfies the relations in the first conclusion of the lemma. The second conclusion is a direct consequence of the definition of the hypercube.
                                                                                                                q.e.d.

As a direct consequence of Lemma 2.2, we have

Corollary 2.3 *For any vertex v of the hypercube, there exists a unique vertex v' (called the complement of v) such that $\delta(v,v') = d$. Moreover, for any two vertex pair v_1, v_2, we have*

$$\delta(v_1, v_2) = \delta(v'_1, v'_2).$$

For convenience, we also introduce the complement of a code c.

Definition 2.4 *let c be a binary code of length d. We call the binary code c' the complement of c if the Hamming distance between c and c' equal d, i.e., every component in c' is generated by switching 0 to 1 and verse vice for the component of c at the same position.*

We now discuss the variety of the optimal solutions of the underlying QAP in this paper. We have

Theorem 2.5 *Suppose that B is the adjacency matrix of a hypercube in a suitable space. Then, for any coefficient matrix A, there exist at least n different optimal solutions to problem (1).*

Proof: Let us assume that X^* is an optimal solution of problem (1). Let us define $B^* = (X^*)^T B X^*$. Then problem (1) can be equivalently stated as

$$\min_{X \in \mathscr{X}} \mathrm{Tr}(AX^T B^* X). \tag{4}$$

The optimality of X^* implies that the identify matrix I is an optimal solution of problem (4). Now let us recall the fact that B^* is the adjacency matrix of a hypercube, i.e., for any index i, j, there is a corresponding vertex pair v_i, v_j such that $b_{ij}^* = 1$ if and only if $\delta(v_i, v_j) = 1$. Now recall Corollary 2.3, we know that for every index i, there exists another index i' such that $\delta(v_i, v_{i'}) = d$. Moreover, by Corollary 2.3, for any vertex pair (v_i, v_j), we have $\delta(v_i, v_j) = \delta(v_{i'}, v_{j'})$. This implies, if we choose the values of the elements of P in (4) by the following rule

$$x_{ij} = \begin{cases} 1 & \text{if } j = i'; \\ 0 & \text{otherwise}; \end{cases}$$

then it must hold $X^T B^* X = B^*$.

Now let us recall the fact that for every vertex v, there is an associated binary code c_v of d digits. Let I be an index set $I \subseteq \{1, 2, \cdots, d\}$ and its complement $\bar{I}$ defined by $\bar{I} = \{1, 2, \cdots, d\} - I$. We can separate the code corresponding to every vertex into two parts c_v^I and $c_v^{\bar{I}}$ respects to the index sets I and $\bar{I}$, i.e.,

$$c_v = c_v^I \oplus c_v^{\bar{I}}.$$

68

It is straightforward to see that for any two vertices v_i, v_j, we have

$$\delta(v_i, v_j) = \delta(c_{v_i}^I, c_{v_j}^I) + \delta(c_{v_i}^{\bar{I}}, c_{v_j}^{\bar{I}}).$$

Next we introduce the following transformation

$$c_v = c_v^I \oplus c_v^{\bar{I}} \Longrightarrow c_{v'}^I \oplus c_v^{\bar{I}}, \tag{5}$$

where $c_{v'}^I$ is the complement of c_v^I. By applying Corollary 2.3, one can easily verify that for any vertex pair (v_i, v_j), the following relation holds

$$\delta(v_i, v_j) = \delta(c_{v_i}^I, c_{v_j}^I) + \delta(c_{v_i}^{\bar{I}}, c_{v_j}^{\bar{I}}) = \delta(c_{v_i'}^I, c_{v_j'}^I) + \delta(c_{v_i}^{\bar{I}}, c_{v_j}^{\bar{I}}).$$

Let X be the corresponding permutation matrix induced by the transformation (5). Our above discussion implies that

$$X^T B^* X = B^*.$$

If we perform the above permutation for every index set $I \subseteq \{1, \cdots, d\}$, one can easily see that we will obtain $2^d = n$ different permutation matrices such that

$$X^T B^* X^T = B^*.$$

This shows that problem (4) has at least n different optimal solutions.   q.e.d.

We next discuss how to use the conclusions from Theorem 2.5 to restrict the search space of the underlying QAP.

Theorem 2.6 *Suppose that B is the adjacency matrix of a hypercube in a suitable space. Then, for any coefficient matrix A and any fixed index pair (i, j), there exists a global optimal solution to problem (1) that satisfies $x_{ij} = 1$.*

Proof: We first prove the special case with fixed index pair $i = j = 1$. Without loss of generality, we can assume that B is the adjacency matrix of the hypercube corresponding to the natural labeling, i.e., v_i corresponds to the binary coding of number $i - 1$. Suppose that P^* is a global optimal solution of problem (1) and v is the first labeled vertex in the final solution. Let c_v be the binary code of vertex v and $I \subseteq \{1, \cdots, d\}$ the index set corresponding to the components of c_v with value 1. Now we can apply the transformation (5) to this particular index set I. Let us denote the corresponding induced permutation matrix by P. From (5), we can conclude that

$$v \Longrightarrow v_1,$$

which implies

$$[X^* X]_{11} = 1.$$

The case for general fixed index pairs follows similarly. This finishes the proof of the theorem.   q.e.d.

We remark the fact that there exist multiple optimal solutions to QAPs is not surprising due to the discrete structure of the constrained set. However, to the best of the authors' knowledge, our result is the first one that shows the number of different optimal solutions is at least n, and such a conclusion holds independent of coefficient matrix A. This partially explains why the classical B&B approach for QAPs might take a long process to locate a global optimum. On the other hand, though the constructed optimal permutation matrices for problem (4) are completely different, the corresponding adjacency matrix of the hypercube remains the same. This indicates that if we can use the geometric properties of the hypercube to derive an optimization model equivalent to model (1), then it is possible that the new optimization model won't have multiple optimal solutions. This will definitely help to speed up the B&B approach applied to the new model.

3. A NEW CHARACTERIZATION OF THE HYPERCUBE

In this section we characterize the hypercube via its adjacency matrix. For a given vertex pair (v, v') of a graph, we define the distance between v and v' as the length of the shortest path that links v and v'. For convenience, we introduce the following definition:

Definition 3.1 *For a given graph $G = (V,E)$, we define the indicator cubic matrix $W_{ij}^l \in \Re^{n \times n \times d}$ by*

$$w_{ij}^l = \begin{cases} 1 & \delta(v_i,v_j) = l; \\ 0 & \delta(v_i,v_j) \neq l \end{cases}$$

We also denote by W^l the square matrix $(W^l)_{ij} = w_{ij}^l$ for $l = 1, \cdots, d$.

Based on Definition 3.1, the matrix W^1 is precisely the adjacency matrix of the graph G.

We next explore the properties of the indicator matrix W of a hypercube. Combining Lemma 2.1, Lemma 2.2 and Definition 3.1, we have

Proposition 3.2 *Let $W \in \Re^{n \times n \times d}$ be the cubic matrix defined by (3.1) and G be a hypercube in $\Re^d$. Then one has*

$$\sum_{l=1}^{d} w_{ij}^l = 1, \qquad \forall i \neq j \in \{1, \cdots, n\}, \tag{6}$$

$$\sum_{j=1}^{n} w_{ij}^l = C(d,l), \qquad \forall i \in \{1, \cdots, n\}, l \in \{1, \cdots, d\}. \tag{7}$$

From the second conclusion of Lemma 2.1, we can derive the following

$$w_{ik}^1 + w_{kj}^1 + w_{ij}^1 \leq 2, \forall i,j,k \in \{1, \cdots, n\}$$

which is also called anti-triangular inequalities in graph theory. By using the first conclusion from Lemma 2.1, we can further derive the following enhanced version of the anti-triangular inequalities

$$\sum_{\substack{l \text{ is odd}}} w_{ik}^l + \sum_{\substack{l \text{ is odd}}} w_{kj}^l + \sum_{\substack{l \text{ is odd}}} w_{ij}^l \leq 2 \qquad \forall i,j,k \in \{1, \cdots, n\};$$

$$\sum_{\substack{l \text{ is even}}} w_{ik}^l + \sum_{\substack{l \text{ is even}}} w_{kj}^l + \sum_{\substack{l \text{ is odd}}} w_{ij}^l \leq 2 \qquad \forall i,j,k \in \{1, \cdots, n\}.$$

For any $X \in \Re^{n \times n}$, let us denote $X_{i.}$ the i-th column of X. We have

Theorem 3.3 *Let $W \in \Re^{n \times n \times d}$ be the cubic matrix defined by (3.1) and G be a hypercube. Then it satisfies the following relations*

$$(W_{i.}^1)^T W_{j.}^1 = \begin{cases} 2 & \delta(v_i,v_j) = 2; \\ 0 & \delta(v_i,v_j) \neq 2 \end{cases} \qquad \forall\, i \neq j. \tag{8}$$

Proof: In order to prove the relation (8), we recall

$$(W_{i.}^1)^T W_{j.}^1 = \sum_{k=1}^{n} w_{ik}^1 w_{jk}^1.$$

If $\delta(v_i,v_j) = 2$, it follows from Lemma 2.2 immediately that $(W_{i.}^1)^T W_{j.}^1 = 2$. If $\delta(v_i,v_j) \neq 2$, then for every $k \in \{1, \cdots, n\}$, it must hold $w_{ik}^1 w_{jk}^1 = 0$. Because otherwise we have $w_{ik}^1 = w_{jk}^1 = 1$, which implies $\delta(v_i,v_j) = 2$ by Lemma 2.1. This contradicts to the assumption that $\delta(v_i,v_j) \neq 2$. This concludes the proof of the theorem.   q.e.d.

When G is a hypercube, the relation (8) established a one-to-one map between w_{ij}^2 and the row (or column) pairs of W^1. Unfortunately, such a relation does not exist between w_{ij}^l and the row (or column) pairs from W^1 and W^{l-1} when $l \geq 3$. This can be seen from the following simple example. Suppose that $w_{12}^1 = w_{23}^1 = w_{34}^1 = 1$, then $w_{13}^2 = 1$. Therefore, $(W_{1.}^2)^T W_{2.}^1 > 0$. However, we have $w_{12}^1 = 1, w_{12}^3 = 0$. Nevertheless, we have

Theorem 3.4 *Let $W \in \Re^{n \times n \times d}$ be the cubic matrix defined by (3.1). Then $\delta(v_i,v_j) = l$ if and only if $(W_{i.}^{l-1})^T W_{j.}^1 > 0$ and $\sum_{k=1}^{l-1} w_{ij}^k = 0$.*

Proof: The necessary part of the theorem follows from Definition 3.1. It remains to prove the sufficient part of the theorem. Suppose that $(W_{i.}^{l-1})^T W_{j.}^1 > 0$. Then there exists a path from v_i to v_j whose length is at most l. Because $\sum_{k=1}^{l-1} w_{ij}^k = 0$, it must hold $\delta(v_i, v_j) = l$.

q.e.d.

We next discuss how to characterize a hypercube in $\Re^d$ by making use of its adjacency matrix. The new characterization is based on the results in [29] where the authors characterized the hypercube in a certain space based on the relations between distinct adjacent edges of a graph. For notational convenience, we introduce the following definition.

Definition 3.5 *A graph $G = (V, E)$ is called an exact 4-cycle graph if each pair of distinct adjacent edges lies in exactly one 4-cycle.*

An exact 4-cycle graph shares many common interest properties as the hypercube. In particular, Laborde et'al [29] observed the following interesting link between an exact 4-cycle graph and the hypercube in a certain space.

Theorem 3.6 *[29] Suppose that a graph $G = (V, E)$ is a connected exact 4-cycle graph. If the degree of every vertex of G is d and $|V| = 2^d$, then G is a hypercube.*

Though the above theorem provides a complete geometric characterization of the hypercube in a certain space, it does not provide an algebraic characterization of the hypercube. We next show how to characterize the so-called exact 4-cycle relation in term of the adjacency matrix of a graph. It is worthwhile mentioning that from a viewpoint of computation, the approach based on the relations among different vertices of a graph is more promising since the number of vertices is much smaller than that of the edges.

Recall that the relation (8) describes the relation whether the distance between two vertices is 2. It follows from Definition 3.1 that

$$2w_{ij}^2 = (W_{i.}^1)^T W_{j.}^1, \quad for\ i \neq j = 1, \cdots, n.$$

Moreover, from Definition 3.1, we can easily see that

$$w_{ij}^1 + w_{ij}^2 \leq 1.$$

The above two relations can be used to define the so-called 4-cycle relation of a graph as shown by the following theorem.

Lemma 3.7 *Let $W \in \Re^{n \times n \times d}$ be the indicator matrix of a graph. If it satisfies the following relations*

$$(W_{i.}^1)^T W_{j.}^1 = 2w_{ij}^2, \quad w_{ij}^2 \in \{0, 1\}, \forall\, i \neq j \in \{1, \cdots, n\}; \tag{9}$$

$$w_{ij}^1 + w_{ij}^2 \leq 1, \quad \forall\, i \neq j = 1, \cdots, n. \tag{10}$$

then G is an exact 4-cycle graph.

Proof: Without loss of generality, we can assume that (v_1, v_2) and (v_2, v_3) are two distinct adjacent edges of G. We therefore have $w_{12}^1 = w_{23}^1 = 1$. This implies $w_{13}^1 = 1$ and thus $(W_1^1)^T W_3^1 = 2$. It follows that there exists another index $k \in \{1, \cdots, n\}$ such that $w_{1k}^1 = w_{k3}^1 = 1$ and the set of four points $\{v_1, v_2, v_3, v_k\}$ forms precisely the unique 4-cycle that contains (v_1, v_2) and (v_2, v_3).

q.e.d.

Before closing this section, we present the main theorem in this section which gives a complete characterization of a hypercube in term of its indicator matrix W.

Theorem 3.8 *Let $W \in \Re^{n \times n \times d}$ be the indicator matrix of a graph with $n = 2^d$. If it satisfies the following relations*

C1

$$\sum_{j=1}^n w_{ij}^l = C(d, l), \quad \forall i \in \{1, \cdots, n\}, l \in \{1, \cdots, d\};$$

C2 The matrices W^1 and W^2 satisfy the relation (9);

C3 For $l \geq 3$,

$$w_{ij}^l = 1 \iff (W_{i.}^{l-1})^T W_{j.}^1 > 0 \text{ and } \sum_{m=1}^{l-1} w_{ij}^m = 0;$$

$$\sum_{l=1}^{d} w_{ij}^{l} = 1, \quad i \neq j \in \{1,\cdots,n\},$$

then G is a hypercube in $\Re^d$.

Proof: The condition C1 ensures that the degree of every vertex in G is d. The second and forth conditions guarantee that two distinct adjacent edges of G lie precisely in a 4-cycle as shown by Lemma 3.7. The third and forth conditions ensure that the graph is connected since every pair of vertices is connected. Applying Theorem 3.6, we can conclude the proof. q.e.d.

Though Theorem 3.8 gives a new characterization of the hypercube in terms of its indicator matrix W, it is desirable to further reduce the number of constraints in the characterization. In what follows we discuss how to remove some constraints in Theorem 3.8.

First we observe that Condition C2 implies that

$$\sum_{j=1}^{n} w_{ij}^{2} = C(d,2), \quad \forall i \in \{1,\cdots,n\}.$$

Secondly, we can relax the conditions in the last two items of Theorem 3.8 to the following

C3.1 For $l \in \{3,\cdots,\lceil\frac{d}{2}\rceil\}$,

$$w_{ij}^{l} = 1 \Longleftrightarrow (W_{i.}^{l-1})^{T} W_{j.}^{1} > 0 \text{ and } \sum_{m=1}^{l-1} w_{ij}^{m} = 0;$$

C4.1

$$\sum_{l=1}^{\lceil\frac{d}{2}\rceil} w_{ij}^{l} \leq 1, \quad i \neq j \in \{1,\cdots,n\}$$

We have

Theorem 3.9 *Let $W \in \Re^{n \times n \times d}$ be the indicator matrix of a graph with $n = 2^d$. If it satisfies the relations C1, C2, C3.1 and C4.1, then G is a hypercube in $\Re^d$.*

Proof: Suppose that G is a graph satisfying the assumptions of the theorem. The conditions C2 and C4.1 ensure that two distinct adjacent edges of G lie precisely in a 4-cycle as shown by Lemma 3.7. For any $v \in V$, let us define

$$\mathscr{S}_1(v,l) = \{v' \in V : \delta(v,v') \leq l\}.$$

It follows immediately that

$$|\mathscr{S}_1(v,\lceil\tfrac{d}{2}\rceil)| = 1 + \sum_{l=1}^{\lceil\frac{d}{2}\rceil} C(d,l) > \frac{n}{2}.$$

Now let us consider any vertex pair $v, v' \in V$. We therefore have

$$|\mathscr{S}_1(v,\lceil\tfrac{d}{2}\rceil)| + |\mathscr{S}_1(v,\lceil\tfrac{d}{2}\rceil)| > n.$$

The above relation implies that there exists at least one vertex $\bar{v}$ in the intersection of the two sets $\mathscr{S}_1(v,\lceil\frac{d}{2}\rceil)$ and $\mathscr{S}_1(v',\lceil\frac{d}{2}\rceil)$. From the definition of $\mathscr{S}_1(v,.)$, we can conclude that there exists a path from v to v', i.e., the graph is connected. The theorem follows immediately from Theorem 3.6. q.e.d.

4. NEW ILP FORMULATIONS AND RELAXATIONS

In this section, we describe our new ILP model for the QAPs associated with the hypercube in $\mathfrak{R}^d$. The new model can be viewed as a combination of the geometric structure of the hypercube and the linear linearization technique for ILPs [3].

We start with the following simple LP model

$$\min \quad \mathrm{Tr}(AZ) \tag{11}$$
$$\text{s.t.} \quad z_{ii} = 0, Z = Z^T, z_{ij} \in [0,1], \quad \forall i \neq j \in \{1,\cdots,n\};$$
$$\sum_{j=1}^{n} z_{ij} = d, \qquad \forall i \in \{1,\cdots,n\}.$$

The above problem can be viewed as a special case of the transportation problem and thus it has an integer optimal solution. It is also easy to see that the graph that satisfies the constraints in the above model is a graph whose vertices have the same degree (d). Thus we can cast model (11) as a relaxation of the QAPs associated with a hypercube. Nevertheless, we have

Theorem 4.1 *Let Z^* be the optimal solution of the LP problem (11) and (X^*, Y^*) be the optimal solution of the LP relaxation of problem (3). If the matrix A is nonnegative and B is the adjacency matrix of the hypercube in $\mathfrak{R}^d$, then it must hold*

$$\sum_{i=1}^{n}\sum_{j=1}^{n}\sum_{k=1}^{n}\sum_{l=1}^{n} a_{ik} b_{jl} y_{ijkl}^* \leq \mathrm{Tr}(AZ^*).$$

Proof: To prove the theorem, we first show that from any optimal solution of the LP problem (11), we can construct a feasible solution for the LP relaxation of problem (3). Suppose that Z^* is the optimal solution of the relaxed version of problem (11). Let us define

$$x_{ij} = \frac{1}{d} z_{ij}^*, \quad i, j = \{1, \cdots, n\};$$
$$y_{ijkl} = x_{ij} x_{kl}, \quad i, j, k, l \in \{1, \cdots, n\}.$$

Then it is easy to verify that (X, Y) is a feasible solution of the LP relaxation of problem (3). Let E be the matrix whose elements have value 1. From the definition of X and Y, we have

$$\mathrm{Tr}(AZ^*) = \mathrm{Tr}(XAX^TE) \geq \mathrm{Tr}(XAX^TB) \geq \mathrm{Tr}(X^*A(X^*)^TB),$$

where the first inequality follows from the fact that $b_{ij} \leq 1$ for $i, j \in \{1, \cdots, n\}$. This finishes the proof of the theorem.
q.e.d.

Theorem 4.1 shows that compared with the classical ILP approaches based on model (3), much stronger lower bound can be derived by using the geometric structure of the hypercube for the underlying QAPs in the present work. We next present a result showing that the bound provided by the LP model (11) is weaker than the bound provided by the model in [20].

Theorem 4.2 *Let Z^* be the optimal solution of the LP problem (11) and (X^*, Y^*) be the optimal solution of the LP relaxation of the model in [20]. If the matrix A is nonnegative and B is the adjacency matrix of the hypercube in $\mathfrak{R}^d$, then it must hold*

$$\sum_{i=1}^{n}\sum_{j=1}^{n}\sum_{k=1}^{n}\sum_{l=1}^{n} a_{ik} b_{jl} y_{ijkl}^* \geq \mathrm{Tr}(AZ^*).$$

Proof: To prove the theorem, it suffices to construct a feasible solution to the LP model (11) from a solution of the LP relaxation of the model in [20]. Let (X^*, Y^*) be such a solution, we define

$$z_{ij} = \sum_{k,l=1}^{n} y_{ikjl}^* b_{kl}.$$

Since $b_{kl} \in \{0,1\}, \forall k, l \in \{1, \cdots, n\}$, it follows immediately that

$$0 \leq z_{ij} \leq \sum_{k,l=1}^{n} y_{ikjl}^* = \sum_{l=1}^{n} x_{jl} = 1.$$

Moreover, because $y^*_{ikil} = 0$ for all $k \neq l$ and $b_{ii} = 0$, we can claim that $z_{ii} = 0$ for all $i \in \{1, \cdots, n\}$. On the other hand, because B is the adjacency matrix of the hypercube in $\Re^d$, for any fixed $j \in \{1, \cdots, n\}$, we have

$$\sum_i^n z_{ij} = \sum_{i=1}^n \sum_{k,l=1}^n y^*_{ikjl} b_{kl} = \sum_{k,l}^1 x^*_{jl} b_{kl}$$

$$= \sum_{l=1}^n x^*_{jl} \sum_{k=1}^n b_{kl} = d \sum_{l=1}^n x^*_{jl} = d.$$

Recall that in the model (3), we have $y^*_{ikjl} = y^*_{jlik}$ and B is symmetric. It follows immediately $z_{ij} = z_{ji}, \forall i, j \in \{1, \cdots, n\}$. From the above discussion we can conclude that Z is a feasible solution of the LP model (11). This finishes the proof of the theorem.   q.e.d.

We can further add more constraints to problem (11) to close the gap between the constraint polyhedral set and the hypercube. For example, we can impose the condition that there is no triangles in graph. This will lead to the following model

$$\begin{aligned}
\min \quad & \mathrm{Tr}(AZ) \\
\text{s.t.} \quad & z_{ii} = 0, Z = Z^T, z_{ij} \in [0,1], \quad \forall i \neq j \in \{1, \cdots, n\}; \\
& \textstyle\sum_{j=1}^n z_{ij} = d, \quad \forall i \in \{1, \cdots, n\} \\
& z_{ij} + z_{jk} + z_{ki} \leq 2, \quad \forall i \neq j \neq k \in \{1, \cdots, n\}.
\end{aligned}$$

We now discuss how to add more constraints to model (11) so that the new model defines precisely a hypercube. Let us recall that when the coefficient matrix B in (1) is the adjacency matrix of a hypercube in $\Re^d$, for any permutation matrix P, the matrix $P^T BP$ remains an adjacency matrix of the hypercube with different numbering of the vertices. From Theorem 3.8, we can characterize such an adjacency matrix via the indicator matrix W. We recall that the conditions C1 and C4.1 can be represented by linear constraints. In what follows we discuss how to represent the conditions C2 and C3.1 by using linear equations and inequalities.

In the sequel we show how to replace the nonlinear relation (9) by a set of linear constraints. Since all the variables w^1_{ij} are binary, $(W^1_{i.})^T W^1_{j.} = 2 = 2w^2_{ij}$ if and only if there exist two indices k, l such that

$$w^1_{ik} = w^1_{jk} = w^1_{il} = w^1_{jl} = 1.$$

Let us define [1]

$$t^1_{ijk} = \min\{w^1_{ik}, w^1_{jk}\}, \quad k = 1, \cdots, n. \tag{12}$$

We can rewrite (9) as

$$\sum_{k=1}^n t^1_{ijk} = 2w^2_{ij}.$$

The relation (12) is still nonlinear. However, since all the elements w^1_{ij} are binary, it is easy to see that it can be replaced by the following relations

$$\begin{aligned}
t^1_{ijk} &\leq \min\{w^1_{ik}, w^1_{jk}\}, \quad k = 1, \cdots, n, \\
w^1_{ik} + w^1_{jk} &\leq 1 + t^1_{ijk}, \quad k = 1, \cdots, n.
\end{aligned}$$

We argue that the binary requirement on w^1_{ij} in (9) can be waived if we impose the following linear constraints $1 \geq w^2_{ij} \geq t^1_{ijk} \geq 0$ for all $i \neq j, k \in \{1, \cdots, n\}$. We need to consider only two cases. Suppose that $(W^1_{i.})^T W^1_{j.} > 0$. Then it must hold $\max_{k=1,\cdots,n} t^1_{ijk} = 1$, and thus $w^2_{ij} = 1$. If $(W^1_{i.})^T W^1_{j.} = 0$, then $w^2_{ij} = 0$. This in turn guarantees the condition (9).

---

[1] The artificial variables t^1_{ijk} can also be derived via the linearization technique in [3]. To see this, let us lift model (11) to higher dimension through the transformation $y_{ijkl} = x_{ij}x_{kl}$. We then use the geometric structure of the hypercube to impose constraints on the variables in the lifted space. In our special case, we have $t^1_{ijk} = y_{ikkj}$.

We now progress to characterize the relation in condition C3.1 by a set of linear constraints. For $l \in \{2, \cdots, d\}$, we introduce new artificial variables t_{ijk}^l satisfying the following relations

$$t_{ijk}^l \leq \min\{w_{ik}^l, w_{jk}^1, w_{ij}^{l+1}\}, \qquad k = 1, \cdots, n, \tag{13}$$

$$w_{ik}^l + w_{jk}^1 \leq 1 + t_{ijk}^l + t_{ikj}^{l-1}, \qquad k = 1, \cdots, n, \tag{14}$$

One can further verify that $w_{ij}^{l+1} = 1$ if and only if $t_{ijk}^l = 1$ and $\sum_{m=1}^l w_{ij}^m = 0$. Note that w_{ij}^{l+1} is binary automatically if the linear equalities (13)-(14) are satisfied and $w_{ij}^1, \cdots, w_{ij}^l$ are binary. This gives a complete characterization of condition C3.1 in Theorem 3.9. We thus have the following result.

Theorem 4.3 *Suppose that the matrix B in problem (1) is the adjacency matrix of a hypercube in $\Re^d$ and $n = 2^d$. Then the QAP (1) can be equivalently cast as the following ILP:*

$$\min \quad \mathrm{Tr}(AW^1) \tag{15}$$

$$\begin{aligned}
s.t. \quad & w_{ii}^1 = 0, w_{ij}^1 = w_{ji}^1 \in \{0,1\}, && \forall\, i,j \in \{1, \cdots, n\}; \\
& w_{ii}^l = 0, w_{ij}^l = w_{ji}^l, w_{ij}^l \in [0,1], && \forall i \neq j \in \{1, \cdots, n\}, l \in \{2, \cdots, d\}, \\
& t_{ijk}^0 = 0, u_{ijk}^0 = 0, && \forall\, i \neq j, k \in \{1, \cdots, n\}, l \in \{1, \cdots, d\}; \\
& t_{ijk}^l + t_{ikj}^{l-1} \leq \min\{w_{ik}^l, w_{jk}^1\}, && \forall i, j, k \in \{1, \cdots, n\}, l \in \{1, \cdots, d-1\}, \\
& 0 \leq t_{ijk}^l \leq w_{ij}^{l+1}, && \forall i, j, k \in \{1, \cdots, n\}, l \in \{1, \cdots, d-1\}, \\
& w_{ik}^l + w_{jk}^1 \leq 1 + t_{ijk}^l + t_{ikj}^{l-1}, && \forall i, j, k \in \{1, \cdots, n\}, l \in \{1, \cdots, d-1\}, \\
& (l+1)w_{ij}^{l+1} = \sum_{k=1}^n t_{ijk}^l, && \forall i \neq j \in \{1, \cdots, n\}, l = \{1, \cdots, d-1\}, \\
& u_{ijk} \leq \min\{w_{ik}^{d-1}, w_{jk}^{d-1}\}, && \forall i \neq j, k \in \{1, \cdots, n\}, \\
& t_{ijk}^1 + u_{ijk} \leq w_{ij}^2, && \forall i, j, k \in \{1, \cdots, n\}, \\
& w_{ik}^{d-1} + w_{kj}^{d-1} \leq 1 + u_{ijk}, && \forall i \neq j, k \in \{1, \cdots, n\}; \\
& \sum_{k=1}^n u_{ijk} = 2w_{ij}^2, && \forall i \neq j \in \{1, \cdots, n\}; \\
& \sum_{j=1, j \neq i}^n t_{ijk}^l = (d-l)w_{ik}^l, && \forall i \neq k \in \{1, \cdots, n\}, l \in \{1, \cdots, d-1\}; \\
& \sum_{i=1, i \neq j}^n t_{ijk}^l = C(d-1, l)w_{kj}^1, && \forall j \neq k \in \{1, \cdots, n\}; \\
& \sum_{j=1, j \neq i}^n u_{ijk} = (d-1)w_{ik}^{d-1}, && \forall i \neq k \in \{1, \cdots, n\}; \\
& \sum_{i=1, i \neq j}^n u_{ijk} = (d-1)w_{kj}^{d-1}, && \forall j \neq k \in \{1, \cdots, n\}; \\
& \sum_{l=1}^{d-1} \left(t_{ijk}^l + t_{ikj}^l\right) = w_{jk}^1, && \forall i \neq j, i \neq k \in \{1, \cdots, n\}, \\
& \sum_{l=1}^d w_{ij}^l = 1, && \forall i \neq j \in \{1, \cdots, n\}.
\end{aligned}$$

Proof: By the discussion preceding the theorem, we can conclude that if W is a feasible solution of problem (15), then W^1 must be an adjacency matrix of the hypercube in $\Re^d$. This finishes the proof of the theorem. q.e.d. .
It is straightforward to verify that the above problem has $\frac{1}{2}n(n-1)$ binary variables and roughly $O(n^3 \log n)$ linear constraints.
An alternative model is the following

$$\min \quad \mathrm{Tr}(AW^1) \tag{16}$$

$$\begin{aligned}
s.t. \quad & w_{ii}^1 = 0, w_{ij}^1 = w_{ji}^1 \in \{0,1\}, && \forall\, i,j \in \{1, \cdots, n\}; \\
& w_{ii}^l = 0, w_{ij}^l = w_{ji}^l, w_{ij}^l \in [0,1], && \forall i \neq j \in \{1, \cdots, n\}, l \in \{2, \cdots, d\}, \\
& w_{ij}^0 = 1, w_{ij}^0 = 0, w_{ij}^{d+1} = 0, && \forall i \neq j, k \in \{1, \cdots, n\}, l \in \{1, \cdots, d\}; \\
& 0 \leq t_{ijk} \leq \min\{w_{ik}^1, w_{jk}^1,\}, && \forall i \neq j, k \in \{1, \cdots, n\}, \\
& w_{ik}^1 + w_{jk}^1 \leq 1 + t_{ijk}, && \forall i \neq j, k \in \{1, \cdots, n\},
\end{aligned}$$

75

$$0 \le u_{ijk} \le \min\{w_{ik}^{d-1}, w_{jk}^{d-1}, \}, \qquad \forall i \ne j, k \in \{1, \cdots, n\},$$

$$w_{ik}^{d-1} + w_{kj}^{d-1} \le 1 + u_{ijk}, \qquad \forall i \ne j, k \subset \{1, \cdots, n\};$$

$$t_{ijk} + u_{ijk} \le w_{ij}^2, \qquad \forall i \ne j, k \in \{1, \cdots, n\};$$

$$\sum_{k=1}^{n} t_{ijk} = 2w_{ij}^2, \qquad \forall i \ne j \in \{1, \cdots, n\}, l = \{1, \cdots, d-1\},$$

$$\sum_{k=1}^{n} u_{ijk} = 2w_{ij}^2, \qquad \forall i \ne j \in \{1, \cdots, n\};$$

$$\sum_{j=1, j \ne i}^{n} t_{ijk} = (d-1)w_{ik}^1, \qquad \forall i \ne k \in \{1, \cdots, n\};$$

$$\sum_{i=1, i \ne j}^{n} t_{ijk} = (d-1)w_{kj}^1, \qquad \forall j \ne k \in \{1, \cdots, n\};$$

$$\sum_{j=1, j \ne i}^{n} u_{ijk} = (d-1)w_{ik}^{d-1}, \qquad \forall i \ne k \in \{1, \cdots, n\};$$

$$\sum_{i=1, i \ne j}^{n} u_{ijk} = (d-1)w_{kj}^{d-1}, \qquad \forall j \ne k \in \{1, \cdots, n\};$$

$$w_{ik}^l + w_{kj}^1 \le 1 + w_{ij}^{l+1} + w_{ij}^{l-1}, \qquad i \ne j, k \in \{1, \cdots, n\}, l \in \{2, \cdots, d\},$$

$$w_{ik}^{d-1} + w_{kj}^{d-l} \le 1 + w_{ij}^{l-1} + w_{ij}^{l+1}, \qquad i \ne j, k \in \{1, \cdots, n\}, l \in \{2, \cdots, d\},$$

$$\sum_{j=1}^{n} w_{ij}^l = C(d, l) \qquad \forall i \in \{1, \cdots, n\}, l \in \{1, \cdots, d\},$$

$$\sum_{l=1}^{d} w_{ij}^l = 1, \qquad \forall i \ne j \in \{1, \cdots, n\}.$$

5. NUMERICAL EXPERIMENTS

In this section, we report some numerical results based on our model. As pointed out in Section 2, the theoretical results in our work can indeed be used to improve most QAP solvers for the underlying problems. Therefore, in this section we only compare the lower bounds provided by various LP relaxations. For a fair comparison, we list the results based on our two models (16) and (15), model (3) (denoted by A&J), and the model in [20] (G&H98)[2] and their improved versions (A&J+ and G&H98+) with additional constraints (6)-(7), the GL[18] bound and the bound by the projection method proposed in [19, 6] (denoted by HRW).

We examine the performance of these models for several different choices of the matrix A. The first one is the matrix used in [39] defined by

$$a_{ij} = \frac{\Delta^3 \sqrt{1-\rho^2}}{(2\pi\sigma^6)^{\frac{3}{2}}} \sum_{l=1}^{n} \exp^{-\frac{\delta^2}{2\sigma^2}((1-\rho^2)(i-n_1)^2 + (j-n_1-\rho*(i-n_1))^2 + (l-n_1-\rho*(i-n_1))^2)},$$

where $n_1 = \frac{n+1}{2}$, and Δ is the step size to quantize the source, σ is the variance of the Gaussian Markov source with zero mean and ρ its correlation coefficient. In our experiment, we set the step size $\Delta = 0.4, \sigma = 1$ and $\rho = 0.1, 0.9$ respectively. These two different choices (denoted by Eng1 and Eng9) of ρ represent the scenarios of the source with dense correlation and non-dense correlation. We also include one example from vector quantization (denoted by VQ) provided by the last author of this work.

The other test problems include the so-called Harper code [23] with

$$a_{ij} = |i - j|.$$

and a random positive semi-definite matrix A. We point out that due to the special structure of the matrix A in Harper code, the optimal code can be found in polynomial time.

Our experiments were done on an 2.4GHz 64-bit AMD Opteron. We used CPLEX-10.2 with the barrier method to solve the LP relaxation. To see the quality of these lower bound, we also list the best solutions that have been found with a heuristics. In the second row for each model, we also list the CPU time in seconds to obtain the bound. The results for $d = 4$ resp. $d = 5$ are summarized in Table 1 resp. Table 2.

---

[2] We did not test the high-order lift model in [21] because it requires a great amount of memory. On the other hand, though very strong lower bounds have been obtained in [21], these bounds are obtained based on the dual Lagrangian method combined with some heuristics, which is quite different from the approach in the present paper.

TABLE 1. Bounds for QAPs associated with hypercube in $\Re^4$

	Harper	Random	Eng1	Eng9	VQ
Best Sol	240	64.344	0.0823832	0.0702545	19.7672
(15)	224.87 1592	64.313 1469	0.0823832 1168	0.0592348 2921	18.8316 2199
(16)	215.64 78	64.0187 143	0.0823832 95	0.0576037 109	17.8121 153
A&J	74 78	62.8863 89	0.0749003 109	0.0479947 88	7.07798 87
A&J +	122.02 82	63.4405 64	0.0771238 65	0.0548599 88	9.8353 70
GH98	120 76	62.8863 77	0.0749003 76	0.0479947 67	11.2419 75
GH98 +	160.47 62	63.4405 48	0.0771238 43	0.0548599 58	13.7575 60
GL	106 1	36.2555 1	0.0149548 1	0.0231322 1	10.2583 1
HRW	197.55 1	50.3012 1	0.0483 1	0.0458 1	15.8599 1

TABLE 2. Bounds for QAPs associated with hypercube in $\Re^5$

	Harper	Random	Eng1	Eng9	VQ
Best Sol	992	438.51	.8397e-1	.12402	20.487
(16)	650	292	.1355e-3	.2978e-3	5.5576
A&J	218	285.30	.6566e-4	.1138e-3	2.8149
A&J +	354.76	288.54	.8094e-4	.1927e-3	3.2097
GH98	324	285.30	.6566e-4	.1138e-3	3.9697
GH98 +	450.71	288.54	.8094e-4	.1927e-3	4.4460
GL	304	107.21	.1022e-4	.2424e-4	2.11940
HRW	783.03	211.96	-.2506	-.1533	.5013

As we can see from both Table 1 and 2, the bound by model (15) is the strongest, followed by model (16). However, it requires more memory and takes longer time to solve the LP relaxation of model (15). When $d = 5$, we run out of memory for model (15). The CPU time for solving the LP relaxation of model (16) is slightly longer than that for A&J and GH98 with a comparable memory usage. Not surprising to see that the improved versions (A&J+ and GH98+) require less CPU time, but result in better bounds than their original counterparts (A&J and GH98). We also note that both the GL and HRW bounds are very easy to compute. However, the GL bound is much weaker than other lower bounds in general. We also note that the HRW model can find very good bounds for the Harper code and the random problem very efficiently. However, as we can see from Table 2, for Eng1 and Eng9, the lower bound when $d = 5$ turns out be useless. Overall, the numerical effort is very comparable as for $d = 4$ and since our emphasis is on bounds, we do not list CPU times for $d = 5$.

6. CONCLUSIONS

In this paper, we have studied the QAPs associated with a hypercube in a finite space. By using the geometric structure of the hypercube, we show that there exist at least n different optimal permutations. Moreover, the symmetries in the hypercube allow us to restrict the search space and thus improve many existing QAP solvers for the underlying problem. Numerical experiments confirm our theoretical conclusions.

New ILP models for the underlying QAP are suggested. Compared with the existing ILP formulations for the underlying problem, our new model has the smallest number of binary variables and its LP relaxations can provide tighter lower bounds as verified by numerical experiments.

There are several different ways to extend our results. First of all, the LP relaxation of our new models still have $O(n^3 \log(n))$ constraints, which implies the model can not be applied to large-scale problems. One possible way to deal with this issue is to use relaxation based on other convex optimization models such as the LP over the second-order cone. To see this, let us recall that the adjacency matrix of a hypercube satisfies the condition

$$(W_{i.}^1)^T W_{j.}^1 \in \{0, 2\}, \quad \forall\, i \neq j$$

which can be relaxed to

$$(W_{i.}^1)^T W_{j.}^1 \leq 2, \quad \forall\, i \neq j,$$

which equals to

$$\|W_{i.}^1 + W_{j.}^1\|^2 \leq 2d + 4, \quad \forall\, i \neq j.$$

In this way, we can relax (15) to a LP over the second-order cone. More study is needed to explore whether such an approach can lead to computational advantages.

Another way to extend our approach is to develop new ILP models for other QAPs with structured coefficient matrices such as the adjacency matrix of a tree and Manhattan distances of rectangular grids.

REFERENCES

1. W.P. Adams and T.A. Johnson, "Improved Linear Programming-Based Lower Bounds for the Quadratic Assignment Problem", In *P.M. Pardalos and H. Wolkowicz, editors, Quadratic Assignment and Related Problems*, DIMACS 25 Series in Discrete Mathematics and Theoretical Computer Science, 1994, pp. 43-75.
2. W.P. Adams and H.D. Sherali, "A tight linearization and an algorithm for zero-one quadrtic programming problems", *Management Science*, **32**, 1274-1290 (1986).
3. W.P. Adams and H.D. Sherali, "A hierarchy of relaxations between continuous and convex hull representations for zero-one programming problems". *SIAM J. Discrete Applied Mathematics* **52**, 83-106 (1990).
4. K.M. Anstreicher and N.W. Brixius, "A new bound for the quadratic assignment problem based on convex quadratic programming", *Mathematical Programming* **80**, 341-357 (2001).
5. K.M. Anstreicher, N.W. Brixius, J.-P. Goux, and J.Linderoth. "Solving large quadratic assignment problems on computational grids", *Mathematical Programming* **91**, 563-588 (2002).
6. G. Ben-David and D. Malah, "Bounds on the performance of vector-quantizers under channel errors". *IEEE Transactions on Information Theory*, **51**, 2227–2235 (2005).
7. R. E. Burkard, "Efficiently Solvable Special Cases of Hard Combinatorial Optimization Problems", *Mathematical Programming* **79**, 55–69 (1997).
8. R. E. Burkard, E. Çela, G. Rote and G.J. Woeginger, "The Quadratic Assignment Problem with a Monotone Anti-Monge Matrix and a Symmetric Toeplitz Matrix: Easy and Hard Cases", *Mathematical Programming*, **82**, 125–158 (1998).
9. R.E. Burkard and J. Offermann. "Entwurf von Schreibmaschinentastaturen mittels quadratischer Zuordnungsprobleme". *Zeitschrift für Operations Research* **21**, B121-B132 (1977).
10. R.E. Burkard, E. Çela, S.E. Karisch and F. Rendl, "QAPLIB-A quadratic assignment problem library". *J. Global Optimization*, **10**, 391–403 (1997).
11. N.T. Cheng and N.K. Kingsbury, "Robust zero-redundancy vector quantization for noisy channels". In *Proceedings of the International Conference on Communications*, Boston, 1989, pp. 1338–1342.
12. N. Christofides and E. Benavent, "An exact algorithm for the quadratic assignment problem". *Operations Research*, **37**, 760–768, (1989).
13. N. Christofides and M. Gerrard, "A graph theoretical analysis of bounds for the quadratic assignment problem". In *P. Hansen eds: Studies on Graphs and Discrete programming*, North Holland, 1981, pp. 61-68.
14. Z. Drezner, "Lower bounds based on linear programming for the quadratic assignment problem". *Comput. Optim. & Applic.*, **4**, 159–169 (1995).
15. G. Finke, R.E. Burkard, and F. Rendl, "Quadratic assignment problems". *Annals of Discrete Mathematics*, **31**, 61–82 (1987).
16. A.M. Frieze and J. Yadegar, "On the quadratic assignment problem", *Discrete Applied Mathematics*, **5**, 89–98 (1983).
17. J.W. Gavett and N.V. Plyter, "The optimal assignment of facilities to locations by branch-and-bound". *Operations Research*, **14**, 210-232 (1966).
18. P.C. Gilmore, "Optimal and suboptimal algorithms for the quadratic assignment problem. *SIAM Journal on Applied Mathematics*, **10**, 305–313 (1962).
19. S.W. Hadley, F. Rendl, and H. Wolkowicz, "A new lower bound via projection for the quadratic assignment problem". *Mathematics of Operations Research*, **17**, 727-739 (1992).

20. P. Hahn. and T. Grant, "Lower bounds for the quadratic assignment problem based upon a dual formulation". *Operations Research*, **46**, 912–922 (1998).
21. P. Hahn, W.L. Hightower, T.A. Johnson, M. Guignard-Spielberg and C. Roucairol, "A Lower Bound for the Quadratic Assignment Problem Based on a Level-2 Reformulation- Linearization Technique", *University of Pennsylvania, Department of Systems Engineering*, Technical Report, 2001.
22. M. Hannan and J.M. Kurtzberg, "Placement techniques". In *Design Automation of Digital Systems: Theory and Techniques*, M.A. Breuer Ed., vol(1), Prentice-hall: Englewood Cliffs, 1972, pp. 213–282.
23. L.H. Harper, "Optimal assignments of numbers to vertices". *J. of the Society for Industrial and Applied Mathematics*, **12** , 131–135 (1964).
24. S.E. Karisch, E. Çela, J. Clausen. and T. Espersen, "A dual framework for lower bounds of the quadratic assignment problem based on linearization". *Computing*, **63**, 351-403 (1999).
25. S.E. Karisch and F. Rendl, "Lower bounds for the quadratic assignment problem via triangle decompositions". *Mathematical Programming*, **71**, 137-152 (1995).
26. L. Kaufman and F. Broeckx, "An algorithm for the quadratic assignment problem using Benders decomposition", *European Journal of Operational Research*, **2**, 204–211 (1978).
27. T. Koopmans and M. Beckmann. "Assignment problems and the location of economic activities", *Econometrica*, **25**, 53–76 (1957).
28. E. Lawler, "The quadratic assignment problem". *Management Science*, **9**, 586-599 (1963).
29. J.M. Laborde, S. Prakash and R. Hebbare, "Another characterization of hypercubes". *Discrete Mathematics*, **39**, 161–166 (1982).
30. E.M. Loiola, N.M. Maia de Abreu, P.O. Boaventura-Netto, P. Hahn, and T. Querido, "A survey for the quadratic assignment problem". *European J. Oper. Res.*, **176**, 657–690 (2007).
31. T. Mautor and C. Roucairol. "A new exact algorithm for the solution of quadratic assignment problems". *Discrete Applied Mathematics*, **55**, 281–293 (1994).
32. C.E. Nugent, T.E. Vollman, and J. Ruml, "An experimental comparison of techniques for the assignment of facilities to locations". *Operations Research*, **16**, 150–173 (1968).
33. L.C. Potter and D.M. Chiang, "Minimax nonredundant channel coding". *IEEE Transactions on Communications*, **43**, 804–811 (1995).
34. K.G. Ramakrishnan, M.G.C. Resende, B. Ramachandran. and J.F. Pekny, "Tight QAP bounds via linear programming". In: *Combinatorial and Global Optimization*, Eds by P.M. Pardalos, A. Migdalas, and R.E. Burkard, World Scientific Publishing, Singapore, 2002, pp. 297–303.
35. M.G.C. Resende, K.G. Ramakrishnan and Z. Drezner, "Computing lower bounds for the quadratic assignment problem with an interior point algorithm for linear programming". *Operations Research*, **43**, 781–791 (1995).
36. M. Scriabin and R.C. Vergin, "Comparison of computer algorithms and visual based methods for plant layout". *Management Science*, **22**, 172–187 (1975).
37. J. Skorin-Kapov, "Tabu search applied to the quadratic assingnment problem". *ORSA Journal on Computing*, **2**, 33–45 (1990).
38. E.D. Taillard, "Comparison of iterative searches for the quadratic assingnment problem". *Location Science*, **3**, 87-105 (1995).
39. X. Wang, X. Wu and S. Dumitrescu, "On optimal index assignment for MAP decoding of Markov sequences", Technical report, *Department of Electrical and Computer Engineering*, McMaster University, Ontario, Canada, 2006.
40. K. Zeger and A. Gersho, "Pseudo-Gray Code". *IEEE Trans. on Communications*, **38**, 2147–2158 (1990).
41. Q. Zhao, S.E. Karisch, F. Rendl, and H. Wolkowicz, "Semidefinite programming relaxations for the quadratic assignment problem". *J. Combinatorial Optimization*, **2**, 71–109 (1998).

Integrate System Modeling for Design and Production Planning of High Quality Products Considering Failure Data

Kazuhiro Aoyama and Tsuyoshi Koga

Department of Systems Innovation, School of Engineering, The University of Tokyo
7-3-1 Hongo, Bunkyo-ku, Tokyo, Japan 113-8656

Abstract. Since the product recall problem is recently observed, realizing design and production activities that can prevent occurrences of a product failure has been becoming a serious issue. This paper proposes "the Synthetic design approach of a product and a production process which enables to reduce occurrences of a product failure from the initial stage of a product development". In order to realize this proposed concept as a specific system, the integrated model of the failure information in the design and production is introduced. This paper shows some examples of design and production for a circuit breaker and an automobile with considering a design error and a production failure using developed prototype system.

Keywords: Design Error, Product Failure, Production Failure, FTA, FMEA, System Integration

INTRODUCTION

In these years, the loss of money and reliability of manufacturing industry, caused by product quality related issues, is increasing. For instance, the number of recalls of automobile has been increasing globally in the recent years [1]. Engineers predicted that this recall rate might progressively increase in the future [2]. Realizing design and production activities that can prevent occurrences of a product failure has been becoming a serious issue.

On the other hand, manufacturing industries make efforts to develop and to introduce the PLM (Product Lifecycle Management) system that enable to manage engineering information from the product design to a production, maintenance, and scrap. This PLM is developed based on digital engineering technology, such as CAD and CAE, and this system realizes sharing of the various information that many engineers engaged in a design and production need by using PDM (Product Data Management). In order to prevent occurrences of a product failure, many companies have been introducing several engineering simulation software, which are called CAE. However, it is necessary to execute simulation that we have to know the physical phenomena of a product failure. That is CAE system cannot tell us where should we check and simulate. Before starting the engineering simulation to solve an occurrence of a product failure, we have to understand a product failure.

An information management system is utilized and huge efforts are paid to guarantee a product quality without any failure. However, in the life cycle of a product

CP1146, *Modelling of Engineering and Technological Problems*, edited by A. H. Siddiqi, A. K. Gupta, and M. Brokate
© 2009 American Institute of Physics 978-0-7354-0683-4/09/$25.00

development, there are many reconsidering activities by various failures and the corresponding failures and radical actions to remove failures are demanded.

The computational representation model of knowledge and know-how is highly desired to solve such difficulty in the prevention of quality issues. It is pointed out that the product failures and product quality related issues are very difficult to represent and visualize. Various factors of failures, errors, faults and defects of final product must be modeled and represented in a computer.

Under existing methodologies in engineering design and manufacturing field, there is no unified and general method for representing and sharing failures or the quality related issues. This is one of the reasons for the existing documents of design and production knowledge (ex. PFMEA (Production Failure Modes and Effects Analysis) sheets, QC (Quality Control) Process Chart, and etc.) are not effectively used in the process planning stage.

Hence, it is desired that the product quality related issues are represented in a computer comprehensive way. This representation realizes a computational method to store, share, manage and utilize a huge number of design and production know-how and knowledge in the process planning stage. So, this is the reason the representation for production quality is required.

In this paper, we re-acknowledge a current matter derived from such a situation, we aim at arguing about the structure of information management to get the competitiveness that enable to reduce occurrences of a product failure in manufacturing industries as much as possible, and can respond to QCD (Quality, Cost, Delivery) that are three important demands of a market [3][4].

APPROACH OF FAILURE PREVENTION IN PRODUCT DEVELOPMENT

Failure Prevention in Information System

Recently, in a product design and a production, an information system has very important role. It manages the huge information related to a product design and production appropriately, and new information systems, which can utilize cases of past success, failure and etc., have been developed. The most of them are called a knowledge management system; there are activities for formalizing an engineer's tacit knowledge and enabling to cooperate with design system and production system closely. Although we have various methods for those activities, the mechanism of preventing failure happening etc. is proposed and used by providing effectively the accumulated case information, which is digitalized, to an engineer [5].

Concerning prevention of a product failure before it happens, it is pointed out that an initial development stage has an important role to consider a product. However, there are not many reports that have realized incorporating failure considering from the initial development stage and utilizing the considering result efficiently to the later stage. Although it can be recognized that failures and product functions are inextricably linked, the expression of a product function becomes so difficult according to the earlier development stage that it is very hard to consider a product failure in an earlier development stage [6].

Besides, although there are many proposals concerning failure considering in a production process, there are few examples that have realized systematic description of a quality problem. There are few information models that can express appropriately the relation that exists between the function of a product, and a behavior/structure and a production. It can be understood that synthetic considering of product failures from a design to a production is not realized for the above reason.

Synthetic Management of Failure Information in Design and Production

This paper proposes "the synthetic design approach of a product and a production process that enables to reduce occurrences of a product failure from the initial stage of product development" as one approach for solving the above-mentioned issue. In order to realize this concept as a specific system, the integrated management model of failure information in a design and a production, which is mentioned later, is introduced, and the following items from a) to d) are realized.

a) Stepwise design approach with expression of a designer's thinking process.
b) The failure reduction approach in a product design by a modeling of failure based on a product behavior.
c) The failure reduction approach in a production process design by a modeling of a production process with considering a product quality.
d) The failure knowledge database.

INTEGRATED MODEL FOR OCCURRENCE MECHANISM OF A PRODUCT FAILURE

Occurrence Mechanism Model

In order to discuss on how we can prevent a product failure, we have to understand the occurrence mechanism of a product failure. An occurrence mechanism model of a product failure is proposed to represent design errors and production failures from the product design stage to the production planning stage. An integrated model of a design error and a production failure provides the causality between a design error and a production failure.

Figure 1 shows the overview of Integrated Model for Occurrence Mechanism of a Product Failure. At the following section, we discuss on this Integrated Model.

Product Information on Design Stage for Realizing Integrated Model

Product Structure and Behavior

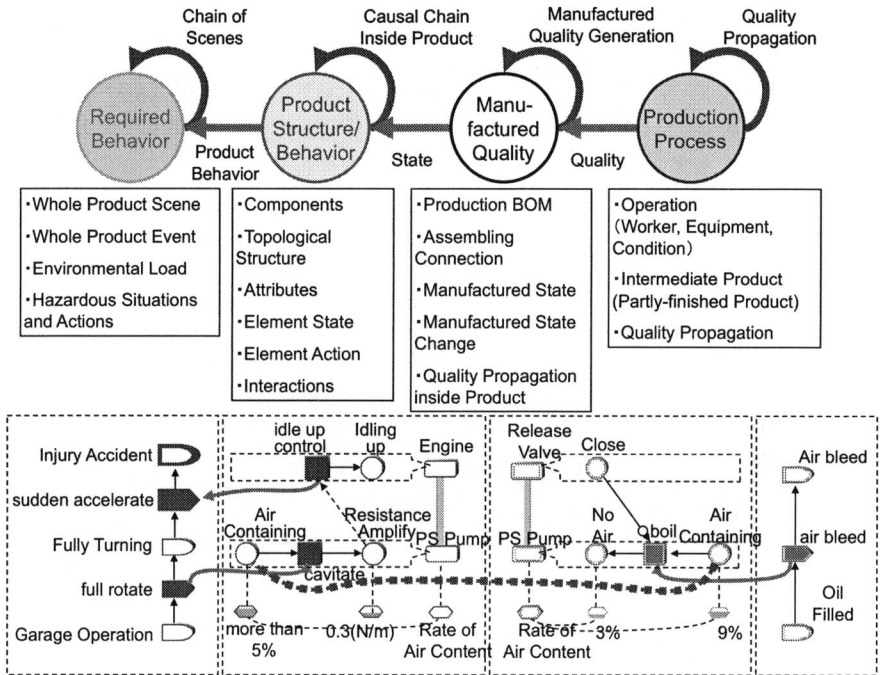

FIGURE 1. Overview of Integrated Model for Occurrence Mechanism of a Product Failure
A product failure model to describe errors and failures on the design stage is proposed using a model of intended and unintended product behavior. 1) Required Product Behavior on Usage Scene 2) Product Structure, Interfaces and State Transition. And more, a product failure model on the production stage is proposed using an operation model and a production quality model. 3) Quality State Transition of a Product in Production Process 4) Relationship between Production Process and Quality State Transition of a Product in Production Process

The following information models are introduced to represent the information of a product structure as E-R model (entity relationship model) [7].

- Entity: E = Component or Parts of a Product.
- Interface: I = The interface between Entities
- Attribute: A = Attributes define and explain Entity. The value of attribute is determined by the current state of the entity

On the other hand, Tomiyama and Umeda proposed FBS (Function, Behavior and Structure) model to describe product functions and product behaviors with a product structure [8][9]. In their research, a meta-model of product functions and behaviors defined by the FBS modeler realize the libraries of product functions.

In this research, referring the FBS model, we introduce new information model for dynamic description of the state and action of a product [10]. Petri-Net is introduced to represent the building blocks of product behavior on the design stage [11][12]. Figure 2 represents the information model of product behavior.

- The State of the Product Component = Place of Petri-Net (Cs; Circular nodes)

- The Action of the Product Component (changes in the values of attributes) = Transition of Petri-Net (Ca; Rectangle nodes)

The product behavior describes all product behavior when we consider the product as a single system. Here, it is necessary to define the representation model of all product behavior. The product behavior is calculated automatically by Reachable Tree

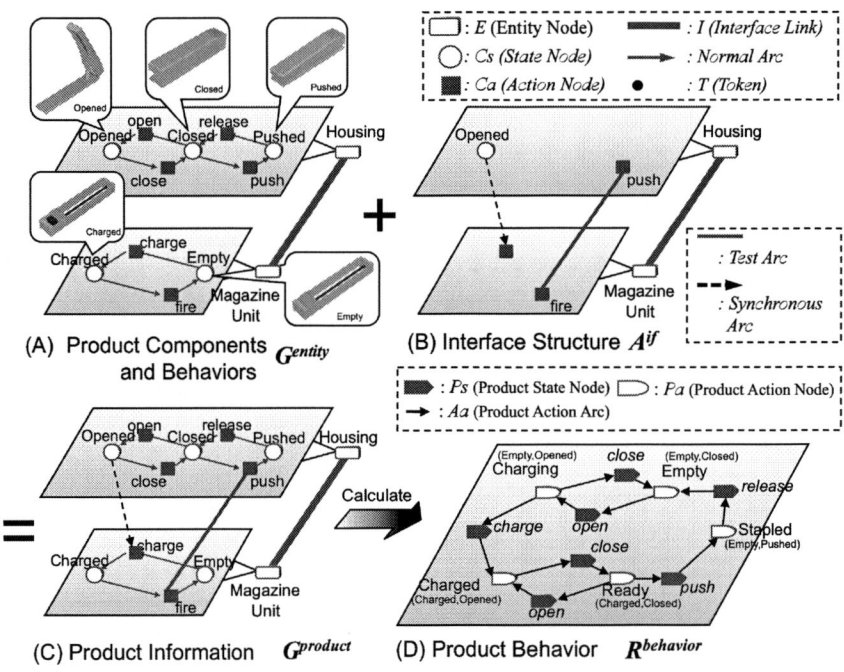

FIGURE 2. An example of the representation model of product behavior

A product, "Stapler" consists of two components (entities), "Housing" and "Magazine Unit". "Housing" has three states, "Opened", "Closed" and "Pushed". In a similar way, "Magazine Unit" has two states: "Charged" and "Empty". Between these states, each entity has transitions (Housing: open, close, push, release; and Magazine Unit: fire, charge). Entities have attributes, the attribute values of which are determined by the state.

For example, "Housing" has the shape attribute. The shape attribute takes the attribute value corresponding to the three states ("Opened", "Closed" and "Pushed"). The component being in the state at a given moment is described as the Token object T (Black circle on the state). In this study, a user or operator of the product is also regarded as one entity that has its behavior. (D) shows the product behavior of the Stapler, which has 5 Product States (Ready, Stapled, Empty, Charging, Charged) and 7 product action arcs (including 1 product action node).

The equation (1), (2), and (3) are definitions of the introduced information model based on Petri-Net model.

$$G^{product} = \left\{ G^{structure}(E;I,A), G^{behavior}(Cs,Ca;Pa) \right\} \dots\dots(1)$$

$$G^{eb} \subseteq G^{behaivor} \dots\dots(2)$$

$$A^{if} = G^{behaivor} - G^{eb} \dots\dots(3)$$

where

$G^{product}$: *Product information,*

$G^{structire}$: *Structure related information of product*

$G^{behaivior}$: *Behavior related information of product*

G^{eb} : *Behaivior of the product components*

A^{if} : *Interactions between product components*

E : *Entities,* I : *Interfaces,* A : *Attributes*

Cs : *Component State,* Ca : *Component Action,*

Pa : *Product Action*

Analysis of behaviors of product components and interactions between these behaviors, which is mentioned below.

Interface Structure and Interaction Between Behaviors

A product consists of many components and parts, which are product elements. Therefore, product behavior is determined by the behavior of these components and their interactions. Those interactions are realized by interfaces between product elements.

This research defines the interactions between components as the effect relationships that the designer does not yet recognize as entities. In the later design stages, such effect relationships can be translated into entity concepts (e.g., cables, linkages or actuators). Although, in the design stage the designer grasps the interface only as the effect relationships, the interface can be regarded as the relationship without concrete entity concepts. The combinations of interactions represent the interface structure.

Product Information on Production Stage

Product in Production Process

The product in production process means product components and parts that are not finished yet. Discussing the issue of a product quality, it is necessary to define a representation model for quality-related issues of a product. As mentioned before,

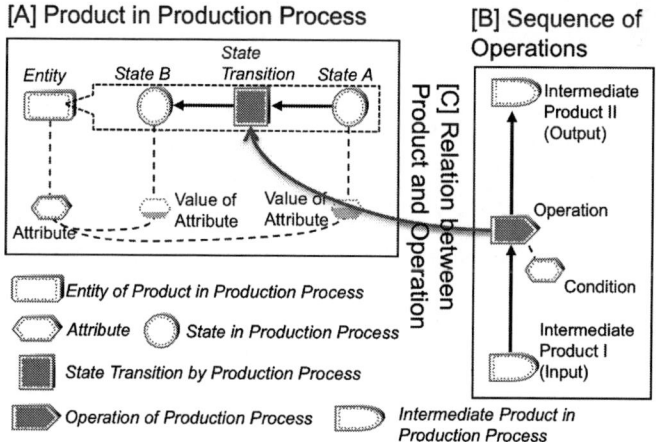

FIGURE 3. A Simple Model for a Production Process
This model consists of the following two types of models and relationships between two models. Model(A) represents Product in Process and Model(B) represents Sequence of Operations. Relationship between Product and Operation is represents by a link. The function of an operation model defined in the model(B) represents a transformation of a product quality state. The sequences of the operations transform the quality state of the product in process from material to a final product.

several kinds of information model are introduced to represent a product structure and a product behavior. Using these models, we propose the representation model for structure and quality of a product in production process. Figure 3(A) shows a simple model for a product in production process.

The product structure is defined as the elements of product (Product Entity in Process: PEP) and as its interfaces between PEP (Product Interface in Process: PIP). The product quality is defined as its quality state. Both of Product Entity and Product Interface have a network of quality states and they changes in production process. They are modeled as the bipartite graph of Quality State (place node) and Quality State Transition (transition node) respectively.

An attribute is defined in the PEP and PIP, and represents the product attribute supposed to be / being manufactured. A token describes an active concept on the time-axis, and represents a current state of the production process. A quality token describes the quality state of product at some point in time axis. A production token describes the progress of production process at some point in time axis.

Production Process Model

The primary function of a production process is the conversion of design information into an actual product. The production process consists of operations that change the physical state of a product.

In this research, we assume that a product is manufactured by several transformations of its form, shape and quality state from material to parts and parts to a final product (assembled product). Basically, this research models a production process as a set of operations, those are done by workers and machines in order to change the product state from material to final product.

This modeling decomposes a production process into several operations with its sequences. The operation is decomposed into its translation function (main task) and translation target (product in process). The product quality is defined as the quality state of the shape and attributes of the product in the production process. Figure 3(B) shows a simple model for the production process in this research. In this research, we introduce the Petri-Net to represent the changes of the product quality state in the production process. The production process model is described based on the Petri-Net.

An operation in a production is defined as the transition node and its function of translation. This node has the main body of its action (worker, machine and equipment). Condition in a production is defined with the operation node, and represents the attribute of equipments that is required for function of translation of quality states. The operation is identified by the input of a product in process and output of a product in process. This operation is represented as a place node in Petri-Net model.

THE SYNTHETIC DESIGN AND PLANNING APPROACH FOR PREVENTING A PRODUCT FAILURE

On the previous chapter, we proposed the integrated model for occurrence mechanism of a product failure. Using this model, we propose the method how to design a product and a production process without a product failure.

- A design process model of product and production process in order to reduce the product failure
- Product failure database system for information sharing and access

Interface Design in order to Achieve Required Behavior

The product failure on the design stage is defined as 'a lack of required behavior' and 'an unexpected behavior' by the introduction of the product behavior concept. Hence, the failure reduction on the product design stage can be defined as a product design with high achievement level of required behavior, and less unexpected and hazardous product behavior.

The interface structure is defined as a combination of interactions between product components in this research. Hence, an interface design method is proposed by comparing two models: 1) required behavior model and 2) product behavior and structure model.

Required Product Behavior on Usage Scene

Synchronization to customer demands is one of the missions of manufacture. Moreover, the starting point of product design is to grasp the demands of the customer. The customer does not always need the product per se, the customer needs the situations or actions that can be performed by using the product. Therefore, the designer generates product information by making assumptions about the scenes and situations in which the product will be used.

We propose a design methodology to propose a product that can achieve the required scenes or situations using the concept of "Product Behavior".

Product Failure due to Errors in Interface Structure

The design of interactions of behaviors of product components, i.e., an interface structure, is a very difficult design problem. If an inappropriate interface structure is designed, fatal failures and accidents may occur due to unexpected product behavior, because the whole product behavior is determined by an interface structure. Tamura et al. noted that many failures and problems in market are caused by known and basic mechanisms. Nonetheless, due to the shortage of assumptions about situations and scenes, these problems still occur.

The interface structure design, that can achieve required behavior and does not involve hazardous behavior, achieves the reduction of product failure on the design stage. A multi-phased interface structure design enables the designer to design the complex product system with less system failure from an early design stage.

87

Design System for Interface Structure

We propose an algorithm that can design the interface structure logically to achieve the required product behavior. The integration between this algorithm and design method of product behavior by multi-stage decomposition enables the designer to reduce the failures and problems caused by the interactions between components from the early design stage.

$$G^{product} = G^{entity} + A^{if} \tag{4}$$

where :

$G^{product}$: Product information, G^{entity} : Product components,

A^{if} : Interface Structure

We propose the new design system of Product Behavior by automatic calculation of the Interface Structure (PBIS). PBIS supports the designer to determine the interface structure interactively based on a list of proposed interface structures that can realize the required product behavior.

A fundamental objective of interface structure design was to design product behavior by estimating and simulating the scenes and situations of product use, and to achieve such product behavior by a product structure that includes as a simple mechanism as possible with no unnecessary motions.

Based on these definitions, we can formalize the Interface Structure Design as following:

Interface Structure Design = The Design Problem of (A^{if}) where (G^{entity}) and ($R^{behavior}$) is defined

Interface Design for Product Quality

Model of Product Behavior (Scene and Event)

The product behavior $R^{behavior}$ describes all product behavior considering the product as a single system. The product behavior can be expressed as:

$$R^{behaivior} = \text{Reachable Tree Analysis}(G^{product}) \tag{5.1}$$

$$R^{behaivior} = \{Ps, Pa, Aa\} \tag{5.2}$$

where :

$G^{product}$: Product information, Ps : Product State,(5.1)

Pa : Product Action, Aa : Product Action Arc

The Product State Node has the associations with the states of product components that describe the product state. When the described product components include the user state, the product state can be called as the scene.

The importance describes the value of scenes that is achieved by the product state. Contrary, the hazard level describes the risk level of unexpected product state, which can derive the failure or unsafe scene.

Aa means the shifting between product states and is described by the arc between product states.

If there is a concrete action event, the product action arc has the Product Action Node (*Pa*) in center of the arc as the ellipse node. The Product Action Node has associations with the action of the product components *Ca* that describe the product actions.

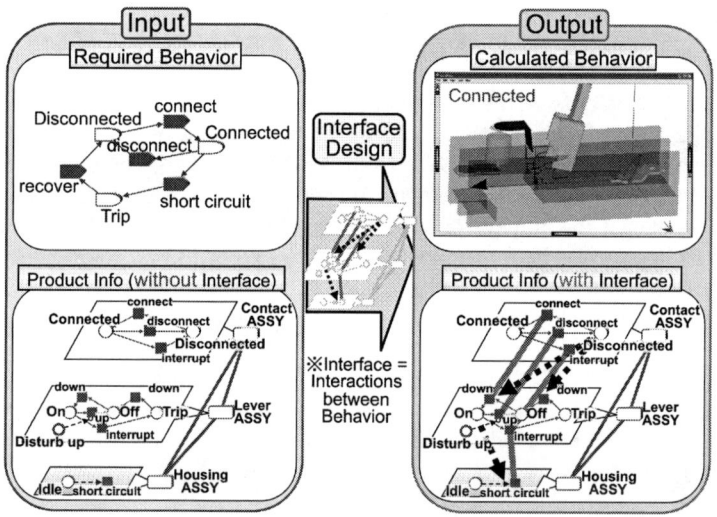

FIGURE 4. Interface Structure Design

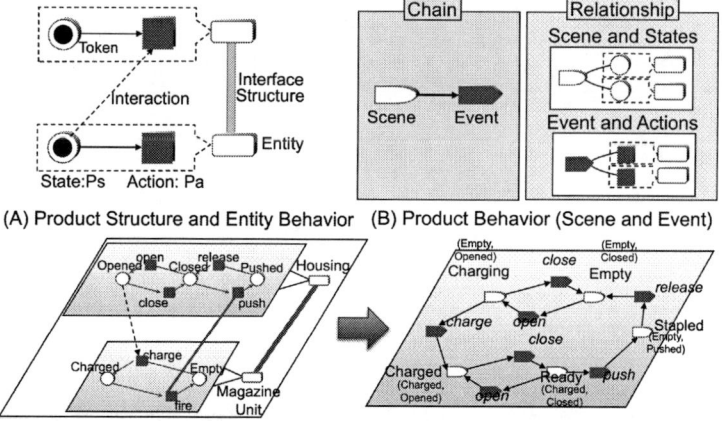

FIGURE 5. Product Behavior describes all product behavior considering the product as a single system. Product Behavior is generated from Entity Behavior and Interactions between behaviors. This consists of Product State and Action. They associates with Components state and actions.

The Product Action Arc and Product Action Node have the same scheme that consists of the name, associations with the product components, importance and hazard level.

Specify the Required Product Behavior

There can be many-to-one associations between the component states & actions and the product states & actions. Such optionally is also part of the designer's knowledge. Therefore, we developed an association library of Product States & Actions (*Cs&Ca*) and Component States & Actions (*Ps&Pa*), and use it for comparison between the calculated product behavior and the required product behavior.

The interface design problem can result in the selection problem of an element in the space of interface structures. In this approach, the space of interface structures is calculated automatically from the interface candidates.

All combinations of the interface candidates (representing the interface structure) can be calculated. It is necessary for the design of interface structure to select the element of the space of the interface structure that can achieve the required product behavior. However, if the number of interface candidates increases, the size of the set of interface structures will become exponentially large. For this reason, the exhaustive search cannot be applied to the large-scale design problem (NP hard). Therefore, the heuristic algorithm to search the solution space is required.

Heuristic Search for the Interface Structure

In this research, we assumed that the heuristic search algorithm by reinforcement learning work effectively on this design problem. The learning process is based on

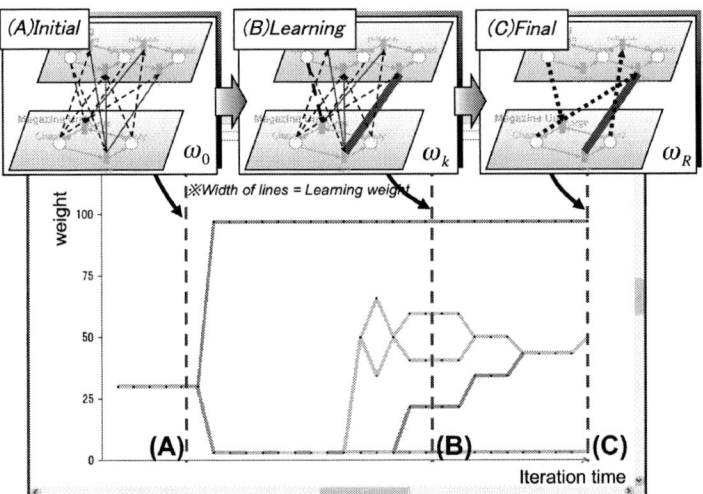

FIGURE 6. The heuristic search algorithm by reinforcement learning is introduced to search interface structure. This chart shows learning process with the evaluated score of interface structure (performance, feasibility and differential vector) as the importance of each interface.

feedback of the evaluated score of interface structure (performance, feasibility and differential vector) as the importance of each interface. This algorithm can propose the solution by following four steps (Step-1 to Step-4):

Step-1: Generate the interface structure at random

First, the importance of each interface candidate w_i is defined (initial value is assigned by constant value: w_0). The importance of each interface is shown as the line width. Based on this importance, the interface structure is generated at random. An important interface candidate is selected by high probability.

Step-2:Calculate the product behavior

The product information is determined when such a generated interface structure is connected to the component behavior. Therefore, we can calculate the product behavior by generating the reachable tree of product information.

Step-3:Evaluate the calculated product behavior

The evaluation method mentioned above enables the designer to evaluate the calculated product behavior. Therefore, this algorithm compares the graph of the calculated product behavior with the graph of the required product behavior. Such a comparison calculates the performance, differential vector, and classification.

Step-4 Feedback Learning

This algorithm changes the importance of the interface candidate w_i+1 based on the results of evaluation. Feedback learning is calculated by increasing the importance of the interface candidates that are included in the high-scoring interface structure, and by reducing the importance of those included in the low-scoring structure.

In the processes from Step-1 to Step-4, the learning process of importance is as shown in Figure 6. The initial value of each importance w_0 can change gradually by the cycles of learning (w_k). Finally important interface candidates and unnecessary interface candidates can be visualized (w_n).

The importance of the interface candidate is changed by each calculation cycle, and finally reaches the local solution. The type of solution calculated by this learning process is dependent on the random values of early calculation iterations, and so it is necessary that this calculation loop (Step-1 to Step-4) is calculated more than once.

Failure Reduction on Production Process Planning Stage

The production process can be represented as the translation process of product quality by a sequence of the operations [5]. Based on the model of the production process, the production failure is modeled as the causal chain between a production process and a product on process. The operation changes a product quality state. The production failure is caused by bad operations, bad operation sequence, and bad material. The production failure model represents the whole state change of production process based on a production model.

A planner of production process must design a sequence of operations with less production failure. Based on a production failure model, computational method of planning system that does not involve a production failure is developed. The input of planning system is a required state of product quality, and the output of planning system is a sequence of operation that does not result in the quality problem. Integration between the multi-staged product behavior design and the process planning method provides the designer the product development with less failure.

Product Failure Model in Production

A failure model of product in production process is proposed based on the production process model. The failure of production process is defined as the generation of failure quality state of product. The failure quality state is defined as a quality state of final product that derives loss of product function. Main factors of this failure are categorized as two types:

Type-A: A product failure caused by a product
This failure is derived from product itself, e.g. an inadequate material, defect of additives or a causal chain inside a product structure.

Type-B: A product failure caused by an operation
This failure is derived from an operation, e.g. worker's or machine's errors.
The expression of 'Good' or 'Bad' of an operation cannot be defined locally. 'Bad' can be only defined as the final quality state of product. Whether bad quality is generated or not depends on the whole production process. This is the reason that the total quality design system considering both qualities related good factors and bad factors is required.

Design Method of Production Quality

Overview of Design Method

This section discusses the algorithm for production planning to reduce the product quality issues. The designer has to define the information of operations and its sequences to avoid and prevent failure quality states. The computational product quality design method of production process is shown in Figure 7.

This research recognizes the design of the production process as the determination procedures of operations and its processes. In order to reduce the product failure quality states, the product quality related issues must be considered in the planning stage of operations and its processes.

The overview of the design method of the product quality in the production process is shown in Figure 7. This figure shows the whole model of the production process and the product quality in the method. The design method operates following three models:

a) Integrated model of product-process-quality   (Figure 7 (a))
b) Space of quality state   (Figure 7 (b))
c) Sequence of Operations   (Figure 7 (c))

The quality design algorithm is defined as the information processing procedure between models a), b) and c).

Flow of Product Design considering a Product Failure

The following steps represent the information processing flow of the design for the product quality in the production process.

Step-1a: Decomposing product structure

The Entity object defined on a component level is decomposed into some entity objects on the parts level. The product structure is defined by the design of interfaces.

Step-1b: Listing feasible operations

The designer lists the feasible operations that an organization or a factory has. The product quality state changes corresponding to feasible operations are described in the product quality information.

Step-1c: Listing possible failures

The designer lists the possible failures related to the applied operations.

Step-2: Generating Integrated model of product-process-quality

By integration of three models (obtained by Step-1a,1b,1c), an integrated model of product-process-quality is generated by design system.

Step-3: Generating space of quality state

The design system calculates the space of quality state of the production process from the integrated model of product-process-quality. The integrated model is defined based on the PN model. A reachable tree calculated from the PN model of the

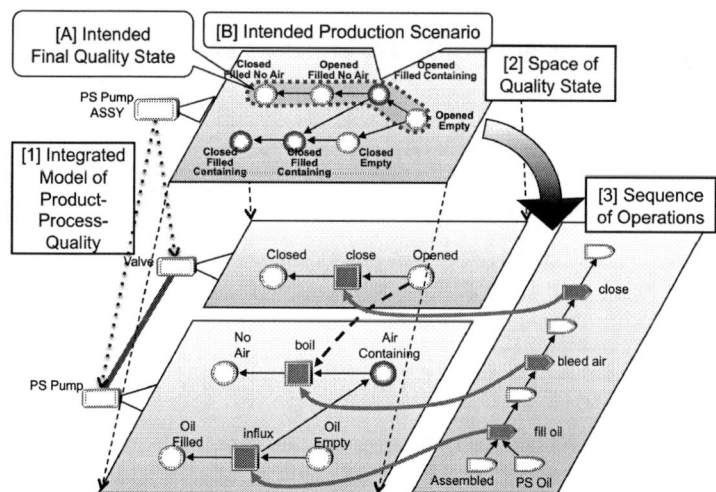

FIGURE 7. Design Method of Production Quality

Z-axis means the hierarchical relationships between an assembled product layer and a parts layer. 'Pump ASSY' consists of two sub-components, 'Valve' and 'Pump'. The space of product quality state of 'Pump ASSY' is represented by a combination of quality states of sub-components, 'Valve' and 'Pump'. The production process can be so complex that the hierarchical representation is introduced to enable a multistage decomposition approach of the quality design of the production process.

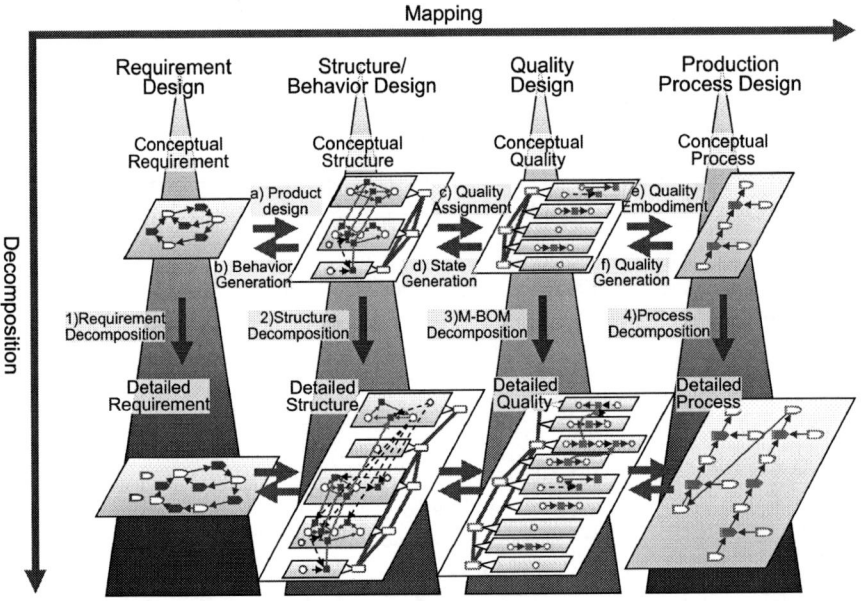

FIGURE 8. shows the failure reduction concept by stepwise refinement of this research. The information processing procedures can be categorized into two axes: 1) mapping from the design stage to the production stage, and 2) detailing process by the decomposition. The zigzagging design process enables a designer to design a product and a production process with full-time comparison of a requirement by multi-staged approach. Stepwise Refinement Model for design a product and production process without a product failure

integrated model means the space of product quality state in a production process.

Step-4: Selecting final quality and scenario

The designer selects a final product quality state and its scenario from the calculated space of product quality state.

Step-5: Output production process

The selected scenario of quality state change includes the operations and its processes. The design system proposes the production process from the operations.

Failure Information Database

The model of a product behavior and a production process provides the meta-model of design error and production failure. A failure information database can be developed based on the meta-model. The failure database enables a designer to share and access on each stage of a product design and a production process planning.

The meta-model based database can be categorized as following four databases: 1)a generalization and specialization database, 2)decomposition and recombination database, 3)causal relationship database and 4)actual data and case database. Those four databases and the stepwise design process enable a designer to reduce a product failure from the early product design stage and the early production planning stage.

Stepwise Refinement Model

The computational management and representation method of design process is necessary to support the product development from early design stage. In order to represent the design process, the model of the designer's decision process is required. This paper assumes that the design process is the embodiment process of the product (and production process). The embodiment process increases the volume of the product information. Hence, this paper assumes that the increasing information model can represent the product information in the design process. Based on the increasing information model, a seamless design support is realized from the initial design stage to the detailed design stage using same product representation model.

Figure 8 shows the failure reduction concept by stepwise refinement of this research. The information processing procedures can be categorized into two axes: 1) mapping from the design stage to the production stage, and 2) detailing process by the decomposition. The zigzagging design process enables a designer to design a product and a production process with full-time comparison of a requirement by multi-staged approach. That means why this design method can reduce the new difficult product failure systematically.

EXECUTE EXAMPLE BY PROTOTYPE SYSTEM

Overview of Proto-type System

Based on the discussions in the previous sections, the prototype system is developed using pure object oriented language Smalltalk (CINCOM VisualWorks 7.4). Figure 9 shows the overview of the proto-type system.

The model view provides the structure of a product behavior model and a production process model. The knowledge view provides the failure knowledge by proposed four databases. The attribute view provides a visual product state in a production process and behavior. The design process view provides the roll-back operation, generating candidates and try & error to avoid a product failure.

The prototype design system can simulate the state change on a product behavior and production process. Based on a production process simulation, the product state is determined. Based on the product behavior simulation, a product failure is calculated. The integrated model and simulation function enables a designer to know the reason of current failure, an impact of process and product change and the better design candidate in the view point of a product failure and its production quality. Hence, this system provides the on-stage check-list and information candidates for PFMEA (Production Failure Modes and Effects Analysis) and FTA (Failure Tree Analysis).

Example: Auto Circuit Breaker

We selected an Auto Circuit Breaker as an example product for confirming the effectiveness of our proposed method. Auto Circuit Breaker has functions: 1) Detecting a thunderbolt, an imprudence, short circuit, and etc. 2) Preventing the destruction by fire of the electric machine used by cutting inside circuit.

Based on the actual information of design and production, we try to show the validity of our method concretely, how we can carry out design a product and a production process stage by stage from the upper design stage, and describe failure information in each stage, and avoid failure. By this example execution, we discuss on the possibility of the realization to decrease failure occurrence and from the initial design stage.

Design the Interface Structure and Production Process at the Initial Design Stage

Design Interface Structure without Failure

On the conceptual design stage of an Auto Circuit Breaker, the system supports the designer to input behavior as the customer requirement (Figure 10(a)), and to output the product information (Figure 10(e)) as following Step-1.1 to Step-1.5. The required

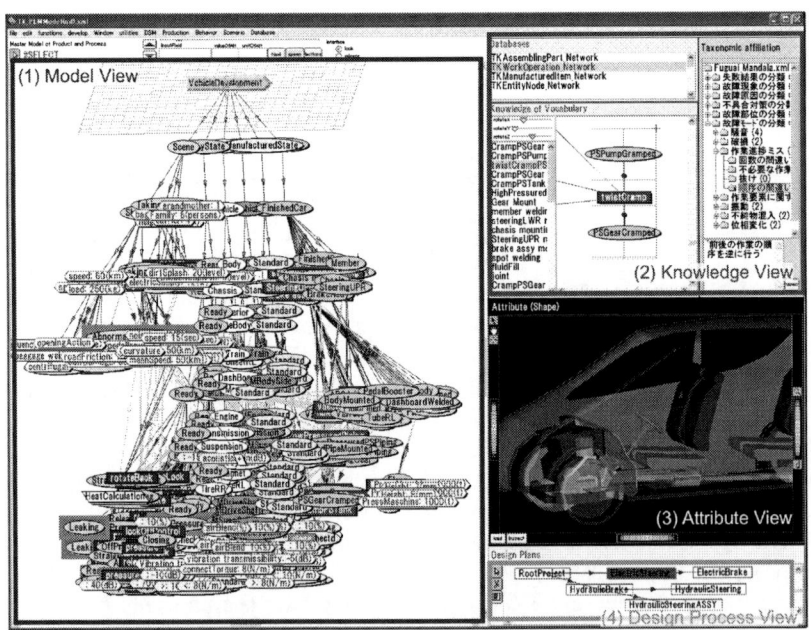

FIGURE 9. Overview of the Proto-type System
The model view provides the structure of a product behavior model and a production process model. The knowledge view provides the failure knowledge by proposed four databases.

96

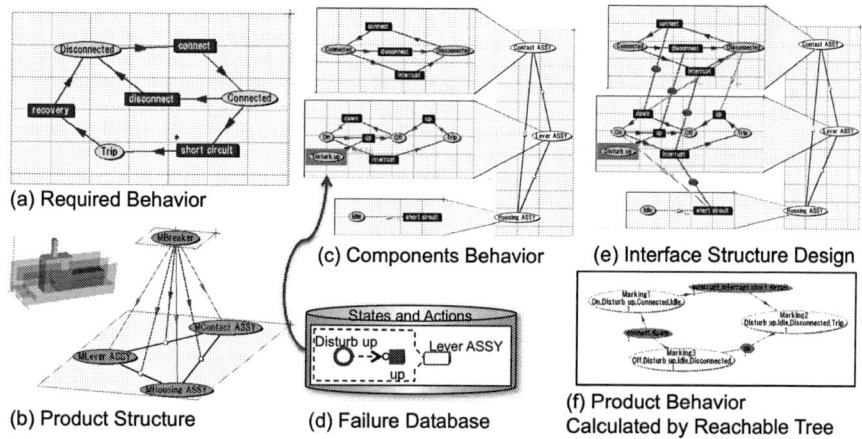

(a) Required Behavior

(c) Components Behavior

(e) Interface Structure Design

(b) Product Structure

(d) Failure Database

(f) Product Behavior
Calculated by Reachable Tree

FIGURE 10. Design the Interface Structure at the Initial Design Stage

behavior is defined as a process that "When it detects short circuit or overload, electrical connection must be separated to break the circuit (Trip). The Auto Circuit Breaker can be recovered to resume normal operation between connected state (Connected) and disconnected state (Disconnected)".

Step-1.1: Considering three product elements (Lever, Housing ASSY, and Contact ASSY), the engineering designer designs behaviors of each element and physical connections between those elements as shown in Figure 10(b).

Step-1.2: The mechanism how it avoids the occurrence of trouble and its interface structure are considered in the usage case when the lever is kept the position "ON" due to some obstacles. So, the usage scene with the state of a lever is represented into product information.

Step-1.3: Calculating the interface structure that enables to avoid failure problems, which is mentioned at Step-1.2. Then, generating the product behavior by reachable tree analysis of Petri-Net.

Step-1.4: Comparing two behaviors those are the required product behavior (Figure 10(a)) and the generated product behavior (Figure 10(f)).

Step-1.5: As sown in Figure 10(e), the interface structure that meets the required product behavior is designed.

Design the Production Process

Using the product information that is designed at the interface structure design (Figure 11(a)), an assemble sequence and design assemble operations for an auto circuit breaker is designed. The following steps show the flow of generating the information of production process that can meet the required product quality.

Step-2.1: Three candidates of assembling sequences are generated form the manufacturing BOM that is generated at the design stage.

Step-2.2: Selecting the feasible assemble operations and the connecting method considering the types of the connections between product elements.

Step-2.3: Describing the information on a product failure into the assemble operations.

Step2.4: Generating the initial plan of production process that has information on product structure, process operation, and product failure (Figure 11(c)).

Step-2.5: Generating the candidates of production process that represents the state transitions of a product structure and a product quality by calculating the reachable tree of Petri-Net (Figure 11(d)), which is generated from production information.

Step-2.6: Selecting the good production scenario without a product failure from the candidates of production process. The selected production process represents an assembling sequence and assemble operations in a production.

Step-2.7: Outputting the final production information from the selected assembling sequence and operations by the describing intermediate products and making relations between quality state transitions in the production (Figure 11(e)).

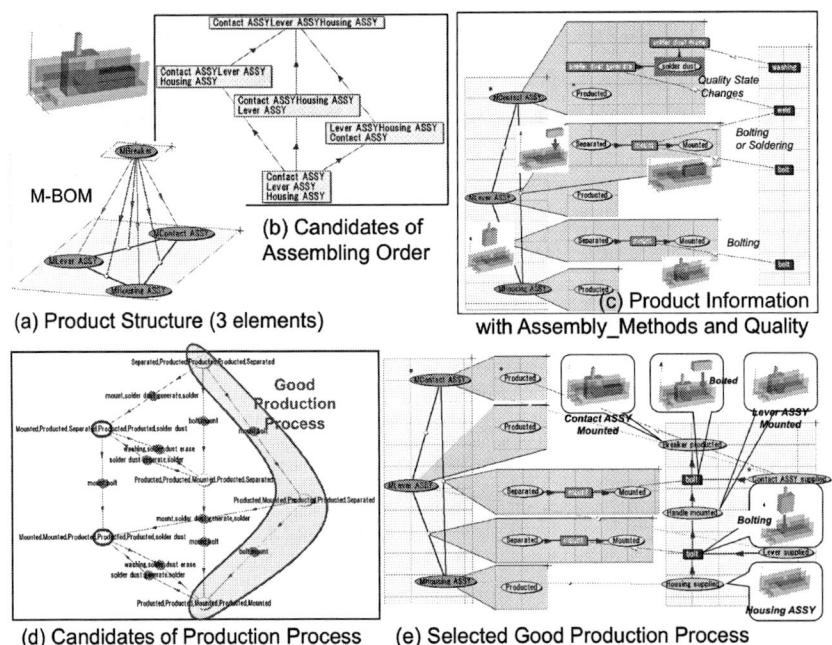

FIGURE 11. Design Production Process at the Initial Design Stage

Design the Interface Structure and Production Process at the Detailed Design Stage

Design Interface Structure without Failure

At the detailed design stage, focusing on the Contact ASSY, which is key assembly component that has many problems of product failures. The Contact ASSY is an important component of the Auto Circuit Breaker, and its function is conduction and insulation of electric current. Usual path of the electric current is from the Contact Plate to the Contact. Because high current derives arcs from the Contact, the Arc Hone is required to avoid the arcs from the Contact, and to realize quick insulation.

Describing many kinds of information that are product function and behavior product quality shows the effectiveness of our proposed method.

Step-3.1: The Contact ASSY is decomposed into three parts to represent the detailed behavior. Figure 12(a)(b) shows the information on a product structure and behaviors of each parts.

Step-3.2: Product failure is described into product information model. The product failure problems are described by same way of initial design. Figure 12(d) shows above-mentioned failure information.

Step-3.3: Calculating the interface structure that enable avoid product failures. Then, generating the product behavior by calculating the reachable analysis of Petri-Net.

Step-3.4: Comparing the required product behavior (Figure 10(a)) and the generated product behavior (Figure 12(e)).

Step-3.5: Figure 12(d) shows the interface structures to meets the required product behavior.

Design Production Process without production failure

Generating the detailed production process information. By the detailed product design, the product structure and role of parts is determined, we can design the production process to realize its structure and roles.

The problems of design the production process of a Contact ASSY in the detailed design means the generating the assembling sequence and planning operations for assembling.

Step-4.1: The two candidates of assembling sequence is generated from the manufacturing BOM considering the connections between product elements.

Step-4.2: Selecting the connecting method considering the connection type between parts. Then, describing product failure into the model of operation generates the initial plan of production process (Figure 13(a)).

Step-4.3: Generating the candidates of production process with product state transitions and quality state transitions by calculating the reachable tree of Petri-Net. Selecting the good production process that has no product failure and cleaning process. This production process represents assembling sequence and operations of assembling (Figure 13(b)).

Step-4.4: Outputting the final production information from the selected assembling sequence and operations by the describing intermediate products and making relations between quality state transitions in the production.

The example of confirmation that change of assembling process drives to lose a product function

Figure 14 show the example output that the detecting the product failure that is derived by change of production process. Before changing process, the quality, which is made by the [a-A]braze operation, generates good quality state that affects no influence to product behavior. It is easy to make sure that the generated behavior meets the requirement. However, the process is changed by some reason. This change is made by swapping operation [a-A] and [a-B].

This swapping of operations produces the bad quality state. The brazing operation after the forming operation produces the quality state [Curved] (Figure 14(b)[b-4]).

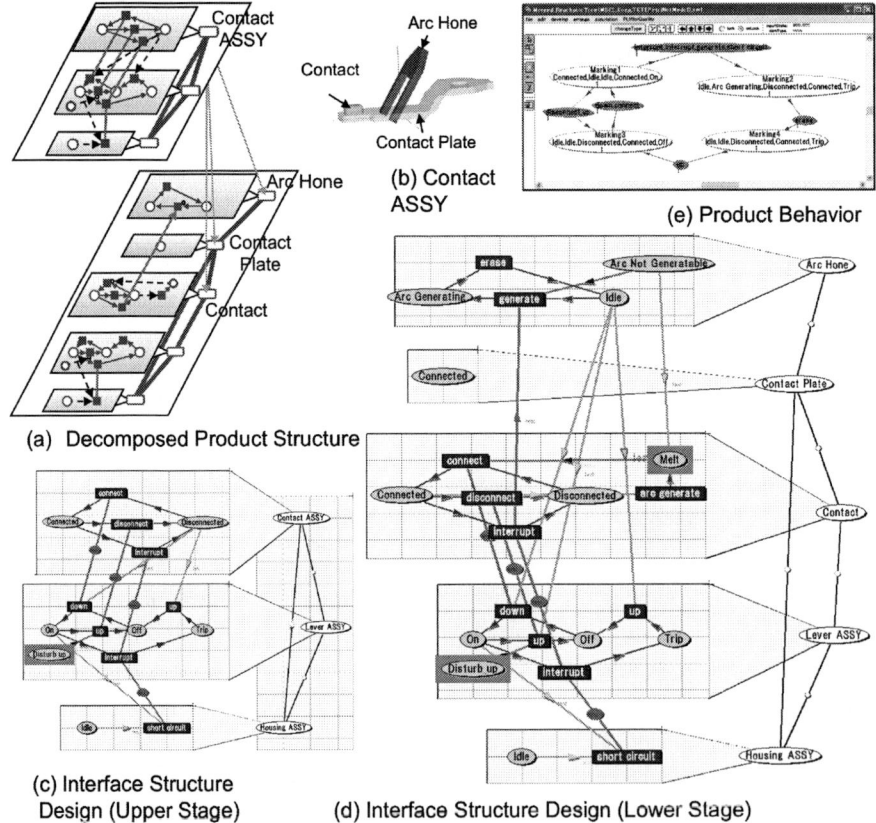

(a) Decomposed Product Structure

(b) Contact ASSY

(e) Product Behavior

(c) Interface Structure Design (Upper Stage)

(d) Interface Structure Design (Lower Stage)

FIGURE 12. Design the Interface Structure at the Detailed Design Stage

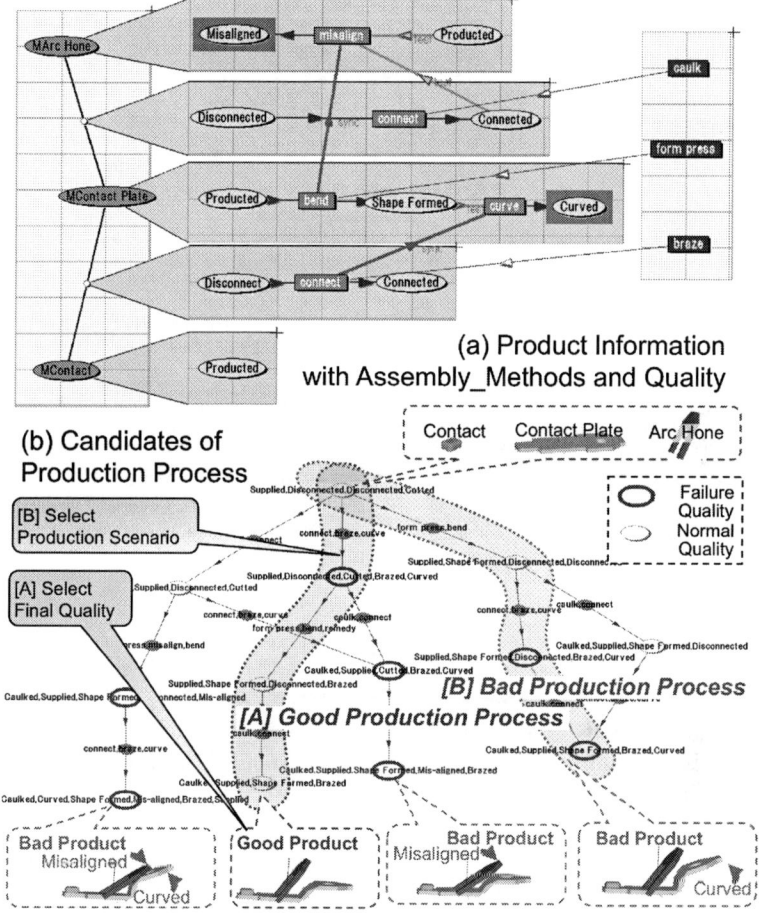

FIGURE 13. Design the Interface Structure and Production Process at the Initial Design Stage

This quality state [Curved] derives the phenomena where there is no arc form Arc-Horn (Figure 15(a)). The long-term using of this product with bad quality derives serious product failure. This is the melting of the contact point and the lost of connection (Figure 15(b)).

CONCLUSION

The design method for a product and production process using failure information is proposed in this paper. The shortening of product development time and reduction of the wasteful design iteration can be reduced based on the detecting and accessing of a quality related issues in a production process that derives critical product behavior.

Focusing on the difficulty of the interface structure design problem, we have developed a methodology for proposition of the interface structure that satisfies the

(a) Good Production Process (Design Result with this research)

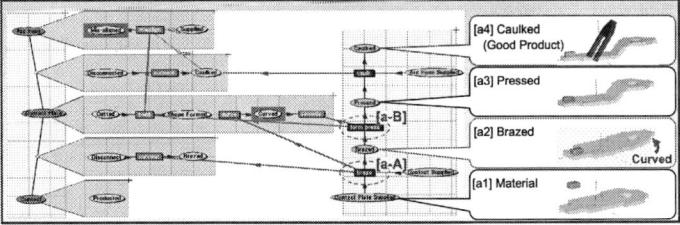

(b) Bad Production Process (Possible Design Result without this research)

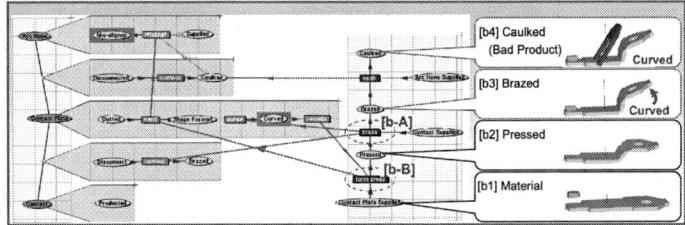

FIGURE 14. The change of assembling process drives to change the product quality state in production

required product behavior by heuristic search algorithm. The representation model of product behavior using Petri-Net formulates the interface structure design. The space of the interface structures is derived from the interface types and its mechanism concepts. By reinforcement learning and evaluation method of the interface structure, an algorithm for the heuristic search was proposed.

We developed prototype system for design Product Behavior by automatic calculation of the Interface Structure. This system enables new design processes by direct mapping of the product behavior to the product components and attributes. This design process is expected to allow the designer to better synchronize with the customer's demands than the legacy design process.

The computational representation model of the production process that represents its quality related issues are proposed. Based on the production process model, the design method of the production process that can achieve required product quality state is developed. This design method enables to plan the production process as one system considering both quality related good factors and bad factors.

The effectiveness of this proposed method is confirmed by applying the actual design problem of the production process of one component of Auto-Circuit Breaker.

The production model and product behavior model proposed in this paper visualize the process of the product quality and product behavior that satisfies the required behavior. This visualization will accelerate the communication between product design engineers and production planning engineers. The designer can know what is going on in the production process. The production-planning engineer can know why a component is required and how to behave through the entire product life cycle.

In the future, based on the knowledge sharing and process integration by the proposed method, we hope engineers provides unprofessional product failure

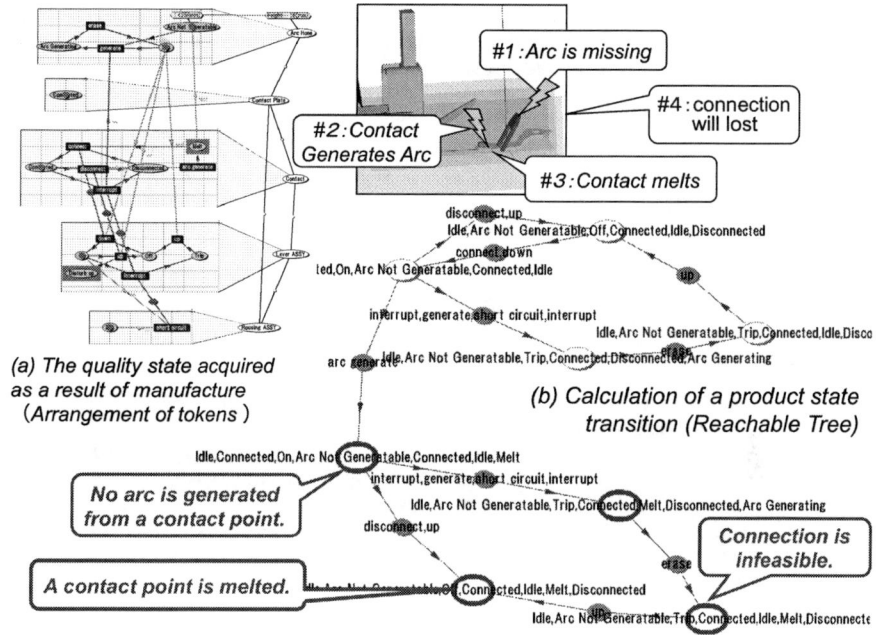

FIGURE 15. Evaluation of the influence of the state of Product
It is find out that the fatal product deficiency "connection is impossible" occurs by influence of the quality of a product.

information with each other, and new technological break-through opportunities will be created by proposed design system.

In these years, large number of product failures in production stage caused by the design change of the production process for the reason of cost-down has been reported. The representation method of production knowledge of sequence of operations or important quality state is highly desired. The proposed production process model and quality design method in this research will contribute to reduce the final product failures in production stage caused by the design change of the production process.

ACKNOWLEDGMENTS

This work and example of the production process of an auto-breaker was partially supported by Fuji Electric Systems Cooperation Co., Ltd. and Fuji Electric Advanced Technology Systems Co., Ltd. Fruitful discussions with and valuable comments by Kenichi Kurotani, Hiroshige Mizutani, and Mitsutoshi Iino, Mitsuhiro Nakamura, Wang Xihong, and Hideyuki Ito are greatly appreciated.

REFERENCES

1. Land Infrastructure and Transportation Ministry, 2002, "The result of analysis on system of automotive recalls," http://www.mlit.go.jp/kisha/kisha03/09/090801_2/01.pdf (in Japanese).

2. Takano, A., 2004, "Troubles are getting closer to you" (in Japanese), Cover story of Nikkei Mechanical, November Issue, Nikkei Business Publications, Inc.(in Japanese)

3. Koga T. and Aoyama K., "A Search Algorithm of Interface Structure to Achieve Required Product Behavior", Proceedings of the ASME 2005 International Design Engineering Technical Conferences & Computers and Information in Engineering Conference, IDETC2005-85051 (CD-ROM), Long Beach, California, USA, September, 2005

4. Koga T. and Aoyama K., "Integrated Product-Process-Causality Model To Prospect Failure Propagation", Proceedings of the ASME 2005 International Design Engineering Technical Conferences & Computers and Information in Engineering Conference, IDETC2005-85459 (CD-ROM) Long Beach, California, USA, September, 2005

5. Tamura, Y. and Iizuka,Y. 2002., "A Study on the Method to Manage Design Knowledge on Failures–Construction of the Knowledge Structure of a Causal Chain of Failures", the Japanese Society for Quality Control, vol. 32, no. 1, pp. 122-135, 2002 (in Japanese).

6. Fuse, M. et al., 2002, "Knowledge Base and Improvement of Production Technology", The Japan Society for Precision Engineering, Vol.68, No.4, pp.507-510, 2002 (in Japanese).

7. Chen, P., 1976, "The Entity Relationship Model-Toward a Unified View of Data," ACM Transactions on Database Systems, 1, (1), pp. 9-36.

8. Umeda, Y., Takeda, H., Tomiyama, T., and Yoshikawa, H., 1990, "Function, Behavior, and Structure," Applications of Artificial Intelligence in Engineering, V. Ed. Gero, Proceedings of the Fifth International Conference, Boston, Massachusetts.

9. Qian, L. and Gero, J. S. (1996). "Function-behaviour-structure paths and their role in analogy-based design", AIEDAM 10:289-312.

10. Koga, T., and Aoyama, K., 2004, "Product Behavior and Topological Structure Design System by Step-by-step Decomposition," ASME 2004 International Design Engineering Technical Conferences & Computers and Information in Engineering Conference, Salt Lake City, Utah, DETC2004/DTM-57513 (CD-ROM).

11. Petri, C. A., 1962, "Kommunikation mit Automaten," Ph.D. thesis, University of Bonn, Bonn, Germany.

12. Murata, T., "Petri Nets: Propaties, Analysis and Applications", Proceedings of the IEEE, 77, 4 (1989), 541.

13. Aoyama K. and Koga T., 2004, "Development of an Artificial Satellite Design Environment by Top-Down oriented Design Supporting System", Proceeding of the 24th International Symposium on Space Technology and Science, Miyazaki, Japan, May 30-June 6, ISTS2004-t-17

On Mathematical Modeling Of Quantum Systems

Achuthan. P [a,b] and Narayanankutty Karuppath [c]

[a] Department of Mathematics, Amrita Vishwa Vidyapeetham,
Coimbatore, India 641 105.
[b] Dept. of Mathematics, Indian Institute of Technology,
Madras, India 600 036.
[c] Dept. of Physics, Amrita Vishwa Vidyapeetham,
Coimbatore, India 641 105.

Abstract. The world of physical systems at the most fundamental levels is replete with efficient, interesting models possessing sufficient ability to represent the reality to a considerable extent. So far, quantum mechanics (QM) forming the basis of almost all natural phenomena, has found beyond doubt its intrinsic ingenuity, capacity and robustness to stand the rigorous tests of validity from and through appropriate calculations and experiments. No serious failures of quantum mechanical predictions have been reported, yet. However, Albert Einstein, the greatest theoretical physicist of the twentieth century and some other eminent men of science have stated firmly and categorically that QM, though successful by and large, is incomplete. There are classical and quantum reality models including those based on *consciousness*. Relativistic quantum theoretical approaches to clearly understand the ultimate nature of matter as well as radiation have still much to accomplish in order to qualify for a final theory of everything (TOE). Mathematical models of better, suitable character as also strength are needed to achieve satisfactory explanation of natural processes and phenomena. We, in this paper, discuss some of these matters with certain apt illustrations as well.

Keywords: EPR, Entanglement, Bell, Modeling, Reality
PACS: 03.65Ud, 03.65Ta

INTRODUCTION

Mathematical modeling is an art of extraordinary power and merit. Also it is a science with great potential and intrinsic coverage over human activities of various kinds. In the realm of microscopic objects it is quantum mechanics (QM) and special relativity which can explain the events and properties[1]. Thanks to the work of several giants of theoretical physics who laid the foundations of modern science, today we are in an elegant and enviable position to understand and appreciate thoroughly vast areas of the scientific and technological world. The integrating ability of mathematics and the modeling processes have to a very large extent enabled man to master and control nature, particularly with the aid and effective application of the efficient computational tools[2].

The much appreciated mathematical methods, techniques and approaches whereby practical problems are tackled have placed at the disposal of man a good deal of

CP1146, *Modelling of Engineering and Technological Problems*, edited by A. H. Siddiqi, A. K. Gupta, and M. Brokate
© 2009 American Institute of Physics 978-0-7354-0683-4/09/$25.00

precious possibilities for probing into the realities of nature at the innermost as well as outermost levels. We would like to stress and point out the efficacy of a combined effort to discover the ultimate truth, possibly with the complementary role of a number of intelligent creations of the human mind and heart including the consciousness[3-5]. Fundamental physics at this point of time stands thoroughly investigated with the active involvement of the high energy accelerators on the one hand and super-telescopes and large distance observational facilities on the other. One is capable of looking at the quarks and quasars with some amount of courage and confidence now with the help of carefully formulated models. It is our main purpose here to emphasize on the rather powerful application of mathematical and physical theories and models to unravel the secrets of the world at all levels of the universe.

WHY MATHEMATICS?

One of the most beautiful creations of the human mind is mathematics. Millennium before the man has started thinking about his whereabouts and started building imaginative models of the inner and environmental aspects of nature, in and out. The geometrical pictures became necessary and the mental construction came handy to represent the observed facts and the conjectured concepts and theoretical explanations. Today mathematics is strong, has spacious structures wherein many attractive, ingenious creative mansions are visible. Thousands of special topics under major headings are listed and pursued at different levels. Abstract as well as application-cum-applied inputs of mathematical edifice are pursued with added vigor. Rigorous proofs are advanced for nice, clever theorems and conjectures. New branches and sub-sections are being discussed. There is almost an exponential growth in the research results. Mathematics has entered into every sphere of the human endeavors. This is, indeed, all for the good of the entire set up known so far.

The advent of super computers has opened up vast vistas and tremendous possibilities for man to reach out higher and deeper in the ever-widening fields of scientific and technical innovations. There is perhaps not a single area currently untouched by mathematical thinking and applications since mathematics does have so much of the enlightening and unifying ability. Even approximate results emerging from mathematical calculations relating to any special practical questions are quite welcome, especially where other avenues are extremely complex and solutions are hard to come by. Optimization with the help of mathematics arrived at to achieve workable results. We are bound to attain perfection in our grasp and command of the world around us sooner than later. For this, mathematics along with other relevant subjects can and will be of great support and help for certain. We consider some of the revolutionary developments in modern theoretical physics that touch the very deep roots of our understanding the very basic nature of reality illustrating the potential of mathematical modeling.

TIME AND RELATIVITY

Relativity theories modify the classical views on space and time quite drastically. According Special Relativity theory (SRT) simultaneity is not absolute but rather dependent on frame of reference. Quantities that have been thought to be invariant (under the Galilean transformation) are no longer so in SRT. The view that time and space are two independent and separate entities is demolished by the requirement of Lorentz invariance of SRT. The confidence about this preference of Lorentz transformation (LT) over the Galilean comes through the Maxwell's electromagnetic equations which are covariant under Lorentz transformations[a]. Both special and General Relativity Theory (GRT) can be considered to be studies of this covariance and invariance under certain mathematical transformations[6]. As we can see time interval (like space interval) is not invariant in SRT as given by

$$\Delta t = \frac{\Delta t_o}{\sqrt{1 - v^2/c^2}} \tag{1}$$

and in GRT gravitational time dilation due the presence of a gravitational field is given by

$$\tau = \tau_o \left(1 + \phi/c^2 \right) \tag{2}$$

where ϕ represents the gravitational potential. But the quantity, ds is called space-time interval which is an absolute entity and therefore has physical significance or reality. In SRT, it is given by

$$ds^2 = dx^2 + dy^2 + dz^2 - c^2 dt^2 \tag{3}$$

And GRT this interval is given by the expression

$$ds^2 = \sum_{i=0}^{3} \sum_{j=0}^{3} g_{ij} dx^i dx^j \tag{4}$$

Being used only in inertial frames (or unaccelerated with no gravity), Lorentz transformation works in the domain of (pseudo) Euclidean space. A generalization of covariance principle into a general type of geometry (non-Euclidean) would yield the equations of GRT. Non-Euclidean geometry describes the so-called space-time continuum in GRT. This requirement is due to the presence of gravity. As a theory of gravity, the most significant force on the cosmic scale, GRT has shaped the Big Bang Cosmology and addressed the problem of the beginning of time. Quantum origin of universe is definitely required to complete the picture. The theories of relativity and

---

[a]   The concept of new displacement current is again almost entirely due to the requirement of vector calculus.

QM are the most successful ones. Nevertheless the modern enigma is that both of these theories stand poles apart and a unification of the two poses problems.

Basically classical theories, SRT and GRT stress the reality of a rigid frame space and time structure as seen by above equations, though we find modifications in each of them. That is, according to this picture every particle follows a trajectory in space and time as endorsed by our every day experience. Every object that we come across in daily life (e.g. bullet, train, planets etc.) moves along a definite path in space. The analog of this in relativity is the world-line through Minkowski space. The Copenhagen view of quantum phenomena challenges this world view. The three most sacrosanct principles of physics like *causality, locality and reality* are being challenged by modern view on QM, or at least many believe so. Einstein and his team fought against these paradigm changes in views afforded by QM which led to the famous debate between Einstein[7] and Bohr[8]. The most important turning points were papers of EPR[7] and Bell[9]. The experiments of Aspect[10] and several others that followed, vindicated the stance that no hidden variable theory can mimic all the results of quantum mechanical systems. To majority of the physicists this seems to have conclusively settled the issue in favor of Bohr's Copenhagen position. Yet doubts about have been raised by some whether it is really so. If it is really true, then the consequences are tremendously bizarre.

MATHEMATICAL MODELING OF REALITY IN QM

Mathematical modeling can reveal the concealed reality. This means that mathematics is not just a language alone. *Basically Mathematics mainly involves exploitation of symmetries in general.* Thus hidden symmetries are revealed as various aspects truth and diverse connections and inter relations revealed through modeling a system. Take for instance, the power of mathematical modeling in QM,

$$H\psi_n = \frac{ih}{2\pi} \frac{\partial}{\partial t} \psi_n \qquad (5)$$

The above is the Schrödinger equation where H, the Hamiltonian in typical and relatively simple cases is given by

$$H = \frac{-h^2}{8\pi^2 m} \nabla^2 + V \qquad (6)$$

and the relativistic Dirac equation is

$$i\frac{h}{2\pi} \frac{\partial}{\partial t} \psi = \left(-\frac{ih}{2\pi} c\alpha.\nabla + \beta mc^2 \right)\psi \qquad (7)$$

$$\alpha, \beta : \text{are matrices.}$$

There are no non-deterministic features in this but Max Born's elucidation of the quantum mechanical[11] wave function as likelihood that a particle may be found between two points, r_1 and r_2 can be articulated as

$$P_{r1,r2} = \int_{r1}^{r2} \psi^* \psi dr \qquad (8)$$

The same can be expressed in terms of vectors in Hilbert space. But there can be several solutions ψ_n to the Schrödinger equation and each one can give a different eigenvalue value to a physical attribute as follows,

$$\hat{O} \psi_n = o_n \psi_n \qquad (9)$$

Unwittingly we are now modeling the uncertainty in the physical observables. From Eq. (8) and (9) it is possible to get the expectation values of a physical attribute a quantum mechanical system as

$$< O(r,p) > = \int_{-\infty}^{+\infty} \psi^* \hat{O} \psi dv \qquad (10)$$

From the above definition of expectation values it follows that the expectation values of position and momentum / velocity do not commute as their operators satisfy

$$< px > - < xp > = \frac{ih}{2\pi} \qquad (11)$$

These facts mathematically lead to the famous Heisenberg's uncertainty relations[12]. One can see that certain extremely important aspect of the *reality* itself is being mathematically modeled here. The most sacrosanct classical idea of determinism gets a jolt by the above simple unassuming mathematical relations. Even the idea of consciousness was conjectured by certain noted scientists. The above piece of illustration of mathematical modeling is totally unparalleled and unprecedented! *It is the concealed symmetry in the situations that is finally revealed and manifested as fragments of truth through coherent mathematical equations.* The relativistic Dirac-Schrödinger equation leads to the spin of electrons and to the inevitability of the existence of antiparticle of electron (positron)[13,14]. This again is from symmetry as preserved by the coherence accorded by mathematical modeling. Similarly the connection between Lorentz invariance and CPT invariance is revealed through purely mathematical logic which is basically the logic of a situation governed by symmetry considerations or similarity of situation. There are so many cases in physics of this type of compelling symmetry considerations which when mathematically formulated would even lead to counter intuitive and surprising revelations and results. QM is fully ridden with such counter-intuitive results.

LOGIC OF MATHEMATICS

The basic symmetries and logic of mathematics pop up in outstanding discoveries. CPT theorem and Lorentz invariance are compelling of such [15]. The prediction of anti-particles by the Dirac equation is due to the compelling symmetry. Spin of electrons and Pauli's exclusion principle are also similar ones but in a more subtle way. These innumerable illustrations in physics are impossible to exhaust. One may observe rather a curiously surprising phenomenon that the imaginary numbers that appear in many places in physics give rise to a new type of logic and weird interpretation that are counter-intuitive. Where ever the factor 'i' (i.e. square root of -1) appears in an equation the situation appears counter intuitive, fuzzy or weird[16]. For instance, the quantum bizarreness can be correlated to the presence of 'i' in the Schrödinger equation. The presence of this imaginary character 'i' is responsible for the non factorizablity in EPR type of pair. The non-factorizablity is well corroborated by several experimental results showing breach of Bell's inequality[17]. Incidentally the surprise over the violations of Bell's inequality still persists in a large scale.

THE PHYSICAL WORLD

The 21st century may primarily belong to biology as claimed by many. May be so, quite justified too, since, man has achieved a high degree of knowledge of the physical world already due to the tremendous progress made in the physical sciences during over the last 100 years and more. Physics is supposed to be the scientific field that helps us to comprehend and master the material world in and around us. Yet one finds words like 'free will', 'information', 'consciousness', 'mind' and the like in physics literature these days. Gross, sturdy structure of matter at chemical level is somewhat well understood thus far. It is the further inner frontiers that are posing immediate challenges to the human intelligence. Perhaps it is only fair to say that in probing the minutest levels of matter and radiation man has reached almost a stage of perfection that does give credit to the human hard work and enthusiastic intellectual efforts of so many brilliant people from all over the globe.

In this scenario one can fairly easily visualize the role of quantum picture which has literally revolutionized the science-oriented human probe into the fundamental truth. We have almost arrived at the most important stage in exploring and expounding the content and character of matter. But light, in spite of the breath-taking contributions of stalwarts like Einstein starting from the photon theme, is an area of continuing work with much thrust forward being experimental. The serious quest for a single theory of everything going beyond the standard model is very much still on. The quark assembly is possibly complete, but still inclusion of gravity in the theory is yet to be accomplished. Certainly, much more research is required to be carried out on all levels, as then and there alone we would have a firm command over the full field as the achievement that is aimed at is to have a complete understanding of the physical world is rather far away. Even so, with good mathematical tools final destination is hopefully not unreachable. Physics aided, pursued and expressed by mathematics ought to reach the coveted destination ultimately. Perhaps, the need of the hour is to proceed with confidence, courage and conviction in knowing the ultimate nature with consciousness at the central focus.

Understanding of nature and truth is the *only major aim* of science, technology being only the spin off, not the target of scientific enterprise. We believe that one can never exclude the existence and reality of *free will* altogether from an ultimate model of the cosmos. Free will can affect the physical cosmos in a tangible and concrete way. The recognition of this is happening among the community of physicists and other scientific circles and researches only recently in a wide spread fashion. The mathematical rigor, required was not available earlier, thanks to the quantum enigma and cosmological requirements. The deterministic philosophy is not completely dead; people are trying to restore it in QM[18]. Of late the role and the existence of free will is also being discussed very much[19]. Absolute causeless free will can be considered as the characteristic of Consciousness[20]. The question of subject-object dichotomy in context of QM was discussed even by the celebrated quantum physicists of earlier generations[21,22]. Physicists and mathematicians are trying to model free will mathematically. Existence of real free will is being debated vigorously by leading scientists. Anthropic principle in cosmology and the quantum riddle have opened the possibility of such considerations rendering the purely classical mechanistic view of the universe naive.

CONCLUDING REMARKS

Maxwell's electromagnetic equations actually revolutionized the use of pure mathematics as a tool for discovery. The introduction of displacement current was a mathematic necessity born out of property of space represented by vector calculus. The development of General Relativity Theory (GRT) by Einstein was exemplary. It was not due to any persuasive observational or experimental evidence but due to purely mathematical compulsion. Dirac's predication of anti-particle and the introduction of spin of electron are fine examples utility and power of mathematical modeling. The value of mathematical modeling cannot be over emphasized. Our contention is that mathematics brings out the inherent symmetry of nature. Interpretation of QM has lead to the question of observer and observed and lead to the extent of throwing away the principles of causality and the abandoning of the space-time picture of reality[19]. QM is a powerful tool that brings out the property of space in physical models leading to surprising discoveries.

Manthanō, the root of mathematics, means to <u>learn</u> and *Scio* the root of science means to <u>know</u>, thus both standing for right knowledge, the greatest gift to and by the mankind. Even though we would very much wish to probe nature up to the deepest and widest extent possible, what is actually possible is only of quite limited in degree and details. Only by means of appropriate, approximate methods and models we can learn about and know anything about the true nature in and around. The truth is that most often the reality is hidden and far away from sight. Even so, we can gain some reasonable insight into the nature of matter, life and the whole dynamics evolving in due course of time via and with the support and through involvement of mathematical/ physical/ intuitive replicas/ models intelligently, carefully and cleverly constructed. In particular, with the quantum mechanical path chosen to knowing the real nature as finally and extensively as allowed, subject to the satisfaction of the uncertainty principle/ relations, we ought to tread cautiously step by step. Only then there can be any hope to achieve the goals set before us in this life and there after.

ACKNOWLEDGMENTS

The Authors express their gratitude to *Sri Mata Amrithanandamayi Devi (Amma)*, Chancellor, Amrita Vishwa Vidyapeetham for unfailing inspirations and opportunity. They also thank *Prof. Dr. Venkat Rangan*, Vice Chancellor for providing opportunity for the work. They wish to thank *Prof. Dr. C.S.Shastry and Prof. Dr. N.N.Pillai* for useful discussions.

REFERENCES

1. P.Achuthan, *"Aspects of Mathematical Modeling and Applications"*, Proceedings, International Conference on Mathematical Modeling in Science and Technology (1988).
2. P.Achuthan, *Aspects of Fundamental Particles and Interactions*, Nobel Prize Week Souvenir AVV, CBE, India (2007).
3. H.P.Stapp, Found.Phys., **12**, 363 (1982).
4. John von Neumann, *Mathematical Foundations of Quantum Mechanics,* New Jersey: Princeton,1955.
5. R.I.Thompson, "The theory of quantum mechanics", *Consciousness and the Laws of Nature,* , Mumbai: Bhaktivedanta Book Trust,1977, pp.20-28.
6. A.Einstein, "The Fundaments of Theoretical Physics", *Ideas and Opinions,* New Delhi: Rupa and Co., 1984, p.329.
7. A. Einstein, B. Podolsky, N. Rosen, Phys. Rev. **47**, 777-780 (1935).
8. N. Bohr, Phys. Rev., **48**, 696 (1935).
9. J.S,.Bell, Physics, **1**,3, 195-200 (1964).
10. A. Aspect, P. Grangier and G. Roger, Phys. Rev. Lett. **49**, 91-94 (1982).
11. *L.I.Schiff, Quantum Mechanics, 3rd edition, Singapore: McGraw-Hill Book Co., 1968, pp.24-25.*
12. *R.P.Feynman, The Feynman Lectures on Physics, Vol. 3, New Delhi: Narosa Publ. House,1998, pp. 1792-1795.*
13. P.A.M. Dirac, *Principles of Quantum Mechanics*, New York: Oxford Univ. Press, 2004, pp. 273-275.
14. Frank Wilczek, "The Dirac Equation*", Proceedings of Dirac Centennial Symposium*, 2002, Edited by Howard Baer and Alexander Belyaev, Singapore: World Scientific, 2003, pp.45-55.
15. P.Achuthan and Narayanankutty Karuppath, Electronic Notes in Discrete Mathematics, **33**, 43-50 (2009).
16. Roger Penrose, *Emperor's New Mind*, New Delhi: Oxford Univ. Press, 2005, pp. 115-127.
17. David Wick, "Testing Bell"*, The Infamous Boundary- Seven Decades of Heresy in Quantum Physics*, New York: Springer Verlag , 1996, pp. 115-128.
18. Gerard 't Hooft, "How does God Play? (Pre-)Determinism at the Planck Scale", arXiv:hep-th/0104219v1(2001).
19. John Conway and Shimon Kohen, Found. Phys., **36**, 10 (2006).
20. P.Achuthan, Narayanankutty Karuppath, 'Aspects of Consciousness and Applications', Science and Spiritual Quest: Third All India Students' Conference, 22-23 Dec. 2007, Tirupati, India
21. E. Schrödinger, *"Quantum Questions-Mystical Writings Of The Worlds Greatest Physicist*s'', Edited by K.Wilber, Boston: Shambala, 2001, pp.92-93.
22. Wolfgang Pauli, *General Principles of Quantum Mechanics,* (Translated from German by P.Achuthan and K.Venkatesan), New Delhi: Allied publishers, 1980, p.1 (The original German edition of this classic work was published under the title, Handbuch Der Physik, Band V Teil 1; Prinzipien der Quantentheorie I, Berlin: Springer-Verlag, 1958).

Visualization and Analysis of Wireless Sensor Network Data for Smart Civil Structure Applications Based On Spatial Correlation Technique

Bhawani Shankar Chowdhry[1,2,3], Neil M White[2], Jai Kumar Jeswani[1],
Khalil Dayo[1], Manorma Rathi[1]

[1] *Mehran University of Engineering & Technology, Jamshoro, Pakistan*
c.bhawani@ieee.org, bsc06v@ecs.soton.ac.uk
[2] *School of Electronics and Computer Science, University of Southampton, UK*
[3] *Regular Associate, The Abdus Salam ICTP, Trieste, ITALY*

Abstract. Disasters affecting infrastructure, such as the 2001 earthquakes in India, 2005 in Pakistan, 2008 in China and the 2004 tsunami in Asia, provide a common need for intelligent buildings and smart civil structures. Now, imagine massive reductions in time to get the infrastructure working again, real-time information on damage to buildings, massive reductions in cost and time to certify that structures are undamaged and can still be operated, reductions in the number of structures to be rebuilt (if they are known not to be damaged). Achieving these ideas would lead to huge, quantifiable, long-term savings to government and industry. Wireless sensor networks (WSNs) can be deployed in buildings to make any civil structure both smart and intelligent. WSNs have recently gained much attention in both public and research communities because they are expected to bring a new paradigm to the interaction between humans, environment, and machines. This paper presents the deployment of WSN nodes in the Top Quality Centralized Instrumentation Centre (TQCIC). We created an *ad hoc* networking application to collect real-time data sensed from the nodes that were randomly distributed throughout the building. If the sensors are relocated, then the application automatically reconfigures itself in the light of the new routing topology. WSNs are event-based systems that rely on the collective effort of several micro-sensor nodes, which are continuously observing a physical phenomenon. WSN applications require spatially dense sensor deployment in order to achieve satisfactory coverage. The degree of spatial correlation increases with the decreasing inter-node separation. Energy consumption is reduced dramatically by having only those sensor nodes with unique readings transmit their data. We report on an algorithm based on a spatial correlation technique that assures high QoS (in terms of SNR) of the network as well as proper utilization of energy, by suppressing redundant data transmission. The visualization and analysis of WSN data are presented in a Windows-based user interface.

Keywords: Wireless Sensor Networks, Spatial Correlation, Smart Civil Infrastructures, QoS, Micro-sensor Node Simulation.

1. INTRODUCTION

One of the biggest and awe-inspiring applications of WSN is claimed to be in intelligent buildings or smart civil structures [1]. Civil infrastructure systems are

CP1146, *Modelling of Engineering and Technological Problems*, edited by A. H. Siddiqi, A. K. Gupta, and M. Brokate
© 2009 American Institute of Physics 978-0-7354-0683-4/09/$25.00

generally the most important assets of a nation. They are large, distributed systems, which are constantly used by public and communities to perform day-to-day activities and commerce. They comprise transportation systems, communication systems, power generation and distribution systems, water supply and sewage treatment systems, and the systems and facilities used for education, medical treatment, and societal governance. The well being and prosperity of a nation is greatly determined by the adequacy, health, safety, and security of its infrastructure systems. To sustain the performance and reliability of such structures, it is essential to have accurate and real-time information about the condition of the structures. Today, information regarding the health of a structure is obtained through scheduled and labor-intensive inspections and analysis, which may not provide the necessary hard information during the critical time before a catastrophic failure strikes, natural or man-made. Similar considerations and arguments apply to vehicles such as airplanes, commuter buses and trains. Exploitation of advanced sensor technologies for rescuing victims after natural disasters and large-scale accidents will contribute a great deal toward realizing safe and secure modern societies [2].

Imagine smart devices in the environment and atmosphere for monitoring resources, in the infrastructures for examining structural health, in industry observing the manufacturing process. That gives some idea of the use of emerging technology of Wireless Sensor Networks (WSNs). The emerging field of wireless sensor networks combines sensing with onboard computation, onboard storage, and wireless communications. WSN is one of the first examples of large-scale computing, where tiny, low-power, smart, cheap, computing nodes with on-board multifunctional sensors, are deployed throughout the physical world of interest providing dense sensing, close to the physical phenomena. This *in situ* sensing, and processing and communicating the information, while coordinating actions with other nodes, results in unexpected opportunities for dealing with the physical environment [3][4]. The development of WSNs was originally motivated by military applications such as battlefield surveillance. However, WSNs are now used in many civilian application areas, including environment and habitat monitoring, healthcare, home automation, traffic control, and many more. WSN is still in its infancy. Work is being done at a rapid pace and the range of potential applications is overwhelming. WSNs will expand people's ability to remotely interact with the physical world. It is predicted that, within this decade, distributed sensing and computing will creep into every home, building, office, factory, car, street, and farm just like the internet [3][4][5].

We have used a Windows-based user interface for visualization and analysis of sensor data, their Network Mapping Topology, Alarm Services, Configuration and Management services. Typical factors to be detected are light, temperature, acceleration, magnetic field and sound. The information generated by these sensors can be helpful in transforming any traditional building into an "Intelligent Building". We use the term *intelligence*, although it is unsatisfactory in an engineering context and no longer has its human or biological meaning. Here, intelligence simply implies that the benefits of buildings for their inhabitants are increased by units that capture the current state of the building and its devices, process those signals, and make appropriate adjustments. Although the intellectual capability of human beings has not been reached by a long way, tremendous progress has been made using smart sensors and control systems.

2. OUR MODEL

We introduce our real focus of interest, which is WIS "Wireless sensor networks for intelligent spaces", and discuss the measurements that we have simulated in the WIS scenario using the Moteview simulator. WIS is expected to judge indoor and outdoor conditions of a building and its devices in order to operate as an integrated system and achieve maximum performance and comfort levels. Two cases of node deployment are considered here: one with MICA2, and the other using MICA2DOT. The nodes are placed so that some of them are near to light, some close to darkness, some close to noisy locations, in order to cover every possibility of noticing any change under the laboratory conditions. Obstructions, such as furniture, PCs and other equipment are also considered, in order to make a realistic environment.

The following hardware components were used for the practical setup.
1. **MIB 510CA** Serial Interface Board
2. **MPR400** MOTE: MICA2
3. **MPR500** MOTE: MICA2DOT
4. **MTS 310CA** Sensor board
5. **MDA500** Data acquisition boards
6. **MTS 510** Flexible Sensor Board
7. Serial **RS 232** cable and power cable

CASE 1: Using MICA2 Nodes
The MICA2 Motes come in three models according to their RF frequency band: the MPR400 (915 MHz), MPR410 (433 MHz), and MPR420 (315 MHz). The Motes use the Chipcon CC1000 FSK modulated radio. All models utilize a powerful Atmega128L micro-controller and a frequency tunable radio with extended range. The MPR4x0 and MPR5x0 radios are compatible and can communicate with each other [8]. (The $x = 0$, 1, or 2 depending on the model / frequency band). Six of these nodes are deployed in TQCIC Lab.

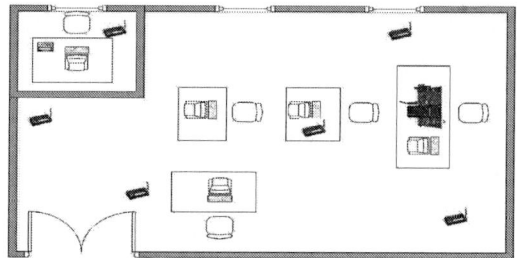

FIGURE1. Deployment of MICA2 nodes in TQCIC Lab

3. VISUALIZATION AND ANALYSIS OF SENSOR NETWORK DATA

Id	voltage	temp	light	accel_x	accel_y	mag_x	mag_y	mic	Time
3	3.03V	-273.15C	920	1.14g	0.74g	25.93mgauss	25.93mgauss	161	10/5/2007 10:10:28 AM
6	3.22V	-273.15C	683	0.36g	1.12g	109.39mgauss	48.89mgauss	171	10/5/2007 10:10:45 AM
5	3.08V	-273.15C	850	0.4g	0.44g	24.71mgauss	24.58mgauss	166	10/5/2007 10:08:15 AM
4	3.22V	-273.15C	725	0.7g	-2.68g	42.81mgauss	24.44mgauss	169	10/5/2007 10:09:47 AM
7	3.3V	-273.15C	825	0.78g	-2.44g	44.43mgauss	24.17mgauss	167	10/5/2007 10:10:26 AM

After installation, it takes a little while for the network to detect nodes and list them, the data values needs to be updated periodically. The refresh time/update duration is user dependent. Here we kept this value 180 seconds. The table shows an extract of five of the sensors reporting on battery voltage, temperature, incident light, acceleration in two horizontal directions, magnetic field likewise, and mic.

The above values detected by the nodes have no particular sequence, but each entry is time-stamped. The voltage value shows the state of the 2 AA batteries attached to each node. The threshold condition for a node to be functional is 2.7 volt. Each node's sensors have added their respective values. The temperature sensor was out of order in our experiment.

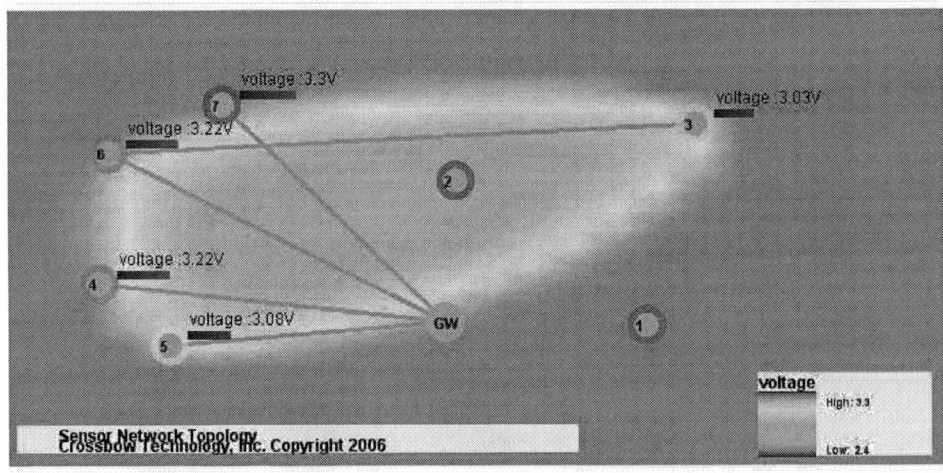

FIGURE 2. Visualization of the sensor network showing voltage levels

This topology reveals the deployment of the nodes. Nodes with ID 1 & ID 2 were isolated from the others and thus they were not detected, because they were out of the range. Though they seem to be close in the diagram, it does not show relative distance. This means a node that seems to be near may not be actually close to the gateway.

An interesting point is that, due to a problem in connectivity, node 3 is not directly connected to the gateway but is connected via node 6. This reveals the "Self-Configuring, Self-Healing" property of WSNs.

CASE 2: Using MICA2DOT Nodes

The MICA2DOT is a Mote designed for applications where physical size is important. Like the MICA2, these are available in three models according to the frequency of the RF transceiver: the MPR500 (915 MHz), MPR510 (433 MHz), and MPR520 (315 MHz). The Motes use the Chipcon CC1000 FSK-modulated radio. All models utilize a powerful ATMega128L microcontroller and a frequency tunable radio with extended range [8]. The MPR4x0 and MPR5x0 radios are compatible and can communicate with each other as long as they use the same frequency. In order to correlate the results obtained in Case 1, another random deployment was carried out using MICA2DOT nodes. The data obtained from these nodes are elaborated in charts shown in Figure 3.

These charts give us peaks and troughs of every parameter by timestamp in a separate window. These give a clear idea of the behavior of any parameter, whether it is smooth or there are rapid fluctuations.

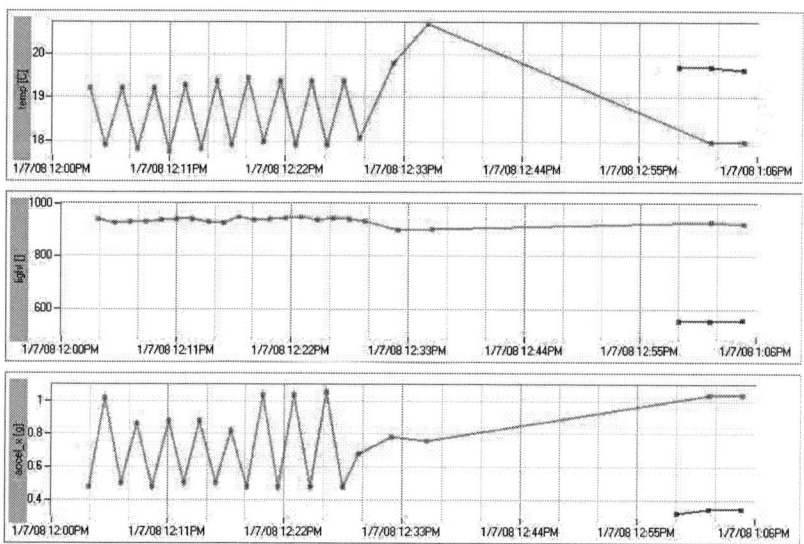

FIGURE 3. Charts of sensor values with one MICA2DOT node

For example, considering temperature for node 2, it starts from 19°C and keeps on fluctuating between 19 and 20. Then it makes a transition from 18 to 21°C and gradually decreases over half an hour, while node 1 is detected and begins to transmit

117

readings around 19.7°C. The light does not fluctuate rapidly and its curve tends linearity.

4. SPATIAL CORRELATION IN WSN

Since WSNs are event-based systems that rely on the collective effort of several micro-sensor nodes, which are continuously observing physical phenomena, WSN applications require spatially-dense sensor deployment in order to achieve satisfactory coverage. As a result, several sensor nodes record information about a single event in a sensor field. If the nodes are close, many of them have approximately same piece of data, which is spatially correlated with each other, subjected to the same event. Only one of the nodes of that small coverage area needs to transmit instead of all. This is known as Spatial Correlation. The degree of Spatial Correlation increases with the decreasing inter-node separation. Allowing only some sensor nodes to transmit their data can reduce energy consumption dramatically [6] [7].

Typical WSN applications require spatially dense sensor deployment in order to achieve satisfactory coverage. Due to high density in the network topology, spatially proximal sensor observations are highly correlated with the degree of correlation increasing with decreasing inter-node separation [6] [7].

Some problems arise when more data is transmitted, such as:

- Loss of packets
- More energy required for transmitting and routing packets
- More bandwidth required for transmitting more data at higher speed
- More computing capabilities for aggregating the data
- More memory is required to store the data.

We have developed an algorithm to cope with these problems. Consider two nodes in the environment with sensing, computation and communication capabilities. After the node has sensed the data, we will run an algorithm before transmission of the data. The algorithm we have developed is based on two factors.

1) We know transmission of data requires a lot more energy than computation, so we start filtering the data at each individual node. The first figure of merit is "Signal-to-Noise Ratio". We have assumed an SNR level in the range 0-1. Each node will generate a random number and if that number is greater than 0.5 then that node will be allowed to transmit its data. Data with SNR<=0.5 will be considered as distortion and its data blocked. There is also the possibility that low SNR data cannot reach the gateway, so allowing that data would ultimately waste network resources.

2) After getting data of the desired SNR, the next task is to identify the set of nodes with similar data or with very close values. The second figure of merit is the "Spatial-Correlation". Again, each node generates a random number in the range 0-1, which we

assume is the sensed value of a parameter (e.g. temperature, pressure), and compare it with the 2nd node's corresponding number. If there is difference of at least 25% or greater, then we will assume that they have different values, otherwise we will consider that they have similar values. Finally, a 'star' will be generated to show values that are correlated or not. A 'Red Star' at zero level shows that correlation is less than threshold and they may be counted as different values, while a 'blue star' at level 1 shows that the values are highly correlated. Now an individual node is in a position to decide which data is to block and which is to send.

4.1 Algorithm

```
clc
SNR_1 = 0
SNR_2 = 0
corr = 0
data_value_node_1 = 0
data_value_node_2 = 0

diff_node_value = 0

SNR_1 = rand()
SNR_2 = rand()
m = 1
n = 1

    if (SNR_1>=0.5)
       data_value_node_1 = rand(2)
    else
       display('SNR_1 is less than the required threshold value')
    end if

    if (SNR_2>=0.5)
       data_value_node_2 = rand(2)
    else
       display('SNR_2 is less than the required threshold value')
    end if

    if (SNR_1>=0.5 && SNR_2<0.5)
       display('SNR_2 is less than threshold so only node 1 will
transmit all its values')
    return
    end if

    if (SNR_1<0.5 && SNR_2>=0.5)
       display('SNR_1 is less than threshold so only node 2 will
transmit all its values')
    return
    end if

if (SNR_1<0.5 && SNR_2<0.5)
    display('both values are below threshold so algorithm would not
run')
    return
```

```
          end if

if (SNR_1>=0.5 && SNR_2>=0.5)
     for i = 1 to 2
        for j = 1 to 2
        i
        j
        data_value_node_1(i,j)

   data_value_node_2(i,j)
diff_node_value = abs(data_value_node_1(i,j) -
data_value_node_2(i,j))

   if diff_node_value<=0.25
      corr = 1
      disp('similiar values')

   elseif diff_node_value>0.25
      corr = 0
      disp('different values')
   end if

   switch(corr)
   case 0
      plot(n,corr,'r*')
      hold on
      n = n + 1
   case 1
      plot(n,corr,'b*')
      hold on
      n = n + 1
   end switch

   data_value_node_1(i,j) = 0 //reset values for next iteration
   data_value_node_2(i,j) = 0 //reset values for next iteration

   axis([-2 20 -3 3])

end
end
display('all the four values are compared, algorithm ends')
end
```

Implementation of the above algorithm assures high QoS (in terms of SNR) of the network, as well as proper utilization of energy by suppressing redundant transmission.

In the Program both nodes Node1 and Node2 have generated SNR values randomly and both values are greater than 0.5 that is our first condition for the nodes to be eligible for second condition. Now after generating SNR values both the nodes generate 4- data values respectively at time slot 1,2,3 and 4. Now we need to calculate whether there is any similarity in the data generated by two nodes or not at individual time slot. Our condition is that if the difference between the data values of the two nodes at single time slot is less than 25% then we assume them similar values otherwise they are different, that's our consideration.

4.2 Interpretation of Simulation Results:

Now at time slot 1: Blue star just as a symbol representing no correlation or zero correlation as the difference between two data values would be >0.25. Interpretation is that both the values should be transmitted by respective nodes. That can be future step on this paper.

Time slot 2: again Blue Star just as a symbol representing no correlation or zero correlation.

Time slot 3: Red star shows correlation value is 1 and difference is <0.25, so both the data values generated by individual nodes are similar. Interpretation is that only single value should be transmitted in order to reduce the amount of energy in such an energy-hunger application. Next step on this should be that which node's value should be selected for transmission.

Time slot 4: again red star which shows correlation value.

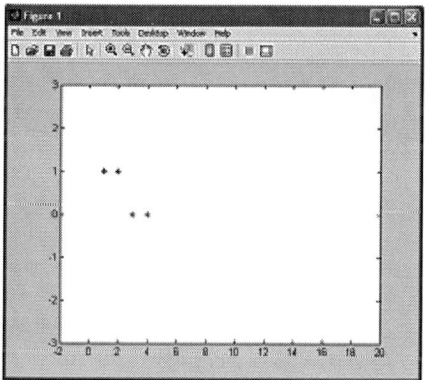

5. CONCLUSION

Sensor nodes are tiny autonomous devices that combine sensing, computing and wireless communication capabilities. These nodes are deeply embedded into the physical surroundings, and gather and process information such as acceleration, temperature, humidity, light characteristics, seismic activities, images, and sound samples from the physical world. A wide variety of applications is expected to use networked systems of sensors. Sensor network deployments and prototype implementations are still not commonplace. However, experiences gained in such deployments are crucial for the sensor network research community. The data obtained from two different deployments of WSN nodes such as MICA2 and MICA2DOT are presented here. A spatial-correlation algorithm is presented which assumes high QOS of network and substantially reduces energy consumption.

These results will allow refining assumptions to be made when designing hardware, software, protocols and mechanisms for sensor networks. With the cost of sensors and monitoring equipment reducing, more and more structures will have provision for embedded sensors for close monitoring. In future, WSNs may be an integral part of our lives, even more than today's PCs. More work is being carried out to make best use of the two or more redundant data values thereby eliminating similarities before transmitting data, representing the aggregate data from multiple nodes.

REFERENCES

1. Chowdhry, BS, Shardha, K, and Rajput, AQK. Development of Wireless Sensor Networks for Smart Civil Structures and Highway Safety, ICSPC 2007, IEEE International Conference on Signal Processing and Communications, Dubai, UAE, 24-27 Nov 2007.

2. Report on the US-Japan Workshop on Sensors, Smart Structures and Mechatronic Systems, University of Tokyo, Tokyo, Japan, 12-13 Nov 2005.

3. Estrin, D, and Srivastava, M. Wireless Sensor Networks, Tutorial, MOBICOM 2002.

4. Kirshnamachari, B. Networking Wireless Sensors, Cambridge University Press, Jan 2006.

5. www.techreview.com

6. Vuran, MC, Akyildiz, IF. Spatial Correlation-based Collaborative Medium Access Control in Wireless Sensor Networks, IEEE/ACM Transactions on Networking archive, 14(2), p 316-329, Apr 2006.

7. Vuran, MC, Akan, OB, Akyildiz, IF. Spatio-temporal correlation: theory and applications for wireless sensor networks, Computer Networks Journal, Elsevier, 45, p 245-259, 2004.

8. MPR/MIB User Manuals & Product References Guide, Crossbow Technology Inc, USA.

A stability result in a weakly damped nonlinear Timoshenko system

Salim A. Messaoudi[1] and Muhammad I. Mustafa[2]
King Fahd University of Petroleum and Minerals
Department of Mathematics and Statistics
Dhahran 31261, Saudi Arabia.
1) E-mail: messaoud@kfupm.edu.sa
2) E-mail: mohmim@kfupm.edu.sa

Abstract

In this paper we consider the following Timoshenko system

$$\varphi_{tt} - (\varphi_x + \psi)_x = 0, \quad (0,1) \times (0, +\infty)$$

$$\psi_{tt} - \psi_{xx} + \varphi_x + \psi + \sigma(t)g(\psi_t) = 0, \quad (0,1) \times (0, +\infty)$$

with Dirichlet boundary conditions where σ is a positive function and g is a nondecreasing function satisfying some growth conditions. We establish a generalized stability result for this system.

Keywords and phrases: Nonlinear stability, Timoshinko system, weakly damped.

AMS Classification: 35B37, 35L55, 74D05, 93D15, 93D20.

1 Introduction

Timoshenko [13] gave the following system of coupled hyperbolic equations

$$\begin{aligned}
\rho u_{tt} &= (K(u_x - \varphi))_x, \quad \text{in } (0, L) \times (0, +\infty) \\
I_\rho \varphi_{tt} &= (EI\varphi_x)_x + K(u_x - \varphi), \quad \text{in } (0, L) \times (0, +\infty),
\end{aligned} \tag{1.1}$$

as a simple model describing the transverse vibration of a beam, where t denotes the time variable and x is the space variable along the beam of length L, in its equilibrium configuration, u is the transverse displacement of the beam and φ is the rotation angle of the filament of the beam. The coefficients ρ, I_ρ, E, I and K are respectively the density (the mass per unit length), the polar moment of inertia of a cross section, Young's modulus of elasticity, the moment of inertia of a cross section, and the shear modulus.

CP1146, *Modelling of Engineering and Technological Problems*, edited by A. H. Siddiqi, A. K. Gupta, and M. Brokate
© 2009 American Institute of Physics 978-0-7354-0683-4/09/$25.00

System (1.1) has been studied by many mathematicians and results concerning existence and asymptotic behavior have been established. Kim and Renardy [3] considered (1.1) together with two linear boundary conditions of the form

$$K\varphi(L,t) - K\frac{\partial u}{\partial x}(L,t) = \alpha\frac{\partial u}{\partial t}(L,t), \quad \forall t \geq 0$$

$$EI\frac{\partial\varphi}{\partial x}(L,t) = -\beta\frac{\partial\varphi}{\partial t}(L,t), \quad \forall t \geq 0$$

and used the multiplier techniques to establish an exponential decay result for the energy of (1.1). They also provided numerical estimates to the eigenvalues of the operator associated with system (1.1). An analogous result was also established by Feng et al. [2], where the stabilization of vibrations in a Timoshenko system was studied. Raposo et al. [6] studied (1.1) with homogeneous Dirichlet boundary conditions and two linear frictional dampings. Precisely, they looked into the following system

$$\begin{aligned}
&\rho_1 u_{tt} - K(u_x - \varphi)_x + u_t = 0, \quad \text{in } (0,L) \times (0,+\infty) \\
&\rho_2\varphi_{tt} - b\varphi_{xx} + K(u_x - \varphi) + \varphi_t = 0, \quad \text{in } (0,L) \times (0,+\infty) \\
&u(0,t) = u(L,t) = \varphi(0,t) = \varphi(L,t) = 0, \quad \forall t > 0
\end{aligned} \tag{1.2}$$

and proved that the energy associated with (1.2) decays exponentially. To obtain their result, they used a method developed by Liu and Zheng [4]. This method is different from the usual ones such as the classical energy method. It mainly uses the semigroup theory. Soufyane and Wehbe [11] showed that it is possible to stabilize uniformly (1.1) by using a unique locally distributed feedback. So, they considered

$$\begin{aligned}
&\rho u_{tt} = (K(u_x - \varphi))_x, \quad \text{in } (0,L) \times (0,+\infty) \\
&I_\rho\varphi_{tt} = (EI\varphi_x)_x + K(u_x - \varphi) - b\varphi_t, \quad \text{in } (0,L) \times (0,+\infty) \\
&u(0,t) = u(L,t) = \varphi(0,t) = \varphi(L,t) = 0, \quad \forall t > 0,
\end{aligned} \tag{1.3}$$

where b is a positive and continuous function, which satisfies

$$b(x) \geq b_0 > 0, \quad \forall x \in [a_0, a_1] \subset [0,L].$$

In fact, they proved that the uniform stability of (1.3) holds if and only if the wave speeds are equal $\left(\frac{K}{\rho} = \frac{EI}{I_\rho}\right)$; otherwise only the asymptotic stability has been proved. This result improves earlier ones by Soufyane [10] and Shi and Feng [8], where an exponential decay of the solution energy of (1.1) together, with two locally distributed feedbacks, had been proved. Xu and Yung [14] studied a system of Timoshenko beams with pointwise feedback controls, sought information about the eigenvalues and eigenfunctions of the system, and used this information to examine the stability of the system. Ammar-Khodja et al. [1] considered a linear Timoshenko-type system with memory of the form

$$\begin{aligned}
&\rho_1\varphi_{tt} - K(\varphi_x + \psi)_x = 0 \\
&\rho_2\psi_{tt} - b\psi_{xx} + \int_0^t g(t-s)\psi_{xx}(s)ds + K(\varphi_x + \psi) = 0
\end{aligned} \tag{1.4}$$

in $(0,L) \times (0,+\infty)$, together with homogeneous boundary conditions. They used the multiplier techniques and proved that the system is uniformly stable if and only if the wave speeds are equal $\left(\frac{K}{\rho_1} = \frac{b}{\rho_2} \right)$ and g decays uniformly. Precisely, they proved an exponential decay if g decays in an exponential rate and polynomially if g decays in a polynomial rate. They also required some extra technical conditions on both g' and g'' to obtain their result. The feedback of memory type has also been used by Santos [7]. He considered a Timoshenko system and showed that the presence of two feedbacks of memory type at a portion of the boundary stabilizes the system uniformly. He also obtained the rate of decay of the energy, which is exactly the rate of decay of the relaxation functions. Shi and Feng [9] investigated a nonuniform Timoshenko beam and showed that, under some locally distributed controls, the vibration of the beam decays exponentially. We refer the reader to [12], [5], and [15] for more results.

In the present work we are concerned with the following nonlinear system

$$\begin{cases} \varphi_{tt} - (\varphi_x + \psi)_x = 0, \quad (0,1) \times \mathbb{R}_+ \\ \psi_{tt} - \psi_{xx} + \varphi_x + \psi + \sigma(t)g(\psi_t) = 0, \quad (0,1) \times \mathbb{R}_+ \\ \varphi(0,t) = \varphi(1,t) = \psi(0,t) = \psi(1,t) = 0, \ t \geq 0 \\ \varphi(x,0) = \varphi_0(x), \ \varphi_t(x,0) = \varphi_1(x), \ x \in (0,1) \\ \psi(x,0) = \psi_0(x), \ \psi_t(x,0) = \psi_1(x), \ x \in (0,1). \end{cases} \quad (1.5)$$

Our aim in this work is to establish a generalized stability result for system (1.5). We should note here that we do not loose the generality by taking the coefficients, appeared in (1.4), to be equal to one and our argument also works for $\frac{K}{\rho_1} = \frac{b}{\rho_2}$. The paper is organized as follows. In section 2, We present some notations and material needed for our work and state our main result. The proof will be given in section 3.

2 Preliminaries

In this section we present some material needed in the proof of our main result. We use the standard Lebesgue space $L^2(0,1)$ and the Sobolev space $H_0^1(0,1)$ with their usual scalar products and norms. We will use c, throughout this paper, to denote a generic positive constant. For σ and g we assume

(H1) $\sigma : \mathbb{R}_+ \to \mathbb{R}_+$ is a bounded differentiable function satisfying

$$\lim_{t \to \infty} \frac{\sigma'(t)}{\sigma(t)} = 0$$

(H2) $g : \mathbb{R} \to \mathbb{R}$ is a nondecreasing differentiable function such that there exist constants $c_1, c_2 > 0$ and $q \geq 1$ for which

$$c_1 \min\{|s|, |s|^q\} \leq |g(s)| \leq c_2 \max\{|s|, |s|^{\frac{1}{q}}\}$$

Remark 2.1. By hypothesis (H1), there exists $t_0 > 0$ such that

$$\left| \frac{\sigma'(t)}{\sigma(t)} \right| \leq 1, \ \forall t \geq t_0. \quad (2.1)$$

Remark 2.2. By hypothesis (H2), we note that

$$sg(s) > 0, \qquad \forall s \neq 0.$$

For completeness we state, without proof, an existence and regularity result.

Proposition 2.1. *Let* (φ_0, φ_1), $(\psi_0, \psi_1) \in H_0^1(0,1) \times L^2(0,1)$ *be given. Assume that (H1) and (H2) are satisfied, then problem (1.5) has a unique global (weak) solution*

$$\varphi, \psi \in C(\mathbb{R}_+; H_0^1(0,1)) \cap C^1(\mathbb{R}_+; L^2(0,1)).$$

Moreover, if

$$(\varphi_0, \varphi_1), (\psi_0, \psi_1) \in (H^2(0,1) \cap H_0^1(0,1)) \times H_0^1(0,1)$$

then the solution satisfies

$$\varphi, \psi \in L^\infty(\mathbb{R}_+; H^2(0,1) \cap H_0^1(0,1)) \cap W^{1,\infty}(\mathbb{R}_+; H_0^1(0,1)) \cap W^{2,\infty}(\mathbb{R}_+; L^2(0,1)).$$

Remark 2.3. This result can be proved using standard arguments such as the semigroup method or the Galerkin method.

Now, we introduce the energy functional

$$E(t) := \frac{1}{2} \int_0^1 \left(\varphi_t^2 + \psi_t^2 + \psi_x^2 + (\varphi_x + \psi)^2 \right) dx \tag{2.2}$$

Our main result is the following

Theorem 2.2. *Let* (φ_0, φ_1), $(\psi_0, \psi_1) \in H_0^1(0,1) \times L^2(0,1)$ *be given. Assume that (H1) and (H2) are satisfied, then there exist two positive constants c and ω such that the solution of problem (1.5) satisfies, for t large,*

$$E(t) \leq ce^{-\omega \int_0^t \sigma(s)ds}, \qquad if \quad q = 1 \tag{2.3}$$

and

$$E(t) \leq c \left(1 + \int_0^t \sigma(s)ds \right)^{-\frac{2}{q-1}}, \qquad if \quad q > 1 \tag{2.4}$$

3 Proof of the main result

In this section we prove our main result. For this purpose we will establish several lemmas.

Lemma 3.1. *Let* (φ, ψ) *be the solution of (1.5). Then the energy functional satisfies*

$$E'(t) = -\sigma(t) \int_0^1 \psi_t g(\psi_t) dx \leq 0. \tag{3.1}$$

Proof. By multiplying equations in (1.5) by φ_t and ψ_t respectively and integrating over $(0,1)$, using integration by parts, hypotheses (H1)-(H2) and some manipulations, we obtain (3.1) for any regular solution. This equality remains valid for weak solutions by a simple density argument.

Now we are going to construct a Lyapunov functional $\mathcal{L}$ equivalent to E, with which we can show the desired result. For this, we define several functionals which allow us to obtain the needed estimates

Lemma 3.2. *Under the assumptions (H1) and (H2), the functional F_1 defined by*

$$F_1(t) := -\sigma(t) \int_0^1 (\psi\psi_t + \varphi\varphi_t)dx$$

satisfies, along the solution, the estimate

$$F_1'(t) \leq \sigma(t)\left[-\frac{1}{2}\int_0^1 (\psi_t^2 + \varphi_t^2)dx + c\int_0^1 (\psi + \varphi_x)^2 dx + c\int_0^1 \psi_x^2 dx + c\int_0^1 g^2(\psi_t)dx \right], \tag{3.2}$$

for any $t \geq t_0$.

Proof. By exploiting equations (1.5) and the fact that

$$\varphi_x^2 \leq 2(\psi + \varphi_x)^2 + 2\psi^2$$

and using Young's and Poincaré's inequalities and (2.1), we have, for $t \geq t_0$,

$$
\begin{aligned}
F_1'(t) &= \sigma(t)\left[-\int_0^1 (\psi_t^2 + \varphi_t^2)dx - \int_0^1 \varphi(\psi + \varphi_x)_x dx - \int_0^1 \psi[\psi_{xx} - \varphi_x - \psi - \sigma(t)g(t)]dx \right] \\
&\quad -\sigma'(t)\int_0^1 (\psi\psi_t + \varphi\varphi_t)dx \\
&= \sigma(t)\left[-\int_0^1 (\psi_t^2 + \varphi_t^2)dx + \int_0^1 \psi_x^2 dx + \int_0^1 \sigma(t)\psi g(\psi_t)dx + \int_0^1 (\psi + \varphi_x)^2 dx \right] \\
&\quad -\sigma'(t)\int_0^1 (\psi\psi_t + \varphi\varphi_t)dx \\
&\leq \sigma(t)\left[-\int_0^1 (\psi_t^2 + \varphi_t^2)dx + \int_0^1 (\psi + \varphi_x)^2 dx + c\int_0^1 \psi_x^2 dx + c\int_0^1 g^2(\psi_t)dx \right] \\
&\quad +\sigma(t)\left[\delta\int_0^1 \psi_t^2 dx + \frac{c}{\delta}\int_0^1 \psi_x^2 dx + \delta\int_0^1 \varphi_t^2 dx + \frac{c}{\delta}\int_0^1 \varphi_x^2 dx \right] \\
&\leq \sigma(t)\left[-\int_0^1 (\psi_t^2 + \varphi_t^2)dx + \int_0^1 (\psi + \varphi_x)^2 dx + c\int_0^1 \psi_x^2 dx + c\int_0^1 g^2(\psi_t)dx \right] \\
&\quad +\sigma(t)\left[\delta\int_0^1 \psi_t^2 dx + \frac{c}{\delta}\int_0^1 \psi_x^2 dx + \delta\int_0^1 \varphi_t^2 dx + \frac{c}{\delta}\int_0^1 (\psi + \varphi_x)^2 dx \right]
\end{aligned}
$$

By choosing $\delta = 1/2$, estimate (3.2) is established.

Lemma 3.3. *Assume that (H1) and (H2) hold. Then, the functional F_2 defined by*

$$F_2(t) := \sigma(t)\left[\int_0^1 \psi_t(\psi + \varphi_x)dx + \int_0^1 \psi_x\varphi_t dx \right]$$

satisfies, along the solution, the estimate

$$F_2'(t) \leq \sigma(t) \left[[\psi_x \varphi_x]_{x=0}^{x=1} - (1 - \varepsilon c) \int_0^1 (\psi + \varphi_x)^2 dx + \varepsilon c \int_0^1 \varphi_t^2 dx \right. \tag{3.3}$$

$$\left. + \frac{c}{\varepsilon} \int_0^1 \psi_x^2 dx + \frac{c}{\varepsilon} \int_0^1 \psi_t^2 dx + \frac{c}{\varepsilon} \int_0^1 g^2(\psi_t) dx \right]$$

for any $t \geq t_0$ and $0 < \varepsilon < 1$.

Proof. By exploiting equations (1.5), and repeating the same procedure as in above, we have, for $t \geq t_0$,

$$F_2'(t) = \sigma(t) \left[\int_0^1 (\varphi_x + \psi)[\psi_{xx} - \varphi_x - \psi - \sigma(t)g(\psi_t)] dx + \int_0^1 (\varphi_{xt} + \psi_t) \psi_t dx \right.$$

$$\left. + \int_0^1 \psi_{xt} \varphi_t dx + \int_0^1 \psi_x (\varphi_x + \psi)_x dx \right] + \sigma'(t) \left[\int_0^1 \psi_t (\psi + \varphi_x) dx + \int_0^1 \psi_x \varphi_t dx \right]$$

$$\leq \sigma(t) \left[[(\psi_x \varphi_x]_{x=0}^{x=1} - \int_0^1 (\psi + \varphi_x)^2 dx + \int_0^1 \psi_t^2 dx - \int_0^1 \sigma(t)(\psi + \varphi_x)g(\psi_t) dx \right]$$

$$+ \sigma(t) \left| \int_0^1 \psi_t (\psi + \varphi_x) dx + \int_0^1 \psi_x \varphi_t dx \right|$$

By using Young's inequality, (3.3) is established.

Lemma 3.4. *Assume that (H1) and (H2) hold. Let $m \in C^1([0,1])$ be a function satisfying $m(0) = -m(1) = 2$. Then there exists $c > 0$ such that for any $0 < \varepsilon < 1$ the functional F_3 defined by*

$$F_3(t) := \sigma(t) \left[\frac{1}{4\varepsilon} \int_0^1 m(x) \psi_t \psi_x dx + \varepsilon \int_0^1 m(x) \varphi_t \varphi_x dx \right]$$

satisfies, along the solution, the estimate

$$F_3'(t) \leq \sigma(t) \left[-\frac{1}{4\varepsilon} \left((\psi_x(1,t))^2 + (\psi_x(0,t))^2 \right) - \varepsilon \left(\varphi_x^2(1,t) + \varphi_x^2(0,t) \right) \right.$$

$$+ (\frac{1}{4} + \varepsilon c) \int_0^1 (\psi + \varphi_x)^2 dx + \varepsilon c \int_0^1 \varphi_t^2 dx \tag{3.4}$$

$$\left. + \frac{c}{\varepsilon^2} \int_0^1 \psi_x^2 dx + \frac{c}{\varepsilon} \int_0^1 \psi_t^2 dx + c \int_0^1 g^2(\psi_t) dx \right], \qquad \forall t \geq t_0.$$

Proof. By exploiting equations (1.5) and repeating the same procedure as in above, we have, for $t \geq t_0$,

$$F_3'(t) = \sigma(t) \left[\frac{1}{4\varepsilon} \int_0^1 m(x) \psi_x [\psi_{xx} - \psi - \varphi_x - \sigma(t)g(\psi_t)] dx + \frac{1}{4\varepsilon} \int_0^1 m(x) \psi_t \psi_{tx} \right.$$

$$\left. + \varepsilon \int_0^1 m(x)(\psi + \varphi_x) \varphi_x dx + \varepsilon \int_0^1 m(x) \varphi_t \varphi_{tx} dx \right]$$

$$+ \sigma'(t) \left[\frac{1}{4\varepsilon} \int_0^1 m(x) \psi_t \psi_x dx + \varepsilon \int_0^1 m(x) \varphi_t \varphi_x dx \right]$$

$$\leq \frac{1}{4\varepsilon}\sigma(t)\left[-\left((\psi_x(1,t))^2 + (\psi_x(0,t))^2\right) - \frac{1}{2}\int_0^1 m'(x)\psi_x^2 dx\right.$$

$$\left. - \int_0^1 m(x)\psi_x(\varphi_x + \psi)dx - \int_0^1 m(x)\psi_x\sigma(t)g(\psi_t)dx - \frac{1}{2}\int_0^1 m'(x)\psi_t^2 dx\right]$$

$$+\varepsilon\sigma(t)\left[\int_0^1 m(x)\psi_x\varphi_x dx - \left(\varphi_x^2(1,t) + \varphi_x^2(0,t)\right) - \frac{1}{2}\int_0^1 m'(x)\varphi_x^2 dx - \frac{1}{2}\int_0^1 m'(x)\varphi_t^2 dx\right]$$

$$\sigma(t)\left|\frac{1}{4\varepsilon}\int_0^1 m(x)\psi_t\psi_x dx + \varepsilon\int_0^1 m(x)\varphi_t\varphi_x dx\right|$$

By using Young's and Poincaré's inequalities, and the fact that

$$\varphi_x^2 \leq 2(\psi + \varphi_x)^2 + 2\psi^2$$

we obtain (3.4).

Lemma 3.5. *Assume that (H1) and (H2) hold. Then, after fixing ε small enough, the functional F defined by*

$$F(t) := 6c\varepsilon F_1(t) + F_2(t) + F_3(t)$$

satisfies, along the solution, the estimate

$$F'(t) \leq \sigma(t)\left[-\frac{1}{2}\int_0^1 (\psi + \varphi_x)^2 dx - \tau\int_0^1 \varphi_t^2 dx \right. \tag{3.5}$$

$$\left. +c\int_0^1 \psi_t^2 dx + c\int_0^1 \psi_x^2 dx + c\int_0^1 g^2(\psi_t)dx\right], \qquad \forall t \geq t_0.$$

where $\tau = c\varepsilon$.

Proof. By using Lemma 3.2, Lemma 3.3,Lemma 3.4, and the fact that

$$\psi_x\varphi_x \leq \varepsilon\varphi_x^2 + \frac{1}{4\varepsilon}\psi_x^2$$

we obtain (3.5).

As in [1], we use the multiplier w given by the solution of

$$-w_{xx} = \psi_x, \quad w(0) = w(1) = 0. \tag{3.6}$$

Lemma 3.6. *The solution of (3.6) satisfies*

$$\int_0^1 w_x^2 dx \leq \int_0^1 \psi^2 dx$$

and

$$\int_0^1 w_t^2 dx \leq \int_0^1 \psi_t^2 dx.$$

Proof. We multiply equation (3.6) by w, integrate by parts, and use the Cauchy-Schwarz inequality, to get

$$\int_0^1 w_x^2 dx \leq \int_0^1 \psi^2 dx.$$

129

Next, we differentiate (3.6) with respect to t to obtain, by similar calculations,

$$\int_0^1 w_{xt}^2 dx \leq \int_0^1 \psi_t^2 dx.$$

Poincaré's inequality, then yields

$$\int_0^1 w_t^2 dx \leq \int_0^1 \psi_t^2 dx.$$

This completes the proof of Lemma 3.6.

Lemma 3.7. *Assume that (H1) and (H2) hold. Then, for any $0 < \varepsilon_1 < 1$ there exists $t_1 \geq t_0$ such that the functional K defined by*

$$K(t) := \sigma(t) \int_0^1 (\psi \psi_t + w \varphi_t) dx$$

satisfies, along the solution, the estimate

$$K'(t) \leq \sigma(t) \left[-\frac{1}{4} \int_0^1 \psi_x^2 dx + \frac{c}{\varepsilon_1} \int_0^1 \psi_t^2 dx + c\varepsilon_1 \int_0^1 \varphi_t^2 dx + c \int_0^1 g^2(\psi_t) dx \right] \qquad (3.7)$$

for all $t \geq t_1$.

Proof. Since $\lim\limits_{t \to \infty} \frac{\sigma'(t)}{\sigma(t)} = 0$, then for any $0 < \varepsilon_1 < 1$ there exists $t_1 \geq t_0$ such that

$$\left| \frac{\sigma'(t)}{\sigma(t)} \right| \leq \varepsilon_1, \qquad \forall t \geq t_1.$$

Using this, equations (1.5) and integrating by parts, we have

$$K'(t) = \sigma(t) \left[\int_0^1 (\psi_t^2 - \psi_x^2) dx - \int_0^1 \psi(\psi + \varphi_x) dx - \int_0^1 \sigma(t) \psi g(\psi_t) dx \right.$$

$$\left. + \int_0^1 w(\psi_x + \varphi_{xx}) dx + \int_0^1 w_t \varphi_t dx \right] + \sigma'(t) \int_0^1 (\psi \psi_t + w \varphi_t) dx$$

$$\leq \quad \sigma(t) \left[\int_0^1 \psi_t^2 dx - \frac{1}{2} \int_0^1 \psi_x^2 dx + c \int_0^1 g^2(\psi_t) dx \right.$$

$$\left. + \int_0^1 (w_x^2 - \psi^2) dx + \varepsilon_1 \int_0^1 \varphi_t^2 dx + \frac{1}{4\varepsilon_1} \int_0^1 w_t^2 dx \right]$$

$$+ \varepsilon_1 \sigma(t) \left[\frac{1}{4\varepsilon_1} \int_0^1 \psi^2 dx + \varepsilon_1 \int_0^1 \psi_t^2 dx + \frac{1}{4\varepsilon_1} \int_0^1 w^2 dx + \varepsilon_1 \int_0^1 \varphi_t^2 dx \right], \qquad \forall t \geq t_1.$$

Then Poincaré's inequality and Lemma 3.6 give the desired result.

We are now ready to prove our main result.

Proof of Theorem 2.2.

For $N_1, N_2 > 0$, let

$$\mathcal{L}(t) := N_1 E(t) + N_2 K(t) + F(t)$$

By combining (3.1), (3.5), (3.7), then for any $0 < \varepsilon_1 < 1$ there exists $t_1 \geq t_0$ such that

$$\mathcal{L}'(t) \leq \sigma(t) \left[-N_1 \int_0^1 \psi_t g(\psi_t) dx - (\frac{N_2}{4} - c) \int_0^1 \psi_x^2 dx - (\tau - c\varepsilon_1 N_2) \int_0^1 \varphi_t^2 dx \right.$$
$$\left. -\frac{1}{2} \int_0^1 (\psi + \varphi_x)^2 dx + (c\frac{N_2}{\varepsilon_1} + c) \int_0^1 \psi_t^2 dx + (cN_2 + c) \int_0^1 g^2(\psi_t) dx \right]$$

for all $t \geq t_1$.

Now, we choose N_2 large enough so that

$$k_1 := (\frac{N_2}{4} - c) > 0$$

then ε_1 small enough so that

$$k_2 := (\tau - c\varepsilon_1 N_2) > 0$$

We fix $t_1 \geq t_0$, depending on ε_1, then

$$\mathcal{L}'(t) \leq \sigma(t) \left[-N_1 \int_0^1 \psi_t g(\psi_t) dx - k_1 \int_0^1 \psi_x^2 dx - k_2 \int_0^1 \varphi_t^2 dx \right. \quad (3.8)$$
$$\left. -\frac{1}{2} \int_0^1 (\psi + \varphi_x)^2 dx + c \int_0^1 \psi_t^2 dx + c \int_0^1 g^2(\psi_t) dx \right]$$

for all $t \geq t_1$.

Case 1: q = 1

In this case, using hypothesis (H2), (3.8) can be written as

$$\mathcal{L}'(t) \leq \sigma(t) \left[-(N_1 c_1 - c) \int_0^1 \psi_t^2 dx - k_1 \int_0^1 \psi_x^2 dx - k_2 \int_0^1 \varphi_t^2 dx - \frac{1}{2} \int_0^1 (\psi + \varphi_x)^2 dx \right]$$
$$(3.9)$$

for all $t \geq t_1$. By choosing N_1 large enough such that

$$k_3 := (N_1 c_1 - c) > 0,$$

we arrive at

$$\mathcal{L}'(t) \leq \sigma(t) \left[-k_3 \int_0^1 \psi_t^2 dx - k_1 \int_0^1 \psi_x^2 dx - k_2 \int_0^1 \varphi_t^2 dx - \frac{1}{2} \int_0^1 (\psi + \varphi_x)^2 dx \right]$$
$$\leq -c\sigma(t) E(t) \quad (3.10)$$

On the other hand, we can choose N_1 even larger (if needed) so that

$$\mathcal{L}(t) \sim E(t). \quad (3.11)$$

Therefore, by using (3.10) and (3.11), we obtain, for a positive constant ω,

$$\mathcal{L}'(t) \leq -\omega\sigma(t)\mathcal{L}(t), \quad t \geq t_1.$$

A simple integration over (t_1, t), leads to

$$\mathcal{L}(t) \leq \mathcal{L}(t_1)e^{-\omega \int_{t_1}^{t} \sigma(s)ds}, \quad t \geq t_1.$$

Consequently, (2.3) is established by virtue of (3.11) and boundedness of E and σ.

Case 2 : q > 1

We consider the following partition of $(0,1)$

$$\Omega^+ = \{x \in (0,1) : |\psi_t| > 1\} \quad and \quad \Omega^- = \{x \in (0,1) : |\psi_t| \leq 1\}$$

From hypothesis (H2) and Holder's inequality, we easily show that

$$\int_{\Omega^+} (\psi_t^2 + g^2(\psi_t))dx \leq c \int_{\Omega^+} \psi_t g(\psi_t)dx \leq c \int_0^1 \psi_t g(\psi_t)dx \qquad (3.12)$$

$$\int_{\Omega^-} (\psi_t^2 + g^2(\psi_t))dx \leq c \int_{\Omega^-} (\psi_t g(\psi_t))^{\frac{2}{q+1}} dx \leq c \left(\int_0^1 \psi_t g(\psi_t)dx \right)^{\frac{2}{q+1}} \qquad (3.13)$$

Therefore, using (3.8), (3.12), (3.13), we deduce that

$$\mathcal{L}'(t) \leq \sigma(t)\left[-k_1 \int_0^1 \psi_x^2 dx - k_2 \int_0^1 \varphi_t^2 dx - \frac{1}{2} \int_0^1 (\psi + \varphi_x)^2 dx - \int_0^1 \psi_t^2 dx \right]$$
$$+\sigma(t)\left[-N_1 \int_0^1 \psi_t g(\psi_t)dx + c \int_0^1 \psi_t g(\psi_t)dx + c \left(\int_0^1 \psi_t g(\psi_t)dx \right)^{\frac{2}{q+1}} \right].$$

Then, by choosing $N_1 = c$, we obtain

$$\mathcal{L}'(t) \leq -c\sigma(t)E(t) + c\sigma(t)\left(\int_0^1 \psi_t g(\psi_t)dx \right)^{\frac{2}{q+1}} \qquad \forall t \geq t_1$$

Using (3.1), we find that

$$\mathcal{L}'(t) \leq -c\sigma(t)E(t) + c\sigma(t)^{\frac{q-1}{q+1}} (-E'(t))^{\frac{2}{q+1}} \qquad \forall t \geq t_1$$

Multiplying the last inequality by $E^{\frac{q-1}{2}}$ yields

$$\mathcal{L}'(t)E^{\frac{q-1}{2}} \leq -c\sigma(t)E(t)^{\frac{q+1}{2}} + c\sigma(t)^{\frac{q-1}{q+1}} E^{\frac{q-1}{2}} (-E'(t))^{\frac{2}{q+1}} \qquad \forall t \geq t_1$$

By using Young's inequality and (3.11), we obtain, for all $\varepsilon > 0$,

$$\left(\mathcal{L}(t)E^{\frac{q-1}{2}}(t) \right)' \leq \frac{q-1}{2}\mathcal{L}(t)E^{\frac{q-3}{2}}(t)E'(t) - c\sigma(t)E(t)^{\frac{q+1}{2}} + \varepsilon\sigma(t)E^{\frac{q+1}{2}}(t) - c_\varepsilon E'(t)$$
$$\leq -(c - \varepsilon)\sigma(t)E^{\frac{q+1}{2}}(t) - c_\varepsilon E'(t) \qquad \forall t \geq t_1$$

By choosing ε small enough, we obtain

$$\left(\mathcal{L}(t)E^{\frac{q-1}{2}}(t) + cE(t) \right)' \leq -c\sigma(t)E^{\frac{q+1}{2}}(t) \qquad \forall t \geq t_1$$

132

We put
$$I(t) = \mathcal{L}(t)E^{\frac{q-1}{2}}(t) + cE(t)$$
and we use (3.11), to deduce that
$$I(t) \sim E(t) \tag{3.14}$$
hence
$$I'(t) \leq -c\sigma(t)I^{\frac{q+1}{2}}(t) \qquad \forall t \geq t_1$$
Asimple integration then leads to
$$I(t) \leq c\left(1 + \int_{t_1}^{t} \sigma(s)ds\right)^{\frac{-2}{q-1}} \qquad \forall t \geq t_1$$
Then, by (3.14), we have
$$E(t) \leq c\left(1 + \int_{t_1}^{t} \sigma(s)ds\right)^{\frac{-2}{q-1}} \qquad \forall t \geq t_1,$$
which gives (2.4).

Examples. We give some examples to illustrate the energy decay rates obtained by Theorem 2.2.
(1) If $\sigma \equiv 1$, then
$$
\begin{aligned}
E(t) &\leq ce^{-\omega t}, &&\text{if } q = 1 \\
E(t) &\leq c(1+t)^{-\frac{2}{q-1}}, &&\text{if } q > 1.
\end{aligned}
$$
(2) If $\sigma(t) = \frac{1}{1+t}$, then
$$
\begin{aligned}
E(t) &\leq \frac{c}{(1+t)^\omega}, &&\text{if } q = 1 \\
E(t) &\leq \frac{c}{[1 + \ln(1+t)]^{\frac{2}{q-1}}}, &&\text{if } q > 1.
\end{aligned}
$$
(3) If $\sigma(t) = \frac{1}{(e+t)\ln(e+t)}$, then
$$
\begin{aligned}
E(t) &\leq \frac{c}{[\ln(e+t)]^\omega}, &&\text{if } q = 1 \\
E(t) &\leq \frac{c}{[1 + \ln(\ln(e+t))]^{\frac{2}{q-1}}}, &&\text{if } q > 1.
\end{aligned}
$$
Remark 3.1. If the hypothesis (H1) is replaced by
$$\sigma \in C^0(\mathbb{R}_+) \quad \text{such that} \quad 0 < a_0 \leq \sigma(t) \leq a_1, \ \forall t \geq 0,$$
then using the same functionals, but not multiplied by $\sigma(t)$, and repeating similar procedure give
$$E(t) \leq ce^{-\omega t}, \qquad \text{if } q = 1$$
and
$$E(t) \leq c(1+t)^{-\frac{2}{q-1}}, \qquad \text{if } q > 1.$$

4 References

1. Ammar-Khodja F., Benabdallah A. Muñoz Rivera J. E. and Racke R., *Energy decay for Timoshenko systems of memory type.* J. Differential Equations **194** no. **1** (2003), 82–115.

2. Feng D-X, Shi D-H, and Zhang W., *Boundary feedback stabilization of Timoshenko beam with boundary dissipation.* Sci. China Ser. **A 41** no. **5** (1998), 483–490.

3. Kim J.U. and Renardy Y., *Boundary control of the Timoshenko beam*, SIAM J. Control Optim. **25** no. **6** (1987),1417–1429.

4. Liu Z. and Zheng S., *Semigroups associated with dissipative systems*, Chapman & Hall/CRC, 1999.

5. Muñoz Rivera J.E. and Racke R., *Global stability for damped Timoshenko systems*, Discrete Contin. Dyn. Syst. **9** no. **6** (2003), 1625–1639.

6. Raposo C.A., Ferreira J., Santos M.L., and Castro N.N.O, *Expoenetial stability for the Timoshenko system with two weak dampings*, Applied Math Letters 18 (2005), 535-541.

7. Santos M., *Decay rates for solutions of a Timoshenko system with a memory condition at the boundary*, Abstr. Appl. Anal. **7** no. **10** (2002), 531–546.

8. Shi D-H and Feng D-X, *Exponential decay of Timoshenko beam with locally distributed feedback*, IMA J. Math. Control Inform. **18** no. **3** (2001), 395–403.

9. Shi D-H and Feng D-X, *Exponential decay rate of the energy of a Timoshenko beam with locally distributed feedback*, ANZIAM J. **44** no. **2** (2002), 205-220.

10. Soufyane A. and Wehbe A., *Uniform stabilization for the Timoshenko beam by a locally distributed damping*, Electron. J. Differential Equations no. **29** (2003), 1-14.

11. Soufyane A., *Stabilisation de la poutre de Timoshenko*, C. R. Acad. Sci. Paris Sér. I Math. **328** no. **8** (1999), 731–734.

12. Taylor S.W., *A smoothing property of a hyperbolic system and boundary controllability*, J. Comput. Appl. Math. **114** (2000), 23–40.

13. Timoshenko S., *On the correction for shear of the differential equation for transverse vibrations of prismaticbars*, Philisophical magazine **41** (1921), 744-746.

14. Xu G-Q and Yung S-P, *Stabilization of Timoshenko beam by means of pointwise controls*, ESAIM, Control Optim. Calc. Var. **9** (2003), 579-600.

15. Yan Q-X, Chen Z. and Feng D-X, *Exponential stability of nonuniform Timoshenko beam with coupled locally distributed feedbacks,* Acta Anal. Funct. Appl. **5** no. **2** (2003), 156-164.

Rainfall Spatial Variability and its Impact on Important Environment Processes in India

N.A. Sontakke, H.N. Singh, and Nityanand Singh

Indian Institute of Tropical Meteorology,
Dr. Homi Bhabha Road, Pashan, Pune – 411008 (INDIA)

Abstract. Desertification, River channel changes and rising trend in surface air temperature are environmental problems of serious concern in India. Elaborate analyses of spatial variation of annual, seasonal and monthly rainfall over the country have been carried out to understand the role of spatial variability on important environmental processes. Spatial variation of annual rainfall expansion and contraction of by examining expansion and contraction of the moisture regions and that of seasonal and monthly rainfall from expansion and contraction of respective period dry and wet zones using highly quality-controlled rainfall data from 316 locations for the period 1871-2006. Impacts of rainfall spatial variability on important environmental problems are discussed.

Keywords: Rainfall Spatial Variability, Moisture Regions, Dry and Wet Zones, Desertification, River Channel Changes, Rising Trend in Temperature.
PACS: 01.30.Cc

INTRODUCTION

All the climate parameters are spatially variable but precipitation is highest due to discontinuity both in space and time. The average annual precipitation of the entire surface of our planet is estimated to be about 1050 millimeters per year or approximately 88 millimeters per month by Global Precipitation Climatology Project (GPCP) established by the World Climatic Research Project (WCRP) in 1986 with the major goal to develop a more complete understanding of the spatial and temporal patterns of global precipitation.. However, actual values vary spatially from less than 10 millimeters per month to a maximum of more than 300 millimeters per month depending on location [1]. Precipitation decreases poleward away from the equator as one moves toward the subtropical regions. The largest annual precipitation totals straddle the equator while the driest regions on Earth lie near the Tropic of Cancer. In addition, precipitation becomes more seasonal as one move away from the Equator due to the shifting locations of global wind and pressure systems.

Over India precipitation varies spatially largely on annual, seasonal and monthly scale. On annual scale, it ranges from 136 mm (Gilgit) to 11079 mm (Cherrapunji). Large part of the country in the northwest in the subtropical region comes under dry region. Location and size of the dry provinces over the five continents represents the average state of the general circulation of the atmosphere. Boundaries of these regions

CP1146, *Modelling of Engineering and Technological Problems*, edited by A. H. Siddiqi, A. K. Gupta, and M. Brokate
© 2009 American Institute of Physics 978-0-7354-0683-4/09/$25.00

expands and contracts enormously year to year and provide reliable information for variability of the general circulation of the atmosphere. Spatial variability of dry regions provides useful information for spatial climate variability on regional scales. Expansion and contraction of dry and wet areas demarcates the region affected by the climatic variations.

The area-averaged rainfall series is very smooth and does not take into account the variance accounted by different stations involved in it. Singh et al. [2] had prepared annual rainfall series over India by nine different methods to bring out the best possible spatially representative series. The Index of Areal Representativeness (IAR) has varied between 10.78% and 15.16% showing the limitations of area-averaged series representativeness prepared by any method. The correlation between all-India summer monsoon rainfall and individual gauges varies largely from 0.6 to -0.1.

Variability in climatic parameters, chief amongst precipitation very largely governs the inherent natural dynamism of dryland ecosystems [3]. Charney [4] in his 'biophysical feedback model has argued that reduction in vegetation increases the albedo, causes sinking motion, additional drying and perpetuate arid conditions. Numerical integration with the GCM has shown that increasing albedo north of Inter Tropical Convergence Zone (ITCZ) from 14% to 35% had the effect of shifting the ITCZ several degrees latitude south and decreasing rainfall in Sahel about 40% during the rainy season. Rodwell and Hoskins [5] have suggested monsoon-desert mechanism to explain for desertification.

Temporarily on annual to monthly scale climate shows large variability and isolines of specific amount fluctuate, shift and expand/contract in time span. Tucker et al. [6] have studied expansion and contraction of Sahara desert over the Sahara-Sahelian boundary using satellite derived NDVI during 1980-1990. They found north-south fluctuation in the boundary between the climatic zones. Overall there was southward shift in the boundary by 130 kilometers compared to its position in 1980. Yang et al. [7] have studied fluctuation of semi-arid region over China during 1951-1999 using annual precipitation and measured evaporation of well spread 295 meteorological stations. The semi-arid region lies along the border of the monsoon circulation and is highly susceptible to environmental changes. Semi-arid zone between humid and arid fluctuates substantially, differ significantly regionally and have the shifting features in the same direction in others shifting features in the same direction in some areas and in opposite direction in others shifting the entire semi-arid belt. The climate in the northeastern region has been warming and becoming wetter over the past 50 years whereas in the central and eastern region it has been warming and becoming drier. Yan et al. [8] have studied climatic fluctuations over northwest China by examining spatial variation of 250mm isohyet (climatic boundary line between arid and semiarid regions) during 1950s to 1990s. They found that the 250mm isohyet has shifted southward about 140 to 240 km between wet year 1964 and the dry year 1980, and concluded decrease in precipitation over the area during the last 40 years. Nesje [9] studied spatial fluctuation of mountain glaciers and ice-caps in western Norway during 1900-2004 and found that they were dependent on winter precipitation and summer temperature. They observed maximum annual retreat of Briksdalsbreen glacier by 130 meters in 2003/2004 and suggested this could be regarded as a 'modern analog' for other Holocene events. The two main objectives of this study are:

- To document climatology, distribution and fluctuations of different moisture regions and seasonal and monthly dry and wet zones over India during the period 1871-2006; and

- To investigate impact of rainfall spatial variability on important environmental problems such as desertification, river channel changes and rising trend in surface air temperature.

RAINFALL DATA USED

Monthly rainfall data from well spread network of 316 raingauge stations over contiguous India for the period 1871-2006 has been used in the present study [10]. Sikkim State and Arunachal Pradesh State in the northeast, northern part of Jammu and Kashmir State, Andaman & Nicobar Islands in the Bay of Bengal, Lakshadweep islands in the Arabian Sea are not considered owing to non-availability of long period rainfall data.

On the mean annual rainfall chart the isohyets of 560 mm, 1040 mm, 1420 mm, 1630 mm and 2450 mm can be used to divide the country into six Thornthwaite's type climatic moisture regions (arid, semiarid, dry subhumid, moist subhumid, humid and perhumid). Climatic moisture regions and yearly maps for latest five years (2002-2006) are shown in Figure 1. Area under different moisture conditions has been calculated using map of India (Administrative) prepared by the National Atlas & Thematic Mapping Organization [11] on "Albers Equal Area Projection with Two Standard Parallels, 15°N and 30°N" projection system and 1:6M scale showing the international boundary, states and 593 districts. This map is digitized in three layers, i) all-India boundary; ii) boundaries of states and union territories; and iii) location of 316 raingauge stations. Climatological and yearly (1871-2006) area of the country under different moisture condition is obtained using GeoMedia Professional 5.1 Geographic Information System (GIS) software package.

On the mean isohyetal chart of particular season/month, one-fourth area of the country with lowest rainfall is identified as very dry and one-fourth with highest rainfall as very wet. The remaining area between these two extremes is divided into two equal halves and designated as dry and wet zones. The climatological rainfall value has been applied to delineate very dry, dry, wet and very wet on yearly basis and the area of each zone is calculated using the GIS. For four seasons climatological and yearly maps for latest five years (2002-2006) are shown in Figures 2-5 in order.

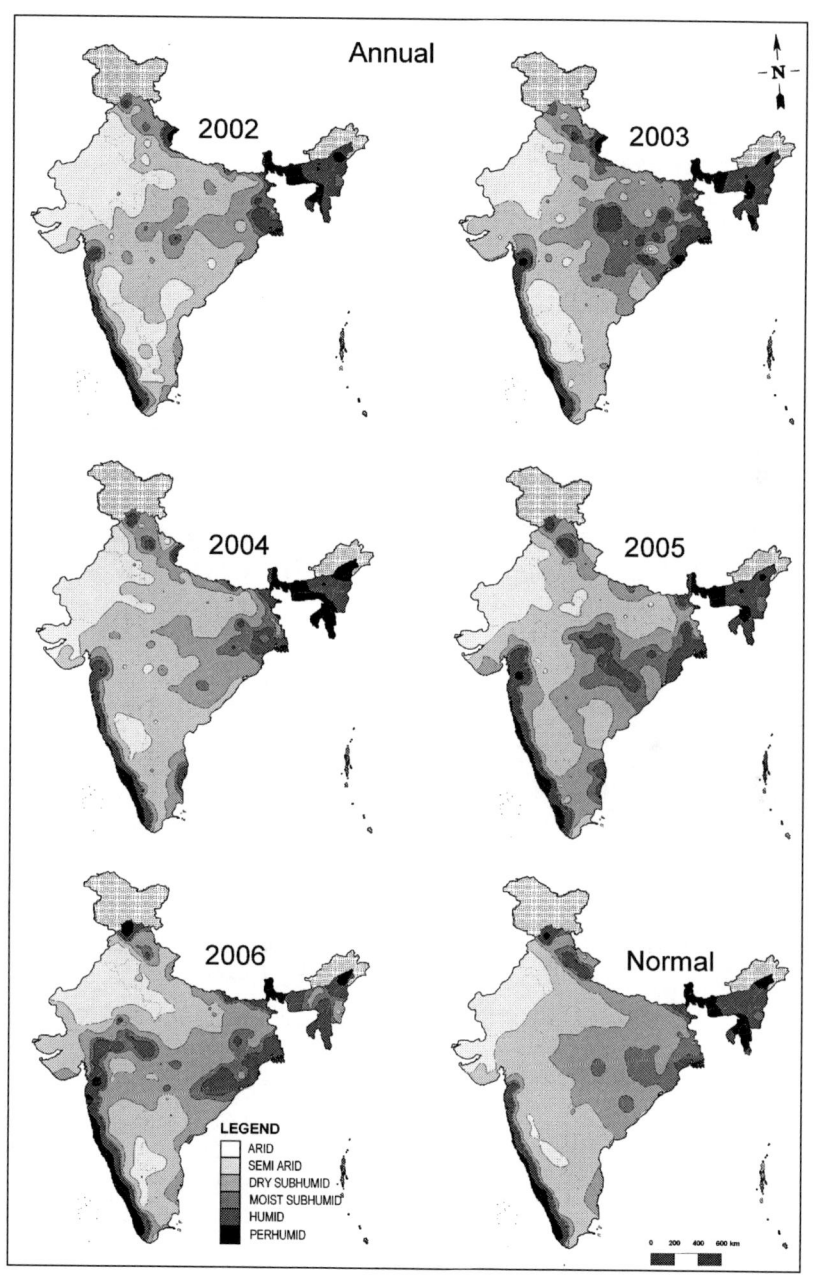

FIGURE 1. Maps of the moisture regions for the 6 latest years 2002-2006 and normal (climatology).

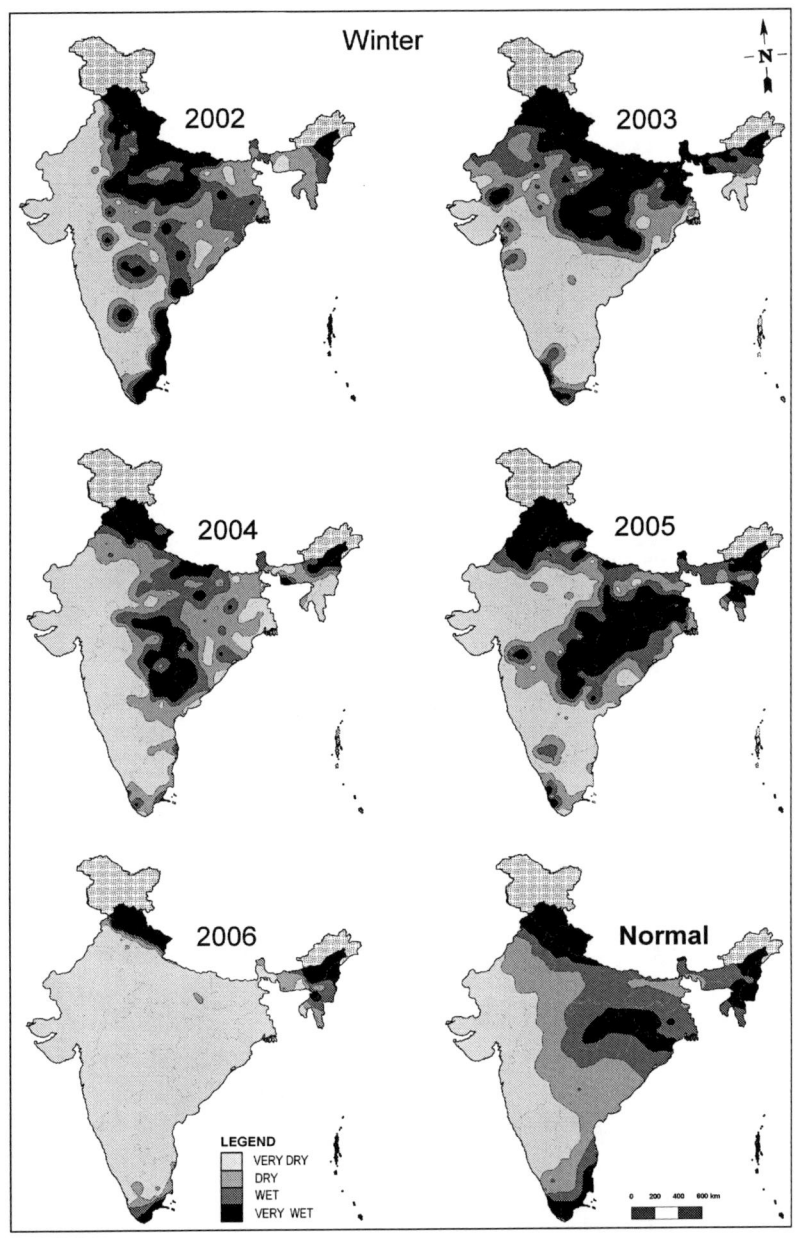

FIGURE 2. Maps of the very dry, dry, wet and very wet zones for the 6 latest years 2002-2006 and normal (climatology).

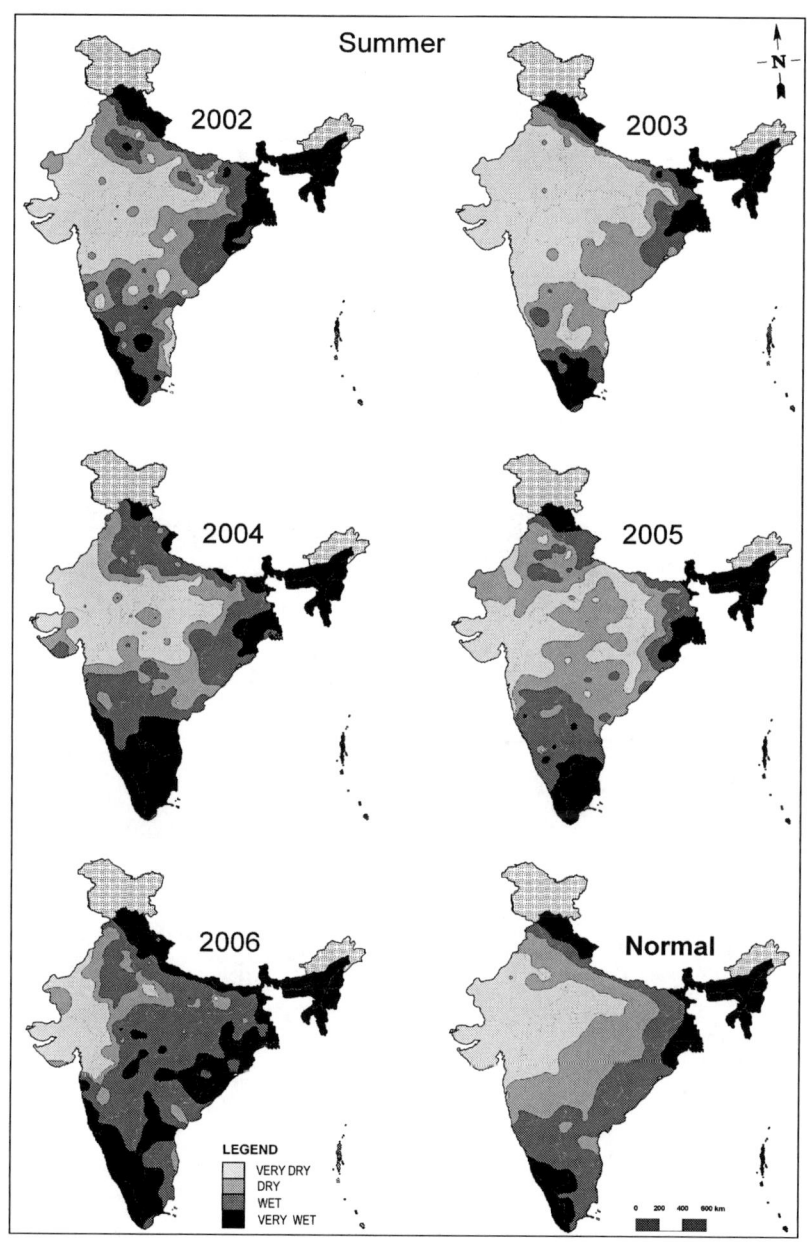

FIGURE 3. Maps of the very dry, dry, wet and very wet zones for the 6 latest years 2002-2006 and normal (climatology).

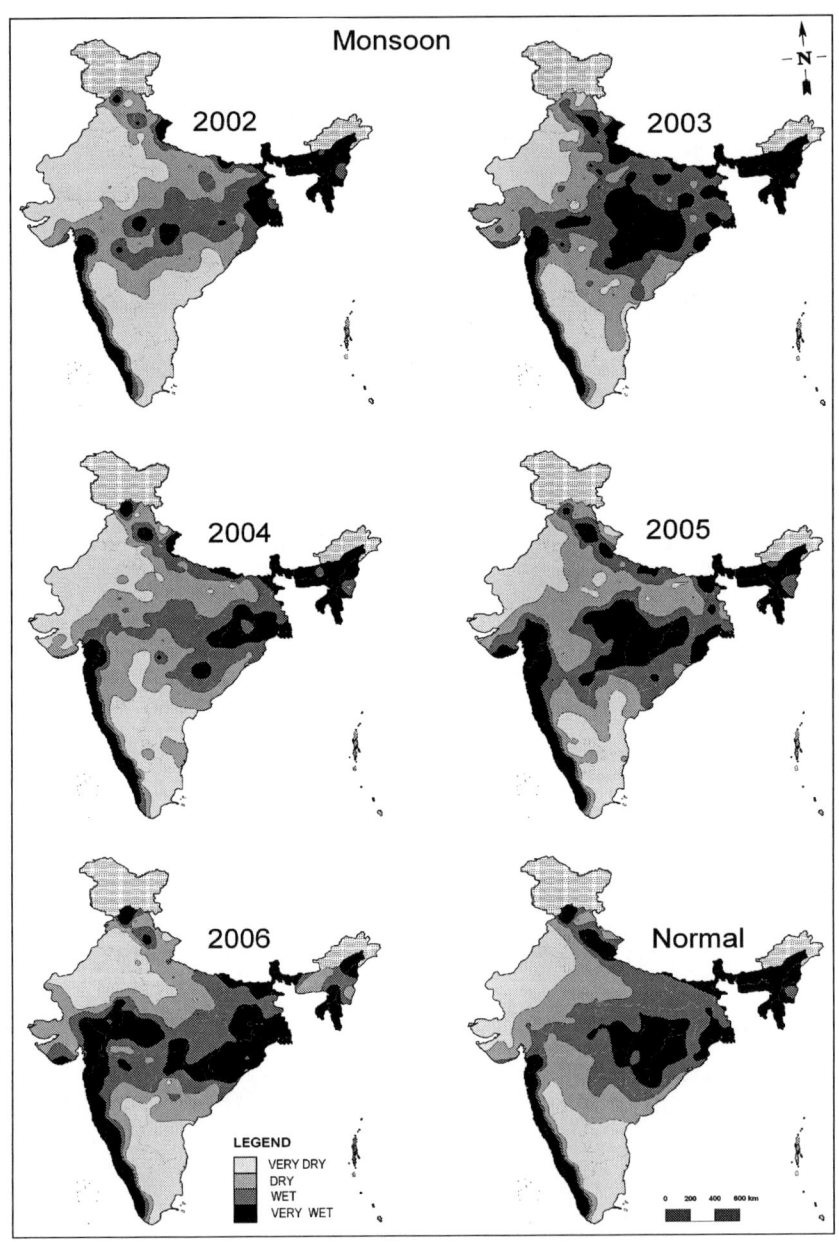

FIGURE 4. Maps of the very dry, dry, wet and very wet zones for the 6 latest years 2002-2006 and normal (climatology).

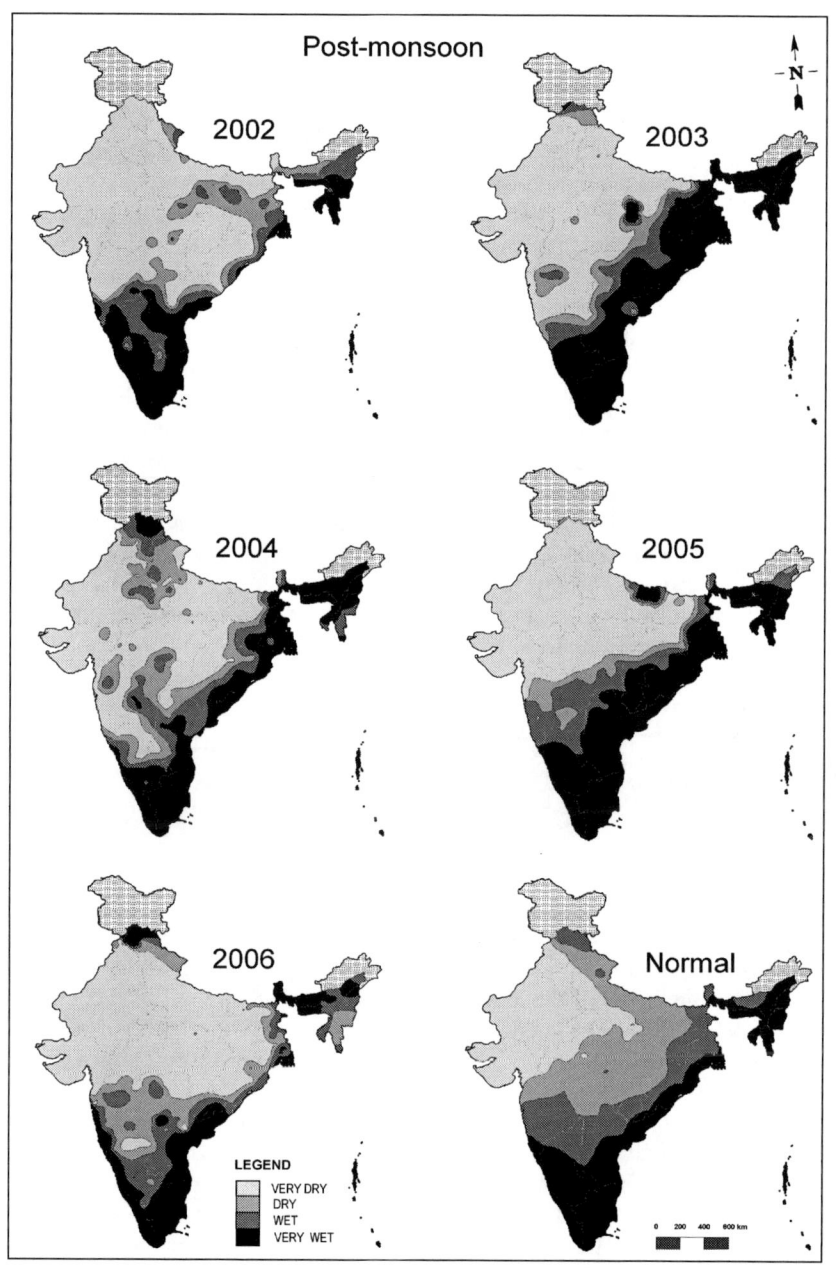

FIGURE 5. Maps of the very dry, dry, wet and very wet zones for the 6 latest years 2002-2006 and normal (climatology).

PROBABILITY DISTRIBUTION USING AVERAGE PROBAILITY DIAGRAM

Probability distribution of interannual variation of annual, seasonal and monthly rainfall at stations over India is simple linear function of respective period mean rainfall. The diagram presenting probability distribution of point rainfall as a function of spatial variation in the respective period mean rainfall is referred to as 'average probability diagram (APD)' [12]. The APD of annual rainfall of the country has been prepared using a long period (1871-2000) data of 316 well spread stations. Rainfall amount expected with any specified exceedance/non-exceedance probability increases linearly with the spatial increase in the respective period mean rainfall. The relationship is presented in a simple diagram for 1%, 5%, 10%, 20%, 30%, 40%, 50%, 60%, 70%, 80%, 90%, 95%, 99% exceedance/non-exceedance probabilities. Finally, envelope curve on highest observed and lowest observed rainfall is also added in the diagram (APD) to get an idea of rainfall expected at the station with known mean rainfall in extreme situation.

For probabilistic description, a line at 560 mm, 1040 mm, 1420 mm, 1630 mm and 2450 mm along the ordinate of the annual rainfall Average Probability Diagram is drawn parallel to abscissa. From the point of intersection of this line and the line, showing average-probabilistic estimated rainfall relationship vertical projection is drawn on the abscissa, and the mean rainfall is estimated using average-probability of the places where specified amount of annual rainfall is expected with specified exceedance/non-exceedance probabilities.

CLIMATOLOGY, DISTRIBUTION AND FLUCTUATION OF THE MOISTURE REGIONS

Arid Region

Normally 15.32% (500,282.5 km^2) area of the country experiences dry condition. The standard deviation (SD) of the arid area variation is 189498.9 km^2 and coefficient of variation (CV) 37.88%. The smallest arid area occurred during 1917 (61,475 km^2; 1.88% of GAOI-geographical area of India) and largest during 1899 (1,320,450 km^2; 40.45% of GAOI). Probability distribution of the arid region suggests that the areas of the country with mean annual rainfall (MAR) less than 320mm would experience arid conditions with exceedance probability 90%, with MAR less than 472 mm with exceedance probability 70% and with MAR less than 582 mm with exceedance probability 50%. The aridity can spread up to mean annual isohyets of 692 mm thrice in 10 years, up to mean annual isohyets of 870 mm once in 10 years and up to mean annual isohyets of 1203 mm once in 100 years.

The prominent dry epochs in the arid area fluctuations are 1896-1925, 1934-1942, 1965-1974, 1984-1987 and 1999-2004, and wet epochs 1871-1895, 1926-1933, 1943-1964, 1975-1983 and 1988-1998. Compared to 1895-1942 during 1943-2006 the arid region shrunk by 18.14%.

Semiarid Region

North-south elongated semiarid region occupies 33.79% (1103236.3 km^2) area in the central parts of the country. The SD and CV of the semiarid region is 1822832.2 km^2 and 16.52%. The smallest semiarid occurred during 1871 (734,150 km^2; 22.49% of GAOI) and largest during 1979 (1,686,025 km^2; 51.65% of GAOI). Mean annual rainfall between 582 mm and 1065 mm experiences semiarid condition with probability exceeding 50%. In the central portion of this area with mean rainfall between 762 mm and 841 mm, the occurrence probability of semiarid condition is 80%, which is highest. Towards the areas of lesser rainfall, the boundary can spread up to 472 mm isohyets with 30% probability, up to 320 mm with 10% and up to 152 mm with 1% probability. Towards wetter side, the boundary can spread up to 1214 mm isohyets with 30% probability, up to 1447 mm with 10% probability and up to 1937 mm with 1% probability. In extreme cases, half of the country can be under the influence of semiarid condition.

The prominent dry epochs in the semiarid area fluctuations are 1900-1912, 1950-1957 and 1964-2004 and the wet epochs 1871-1899, 1913-1949 and 1958-1963. Compared to 1913-1963 during 1964-2006 the semiarid area has expanded by 12.95%.

Dry Subhumid Region

Normally, 793,137 km^2 (24.3% of GAOI) experience dry subhumid condition. The standard deviation (SD) of the arid area variation is 150,177.2 km^2 and coefficient of variation (CV) 18.93%.The area is found along the slope of the topographic regions such as central highlands, leeside of the Sahyadri Range (or Western Ghats), Western Himalaya and eastern India. The smallest dry subhumid region occurred during 1936 (436,450 km^2; 13.37%) and largest during 1916 (1,217,075 km^2; 37.28%). The occurrence probability of dry subhumid conditions exceeds 70% over the area with MAR between 1214 mm and 1280 mm, which is the highest. Towards drier side, the boundary spreads up to 923 mm annual isohyets with 30%, up to 727 mm with 10% and up to 461 mm isohyets with 1% probability. Excluding northwestern part, high reaches of Sahayadri Range, Western Himalaya and eastern parts about 24.3% area of the country can be under the influence of dry subhumid conditions once in 100 years.

The prominent dry epochs in the dry subhumid area fluctuations are 1914-1917 and 1953-1983 and wet epochs 1871-1913, 1918-1952 and 1984-2006. Compared to 1948-1983 during 1984-2006 the dry subhumid region has shrunk by 9.15%.

Moist Subhumid Region

Normally, 289,239 km^2 (8.86% of GAOI) area of the country experience moist sub-humid conditions. The standard deviation (SD) of the arid area variation is 91,901.0 km^2 and coefficient of variation (CV) 31.76%.The smallest moist subhumid region occurred during 1974 (84,450 km^2; 2.59% of GAOI) and largest during 1990 (481,775 km^2; 14.76% of GAOI). The moist subhumid condition occurs with probability greater than 50% where MAR is between 1447 mm and 1658 mm. In the core region of this area with MAR between 1280 mm and 1855 mm, the probability exceeds 70%. Towards wetter side, the boundary spreads up to 1855 mm with 30% probability, up to 2157 mm with 10% and up to 2841 mm isohyets with 1% probability.

The prominent dry epochs in the moist subhumid area fluctuations are 1876-1883, 1952-1989 and 1999-2004 and wet epochs 1884-1951 and 1990-1998. Compared to 1884-1951 during 1952-2006 the moist subhumid region has shrunk by 15.47%.

Humid Region

Normally, 404,151 km^2 (12.38% of GAOI) area of the country experience humid conditions. The standard deviation (SD) of the arid area variation is 131,700.0 km^2 and coefficient of variation (CV) 32.58%.The smallest humid region occurred during 1974 (194,900 km^2; 5.97% of GAOI) and largest during 1893 (934,975 km^2; 28.64% of GAOI). The humid conditions occur where MAR is between 1658 mm and 2482 mm with probability greater than 50%. In the core region of this area with MAR between 1976 mm and 2113 mm, the probability exceeds 80%, which is highest. Towards drier side, the humid condition spreads up to 1478 mm with 30% probability, up to 1227 mm isohyets with 10% and up to 841 mm with 1% probability. Towards wetter side, the humid conditions spread up to 2746 mm isohyets with 30% probability, up to 3143 mm with 10% and up to 4096 mm with 1% probability.

The prominent dry epochs in the humid area fluctuations are 1899-1913, 1949-1953, 1964-1989 and 1998-2004 and wet epochs 1871-1898, 1914-1948, 1954-1963 and 1990-1997. Compared to 1911-1959 during 1960-2006 the humid region has shrunk by 9.11%.

Perhumid Region

Normally, 174,327 km^2 (5.34% of GAOI) area of the country experiences perhumid conditions. The standard deviation (SD) of the arid area variation is 48,213.5 km^2 and coefficient of variation (CV) 27.66%.The smallest perhumid region occurred during 1873 (35,925 km^2; 1.10% of GAOI) and largest during 1879 (439,950 km^2; 13.48% of GAOI). Over upper reaches of the orographic regions with MAR greater than 2746 mm, the occurrence probability of perhumid conditions exceeds 70%, with MAR greater than 3143 mm 90%, and with MAR greater than 4096 mm 99%. Towards drier side, the boundary of perhumid region can spread up to 2248 mm isohyets with 30%, up to 1922 mm with 10% and up to 1370 mm isohyets with 1% probability.

The prominent dry epochs in the perhumid area fluctuations are 1898-1910, 1935-1942, 1960-1987 and 1995-2006 and wet epochs 1871-1897, 1911-1934, 1943-1959 and 1988-1994. Compared to 1911-1959 during 1960-2006 the perhumid region has shrunk by 17.25%.

In summary, the semiarid region shows expanding tendency while others show shrinking tendency in recent period. The semiarid area, which occupies large portion in the central part of the country, is expanding significantly over other regions. Overall effect is that country is experiencing dry climate in the recent period.

CLIMATOLOGY AND DISTRIBUTION OF SEASONAL AND MONTHLY DRY AND WET ZONES

On each of the mean seasonal and monthly rainfall chart prepared using 316 stations' rainfall data, the area of the country is divided into four equal parts and is identified as very dry, dry, wet and very wet The rainfall threshold values derived for four seasons and 12 months are given Table 1. Based on these threshold values for respective months and seasons, the thematic maps have been prepared showing very dry, dry, wet and very wet areas for each year from 1871-2006 of different seasons and months. The area under each rainfall condition is calculated using GeoMedia Professional 5.1.

In order to understand the combined effect of spatial variations of the very dry, dry, wet and very wet rainfall conditions over the country a Rainfall Spatial Distribution Index (RSDI) is developed, which is defined as [13];

$$RSDI = \frac{1.a_{VD} + 2.a_D + 3.a_W + 4.a_{VW}}{A} \tag{1}$$

where a_{VD} denotes area of the country under very dry (VD) condition, a_D under dry (D), a_W under wet (W) and a_{VW} under very wet condition (VW), and A is the total geographical area of the country. High RSDI value is indicative of enhanced rainfall over large areas, including dry areas and vice versa.

TABLE . Mean rainfall threshold (in mm) for identification of very dry, dry, wet and very wet zones of the different seasons and months in India.

Period	Very Dry	Dry	Wet	Very Wet
Winter	11	11 - 27	28 - 45	45
Summer	26	26 - 61	62 - 152	152
Summer Monsoon	500	500 - 830	831 - 1120	1120
Post-monsoon	53	53 - 89	90 - 150	150
January	5	5 - 12	13 - 20	20
February	5	5 - 13	14 - 26	26
March	6	6 - 11	12 - 23	23
April	5	5 - 16	17 - 42	42
May	12	12 - 31	32 - 73	73
June	68	68 - 126	127 - 192	192
July	162	162 - 286	287 - 362	362
August	140	140 - 264	265 - 340	340
September	121	121 - 170	171 - 216	216
October	32	32 - 54	55 - 111	111
November	8	8 - 16	17 - 30	30
December	4	4 - 7	8 - 12	12

Winter

Very dry: Normally western central part of the country is under very dry condition i.e. rainfall less than 11 mm. The occurrence probability of very dry conditions is greater than 50% where mean winter rainfall is less than 20 mm. It exceeds 60% for area less than 15 mm. Outwards of normal limit the boundary spreads up to 30 mm isohyets with 30% probability, up to 71 mm with 5% probability.

Dry: Normally north-south elongated adjacent part to the very dry condition of the country is under dry condition having rainfall between 11 and 27 mm. The occurrence probability exceeds 50% where the rainfall is between 20 mm and 38 mm. Towards drier side the boundary spreads up to 24 mm isohyets with 60% probability. Towards wetter side, the spreading could be up to 56 mm isohyets with 30% probability, up to 71 mm with 20% probability.

Wet: Normally three parts viz. western northeast India, adjacent eastern India and inner part of southeast peninsula experiences wet condition viz. rainfall between 28 and 45 mm. The wet conditions occur with probability 50% and greater over the area that receive mean winter rainfall between 38 mm and 56 mm. Towards the drier side the wet area boundary spreads up to 24 mm isohyets with 30% probability and up to 6 mm isohyets with 10% probability. Towards higher rainfall, the boundary spreads up to 82 mm isohyets with 30% probability, up to 104 mm with 20% and up to 152 mm isohyets with 10% probability.

Very wet: Normally very wet condition viz. rainfall more than 45 mm is experienced over four isolated parts of the country viz. northernmost India, eastern central India and eastern northeast India. The very wet conditions occur with probability greater than 50% where mean rainfall is more than 55 mm. Occurrence probability exceeds 70% for rainfall greater than 82 mm, 90% for greater than 152 mm and 95% for greater than 220 mm.

January

Very dry: During January, the very dry condition occurs with 40% probability and greater where mean rainfall is less than 16 mm. The occurrence probability is more than 30% where mean January rainfall is less than 6 mm. Outward of normal limit the boundary spreads up to 31 mm isohyets with 20% probability, up to 56 mm with 10% and up to 111 mm with 5% probability.

Dry: The dry condition occurs with probability 50% and greater where mean January rainfall is between 12 mm and 22 mm. Towards drier side with rainfall less than 9 mm the occurrence probability of dry condition is less than 40% . Towards wetter side, the boundary spreads up to 17 mm isohyets with 40% probability.

Wet: The area with mean January rainfall between 22 mm and 30 mm experiences wet conditions with 50% probability and greater. Towards the area of lesser rainfall, the boundary spreads up to 12 mm isohyets with 30% probability and up to 2 mm isohyets with 10% probability.

Very wet: The very wet condition during January occurs with probability 50% and greater where mean rainfall is greater than 30 mm. The occurrence probability exceeds 70% for rainfall greater than 54 mm, 90% for greater than 186 mm, 95% for rainfall greater than 405 mm. Towards drier side, the boundary spreads up to 18 mm isohyets with 30% probability, up to 5 mm with 10% probability.

February

Very dry: During February, the very dry condition occurs with 30% probability and greater where the mean rainfall is less than 20 mm. Outward of normal limit the boundary spreads up to 31 mm isohyets with 20% probability, up to 52 mm with 10%, up to 105 mm with 5% and up to 1074 mm isohyets with 1% probability.

Dry: The area with mean February rainfall between 12 mm and 23 mm experiences dry conditions with 50% probability and greater. Towards drier side, the probability is less than 30% where rainfall is more than 7 mm and up to 3 with 20% probability. Towards wetter side, the boundary spreads up to 18 mm isohyets with 40% probability.

Wet: The area with mean February rainfall between 23 mm and 36 mm experiences wet conditions with 50% probability and greater. Towards the area of lesser rainfall, the boundary spreads up to 13 mm isohyets with 30% probability and up to 2 mm with 10% probability.

Very wet: The very wet condition during February occurs with probability 50% and greater where mean rainfall is more than 36 mm. The occurrence probability exceeds 70% for rainfall more than 66 mm, 90% for more than 220 mm and 95% for more than 492 mm. Towards drier side, the boundary spreads up to 22 mm isohyets with 30% probability, up to 16 mm with 20% and up to 2 mm isohyets with 5% probability.

Summer

Very dry: Normally lower/south northwest India experiences very dry condition i.e. rainfall less than 26 mm in summer. During summer season, the areas with mean rainfall less than 36 mm experiences very dry conditions with probability exceeding 50%. The probability exceeds 70% where mean rainfall is less than 20 mm. Outward of normal limit the boundary spreads up to 53 mm isohyets with 30% probability, up to 80 mm with 10%, up to 94 mm with 5% and up to 144 mm isohyets with 1% probability.

Dry: East to very dry area is occupied by dry condition i.e. rainfall between 26 and 61mm. Occurrence probability of dry condition is greater than 50% where mean rainfall is between 36 mm and 74 mm. Towards drier side, the boundary spreads up to 20 mm isohyets with 30% probability and up to 10 mm with 20% probability. Towards wetter side, the boundary spreads up to 63 mm isohyets with 40% probability.

Wet: Inner south peninsula, east coast and border of north India experiences wet condition i.e. rainfall between 28 and 45 mm. The wet conditions occur with probability greater than 50% where mean summer rainfall is between 74 mm and 168 mm. The occurrence probability exceeds 70% with rainfall between 98 mm and 210 mm. Towards the lesser rainfall side, the boundary spreads up to 52 mm isohyets with 30% probability, up to 20 mm with 10% and up to 3 mm isohyets with 5% probability. Towards higher rainfall, the boundary spreads up to 130 mm isohyets with 30% probability.

Very wet: Southwest peninsula, northeast India and northernmost India comes under very wet condition i.e. rainfall more than 45 mm. The very wet condition prevails with 50% probability and greater where mean rainfall is more than 168 mm. The probability exceeds 70% where mean rainfall is more than 211 mm, 90% for more than 295 mm, 95% for more than 347 mm and 99% for more than 598 mm. Towards drier side, the boundary spreads up to 81 mm isohyets with 10% probability and up to 7 mm isohyets with 1% probability.

March

Very dry: During March, the very dry condition occurs with 50% probability and greater where mean rainfall is less than 14 mm. Outward of normal limit the boundary spreads up to 21 mm isohyets with 30% probability, up to 29 mm with 20%, up to 45 mm with 10%, upto 78 mm with 5% and up to 474 mm with 1% probability.

Dry: The dry condition occurs with 50% probability and greater where mean March rainfall is between 14 mm and 21 mm. Towards drier side, the probability is less than 30% where rainfall is less than 7 mm and up to 3 with 20% probability.

Wet: The area with mean March rainfall between 21 mm and 34 mm experiences wet conditions with 50% probability and greater. Towards the area of lesser rainfall, the boundary spreads up to 12 mm isohyets with 30% probability and up to 1 with 10% probability. Towards higher rainfall, the boundary spreads up to 27 mm isohyets with 40% probability.

Very wet: The very wet condition during March occurs with probability 50% and greater where mean rainfall is greater than 34 mm. The occurrence probability exceeds 70% for rainfall more than 55 mm, 90% for more than 139 mm, 95% for more than 265 mm and 99% for more than 1782 mm. Towards drier side, the boundary spreads up to 20 mm isohyets with 30% probability and up to 6 mm isohyets with 10% probability.

April

Very dry: During April, the very dry condition occurs with 50% probability and greater where mean rainfall less than 11 mm. Outward of normal limit the boundary spreads up to 21 mm isohyets with 20% probability, up to 37 mm with 5% and up to 87 mm with 1% probability.

Dry: In the core region with rainfall between 11 mm and 24 mm the occurrence probability exceeds 50%. Towards drier side, the probability is less than 40% where rainfall is less than 8 mm and probability 20% where rainfall is less than 2 mm.

Towards wetter side, the boundary spreads up to 20 mm isohyets with 40% probability.

Wet: The area with mean April rainfall between 24 mm and 51 mm experiences wet conditions with 50% probability and greater. Towards the area of lesser rainfall, the boundary spreads up to 15 mm isohyets with 30% probability and up to 3 mm with 10% probability. Towards higher rainfall, the boundary spreads up to 36 mm isohyets with 30% probability.

Very wet: The very wet condition during April occurs with probability 50% and greater where mean rainfall is greater than 51 mm. Occurrence probability exceeds 70% for rainfall greater than 73 mm, 90% for greater than 129 mm, 95% for greater than 198 mm and 99% for greater than 596 mm. Towards drier side, the boundary can spread up to 36 mm isohyets with 30% probability and up to 9 mm isohyets with 5% probability.

May

Very dry: During May, the very dry condition occurs with 50% probability and greater where mean May rainfall is less than 22 mm. Outward of normal limit the boundary spreads up to 33 mm isohyets with 30% probability, up to 51 mm with 10%, up to 62 mm with 5% and up to 104 mm isohyets with 1% probability.

Dry: The dry condition occurs with 50% probability and greater where mean May rainfall is between 22 mm and 44 mm. In the core region with rainfall between 27 mm and 52 mm the occurrence probability exceeds 60%. Towards drier side, the probability is less than 20% where mean rainfall is less than 5 mm. Towards wetter side, the boundary spreads up to 36 mm isohyets with 40% probability.

Wet: The area with mean May rainfall between 44 mm and 89 mm experiences wet conditions with 50% probability and greater. Towards the area of lesser rainfall, the boundary spreads up to 28 mm isohyets with 30% probability and up to 7 mm with 10%. Towards higher rainfall, the boundary spreads up to 76 mm isohyets with 40% probability.

Very wet: The very wet condition during May occurs with 50% probability and greater where mean rainfall is more than 89 mm. The occurrence probability exceeds 70% for rainfall more than 121 mm, 90% for more than 193 mm, 95% for more than 247 mm and 99% for more than 471 mm. Towards drier side, the boundary spread up to 62 mm isohyets with 30% probability, up to 31 mm with 10% and up to 17 mm isohyets with 5% probability.

Summer Monsoon

Very dry: Normally in two isolated areas viz. northwest India and south peninsula except west coast very dry condition i.e. rainfall less than 500mm.occurs in monsoon. The very dry condition occurs with probability 50% and greater where mean rainfall is less than 521 mm. The occurrence probability exceeds 70% where mean monsoon rainfall is less than 426 mm, 90% for less than 294 mm, 95% for less than 234 mm and 99% for less than 134 mm. Outward of normal limit the boundary spreads up to 622 mm isohyets with 30% probability, up to 790 mm with 10%, up to 882 mm with 5% and up to 1160 mm isohyets with 1% probability.

Dry: Dry condition i.e. rainfall between 500 and 830 mm is experienced in adjacent eastern part of very dry area in northwest India and north of very dry area in south peninsula. The dry condition occurs with probability 50% and greater where mean monsoon rainfall is between 521 mm and 854 mm. In the core region with rainfall between 622 mm and 987 mm, the occurrence probability exceeds 70%. Towards drier side, the boundary spreads up to 426 mm isohyets with 30% probability, up to 294 mm with 10% and up to 134 mm isohyets with 1% probability. Towards wetter side, the boundary spreads up to 731 mm isohyets with 30% probability.

Wet: Normally large part of eastern central India and adjacent north India is under wet condition i.e. rainfall between 831 and 1120mm. The area with mean monsoon rainfall between 854 mm and 1144 mm experiences wet conditions with probability 50% and greater. The probability exceeds 60% in the core region with rainfall between 917 mm and 1221 mm. Towards the area of lesser rainfall, the boundary spreads up to 731 mm isohyets with 30% probability, up to 564 mm with 10% and up to 337 mm isohyets with 1% probability. Towards higher rainfall, the boundary spreads up to 997 mm isohyets with 30% probability.

Very wet: Very wet i.e. rainfall more than 1120mm is experienced over areas mainly viz. west coast, northeast India, central and east part of central India and part of northernmost India. The very wet condition during monsoon season occurs with 50% probability and greater where mean rainfall is more than 1144 mm. It occurs with probability greater than 70% where rainfall is more than 1306 mm, 90% where rainfall is more than 1581 mm, 95% where rainfall is more than 1745 mm and 99% where rainfall is more than 2308 mm. Towards drier side, the boundary spreads up to 997 mm isohyets with 30% probability, up to 799 mm with 10%, up to 706 mm with 5% probability and up to 515 mm isohyets with 1% probability.

June

Very dry: Normally two isolated areas viz. northwest India and south peninsula except west coast experiences very dry condition i.e. rainfall less than 68mm. in June. The very dry condition occurs with 50% probability and greater where mean June rainfall is less than 86 mm. The occurrence probability exceeds 80% where mean June rainfall is less than 31 mm. Outward of normal limit the boundary spreads up to 121 mm isohyets with 30% probability, up to 177 mm with 10%, up to 212 mm with 5% and up to 331 mm isohyets with 1% probability.

Dry: Dry condition i.e. rainfall between 68 and 126 mm is experienced in adjacent eastern part of very dry area in northwest India and north of very dry area in south peninsula. The area with mean June rainfall between 86 mm and 146 mm experiences wet condition with 50% probability and greater. The probability exceeds 60% where mean rainfall is between 102 mm and 167 mm. Towards the area of lesser rainfall the boundary spreads up to 52 mm isohyets with 30% probability, up to 2 mm with 10%

probability. Towards higher rainfall, the boundary spreads up to 124 mm isohyets with 40% probability.

Wet: Normally large part of central India and border area of north India is under wet condition i.e. rainfall between 127 and 192 mm. The dry condition occurs with 50% probability and greater where mean June rainfall is between 146 mm and 212 mm. In the core region with rainfall between 167 mm and 238 mm, the occurrence probability exceeds 60%. Towards drier side, the boundary spreads up to 124 isohyets with 40% probability, up to 78 mm isohyets with 20% probability and up to 17 mm isohyets with 5% probability. Towards wetter side, the boundary spreads up to 185 mm isohyets with 40% probability.

Very wet: Very wet condition i.e. rainfall more than 192mm is experienced over areas mainly viz. west coast, northeast India, large part of east peninsula and small part of northernmost India. The very wet condition during June occurs with 50% probability and greater where mean rainfall is more than 212 mm. The occurrence probability exceeds 70% where rainfall is more than 269 mm, 90% for more than 376 mm, 95% for more than 458 mm and 99% for more than 768 mm. Towards drier side, the boundary spreads up to 159 mm isohyets with 30% probability, up to 91 mm with 10% probability and up to 60 mm isohyets with 5% probability.

July

Very dry: Normally two isolated areas viz. northwest India and south peninsula except west coast experiences very dry condition i.e. rainfall less than 162mm. in July. The very dry condition occurs with 50% probability and greater where mean July rainfall is less than 180 mm. The occurrence probability exceeds 95% where July rainfall is less than 41 mm. Outward of normal limit the boundary spreads up to 234 mm isohyets with 30% probability, up to 343 mm with 10%, up to 424 mm with 5% and up to 807 mm with 1% probability.

Dry: Dry condition i.e. rainfall between 162 and 286 mm is experienced in adjacent eastern part of very dry area in northwest India and north of very dry area in south peninsula. The dry condition occurs with 50% probability and greater where mean July rainfall is between 180 mm and 306 mm. In the core region with mean rainfall between 205 mm and 341 mm, the occurrence probability exceeds 60%. Towards drier side, the boundary spreads up to 133 isohyets with 30% probability and up to 41 mm isohyets with 5% probability. Towards wetter side, the boundary spreads up to 241 mm isohyets with 30% probability.

Wet: Normally extended central India and contiguous border area of north India is under wet condition i.e. rainfall between 287 and 362 mm. The area with mean July rainfall between 306 mm and 381 mm experiences wet conditions with 50% probability and greater. Towards the area of lesser rainfall, the boundary spreads up to 241 mm isohyets with 30% probability, up to 158 mm with 10% and up to 43 mm with 1% probability. Towards higher rainfall, the boundary spreads up to 343 mm isohyets with 40% probability.

Very wet: Very wet condition i.e. rainfall more than 362mm is experienced over areas mainly viz. west coast, large part of northeast India, central India and part of northernmost India. The very wet condition during July occurs with 50% probability and greater where mean rainfall is greater than 381 mm. The occurrence probability exceeds 70% where rainfall is more than 475 mm, 90% for more than 680 mm, 95% for more than 841 mm and 99% for more than 1677 mm. Towards drier side, the boundary spreads up to 306 mm isohyets with 30% probability, up to 211 mm with 10%, up to 169 mm with 5% and up to 81 mm isohyets with 1% probability.

August

Very dry: Normally two isolated areas viz. western northwest India and south peninsula except west coast experiences very dry condition i.e. rainfall less than 140mm. in August. During August, the very dry condition occurs with probability 50% and more where mean rainfall is less than 157 mm. The occurrence probability exceeds 70% where rainfall is less than 113 mm, 90% with less than 54 mm and 95% with rainfall less than 27 mm. Outward of normal limit the boundary spreads up to 211 mm isohyets with 30% probability, up to 307 mm with 10%, up to 366 mm with 5% and up to 600 mm isohyets with 1% probability.

Dry: Dry condition i.e. rainfall between 140 and 264 mm is experienced in adjacent eastern part of very dry area in northwest India and north of very dry area in south peninsula. Dry condition occurs with 50% probability and greater where mean August rainfall is between 157 mm and 287 mm. The occurrence probability exceeds 60% where mean rainfall is between 182 mm and 324 mm. Towards drier side, the boundary spreads up to 113 isohyets with 30% probability, up to 54 mm with 10% and up to 27 mm isohyets with 5% probability. Towards wetter side, the boundary spreads up to 221 mm isohyets with 30% probability.

Wet: Normally central India, contiguous border area of north India and central part of northeast India is under wet condition i.e. rainfall between 265 and 340 mm. The area with mean August rainfall between 287 mm and 364 mm experiences wet conditions with 50% probability and greater. Towards the area of lesser rainfall, the boundary spreads up to 221 mm isohyets with 30% probability, up to 141 mm with 10% and up to 38 mm isohyets with 1% probability.

Very wet: Very wet condition i.e. rainfall more than 340mm is experienced over areas mainly viz. west coast, large part of northeast India, central India and part of northernmost India. The very wet condition during August occurs with 50% probability and greater where mean rainfall is more than 364 mm. The occurrence probability exceeds 70% where mean rainfall is more than 457 mm, 90% for greater than 639 mm, 95% for greater than 759 mm and 99% for greater than 1305 mm. Towards drier side the probability is less than 30% where rainfall is less than 286 mm, 10% for rainfall less 193 mm and 1% for less than 74 mm.

September

Very dry: Northwest India, extreme southeast India, inner middle area of west coast and some isolated areas experience very dry condition i.e. rainfall less than 121 mm. during September. The very dry condition occurs with 50% probability and greater where mean September rainfall is less than 144 mm. The occurrence probability exceeds 90% where mean September rainfall is less than 32 mm. Outward of normal limit the boundary spreads up to 200 mm isohyets with 30% probability, up to 304 mm with 10%, up to 379 mm with 5% and up to 706 mm with 1% probability.

Dry: Normally, dry condition i.e. rainfall between 121 mm and 170 mm. occurs over adjacent eastern part of very dry area in northwest India and north of very dry area in south peninsula. The area with mean September rainfall between 144 mm and 194 mm experiences wet conditions with 50% probability and greater. Towards drier side, the boundary spreads up to 95 mm isohyets with 30% probability, up to 32 mm isohyets with 10% probability and up to 2 mm isohyets with 1% probability.

Wet: Central India, adjacent part of north India border and inner area of west coast mainly experiences wet condition i.e. rainfall between 171 and 216 mm. Towards the area of lesser rainfall, the boundary spreads up to 137 mm isohyets with 30% probability, up to 66 mm with 10% and up to 32 mm isohyets with 5% probability.

Very wet: Very wet condition i.e. rainfall more than 216 mm is experienced over areas mainly viz. west coast, northeast India and adjacent east peninsula, and parts of northern border of India. The very wet condition during September occurs with 50% probability and greater where mean rainfall is more than 275 mm. The occurrence probability exceeds 70% where mean rainfall is more than 375 mm, 90% for more than 603 mm and 95% for more than 1171 mm. Towards drier side, the boundary spreads up to 207 mm isohyets with 30% probability, up to 142 mm with 10% and up to 60 mm isohyets with 1% probability.

Post-Monsoon

Very dry: Extended northwest India occupies normally very dry area i.e. rainfall less than 53 mm. During post-monsoon season, the very dry condition occurs with 50% probability and greater where mean rainfall is less than 72 mm. The occurrence probability exceeds 70% where the mean rainfall is less than 44 mm, 80% where the mean rainfall is less than 25 mm. Outward of normal limit the boundary spreads up to 102 mm isohyets with 30% probability, up to 156 mm with 10%, up to 190 mm with 5% and up to 331 mm isohyets with 1% probability.

Dry: Adjacent to very dry area in southeast direction covering central and northern border of India is the dry area i.e. rainfall between 53 and 89 mm. The dry condition occurs with probability 50% and greater where mean post-monsoon rainfall is between 72 mm and 109 mm. In the core region with rainfall between 87 mm and 128 mm, the probability exceeds 60%. Towards drier side, the boundary spreads up to 44 isohyets with 30% probability and up to 25 mm isohyets with 20% probability. Towards wetter side, the boundary spreads up to 92 mm isohyets with 40% probability.

Wet: Wet area i.e. rainfall between 90 and 150 mm. is southeast to dry area covering middle peninsula and inner east India. The area with mean post-monsoon rainfall between 109 mm and 169 mm experiences wet conditions with probability 50% and greater. The probability exceeds 60% in the core region with rainfall between 128 mm and 194 mm. Towards the area of lesser rainfall, the boundary spreads up to 75 mm isohyets with 30% probability, up to 53 mm with 20%, up to 23 mm isohyets with 10% probability. Towards higher rainfall, the boundary spreads up to 147 mm isohyets with 40% probability.

Very wet: South peninsula, east coast and large part of northeast India experiences very wet condition i.e. rainfall more than 150mm. The very wet condition during post-monsoon season occurs with 50% probability and greater where mean rainfall is more than 169 mm. It occurs with probability greater than 70% where mean rainfall is more than 224 mm, 90% where mean rainfall is more than 342 mm, 95% where mean rainfall is more than 425 mm and 99% where mean rainfall is more than 811 mm. Towards drier side, the boundary spreads up to 125 mm isohyets with 30% probability, up to 64 mm with 10%, up to 31 mm with 5% probability.

October

Very dry: During October, the very dry condition occurs with 50% probability and greater where mean rainfall is less than 53 mm. The occurrence probability exceeds 70% where the mean rainfall is less than 31 mm, 80% where the mean rainfall is less than 16 mm. Outward of normal limit the boundary spreads up to 75 mm isohyets with 30% probability, up to 125 mm with 10%, up to 164 mm with 5% and up to 389 mm isohyets with 1% probability.

Dry: The dry condition occurs with probability 50% and greater where mean October rainfall is between 53 mm and 76 mm. Towards drier side, the boundary spreads up to 31 isohyets with 30% probability and up to 16 mm isohyets with 20% probability.

Wet: The area with mean October rainfall between 76 mm and 133 mm experiences wet conditions with 50% probability and greater. The occurrence probability exceeds 60% where mean rainfall is between 90 mm and 157 mm. Towards higher rainfall, the boundary spreads up to 112 mm isohyets with 40% probability. Towards drier side, the boundary spreads up to 48 isohyets with 30% probability and up to 7 mm isohyets with 10% probability

Very wet: The very wet condition during October occurs with 50% probability and greater where mean rainfall is more than 157 mm. The occurrence probability exceeds 70% where rainfall is more than 233 mm, 90% for more than 461 mm and 95% for more than 1225 mm. Towards drier side, the boundary spreads up to 92 mm isohyets with 20% probability, up to 41 mm with 5% and up to 15 mm isohyets with 1% probability.

November

Very dry: During November, the very dry condition occurs with 50% probability and greater where mean rainfall is less than 20 mm. The occurrence probability exceeds 70% where the mean rainfall is less than 12 mm and 80% where the mean rainfall is less than 5 mm. Outward of normal limit the boundary spreads up to 28 mm isohyets with 30% probability, up to 50 mm with 10%, up to 80 mm with 5% and up to 326 mm isohyets with 1% probability.

Dry: The dry condition occurs with 50% probability and greater where mean November rainfall is between 20 mm and 30 mm. Towards drier side, the boundary spreads up to 12 mm isohyets with 30% probability and up to 5 mm with 20% probability.

Wet: The area with mean November rainfall between 30 mm and 44 mm experiences wet conditions with 50% probability and greater. Towards the area of lesser rainfall, the boundary spreads up to 25 mm isohyets with 40% probability, up to 19 mm isohyets with 30% probability and up to 11 mm isohyets with 20% probability.

Very wet: The very wet condition during November occurs with 50% probability and greater where mean rainfall is more than 44 mm. The occurrence probability exceeds 70% where rainfall is more than 66 mm, 90% for more than 142 mm, 95% for more than 251 mm and 99% for more than 1172 mm. On the drier side, the boundary spreads up to 29 mm isohyets with 30% and up to 6 mm isohyets with 10% probability.

December

Very dry: During December, the very dry condition occurs with 50% probability and greater where mean rainfall is less than 12 mm. The occurrence probability exceeds 70% where the mean rainfall is less than 7 mm and 80% where the mean rainfall is less than 4 mm. Outward of normal limit the boundary spreads up to 21 mm isohyets with 30% probability, up to 70 mm with 10% and up to 238 mm with 5% probability.

Dry: The dry condition occurs with 50% probability and greater where mean December rainfall is between 12 mm and 18 mm. Towards drier side, the boundary spreads up to 7 mm isohyets with 30% probability, and up to 4 mm isohyets with 20% probability.

Wet: The area with mean December rainfall between 18 mm and 23 mm experiences wet conditions with 50% probability and greater. Towards drier side, the boundary spreads up to 14 mm isohyets with 30% probability and up to 1 mm with 10% probability.

Very wet: The very wet condition during December occurs with 50% probability and greater where mean rainfall is more than 23 mm. The occurrence probability exceeds 70% where mean December rainfall is more than 46 mm, 90% for more than 193 mm and 95% for more than 695 mm. Towards drier side, the boundary spreads up to 13 mm isohyets with 30% probability and up to 2 mm isohyets with 10% probability.

CHIEF FEATURES OF FLUCTUATIONS OF SEASONAL AND MONTHLY DRY AND WET ZONES

Winter

In RSDI fluctuations, the dry epochs are 1872-1900, 1949-77 and 1987-2006 and the wet epochs 1901-48 and 1978-86. In very dry area fluctuations the dry epochs are 1872-1904 and 1949-2001 and the wet epochs 1905-48 and 2002-05; in dry area the dry epochs are 1916-24, 1938-53 and 1975-81 and the wet epochs 1876-1915, 1925-37, 1954-74 and 1982-2006; in wet area the dry epochs are 1872-1914, 1930-34, 1954-77 and 1988-2006 and the wet epochs 1915-29, 1935-53 and 1978-87; and in very wet area the dry epochs are 1872-1897, 1949-81, 1987-93 and 2001-06 and the wet epochs are 1898-1948, 1982-86 and 1994-2000.

The winter RSDI increased up to 1943, and then decreased up to 2006. During 1949-2006, the RSDI was 9% below compared to 1901-1948 (significant at 0.1%). The recent decrease in winter RSDI is due to expansion of very dry area and contraction/shrinking of wet and very wet areas. The very dry area was 25% larger during 1949-2001 compared to 1905-1948 (significant at 0.1%), the wet area was 16% smaller during 1954-2006 compared to 1915-1953 (significant at 5%) and the very wet area was 19% smaller during 1949-2006 compared to 1898-1948 (significant at 0.1%). The dry area was however 22% smaller during 1982-2006 compared to 1975-1981 (significant at 10%).

January

In RSDI fluctuation, the dry epochs are 1871-86, 1896-1918, 1930-40, 1962-79 and 1988-2003 and the wet epochs 1887-95, 1919-29, 1941-61 and 1980-87. In very dry area variation the epochal patterns: dry- 1872-86, 1894-1900, 1909-18, 1927-40, 1960-78, 1988-93 and 1998-2006, and the wet- 1887-93, 1901-08, 1919-26, 1941-59, 1979-87 and 1994-97; in dry area the epochs are: dry- 1888-92, 1902-15 and 1961-2005, and wet- 1871-81, 1893-1901 and 1916-60; in wet area variation the epochs: dry- 1872-86, 1894-1916, 1927-51 and largely fluctuating thereafter, and wet- 1887-93 and 1917-26; and in very wet area variation: dry- 1872-86, 1896-1916, 1931-37 and 1962-2006, and wet- 1887-95, 1917-30 and 1938-61.

The RSDI of the month was 10% lower during 1962-2003 compared to 1919-1961 (significant at 5%), due to larger than normal dry area and smaller than normal very wet area. The dry area is 18% larger during 1961-2005 compared to 1916-1960 (significant at 5%) and the very wet area 23% smaller during 1962- 2006 compared to 1917-1961(significant at 5%). The very dry and wet areas have displayed highly fluctuating behaviour.

February

In RSDI fluctuation, the dry epochs are 1876-92, 1909-26 and 1949-78, and the wet epochs 1893-1908, 1927-48 and 1979-2003. In very dry area variation the epochal patterns are: dry- 1881-90, 1908-26, 1952-80 and 1992-2006, and wet- 1872-80, 1891-1907, 1927-49 and 1981-91; in the dry area: dry- 1871-79, 1890-1900, 1937-53 and 1961-83, and wet- 1880-89, 1901-36, 1954-60 and 1984-2004; in the wet area: dry- 1881-91, 1906-30, 1953-72 and largely fluctuating thereafter, and wet- 1872-80, 1892-1905 and 1931-52; and in the very wet area: dry- 1871-95, 1913-25 and 1948-80, and wet- 1896-1912, 1926-47 and 1981-2003.

The RSDI showed rising trend from 1957 and was 13% higher during 1979-2003 compared to 1949-1978 (significant at 5%). The very dry area showed moderately shrinking tendency since 1957, the dry area showed shrinking tendency since 1979 and the very wet area sharply increasing since 1957, but, the wet area fluctuated around the long term mean. The dry area was 16% smaller during 1984-2004 compared to 1961-1983 (significant at 1%) and the very wet area 32% larger during 1981- 2003 compared to 1948-1980 (significant at 1%).

In summary, the winter rainfall was below normal since 1949 mainly due to expansion of very dry area and shrinkage of wet and very wet areas. The January rainfall was also mostly below normal since 1962 due to expansion of dry area and shrinkage of very wet area. The February rainfall however showed sharp increase since 1957 and was above normal since 1983 due to shrinkage of dry area and expansion of very wet area.

Summer

In RSDI fluctuation, the dry epochs are 1882-1912, 1921-35, 1947-61 and 1972-76, and the wet epochs 1877-81, 1913-20, 1936-46, 1962-71 and 1977-2006. In the very dry area variation the epochal patterns are: dry- 1874-1912, 1921-76 and 1988-96, and wet- 1913-20, 1977-87 and 1997-2006; in the dry area: dry- 1871-89, 1903-13, 1927-36 and 1976-2006, and wet- 1890-1902, 1914-26 and 1937-75; in the wet area: dry- 1871-77, 1887-1906, 1921-35, 1947-55, 1972-80 and 1987-96, and wet- 1878-86, 1907-20, 1936-46, 1956-71, 1981-86 and 1997-2006; and in the very wet area: dry- 1871-1932 and 1957-76, and wet- 1933-56 and 1977-2006.

The summer RSDI showed moderate increasing tendency since 1922 and became sharp since 1974 with values 5% higher during 1977-2006 compared to 1921-1976 (not significant). Increase in the RSDI was due to mild decrease in very dry area since 1922 and increasing tendency in dry, wet and very wet areas since 1954, 1974 and 1966 respectively. The very dry area shrank during 1977-2006 by 16% compared to 1921-1976, the dry area was larger by 17% during 1976-2006 compared to 1937-1975 (significant at 1%), the wet area was larger by 36% in recent years since 1997 compared to 1987-1996 (significant at 1%) and the very wet area expanded by 10% since 1977 compared to 1957-1976 (significant at 0.1%).

March

In RSDI fluctuation, the dry epochs are 1871-1903, 1921-35, 1971-77, 1983-86 and 1994-2004, and the wet epochs 1904-20, 1936-70, 1978-82 and 1987-93. In the very dry area variation the epochal patterns: dry- 1871-1903, 1916-35, 1968-76, 1983-86 and 1991-2004, and the wet- 1904-15, 1936-67, 1977-82 and 1987-1990; in the dry area: dry- 1928-41, 1954-63 and 1974-93, and the wet- 1872-1927, 1942-53, 1964-73 and 1994-2003; in the wet area: dry- 1871-1902, 1920-30, 1942-49, 1964-74 and 1994-2002, and wet- 1903-19, 1931-41, 1950-63, 1975-93 and 2003-06; and in the very wet area: dry- 1871-89, 1894-1903, 1918-38, 1953-62 and 1971-2004, and the wet- 1890-93, 1904-17, 1939-52 and 1963-70.

The March RSDI showed mild increasing tendency up to 1967 and mild decreasing tendency thereafter. During 1971-2004 the rainfall was 6% less compared to 1936-1970 (not significant). The very dry area was spreading from 1989, the dry area shrinking from 1979, the wet area highly fluctuating and the very wet area shrinking from 1967.

April

In RSDI fluctuation, the dry epochs are 1871-94, 1916-24, 1947-57, 1966-75 and 1989-95, and the wet epochs 1895-1915, 1925-46, 1958-65, 1976-88 and 1996-2006. In the very dry area variation the epochal patterns are: dry- 1873-96, 1903-27, 1947-57 and 1966-75, and wet- 1897-1902, 1928-46, 1958-65 and 1976-2006; in the dry area: dry- 1881-95, 1914-47, 1958-65 and 1976-2005, and wet- 1873-80, 1896-1913, 1948-57 and 1966-75; in the wet area: dry- 1873-80, 1890-96, 1901-25 and 1971-75, and wet- 1881-89, 1897-1900, 1926-70 and 1976-2001; and in the very wet area: dry- 1871-94, 1915-24, 1947-75 and 1989-95, and wet- 1895-1914, 1925-46, 1976-88 and 1996-2006.

The April RSDI showed rising trend during 1885-1944, falling trend during 1945-1974 and again falling trend during 1975-2006. From 1996 onwards, the RSDI was 9% higher (not significant) compared to 1989-1995. The very dry area showed falling trend from 1974, the dry area rising trend from 1955, the wet area highly fluctuating around long term mean and the very wet area rising trend from 1980. The very dry area was 18% lower during 1976-2006 compared to 1966-1975 (significant at 5%). As a complimentary compensation the dry area was 16% larger during 1976- 2005 compared to 1966-1975 (is significant at 1%), the wet area 13% larger during 1926-2001 compared to 1901-1925 (significant at 5%), and the very wet area 21% larger from 1996 compared to 1989-1995 (significant at 5%).

May

In RSDI fluctuation, the dry epochs are 1890-1912, 1921-68 and the wet epochs 173-89, 1913-20 and 1969-2006. In the very dry area variation the epochal patterns are: dry- 1887-1911 and 1921-68, and wet- 1871-1886, 1912-20 and 1969-2006; in the dry area: dry- 1871-1901, 1913-27, 1938-45, 1974-88 and 1996-2005, and wet- 1902-12, 1928-37, 1946-73 and 1989-95; in the wet area: dry- 1874-82, 1890-1909,

1921-44, 1952-65 and 1989-96, and wet- 1883-89, 1910-20, 1945-51, 1966-88 and 1997-2002; and in the very wet area: dry- 1886-1900, 1906-12, 1921-48 and 1962-73, and wet- 1876-85, 1901-05, 1913-20, 1949-61 and 1974-2006.

The May RSDI showed increasing trend from 1936 onwards; the values are higher by 8% since 1969 compared to 1921-1968 (significant at 5%). The very dry area showed shrinking tendency from 1936 onwards, and it has contracted since 1969 by 20% compared to 1921-1968 (significant at 1%). The dry, the wet and the very wet areas showed spreading tendency from 1955, 1954 and 1967 respectively. The dry area expanded during 1974-2005 by 23% compared to 1946-1973 (significant at 0.1%), the wet area expanded during 1997-2002 by 65% compared to 1989-1996 (significant at 0.1%) and the very wet area expanded since 1974 by 18% compared to 1962-1973 (significant at 5%).

The summer rainfall showed increasing trend since 1922; the increase was sharp since 1974 as the very dry area started contracting since 1922 and the dry and the very wet areas started expanding since 1954 and 1966 respectively. The increase in summer rainfall was contributed by April and May rainfall but the March rainfall showed slight decrease. Increase in April and May rainfall is due to shrinking of the very dry area and expansion of the dry, the wet and the very wet areas.

Summer monsoon

In RSDI fluctuation, the dry epochs are 1898-1925 and 1974-2006, and the wet epochs 1871-97 and 1926-73. In the very dry area variation the epochs are: dry- 1895-1941, 1965-74, 1979-87 and 1999-2006, and wet- 1872-94, 1942-64, 1975-78 and 1988-98; in the dry area: dry- 1898-1912 and 1961-2006, and wet- 1871-97 and 1913-60; in the wet area: dry- 1874-1948 and 1984-2005, and wet- 1949-83; and in the very wet area: dry- 1899-1915 and 1962-2006, and wet-1874-98 and 1916-61.

The summer monsoon RSDI decreased during 1871-1900, increased during 1901-1961 and showed decreasing trend since 1962. The recent decrease in RSDI was due to increase in the very dry and the dry areas since 1958 and 1938 respectively, and compensatory decrease in the wet and the very wet areas since 1963 and 1962 respectively. Since 1974, the RSDI values were lower compared to 1926-1973 by 4% (significant at 5% level). The very dry area was 31% larger during 1999-2006 compared to 1988-1998 (not significant). The dry area was larger since 1961 compared to 1913-1960 by 16% (significant at 0.1% level) and compared to 1871-1960 by 15% (significant at 0.1% level). The wet area was 11% smaller during 1984-2005 compared to 1949-1983 (not significant). And the very wet area was larger since 1962 compared to 1916-1961 by 18% (significant at 0.1% level) and compared to 1871-1961 by 19% (significant at 0.1% level). During 1965-2006 the monsoon rainfall was less by 4.72% compared to the period 1931-1964.

June

In RSDI fluctuation, the dry epochs are 1900-08, 1923-32 and 1943-74 and the wet epochs 1871-99, 1909-22, 1933-42 and 1975-2006. In the very dry area variation the epochal patterns are: dry- 1900-08, 1918-32 and 1945-69, and wet- 1871-99, 1909-17,

161

1933-44 and 1970-2006; in the dry area: dry- 1939-2006, and wet- 1876-1938; in the wet area: dry- 1900-08, 1918-27, 1958-74 and 1982-90, and wet- 1871-99, 1909-17, 1928-57, 1975-81 and 1991-2005; and in the very wet area: dry- 1900-05, 1923-32 and 1942-76, and the wet- 1871-99, 1906-22, 1933-41 and 1977-2006.

The June RSDI showed slight increasing trend since 1902 and a prominent trend since 1966. The very dry area showed shrinking tendency since 1925, the dry area spreading tendency from 1892 onwards and the very wet area spreading tendency since 1974 whereas the wet area fluctuated around the long term mean. The RSDI values were 7% higher since 1975 compared to 1943-1974 (significant at 5% level). The very dry area contracted by 18% since 1970 compared to 1945-1969 (significant at 0.1% level), the dry area expanded by 22% since 1939 compared to 1876-1938 (significant at 0.1% level), the wet area expanded by 12% (not significant) since 1991 compared to 1982-1990 and the very wet area expanded by 19% since 1977 compared to 1941-1976 (significant at 10% level). During 1995-2006, the June rainfall was 4.81% less compared to the period 1931-1964.

July

In RSDI fluctuation, the dry epochs are 1895-1921 and 1970-2006 and the wet epochs 1871-94 and 1922-69. In the very dry area variation the epochal patterns area: dry- 1899-1918, 1936-41, 1970-74 and 1982-2006, and wet- 1871-98, 1919-35, 1942-69 and 1975-82; in the dry area: dry- 1891-1910 and 1954-2001, and wet- 1871-90, 1911-53 and 2002-06; in the wet area: dry- 1878-1903, 1946-69 and 1975-86, and wet- 1871-77, 1904-45, 1970-74 and 1987-2006; and in the very wet area: dry- 1895-1921 and 1965-2006, and wet- 1871-94 and 1922-64.

The July RSDI showed sharp fall from 1945 onwards and the values were lesser since 1970 by 6% compared to 1922-1969 (significant at 0.1% level). The wet area was shrinking since 1977 and the very wet area fast shrinking since 1945. As a result, the very dry area was spreading fast since 1960 and the dry area during 1927-1982. The very dry expanded since 1970 by 15% compared to 1919-1969 (significant at 10% level), and the dry area during 1954-2001 by 14% compared to 1911-1953 (significant at 1% level). The wet area reduced since 1987 by 12% compared to 1975-1986 (not significant) and the very wet area reduced since 1965 by 19% compared to 1922-1964 (significant at 0.1% level). During 1965-2006 the July rainfall was less by 6.93% compared to 1931-1964.

August

In RSDI fluctuation, the dry epochs are 1898-1937 and 1979-2006 and the wet epochs 1871-97 and 1938-78. In the very dry area variations the epochal patterns are: dry- 1871-81, 1895-1943 and 1999-2006, and wet- 1882-94 and 1944-98; in the dry area: dry- 1902-17 and 1962-2006, and wet- 1871-1901 and 1918-61; in the wet area: dry- 1875-1929, 1939-49, 1964-72 and 1999-2006, and wet- 1930-38, 1950-63 and 1973-98; and in the very wet area: dry- 1902-39 and 1985-2006, and wet- 1871-1901 and 1940-84.

The August RSDI increased during 1906-1977 but decreased sharply since 1979 onwards; the values were 6% smaller compared to 1938-1978 (significant at 5% level). The very dry and the dry areas expanded since 1977 and 1942 onwards respectively. The wet area showed shrinking tendency since 1958 and sharply since 1989 while the very wet area showed sharp shrinking tendency since 1943. The very dry area was 30% larger since 1999 compared to 1944-1998 (significant at 10% level), the dry area larger by 13% since 1962 compared to 1918-1961 (significant at 5% level), while the wet area was smaller by 21% since 1999 compared to 1973-1998 (not significant) and the very wet area smaller by 16% since 1985 compared to 1940-1984 (significant at 10% level). During 1965-2006 the August rainfall was less by 3.36% compared to 1931-1964.

September

In RSDI fluctuation, the dry epochs are 1899-1909, 1925-30 and 1965-2006 and the wet epochs 1871-98, 1910-24 and 1931-64. In the very dry area variations the epochal patterns are: dry- 1895-1913, 1927-30 and 1972-2006, and wet- 1871-94, 1914-26 and 1931-71; in the dry area: dry- 1873-78, 1882-86, 1951-59, 1966-77 and 1988-2006, and wet- 1879-81, 1887-50, 1960-65 and 1978-87; in the wet area: dry-1943-49 and 1972-2006, and wet-1871-1942 and 1950-71; and in the very wet area: dry- 1904-09, 1927-30 and 1965-2006, and wet- 1871-1903, 1910-26 and 1931-64.

The September RSDI showed sharp decreasing trend since 1962. The decrease was due to fast spread of the very dry area since 1961, mild reduction in wet area since 1963 and fast reduction in very wet area since 1961. The dry area showed gentle spreading trend from 1900 onwards. The RSDI values were 8% below since 1965 compared to 1931-1964 (significant at 1% level). The very dry area expanded by 18% since 1972 compared to 1931- 1971 (significant at 5 % level), the dry area expanded by 28% since 1988 compared to 1978-1987 (significant at 5% level), the wet area shrunk by 14% since 1972 compared to 1950-1971 (significant at 5% level), and the very wet area shrunk by 19% since 1965 compared to 1931-1964 (significant at 5% level). During 1965-2006 the September rainfall was less by 8.92% compared to 1931-1964.

The summer monsoon rainfall showed decreasing trend since 1962 as the very dry, the dry areas expanded, the wet, and the very wet areas shrunk since 1963 and 1945 respectively. July, August and September rainfalls contribute the decrease in monsoon rainfall.

Post-monsoon

In RSDI fluctuation, the dry epochs are 1871-76, 1895-1926, 1938-43, 1964-72, 1988-93 and 2000-06, and the wet epochs 1877-94, 1927-37, 1944-63, 1973-87 and 1994-99. In the very dry area variations the epochal patterns are: dry- 1871-78, 1888-1923, 1938-54, 1964-70, 1988-94 and 2000-06, and wet- 1879-87, 1924-37, 1955-63, 1971-87 and 1995-99; in the dry area: dry- 1874-84, 1922-34 and 1958-83, and wet-1885-1921, 1935-57 and 1984-2006; in the wet area: dry- 1871-76, 1888-1922, 1938-54, 1964-75, 1986-94 and 1999-2006, and wet- 1877-87, 1923-37, 1955-63, 1976-85 and 1995-98; and in the very wet area: dry- 1871-80, 1895-1926, 1938-43, 1964-72 and 1980-84, and wet- 1881-94, 1927-37, 1944-63, 1973-79 and 1985-2003.

The RSDI showed sharp rising trend during 1907-1957; mainly due to decrease in very dry area and increase in very wet area. The RSDI values were higher by 12% during 1927-1999 compared to 1895-1926 (significant at 1% level). The very dry area was 20% smaller during 1924-1999 compared to 1888-1923 (significant at 1% level) while the very wet area was 23% larger during 1927-2003 compared to 1895-1926 (significant at 5% level). In the latest years, however, since 2000 the RSDI values were 18% below compared to 1994-1999 (significant at 5% level). The very dry area was 89% larger since 2000 compared to 1995-1999 (significant at 5% level), the dry area 26% smaller since 1984 compared to 1958-1983 (significant at 1% level), the wet area 40% smaller during 1999-2006 compared to 1995-1998 (significant at 5% level) and the very wet area 50% larger during 1985-2003 compared to 1980-1984 (significant at 0.1% level). Compared to 1931-1964, the post-monsoon rainfall was less by 3.14% during 1965-2006, but 5.1% more during 1995-2006.

October

In RSDI fluctuation, the dry epochs are 1871-78, 1895-1927, 1940-54, 1964-70 and 1976-82, and the wet epochs1879-94, 1928-39, 1955-63, 1971-75 and 1983-2006. In the very dry area variations the epochal patterns are: dry- 1871-78, 1892-1926, 1934-53 and 1964-81, and wet- 1879-91, 1927-33, 1954-63 and 1982-2003; in the dry area: dry- 1878-88, 1919-35 and 1966-2002, and wet- 1871-77, 1889-1918, 1936-65 and 2003-06; in the wet area: dry 1871-74, 1888-1914, 1934-54, 1962-81, 1988-91 and 2003-06, and wet- 1875-87, 1915-33, 1955-61, 1982-87 and 1992-2002; and in the very wet area: dry- 1871-83, 1899-1927, 1934-42, 1964-69 and 1976-84, and wet- 1884-98, 1928-33, 1943-63, 1970-75 and 1985-2002.

The October RSDI showed increasing tendency from 1907, and in recent period since 1983 the RSDI values were 25% higher compared to 1976-1982 (significant at 1% level). The very dry area showed shrinking tendency since 1907 but the dry area spreading tendency from 1871 onwards. The very wet area expanded during 1907-1957 and from 1990 onwards while the wet area fluctuated around the long-term mean. The very dry area was 20% smaller during 1982-2003 compared to 1964-1981 (significant at 5% level) and the dry area 24% larger during 1966-2002 compared to 1936-1965 (significant at 1% level). On the other hand the very wet area was 66% larger during 1985-2002 compared to 1976-1984 (significant at 0.1% level). Compared to 1931-1964, the October rainfall was less by 10.58% during 1965-2006, but 7.23% more during 1995-2006.

November

In RSDI fluctuation, the dry epochs are 1871-81, 1895-1909, 1937-71 and 1983-2006, and the wet epochs 1882-94, 1910-36 and 1972-82. In the very dry area variations the epochal patterns are: dry- 1871-81, 1897-1909, 1935-75 and 1983-2006, and wet- 1882-96, 1910-34 and 1976-82; in the dry area: dry- 1871-96, 1910-16, 1928-40, 1958-83 and 1990-2003, and wet- 1897-1909, 1917-27, 1941-57 and 1984-

164

89; in the wet area: dry- 1871-80, 1900-09, 1920-55 and 1983-2006, and wet- 1881-99, 1910-19 and 1956-82; and in the very wet area: dry- 1871-81, 1895-1909, 1937-75 and 1983-2006, and wet- 1882-94, 1910-36 and 1976-82.

The November RSDI showed overall decreasing tendency from 1912 and a steep fall since 1978. The RSDI values were 19% below from 1983 compared to 1972-1982 (significant at 5% level). The very dry area showed increasing trend since 1978, and was 80% larger since 1983 compared to 1976-1982 (significant at 5% level). The dry area was 55% larger during 1990-2003 compared to 1984-1989 (significant at 5% level). The wet area decreased since 1962, and was 27% less since 1983 compared to 1956-1982 (significant at 1% level). The very wet area was 49% less since 1983 compared to 1976-1982 (significant at 5% level). Compared to 1931-1964, the November rainfall was higher by 6.54% during 1965-2006, but 3.23% less during 1995-2006.

December

In RSDI fluctuation, the dry epochs are 1891-1903, 1910-21, 1939-60, 1968-76 and 1992-2006 and the wet epochs 1871-90, 1904-09, 1922-38, 1961-67 and 1977-91. In the very dry area variations the epochal patterns are: dry- 1891-1921, 1938-60, 1968-76 and 1992-2006, and wet- 1871-90, 1922-37, 1961-67 and 1977-91; in the dry area: dry- 1871-1900, 1922-40 and 1972-90, and wet- 1901-21, 1941-71 and 1991-2006; in the wet area: dry- 1903-21, 1943-76 and 1992-2006, and wet- 1871-1902, 1922-42 and 1977-91; and in the very wet area: dry- 1871-79, 1891-1921, 1938-61, 1968-76 and 1992-2006, and wet- 1880-90, 1922-37, 1962-67 and 1977-91.

The December RSDI showed sharp decreasing tendency since 1988; the very dry area was spreading since then. The dry, the wet and the very wet areas showed shrinking tendency since 1980. The RSDI was 24% less during 1992-2006 compared to 1977-1991 (significant at 0.1% level). The very dry area was 46% above since 1992 compared to 1977-1991 (significant at 0.1% level), the dry area was below by 44% from 1991 compared to 1972-1990 (significant at 0.1% level), the wet area was below since 1992 compared to 1977-1991 by 51% (significant at 0.1% level), and the very wet area was 39% below since 1992 compared to 1977-1991 (significant at 0.1% level).

IMPACT OF RAINFALL VARIABILITY ON ENVIRONMENT

Shrinking Desert Area of the Rajasthan State

In 1973, British Geographer-turned-Climatologist D. Winstanley wrote an article [14] in which he predicted that by the year 2030 the Sahara and Rajaputana Deserts will have advanced some 100 km further south. The study created wide spread stir. American Historian-turned-Environmental Scientist Edward S. Haynes [15] compiled the Land use/cover change (LUCC) data of Rajasthan State for the period 1860-1980 by extracting information from variety of sources such as reports of revenue, administrative and forest departments, district gazetteer, historical evidences,

165

travelers' diary and satellite imageries. Over the state since 1860, expanding trend is seen in arable land at the rate of 87.88 km²/year and in settled and built-up areas 10.77 km²/year; while shrinking trend is seen in forest cover at the rate of 9.44 km²/year, in interrupted woods 25.94 km²/year, in grass/shrubs complexes 17.99 km²/year, in arid and unvegetated (desert) area 44.77 km²/year and in herbaceous wetlands 0.2 km²/year. Practically there is no change in surface water cover. Here important to note is the contracting tendency in the desert area based on LUCC data representing the ground realities which contradicts the popular notion of 'desert is marching' based on hearsay. In western Rajasthan cultivation is expanding at the expanse of desert areas due to improved rainfall conditions (35 mm/year), to the detriment of grasslands and shrub woods in the Aravallis and at the expense of denser wooded areas on the eastern and southern fringes of the province.

River Channel Changes over Indo-Gangetic Plains

River channel changes over the Indo-Gangetic Plains (IGPs) are natural environmental hazards of serious concern. Three forgings have been identified causing the river channel changes: tectonic disturbances (resulting into geological processes like folding, faulting, fracturing, rupturing, earthquake, volcano etc.), hydrologic phenomena (depending upon ground slope, rate and depth of flow, width and size of the flood plain, sediment transport etc.) and meteorologic factors (Coriolis Effect, spatial shift in rainfall/climatic belt and strength and direction of surface winds). During cooler northern hemisphere atmosphere, monsoon circulation will be weaker and low-level westerlies over the IGPs stronger, consequent upon which rainfall activities and river courses would shift eastward. While opposite will be true with warmer atmosphere- monsoon circulation and low-level easterlies over the IGPs will be stronger, consequent upon which rainfall activities and river courses would shift westward. However, in the recent complex situation with warmer lower troposphere (2.2°C/100-year) and cooler upper troposphere and lower stratosphere (2.2°C/100-year) the monsoon is likely to be highly variable. In the random occurrence if the entire depth of troposphere is warmer, the monsoon will be intense over northwest India and the river channels of the IGPs would shift westward, and if the full-depth of troposphere is cooler, the monsoon will be weaker and the rivers will shift eastward.

RISING TREND IN SURFACE AIR TEMPERATURE OVER WHOLE INDIA

Since 1850 the mean annual surface air temperature (SAT) of the global atmosphere has increased by about 1°C. The annual SAT over India also showed rising trend at the rate of 0.5°C/100-year during 1901-2006. The Indian temperature trends are generally considered as regional manifestation of global trends. Robust anti-correlation between rainfall and SAT is seen on annual, seasonal and monthly scales over the country during the period 1901-2006. Between maximum temperature and rainfall the relationship was found to vary between -0.5 to -0.8. Similar relationship was found between maximum temperature and dry area (very dry and dry zones). On the other hand, minimum temperature and wet area (wet and very wet zones) were found weakly negatively correlated for the months March through July and highly positively correlated for the months January, February and August through December. In general, maximum and mean temperatures fluctuations over the country are highly negatively correlated with rainfall; recent sharp rising trend in temperature is essentially due to large-scale decreasing trend in rainfall.

THE END NOTE

In majority of cases, the dry provinces of the country showed spreading tendency in recent years while wet provinces shrinking tendency. The integrated effect is large-scale decline in rainfall. Earliest possible action may be initiated to store the available surplus rainfall to the maximum possible extent in order to have stable food grain production and to meet drinking water supply and hydroelectric power generation.

ACKNOWLEDGEMENTS

The authors are extremely grateful to Prof. B. N. Goswami, Director, Indian Institute of Tropical Meteorology, Pune for necessary facilities to pursue this study. The rainfall data used in this study is provided by the India Meteorological Department, Pune and is thankfully acknowledged. Part of this research was supported by Department of Science and Technology (DST No. ES/48/ICRP/003/2000), Government of India.

REFERENCES

1. Pidwirny, Michael, "Global distribution of precipitation." in *Encyclopedia of Earth,* edited by Cutler J. Cleveland, Washington, D.C.: Environmental Information Coalition, National Council for Science and the Environment, 2006 (http://www.eoearth.org/article/Global_distribution_of_precipitation)
2. Singh N., Pant, G. B. and Mulye, S. S., "A statistical package for constructing representative area-averaged rainfall series and its updating using selected stations data" in *Use of Computers of computers in Hydrology and Water Resources,* National Seminar Proceedings, Indian Institute of Technology, Delhi, 1991, **Vol. II**, pp. 122-134.
3. Agnew Clive Agnew and Anderson Ewan W., *Water Resources in the Arid Realm,* Published by Routledge, ISBN 0415043468, 9780415043465, 1992, 329 pp.
4. Charney J.G., *Quarterly Journal of the Royal meteorological Society*, 101, 193-202 (1975).
5. Rodwell and Hoskins., *Quarterly Journal of the Royal meteorological Society*, 122, 1385-1404 (1996).
6. Tucker C. J., H. E. Dregne, W. W. Newcomb, *Science*, 253, 299-301 (1991)
7. Yang J., Ding Y., Chen R., and Liu Lianyou, *Climate Change*, 72, 171-188 (2005)
8. Yan Li Xiao, Wang XueQuan, and Gao QianZhao, *Journal of Food, Agriculture & Environment*, 1, (3/4), 229-238 (2003)
9. Nesje Atle, *The Holocene*, 15, 8, 1245-1252 (2005)
10. Sontakke N.A., Singh Nityanand and Singh H.N., *The Holocene*, 18, 7, 1055-1066 (2008)
11. Anonymous, National Atlas of India Physiographic Regions of India, Kolkata, India: Director, the National Atlas & Thematic Mapping Organization, 1986, Plate 41.
12. Singh N. and Mulye S.S., Theor. Appl. Climatol., 44, 3-4, 209-221 (1991)
13. Singh, N., Sontakke, N.A., Singh, H.N. and Pandey, A.K., "Recent trends in spatiotemporal variation of rainfall over India- an investigation into basin-scale rainfall fluctuations" in *Regional Hydrological Impacts of Climate Change- Hydroclimatic Variability,* edited by Franks, S.,

Wagener, T., Bogh, E., Guptya, H.V., Bastidas, L., Nobre, C. and Galvao, C. de O., Wallingford, Oxfordshire, UK: IAHS, 2005, **296**, pp. 273-282.

14. Winstanley, D., *Nature*, **245**, 190-194 (1973)

15. Haynes, Edward, S., "Land use, Natural resources and the Rajput State, 1780-1980" in *Desert, Drought and Development: Studies in Resource Management and Sustainability,* edited by R. Hooja and R. Joshi, Jaipur: Rawat Publications, 1999, pp 53-119.

Modelling Of Displacement Washing Of Pulp Bed Using Orthogonal Collocation On Finite Elements

Shelly Arora[a*], František Potůček[b], S.S. Dhaliwal[c], V.K. Kukreja[c]

[a] Department of Mathematics, Punjabi University, Patiala, India
[b] Department of Wood Pulp and Paper, University of Pardubice, Czech Republic
[c] Department of Mathematics, SLIET, Longowal, India

Abstract. Mechanism of displacement washing of packed bed of porous, compressible and cylindrical particles, e.g., fibers is presented with the help of an axial dispersion model involving Peclet number (Pe) and Biot number (Bi). Bulk fluid concentration and intra-pore solute concentration are related by Langmuir adsorption isotherm. Model equations have been solved using orthogonal collocation on finite elements using Lagrangian interpolating polynomials as base functions. Displacement washing has been simulated using a laboratory washing cell and experiments have been performed on pulp beds formed from unbeaten, unbleached kraft fibers. Model predicted values have been compared with experimental values to check the applicability of the method.

Keywords: Peclet number, Biot number, Lagrangian interpolation polynomials, orthogonal collocation, finite elements.
PACS : 81.05Ni, 81.05Rm

INTRODUCTION

Problem of heat and mass transfer during flow through packed beds of porous and compressible particles such as fibers have gained momentum in the field of mathematical modelling for the past few years. A variety of models related to adsorption-desorption and diffusion-dispersion phenomenon [1-7] have been developed.

In present study displacement washing of fibers which are porous in nature is presented through a dynamic model using Peclet number, Pe and Biot number, Bi. Peclet number is ratio of advection to dispersion whereas Biot number relates mass transfer resistances inside and on the surface of body.

Due to porous nature of fibers, solute present in pores of fibers diffuse out when it comes in contact with external fluid. Therefore, the mechanism of axial dispersion in the continuous phase between particles and concentration of fluid within and outside the particles play an important role. Due to adsorption of solute on fiber surface, mass transfer takes place through internal and external fiber surface and from fiber surface to the external fluid.

CP1146, *Modelling of Engineering and Technological Problems*, edited by A. H. Siddiqi, A. K. Gupta, and M. Brokate
© 2009 American Institute of Physics 978-0-7354-0683-4/09/$25.00

MODEL FORMULATION

Basically there are two types of mechanism related to fluid concentration in packed bed of fibers. First one is related to the transfer rate of material between fluid and fibers. This type of isotherm has been discussed widely for both linear and non-linear isotherms. The second type of rate mechanism is one in which solid phase diffusion into the interior of the bed particles is an important rate step and the rate equation is replaced by an integral equation.

In this paper the first type of rate mechanism is followed for Langmuir adsorption isotherm to study the displacement washing of fluid flow through the bed of packed fibers.

Model Equations For Particle Phase

In particle phase, the solute present in the pores of fibers diffuse out when comes in contact with external fluid. Movement of solute within fiber pores is defined by Fick's law. Mass transfer through the stagnant layer between fiber and external fluid is controlled by film resistance mass transfer coefficient k_f.

$$\frac{\partial q}{\partial t} + \frac{1-\beta}{\beta} \frac{\partial n}{\partial t} = \frac{k_f}{KR}(c-q) \tag{1}$$

Adsorption Isotherm

Langmuir type adsorption isotherm has been followed to relate the concentration of solute adsorbed on fiber surface and intrapore concentration.

$$\frac{\partial n}{\partial t} = \frac{qk_1}{C_0}(N_0 - n) - k_2 n \tag{2}$$

Model Equations For Bulk Fluid Phase

In this phase the impurities adsorbed on fiber surface and within fiber pores are washed with the aid of external fluid. These impurities detach from fiber surface and mixes with the external fluid by dispersion defined by axial dispersion coefficient D_L and is independent of axial distance L.

$$u\frac{\partial c}{\partial z} + \frac{\partial c}{\partial t} + \frac{2(1-\varepsilon)k_f}{\varepsilon KR}(c-q) = D_L \frac{\partial^2 c}{\partial z^2} \tag{3}$$

Boundary conditions: $uc - D_L \dfrac{\partial c}{\partial z} = 0$ at $z = 0$ (4)

$$\frac{\partial c}{\partial z} = 0 \qquad\qquad \text{at } z = L \tag{5}$$

Initial condition: $C = q = C_0$ and $n = N_0$ at $t = 0$ (6)

170

The set of equations (1) to (6) are converted into following dimensionless form using non dimensional variables given in nomenclature.

$$\frac{\partial Q}{\partial \tau} + \mu N' \frac{\partial N}{\partial \tau} = Bi(C - Q) \tag{7}$$

$$\frac{\partial N}{\partial \tau} = P^* \left[Q(1 - N) - \left(\frac{N}{k^*} \right) \right] \tag{8}$$

$$\frac{\partial C}{\partial \tau} = \frac{1}{Pe} \frac{\partial^2 C}{\partial \xi^2} - \frac{\partial C}{\partial \xi} - \theta Bi(C - Q) \tag{9}$$

Boundary conditions: $C - \dfrac{1}{Pe} \dfrac{\partial C}{\partial \xi} = 0 \qquad$ at $\xi = 0$ (10)

$$\frac{\partial C}{\partial \xi} = 0 \qquad \text{at } \xi = 1 \tag{11}$$

Initial condition: $\quad C = Q = N = 1 \qquad$ at $\tau = 0$ (12)

ORTHOGONAL COLLOCATION ON FINITE ELEMENTS

Method of orthogonal collocation on finite elements (OCFE) has been followed to solve the system of partial differential equations (7) to (12). In this method axial domain is divided into small sub-domains called elements. The global variable is transformed to local variable in ℓ^{th} element using the transformation $u = \dfrac{x - x_\ell}{\Delta x \ell}$; where $\Delta x_\ell = x_\ell - x_{\ell+1}$. Orthogonal collocation is applied on local variable with in each element. Zeros of Legendre polynomials have been taken as collocation points. The approximate function is discretized in terms of Lagrangian interpolation polynomials in ℓ^{th} element as:

$$C^\ell = \sum_{i=1}^{m} l_i(x) C_i^\ell \; ; \; \ell = 1, 2, 3, \dots, r$$

Convergence of numerical solutions does not depend upon number of collocation points rather it depends upon the number of elements to be taken in the domain of interest.

NUMERICAL SIMULATION

OCFE is applied on the set of partial differential equations (PDE's) (7) to (12) which convert these PDE's into a set of differential algebraic equations (DAE's) as follows:

$$\frac{\partial Q_j^\ell}{\partial \tau} + \frac{1-\beta}{\beta}\frac{\partial N_j^\ell}{\partial \tau} = Bi(C_j^\ell - Q_j^\ell); \quad j = 2,3,\ldots,m \text{ and } \ell = 1,2,3,\ldots,r \tag{13}$$

$$\frac{\partial N_j^\ell}{\partial \tau} = P^*\left[Q_j^\ell(1-N_j^\ell) - k^{*-1}N_j^\ell\right]; \quad j = 2,3,\ldots,m \text{ and } \ell = 1,2,3,\ldots,r \tag{14}$$

$$\frac{\partial C_j^\ell}{\partial \tau} = \frac{1}{Pe}\sum_{i=1}^{m+1}B_{ji}C_i^\ell - \sum_{i=1}^{m+1}A_{ji}C_i^\ell - \theta Bi(C_j^\ell - Q_j^\ell); j = 2,3,\ldots,m \text{ and } \ell = 1,2,3,\ldots,r \tag{15}$$

$$C_1^1 - \frac{1}{Pe}\sum_{i=1}^{m+1}A_{1i}C_i^1 = 0, \text{ at } \xi = 0 \tag{16}$$

$$\sum_{i=1}^{m+1}A_{m+1i}C_i^r = 0, \text{ at } \xi = 1 \tag{17}$$

The resulting stiff system of $r(3m+1)+1$ DAE's is solved using MATLAB with ode15s system solver. Here the set of differential algebraic equations is not converted into the set of ordinary differential equations and so the computational time is saved.

EXPERIMENTAL

The stimulus-response experiments, using a step input, have been carried out in the displacement washing cell consisting of a vertical glass cylinder 110 mm high, 36.4 mm inside diameter, and closed at the lower end by a permeable septum.

Pulp beds were formed from a dilute suspension of unbeaten, unbleached kraft pulp in black liquor. In all runs, the beds were compressed to a final desired thickness of 30 mm. The Schopper-Riegler freeness had a value of 13 SR. To start the experiment, distilled water maintained at a temperature of about 20 °C was distributed uniformly via piston to the top of bed, approximating a step change in concentration. Depending on the permeability of bed, wash liquid was forced through the pulp bed under the pressure drop up to 7 kPa. Samples of the washing effluent taking at different timed intervals until the effluent was colourless were analyzed for alkali lignin using an UV spectrophotometer operating at a wavelength of 295 nm. The initial bed liquor concentration was 50 g of alkali lignin per litre.

RESULTS AND DISCUSSION

In washing process, axial dispersion, interstitial velocity, cake thickness plays an important role as well as the bed preparation. Both Peclet number and Biot number depends upon interstitial wash liquid velocity and axial dispersion coefficient. Interstitial velocity as given by Darcy's law depends upon bed porosity which is determined with the help of effective specific volume of fibers and bed consistency.

172

Effect of Pe On Exit Solute Concentration

In Fig. 1 theoretical behavior of exit solute concentration profiles is shown for *Bi* 10 and for different values of *Pe*. It is observed that for *Pe* = 10 solution profile take a long time to converge to steady state condition as compared to *Pe* = 30. However, no significant effect is observed on solution profiles for *Pe* ≥ 30. [1] has mentioned that for displacement washing of porous structure of particles efficient washing operations can be achieved for *Pe* ≤ 40. In our case bed thickness is comparatively small, therefore time taken by the solute to leach from fiber surface is comparatively large which contributes to small *Pe*.

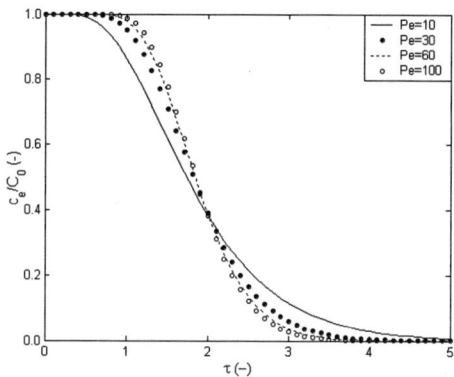

FIGURE 1: Behavior of concentration profiles for different values of Peclet number.

Effect of Bi on Exit Solute Concentration

Biot number signifies mass transfer resistance inside and on the surface of body. In Fig. 2 theoretical behavior of exit solute concentration is shown for *Pe* = 30 and for different values of *Bi*. It is observed from this figure that no remarkable change occur in solution profiles for *Bi* ≥ 10. For *Bi* = 50 and *Bi* = 100 solution profiles almost overlap each other. Mathematically, it may be due to the fact that increase in *Bi* reduces the effect of dispersion term which ultimately affects the solution profiles and therefore, no significant change occurs for *Bi* ≥ 10. Under practical conditions, [8] has shown that for fixed bed adsorbers, *Bi* ≤ 10 is of practical importance due to large volume equilibrium constant.

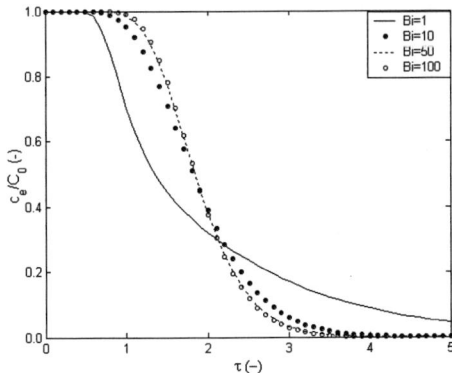

FIGURE 2: Behavior of concentration profiles for different values of Biot number.

Effect of *Pe* And *Bi* on Exit Solute Concentration

In Fig. 3, solution profiles are plotted for different values of *Pe* and *Bi* on the basis of experiments performed on a washing cell. It is observed that better washing operations are achieved for *Pe* = 21 and *Bi* = 15 as compared to *Pe* = 8 and *Bi* = 2.3. It is due to the fact that for large *Pe*, axial dispersion is small causing less back mixing. Similarly for large *Bi*, *L/u* ratio increases as well as the k_f. However, for small velocity profiles the effect of k_f is negligible. Both *Pe* and *Bi* are affected by velocity profiles and also the efficient washing operations can be achieved for small velocity profiles. Therefore, from theoretical point of view breakthrough curves agree well with the theory that efficient washing operations can be achieved for large *Pe* and *Bi* as compared to smaller ones.

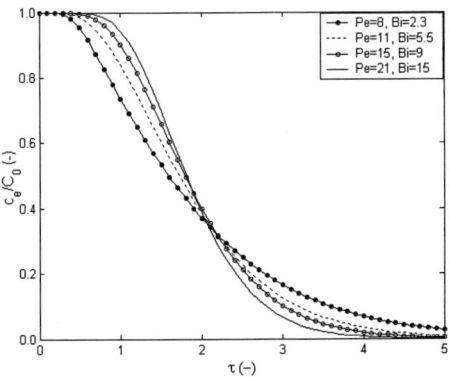

FIGURE 3: Behavior of solution profiles for different values of Peclet number and Biot number.

In Table 1 the relative error between experimental and calculated values is presented for a set of washing run. A good agreement is found between experimental and calculated values as the relative error is found to be less than 1%. It signifies the applicability of the model on washing cell.

174

Table 1: Comparison of experimental and calculated values
For $Pe = 12.25$ and $Bi = 7.4$

Experimental	Calculated	Relative Error (%)
9.5080×10^{-1}	9.5142×10^{-1}	0.0652
8.4810×10^{-1}	8.4652×10^{-1}	0.1863
6.5190×10^{-1}	6.4979×10^{-1}	0.3237
3.0720×10^{-1}	3.0796×10^{-1}	0.2474
9.7270×10^{-3}	9.7904×10^{-3}	0.6518
2.0310×10^{-3}	2.0378×10^{-3}	0.3348
8.1400×10^{-4}	8.1385×10^{-4}	0.0184

Effect of *Pe* and *Bi* on Concentration of Solute Adsorbed on Fiber Surface

In Fig. 4, the theoretical behavior of solute concentration adsorbed on fiber surface is shown for different values of *Pe* and *Bi*. It is observed that for $Pe = 11$ and $Bi = 5.5$, the retention time for the solute adsorbed on fiber surface is higher as compared to $Pe = 21$ and $Bi = 15$. It signifies the fact that solute adsorbed on fiber surface takes a long time to leach out from the fiber surface for small *Pe* and *Bi*. It is due to the reason that for small *Pe* more back mixing occurs causing more dispersion of solute and small *Bi* reduces the equilibrium constant. Also [2] has mentioned in his studies that solute adsorbed on porous particles leaches out rapidly for *Pe* greater than 10.

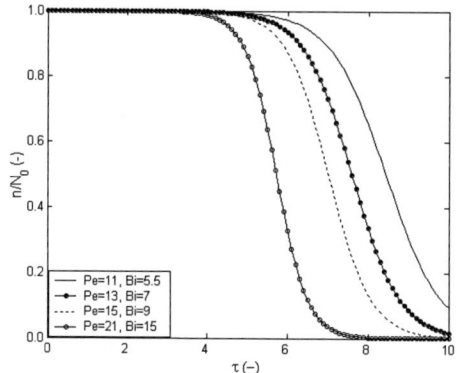

FIGURE 4: Behavior of solution profile for concentration of solute adsorbed on fiber surface for different values of Peclet number and Biot number.

CONCLUSIONS

In this paper a mathematical model for a displacement washing cell is proposed consisting of both the Peclet number and Biot number. It was observed that both Peclet number and Biot number effects the washing process. Experimental values of exit solute concentration were compared to the calculated ones and the relative error was found to be in the range of $\pm 1\%$. It was also observed from breakthrough curves that washing process is significantly effected by Peclet number for $Pe \leq 30$ and Biot number for $Bi \leq 10$.

NOMENCLATURE

B_i	=	Biot number, $= (k_f L\beta)/(KRu)$, dimensionless
c	=	Concentration of solute in liquor, kg /m^3
C	=	Dimensionless concentration, ($= c / C_0$)
C_0	=	Initial solute concentration, kg /m^3
D_L	=	Axial dispersion coefficient, m^2/s
k^*	=	Dimensionless parameter, ($= k_1 / k_2$)
k_1, k_2	=	Mass transfer coefficient, 1/s
k_f	=	Film resistance mass transfer coefficient, m/s
K	=	Volume equilibrium constant, dimensionless
L	=	Thickness of the bed, m
n	=	Concentration of solute adsorbed on the fibers, kg /m^3
N	=	Dimensionless concentration of solute adsorbed on fibers, ($= n / N_0$)
N'	=	Dimensionless parameter, ($= N_0 / C_0$)
N_0	=	Initial concentration of solute adsorbed on the fibers, kg /m^3
P^*	=	Dimensionless parameter, $= k_1 L / u$
Pe	=	Peclet number, $= uL / D_L$, dimensionless
q	=	Intrapore solute concentration, kg /m^3
Q	=	Dimensionless intrapore solute concentration, ($= q / C_0$)
R	=	Pore radius of fibers, m
t	=	Time, s
u	=	Interstitial wash liquid velocity through bed, m /s
z	=	Distance from point of introduction of solvent, m

Greek Symbols

β	=	Fiber porosity, dimensionless
ε	=	Porosity of bed, dimensionless
θ	=	Dimensionless parameter, $= 2(1-\varepsilon)/\varepsilon$, dimensionless
ξ	=	Dimensionless axial distance, $= z / L$, dimensionless
τ	=	Dimensionless time, $= tu / L$, dimensionless

REFERENCES

1. M. Al-Jabari, A.R.P. Van Heiningen and T.G.M. Van De Ven, *Journal of Pulp and Paper Science*, **20** (9), J249-J253 (1994).
2. S. Arora, S.S. Dhaliwal and V.K. Kukreja, *Comp. & Chem. Engr.*, **30** (6-7), 1054-1060 (2006).
3. L.P. Ding and S.K. Bhatia, *A.I.Ch.E*, **49** (5), 883-895 (2003).
4. G. Eriksson, A. Rasmuson and H. Theliander, *Seperation Technology*, **6**, 201-210 (1996).
5. F. Liu and S.K. Bhatia, *Chem. Engr. Sci.*, **56**, 3727-3735 (2001).
6. F. Potucek, *Coll. of Czech. Chem. Comm.*, **62**, 626 – 644 (1997).
7. N.S. Raghvan and D.M. Ruthven, *A.I.Ch.E*, **29** (6), 922-925 (1983).
8. P. Sridhar, N.V.S. Sastri, J.M. Modak, and A.K. Mukherjee, *Chem. Engr. & Tech.*, **17**, 422-429 (1994).

A New Differential Evolution Algorithm and Its Application to Real Life Problems

Millie Pant, Musrrat Ali and V. P. Singh

Department of Paper Technology,
Indian Institute of Technology Roorkee, Saharanpur Campus, Saharanpur – 247001, India.
millifpt@iitr.ernet.in,   musrrat.iitr@gmail.com ,   singhvp2@yahoo.co.in

Abstract. Most of the real life problems occurring in various disciplines of science and engineering can be modeled as optimization problems. Also, most of these problems are nonlinear in nature which requires a suitable and efficient optimization algorithm to reach to an optimum value. In the past few years various algorithms has been proposed to deal with nonlinear optimization problems. Differential Evolution (DE) is a stochastic, population based search technique, which can be classified as an Evolutionary Algorithm (EA) using the concepts of selection crossover and reproduction to guide the search. It has emerged as a powerful tool for solving optimization problems in the past few years. However, the convergence rate of DE still does not meet all the requirements, and attempts to speed up differential evolution are considered necessary. In order to improve the performance of DE, we propose a modified DE algorithm called DEPCX which uses parent centric approach to manipulate the solution vectors. The performance of DEPCX is validated on a test bed of five benchmark functions and five real life engineering design problems. Numerical results are compared with original differential evolution (DE) and with TDE, another recently modified version of DE. Empirical analysis of the results clearly indicates the competence and efficiency of the proposed DEPCX algorithm for solving benchmark as well as real life problems with a good convergence rate.

Keywords: Stochastic optimization, differential evolution, mutation operation, crossover.

1. INTRODUCTION

Most of the real life problems occurring in the various disciplines of science and engineering are modeled as non-linear optimization problems. These non-linear optimization problems may further be classified broadly as unimodal and multimodal. Multimodal problems are generally considered more difficult to solve because unlike unimodal problems they are assisted with several local and global optima. Ideally, the user is interested in determining the global optimal solution because it gives the true objective function value. Most of the traditional methods available in literature are gradient based search techniques which require auxiliary properties like differentiability and continuity of the objective function and also they do not guarantee to obtain the global optimum. In the past few years scientist and researchers have laid emphasis on nontraditional optimization techniques which can obtain the global optimum solution and are also independent of the nature of the problems.

Evolutionary Algorithms (EAs) are general-purpose stochastic search methods imitating the phenomena of biological evolution. These are also known as population based search techniques as they work with a population of solutions rather than a single solution, which makes them different from optimization methods, such as hill-climbing [1] and simulated annealing [2].

CP1146, *Modelling of Engineering and Technological Problems,* edited by A. H. Siddiqi, A. K. Gupta, and M. Brokate
© 2009 American Institute of Physics 978-0-7354-0683-4/09/$25.00

One of the reasons of the success of EAs is their population based strategy which prevents them from getting trapped in a local optimal solution and consequently increases their probability of finding a global optimal solution. Thus, EAs can be viewed as global optimization algorithms. Some frequently used EAs include Evolutionary Programming (EP) [3], Evolution Strategies (ES) [4] and Genetic Algorithms (GA) [5].

DE, proposed by Storn and Price [6], is comparatively a newer addition to the class of EAs. DE is similar to GAs in the sense that it uses same evolutionary operators like mutation and crossover for guiding the particles towards the optimum solution. Nevertheless, it's the application of these operators that makes DE different from GA. The main difference between GAs and DE is that; in GAs, mutation is the result of small perturbations to the genes of an individual while in DE mutation is the result of arithmetic combinations of individuals. Also in DE, mutation plays a prominent role whereas, in GA, crossover is the major operator. At the beginning of the evolution process, the mutation operator of DE favors exploration. As evolution progresses, the mutation operator favors exploitation. Hence, DE automatically adapts the mutation increments (i.e. search step) to the best value based on the stage of the evolutionary process. Mutation in DE is therefore not based on a predefined probability density function.

DE is easy to implement, requires little parameter tuning and exhibits fast convergence. It has been successfully applied to solve a wide range of optimization problems such as clustering [7], unsupervised image classification [8], digital filter design [6], optimization of non-linear functions [9], global optimization of non-linear chemical engineering processes [10] and multi-objective optimization [11] etc.

Although, DE has given good results in most of the real life and test problems, researchers have observed that its convergence rate do not meet the desired expectations in some of the cases. In order to overcome this drawback of DE, the authors proposed a modified version called DEPCX [12], which uses parent centric approach to generate new solution vectors. This modification provides a measure to tune the balance between the convergence rate and the robustness of the algorithm. It accelerates the convergence velocity of DE algorithm locally without compromising the quality of solution. Encouraged by the preliminary results, in the present study we have extended the DEPCX for solving real life problems and have discussed the results in more detail.

Remaining of the paper is organized as follows; in Section 2, we give a brief description of DE. Section 3, describes the PCX operator. In section 4, we explain the proposed DEPCX algorithm. Benchmark problems and real life problems are given in section 5. Section 6 deals with experimental settings and parameter selection. Section 7 makes comparison of algorithms. The paper finally concludes with section 8.

2. DIFFERENTIAL EVOLUTION (DE)

Standard In this section, we briefly describe the classical DE algorithm as given by Storn and Price. DE attempts to replace each point in population set S with a new better point. Therefore, in each generation, NP (population size) competitions are held to determine the members of S for the next generation. The ith (i = 1, 2, . . . , NP) competition is held to replace $X_{i,G}$ in S. Considering $X_{i,G}$ as the target point, a trial point $U_{i,G+1}$ is found from two points (parents), the point $X_{i,G}$, i.e., the target point and the mutated point $V_{i,G+1}$ are determined by the mutation operation. In its mutation phase, DE randomly selects three distinct points from the population. The ith perturbed individual, $V_{i,G+1}$, is therefore generated based on the three chosen individuals as follows:

$$V_{i,G+1} = X_{r3,G} + F * (X_{r1,G} - X_{r2,G}) \qquad (1)$$

Where, $i = 1 \ldots$ NP, r1, r2, r3 $\in \{1, \ldots, NP\}$ are randomly selected such that $r_1 \neq r_2 \neq r_3 \neq i$. Scaling factor F (F $\in [0, 1.2]$) is a control parameter of the DE algorithm. The perturbed individual, $V_{i,G+1} = (v_{1,i,G+1}, \ldots, v_{D,i,G+1})$, and the current population member, $X_{i,G} = (x_{1,i,G}, \ldots, x_{D,i,G})$, are then subject to the crossover operation, that finally generates the population of candidates, or "trial" vectors, $U_{i,G+1} = (u_{1,i,G+1}, \ldots, u_{D,i,G+1})$, as follows:

$$u_{j,i,G+1} = \begin{cases} v_{j,i,G+1} & \text{if } rand_j \leq C_r \vee j = k \\ x_{j,i,G} & \text{otherwise} \end{cases} \qquad (2)$$

Where j, k $\in \{1,\ldots, D\}$ k is a random parameter index, chosen once for each i, Cr is crossover factor $\in [0, 1]$.

The population for the next generation is selected from the individuals in current population and its corresponding trial vector according to the following rule:

$$X_{i,G+1} = \begin{cases} U_{i,G+1} & \text{if } f(U_{i,G+1}) \leq f(X_{i,G}) \\ X_{i,G} & \text{otherwise} \end{cases} \qquad (3)$$

Thus, each individual of the temporary (trial) population is compared with its counterpart in the current population. The one with the lower objective function value survives the tournament selection and go to the next generation. As a result, all the individuals of the next generation are as good or better than their counterparts in the current generation. A notable point in DE's selection scheme is that a trial vector is not compared against all the individuals in the current generation, but only against one individual, its counterpart, in the current generation.

3. PARENT CENTRIC CROSSOVER

This Parent centric crossover (PCX) was first given by Deb [13, 14]. It is a multiparent operator consisting of more than one parent. Its working may be described as follows: in the beginning μ (say three) parents are selected for which the mean vector $\vec{g}$ is computed. From the three chosen parents, one parent $\vec{x}_{p,G}$ is selected at random to generate offspring and direction vector $\vec{d}_p = \vec{x}_{p,G} - \vec{g}$ is calculated. From the remaining (μ-1) parents perpendicular distances D_i to the line $\vec{d}_p$ are computed and their average $\overline{D}$ is found. The offspring is created as follows:

$$V_{i,G+1} = \vec{x}_{p,G} + w_\varsigma \vec{d}_p + \sum_{i=1,i\neq p}^{\mu} w_\eta \overline{D} \vec{e}_i \qquad (4)$$

Where $\vec{e_i}$ are the (μ-1) orthonormal bases that span the sub space perpendicular to $\vec{d_p}$. The parameters w_ς and w_η are zero mean normally distributed variables with variances σ_ς^2 and σ_η^2 respectively.

4. PROPOSED DEPCX

The proposed DEPCX is a simple and modified version of the original DE algorithm. The PCX operator described in section 3 creates new candidate solutions (children) near by the parent vectors. In [13, 14], PCX is a normal crossover operator, whereas in DEPCX it is treated as a mutant vector and is applied to each candidate solution vector. Thus DEPCX differs from original DE only in mutation phase. Crossover and selection scheme for DEPCX is same as in original DE. The presence of PCX operator in DE helps in enhancing its performance by giving it a chance of exploring its neighborhood more efficiently. The pseudo code of DEPCX algorithm is given here:

```
1. Randomly initialize the population Pɢ
2. REPEAT Until search converged
3.         REPEAT for each individual I∈Pɢ
4.              Mutation operation by equation (4).
5.              Crossover operation.
6.              Evaluation of the function.
7.              Selection.
8.         ENDREPEAT
9. ENDREPEAT.
```

5. BENCHMARK and REAL LIFE PROBLEMS

5.1 Benchmark Problems

The performance of proposed DEPCX is evaluated on a test bed of five standard, unconstrained benchmark problems taken from the literature [15]. All the test problems are scalable and are tested for 10, 50 and 100 number of variables. Except for the first function f1, all the other test functions are highly multimodal in nature. The degree of multimodality for functions f2 – f5, keeps on increasing with the increase in the number of variables thereby, increasing the complexity of the problem. Function f4 is a noisy function due to the presence of a uniformly distributed random noise.

5.2 Real Life Problems

The credibility of an optimization algorithm can be validated only if it is suitable for solving real life problems along with the test problems. In the present study we considered five problems that are common in various fields of engineering designs.

All the problems are highly nonlinear in nature. Mathematical models of the problems are given below:

TABLE 1: benchmark problems

Function	Dim	Ranges	Min value
$f_1(X) = \sum_{i=1}^{n} x_i^2$	10/ 50/ 100	[-5.12,5.12]	0
$f_2(X) = -20 * \exp\left(-.2\sqrt{1/n \sum_{i=1}^{n} x_i^2}\right) - \exp\left(1/n \sum_{i=1}^{n} \cos(2\pi x_i)\right) + 20 + e$	10/ 50/ 100	[-32, 32]	0
$f_3(X) = \sum_{i=1}^{n}\left(-x_i \sin\left(\sqrt{|x_i|}\right)\right)$	10/ 50/ 100	[-500, 500]	- 418.9829*n
$f_4(X) = \left(\sum_{i=1}^{n} i \times x_i^4\right) + rand[0,1[$	10/ 50/ 100	[-1.28,1.28]	0
$f_5(X) = \sum_{i=1}^{n}\left(x_i^2 - 10\cos(2\pi x_i) + 10\right)$	10/ 50/ 100	[-5.12,5.12]	0

1. *Gas transmission compressor design [16].*

Min $f(x) = 8.61 \times 10^5 \times x_1^{1/2} x_2 x_3^{-2/3} (x_2^2 - 1)^{-1/2} + 3.69 \times 10^4 \times x_3 + 7.72 \times 10^8 \times x_1^{-1} x_2^{0.219}$
$-765.43 \times 10^6 \times x_1^{-1}$

Bounds: $10 \le x_1 \le 55$, $1.1 \le x_2 \le 2$, $10 \le x_3 \le 40$

2. *Optimal capacity of gas production facilities [16].*

Min $f(x) = 61.8 + 5.72x_1 + 0.2623[(40 - x_1)\ln(\frac{x_2}{200})]^{-0.85}$

$+ 0.087(40 - x_1)\ln(\frac{x_2}{200}) + 700.23x_2^{-0.75}$

Subject to: $x_1 \ge 17.5$, $x_2 \ge 200$.

Bounds: $17.5 \le x_1 \le 40$, $300 \le x_2 \le 600$.

3. *Design of a gear train [17].*

The compound gear train problem shown in Fig. 1. It is to be designed such that the gear ratio is as close as possible to 1/6.931. For each gear the number of teeth must be between 12 and 60. Since the number of teeth is to be an integer, the variables must be integers.

The mathematical model of gear train design is given by,

Min $f = \left\{\dfrac{1}{6.931} - \dfrac{T_d T_b}{T_a T_f}\right\}^2 = \left\{\dfrac{1}{6.931} - \dfrac{x_1 x_2}{x_3 x_4}\right\}^2$

Subject to: $12 \le x_i \le 60$ $i = 1,2,3,4$

$[x_1, x_2, x_3, x_4] = [T_d, T_b, T_a, T_f]$, x_i's should be integers.

Where T_a, T_b, T_d, and T_f are the number of teeth on gears A, B, D and F respectively.

181

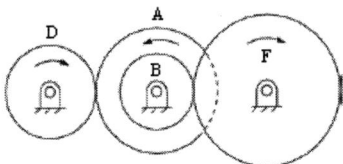

FIGURE 1. Compound Gear Train

4. *Optimal thermohydralic performance of an artificially roughened air heater [18].*

$$\text{Max } L = 2.51 * \ln e^{+} + 5.5 - 0.1 R_M - G_H$$

Where $R_M = 0.95 x_2^{0.53}$; $GH = 4.5(e^{+})^{0.28}(0.7)^{0.57}$; $e^{+} = x_1 x_3 (\bar{f}/2)^{1/2}$; $\bar{f} = (f_s + f_r)/2$

$f_s = 0.079 x_3^{-0.25}$; $f_r = 2(0.95 x_3^{0.53} + 2.5 * \ln(1/2x_1)^2 - 3.75)^{-2}$;

Bounds: $0.02 \le x_1 \le 0.8$, $10 \le x_2 \le 40$, $3000 \le x_3 \le 20000$

5. *Frequency modulation sound parameter identification [19].*

The problem is to satisfy six parameters $a_1, w_1, a_2, w_2, a_3, w_3$ of the frequency modulation sound model is given here.

$$y(t) = a_1 \times \sin\left(w_1 \times t \times \theta + a_2 \times \sin\left(w_2 \times t \times \theta + a_3 \times \sin\left(w_3 \times t \times \theta\right)\right)\right)$$

With $\theta = (2.\pi/100)$. The fitness function is defined as the sum of square error between the evolved data and the model data as follows:

$$f\left(a_1, w_1, a_2, w_2, a_3, w_3\right) = \sum_{t=0}^{100} \left(y(t) - y_0(t)\right)^2$$

Where the model data are given by the following equation.

$$y_0(t) = 1.0 \times \sin\left(5.0 \times t \times \theta + 1.5 \times \sin\left(4.8 \times t \times \theta + 2.0 \times \sin\left(4.9 \times t \times \theta\right)\right)\right)$$

Each parameter is in the range (-6.4, 6.35). This is highly complex multimodal problem with minimum value zero.

6. EXPERIMENTAL SETTINGS and PARAMETER SELECTION

Population size for all the algorithms is taken as NP (=3*D), where D is the dimension of the problem. Crossover rate (Cr) and scaling factor (F) are fixed at 0.01 and 0.5 respectively; for TDE, the trigonometric mutation probability Pt is 0.01; Maximum generations for all the algorithms are fixed at 8000; PCX parameters σ_ς and σ_η are taken as 0.1 each. All the algorithms were run 30 times for each of the test problems to determine the mean fitness and diversity. In every case, a run was terminated when the function values of all points in population S were identical to an accuracy of four decimal places, i.e. $|f_{max} - f_{min}| \le \varepsilon = 10^{-4}$ or when the maximum generation was reached. We do not claim these values to be the optimal for any given problem, in general, but empirically they are good values to choose. All the algorithms are executed on a Pentium IV PC using Dev C++.

7. NUMERICAL RESULTS and COMPARISONS of ALGORITHMS

The performance is DEPCX is compared with the classical DE and with Trigonometric Differential Evolution TDE [20]. TDE is one of the latest versions of DE and has reportedly given better performance than the original DE.

Numerical results of benchmark and real life problems are reported in Table 2 and Table 3 respectively. For each function the results are arranged as mean fitness and average number of generations of thirty runs and dimension ten, fifty and hundred. In figure 2, the graphs represent the mean of the best evaluation in thirty runs for the functions of fifty dimensions. Because of limited space, only some representative graphs for different functions are presented.

From the numerical results given in Table 2, it can be observed that out of the 15 test cases considered, DEPCX gave superior solution in 7 cases (success rate = 46.2%) in comparison to TDE and DE. TDE and DE gave good performances in 2 and 3 cases respectively. TDE gave better performance than DEPCX and DE for 100 variables problems. Surprisingly, for f4 which is a noisy function, DE gave best results followed closely by DEPCX and TDE. Also we would like to add that in terms of fitness function values, the performance of all the algorithms vary from each other only marginally. However, the performance comparison of the algorithms becomes more visible when the convergence rate is considered. It can be seen that for all the test problems, except for f4, DEPCX gave a visibly better convergence rate. Similar observations can be made from Table 3, where the results of real life problems are given. For all the real life problems DEPCX converged at a much faster rate in comparison to the other algorithms.

TABLE 2: Mean of fitness and generation for all algorithms on benchmark problems of dimensionality 10, 50,100(mean of 30 runs) for each problem.

Fun	Dim.	DE		DEPCX		TDE	
		Fitness	Gen	Fitness	Gen	Fitness	Gen
f_1	10	2.43176e-006	173	**7.15463e-008**	**125**	3.00576e-006	161
	50	7.80355e-006	925	**3.23314e-006**	**750**	5.55495e-006	879
	100	**3.0074e-005**	1175	4.60153e-005	1550	3.42414e-005	1550
$f2$	10	8.91482e-006	339	**3.32626e-006**	**200**	1.17702e-005	300
	50	3.29725e-005	1575	**1.99416e-005**	**1300**	2.6708e-005	1466
	100	0.000124886	2800	8.60006e-005	**2600**	**7.28019e-005**	2550
$f3$	10	-4189.83	275	-4189.83	**175**	-4189.83	265
	50	-20949.1	1350	-20949.1	1475	-20949.1	**1300**
	100	-41898.3	2750	-41898.3	2525	-41898.3	2550
$f4$	10	**0.00065363**	8000	0.00155938	8000	0.000671257	8000
	50	0.022227	8000	**0.0136434**	8000	0.0161357	8000
	100	**0.0525307**	8000	0.0538623	8000	0.0582792	8000
$f5$	10	1.17102e-006	275	**5.42596e-008**	**250**	6.52855e-007	281
	50	6.72631e-006	1600	**3.06816e-006**	**1389**	5.91064e-006	1425
	100	3.46447e-005	3250	3.90783e-005	**2830**	**1.05083e-005**	3050

TABLE 3: Mean of fitness and Generation for all algorithms for real life problems

Fun	Dim.	DE		DEPCX		TDE	
		Fitness	Mean Gen.	Fitness	Mean Gen.	Fitness	Mean Gen.
f_1	3	2.96438e+006	937.6	2.97765e+06	**98**	2.96438e+06	571.8
$f2$	2	169.844	51	**169.846**	**38**	169.847	67.4
$f3$	4	1.76382e-008	38.6	6.5688e-009	**26**	1.40671e-09	37
$f4$	3	**4.21422**	93.2	4.20779	**50.2**	4.21422	75
$f5$	6	3.01253	8000	**5.35737**	8000	3.95712	8000

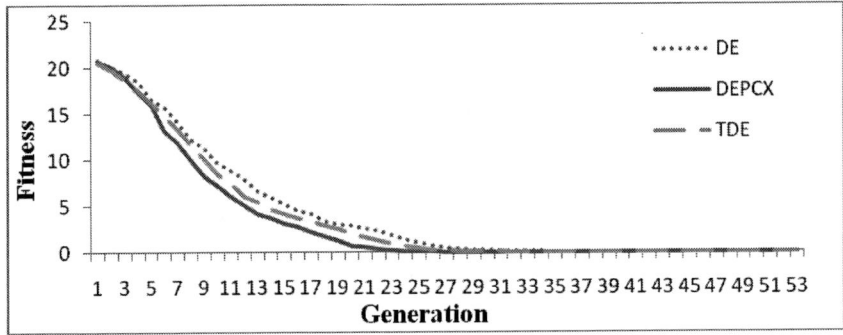

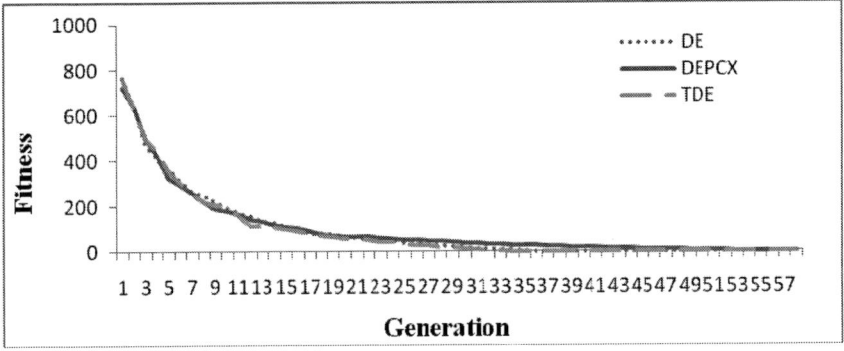

FIGURE 2: fitness Vs generation for function f2 and f5 of fifty dimension.

8. CONCLUSION

A parent centric approach to DE is proposed in this paper in order to improve the convergence rate without compromising with the quality of the solution. The proposed DEPCX algorithm helps in attaining the better convergence rate in comparison to DE and TDE with a good solution quality. The parent centric approach i.e. getting a solution nearby the parent helps in exploration of the neighborhood in order to get a better candidate for the next generation. We would like to maintain that although in terms of solution quality, no algorithm can be claimed as a clear winner but in terms of convergence rate DEPCX definitely gave a better performance. In future we shall apply DEPCX for solving more complex problems and also we will extend it for solving constrained problems.

REFERENCES

[1] Z. Michalewicz, D. Fogel, "How to Solve It: Modern Heuristics". Springer, Berlin. 2000.

[2] Van Laarhoven, P. Aarts,"Simulated Annealing: Theory and Applications." Kluwer Academic Publishers 1987.

[3] L.Fogel," Evolutionary programming in perspective: The top-down view. In: Zurada, J.M., Marks, R. Jr., Robinson, C. (Eds.), Computational Intelligence: Imitating Life. IEEE Press, Piscataway, NJ, USA. 1994.

[4] T.Back, F. Hoffmeister, H.Schwefel, "A survey of evolution strategies." In: Proceedings of the Fourth International Conference on Genetic Algorithms and their Applications, 1991.pp. 2–9.

[5] D. Goldberg, "Genetic Algorithms in Search Optimization and Machine Learning." Addison-Wesley. 1989.

[6] R.Storn,"Differential evolution design for an IIR-filter with requirements for magnitude and group delay". Technical Report TR-95-026, International Computer Science Institute, Berkeley, CA 1995.

[7] S.Paterlini, T.Krink, "High performance clustering with differential evolution." In: Proceedings of the IEEE Congress on Evolutionary Computation, vol. 2, 2004, pp. 2004–2011.

[8] M.Omran, A.Engelbrecht, A.Salman, Differential evolution methods for unsupervised image classification." In: Proceedings of the IEEE Congress on Evolutionary Computation, vol. 2, 2005a, pp. 966–973.

[9] B. Babu, R.Angira, "Optimization of non-linear functions using evolutionary computation". In: Proceedings of the 12th ISME International Conference on Mechanical Engineering, India, 2001, pp. 153–157.

[10] R.Angira, B.Babu, "Evolutionary computation for global optimization of non-linear chemical engineering processes." In: Proceedings of International Symposium on Process Systems Engineering and Control, Mumbai, 2003,pp. 87–91.

[11] H. Abbass, "A memetic pareto evolutionary approach to artificial neural networks" Lecture Notes in Artificial Intelligence, vol. 2256. Springer, 2002a, pp. 1–12.

[12] M.Pant, M.Ali, V.P.Singh, "Differential evolution with parent centric crossover" Second UKSIM European symposium on computer modeling and simulation, Liverpool UK, 2008, pp 141-146.

[13] K Deb, A.Anand D Joshi "A computationally efficient evolutionary algorithm for real-parameter optimization." Evol Comput J 2002, 10(4): 371–395.

[14] K Deb, "A population-based algorithm-generator for real-parameter optimization" .soft Comput J 2005, 9: 236–253.

[15] X.Yao, Y.Liu, "Fast evolutionary programming" in L.J.Fogel, P.J.Angeline and T.Back, editors, Proceeding of the 5th Annual Conference on Evolutionary Programming MIT Press, 1996, pp 451-460.

[16] C.S.Beightler, D.T.Phillips, " Applied geometric programming", john wiley and sons, new York 1976.

[17] B.V.Babu, "new optimization techniques in engineering", springer-verlag, berlin Heidelberg, 2004.

[18] B.N.Prasad, J.S.Saini, "Optimal thermo hydraulic performance of artificially roughened solar air heater", journal solar e, butergy, 1991 (47), pp 91-96.

[19] C.G.Martinez, M.Lozano, F. Herrera, D.Molina, A.M. Sanchez, " Global and local real coded genetic algorithms based on parent centric crossover operators", European journal of operational research 185, 2008, pp 1088-1113.

[20] Hui-Yuan Fan, Jouni Lampinen, "A Trigonometric Mutation Operation to Differential Evolution," Journal of Global Optimization 2003, 27:105-129.

SECTION C:
WAVELETS ANALYSIS WITH APPLICATIONS

Banach Gelfand Triples for Applications in Physics and Engineering

Hans G. Feichtinger

Faculty of Mathematics, NuHAG, University Vienna, AUSTRIA

Abstract.
The principle of extension is widespread within mathematics. Starting from simple objects one constructs more sophisticated ones, with a kind of natural embedding from the set of old objects to the new, enlarged set. Usually a set of operations on the old set can still be carried out, but maybe also some new ones. Done properly one obtains more completed objects of a *similar kind*, with additional useful properties. Let us give a simple example: While multiplication and addition can be done exactly and perfectly in the setting of $\mathbb{Q}$, the rational numbers, the field $\mathbb{R}$ of real numbers has the advantage of being complete (Cauchy sequences have a limit ...) and hence allowing for numbers like π or $\sqrt{2}$. Finally the even "more complicated" field $\mathbb{C}$ of complex numbers allows to find solutions to equations like $z^2 = -1$. The chain of inclusions of fields, $\mathbb{Q} \subset \mathbb{R} \subset \mathbb{C}$ is a good motivating example in the domain of "numbers".

The main subject of the present survey-type article is a new theory of Banach Gelfand triples (BGTs), providing a similar setting in the context of (generalized) functions. Test functions are the simple objects, elements of the Hilbert space $L^2(\mathbb{R}^d)$ are well suited in order to describe concepts of orthogonality, and they can be approximated to any given precision (in the $\|\cdot\|_2$-norm) by test functions. Finally one needs an even larger (Banach) space of *generalized functions* resp. *distributions*, containing among others pure frequencies and Dirac measures in order to describe various mappings between such Banach Gelfand triples in terms of the most important "elementary building blocks", in a clear analogy to the finite/discrete setting (where Dirac measures correspond to unit vectors).

Our concrete Banach Gelfand triple is based on the Segal algebra $S_0(\mathbb{R}^d)$, which coincides with the *modulation space* $M^1(\mathbb{R}^d) = M_0^{1,1}(\mathbb{R}^d)$, and plays a very important and natural role for time-frequency analysis. We will point out that it provides the appropriate setting for a description of many problems in engineering or physics, including the classical Fourier transform or the Kohn-Nirenberg or Weyl calculus for pseudo-differential operators. Particular emphasis will be given to the concept of w^*-convergence and w^*-continuity of operators which allows to prove conceptual uniqueness results, and to give a correct interpretation to certain formal expressions coming up in various versions of the Dirac formalism.

Keywords: Keywords: Banach Gelfand triples, Fourier transform, Kohn-Nirenberg Symbol, w^*-convergence, spectrogram
PACS: 02.30.Sa Functional analysis, 02.30.Tb Operator theory, 02.30.Nw Fourier analysis, 02.30.Jr Partial differential equations, 03.65.Ca QM Formalism

CP1146, *Modelling of Engineering and Technological Problems*, edited by A. H. Siddiqi, A. K. Gupta, and M. Brokate
© 2009 American Institute of Physics 978-0-7354-0683-4/09/$25.00

MOTIVATION AND INTRODUCTION

Although *Gelfand triples* such as $(\mathscr{S}, L^2, \mathscr{S}')$ resp. so-called *rigged Hilbert spaces* (Hilbert spaces endowed with an extra structure of surrounding spaces) have a certain tradition, mostly within theoretical physics, not much systematic mathematical investigation of this concept has been made. It is the purpose of the present paper to bring the advantages of the concept of *Banach Gelfand triples* to the attention of a wider community, to exhibit a concrete, simple and versatile example, coming from time-frequency analysis, and to show how natural it is. The concrete content of these notes is only indicative for the potential, both for the strict derivation of vague but somehow valid claims, but also for teaching purposes, in a context where not the full power of Lebesgue integration or the theory of nuclear topological vector spaces is available. In fact, we even believe that some of the involved mathematical concepts can be replaced by more natural and hence more simple ones.

We address physicists and engineers and mathematicians interested in applications or who have to teach students from the above community. While the applied scientists are often using symbolic expressions and derive in this ways valid identities the more strict mathematical view-point requires to have solid mathematical definitions, clear rules and valid logical concatenations of arguments, step by step. By suggesting the concept of Banach Gelfand triples (BGTs), which somehow extend the idea of rigged Hilbert spaces, we hope to offer a quite natural but very powerful tool, which allows to validate some of these heuristic ideas. One of the specific points emphasized is the relevance of w^*-convergence of sequences of generalized functions and $w^* - w^*$-continuity of operators. Intuitively this can be explained to an audio engineer as follows: A sequence σ_n of distributions converges to σ_0 in the w^*-sense if (and only if) the spectrum (the short-time Fourier transform) of σ_n with respect to any reasonable (say Gaussian) window is going to look more and more like the spectrum of σ_0 over larger and larger parts of the time-frequency plane.

Let us mention that this is a written realization of explanations and statements given at various occasions in talks on this subject during the last four years[1]. The material will be covered in much more detail in a forth-coming book publication by the author (jointly with G. Zimmermann, for Birkhäuser's NAHA series).

We also view this as a part of a series of publications, showing how to get from basic linear algebra concepts to time-frequency analysis, in particular to Gabor analysis ([65], the discretized and computationally relevant version of time-frequency analysis). It starts with the "Guided Tour from Linear Algebra to the Foundations of Gabor Analysis" ([56]), where the basic algebraic principles are explained using the standard concepts of *linear independence* and *generating systems of vectors*. It uses linear algebra terminology, and works in the setting of finite dimensional vector spaces (cf. e.g. [108]). In fact, finite vectors are understood as functions on the cyclic group $\mathbb{Z}_N$ of unit roots of order N, and the properties of involved matrices (e.g. PINV-matrices) show how to

---

[1] Most of them are downloadable from NuHAG Talk server
http://www.univie.ac.at/nuhag-php/program/talks.php

obtain implementations in an efficient way[2]. The algebraic theory is then pursued in [54] in the setting of general finite Abelian groups. Based on this the papers [55, 24] provide a refined view on the tools needed to handle the continuous case. Basic facts had been already presented in [53, 61] and above all in the book "Foundations of Time-Frequency Analysis" by K. Gröchenig ([69]).

Linear Algebra and Matrix Analysis

Coming from linear algebra we have learned to focus on *bases*, i.e. *coordinate systems* which allow to express *any vector* in a *unique way* as a *finite linear combination* of the elements of a basis. In matrix terminology this boils down to concentrate on *invertible* matrices A, which have the pleasant property of allowing for *every right hand side* b a *unique solution* x of the linear equation expressed[3] as $A * x = b$, or equivalently write b as a linear combination of column vectors of A. Solving for x is then possible in various ways, e.g. using Gauss elimination, but in MATLAB^{TM} one could simply use the command $x = inv(A) * b$. If one makes use of a scalar product on $\mathbb{R}^d$ or $\mathbb{C}^d$ one finds that some bases are much more convenient than others, because they allow for an effortless calculation of the coefficients of a vector, by calculating scalar products. Let $(u_k)_{k=1}^d$ in $\mathbb{C}^d$ be such an *orthonormal basis*, then we can form a matrix U, with these vectors as column vectors. The fact that $x = \sum_{k=1}^d \langle x, u_k \rangle u_k$ for all $x \in \mathbb{C}^d$ is then equivalent to the fact that [4] $U * U' = Id$, the unit matrix of size $n \times n$, or equivalently:

$$Id = \sum_{k=1}^d P_k, \quad \text{where} \quad P_k = u_k * u_k'. \tag{1}$$

Since for the case of square matrices any right inverse matrix is also a left inverse matrix this good property is indeed equivalent to $U' * U = Id$. This is compactly expressing the fact that the columns of U (and hence in fact also the rows) form an orthonormal set, or in terms of the individual elements of the Gramian matrix $G = U' * U$ and using Kronecker's δ-symbol:

$$\langle u_k, u_j \rangle = \delta_{k,j}. \tag{2}$$

Much of this *spirit of doing linear algebra*, i.e. to work in the setting of finite dimensional vector spaces, using bases to expand vectors, or matrices in order to describe linear mappings, is simulated in the *bra-ket formalism* going back to Paul Dirac. This allows for continuous integrals instead of (finite or infinite) sums, keeping in mind the dual use of vectors, either as building blocks for *synthesis* (as with matrix multiplication $A \mapsto A * x$, building linear combinations of the column vectors of A) or *analysis*, taking scalar products with the same set of vectors, by forming $y \mapsto A' * y$. Unfortunately this freedom makes things occasionally quite vague, due to a couple of new problems:

---

[2] E.g. within the MATLAB^{TM} software.

[3] Here $*$ denotes matrix multiplication, following the convention used by MATLAB.

[4] We use the MATLAB convention of writing U' for the transpose, conjugate matrix of U.

1. vectors and operators are expanded as a *continuous superposition* (in terms of integrals) of certain building blocks instead of a series or sum;
2. the meaning of these integrals is not obvious (Riemann, Lebesgue);
3. the building blocks may not belong to the Hilbert space anymore;
4. hence scalar products between two such elements are not meaningful a priori;
5. one may even have problems with the domain of the rank-one operators;
6. as in the finite-dimensional case, one may have orthonormality *without completeness* (and vice versa); however, in the infinite dimensional setting one cannot argue with dimensions.

Frames and Riesz Bases in Hilbert Spaces

Let us therefore describe an intermediate step, where we have collections of vectors in a Hilbert space $\mathscr{H}$, for which the synthesis and/or the analysis mapping make sense, as bounded linear mappings between $\mathscr{H}$ and $\ell^2 = \ell^2(I)$ for some (countable) index sequence I. We will see concrete examples (Gabor families) in a moment.

Definition 1. A family $(g_i)_{i \in I}$ in a Hilbert space $\mathscr{H}$ is called a *Bessel family* if the *analysis* mapping $\mathbf{C} : f \mapsto (\langle f, g_i \rangle)_{i \in I}$ is bounded from $\mathscr{H}$ into $\ell^2(I)$, i.e. if and only if there exists some positive constant $B > 0$ such that

$$\|\mathbf{C}f\|^2_{\ell^2(I)} = \sum_{i \in I} |\langle f, g_i \rangle|^2 \leq B\|f\|^2_{\mathscr{H}} \quad \text{for all } f \in \mathscr{H}. \tag{3}$$

By adjointness this is the case if and only if the corresponding *synthesis mapping* $\mathbf{R} : \mathbf{c} = (c_i)_{i \in I} \mapsto \sum_{i \in I} c_i g_i$ is bounded. Using standard terminology known from O. Christensen's book ([19]) one defines:

Definition 2. A family $(g_i)_{i \in I}$ in a Hilbert space $\mathscr{H}$ is called a *frame* if there exist constants $A, B > 0$ such that for all $f \in \mathscr{H}$

$$A\|f\|^2 \leq \sum_{i \in I} |\langle f, g_i \rangle|^2 \leq B\|f\|^2. \tag{4}$$

Definition 3. A family $(g_i)_{i \in I}$ in $\mathscr{H}$ is called a *Riesz (basic) sequence* if $\sum_{i \in I} c_i g_i$ has a Hilbert space norm equivalent to the ℓ^2-norm of the sequence $(c_i)_{i \in I}$, i.e. if there exist constants $C, D > 0$ such that

$$C \cdot \|\mathbf{c}\|^2_{\ell^2} \leq \|\sum_{i \in I} c_i g_i\|^2_{\mathscr{H}} \leq D\|\mathbf{c}\|^2_{\ell^2} \quad \text{for all } \mathbf{c} \in \ell^2. \tag{5}$$

If $(g_i)_{i \in I}$ is a frame respectively Riesz sequence then the analysis mapping $\mathbf{C}$ respectively synthesis mapping $\mathbf{R}$ establishes an isomorphism between its domain Hilbert space and its closed(!) range within its target Hilbert space.

One easily shows that a family $(g_i)_{i \in I}$ in a Hilbert space $\mathscr{H}$ is a frame if and only if the so-called *frame operator* $\mathbf{S} := \mathbf{R} \circ \mathbf{C}$ is bounded and invertible, with bounded inverse.

Analogous results apply for Riesz basic sequences, with $\mathbf{C} \circ \mathbf{R}$ instead of $\mathbf{R} \circ \mathbf{C}$. Note that the fact that the composition $\mathbf{R} \circ \mathbf{C}$ (in the case of a frame) or $\mathbf{C} \circ \mathbf{R}$ (for a Riesz basic sequence) is invertible does *not* imply that $\mathbf{C}$ or $\mathbf{R}$ is invertible. However, if this is indeed the case we have: A family $(g_i)_{i \in I}$ is called a *Riesz basis* for $\mathcal{H}$ if it is both a frame and a Riesz sequence. In that case of course both $\mathbf{C}$ and $\mathbf{R}$ establish isomorphisms between $\mathcal{H}$ and $\ell^2(I)$.

It is not surprising that many of the concepts known from linear algebra extend first in a very natural way to (separable) Hilbert spaces $\mathcal{H}$ such as $L^2(\mathbb{R}^d)$. Instead of finite sequences of vectors (resp. functionals) one deals with infinite sequences and makes corresponding boundedness assumptions, which allow to establish a rather complete analogy to the *four spaces* concept proposed by G. Strang ([108]): The closed range of an operator and its adjoint together with the corresponding null-spaces can be used to have a full geometric understanding of the situation.

For the case of a frame, the situation is best described by the following diagram, where $\mathbf{C}$ is the analysis mapping $f \mapsto (\langle f, g_i \rangle)$, the operator $\tilde{\mathbf{R}}$ is defined as $\tilde{\mathbf{R}} := \mathbf{S}^{-1} \circ \mathbf{R}$, and $\mathbf{P} := \mathbf{C} \circ \tilde{\mathbf{R}}$ is the orthogonal projection from $\ell^2(I)$ onto the range of $\mathbf{C}$.

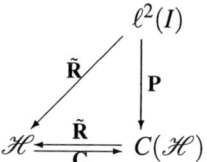

Hence $\tilde{\mathbf{R}} \circ \mathbf{C} = Id_{\mathcal{H}}$, i.e. $\tilde{\mathbf{R}}$ is a left-inverse to $\mathbf{C}$, the so called *Moore-Penrose inverse* to $\mathbf{C}$ (realized as PINV in MATLAB). Explicitly one finds for $f \in \mathcal{H}$

$$
\begin{aligned}
f &= Id_{\mathcal{H}} f = (\tilde{\mathbf{R}} \circ \mathbf{C}) f = (\mathbf{S}^{-1} \circ \mathbf{R} \circ \mathbf{C}) f \quad &(6) \\
&= \mathbf{S}^{-1} (\sum_{i \in I} \langle f, g_i \rangle g_i) = \sum_{i \in I} \langle f, g_i \rangle \mathbf{S}^{-1} g_i. \quad &(7)
\end{aligned}
$$

This motivates the definition of the so-called *dual frame* $(\tilde{g}_i)_{i \in I}$ by $\tilde{g}_i := \mathbf{S}^{-1} g_i$. Using the dual frame one can reconstruct any $f \in \mathcal{H}$ from its coefficients $\mathbf{C} f = (\langle f, g_i \rangle)$ as

$$
f = \sum_{i \in I} \langle f, g_i \rangle \tilde{g}_i, \quad (8)
$$

i.e. $\tilde{\mathbf{R}}$ is just the synthesis operator with respect to the dual frame $(\tilde{g}_i)_{i \in I}$.

For details concerning *frames* resp. *Riesz basic sequences* we refer to O. Christensen's book ([19]) or [77]. The definition of a frame can be generalized to also cover *continuous frames*, e.g. *coherent frames* obtained by the action of a continuous group on some reference vector. Instead of a discrete (typically countable) index set a measure space Ω is used, the mapping $\mathbf{C}$ is now an injective mapping into $L^2(\Omega)$ with closed range, and hence the same kind of diagram is still valid. This concept has made early appearence in the work of G. Kaiser ([86, 85]), and S.T. Ali, J.P. Antoine and J.P. Gazeau ([3, 4]). There are more recent papers on this subject by J.P. Gabardo and D. Han [64] or M. Fornasier and H. Rauhut in [63], discussing the transition from a (redudant) continuous to a (typically still redundant) discrete frame. Their work has been certainly inspired by the papers on coorbit theory by Feichtinger/Gröchenig ([48, 49]), which are also the basis for the first appearance of Banach frames in [67]. In this setting (the so-called *coorbit spaces*) concrete continuous frames appear in the context of irreducible, square-integrable group representations. Further generalizations of coorbit theory and continuous frames are treated in the work of S. Dahlke and his coauthors, [31, 30, 29].

Going beyond Frame Theory, towards Dirac

The Fourier transform is an important tool for both physics and engineering, making use of the "pure frequencies". What makes them so important is the fact that they are *eigenvectors* for the translation operators. Mathematicians like to consider the functions $\chi_s(t) = e^{2\pi i s \cdot t}$ as characters of the group $\mathbb{R}^d$, viewed as a LCA (= locally compact Abelian) group, with respect to addition of vectors. The exponential law implies that $\chi_s(x+y) = \chi_s(x) \cdot \chi_s(y)$, $x, y \in \mathbb{R}^d$. Since we have the pointwise relation $\chi_r \cdot \chi_s = \chi_{r+s}$ we find that the *dual group*, or *frequency domain* is just

$$\widehat{\mathbb{R}^d} = \{\chi_s \,|\, s \in \mathbb{R}^d\}. \tag{9}$$

Spectral synthesis and *spectral analysis* (or Fourier analysis, or harmonic analysis in more general terms, see [102, 11]) address the question whether one can *compose* any signal, function, distribution f from this (continuous) family of "elementary building blocks" by superposition (since we have a continuous parameter it is natural to think of an integral representation), and on the other hand, wether and how one can identify the required coefficients (amplitudes/spectral components) from the signal f.

As in linear algebra, one has to settle the problem whether *every function*, or more precisely, every element from a given (topological) vector space can be represented, and secondly whether the representation is unique. As we will see, the setting of BGTs will also allow to differentiate and decide which one of the objects (of different complexity) can be composed or decomposed in which concrete way, e.g. through integral representation, in the weak sense in the case of the Hilbert space $\mathscr{H} = L^2(\mathbb{R}^d)$ or in the w^*-sense within the dual space $S_0'(\mathbb{R}^d)$.

Although it would be more natural from the linear algebra view-point described above to start with the synthesis problem, we find it more natural (in the Fourier context) to start with the analysis part. After all, according to our philosophy the two operations are mutually adjoint to each other.

Following the usual path the Fourier transform $\mathscr{F}$ is defined on $L^1(\mathbb{R}^d)$, the Banach space of all absolutely Lebesgue-integrable functions (modulo null-functions) as an integral transform as follows:

$$(\mathscr{F}f)(s) \equiv \hat{f}(s) = \int_{\mathbb{R}^d} f(t) \overline{\chi_s(t)} \, dt = \langle f, \chi_s \rangle. \tag{10}$$

We will see later that it suffices to know it on some smaller spaces (such as the Schwartz space $\mathscr{S}(\mathbb{R}^d)$ or $S_0(\mathbb{R}^d)$, where it is enough to use the ordinary Riemannian integral).

Together with $\mathscr{F}$ we have to consider the adjoint mapping, i.e. *Fourier synthesis*, which in analogy to the situation in matrix analysis is given, at least for nice functions h, by

$$\mathscr{F}^* h = \int_{\mathbb{R}^d} h(s) \chi_s \, ds. \qquad (11)$$

This integral can be understood in the following sense: Each χ_s is a bounded and continuous function with $|\chi_s(t)| = 1$ for $t, s \in \mathbb{R}^d$, hence we have a pointwise well-defined function $(\mathscr{F}^* h)(t)$ if h is Riemann-integrable.

Although it is well known how to extend[5] to a unitary automorphism of $\left(L^2(\mathbb{R}^d), \|\cdot\|_2\right)$, thanks to the fundamental identity of the Fourier transform

$$\int_{\mathbb{R}^d} \hat{g}(y) f(y) \, dy = \int_{\mathbb{R}^d} \hat{f}(s) g(s) \, ds, \qquad (12)$$

it is clear that one has to expect a lot of trouble with the domains of $\mathscr{F}$ and $\mathscr{F}^*$, because the different domains do not fit (a typical element $\hat{f} \in \mathscr{F}L^1(\mathbb{R}^d)$ may not be integrable itself, e.g. if f has discontinuities) and because the elementary building blocks, the pure frequencies $(\chi_s)_{s \in \mathbb{R}^d}$, do *not belong* to the Hilbert space $L^2(\mathbb{R}^d)$. On the other hand it is tempting to describe this continuous family (as Dirac did in some sense) as a "continuous coordinate system", satisfying a kind of (distributional) orthogonality relation as well as a decomposition of the identity operator as a continuous integral of rank-one operators comparable to the pair (2) and (1). We will provide arguments towards a meaningful interpretation of such claims in the context of *Banach Gelfand triples*.

THE BANACH GELFAND TRIPLE (S_0, L^2, S_0')

The above observation already calls for a unified treatment of the Fourier transform in the finite as well as in the Euclidean setting, or even (according to A. Weil) in the setting of LCA (locally compact Abelian) groups, including the field of p-adic numbers (see [116, 83, 102]). It is also clear that one cannot - despite its importance - stay within the Hilbert space $L^2(\mathbb{R}^d)$ anymore. We will try to convince the reader that the concept of BGTs (= Banach Gelfand triples) is a good way out of this problem.

Banach Gelfand Triples and their Morphisms

Recall the famous formula $e^{2\pi i} = 1$. It would not have made sense to a Greek mathematician, even if he had perfect knowledge of the field $\mathbb{Q}$ of rationals. One has to be able to create irrational numbers such as π beforehand, and one also has to be able to extend addition and multiplication to the larger domain of complex numbers. Finally one has to have a canonical way to give a meaning to the power series expression (adding up

---

[5] By arguments quite similar to those used in the extension of multiplication of rational numbers to the domain $\mathbb{R}$.

infinitely many of those numbers) to reach the perfect calculus of exponential functions in the complex domain. We will follow a similar path, with the Hilbert space $\mathscr{H}$ (typically $L^2(\mathbb{R}^d)$) playing the role of $\mathbb{R}^d$ and the generalized functions being the analogue of the complex numbers. Before going to the concrete BGT (S_0, L^2, S_0') let us introduce the concept of Banach Gelfand triples in full generality (for the sake of simplicity we restrict our discussions to the case of *separable* Banach spaces).

Definition 4. We call a triple of vector spaces $(B, \mathscr{H}, B')$ a *Banach Gelfand triple* if $(B, \|\cdot\|_B)$ is Banach space, which is dense in some Hilbert space $\mathscr{H}$, and which in turn is contained in B', the dual of B.

There are many examples, and the basic fact is a natural embedding of the elements of B (usually the space of *test functions*) into its dual space B', the space of *generalized functions* or *distributions*[6].

Although the idea of *rigged Hilbert spaces* ([103, 6, 35, 120, 36, 2, 7]) is very close to our BGT concept there are two important differences First of all it is clear that we allow for Banach spaces instead of a Hilbert spaces of dual Hilbert spaces "surrounding" the central Hilbert space, nor any *nuclear topological* vector space, such as $\mathscr{S}(\mathbb{R}^d)$. The concrete example, starting from the space $S_0(\mathbb{R}^d)$ allows to obtain nevertheless a kernel theorem. We are not aware of any kernel theorem for rigged Hilbert spaces other than those using nuclear (hence not Banach or Hilbert) spaces. One can trace the validity of the kernel theorem back to the tensor product factorization property (Lemma 4), which in turn has to do with the "separation of variables" property in the Fourier algebra, which has been historically one of the highlights of J.B. Fourier's concept.

It has been expressed by several authors (cf. [36, 37, 66?]) that rigged Hilbert spaces (a triple of Hilbert spaces, forming a BGT in our sense) allows to describe valid identities which cannot be formulated in the Hilbert space setting alone[7] We take the same view-point, but emphasize the close connection between the inner Banach space and its dual by working with *four topologies*, i.e. by giving the (natural) w^*-topology on B' a prominent role. Note that the dual space for S_0' endowed with the w^*-topology (often denoted as the weak $\sigma(B', B)$ topology) is just B itself, and hence one has a kind of Riesz-representation theory for BGTs in the background. Furthermore it is helpful to recall that bounded (closed) subsets in B' are compact in this topology according to the theorem of Banach-Alaoglou ([106], section 3.15).

The *prototype* of a Banach Gelfand triple is $(\ell^1, \ell^2, \ell^\infty)(\mathbb{Z})$, where the w^*-topology describes coordinate-wise convergence, i.e. views ℓ^∞ as subset of $\mathbb{R}^{\mathbb{Z}}$ with the product topology in the sense of Tychonoff.

In fact, one may view Banach Gelfand triples as a new *category* in the spirit of MacLane ([96]), where the *morphisms* are the "structure preserving mappings", i.e.

---

[6] While smaller spaces of test functions give larger space of bounded linear functionals on them, one has to keep in mind that B is not degenerating, because then this construction breaks down. So for our purpose one may think of a situation where $\mathscr{S}(\mathbb{R}^d) \subset B$.

[7] For example, point evaluations do not make sense on $L^2(\mathbb{R}^d)$ while they make perfect sense on a Sobolev space, once the smoothness parameter satisfies $s > d/2$, according to Sobolev's embedding theorem.

linear mappings which are continuous with respect to each of the *four* (!) topologies.

Definition 5. If $(B_1, \mathcal{H}_1, B_1')$ and $(B_2, \mathcal{H}_2, B_2')$ are Gelfand triples then a linear operator T is called a [unitary] Gelfand triple isomorphism if

1. A is an isomorphism between B_1 and B_2.
2. A is [a unitary operator resp.] an isomorphism between $\mathcal{H}_1$ and $\mathcal{H}_2$.
3. A extends to a weak* isomorphism as well as a norm-to-norm continuous isomorphism between B_1' and B_2'.

In principle every ONB (= *orthonormal basis*) $\Psi = (\psi_i)_{i \in I}$ for a given Hilbert space $\mathcal{H}$ can be used to establish such a unitary isomorphism, by choosing as B the space of elements within $\mathcal{H}$ which have an absolutely convergent expansion, i.e. satisfy $\sum_{i \in I} |\langle x, \psi_i \rangle| < \infty$. Of course, this space, which deserves perhaps the symbol A_Ψ, depends on the choice of the orthonormal basis Ψ, but of course one has many *equivalent* bases describing the same space.

For the case of the perhaps most important ONB for $\mathcal{H} = L^2([0,1])$, i.e. for the trigonometric system, the corresponding definition is already around since the times of N. Wiener, who suggested to consider specifically $A(\mathbb{T})$, the space of absolutely continuous Fourier series, because it has very good and useful properties (compared to the Lebesgue space $(L^1(\mathbb{T}), \|\cdot\|_1)$, where e.g. the Fourier inversion is a non-trivial matter). It is also not surprising in retrospect to see that in the discussion the dual space $PM(\mathbb{T}) = A(\mathbb{T})'$ came up, the space of *pseudo-measures*. One can extend the Fourier transform to this space, and in fact interpret this extended mapping, in conjunction with the classical Plancherel theorem as the first unitary Banach Gelfand triple isomorphism, between $(A, L^2, PM)(\mathbb{T})$ and $(\ell^1, \ell^2, \ell^\infty)(\mathbb{Z})$. [8]

It is the main goal of this article to show how the use of the Banach algebra $S_0(\mathbb{R}^d)$ allows to have a similar interpretation of the Fourier transform (and many other mappings relevant for physics, engineering, or mathematical considerations in time-frequency analysis), how to make use of the w^*-concept and how to re-interpret the Dirac formalism in this context.

Having expounded the general theory of Banach Gelfand triples, we are now ready to introduce the constituents of a particularly useful example, namely the Banach Gelfand triple (S_0, L^2, S_0').

Modulation Spaces

The Banach space $(S_0(\mathbb{R}^d), \|\cdot\|_{S_0})$ of test functions to be used in the following is a particular instance of a class of function spaces studied in time-frequency analysis (TF-analysis), called *modulation spaces*. In order to define these spaces we have to recall some concepts from that field. The basic tools in TF-analysis are *time*- and *frequency shifts* (TF-shift) given by $T_x f(t) = f(t - x)$ and $M_\omega f(t) = e^{2\pi i \omega \cdot t} f(t)$, for functions f

---

[8] The Segal algebra $S_0(G)$, defined for general LCA (= locally compact Abelian) groups is in fact a generalization of this construction, i.e. $S_0(\mathbb{T}) = A(\mathbb{T})$.

on $\mathbb{R}^d$. They are combined to (unitary) *time-frequency shift* operators

$$\pi(\lambda) = \pi(x, \omega) = M_\omega T_x, \quad \text{for } \lambda = (x, \omega) \in \mathbb{R}^d \times \widehat{\mathbb{R}}^d. \tag{13}$$

Using these operators one defines (e.g. on $L^2(\mathbb{R}^d)$ or for continuous and absolutely Riemann-integrable functions) the *Short-Time Fourier Transform* as a function on the *time-frequency plane* ([69]) resp. *phase space* ([62]) in the following way:

$$V_g f(\lambda) = V_g f(x, \omega) = \langle f, M_\omega T_x g \rangle = \langle f, \pi(\lambda) g \rangle \text{ for } \lambda = (x, \omega) \in \mathbb{R}^d \times \widehat{\mathbb{R}}^d. \tag{14}$$

Modulation spaces occur in the study of the concentration of a function in the time-frequency plane, described in terms of function spaces over $\mathbb{R}^d \times \widehat{\mathbb{R}}^d$. The classical ones are defined as follows: Let $g \in \mathscr{S}(\mathbb{R}^d)$ be a Schwartz function, $1 \leq p, q < \infty, s \in \mathbb{R}$, then

$$M_s^{p,q}(\mathbb{R}^d) = \{f \in \mathscr{S}'(\mathbb{R}^d) : \|f\|_{M_s^{p,q}} < \infty\}, \tag{15}$$

where the norm $\|f\|_{M_s^{p,q}}$ on $M_s^{p,q}(\mathbb{R}^d)$ is given as

$$\|f\|_{M_s^{p,q}} := \left(\int \left(\int |\langle f, M_\omega T_x g \rangle|^p \, dx \right)^{q/p} (1 + |\omega|)^{sq} \, d\omega \right)^{1/q}, \tag{16}$$

i.e. for which $V_g f$ belongs to some weighted mixed-norm space over phase space. In the "classical" case the weight depends only on frequency, hence the spaces are isometrically translation invariant. The only important facts about the constraint imposed on $V_g f$ is the membership in a *solid*[9] and *translation invariant* Banach space over $\mathbb{R}^{2d}$. We use the abbreviations $M_s^p := M_s^{p,p}$ and $M^p := M_0^{p,p}$.

The modulation space $M_s^{p,q}(\mathbb{R}^d)$ is a Banach space of tempered distributions, the definition is independent of the analyzing function g, and different g's yield equivalent norms on these spaces. The Gauss function is a good choice. Among the modulation spaces are the following important function spaces:

(a) the space $S_0(\mathbb{R}^d)$ we are after is just $M_0^{1,1}(\mathbb{R}^d) = M^1(\mathbb{R}^d)$;

(b) $L^2(\mathbb{R}^d) = M_0^{2,2}(\mathbb{R}^d)$;

(c) the Bessel potential spaces $\mathscr{H}_s(\mathbb{R}^d)$, defined via the Fourier transform by

$$\mathscr{H}_s(\mathbb{R}^d) = \{f \in \mathscr{S}' : \int |\hat{f}(\omega)|^2 (1 + |\omega|)^{2s} \, d\omega < \infty\} \tag{17}$$

coincide with the modulation spaces $M_s^{2,2}(\mathbb{R}^d)$;

(d) the Shubin classes $Q_s(\mathbb{R}^d)$, which can be characterized by a weighted $L^2(\mathbb{R}^{2d})$-condition with respect to the radial symmetric weight over phase-space of the form $v_s(\lambda) = (1 + |\lambda|^2)^{s/2}$ instead of the usual weight $w_s(\omega) := (1 + |\omega|)^s$ resp. $(1 + |\omega|^2)^{s/2}$.

---

[9] In a solid space the norm behaves monotonically, i.e. $|F(x)| \leq |G(x)|$ for all $x \in \mathbb{R}^d$ implies that the norm of F is smaller than the norm of G.

A lot of details on these spaces can be found in the book of Gröchenig, and in the survey note [46] (written in 1983 and published in 2003).

The original description of the modulation spaces was in terms of generalized Wiener amalgams, on the Fourier transform side:

$$M_s^{pq}(\mathbb{R}^d) = \mathscr{F}^{-1}[W(\mathscr{F}L^p, \ell_{w_s}^q)(\mathbb{R}^d)],$$

or equivalently, Banach spaces of distributions obtained using BUPUs (bounded uniform partitions of unity, such as a collection of shifted B-splines) (cf. [43, 78]).

The Banach space $S_0(\mathbb{R}^d)$ and its Various Descriptions

In the following we will establish the basic properties of the Banach space of test functions on which our BGT-approach will be based[10]. $S_0(\mathbb{R}^d)$ can be described in many ways and many equivalent norms can be used to characterize this space. Originally (see [42]) it was introduced as the *Wiener amalgam space* $W(\mathscr{F}L^1, \ell^1)$ (see [43] for generalities of this concept), but the equivalence between discrete and "continuous" norms (using control functions) can be used to show that it coincides with the coorbit space (as developed in full generality in [48]) or with the *modulation space* $(M^1(\mathbb{R}^d), \|\cdot\|_{M^1})$ (see the book [69] for a good introduction to the subject in the context of time-frequency analysis). We will follow the description given there, going back to [45], published in 1989.

According to the description above we can define $S_0(\mathbb{R}^d) := M^1(\mathbb{R}^d)$ by means of the STFT with respect to the Gaussian window $g_0(t) = e^{-\pi|t|^2}$. This choice has the advantage that Fourier invariance of this space is easily verified. It is also not difficult to check that $M^1(\mathbb{R}^d) \subset L^1 \cap C_0(\mathbb{R}^d)$[11]. The following is an alternative definition not making reference to Lebesgue integrals (and thus suitable for applied courses):

Definition 6. $S_0(\mathbb{R}^d) := \{f \in C_0(\mathbb{R}^d) : f$ absolutely Riemann-integrable over $\mathbb{R}^d$, $V_{g_0}f$ absolutely Riemann-integrable over $\mathbb{R}^d \times \widehat{\mathbb{R}}^d\}$, with the norm $\|f\|_{S_0} := \|V_{g_0}f\|_{L^1}$.

An *atomic* characterization[12] also used by H. Reiter (see [101]) is

Theorem 1. *We call a function $f \in \mathscr{F}L^1(\mathbb{R}^d)$ an atom (on the time-side) if* $\operatorname{supp}(f) \subseteq B_1(0)$ *for some $x \in \mathbb{R}^d$. Then $S_0(\mathbb{R}^d)$ consists of all absolutely convergent sums of atoms, i.e. $f \in S_0(\mathbb{R}^d)$ if and only it has a representation as*

$$f = \sum_{n \geq 1} T_{x_n} f_n, \quad with \quad \sum_{n \geq 1} \|\widehat{f_n}\|_{L^1} < \infty. \tag{18}$$

---

[10] It is occasionally referred to as *Feichtinger's algebra* in the literature, see [102].

[11] We write $(C_0(\mathbb{R}^d), \|\cdot\|_\infty)$ for the space of continuous, complex-valued functions, vanishing at infinity, i.e. with $\lim_{|x|\to\infty} f(x) = 0$, endowed with the sup-norm $\|f\|_\infty := \sup_{x \in \mathbb{R}^d} |f(x)|$.

[12] This atomic characterization should be reminiscent of the atomic characterization of Hardy spaces, given by Coifman and Weiss ([23]).

Endowed with the natural norm, i.e.

$$\|f\|_{S_0} := \inf\left\{\sum \|f_n\|_{\mathscr{F}L^1(\mathbb{R}^d)} : f = \sum T_{x_n} f_n, \sum \|f_n\|_{\mathscr{F}L^1(\mathbb{R}^d)} < \infty\right\}. \qquad (19)$$

$S_0(\mathbb{R}^d)$ *is the smallest (non-trivial) Banach space* $(B, \|\cdot\|_B)$ *with the property* $\|\pi(\lambda)f\|_B = \|f\|_B$ *for all* $f \in B, \lambda \in \mathbb{R}^d \times \widehat{\mathbb{R}}^d$, *i.e. it is continuously embedded into any other space with this property*[13]. *Moreover* $S_0(\mathbb{R}^d)$ *is invariant under the Fourier transform.*

This is [41], Thm.1. See [94, 95] for further characterizations of $S_0(\mathbb{R}^d)$, and of course the book [69]. It was the clue for many other interesting properties of $S_0(\mathbb{R}^d)$, which are nowadays proved using TF-arguments. Among others one has the following characterization. Since $\hat{g}_0 = g_0$ it sheds some light on the Fourier invariance of $S_0(\mathbb{R}^d)$. Due to the Fourier invariance one can also avoid the $\mathscr{F}L^1$ norm $\|h\|_{\mathscr{F}L^1} := \|f\|_{L^1}$ for $h = \hat{f}$ by doing the decomposition into pieces of equal size on the Fourier transform side. In this way one achieves a description of $f \in S_0(\mathbb{R}^d)$ as a sum of band-pass signals. This is what Hans Reiter really used, in [100, 101].

Lemma 1. *All absolutely convergent series of time-frequency shifts of g_0 are contained in $S_0(\mathbb{R}^d)$, and even make up all of $S_0(\mathbb{R}^d)$, i.e.*

$$S_0(\mathbb{R}^d) = \left\{\sum_{n\in\mathbb{N}} a_n M_{\xi_n} T_{x_n} g_0 : (x_n, \xi_n) \in \mathbb{R}^d \times \widehat{\mathbb{R}}^d, (a_n)_{n\in\mathbb{N}} \in \ell^1(\mathbb{N})\right\}. \qquad (20)$$

Since the choice of the window in the definition of modulation spaces gives these definition some smell of arbitrariness, some people prefer the characterization of $S_0(\mathbb{R}^d)$ using the (quadratic) Wigner distribution as a suitable alternatively, despite the fact that from the description below it is a-priori not clear why $S_0(\mathbb{R}^d)$ should be a linear manifold. Let us recall the definition of the cross-Wigner distribution (see [20, 21, 22, 79, 80] for $f, g \in L^2(\mathbb{R}^d)$ first, with $z = (x, \xi)$:

$$W(f, g)(z) = \int_{\mathbb{R}^d} e^{-2\pi i \xi \cdot y} f(x + \tfrac{1}{2}y)\overline{g(x - \tfrac{1}{2}y)}\, dy. \qquad (21)$$

Lemma 2. $f \in S_0(\mathbb{R}^d)$ *if and only if the Wigner function $W(f, f) \in L^1(\mathbb{R}^{2d})$.*

Whereas some basic invariance properties of $S_0(\mathbb{R}^d)$, or properties like the restriction to subgroups or integration along subgroups can be derived quite easily (cf. [40]) the last criterion is the most useful for the derivation of metaplectic invariance (cf. last section).

The dual space $(S_0'(\mathbb{R}^d), \|\cdot\|_{S_0'})$

Together with the space $S_0(\mathbb{R}^d)$ of test functions we will have to consider its dual space, the collection of all bounded linear functionals on $(S_0(\mathbb{R}^d), \|\cdot\|_{S_0})$. Since $S_0(\mathbb{R}^d)$

---

[13] See [102, 41, 99, 101] for background on Segal algebras resp. the Segal algebra viewpoint on $S_0(\mathbb{R}^d)$. It is also the smallest strongly character invariant Segal algebra.

is the smallest isometrically TF-invariant Banach space its dual is essentially the largest space (of distributions) with this property. We will make use of this space, endowed with its standard norm respectively the w^*-topology.

We use the symbol $(S_0'(\mathbb{R}^d), \|\cdot\|_{S_0'})$ for the space of bounded linear functionals on $\left(S_0(\mathbb{R}^d), \|\cdot\|_{S_0}\right)$, where the norm for $\sigma \in S_0'(\mathbb{R}^d)$ is defined as usual by

$$\|\sigma\|_{S_0'} := \sup\{|\sigma(f)| : f \in S_0(\mathbb{R}^d), \|f\|_{S_0} \le 1\}. \tag{22}$$

Definition 7. A distribution $\sigma \in S_0'$ is *regular*, if there exists a locally integrable function $\varphi \in L^1{}_{\text{loc}}(\mathbb{R}^d)$ such that

$$\sigma(f) = \int_{\mathbb{R}^d} \varphi(t) f(t)\, dt \quad \text{for all } f \in S_0(\mathbb{R}^d). \tag{23}$$

In this case, we write $\sigma =: \sigma_\varphi$.

Here, we have $\sigma_\varphi = \sigma_\psi$ if and only if $\varphi(t) = \psi(t)$ almost everywhere. Regularity of a distribution does not necessarily imply that the integral in (23) is absolutely convergent for all $f \in S_0$. This holds for φ in an appropriate Wiener amalgam space, though.

Proposition 1. *For $\varphi \in W(L^1, \ell^\infty)$, we have that $\sigma_\varphi \in S_0'$ with $\|\sigma_\varphi\|_{S_0'} \le \|\varphi\|_{W(L^1,\ell^\infty)}$, and the integral in (23) is absolutely convergent for all $f \in S_0$.*

In particular, we see that spaces like S_0, L^p, $W(L^{p_1}, \ell^{p_2})$ are continuously embedded in S_0' for $1 \le p, p_1, p_2 \le \infty$, in the sense that for an element φ of one of these spaces, we have $\sigma_\varphi \in S_0'$, and the norm of σ_φ can be estimated from above by the respective norm of φ. For $p_2 = \infty$ this argument implies that we can even consider periodic functions $L^p(\mathbb{T}^d)$ as subspaces of $W(L^p, \ell^\infty) \subset S_0'(\mathbb{R}^d)$. In a similar way every bounded measure $\mu \in M_b(\mathbb{R}^d)$ can be identified with $\sigma_\mu \in S_0'$ via

$$\sigma_\mu(f) = \int_{\mathbb{R}^d} f(t)\, d\mu(t) \quad \text{for all } f \in S_0,$$

with $\|\sigma_\mu\|_{S_0'} \le C \|\mu\|_{M_b}$. In particular, all finite discrete measures define elements of S_0'. But there are also many other (unbounded) measures within $S_0'(\mathbb{R}^d)$, since the space of *translation-bounded* measures $W(M_b, \ell^\infty)$ is contained S_0'. For example, $\sqcup\!\sqcup_\Lambda :=$ $\sum_{\lambda \in \Lambda} \delta_\lambda \in S_0'(\mathbb{R})$ for any lattice $\Lambda \lhd \mathbb{R}^d$.

The standard methods for Wiener amalgam spaces (cf. [43, 78]) imply that $S_0'(\mathbb{R}^d)$ can be characterized as $W(\mathscr{F}L^\infty, \ell^\infty)(\mathbb{R}^d)$, the space of *translation bounded* quasi-measures, because $\mathscr{F}L^\infty(\mathbb{R}^d) = PM(\mathbb{R}^d) := \{\sigma = \mathscr{F}^{-1}h, \text{ for some } h \in L^\infty(\mathbb{R}^d)\}$, the space of *pseudo-measures*, coincides locally with the space of *quasi-measures* ([92, 39]).

There are also quite useful convolution relations, such as

$$S_0 * S_0' \subseteq W(\mathscr{F}L^1, \ell^\infty) = \mathscr{M}(S_0)(\mathbb{R}^d), \tag{24}$$

where $\mathscr{M}(S_0)$ are the pointwise multipliers of $S_0(\mathbb{R}^d)$. However, $S_0(\mathbb{R}^d)$ is not dense in $S_0'(\mathbb{R}^d)$ with respect to the norm topology and therefore we have to invoke a second, weaker topology on this dual space.

w^*-convergence in $S_0'(\mathbb{R}^d)$

A sequence $(\sigma_n)_{n\in\mathbb{N}}$ in $S_0'(\mathbb{R}^d)$ is w^*-convergent[14] to $\sigma_0 \in S_0'(\mathbb{R}^d)$, in symbols

$$\sigma_0 = w^* - \lim_n \sigma_n \tag{25}$$

if for every test function $f \in S_0(\mathbb{R}^d)$ one has

$$\lim_n \sigma_n(f) \to \sigma_0(f), \tag{26}$$

i.e. pointwise convergence of the sequence (σ_n) to some limit σ_0. The following equivalent characterization is valid for arbitrary Banach spaces:

Lemma 3. *A (bounded) sequence $(\sigma_n)_{n\in\mathbb{N}}$ in $S_0'(\mathbb{R}^d)$ is w^*-convergent to σ_0 if and only if for every compact $M \subset S_0(\mathbb{R}^d)$ and every $\varepsilon > 0$ one has: There exists some index n_0 such that $n \geq n_0$ implies*

$$|\sigma_n(f) - \sigma_0(f)| \leq \varepsilon, \text{ for all } f \in M. \tag{27}$$

Since the atomic characterization of $\left(S_0(\mathbb{R}^d), \|\cdot\|_{S_0}\right)$ implies that for any non-zero $g \in S_0(\mathbb{R}^d)$ the set of all TF-shifted copies of g, i.e. the family $\{\pi(\lambda)g \,|\, \lambda \in \mathbb{R}^d \times \widehat{\mathbb{R}}^d\}$ is total in $\left(S_0(\mathbb{R}^d), \|\cdot\|_{S_0}\right)$, we arrive at the following characterization of w^*-convergence:

Theorem 2. *A bounded sequence $(\sigma_n)_{n\in\mathbb{N}}$ is w^*-convergent to $\sigma_0 \in S_0'$ if and only if for some (and therefore for any) non-zero $g \in S_0(\mathbb{R}^d)$ one has pointwise, or equivalently uniform convergence over compact sets of the TF-plane of $V_g\sigma_n$ to $V_g\sigma_0$. More explicitely: For every $R > 0$ and $\varepsilon > 0$ there exists some index n_0 such that*

$$|V_g(\sigma_n)(\lambda) - V_g(\sigma_0)(\lambda)| \leq \varepsilon \quad \forall n \geq n_0, \lambda \quad \text{with} \quad |\lambda| \leq R. \tag{28}$$

A verbal description of this situation is to say that the spectrograms of σ_n look more and more similar to the spectrogram of σ_0 over larger and larger parts of phase space.

There are of course many important examples where w^*-convergence is valid, while in contrast we do not have norm convergence, even for some simple examples as

1. if $x_n \to x_0$, then $\delta_{x_0} = w^* - \lim_n \delta_{x_n}$, while $\|\delta_x - \delta_y\|_{S_0'} = 2$ for $x \neq y$.
2. $\chi_{s_n} \to \chi_{s_0}$ in the w^*-sense if and only if $s_n \to s_0$;
3. $\sqcup\!\sqcup_r = \sum_{k\in\mathbb{Z}^d}\delta_{rk} \to \delta_0$ for $r \to \infty$;
4. $(St_\rho g)_{\rho\to 0} \to \delta_0$ in the w^*-topology, for $\rho \to 0$, if $\int_{\mathbb{R}^d} g(x)dx = 1$, where $St_\rho g(x) = \rho^{-d}g(x/\rho)$ is the L^1-normalized, dilated version of g;
5. $h\sqcup\!\sqcup_h \to \mathbf{1} = \sigma_1$ for $h \to 0$ (Riemannian integrals definition for $f \in S_0(\mathbb{R}^d)$).

---

[14] The study of this convergence goes back to [17], where *relative completions* have been introduced for the study of multiplier spaces.

For later use let us describe explicitly what it means that a linear mapping T on $S_0'(\mathbb{R}^d)$ is $w^* - w^*$-continuous using bounded and w^*-convergent sequences:

$$\sigma_n(f) \to \sigma_0(f) \quad \forall f \in S_0(\mathbb{R}^d) \quad \Rightarrow \quad T(\sigma_n)(g) \to T(\sigma_0)(g) \quad \forall g \in S_0(\mathbb{R}^d).$$

Under the boundedness assumption it is enough to test convergence on total subsets of $S_0(\mathbb{R}^d)$ only, e.g. on the set of atoms (or coherent states) $(\pi(\lambda)g)_{\lambda \in \mathbb{R}^d \times \widehat{\mathbb{R}}^d}$.

Later on we will see that the usually vague and heuristic argument, exhibiting the Fourier transform as a limit of Fourier series expansions, can be made precise in such a context. In fact, the Fourier transform $\hat{f}$ of $f \in L^1(\mathbb{R}^d)$ can be viewed as the w^*-limit of the Fourier transforms of the correspondingly periodized version of f (in fact classical Fourier series expansions), with the period length going to infinity.

Practically all the invariance properties of $S_0(\mathbb{R}^d)$, including its invariance under the Fourier transform, can be extended to invariance properties for $S_0'(\mathbb{R}^d)$. One possible explanation for this fact is the w^*-density of $S_0(\mathbb{R}^d)$ in $S_0'(\mathbb{R}^d)$. From the point of view of introducing the extended operators it is more convenient to use adjointness relations, which we will do later on, using Banach Gelfand triples.

The Fourier Transform on (S_0, L^2, S_0')

We now come back to the question we started with, namely to define a convenient setting for the Fourier transform. Using our Banach Gelfand triple (S_0, L^2, S_0'), we find the following, satisfactory answer. It is a perfect demonstration example for the power of *unitary Banach Gelfand triple automorphisms*.

Theorem 3. *The Fourier transform, defined in the usual way via*

$$\hat{f}(s) = \int_{\mathbb{R}^d} f(t) e^{2\pi i s \cdot t} dt \quad for \quad f \in S_0(\mathbb{R}^d) \tag{29}$$

extends in a unique way to a (unitary) Banach Gelfand triple automorphism, based on the definition

$$\hat{\sigma}(f) := \sigma(\hat{f}), \quad for \quad \sigma \in S_0'(\mathbb{R}^d), f \in S_0(\mathbb{R}^d). \tag{30}$$

It is also characterized by the fact that it is mapping the pure frequencies χ_s are mapped on the corresponding Dirac measures δ_s.

The direct statement is based on the Fourier invariance of $S_0(\mathbb{R}^d)$, while the uniqueness follows from the w^*-density of $S_0(\mathbb{R}^d)$ respectively trigonometric polynomials in $S_0'(\mathbb{R}^d)$.

Gabor characterization of (S_0, L^2, S_0')

The space $\left(S_0(\mathbb{R}^d), \|\cdot\|_{S_0}\right)$ has a number of further equivalent properties, some of them are quite convenient for various purposes. We will use Weyl-Heisenberg families, indexed by *lattices* $\Lambda = A\mathbb{Z}^{2d} \lhd \mathbb{R}^{2d}$, for some non-singular $2d \times 2d$ matrix A:

203

Definition 8. A family $(\pi(\lambda)g)_{\lambda\in\Lambda}$ is called a *Weyl-Heisenberg family*. It will be convenient to write simply $(g_\lambda)_{\lambda\in\Lambda}$.

A WH-family is also called a *Gabor family*, cf. [65]. If a WH-family is a frame or Riesz basis we will speak of a *Gabor frame* or *Gabor Riesz basis* (for its closed linear span). D. Gabor suggested to use the Gauss-function $g = g_0$, and the ("critical") von Neumann lattice $\Lambda = \mathbb{Z}^{2d}$. Despite the perfect time-frequency localization this family is not a Riesz basis for $\mathscr{H} = L^2(\mathbb{R}^d)$, and the so-called dual γ_{Bast} proposed by M. Bastiaans ([10]) is not in $L^2(\mathbb{R}^d)$. There are two other important results to be mentioned here. For their description we recall the *adjoint lattice* Λ°, which consists of those elements in $\mathbb{R}^{2d}$ which satisfy the commutation property

$$\pi(\lambda^\circ)\pi(\lambda) = \pi(\lambda)\pi(\lambda^\circ) \quad \text{for all } \lambda\in\Lambda. \tag{31}$$

The so-called Wexler-Raz principle (see [118, 33, 82, 53]) says that a WH-family $(g_\lambda)_{\lambda\in\Lambda}$ is a Gabor frame if and only the Gabor frame operator $S : f \mapsto \sum_{\lambda\in\Lambda}\langle f, g_\lambda\rangle g_\lambda$ is invertible, or if and only if there exists a dual WH-family of the form $(\tilde{g}_\lambda)_{\lambda\in\Lambda}$ with a generator $\tilde{g}$, characterized either as the solution of the frame equation $S\tilde{g} = g$, or equivalently $\tilde{g} = S^{-1}g$. There are many other possible (non-canonical) dual functions γ, yielding perfect reconstruction, which are characterized according to [118] by the so-called bi-orthogonality relation.

$$V_g\gamma(\lambda^\circ - \mu^\circ) = \langle \pi(\mu^\circ)\gamma, \pi(\lambda^\circ)g \rangle = \delta_{\lambda^\circ,\mu^\circ}, \quad \text{for} \quad \lambda^\circ,\mu^\circ \in \Lambda^\circ. \tag{32}$$

The so-called *Ron-Shen duality* gives more detailed information (cf. [105, 53]): The condition number of the Gabor frame $(g_\lambda)_{\lambda\in\Lambda}$ is the same as condition number of the Gabor Riesz (basic) sequence $(g_{\lambda^\circ})_{\lambda^\circ\in\Lambda^\circ}$, with explicit constants (going back to a symplectic version of Poisson's formula) relating upper and lower frame bounds. This result has a great impact for applications in communication theory. While one tries to use (preferably tight and) *low redundancy* Gabor frames with good localization properties in order to expand signals, avoiding the storage of too many coefficients for the Gabor expansion, one is interested to use Gaborian Riesz bases for the transmission of data, because the well chosen Gabor atoms (obtained using beam-shaping) g ensure that the family $(g_\lambda)_{\lambda\in\Lambda}$ consists of *joint approximate* eigenvectors to all *underspread* resp. *slowly varying* linear systems, i.e. linear operators which have a spreading function supported by a small rectangle in $\mathbb{R}^d \times \widehat{\mathbb{R}}^d$, determined by the maximal time-delay and Doppler shift respectively (see [74]). Ground breaking work in this direction has been done in the PhD thesis [90] of W. Kozek; the link to (S_0, L^2, S_0') has been established in [53].

It is one of the striking recent results due to Gröchenig and Leinert ([75], following the rational case in [50]) to show that $g \in S_0(\mathbb{R}^d)$ implies also that the canonical dual $\tilde{g}$ is in $S_0(\mathbb{R}^d)$, or equivalently (because the frame operator associated with the Gabor system $(\pi(\lambda)\tilde{g})_{\lambda\in\Lambda}$ is just S^{-1}, the inverse of the frame operator for the WH-family $(\pi(\lambda)g)_{\lambda\in\Lambda}$. Expressed in terms of BGT-morphisms their result can be rephrased as follows. The boundedness part of the theorem below is given in detail in [61]:

Theorem 4. *Assume that* $(g_\lambda)_{\lambda \in \Lambda}$ *be a Gabor frame with* $g \in S_0(\mathbb{R}^d)$, *and hence* $S = S_{g,\Lambda} : f \mapsto \sum_{\lambda \in \Lambda} \langle f, \pi(\lambda)g \rangle g_\lambda$ *is a a BGT-morphism on* (S_0, L^2, S_0') *into itself. If* S *is invertible at the* L^2-*level, then it is already BGT-isomorphism.*
In particular, $\tilde{g} = S^{-1}(g)$ *is in* $S_0(\mathbb{R}^d)$ *in this case and*

$$f = \sum_{\lambda \in \Lambda} V_{\tilde{g}} f(\lambda) g_\lambda = \sum_{\lambda \in \Lambda} V_g f(\lambda) \tilde{g}_\lambda. \tag{33}$$

We will call the corresponding families S_0-Gabor families. Another result where the BGT-spirit comes through and the relevance of considering Gabor problems at all three levels is evident can be found in [71] on "Gabor frames without inequalities".

With this background we can give a characterization of elements in each of the levels of (S_0, L^2, S_0') in terms of Gabor coefficients:

Theorem 5. *Let* $g \in \mathscr{S}(\mathbb{R}^d)$ *be given such that* $(g_\lambda)_{\lambda \in \Lambda}$ *is a Gabor frame with canonical dual* $(\tilde{g}_\lambda)_{\lambda \in \Lambda}$ *(also in* $\mathscr{S}(\mathbb{R}^d)$*). Then one has: A tempered distribution* $f \in \mathscr{S}'(\mathbb{R}^d)$ *belongs to* (S_0, L^2, S_0') *if and only if the following (equivalent!) conditions are satisfied:*

1. *f has a representation of the form[15] $f = \sum_{\lambda \in \Lambda} c_\lambda g_\lambda$, with $(c_\lambda)_{\lambda \in \Lambda}$ from $(\ell^1, \ell^2, \ell^\infty)(\Lambda)$;*
2. *The canonical coefficients $(V_{\tilde{g}} f(\lambda))_{\lambda \in \Lambda} \in (\ell^1, \ell^2, \ell^\infty)(\Lambda)$;*
3. *The sampled STFT with window g satisfies: $(V_g f(\lambda))_{\lambda \in \Lambda} \in (\ell^1, \ell^2, \ell^\infty)(\Lambda)$;*

Overall this can be expressed by the fact that the reconstruction mapping $\mathbf{R} : (c_\lambda)_{\lambda \in \Lambda} \mapsto \sum_{\lambda \in \Lambda} c_\lambda \tilde{g}_\lambda$ completes the following diagram:

$$
\begin{array}{ccc}
 & & (\ell^1, \ell^2, \ell^\infty)(\Lambda) \\
 & {}^{\mathbf{R}}\diagup & \downarrow {\scriptstyle \mathbf{P}} \\
(S_0, L^2, S_0') & \underset{V_g|_\Lambda}{\overset{\mathbf{R}}{\rightleftarrows}} & V_g((S_0, L^2, S_0'))
\end{array}
$$

Poisson's Formula, Sampling and Periodization

Using $S_0(\mathbb{R}^d)$ the classical Poisson's formula can be formulated as follows:

Theorem 6. *For $f \in S_0(\mathbb{R}^d)$ one has*

$$\sum_{k \in \mathbb{Z}^d} f(k) = \sum_{n \in \mathbb{Z}^d} \hat{f}(n), \tag{34}$$

the sum being absolutely convergent on both sides.

---

[15] Here $*$ denotes convolution, in contrast to the use of $*$ earlier on, where it was representing matrix multiplication, using MATLAB conventions.

This formula does not hold for arbitrary functions, even if both the left hand side and the right hand side are absolutely convergent, as has been described in the book of Katznelson ([87, 88]). Most of the usual conditions on f which are sufficient for the validity of (34) can be interpreted as sufficient conditions for f to belong to $S_0(\mathbb{R}^d)$ (cf. [84, 68]). The symplectic version of Poisson relation is also highly relevant for Gabor analysis ([8, 112]).

The key properties of $S_0(\mathbb{R}^d)$ needed to verify Thm. 6 are the fact that the restriction of a function $f \in S_0(\mathbb{R}^d)$ is in $\ell^1(\mathbb{Z}^d)$, that the $\mathbb{Z}^d$-periodization of f is uniformly convergent, and the fact that the periodized function f_{per} has as its Fourier coefficients just the samples $(\hat{f}(n))$, which are again in $\ell^1(\mathbb{Z}^d)$, due to the Fourier invariance of $S_0(\mathbb{R}^d)$.

It is an easy exercise to translate the Poisson formula into a statement about the Fourier invariance of the so-called Shah-distribution $\sqcup\!\!\sqcup_{\mathbb{Z}^d}$ (also called Dirac Comb, etc.):

Theorem 7. *The Shah-distribution $\sqcup\!\!\sqcup_{\mathbb{Z}^d}$ belongs to $S_0'(\mathbb{R}^d)$, and*

$$\widehat{\sqcup\!\!\sqcup}_{\mathbb{Z}^d} = \sqcup\!\!\sqcup_{\mathbb{Z}^d}$$

Using the invariance of $S_0(\mathbb{R}^d)$ under transformation of the argument it is easily extended to other lattices of the form $\Lambda = A * \mathbb{Z}^d \lhd \mathbb{R}^d$, where $det(A) \neq 0$. For the sake of simplicity we will use ordinary dilation, which gives then $\sqcup\!\!\sqcup_a = \sum_{k \in \mathbb{Z}^d} \delta_{ak}$, which has as its (generalized) Fourier transform $b\sqcup\!\!\sqcup_b$, with $b = 1/a$.

One of the most important principles in harmonic analysis is the idea that sampling on the "time-side" corresponds to periodization on the frequency side. The most important consequence of this principle is the so-called Shannon sampling theorem, according to which a band-limited signal can be recovered from its regular or *equidistant* samples. Again, we do the detailed discussion only for the normalized case, i.e. for the case that the *Nyquist sampling rate* is the sampling over the integer lattice $\mathbb{Z}^d$, or in other words, that the spectrum (the support of the Fourier transform of $\hat{f}$ under discussion) is contained in the cube $Q := [-1/2, 1/2]^d$. We write $\mathbf{1}_Q$ for the indicator function of Q and define $\mathrm{SINC} = \mathscr{F}^{-1}(\mathbf{1}_Q)$.

Theorem 8. *[Shannon Sampling Theorem]*
For any $f \in L^2(\mathbb{R}^d)$ with $\mathrm{supp}(f) \subseteq Q$ one has

$$f(t) = \sum_{n \in \mathbb{Z}^d} f(n) T_n \mathrm{SINC}(t), \qquad (35)$$

with absolute and uniform convergence of the series and norm convergence in $L^2(\mathbb{R}^d)$.

The proof is based on the observation that the family $(T_n \mathrm{SINC})_{n \in \mathbb{Z}^d}$ is an orthonormal basis for the closed subspace $B^Q := \{f \in L^2(\mathbb{R}^d) \mid \mathrm{supp}(\hat{f}) \subseteq Q\}$ of $(L^2(\mathbb{R}^d), \|\cdot\|_2)$. In fact, one has convergence with respect to the Wiener amalgam norm $W(C_0, \ell^2)(\mathbb{R}^d)$, which implies both uniform and L^2-convergence. The fact that the SINC function is the (inverse) Fourier transform of the indicator function $\mathbf{1}_Q$, which is a Fourier multiplier for

$1 < p < \infty$ also implies, that a similar statement holds true for band-limited functions in $L^p(\mathbb{R}^d)$, for the same range of parameters.

Proof. Given the sampling values $(f(n))_{n \in \mathbb{Z}^d}$ we have all the information in our hands to describe $\sqcup\!\sqcup_{\mathbb{Z}^d} \cdot f = \sum_{n \in \mathbb{Z}^d} f(n) \delta_n$. This is a well defined (unbounded but translation-bounded) discrete measure in $S_0'(\mathbb{R}^d)$ which has a Fourier transform of the form

$$\widehat{\sqcup\!\sqcup_{\mathbb{Z}^d} \cdot f} = \widehat{\sqcup\!\sqcup_{\mathbb{Z}^d}} * \hat{f} = \sum_{k \in \mathbb{Z}^d} \delta_n * \hat{f} = \sum_{k \in \mathbb{Z}^d} T_n \hat{f} \tag{36}$$

which is nothing but the $\mathbb{Z}^d$-periodic version of $\hat{f}$. We can now use the fact that Q is a fundamental domain for the lattice $\mathbb{Z}^d \lhd \mathbb{R}^d$, hence $|n + Q \cap k + Q| = 0$ for $n \neq k$. Multiplying this periodic version by $\mathbf{1}_Q$ gives us exactly the original basic period, which is $\hat{f}$, or back on the time domain

$$f = (\sqcup\!\sqcup_{\mathbb{Z}^d} \cdot f) * \mathscr{F}^{-1}(\mathbf{1}_Q) = \sum_{n \in \mathbb{Z}^d} f(n) \delta_n * \text{SINC} = \sum_{n \in \mathbb{Z}^d} f(n) T_n \text{SINC}. \tag{37}$$

This series is convergent in $L^2(\mathbb{R}^d)$ because on the Fourier transform side we just have the Fourier expansion of $\hat{f}$ (taken as a periodic function on $\mathbb{R}^d$). On the other hand SINC belongs to $L^2(\mathbb{R}^d)$ and even the Wiener amalgam space $W(C_0, \ell^2)(\mathbb{R}^d)$, which implies uniform and pointwise absolute convergence. $\square$

BANACH GELFAND TRIPLES AND OPERATORS

In this section we will indicate the role of BGTs for the description of operators. The same role which is played by the pure frequencies for Fourier analysis (they are perfect building blocks forming an orthonormal basis in the case of finite Abelian groups but fail to belong to the natural Hilbert space) is now taken by other systems of natural objects. From the point of time-frequency analysis of course the collection of $(\pi(\lambda))_{\lambda \in \mathbb{R}^d \times \widehat{\mathbb{R}}^d}$ is a very natural choice, but again they are *not* in the natural Hilbert space, now $\mathscr{H}\mathscr{S}$, the space of all Hilbert Schmidt operators on $L^2(\mathbb{R}^d)$, endowed with the $\mathscr{H}\mathscr{S}$-scalar product $\langle T, S \rangle_{\mathscr{H}\mathscr{S}} := \text{Tr}(TS^*)$.

On the other hand one of the most exciting developments in the field is the realization, that pseudo-differential operators have a very natural description in terms of time-frequency expressions. To give an example: modulation spaces turn out to be the most natural spaces in order to describe *slowly varying channels*, i.e. convolution operators with a *time-variant kernel* (in an engineering terminology), resp. certain classes of pseudo-differential operators. These are the systems which preserve localization in the TF-sense and hence have a matrix representaion which is mostly concentrated along the diagonal. In the extreme case on has Gabor multipliers, i.e. operators which are factorized through a diagonal matrix, acting on the Gabor coefficients.

There is a large number of papers on Gabor multipliers, such as [57, 12, 38, 5] and a self-contained survey (master thesis) by K. Schnass [107] from 2004 or the PhD thesis of P. Balazs ([9]).

AntiWick operators are operators which are defined as STFT-multipliers. They make use of the inversion formula for the STFT (which in turn based on the isometric properties of $f \mapsto V_g(f)$ from $L^2(\mathbb{R}^d)$ into $L^2(\mathbb{R}^{2d})$):

$$\int_{\mathbb{R}^d \times \widehat{\mathbb{R}}^d} (V_{g_1} f)(x, \omega) M_\omega T_x g_2 \, dx \, d\omega = \langle g_2, g_1 \rangle f \qquad (38)$$

So one can use any $g_2 \in L^2(\mathbb{R}^d)$ for reconstruction of f from V_{g_1} as long as it is not orthogonal to g_1. Usually the integral has to be understood in the *weak sense* for $g_1, g_2 \in L^2(\mathbb{R}^d)$, but if both of them are in $S_0(\mathbb{R}^d)$ (cf. [117]) then one can even read the above integral as limit of vector-valued Riemannian integrals, which are norm convergent in $\mathscr{H} = L^2(\mathbb{R}^d)$. Due to the good local properties of functions in the range of the STFT one can even work with rough symbols (see [93, 13, 28])

Adjointness Relations

First of all let us mention some principles that allow us to extend bounded linear mappings betwenn S_0-spaces to BGT-morphisms. The following principle is quite useful in order to "automatically extend" a mapping between the "inner spaces" to their dual spaces.

Theorem 9. *Let T be a BGT-homomorphism from $(B_1, \mathscr{H}_1, B_1')$ into $(B_2, \mathscr{H}_2, B_2')$, i.e. a bounded linear mapping which is bounded on all three layers, as well as $w^* - w^*$-continuous. Then there exists a unique adjoint GT-homomorphism, i.e. another BGT-homomorphism (denoted by) T^* from $(B_2, \mathscr{H}_2, B_2')$ into $(B_1, \mathscr{H}_1, B_1')$, such that $T^* : \mathscr{H}_2 \mapsto \mathscr{H}_1$ is the adjoint operator, which extends to a GT-morphism in a unique way. Therefore we have the identity*

$$\langle T f, g \rangle_{(B_2, \mathscr{H}_2, B_2')} = \langle f, T^* g \rangle_{(B_1, \mathscr{H}_1, B_1')} \qquad (39)$$

whenever the pairing makes sense. Moreover $T^{**} = T$, i.e. in this sense any BGT-morphism is the adjoint of another (uniquely determined) adjoint BGT-morphism, denoted by T^*.

The case of unitary operators has been discussed already in [53], Thm.7.3.3 (Extension of Unitary Gelfand Triple Isomorphism), p. 239.

Theorem 10. *A unitary mapping U acting from $L^2(\mathbb{R}^d)$ to $L^2(\mathbb{R}^d)$ extends to an isomorphism between the Gelfand triples $(S_0, L^2, S_0')(\mathbb{R}^d)$ to $(S_0, L^2, S_0')(\mathbb{R}^d)$ if and only if the restriction of U and also of its adjoint U^* are bounded linear operators from $S_0(\mathbb{R}^d)$ to $S_0(\mathbb{R}^d)$.*

Remark 1. There are good reasons why the "central" Hilbert space $\mathscr{H} = L^2(\mathbb{R}^d)$ usually plays the dominant rule, just think of Plancherel's theorem as the central property of the Fourier transform, describing it as a *unitary mapping* on $L^2(\mathbb{R}^d)$. However, from an abstract point of view it is not so important, and in most cases the isomorphism property at the S_0 and S_0'-level (both with the norm and the w^*-topology) implies already that one has the full continuity claim for the Hilbert space automatically, becausean application of complex interpolation between the dual pair $S_0(G)$ and $S_0'(G)$ yields[16]

$$[S_0, S_0']_{[1/2]} = L^2.$$

Theorem 11. *For any group $(T_i)_{i \in I}$ of unitary operators leaving $S_0(\mathbb{R}^d)$ invariant, which satisfy*

$$\langle f, g \rangle = \langle T_i f, T_i g \rangle \quad \forall f, g \in S_0(\mathbb{R}^d), \tag{40}$$

one has: The action of $(T_i)_{i \in I}$ extends in a unique way to a unitary Banach Gelfand triple automorphism of $(S_0, L^2, S_0')(\mathbb{R}^d)$.

Proof. The assumption (40) implies (just in the same way as Plancherel's theorem is usually proved) that it is well defined and isometric on $S_0(\mathbb{R}^d)$ with respect to the L^2-norm. Due to the density of $S_0(\mathbb{R}^d)$ in $L^2(\mathbb{R}^d)$ it can be uniquely extended to an isometric and in fact unitary automorphism on $L^2(\mathbb{R}^d)$. $\qquad\square$

Kernel Theorems and Gelfand Triples

The *nuclear* Frechet space $\mathscr{S}(\mathbb{R}^d)$ and its dual, the space of $\mathscr{S}'(\mathbb{R}^d)$ of tempered distributions are the prototype of function spaces for which on can prove a so-called kernel theorem, a continuous analogue of the existence of a matrix, completely describing the operator. We next prepare a similar principle for our BGT-setting.

Given two functions f^1 and f^2 on $\mathbb{R}^d$ respectively, we set $f^1 \otimes f^2$

$$f^1 \otimes f^2(x_1, x_2) = f^1(x_1) f^2(x_2), \ x_1, x_2 \in \mathbb{R}^d. \tag{41}$$

Given two Banach spaces B_1 and B_2 embedded into $\mathscr{S}'(\mathbb{R}^d)$, $B_1 \hat{\otimes} B_2$ denotes their *projective tensor product*, i.e.

$$\left\{ f \mid f = \sum f_n^1 \otimes f_n^2, \ \sum \|f_n^1\|_{B_1} \|f_n^2\|_{B_2} < \infty \right\}. \tag{42}$$

It is easy to show that this defines a Banach space of tempered distributions on $\mathbb{R}^{2d}$ with respect to the (quotient) norm:

$$\|f\|_{\hat{\otimes}} := \inf \left\{ \sum \|f_n^1\|_{B^1} \|f_n^2\|_{B^2}, \dots \right\}, \tag{43}$$

where the infimum is taken over all admissible representations.

One of the most important properties of $S_0(\mathbb{R}^d)$ (leading to a characterization given by V. Losert, [94]) is the tensor-product factorization:

Lemma 4.

$$S_0(\mathbb{R}^k) \hat{\otimes} S_0(\mathbb{R}^n) \cong S_0(\mathbb{R}^{k+n}). \tag{44}$$

---

[16] One way to understand/accept this fact is to invoke the fact that the Wilson bases establish, at least for elementary locally compact Abelian groups, a BGT-isomorphism between (S_0, L^2, S_0') and $(\ell^1, \ell^2, \ell^\infty)$.

The easiest way to realize this relationship is to make use of the atomic decomposition of $S_0(\mathbb{R}^m)$, observing that both time- and frequency shifts, but also the multi-dimensional Gaussian function factorize into lower dimensional partial ingredients. This tensor product property of S_0 is on the other hand the basis for the realization of the so called kernel theorem (see [53], Chap.7.4).

The setting of BGTs is also well suited for the description of linear operators. The so-called *kernel theorem* shows how essentially every reasonable[17] operator T can be interpreted as a kind of *integral operator*, also called *kernel* from a suitable class of generalized functions. One may expect similarity to the finite discrete case (= matrix multiplication), but now with continuous variables:

$$Tf(x) = \int_{\mathbb{R}^d} K(x,y)f(y)dy \qquad (45)$$

in analogy to the description of $u = Tf \simeq u = A * z \in \mathbb{C}^m$ via coordinates

$$(Tz)_s = \sum_{k=1}^n a_{s,k} z_k, \quad \text{for} \quad s = 1,\ldots n. \qquad (46)$$

The usual way of finding the appropriate $m \times n$ matrix A for a linear mapping from $\mathbb{C}^n$ to $\mathbb{C}^m$ is easy: recall that one obtains coordinate number s of the vector u via scalar product with e_s, that the k-th column of A has to correspond to $T(e_k)$ if (46) is supposed to be valid, hence the individual entry must be

$$a_{s,k} = A[s,k] = \langle T(e_k, e_s) \rangle. \qquad (47)$$

Viewing $A = (A(n,k))$ as a function over the product index set one can say that has to take the scalar product of A (in the sense of the Euclidean space $\mathbb{C}^{mn}$) with the unit matrix $e_s * \otimes e_k$, which can also be expressed as via a trace formula of operators:

$$A[s,k] = \langle A, e_s \otimes e_k^* \rangle = \text{Tr}(A * (e_k * e_s^*)).^{[18]} \qquad (48)$$

There is a lot of literature about Dirac's formalism. On the one hand it is very intuitive, on the other hand it has created a lot of discussion concerning the strict mathematical interpretation of these formal symbols. Even engineers are by now aware of the fact δ_0 is no just another function, which is zero everywhere except at 0, but "so strongly infinite" that the integral equals 1. Nowadays it is well known that the Dirac *measure* $\delta_x : f \mapsto f(x)$ is a good way to formalize this procedure, but this still does not explain what the connection between Kronecker's δ, usually written as $\delta_{i,j}$ and Dirac's symbol which is nowadays just a distribution (of *one* variable, so to say) applied on a test function, while in many early interpretations of what Dirac might have had in mind with his symbol the idea of using the symbols he had introduced often comes with the recommendation of only using it within an integral, and not as an individual object. In

---

[17] E.g. T is a bounded linear operator from some L^q-space to another L^p−space.
[18] Where $*$ corresponds to matrix multiplication in a MATLAB setting.

fact, we will argue that one should consider δ_{x-y} as a distributional kernel representing the identity mapping, so simply the continuous analogue of Kronecker's symbol.

Moreover it appears as a way to express a kind of orthogonality relations between "pure frequencies" (and because they are not square-integrable the scalar product between $\langle \chi_s, \chi_s \rangle$ has to be $+\infty$) which allows to do derivations of very useful formulas[19].

Sequences of regularizing operators

Once we have continuous variables one comes into a world where finite dimensional arguments break down, where one may have unbounded operators, and even point evaluations are not always possible, i.e. the use of e.g. Dirac distributions is required. Nevertheless one has a number of different products, which often can be written as integrals (convolution, twisted convolution) or using point values (defining the "ordinary pointwise product"), and sometimes such products immediately make sense, in some other case on has first approximate the involved ingredients before applying the operation, using a regularized version of one or both partners involved, and then let the regularization parameter tend to 0 or ∞, as appropriate.

Such a principle is not really new, as many special cases can be located in the literature. The definition of the Fourier transform is one case, where one has to "push" general $L^2(\mathbb{R}^d)$-functions into $L^1(\mathbb{R}^d)$ (in case they are not already within $L^1 \cap L^2(\mathbb{R}^d)$, which is fortunately a dense subspace of $(L^2(\mathbb{R}^d), \|\cdot\|_2)$), e.g. via pointwise multiplication with the indicator function $\mathbf{1}_{[-N,N]}$, for $N \to \infty$. For the inversion of the Fourier transform a similar strategy can be applied, now by doing a pointwise multiplication with some suitable summability kernel. Although it would again be enough to use any localizing function, it has been realized that a sharp frequency cut-off is not a good way, since $\mathscr{F}^{-1}\mathbf{1}_{[-N,N]} \notin L^1(\mathbb{R}^d)$. Choosing a summability kernel from $S_0(\mathbb{R}^d)$ will help and ensure that its inverse Fourier transform is in $L^1(\mathbb{R}^d)$ as well. Since stretching in Fourier space is the same as L^1-norm preserving dilation the resulting sequence of Dirac-like convolution kernels is an approximate identity for the Banach Gelfand triple (S_0, L^2, S_0'), while the SINC-function is not having this good property.

Wiener amalgam convolution and pointwise multiplier results ([78]) imply that

$$S_0(\mathbb{R}^d) \cdot (S_0'(\mathbb{R}^d) * S_0(\mathbb{R}^d)) \subseteq S_0(\mathbb{R}^d), \quad S_0(\mathbb{R}^d) * (S_0'(\mathbb{R}^d) \cdot S_0(\mathbb{R}^d)) \subseteq S_0(\mathbb{R}^d) \quad (49)$$

Proof. The key arguments for both of these regularization procedures, be it convolution followed by pointwise multiplication (a so-called *product-convolution operator*, for short PC-operator), or corresponding CP-operators, are based on the pointwise and convolutive behaviour of generalized Wiener amalgam spaces, such as the relation $S_0(\mathbb{R}^d) * S_0'(\mathbb{R}^d) = W(\mathscr{F}L^1, \ell^1) * W(\mathscr{F}L^\infty, \ell^\infty) \subseteq W(\mathscr{F}L^1, \ell^\infty)$. $\quad\square$

---

[19] We suggest to view the well known identity $e^{2\pi i} = 1$ in a similar way, as an extremely useful formula which makes use of the complex numbers, the irrational number π, which is never explicitly and constructively realized, let alone the power series expression of the exponential function.

Let now $h \in \mathscr{F}L^1(\mathbb{R}^d)$ be given with $h(0) = 1$. Then the dilated version $h_n(t) = h(t/n)$ are a uniformly bounded family of multipliers on (S_0, L^2, S_0'), tending to the identity operator in a suitable way. Similarly, the usual Dirac sequences, obtained by compressing a function $g \in L^1(\mathbb{R}^d)$ with $\int_{\mathbb{R}^d} g(x)dx = 1$ are showing a similar behavior: $g_n(t) = n \cdot g(nt)$

Following the above rules the combination of the two procedures, i.e. product-convolution or convolution-product operators of the form provide suitable regularizers: $A_n f = g_n * (h_n \cdot f)$ or $B_n f = h_n \cdot (g_n * f)$.

Following Theorem 5 we know that elements in $f \in (S_0, L^2, S_0')$ can be characterized (among others) by their minimal norm coefficients, given in the form $(V_{\tilde{g}} f(\lambda))_{\lambda \in \Lambda}$. It is therefore clear that the partial sum operators for this canonical Gabor expansions, such as

$$A_N f := \sum_{max(|\lambda_1|,|\lambda_2|) \leq N} V_{\tilde{g}} f(\lambda) g_\lambda \tag{50}$$

is mapping $S_0'(\mathbb{R}^d)$ into $S_0(\mathbb{R}^d)$, while on the other hand one has obviously that $A_N f \to f$ as $N \to \infty$ for any f in $S_0(\mathbb{R}^d)$ or $L^2(\mathbb{R}^d)$ respectively, in the corresponding norm, while the convergence occurs in the w^*-sense, for all $f \in S_0'(\mathbb{R}^d)$.

Similar statements can be made for rectangular or any other kind of 'exhausting partial sums, also with respect to Wilson basis. The better the building blocks are (in terms of time-frequency localization, typically expressed using membership in the modulation spaces $M_{v_s}^1(\mathbb{R}^d)$) the more can be said about the rate of approximation, given the quality of the signal, i.e. speed of approximation of f in some Shubin class $Q_s(\mathbb{R}^d)$, measured in the L^2-norm.

Various types of regularizations are also used in the discussion about the most general definition of convolution between distributions, see the work of M. Oberguggenberger ([1, 18]). In fact, one can say, that the basic idea is to assume the the limit of $A_N \sigma_1 * A_N \sigma_2$ exists (for a sufficiently rich class of regularization operators, implying that this limit is then independent of the particular choice of the sequence (A_N)).

Kernel Theorem for $S_0(\mathbb{R}^d)$

There are many different ways to show that the space of test functions $S_0(\mathbb{R}^d)$ is w^*-dense in $S_0'(\mathbb{R}^d)$. One very important and natural way (also valid in a similar way for the space of Schwartz test functions from the space $\mathscr{S}(\mathbb{R}^d)$ of rapidly decreasing functions and its corresponding dual space, $\mathscr{S}'(\mathbb{R}^d)$ of tempered distributions)

Theorem 12. *If K is a bounded operator from $S_0(\mathbb{R}^d)$ to $S_0'(\mathbb{R}^d)$, then there exists a unique kernel $k \in S_0'(\mathbb{R}^{2d})$ such that $\langle Kf, g \rangle = \langle k, g \otimes f \rangle$ for $f, g \in S_0(\mathbb{R}^d)$, where $g \otimes f(x, y) = g(x)f(y)$.*

Formally sometimes one writes by "abuse of language"

$$Kf(x) = \int_{\mathbb{R}^d} k(x, y) f(y) dy \tag{51}$$

with the understanding that one can define the action of the functional $Kf \in S_0'(\mathbb{R}^d)$ as

$$Kf(g) = \int_{\mathbb{R}^d} \int_{\mathbb{R}^d} k(x,y)f(y)dyg(x)dx = \int_{\mathbb{R}^d} \int_{\mathbb{R}^d} k(x,y)g(x)f(y)dxdy. \qquad (52)$$

This result is the "outer shell of the Gelfand triple isomorphism, which corresponds to the well-known result that Hilbert Schmidt operators on $L^2(\mathbb{R}^d)$ are just those compact operators which arise as integral operators with $L^2(\mathbb{R}^{2d})$-kernels. The complete picture can again be expressed by a unitary Gelfand triple isomorphism. Let us start with the classical setting: The Hilbert space $\mathcal{HS}$ of Hilbert Schmidt operators on $L^2(\mathbb{R}^d)$ is important, because the classical *kernel theorem* establishes a unitary mapping between operators $T \in \mathcal{HS}$ and their kernels K. The scalar product of $\mathcal{HS}$−operators is given by $\langle T, S \rangle_{\mathcal{HS}} = \text{Tr}(T * S')$ and turns $\mathcal{HS}$ into a Hilbert space.

Theorem 13 (Kernel Theorem for S_0). *Let $T \in \mathcal{HS}$ be given, with kernel $K \in L^2(\mathbb{R}^{2d})$. Such an operator has a kernel in $S_0(\mathbb{R}^{2d})$ if and only if it maps bounded, w^*−convergent sequences in $S_0'(\mathbb{R}^d)$ into norm convergent in $S_0(\mathbb{R}^{2d})$. The most general operators from $\mathcal{L}(S_0, S_0')$ are in a one-to-one correspondence with $S_0'(\mathbb{R}^{2d})$.*

Overall the kernel theorem allows us to establish a unitary BGT isomophism between the BGT $(\mathcal{L}(S_0', S_0), \mathcal{HS}, \mathcal{L}(S_0, S_0'))$ of operator spaces and the corresponding kernels in $(S_0, L^2, S_0')(\mathbb{R}^{2d})$.

Remark 2. Note that for *regularizing* kernels in $S_0(\mathbb{R}^{2d})$ the usual identification (recall that the entry of a matrix $a_{n,k}$ is the coordinate number n of the image of the k−th unit vector under that action of the matrix $A = (a_{n,k})$) holds:

$$k(x,y) = K(\delta_y)(x) = \delta_x(K(\delta_y)). \qquad (53)$$

Since $\delta_y \in S_0'(\mathbb{R}^d)$ and thus $K(\delta_y) \in S_0(\mathbb{R}^d)$ the pointwise evaluation makes sense.

Remark 3. It is of course interesting to ask how the w^*-topology can be transferred to the operator level. Here again a characterization of general linear operators using Gabor expansions comes into the picture:

Definition 9. Assume that $(g_\lambda)_{\lambda \in \Lambda}$ and $(\tilde{g}_\lambda)_{\lambda \in \Lambda}$ is a dual pair of Gabor frames, with $g, \tilde{g} \in S_0(\mathbb{R}^d)$, and assume $T \in \mathcal{L}(S_0, S_0')$, i.e. that T is a bounded linear operator from $S_0(\mathbb{R}^d)$ into $S_0'(\mathbb{R}^d)$. Then the *matrix elements* of T with respect to the Gabor frame are

$$a_{\lambda,\lambda'} := \langle Tg_\lambda, \tilde{g}_{\lambda'} \rangle, \quad \lambda, \lambda' \in \Lambda. \qquad (54)$$

Using these matrix coefficients (one can use either g or $\tilde{g}$), both in the first or the second place, on obtains

Lemma 5. *Let T_n be a sequence of operators from $S_0(\mathbb{R}^d)$ into $S_0'(\mathbb{R}^d)$, such that the corresponding kernels K^n form a bounded sequence in $S_0'(\mathbb{R}^{2d})$, convergent to K^0 in the w^*-sense. Then T_nf is w^*-convergent for every $f \in S_0(\mathbb{R}^d)$ to some limiting operator $T^0(f) = \lim_n T_nf$ and conversely. In particular, $K^0 = w^* - \lim_n K_n$ if and only if all the matrix coefficients converge pointwise, i.e. for each pair $(\lambda, \lambda') \in \Lambda \times \Lambda$ one has*

$$a_{\lambda,\lambda'}^n \rightarrow a_{\lambda,\lambda'}^0 \quad for \ n \rightarrow \infty.$$

Kohn-Nirenberg Symbol and Spreading Function

The *Kohn-Nirenberg symbol* of an operator (respectively its *symplectic* Fourier transform, the so-called *spreading symbol*) can be obtained from the kernel by applying suitable coordinate transforms (automorphisms) and partial Fourier transforms [20]. Hence they define functions or distributions over $\mathbb{R}^d \times \widehat{\mathbb{R}}^d$. Since all these ingredients are unitary BGT isomorphisms of (S_0, L^2, S_0') the known correspondences at the level of $\mathscr{H}\mathscr{S}$-operators can be extended to BGT isomorphisms.

Theorem 14. *The correspondence between an operator T with kernel K from the Banach Gelfand triple $(\mathscr{L}(S_0', S_0), \mathscr{H}\mathscr{S}, \mathscr{L}(S_0, S_0'))$ and the corresponding spreading distribution $\eta(T)$ in $S_0'(\mathbb{R}^{2d})$ is the uniquely defined Gelfand triple isomorphism between $(\mathscr{L}(S_0', S_0), \mathscr{H}\mathscr{S}, \mathscr{L}(S_0, S_0'))$ and $(S_0, L^2, S_0')(\mathbb{R}^d \times \widehat{\mathbb{R}}^d)$ which maps the time-frequency shift operators $M_y \circ T_x$ onto the Dirac measure $\delta_{(x,y)}$.*

The w^*-continuity of this mapping allows among others to calculate (in the sense of approximate) $\eta(T)$ by first dealing with regularizing operators from $\mathscr{L}(S_0', S_0)$ with kernels and symbols in S_0. For this "core" space one can apply transformations and partial Fourier transform in a direct way, while more general case is realized either by taking w^*-limits of using an adjointness argument.

The Kohn-Nirenberg description of operators is particularly interesting in the discussion of *Gabor multipliers*, i.e. of operators of the form

$$Tf = \sum_{\lambda \in \Lambda} m_\lambda \langle f, \pi(\lambda)g\rangle g_\lambda = \sum_{\lambda \in \Lambda} m_\lambda P_\lambda(f), \tag{55}$$

where P_λ is the projection of f onto the one-dimensional space generated by g_λ. Equivalently, $P_\lambda = \pi(\lambda) \circ P_0 \pi(\lambda)'$. The mapping $H \mapsto \pi(\lambda) \circ P_0 \pi(\lambda)'$ is a unitary group representation of the additive group $\mathbb{R}^{2d}$ on the Hilbert space $\mathscr{H}\mathscr{S}$, and one crucial facts is the relation

$$\kappa[\pi(\lambda) \circ P_0 \pi(\lambda)'] = T_\lambda \kappa(H), \quad \lambda \in \Lambda. \tag{56}$$

Composition of Operators

Given the kernel representation (or whatever other form of "symbol", from Weyl- to Kohn-Nirenberg or spreading representation) it is clear that the composition of operators corresponds to some kind of composition rule at the level of symbols. For the case of matrices we *know* that we have to perform matrix multiplication, i.e. the *matrix-product* $C := A * B$ is given (coefficientwise) by the rule

$$c_{k,l} = \sum_{s=1}^{n} a_{k,s} b_{s,l} \tag{57}$$

---

[20] i.e. Fourier transforms with respect to one variable only with $\mathbb{R}^d \times \mathbb{R}^d$.

whenever the matrix product is possible, resp. whenever the composition of operators is possible (the range of B has to be equal to the domain of A, in our example $\mathbb{C}^m$).

Of course the situation is - from a purely technical point - much more delicate in the case of (perhaps even unbounded) linear mappings between (infinite) dimensional vector spaces (typically Hilbert spaces or Banach spaces), and even if we are next discussing the composition of BGT-morphism it is not absolutely clear how to interpret their composition (which is kind of obvious from the point of view of operators).

Let us therefore consider first the composition of two simple integral operators, with the corresponding kernels $K_2(x,s)$ and $K_1(s,y)$ in $S_0(\mathbb{R}^{2d})$. It is not difficult to verify that one has in such a case, in complete analogy to case of matrix multiplication:

Lemma 6. *The composition of two operators $T_2 \circ T_1$, both of which have a kernel representation with $S_0(\mathbb{R}^{2d})$-kernels $K_2(x,s)$ and $K_1(s,y)$ respectively, has a kernel in $S_0(\mathbb{R}^{2d})$ of the form*

$$K(x,y) = \int_{\mathbb{R}^d} K_2(x,s)K_1(s,y)\,ds \qquad (58)$$

This formula is also valid if one of the kernels belongs to $L^\infty(\mathbb{R}^{2d}) \subset S_0'(\mathbb{R}^{2d})$[21].

Proof. The kernels in K_1 and K_2 define a bounded linear operator from $S_0'(\mathbb{R}^d)$ space into $S_0(\mathbb{R}^d)$, converting w^*-convergent sequences into norm convergent sequences. Hence one can compose the operators, but also verify without difficulties (under the L^∞-condition) the existence of the corresponding integrals in (58)[22]. □

For more general cases, e.g. for the composition of general bounded linear operators on $\mathscr{H} = L^2(\mathbb{R}^d)$ it turns out that a composition rule like the simple integral composition of (58) may become questionable. Among others, because it is known to be hard to characterize the L^2-boundedness of the operator T in terms of the kernel $K(x,s)$. Therefore one has to use the approximation of operators by "good" ones before calculating the "product-kernel", i.e. the (distributional) kernel of the composite linear mapping. In order to realize this in a systematic way (admitting that there are many other ways of doing it) we formulate an auxiliary result. It is based on the use of sequences of regularzing operators with kernels in $S_0(\mathbb{R}^{2d})$, i.e. a bounded sequence A_n of BGT-morphism with kernels $K^n(s,u) \in S_0(\mathbb{R}^{2d})$ such that the sequence (as well as its adjoint) acts as an approximation to the identity operator on $S_0'(\mathbb{R}^d)$ (hence on the larger spaces), i.e. satisfies $\|A_n f - f\|_{S_0} \to 0$ for $n \to \infty$, for each $f \in S_0(\mathbb{R}^d)$.

Lemma 7. *For each regularizing sequence $A_n, n \geq 1$ and linear mappings T_1 and T_2 one finds that $A_n \circ T_1$ resp. $T_2 \circ A_n$ are regularizing operators in $\mathscr{L}(S_0', S_0)(\mathbb{R}^d)$. Hence their kernel of composite mappings such as $T_2 \circ A_n \circ A_n \circ T_1$ can be composed according to formula (58), and the product kernels K^n obtained in this way are w^*-convergent to the*

---

[21] Various properties of the kernel of the composite mapping can be derived from the properties of the resulting product operators. The composition itself need not be "well-defined" in the sense of Lebesgue-integrability almost everywhere. This problem can be overcome using regularization techniques described below.

[22] Here is a warning in place: even if the kernels are given as bounded and continuous functions we do not claim in the most general case that the integration has to make sense in the Lebesgue sense!

kernel of $T_2 \circ T_1$. At the same time the corresponding operators are convergent (in the sense of pointwise convergence) to the action of $T_2 \circ T_1$, because

$$\|T_2(T_1 f) - T_2(A_n(T_1 f))\|_{S_0'} \to 0 \quad \text{for all} \quad f \in S_0(\mathbb{R}^d). \qquad (59)$$

There are of course many variations of this principle, and the concrete form of the regularizing operator can vary from case to case. Again the Fourier transform is a perfect example. We know that the Fourier transform, viewed as an *integral transform* on $\left(L^1(\mathbb{R}^d), \|\cdot\|_1\right)$, maps into $\left(\mathscr{F}L^1(\mathbb{R}^d), \|\cdot\|_{\mathscr{F}L^1}\right) \hookrightarrow \left(C_0(\mathbb{R}^d), \|\cdot\|_\infty\right)$, according to the Riemann-Lebesgue Lemma, as a proper but dense subspace. The problem with Fourier inversion on $\mathscr{F}L^1(\mathbb{R}^d)$ is not the roughness of those functions, but their lack of decay, since they need not be integrable. Since $L^1(\mathbb{R}^d) * S_0(\mathbb{R}^d) \subseteq S_0(\mathbb{R}^d)$ we have the pointwise relationship $\mathscr{F}L^1(\mathbb{R}^d) \cdot S_0(\mathbb{R}^d) \subseteq S_0(\mathbb{R}^d)$, or $\mathscr{F}L^1(\mathbb{R}^d) \hookrightarrow \mathscr{M}(S_0)(\mathbb{R}^d)$ (the pointwise multipliers of $S_0(\mathbb{R}^d)$). Hence it is enough that regularization takes place in the form of pointwise multiplication with any function $h \in S_0(\mathbb{R}^d)$, typically $h_n(t) = h(t/n)$ for $n \to \infty$, with $h(0) = 1$. That indeed all known classical summability kernels are in fact elements of $S_0(\mathbb{R}^d)$ has been investigated in some detail in joint work with F. Weisz ([58, 59, 60]). Of course choices such as the Gauss-Weierstrass kernel $g(t) = e^{-\pi|t|^2}$, the inverse exponential $h(t) = exp(-|t|)$ or $h(t) = 1/(1+t^2)$ on $\mathbb{R}$ come to mind.

FURTHER APPLICATIONS, COMMENTS, OUTLOOK

So far we have only outlined some general principles where the setting of Banach Gelfand triples, and specifically the S_0-BGT come very handy and natural. In the rest of this paper let us just give some indications about further areas where such a setting appears to be quite natural.

Generalized stochastic processes

Already the PhD thesis of A.J.E.M. Janssen [81] indicates that generalized stochastic processes can be modeled appropriately using distribution theoretic methods. His space of test functions did not allow for compactly supported elements, hence he could not define the support of linear functionals in his setting. In this respect the setting of the BGT (S_0, L^2, S_0') is more suitable for a treatment of generalized stochastic processes. We can give only a quick indication of how this works (up to the topic of "spectral representations" of stationary stochastic processes), see [52].

First of all we view a *generalized stochastic process* as a generalization of an ordinary stochastic process, in the sense that an ordinary stochastic process assigns to each $x \in \mathbb{R}^d$ some random variable, abstractly speaking some element $\rho(x)$ in some Hilbert space[23] [of L^2-functions on some measure space, usually with expected value $E(X) = 0$]. As in

---

[23] Usually it is the set of all square-integrable functions random variables with zero expectation over some probability space, but this does not matter in our description of basic concepts.

the case of regular distributions one can integrate against a test function, i.e. extend the mapping $x \mapsto \rho(x) \in \mathcal{H}$ to a linear mapping

$$k \mapsto \rho(k) := \int_{\mathbb{R}^d} k(x)\rho(x)dx,$$

which is well defined at least for $C_c(\mathbb{R}^d)$, the space of compactly supported, continuous and complex-valued functions on $\mathbb{R}^d$. For us $S_0(\mathbb{R}^d)$ is more attractive as a (Banach) space of test functions and therefore we give the following definition:

Definition 10. We call a bounded linear mapping $\rho : f \mapsto \rho(f)$ from $S_0(\mathbb{R}^d)$ into some Hilbert space $\mathcal{H}$ a *generalized stochastic process*, for short a GSP.

In the standard approach to stochastic processe it is quite cumbersome, at least from the technical point of view, to check the existence of an autocorrelation function (resp. distribution) or to provide the spectral representation of a GSP, using vector-valued measures, such things become quite smooth and natural in our setting:

Definition 11. For any GSP ρ one defines its Fourier transform $\hat{\rho}$ via

$$\hat{\rho}(f) := \rho(\hat{f}), \quad \forall f \in S_0(\mathbb{R}^d). \tag{60}$$

Obviously the inverse Fourier transform of a GSP is defined in an analogous manner, and thus every GSP has a spectral representation in this sense. An important object for GSPs is the autocorrelation of such a process, which is given as follows:

Definition 12. Let ρ be a GSP. The **autocovariance** is characterized via

$$\langle \sigma_\rho, f \otimes g \rangle := \langle \rho(f) | \rho(\bar{g}) \rangle \; \forall f, g \in S_0(\mathbb{R}^d). \tag{61}$$

Theorem 15. *For a GSP ρ the following properties are equivalent:*
a) ρ stationary $\iff$ σ_ρ diagonally invariant, i.e. $L_{(x,x)}\sigma_\rho = \sigma_\rho \; \forall x \in \mathbb{R}^d$;
b) ρ bounded $\iff$ σ_ρ extends in a unique way to a bimeasure on $\mathbb{R}^d \times \mathbb{R}^d$;
c) ρ orthogonally scattered
  $\iff$ σ_ρ has support on the diagonal, i.e. $supp(\sigma_\rho) \subseteq \Delta_{\mathbb{R}^d} := \{(x,x) \mid x \in \mathbb{R}^d\}$;
  $\iff$ there exists a positive and translation bounded measure τ_ρ with:

$$\langle \sigma_\rho, f \otimes g \rangle = \langle \tau_\rho, fg \rangle \; \forall f, g \in S_0(\mathbb{R}^d).$$

Corollary 1. *A GSP ρ is bounded and orthogonally scattered if and only if there exists a bounded measure μ_ρ on $\mathbb{R}^d$ such that*

$$\langle \sigma_\rho, f \otimes g \rangle = \langle \mu_\rho, fg \rangle = \int_{\mathbb{R}^d} f(x)g(x)d\mu_\rho(x) \quad \forall f, g \in S_0(\mathbb{R}^d) \tag{62}$$

These statements should only indicate that the BGT (S_0, L^2, S_0') is also very helpful in this context, and therefore likely to be useful in the context of stochastic signal processing, where most often differentiation does not play any role (which in turn would justify using the Schwartz space $\mathscr{S}(\mathbb{R}^d)$ instead), cf. also the PhD thesis of B. Keville ([89]). There is more recent work using such tools by P. Wahlberg ([114, 115]).

Modulation spaces and Coorbit Theory

The Banach Gelfand triple (S_0, L^2, S_0') is just a prototype of the much more general family of modulation spaces, introduced in the early 80's (see ([44, 46])). The by now classical modulation spaces $M_s^{p,q}(\mathbb{R}^d)$ have been modeled in similarity to the family of (inhomogenous) Besov spaces and have similar properties. Moreover, the classical L^2–Sobolev spaces belong to both families (by choosing $p = 2 = q$). The parameter $s \in \mathbb{R}$ is the most important one, describing the smoothness. The family of modulation spaces is closed under duality (at least for finite parameters) or complex interpolation. A summary of the state of the art is given in the survey article ([47], in the special issue of STSIP on modulation spaces).

In the last few years these spaces have found a lot of interest both as a family of Banach spaces of (tempered) distributions of its own right, but above all as a natural tool to describe pseudo-differential operators ([109, 104, 110, 72, 91, 73, 70, 111] and many others, or Chap. 14 of ([69]).)

At the beginning there was the impression that the defining property of a modulation space is the fact that it is a *Wiener amalgam space* (see ([43, 78]) on the Fourier transform side, meaning that it is characterized by *uniform* decomposition of $\hat{f}$, for $f \in \mathscr{S}'(\mathbb{R}^d)$ (as opposed to standard dyadic decompositions used for Besov spaces, see ([97, 113])), or perhaps because a mixed norm-space was used over the TF-plane $\mathbb{R}^d \times \widehat{\mathbb{R}}^d$, with a specific order of integration (first along the time axis, with respect to the L^p-norm, and then in the frequency direction, using an L^q-norm with polynomial weight m), be it a partially discrete or continuous norm of the form

$$\|f\|_{M_m^{p,q}} = \|V_g f\|_{L_m^{p,q}} = \left(\int_{\mathbb{R}^d} \left(\int_{\mathbb{R}^d} |V_g f(x, \omega)|^p m(x, \omega)^p \, dx \right)^{q/p} d\omega \right)^{1/q}, \qquad (63)$$

However, soon the time-frequency point of view suggested to make also use of radial symmetric weights in phase space, not only of weights depending only on the frequency parameter, but rather on polynomial weights of the form $v_s(x, \omega) := (1 + |x|^2 + |\omega|^2)^{s/2}$. The advantage of the corresponding space $M_{v_s}^p(\mathbb{R}^d)$, defined via a weighted L^p-condition with weight $v_s, s \in \mathbb{R}$, on the short-time Fourier transform $V_{g_0} f$ (with respect to any non-zero window g_0, say the Gauss function) is the fact, that they are invariant, not only under the Fourier transform, but even under fractional Fourier transform (and even the whole metaplectic group, cf. ([69]), Chap.9.4).

The realization that modulation spaces and the classical family of Besov-Triebel-Lizorkin spaces have a lot in common, namely the fact that they can be described using so-called representation coefficients of (square-) integrable and irreducible group representations[24] had a great impact on the view on modulation spaces. They appear now as a special example of a more general principle, which is described through the theory of so-called coorbit spaces $\mathscr{C}o(Y)$ (see ([48])). From this point of view modulation spaces are those spaces which are described by the (global) behaviour of the STFT of its

---

[24] For the affine $ax + b$-group one obtains the continuous wavelet transform, while one has the STFT in the case of the Heisenberg group $\mathbb{R}^d \times \widehat{\mathbb{R}}^d \times \mathbb{T}$, using the Schrödinger representation on $\mathscr{H} = L^2(\mathbb{R}^d)$.

elements, expressed by some solid and translation invariant Banach space of functions over phase space $\mathbb{R}^d \times \widehat{\mathbb{R}}^d$.

METAPLECTIC OPERATORS AND SCHRÖDINGER EQUATION

In this last section we try to indicate that the invariance properties of the BGT (S_0, L^2, S_0') can be used to describe in the case of quadratic Hamiltonians the properties of solutions of the Schrödinger equation in the BGT setting.

Metaplectic and Heisenberg–Weyl invariance properties

Recall that the metaplectic group $\mathrm{Mp}(2d, \mathbb{R})$ is the unitary representation of the connected double covering of the symplectic group $\mathrm{Sp}(2d, \mathbb{R})$ (see e.g. [62]). The metaplectic group is generated by the following elementary unitary operators:

- The Fourier transform $\widehat{J} = i^{-n/2} F$, that is

$$\widehat{J}\psi(x) = \int_{\mathbb{R}^d} e^{2\pi i x \cdot x'} \psi(x') dx' \qquad (64)$$

whose projection on $\mathrm{Sp}(2d, \mathbb{R})$ is the standard symplectic matrix $J = \begin{pmatrix} 0 & I \\ -I & 0 \end{pmatrix}$;

- The "chirps" $\widehat{V_{-P}}$ defined, for $P = P^T$, by

$$\widehat{V_{-P}}\psi(x) = e^{2\pi i P x \cdot x} \psi(x) \qquad (65)$$

and whose projection on $\mathrm{Sp}(2, \mathbb{R})$ are the symplectic shears $\begin{pmatrix} I & 0 \\ P & I \end{pmatrix}$;

- The unitary changes of variables, defined for invertible L, by

$$\widehat{M_{L,m}}\psi(x) = i^m \sqrt{|\det L|}\, \psi(Lx) \qquad (66)$$

where the integer m corresponds to a choice of $\arg \det L$; the projection of $\widehat{M_{L,m}}$ on $\mathrm{Sp}(2d, \mathbb{R})$ is $\begin{pmatrix} L^{-1} & 0 \\ 0 & L^T \end{pmatrix}$.

Proposition 2. *The Segal algebra $S_0(\mathbb{R}^d)$ is invariant under the action of $\mathrm{Mp}(2d, \mathbb{R})$; in particular $\psi \in S_0(\mathbb{R}^d)$ if and only if $F\psi \in S_0(\mathbb{R}^d)$.*

Proof. This is an immediate property of the metaplectic covariance property

$$W(\widehat{S}\psi)(z) = W\psi(S^{-1}z) \qquad (67)$$

of the Wigner distribution ($S \in \mathrm{Sp}(2d, \mathbb{R})$ the projection of $\widehat{S} \in \mathrm{Mp}(2d, \mathbb{R})$) and of the characterization given in Lemma (2). $\qquad \square$

Further references are [40, 43, 44, 62, 34]. Finally let us mention that $S_0(\mathbb{R}^d)$ can be charactized via Wilson bases (see [53]) and local Fourier bases (see [76]) . One can show that $f \in L^2(\mathbb{R}^d)$ is in $S_0(\mathbb{R}^d)$ if and only if it has Wilson coefficients in $\ell^1(I)$, where I is essentially a half-space in $\mathbb{Z}^d \times \mathbb{Z}^d$. In this sense Wilson bases over $\mathbb{R}^d$ are like the Fourier basis (defining $A(\mathbb{T})$ and $PM(\mathbb{T})$) for the torus group.

The construction of Wilson bases was published by Daubechies/Jaffard/Journe in [32]. This author learned about Wilson bases from I. Daubechies already in 1989 it was possible to publish the follow-up result (connecting it with modulation spaces) already one year later in [51]. Wilson bases in the discrete domain are given in [14, 15, 16]. They have also been used to prove the kernel theorem in [53].

The Schrödinger equation for quadratic Hamiltonians

The metaplectic group $\mathrm{Mp}(2d, \mathbb{R})$ plays a crucial role in quantum mechanics because of the following property. Consider a Hamiltonian function H which is quadratic in the x_j, p_k variables:

$$H(x, p) = \frac{1}{2}(x, p)M(x, p)^T \tag{68}$$

(M a real symmetric $2d \times 2d$ matrix). Such Hamiltonians generalize the "harmonic oscillator"

$$H(x, p) = \frac{1}{2m}(|p|^2 + m^2 \omega^2 |x|^2) \tag{69}$$

familiar from elementary physics. The solution of the Hamilton equations

$$\frac{dx}{dt} = \frac{\partial H}{\partial p} \ , \ \frac{dp}{dt} = -\frac{\partial H}{\partial x} \tag{70}$$

is explicitly given by

$$(x(t), p(t)) = S_t(x(0), p(0))$$
$$\text{with } S_t = \exp(tJM);$$

since JM is in the Lie algebra of $\mathrm{Sp}(2d, \mathbb{R})$ we have $S_t \in \mathrm{Sp}(2d, \mathbb{R})$ for every $t \in \mathbb{R}$. Now, when t varies the symplectic matrices S_t describe a differentiable curve in $\mathrm{Sp}(2d, \mathbb{R})$ passing through the identity at time $T = 0$ (in fact, (S_t) is a one-parameter subgroup). It follows from a classical result from the theory of covering spaces (the "unique path lifting property") that there exists a unique path $t \longmapsto \widehat{S}_t$ in $\mathrm{Mp}(2d, \mathbb{R})$ whose projection is precisely the path $t \longmapsto S_t$; in particular $\widehat{S}_0$ is the identity in $\mathrm{Mp}(2d, \mathbb{R})$. The interest of these considerations comes from the following well-known result, whose second part trivially follows from Proposition 2 above:

Proposition 3. *(i) Consider the Schrödinger equation*

$$i\hbar \frac{\partial \psi}{\partial t} = H(x, -i\hbar \partial_x)\psi \tag{71}$$

where $H(x, -i\hbar\partial_x)$ is the partial differential operator obtained by Weyl quantization from the quadratic Hamiltonian H. Its solution is given by the formula

$$\psi(x,t) = \widehat{S}_t \psi(x,0). \tag{72}$$

(ii) Thus, if $\psi(\cdot,0) \in S_0(\mathbb{R}^d)$ then $\psi \in S_0(\mathbb{R}^d)$ and the solution depends continuously in the S_0-norm on the time parameter t.

Part (i) has been known for a very long time, it has been implicit in the early work of Hermann Weyl ([119]), and proofs can be found in [62].

We said above that there is an alternative description of the metaplectic group $Mp(2d,\mathbb{R})$ in terms of generators. We set

$$W(x,x') = \frac{1}{2}Px \cdot x - Lx \cdot x' + \frac{1}{2}Qx' \cdot x' \tag{73}$$

where P (resp. Q) and L are as above, and consider the Fourier integral operator $\widehat{S}_{W,m}$ defined by

$$\widehat{S}_{W,m}\psi(x) = \left(\frac{1}{2\pi i\hbar}\right)^{n/2} \Delta(W) \int_{\mathbb{R}^d} e^{\frac{i}{\hbar}W(x,x')}\psi(x')dx'$$
$$\Delta(W) = i^m \sqrt{|\det L|}.$$

One verifies, by simple inspection, that $\widehat{S}_{W,m}$ is easily expressed in terms of the elementary generators of $Mp(2d,\mathbb{R})$, in fact:

$$\widehat{S}_{W,m} = \widehat{V_{-P}} \widehat{M_{L,m}} \widehat{JV_{-Q}}. \tag{74}$$

It follows that $\widehat{S}_{W,m} \in Mp(2d,\mathbb{R})$; one proves that the Fourier integral operators $\widehat{S}_{W,m}$ generate the metaplectic group, more precisely: every $\widehat{S} \in Mp(2d,\mathbb{R})$ can be written (non-uniquely) as a product of exactly two such operators: $\widehat{S} = \widehat{S}_{W,m}\widehat{S}_{W',m'}$. In the case of the Schrödinger equation, it turns out that if the Hessian matrix M of the Hamiltonian is non-singular, the operators $\widehat{S}_t$ are, except for a set of exceptional values of t, of the type $\widehat{S}_{W,m}$.

Note that more concrete realizations of this principle allow E. Cordero and coauthors to derive Strichartz-type estimates for the solutions of the Schrödinger equation (see [26, 25, 27]).

A Fresh Look on Dirac's Functional Calculus

In the case of matrices unitary matrices U are the most useful ones. A complex-valued $n \times n$-matrix U is *unitary* if and only one has

$$U * U' = Id_n = U' * U. \tag{75}$$

Sometimes it is also interesting to consider rectangular matrices U of size $m \times n$, satisfying one or the other of these two properties. If $n \leq m$ it is still possible that $U' * U = Id_n$, or equivalently, that the columns of U form an *orthonormal system*. In particular, they are a linear independent system of vectors. Alternatively for $n \geq m$ one can have a "perfect set of generators", or a so called *tight frame*, satisfying $U * U' = Id_m$, or $x = \sum_{k=1}^{n} \langle x, u_k \rangle u_k$ for all $x \in \mathbb{C}^m$ [25]. It is obvious that only for the case $n = m$ (finite dimension!) these two properties are *equivalent*, while in one can have one without the other, also for the case $\mathcal{H} = \ell^2$.

Let us now go for the analogue of these two identities in the case of the Fourier transform. Recall that $\mathscr{F}$ is a unitary Banach Gelfand morphism, which however is typically used in the spirit of an *analysis mapping* $\mathscr{F} : f \mapsto (\langle f, \chi_s \rangle)$ or $(\langle f, \chi_s \rangle)$, while the inverse Fourier transform (*synthesis* of a function or distribution from "pure frequencies") is more in the spirit of the adjoint mapping. Since in both cases the kernel for the corresponding mapping is continuous and bounded (actually smooth), namely $K(x, s) = e^{-2\pi i s \cdot x}$ for $\mathscr{F}$ and $K(s, y) = e^{2\pi i s \cdot x}$ for $\mathscr{F}^{-1}$ the fact that they are corresponding to two mappings which are inverse to each other, i.e. they satisfy

$$\mathscr{F} \circ \mathscr{F}^{-1} = Id_{\mathcal{H}} = \mathscr{F}^{-1} \circ \mathscr{F}, \tag{76}$$

implies obviously that the composition of their symbols according to (58) has to result in the kernel of the identity operator. This brings us to a short discussion of the connection between the *Kronecker's Delta* and *Dirac's Delta*. Modern distribution theory (and in fact the kernel theorem) tell us that the identity mapping (or equivalently multiplication by the constant 1) is given by a kind of δ-distribution concentrated along the diagonal, namely the functional (we just use an "arbitrary symbol" reminding of this idea) δ_Δ, given as an element of $S_0'(\mathbb{R}^{2d})$ via the action $f \mapsto \int_{\mathbb{R}^d} f(t, t) dt$ for $f \in S_0(\mathbb{R}^{2d})$. In the matrix setting we can view the unit matrix Id_n as a collection of unit vectors, which is clearly described in an equivalent way by the Kronecker δ-function Δ_{kron}. Viewed as a matrix kernel we have of course $\Delta_{kron} * x = x$. The continuous analogue of such a situation is a kernel $K(x, y)$ such that

$$\int_{\mathbb{R}^d} K(x, y) f(y) dx = f(x),$$

so somehow one should have for any fixed x that $K(x, y)$ represents δ_x (in the sense of the point measure at x). However, starting from a general distributional kernel $x \in S_0'(\mathbb{R}^{2d})$, even if we write it symbolically in the form the "restriction" to x, i.e. the distribution $K(x, \cdot)$ does not make sense a priori. So we should probably really interpret the Dirac symbol as a continuous analogue of the Kronecker symbol.

In books and papers on *quantum mechanics*, (using slightly different symbols) one often finds relationships such as the following formulas:

$$\langle \chi_s, \chi_t \rangle = \delta(s - t), \, s, t \in \mathbb{R}^d, \tag{77}$$

---

[25] One can show that these systems are nothing else than orthogonal l projections of orthonormal bases in higher dimensions.

as a replacement for the orthogonality relation, mostly in the form

$$\langle \chi_s, \chi_t \rangle = \delta_t(s), \quad s, t \in \mathbb{R}^d, \tag{78}$$

which is view-point similar to the interpretation of the $n \times n$ identity matrix as a collection of unit vectors (cf. (75)). On the other hand one finds expressions such as

$$Id_{\mathscr{H}} = \int_{\mathbb{R}^d} |\chi_s\rangle\langle\chi_s| ds \tag{79}$$

claiming that the identity operator can viewed as a superposition of rank-one operators (cf. [98], Chap.1), evidently expressing *completeness* of the system of characters $\widehat{\mathbb{R}^d}$.

We can say: both formulas can be given *their proper meanings* in different ways. First of all we view the Fourier transform and its inverse (or equivalently its conjugate kernel) as Banach Gelfand triple morphisms. In some cases it is the "how", i.e. the way how the transformation is first defined, at least on the space of test function $S_0(\mathbb{R}^d)$, which catches our intention. One is lead to believe that the Fourier transform is primarily an integral transform, which has the Lebesgue space $L^1(\mathbb{R}^d)$ as natural domain. On the other hand (when we talk about Fourier synthesis) the w^*-convergence is helpful, because it does not make sense to interpret the Fourier inversion formula (we write f as a superposition of pure frequencies) in any other natural topology.

Despite the fact that $\chi_s \in C_b(\mathbb{R}^d) \subset S_0'(\mathbb{R}^d)$ is only applicable (via integration) on test functions from $S_0(\mathbb{R}^d)$, concrete (hard analysis) arguments allow to show that they determine even a *unitary Banach Gelfand Gelfand triple automorphism*. Obviously one has $\mathscr{F} \circ \mathscr{F}^{-1} = Id$ as well as $\mathscr{F}^{-1} \circ \mathscr{F} = Id$. Both identities can be useful, e.g. in order to show that the Fourier transform is injective, or that a given function is the Fourier transform of another function (or distribution) of the same kind. If we try to describe these two mappings through their kernels and try to compose the kernels using the standard composition formula for kernels we end up with exactly the relations (**??**) and (79) respectively. Of course, one can combine these kernels with regularizing operators, in order to have kernels from $S_0(\mathbb{R}^{2d})$, and in this case the composition can be carried out in the usual way, using Riemannian integrals. Their products are then well defined (according to (58)), and then the claim is: in the w^*-sense the limit of these kernels is the (kernel of the) identity operator, or $\delta(t - s)$, which is in standard terminology the tensor product of δ_0 (the usual Dirac measures at zero) with the function constant one, rotated by 45 degrees. From this point of view Dirac's intention might not have been too far away from simply going from the well-known Kronecker symbol with discrete entries to a continuous version. The fact that the Fourier transform is using building blocks from "outside the Hilbert space" gives troubles to anybody who tries to stay within the world of Hilbert spaces, while the viewpoint of Banach Gelfand triples (in our view only a convenient realization of the idea underlying the concept of rigged Hilbert spaces) opens up a new view and a technically sound perspective. As mathematician we suggest therefore to provide in any concrete application the details of the involved BGT-morphism instead of relying on the symbolic calculus per se. Most likely one can overcome the purely technical problems using the idea of approximation by test functions using regularization ideas, while obviously at critical points (where the symbolic manipulations lead to misleading conclusions) such justification will fail for good reasons.

ACKNOWLEDGMENTS

The author gratefully acknowledges support in the preparation of this manuscript by a number of present and earlier coworkers, especially Werner Kozek, Maurice De Gosson, Harald Stockinger, and Franz Luef. He was partially supported by the Marie Curie Excellence Grant project EUCETIFA (FP6-517154). Thanks go also to professor Abul Hasan Siddiqi for inviting the author to the ISIAM Conference in Agra, January 2009. The notes as it was held (with a number of illustrations) can be viewed at `http://univie.ac.at/nuhag-php/dateien/talks/1055_agra09SIAM.pdf`

REFERENCES

1. Extrapolating a band-limited function from its samples taken in a finite. *IEEE Trans. Inf. Theory*, 32:464–470, 1986.
2. S. Albeverio, R. Bozhok, and V. Koshmanenko. The rigged Hilbert spaces approach in singular perturbation Theory. *Rep. Math. Phys.*, 58(2):227–246, 2006.
3. S. T. Ali, J.-P. Antoine, and J. P. Gazeau. Continuous frames in Hilbert space. *Ann. Phys.*, 222(1):1–37, 1993.
4. S. T. Ali, J.-P. Antoine, and J.-P. Gazeau. *Coherent States, Wavelets and their Generalizations.* Graduate Texts in Contemporary Physics. Springer, New York, 2000.
5. A. Amann, H. G. Feichtinger, A. Klotz, G. Kracher, T. Werther, H. Gilly, and M. Baubin. CPR Artefact Removal in ECG Signals Using Gabor Multipliers. *IEEE Trans. Biomedical Engineering*, 2007.
6. J.-P. Antoine. Quantum Mechanics Beyond Hilbert Space. In A. Bohm, H.-D. Doebner, and P. Kileanowski, editors, *Irreversibility and Causality, Semigroups and Rigged Hilbert Spaces. A Selection of Articles Presented at the 21st International Colloquium on Group Theoretical Methods in Physics (ICGTMP) at Goslar, Germany, 16-21 July, 1996.*, volume 504 of *Lecture Notes in Physics*, Berlin, 1998. Springer Verlag.
7. J.-P. Antoine and C. Trapani. Partial Inner Product Spaces - Theory and Applications. 2009.
8. L. Auslander and Y. Meyer. A generalized Poisson summation formula. *Appl. Comput. Harmon. Anal.*, 3(4):372–376, 1996.
9. P. Balazs. *Regular and Irregular Gabor Multiplier With Application To Psychoacoustic Masking.* PhD thesis, University of Vienna, 2005.
10. M. J. Bastiaans. Gabor's expansion of a signal into Gaussian elementary signals. *Proc. IEEE*, 68(4):538–539, 1980.
11. J. J. Benedetto. *Spectral Synthesis.* Academic Press, Inc, Francisco, 1975.
12. J. J. Benedetto and G. E. Pfander. Frame expansions for Gabor multipliers. *Appl. Comput. Harm. Anal.*, 20(1):26–40, 2006.
13. P. Boggiatto, E. Cordero, and K. Gröchenig. Generalized anti-Wick operators with symbols in distributional Sobolev spaces. *Integral Equations Operator Theory*, 48(4):427–442, 2004.
14. H. Bölcskei, H. G. Feichtinger, K. Gröchenig, and F. Hlawatsch. Discrete-time Wilson expansions. In *Proc. IEEE-SP 1996 Int. Sympos. Time-Frequency Time-Scale Analysis*, pages 525–528, jun 1996.
15. H. Bölcskei, K. Gröchenig, F. Hlawatsch, and H. G. Feichtinger. Oversampled Wilson expansions. *IEEE Signal Proc. Letters*, 4(4):106–108, 1997.
16. H. Bölcskei and F. Hlawatsch. Oversampled Wilson-type cosine modulated filter banks with linear phase. In *Asilomar Conf. on Signals, Systems, and Computers*, pages 998–1002, nov 1996.

17. W. Braun and H. G. Feichtinger. Banach spaces of distributions having two module structures. *J. Funct. Anal.*, 51:174–212, 1983.

18. P. L. Butzer, S. Jansche, and R. L. Stens. Functional analytic methods in the solution of the fundamental theorems on best-weighted algebraic approximation. In *Approximation theory, Proc. 6th Southeast. Approximation Theory Conf., Memphis/TN (USA) 1991*, volume 138 of *Lect. Notes Pure Appl. Math.*, pages 151–205, 1992.

19. O. Christensen. *An Introduction to Frames and Riesz Bases*. Applied and Numerical Harmonic Analysis. Birkhäuser, Boston, 2003.

20. T. A. C. M. Claasen and W. F. G. Mecklenbraeuker. The Wigner distribution - a tool for time-frequency signal analysis. I: Continuous-time signals. *Philips J. Res.*, 35:217–250, 1980.

21. T. A. C. M. Claasen and W. F. G. Mecklenbraeuker. The Wigner distribution - a tool for time-frequency signal analysis. II: Discrete-time signals. *Philips J. Res.*, 35:276–300, 1980.

22. T. A. C. M. Claasen and W. F. G. Mecklenbraeuker. The Wigner distribution - a tool for time-frequency signal analysis. III: Relations with other time-frequency signal transformations. *Philips J. Res.*, 35:372–389, 1980.

23. R. R. Coifman and G. Weiss. *Analyse harmonique non-commutative sur certains espaces homog'enes. Etude de certaines int'egrales singuli'eres. (Non-commutative harmonic analysis on certain homogeneous spaces. Study of certain singular integrals.).*, volume 242 of *Lecture Notes in Mathematics*. Springer-Verlag, Berlin-Heidelberg-New York, 1971.

24. E. Cordero, H. G. Feichtinger, and F. Luef. Banach Gelfand triples for Gabor analysis. In *'Pseudo-Differential Operators, Quantization and Signals', C.I.M.E. 2006*, Lecture Notes in Mathematics. Springer, 2007.

25. E. Cordero and F. Nicola. Some new Strichartz estimates for the Schrödinger equation. *J. Differ. Equations*, 245(7):1945–1974, 2008.

26. E. Cordero and F. Nicola. Strichartz estimates in Wiener amalgam spaces for the Schrödinger equation. *Math. Nachr.*, 281(1):25–41, 2008.

27. E. Cordero, F. Nicola, and L. Rodino. Sparsity of Gabor representation of Schrödinger propagators. *Applied and Computational Harmonic Analysis,*, In Press, Corrected Proof,(,):– „ 2008,.

28. E. Cordero and L. Rodino. Wick calculus: a time-frequency approach. *Osaka J. Math.*, 42(1):43–63, 2005.

29. S. Dahlke, M. Fornasier, H. Rauhut, G. Steidl, and G. Teschke. Generalized coorbit theory, Banach frames, and the relation to *alpha*-modulation spaces. *Proc. Lond. Math. Soc. (3)*, 96(2):464–506, 2008.

30. S. Dahlke, G. Steidl, and G. Teschke. Coorbit spaces and Banach frames on homogeneous spaces with applications to the sphere. *Adv. Comput. Math.*, 21(1-2):147–180, 2004.

31. S. Dahlke, G. Steidl, and G. Teschke. Weighted coorbit spaces and Banach frames on homogeneous spaces. *J. Fourier Anal. Appl.*, 10(5):507–539, 2004.

32. I. Daubechies, S. Jaffard, and J. L. Journ'e. A simple Wilson orthonormal basis with exponential decay. *SIAM J.Math. Anal.*, 22:554–573, 1991.

33. I. Daubechies, H. J. Landau, and Z. Landau. Gabor time-frequency lattices and the Wexler-Raz identity. *J. Fourier Anal. Appl.*, 1(4):437–478, 1995.

34. M. A. de Gosson. *Symplectic Geometry and Quantum Mechanics*, volume 166 of *Operator Theory: Advances and Applications. Advances in Partial Differential Equations*. Birkhäuser, Basel, 2006.

35. R. de la Madrid. Rigged Hilbert space approach to the Schrödinger equation. *J. Phys. A, Math. Gen.*, 35(2):319–342, 2002.

36. R. de la Madrid. The role of the rigged Hilbert space in quantum mechanics. *Eur. J. Phys.*, 26(2):277–312, 2005.

37. R. de la Madrid, A. Bohm, and M. Gadella. Rigged Hilbert space treatment of continuous spectrum. *Fortschr. Phys.*, 50(2):185–216, 2002.

38. M. Dörfler and B. Torresani. On the time-frequency representation of operators and generalized Gabor multiplier approximations. *preprint*, 2007.

39. H. G. Feichtinger. Un espace de Banach de distributions temp'er'ees sur les groupes localement compacts ab'eliens. *Compt.Rend.Acad.Sci.Paris, Ser.A*, 290(17):791–794, 1980.

40. H. G. Feichtinger. A characterization of minimal homogeneous Banach spaces. *Proc. Amer. Math. Soc.*, 81:55–61, 1981.

41. H. G. Feichtinger. On a new Segal algebra. *Monatsh. Math.*, 92:269–289, 1981.

42. H. G. Feichtinger. A new family of functional spaces on the Euclidean n-space. In *Proc.Conf. on Theory of Approximation of Functions*, Teor. Priblizh., 1983.

43. H. G. Feichtinger. Banach convolution algebras of Wiener type. In *Proc. Conf. on Functions, Series, Operators, Budapest 1980*, volume 35 of *Colloq. Math. Soc. Janos Bolyai*, pages 509–524. North-Holland, Amsterdam, Eds. B. Sz.-Nagy and J. Szabados. edition, 1983.

44. H. G. Feichtinger. Modulation spaces on locally compact Abelian groups. Technical report, January 1983.

45. H. G. Feichtinger. An elementary approach to the generalized Fourier transform. In T. Rassias, editor, *Topics in Mathematical Analysis*, pages 246–272. World Sci.Pub., 1989.

46. H. G. Feichtinger. Modulation spaces of locally compact Abelian groups. In R. Radha, M. Krishna, and S. Thangavelu, editors, *Proc. Internat. Conf. on Wavelets and Applications*, pages 1–56, Chennai, January 2002, 2003. New Delhi Allied Publishers.

47. H. G. Feichtinger. Modulation Spaces: Looking Back and Ahead. *Sampl. Theory Signal Image Process.*, 5(2):109–140, 2006.

48. H. G. Feichtinger and K. Gröchenig. Banach spaces related to integrable group representations and their atomic decompositions, I. *J. Funct. Anal.*, 86:307–340, 1989.

49. H. G. Feichtinger and K. Gröchenig. Banach spaces related to integrable group representations and their atomic decompositions, II. *Monatsh. Math.*, 108:129–148, 1989.

50. H. G. Feichtinger and K. Gröchenig. Gabor frames and time-frequency analysis of distributions. *J. Funct. Anal.*, 146(2):464–495, 1997.

51. H. G. Feichtinger, K. Gröchenig, and D. F. Walnut. Wilson bases and modulation spaces. *Math. Nachr.*, 155:7–17, 1992.

52. H. G. Feichtinger and W. Hörmann. Harmonic analysis of generalized stochastic processes on locally compact abelian groups, 1990.

53. H. G. Feichtinger and W. Kozek. Quantization of TF lattice-invariant operators on elementary LCA groups. In H. Feichtinger and T. Strohmer, editors, *Gabor Analysis and Algorithms. Theory and Applications.*, Applied and Numerical Harmonic Analysis, pages 233–266, 452–488, Boston, MA, 1998. Birkhäuser Boston.

54. H. G. Feichtinger, W. Kozek, and F. Luef. Gabor Analysis over finite Abelian groups. *Appl. Comput. Harmon. Anal.*, 26(2):230–248, 2009.

55. H. G. Feichtinger and F. Luef. Wiener Amalgam Spaces for the Fundamental Identity of Gabor Analysis. *Collect. Math.*, 57(Extra Volume (2006)):233–253, 2006.

56. H. G. Feichtinger, F. Luef, and T. Werther. A Guided Tour from Linear Algebra to the Foundations of Gabor Analysis. In *Gabor and Wavelet Frames*, volume 10 of *IMS Lecture Notes Series*, pages 1–49. 2007.

57. H. G. Feichtinger and K. Nowak. A first survey of Gabor multipliers. In H. Feichtinger and T. Strohmer, editors, *Advances in Gabor Analysis*, Appl. Numer. Harmon. Anal., pages 99–128. Birkhäuser, 2003.

58. H. G. Feichtinger and F. Weisz. The Segal algebra $S_0(R^d)$ and norm summability of Fourier series and Fourier transforms. *Monatsh. Math.*, 148:333–349, 2006.

59. H. G. Feichtinger and F. Weisz. Wiener amalgams and pointwise summability of Fourier transforms and Fourier series. *Math. Proc. Cambridge Philos. Soc.*, 140:509–536, 2006.

60. H. G. Feichtinger and F. Weisz. Herz spaces and summability of Fourier transforms. *Math. Nachr.*, 281(3):1–16, 2008.

61. H. G. Feichtinger and G. Zimmermann. A Banach space of test functions for Gabor analysis. In H. Feichtinger and T. Strohmer, editors, *Gabor Analysis and Algorithms: Theory and Applications*, Applied and Numerical Harmonic Analysis, pages 123–170, Boston, MA, 1998. Birkhäuser Boston.

62. G. B. Folland. *Harmonic Analysis in Phase Space*. Princeton University Press, Princeton, N.J., 1989.

63. M. Fornasier and H. Rauhut. Continuous Frames, Function Spaces, and the Discretization Problem. *J. Fourier Anal. Appl.*, 11(3):245–287, 2005.

64. J.-P. Gabardo and D. Han. Frames associated with measurable spaces. *Adv. Comput. Math.*, 18(2-4):127–147, 2003.

65. D. Gabor. Theory of communication. *J. IEE*, 93(26):429–457, 1946.

66. M. Gadella and F. Gomez. A unified mathematical formalism for the Dirac formulation of quantum mechanics. *Found. Phys.*, 32(6):815–869, 2002.
67. K. Gröchenig. Describing functions: atomic decompositions versus frames. *Monatsh. Math.*, 112(3):1–41, 1991.
68. K. Gröchenig. An uncertainty principle related to the Poisson summation formula. *Stud. Math.*, 121(1), 1996.
69. K. Gröchenig. *Foundations of Time-Frequency Analysis*. Appl. Numer. Harmon. Anal. Birkhäuser Boston, Boston, MA, 2001.
70. K. Gröchenig. Composition and spectral invariance of pseudodifferential operators on modulation spaces. *J. Anal. Math.*, 98:65–82, 2006.
71. K. Gröchenig. Gabor frames without inequalities. *International Mathematics Research Notices 2007*, (23), 2007.
72. K. Gröchenig and C. Heil. Modulation spaces and pseudodifferential operators. *Integral Equations Operator Theory*, 34(4):439–457, 1999.
73. K. Gröchenig and C. Heil. Modulation spaces as symbol classes for pseudodifferential operators. In R. R. M. Krishna, editor, *Wavelets and Their Applications*, pages 151–170. Allied Publishers, Chennai, 2003.
74. K. Gröchenig, F. Hlawatsch, A. Klotz, G. Matz, and A. Skupch. Advanced Mathematical Models for the Design and Optimization of Low-Interference Wireless Multicarrier Systems. 2006.
75. K. Gröchenig and M. Leinert. Wiener's lemma for twisted convolution and Gabor frames. *J. Amer. Math. Soc.*, 17:1–18, 2004.
76. K. Gröchenig and S. Samarah. Non-linear approximation with local Fourier bases. *Constr. Approx.*, 16(3):317–331, 2000.
77. D. Han, K. Kornelson, D. Larson, and E. Weber. *Frames for Undergraduates.*, volume 40 of *Student Mathematical Library*. American Mathematical Society (AMS), Providence, RI, 2007.
78. C. Heil. An introduction to weighted Wiener amalgams. In M. Krishna, R. Radha, and S. Thangavelu, editors, *Wavelets and their Applications (Chennai, January 2002)*, pages 183–216. Allied Publishers, New Delhi, 2003.
79. F. Hlawatsch and P. Flandrin. The interference structure of the Wigner distribution and related time-frequency signal representations. In F. Hlawatsch and W. Mecklenbräuker, editors, *The Wigner Distribution - Theory and Applications in Signal Processing*, pages 59–133. Elsevier Science, 1997.
80. F. Hlawatsch and W. Kozek. The Wigner distribution of a linear signal space. *IEEE Trans. Signal Processing*, 41:1248–1258, 1993.
81. A. Janssen. *Application of the Wigner Distribution to Harmonic Analysis of Generalized Stochastic Processes. Dissertation.* 1979.
82. A. J. E. M. Janssen. Duality and biorthogonality for Weyl-Heisenberg frames. *J. Fourier Anal. Appl.*, 1(4):403–436, 1995.
83. J. P. Kahane. *Some Random Series of Functions.* 1968.
84. J. P. Kahane and P. G. Lemari'e Rieusset. Remarques sur la formule sommatoire de Poisson. *Studia Math.*, 109:303–316, 1994.
85. G. Kaiser. *A Friendly Guide to Wavelets*. Birkhäuser, Boston, 1994.
86. P. Kasperkovitz. Quasiclassical descriptions of quantum systems based on coherent states: Product formulae. *J. Phys. A, Math. Gen.*, 23(23):5493–5512, 1990.
87. Y. Katznelson. Une remarque concernant la formule de Poisson. *Studia Math.*, 19:107–108, 1967.
88. Y. Katznelson. *An Introduction to Harmonic Analysis. 2nd corr. ed.* Dover Publications Inc, New York, 1976.
89. B. Keville. *Multidimensional Second Order Generalised Stochastic Processes on Locally Compact Abelian Groups.* PhD thesis, Trinity College Dublin, 2003.
90. W. Kozek. *Matched Weyl-Heisenberg Expansions of Nonstationary Environments.* PhD thesis, University of Technology Vienna, 1997.
91. D. Labate. Pseudodifferential operators on modulation spaces. *J. Math. Anal. Appl.*, 262(1):242–255, 2001.
92. R. Larsen. *An Introduction to the Theory of Multipliers.* Springer, 1971.
93. N. Lerner. The Wick calculus of pseudo-differential operators and energy estimates. In J.-M. Bony and et al., editors, *New trends in microlocal analysis.*, pages 23–37. Springer, Tokyo, 1997.

94. V. Losert. A characterization of the minimal strongly character invariant Segal algebra. *Ann. Inst. Fourier*, 30:129–139, 1980.

95. V. Losert. A characterization of groups with the one-sided Wiener property. *J. Reine Angew. Math.*, 331:47–57, 1982.

96. S. MacLane. *Categories for the working mathematician. 4th corrected printing.* Graduate Texts in Mathematics, 5. New York etc.: Springer-Verlag. ix, 1988.

97. J. Peetre. *New Thoughts on Besov Spaces.* Mathematics Department, Duke University, Durham, 1976.

98. A. Perelomov. *Generalized Coherent States and their Applications.* Springer, Berlin, 1986.

99. H. Reiter. L^1-*algebras and Segal Algebras.* Springer, Berlin, Heidelberg, New York, 1971.

100. H. Reiter. Theta functions and symplectic groups. *Monatsh. Math.*, 97:219–232, 1984.

101. H. Reiter. *Metaplectic Groups and Segal Algebras.* Lect. Notes in Mathematics. Springer, Berlin, 1989.

102. H. Reiter and J. D. Stegeman. *Classical Harmonic Analysis and Locally Compact Groups. 2nd ed.* Clarendon Press, Oxford, 2000.

103. J. E. Roberts. Rigged Hilbert spaces in quantum mechanics. *Commun. Math. Phys.*, 3:98–119, 1966.

104. R. Rochberg and K. Tachizawa. Pseudodifferential operators, Gabor frames, and local trigonometric bases. In H. Feichtinger and T. Strohmer, editors, *Gabor Analysis and Algorithms: Theory and Applications*, pages 171–192, 453–488. Birkhäuser, Boston, 1998.

105. A. Ron and Z. Shen. Weyl-Heisenberg frames and Riesz bases in $L_2(\mathbb{R}^d)$. *Duke Math. J.*, 89(2):237–282, 1997.

106. W. Rudin. *Functional Analysis.* McGraw-Hill Book Company, New York, 1973.

107. K. Schnass. Gabor Multipliers. A self-contained survey. Master's thesis, University of Vienna, 2004.

108. G. Strang. *Introduction to Linear Algebra. 3rd ed.* Wellesley-Cambridge Press, Wellesley, MA, 2003.

109. K. Tachizawa. The boundedness of pseudodifferential operators on modulation spaces. *Math. Nachr.*, 168:263–277, 1994.

110. K. Tachizawa. A generalization of Calder'on-Vaillancourt's theorem. *RIMS Kokyuroku*, 1102:64–75, 1999.

111. J. Toft. Continuity properties for modulation spaces, with applications to pseudo-differential calculus. I. *J. Funct. Anal.*, 207(2):399–429, 2004.

112. R. Tolimieri and R. Orr. Poisson summation, the ambiguity function, and the theory of Weyl–Heisenberg frames. *J.Four. Anal. Appl.*, 1:233–247, 1995.

113. H. Triebel. *Theory of Function Spaces.*, volume 78 of *Monographs in Mathematics.* Birkhäuser, Basel, 1983.

114. P. Wahlberg. Vector-valued modulation spaces and localization operators with operator-valued symbols. *Integral Equations Oper. Theory*, 59(1):99–128, 2007.

115. P. Wahlberg and P. Schreier. Gabor discretization of the Weyl product for modulation spaces and filtering of nonstationary stochastic processes. *Appl. Comput. Harmon. Anal.*, 26:97–120, January 2009.

116. A. Weil. *L'integration dans les Groupes Topologiques et ses Applications.* Hermann and Cie, Paris, 1940.

117. F. Weisz. Inversion of the short-time Fourier transform using Riemannian sums. *J. Fourier Anal. Appl.*, 13(3):357–368, 2007.

118. J. Wexler and S. Raz. Discrete Gabor expansions. *Signal Proc.*, 21:207–220, 1990.

119. H. Weyl. *Gruppentheorie und Quantenmechanik.* S. Hirzel, Leipzig, 1928.

120. S. Wickramasekara and A. Bohm. Representation of semigroups in rigged Hilbert spaces: subsemigroups of the Weyl-Heisenberg group. *J. Math. Phys.*, 44(2):930–942, 2003.

Wavelet Analyses of Oil Prices, USD Variations and Impact on Logistics

M. Melek [a], A. Tokgozlu [b] and Z. Aslan [a]

[a] Beykoz Logistic School of Higher Education, 34805, Beykoz, Istanbul, Turkey

[b] Suleyman Demirel University, Faculty of Science, Isparta, Turkey

Abstract. This paper is related with temporal variations of historical oil prices and Dollar and Euro in Turkey. Daily data based on OECD and Central Bank of Turkey records beginning from 1946 has been considered. 1D-continuous wavelets and wavelet packets analysis techniques have been applied on data. Wavelet techniques help to detect abrupt changing's, increasing and decreasing trends of data. Estimation of variables has been presented by using linear regression estimation techniques. The results of this study have been compared with the small and large scale effects. Transportation costs of track show a similar variation with fuel prices. The second part of the paper is related with estimation of imports, exports, costs, total number of vehicles and annual variations by considering temporal variation of oil prices and Dollar currency in Turkey. Wavelet techniques offer a user friendly methodology to interpret some local effects on increasing trend of imports and exports data.

Keywords: Wavelet, crude oil prices, Dollar and Euro variations, imports, exports.

INTRODUCTION

Wavelet methods invented in mid-eighties have attracted attention of engineers, physicists, computer scientists and mathematicians for applications in different fields. Coifman and Daubechies have been given awards of U.S.A. in the year 2000 for their contribution in this field (Siddiqi et al., 2005). Wavelet methods have been applied for developing techniques which may reduce the exploration and production cost besides predicting fluctuation in oil markets. Furati et al., introduce this methodology and describe the results applied to a filed data. A wavelet based prediction procedure is studied and tested with real data. If the information of signal carrying specific information consists of different components the individual contribution of any component to the total signal is different.

DATA

The data from OECD and National Central Bank data (Dollar, EURO, oil prices, annual imports and exports values in Turkey) recorded in different ranges of 1946-2006 is processed. Each data are examined by using MATLAB Wavelet Packet and SPSS Program.

Jean Morlet, French geophysicist introduced the concept of wavelet, meaning small wave and studied wavelet transform as a new tool for seismic signal and analysis in 1982. In 1984, the collaboration of Morlet and Grossmann yielded detailed mathematical study of the continuous wavelet transform and their applications, [1-3]. Wavelet methodology has been applied to study of atmospheric turbulence, ocean wind waves, sea floor bathymetry, seismic data, environmental biology, electrocardiogram data, temperature variation and global warming [4,5]. Wavelet methodology is capable of revealing aspects of data that other signal analysis techniques lack, aspects like trends, break down points, discontinuities in higher derivatives and self similarity [9-12].

CP1146, *Modelling of Engineering and Technological Problems*, edited by A. H. Siddiqi, A. K. Gupta, and M. Brokate
© 2009 American Institute of Physics 978-0-7354-0683-4/09/$25.00

In this paper not only temporal variation of data has been analyzed, but also linear estimation of import and export data has been discussed, [13-16].

METHOD

Wavelet analysis detects all trend of a given signal. A wavelet is one of the waveform of limited duration. Fourier analysis consists of breaking up of a signal into line waves of different frequencies. Wavelet analysis is also breaking of a signal function but into shifted and scaled versions of mother wavelet [6-8].

As a result of this study, it is expected the following outputs: i) to explain the affects of small meso and large scale fluctuations on economical and logistical variables; ii) to get some experience how wavelet methodology is applied on economical problems.

BASICS OF WAVELET ANALYSIS

Wavelet is families of small waves generated from a single functions f(t) which is called mother wavelet. A sufficient condition for a function f(t) to qualify as a mother wavelet is discussed by Meyer and Siddiqi [1,8]

$$\int_{-\infty}^{\infty} |f(t)|^2 \, dt < \infty \tag{1a}$$

The Fourier transform F of f(t) is defined as

$$F(w) = \int_{-\infty}^{\infty} f(t) e^{iwt} \, dt \tag{1b}$$

A function $\psi(t)$ satisfying the following condition is called a continuous wavelet [8];

$$\int_{-\infty}^{\infty} |\psi(t)|^2 \, dt = 1 \tag{2a}$$

and

$$\int_{-\infty}^{\infty} |\psi(t)| \, dt = 0 \tag{2b}$$

Higher order moments may be zero, that is,

$$\int_{-\infty}^{\infty} t^k \psi(t) dt = 0 \qquad \text{for} \quad k=0,....,N-1$$

The wavelet transform of f(t) denoted by $W_f(a,b)$ is defined as:

230

$$W_f(a,b) = \frac{1}{\sqrt{a}} \int_{-\infty}^{\infty} \psi((u-b)/a) f(u) du = \int_{-\infty}^{\infty} f(u) \psi_{a,b}^{(u)} du$$

$$(3)$$

where $\quad \psi_{a,b}^{(u)} = \frac{1}{\sqrt{a}} \psi((u-b)/a)$ $\qquad\qquad\qquad\qquad (4)$

Here "a" is a scaling parameter, b is a location parameter and $\psi_{a,b}^{(u)}$ is often called continuous wavelet (or daughter wavelet) while $\psi(u)$ is the mother wavelet.

In the following sections f(t) will be considered as daily or annual USD and Euro currencies, oil prices, import and export data deviations from mean values in Turkey. The theory mentioned above has been applied on economical and logistical variables in this study. These data are treated like a signal using the graphical interface tools. Wavelet tools of the MATLAB software perform the wavelet analysis. Application of Wavelet Menu on data helps to detect;
- discontinuities and breakdown points,
- long term evolution,
- self similarity
- pure frequencies.

ANALYSIS AND RESULTS

Analysis of USD Variations

In Figure 1, wavelet analysis of Dollar values recorded between January 1950 and June 2006 has been presented. Large scale effects on Dollar variation in 1960, 1976 and 1985 are observed [2].

In 2001 because of budget deficit, increasing trend was higher than the previous period. In 2004, December, decreasing trend has been recorded.

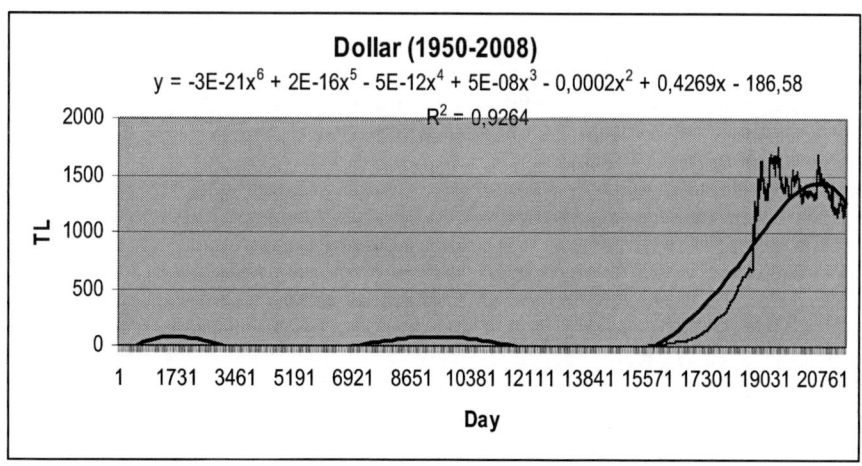

FIGURE 1.a Temporal variation of Dollar Values (January 1950-

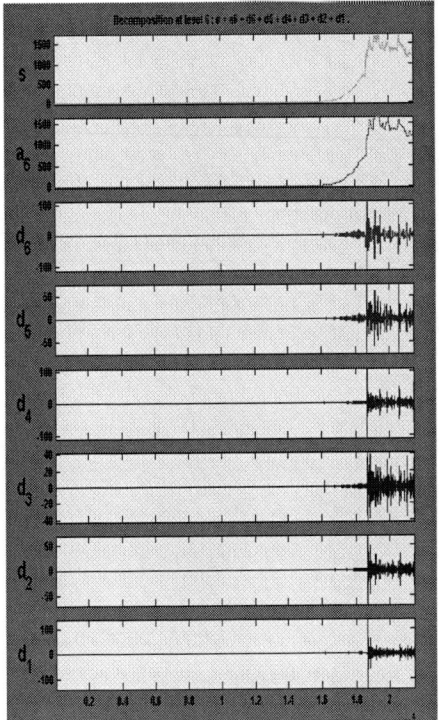

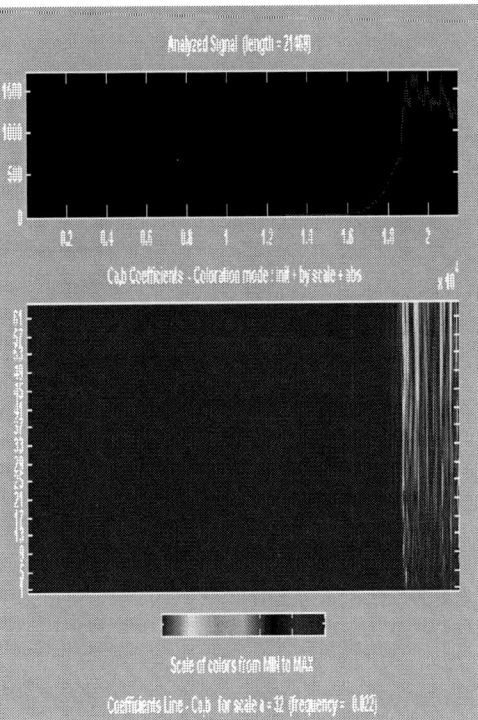

FIGURE 1.b- Db, Level 6, 1D Wavelet USD 1950-2008

FIGURE . c - DMeyer, 1D Continuous Wavelet, Sampling: 1 USD 1950-2008

Figure 1a shows temporal variation of USD values. There is a sufficient evidence of the 6th. order polynomial relation with raw data. The polynomial approximation represents periodic variation of Dollar. The first and second details D1 and D2 show high frequency and the discontinuity most clearly (Figure1b). The long term evolution of wavelet analysis can detect the overall trend of a signal. As the scale increase, the resolution decreases, producing better estimate of the unknown trend (Figure 1b). Figure 1c represents self - similarity. Scale values (days) are on the vertical axis. Large scale influences on variable have been observed in the second part of study period.

Figure 2(a) shows bi - modal distribution (two Gaussian distributions) of USD histogram. D1 and D2 are the first and second details of USD currency between 2005 and 2008 (Figure 2b). They contain the high frequency of the signal. Long –term evolution in level 10 explains the better estimation of the unknown trend. In Figure 2c, self similarity has been detected. Different scale effects on temporal variation of Dollar currency generate a characteristic pattern. In the middle part of 2006, large scale evens play a significant role on it's variation. Moving average curve fitting on data is presented in Figure 2d. This figure shows a decreasing trend.

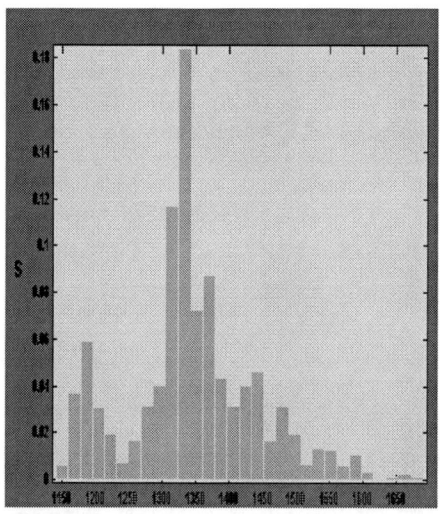

FIGURE 2.a Frequency histogram of USD, 2005-2008

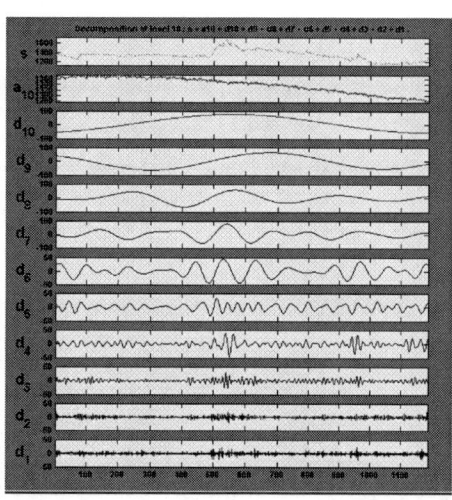

FIGURE 2.b Decomposition of Daily mean USD values using DMeyer with 10 levels, (1D wavelet, USD, 2005-2008)

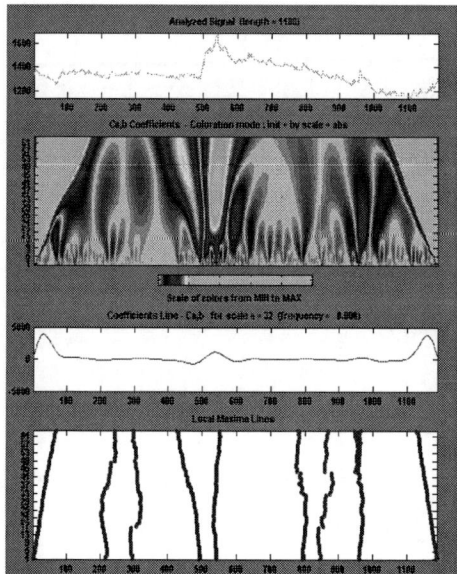

FIGURE 2.c USD Mexh hsv, 1 January 2005- 4 April 2008 (May 2006 max) End of 2006 local scale effects

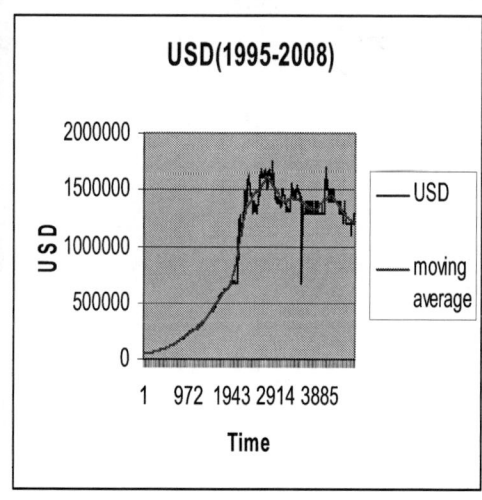

FIGURE 2.d Daily USD variations

233

ANALYSIS OF OIL PRICES

FIGURES 3 (a-c) explain annual average variations of crude oil prices in the range between 1946 and 2006. Increasing trend has been recorded beginning from 1960's. But, the increasing trend which was observed during the last decade is more than the previous period. Self similarity and a repeating pattern in wavelet coefficients plot is characteristic of a signal that looks similar on different scales. Vertical axis (scale values) changes in between 1-16 years. Periodicity of large scale evens decreases and these events have mostly been observed in the second part of the study period.

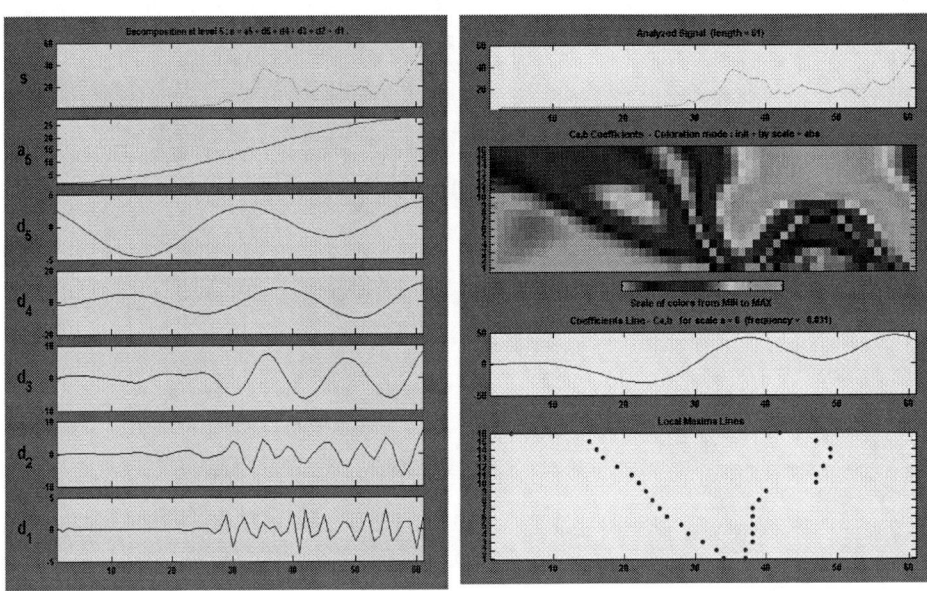

FIGURE 3.a DMeyer wavelet 1-D 1946-2006 oil prices (1981 and 1989 max)

FIGURE 3.b Continuous 1DMexh 1946-2006 oil (1981 and 1989 large scale effects)

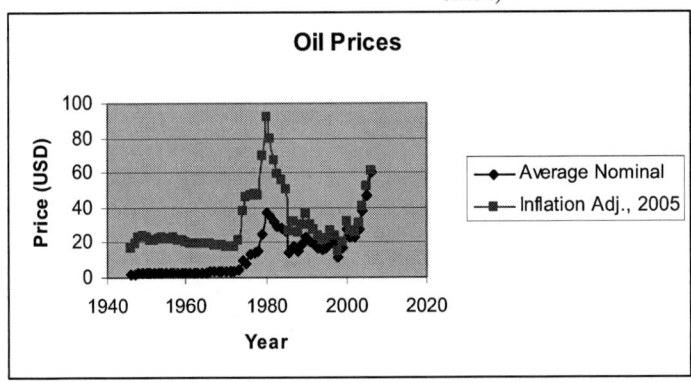

FIGURE 3.c Average Annual Crude Oil prices (average nominal values) and adjusted values (along with inflation rate in 2005) between 1946-2006

234

STATISTICAL ANALYSIS OF LOGISTIC VARIABLES

This part of the paper is related with logistics variables. Temporal variations of logistic variables have been compared with one of the economical variables; USD values.

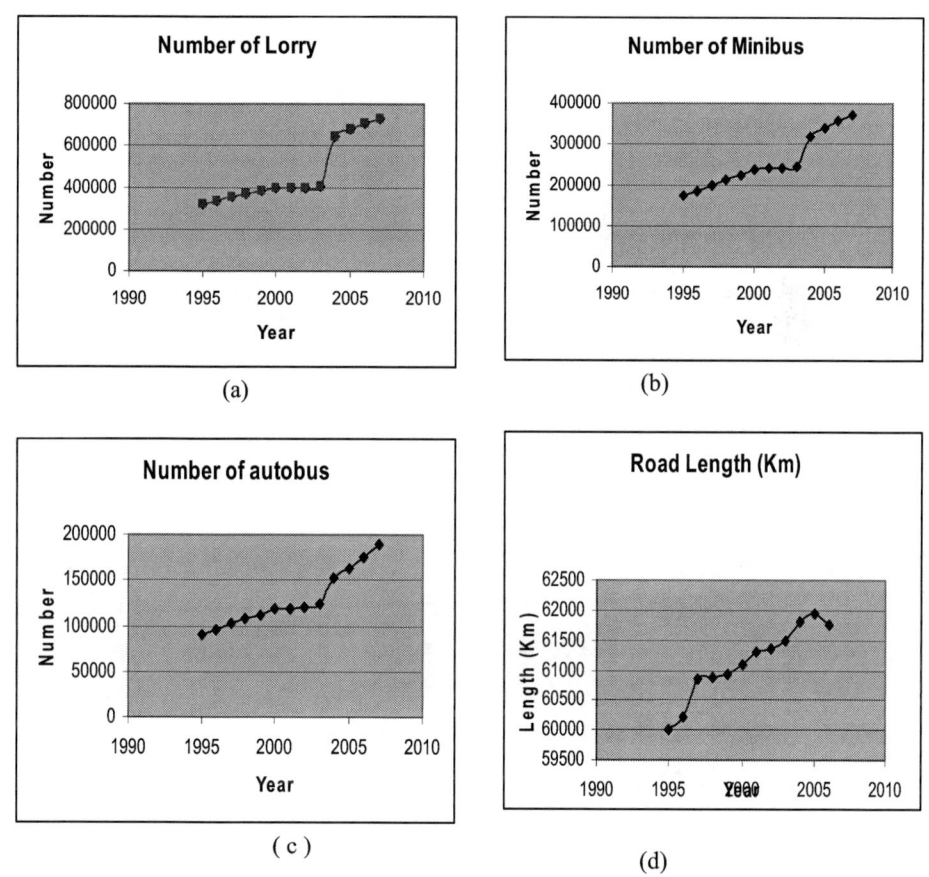

(a)

(b)

(c)

(d)

FIGURE 4(a-d) Variations of logistics variables

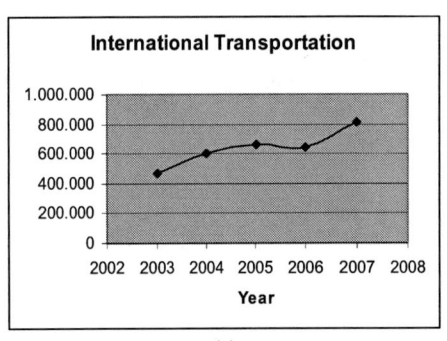

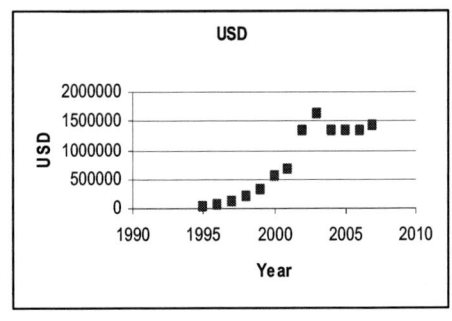

(e)   (f)

FIGURE 4 (e-f) Annual variations of international transportation and USD exchange rate

Annual variations of some logistics variables (numbers of vehicles- lorry, minibus, autobus, total number of international transportation - and road length) have been presented in FIGURES 4 (a-e). All variables show a similar increasing trend with the variation of USD values in recent years.

ANALYSIS OF EXPORT AND IMPORT VALUES

Temporal variations of annual total exports and imports (million Dollars) in Turkey have been discussed in Figs. 5.

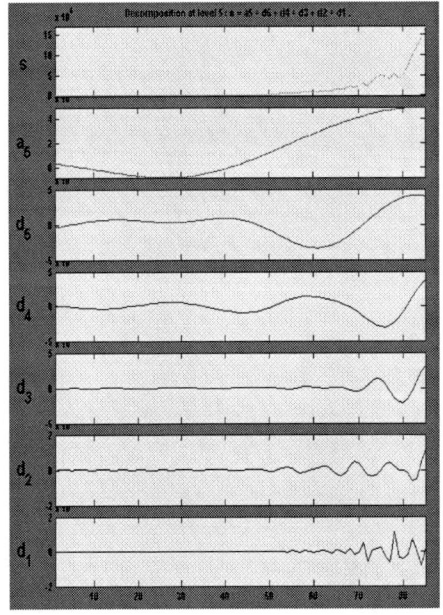

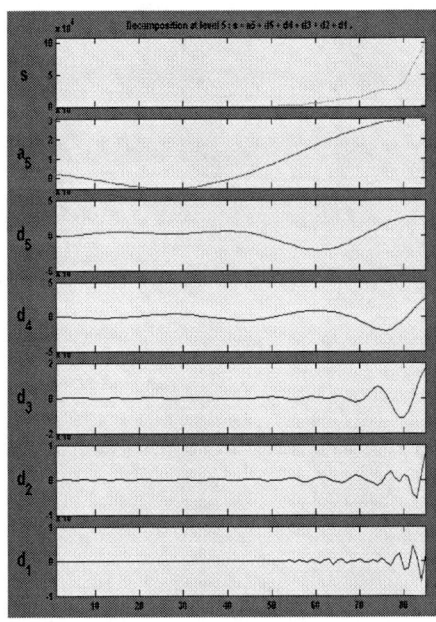

FIGURE 5.a Import, DMeyer, Level 5, 1923-2007   FIGURE 5.b Export, DMeyer, Level 5, 1923-2007

Beginning from 1980's, an increasing trend has been observed. Increasing trend of variables is more in last decade. Similar trends have been observed in annul total imports in Turkey.

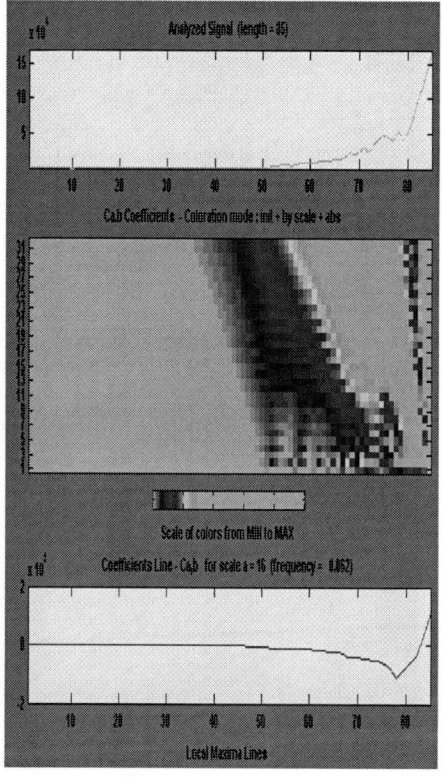 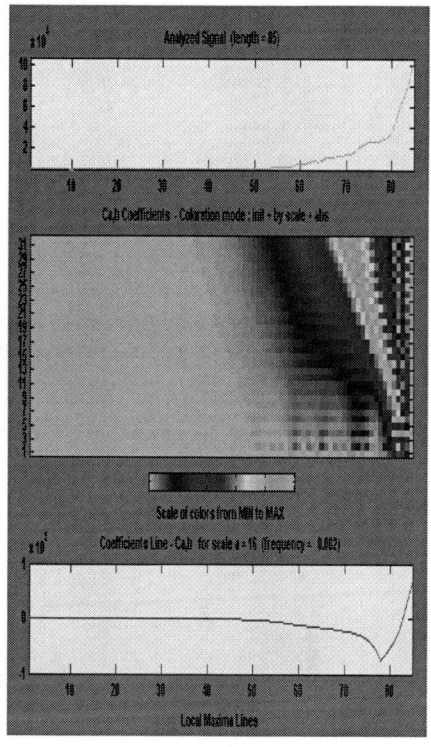

FIGURE 5.c Annual Total Import Values, Db with sampling rate 1, 1D-CW

FIGURE 5.d Annual total Export Values in Turkey, Db with sampling rate 1, 1D CW

In 2000's, $20x10^6$ USD values of annual total imports have been observed every 2-3 years, $40x10^{\ 6}$ import values have been observed every 5-9 years. Large scale effects have been observed at the end of the period. Small scale fluctuations have been observed between 1923 and 1970. Meso and large scale effects have been observed during the last 30 years period. With 4- 5 years periodicity, positive large scale effects (80 million USD) have been observed in 2006 and 2007.

Similar structure has been observed with annual total imports and exports in Turkey. But positive effects of large scale fluctuations are lower (60-70 million USD).

Linear correlation equations are given at Table 1. Positive and negative effects on variations and their significance have been analyzed.

Table 1- Linear Comparison of economical variables

Equation	R^2	Explanation
Export = 1478,4 CO -4936	0,87	There is a sufficient evidence of this relation with $\alpha = 0,01$
Export = 0,03 USD + 18920	0,50	
Export = 1260,2 CO + 0,0008 USD – 5594,2	0,91	There is a sufficient evidence of this relation with $\alpha = 0,01$
Import = 76,4 CO + 5532	0,32	$\alpha = 0.01$
Import = 0,0002 USD + 7842	0,0	$\alpha = 0.01$
Import = 113,1 CO – 0,0014 USD + 5642,5	0,46	There is a sufficient evidence of this relation with $\alpha = 0,10$

ANALYSIS OF DAILY EURO VARIATIONS

Increasing trend of variations of Euro values between 1999 and 2008 is given at Figure 6.a. Daily exchange buying rates of Euro presents three-modal distributions (Figure 6.b).

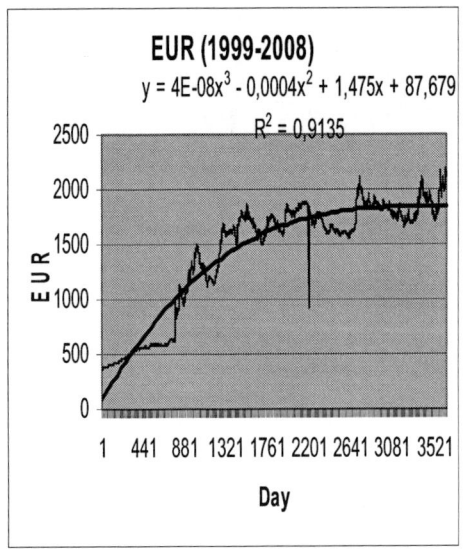

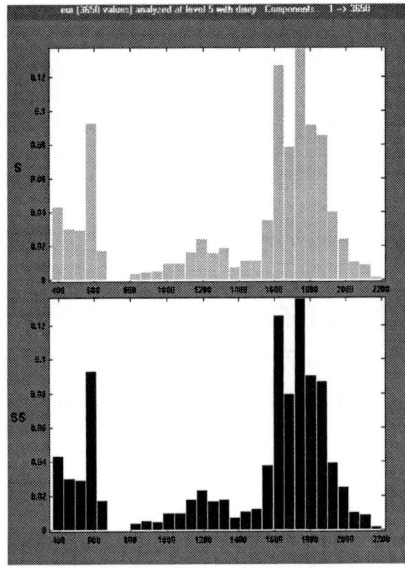

FIGURE 6.a Daily Variations of Euro values between 1999 and 2008.

Figure 6 (b) - Frequency Histogram of Euro Exchange (Tree modal distributions)

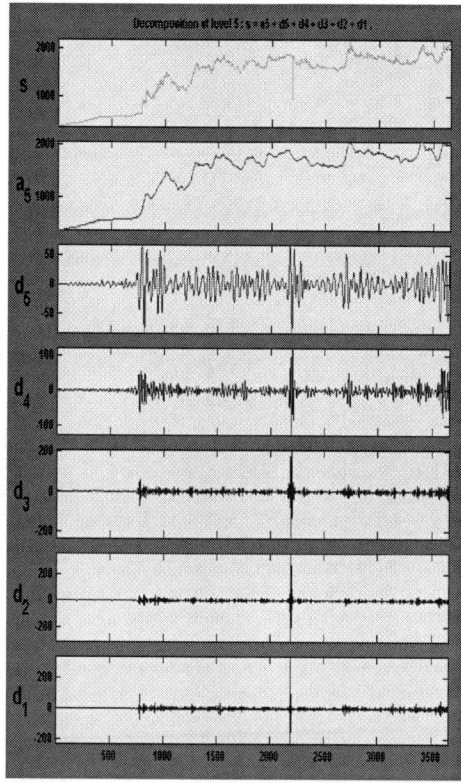

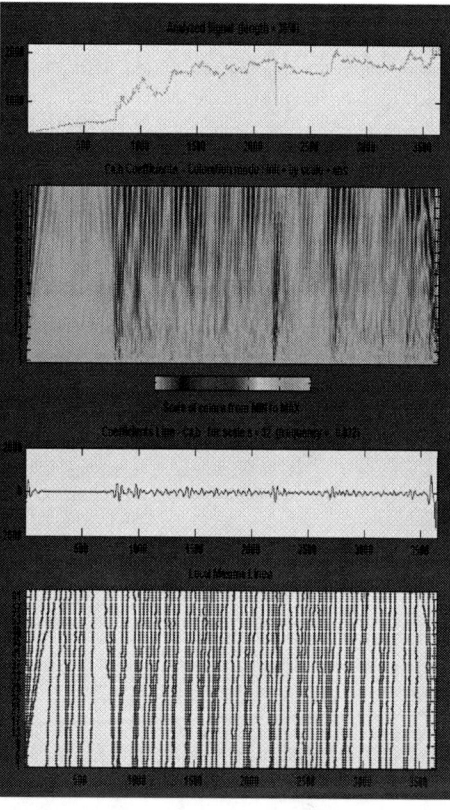

FIGURE 6.c Euro DMeyer with 5 levels
1D Wavelet

FIGURE 6.d EUR DMeyer Sampling
Period1 1D Continuous Wavelet

RESULTS

Analysis of data shows the variation of daily exchange rate between Turkish Lira and buying rates of foreign currencies (Dollar; during 1946-2008 and Euro; during 1999-2008). Meyer, DMeyer and Maxh wavelet functions show that abrupt variation of exchange rates. Bimodal (two Gaussian) distribution has been observed in Dollar variations. But Daily exchange buying rates of Euro presents three-modal distributions. Specific results of this study are listed below:

i)   The buying values of USD increases beginning from 1970's

ii)   Between 2005 and 2008, daily USD variations show a decreasing trend. Frequency distribution functions have been represented by two Gaussian functions.

iii)   Influence of large scale effects on average annual crude oil prices has been recorded.

iv)   Numbers of motor vehicles show an increasing trend.

v)   Annual total imports and exports have been increasing trends.

Increasing trends of buying values of Dollar and Euro, annual crude oil prices, annual total imports and exports in Turkey have been observed in the last part of investigation period. These trends are mostly associated with small scale effects. Increasing demands of logistical variables have also been estimated by EU Development Report until 2010.

REFERENCES

1. Siddiqi, A.H. S. Khan and S. Rehman, (2005): Wind Speed Simulation Using Wavelets, American Journal of Applied Sciences, Vol. 2 (2), p. 557-564.
2. Aslan, Z., H. Gencoglu, (2008): Wavelet Analyses of Oil Prices and USD Variations, Int. Workshop II: mini Symposium on Applications of Wavelets to Real World Problems, IWW2008, ISBN: 978-975-00652-4-8, p. 46-50. May 28-29, Istanbul, Turkey.
3. Can, Z., Z. Aslan, O. Oğuz, and A. H. Siddiqi, (2005): "Wavelet Transforms of Meteorological Parameters and Gravity Waves", Annales Geophysicae, Vol. 23, pp. 1-5.
4. Furati, K. M., P. Manchanda, A. H. Siddiqi, (2008): Wavelet methods in Oil Industry, Int. Workshop II: mini Symposium on Applications of Wavelets to Real World Problems, IWW2008, ISBN: 978-975-00652-4-8, p. 26-36, May 28-29, Istanbul, Turkey.
5. J. J. Puplava, (2004), "The Great Inflation, Part 1, The Nature of Money, September 30, 2004,
6. Manchanda, P., M. R. Islam and A. H. Siddiqi (2005): "Certain Observation on Non-uniform Haar Wavelets", International Workshop on Applications of Wavelets to Real World Problems; Ed(s). Ak. H.Siddiqi, S.Alsan, M. rasulov, O. Oğuz and Z. Aslan, ISBN 975-6516-11-9-ITICU, P. No: 12, Design, Istanbul, pp.85-95.
7. Menna, P., R. Gembi, W. Gillet, G. Deschamps, G. Guiu, R. Ostorn, D. Anderson, H. Schlz, (2005): "European Photovoltaic, RTD and Demonstration Diagram", ISBN 3-936-338-14-0.
8. Meyer Y., (2000): "The Role of Oscillation in some Nonlinear Problems", School on Mathematical Problems in Image Processing, ICTP, SMR1237/4-22, September, 2000.
9. Saatcioglu, C., H. L. Koralp (2005): "The Turkish Broad Money Demand", Istanbul Commerce University, Journal of Social Sciences, Year, 4, No: 7, 2005/1, pp. 139-165.
10. Siddiqi, A. H., (2005): "Trends in Wavelet Applications", International Workshop on Applications of Wavelets to Real World Problems; Ed(s). A. H.Siddiqi, S.Alsan, M. Resulov, O. Oğuz and Z. Aslan, ISBN 975-6516-11-9-ITICU, P. No: 12, Design, Istanbul, pp. 1-26.
11. http://www.oecd.org, (January, 2009).
12. http://www.wrenuk.co.uk/, (January, 2009).
13. http://www.scmr.com/article/ CA6589644.htm, (December, 2008).
14. http://www.cordis.lu/fp6/eoi/instruments/sustainable.htm, (November 2008).
15. http://www.economagic.com, (December, 2008).
16. http: //www.ioga.com, (December, 2008).

Wavelets and Differential Equations–A short review

Mani Mehra*
Department of Mathematics,
Indian Institute of Technology, New Delhi, India

(Dated: April 26, 2009)

PACS numbers: 01.30.Cc
Keywords: Wavelets, Numerical methods

I. INTRODUCTION:

The applications of wavelet theory in numerical methods for solving differential equations are roughly 20 years old. In the early nineties people were very optimistic because it seemed that many nice properties of wavelets would automatically leads to efficient numerical method for differential equations. The reason for this optimism was the fact that many nonlinear partial differential equations (PDEs) have solution containing local phenomena (e.g. formation of shock, hurricanes) and interactions between several scales (e.g. turbulence particularly atmospheric turbulence because there is motion on a continuous range of length scales). Such solutions can be well represented in wavelet bases because of its nice properties few of them like compact support (locality in space) and vanishing moment (locality in scale). Furthermore, this early optimism has been already honored by many authors [1–6] working in this area since then. Nevertheless, there often remains a large gap between a theoretical wavelet paper and the needs of an applied mathematician. This paper is an attempt to bridge this gap by providing a short review on wavelet based numerical methods for differential equations.

Most common numerical methods used for numerical solution of physical problems (mostly leads to partial differential equation) fall in to following classes.

- *Finite difference methods (FDM)*
  The different unknowns are defined by their values on discrete (finite) grid and differential operators are replaced by difference operators using neighboring points. See [7–9] for details.

- *Finite volume methods (FVM)*
  Similar to the finite difference method, values are calculated at discrete places on a meshed geometry. "Finite volume" refers to the small volume surrounding each node point on a mesh. See [10] for details.

- *Finite elements methods (FEM)*
  The unknown solution is approximated by a linear combination of a set of linearly independent test functions, which are piecewise continuous and non vanishing only on the finite number of elements in the domain. Examples of methods that use higher degree piecewise polynomial basis functions are the hp-FEM. See [11] for details.

- *Spectral methods*
  Utilizing basis functions which are infinitely differentiable and non vanishing on the entire domain (global support). See [12, 13] for details.

- *Wavelet methods*
  Utilizing basis functions which are differentiable (according to the requirement) and non vanishing on the compact support. See [14] for details.

Moreover, FDM and FVM are an approximation to the differential equation while other methods are an approximation to its solution. As we noted earlier, spectral bases are infinitely differentiable, but have global support. On the other hand, bases functions used in finite difference or finite element methods have small compact support but poor continuity properties. Conclusively, spectral method have good accuracy, but poor spatial localization, while FDM, FVM and FEM have good spatial localization but poor accuracy. Wavelet based numerical methods seem to combine the advantage (spectral accuracy as well good localization) of all the methods using wavelet bases. Schematically, common wavelet based numerical methods for differential equations can be separated into the following categories:

---

*URL: http://web.iitd.ac.in/~mmehra

CP1146, *Modelling of Engineering and Technological Problems,* edited by A. H. Siddiqi, A. K. Gupta, and M. Brokate
© 2009 American Institute of Physics 978-0-7354-0683-4/09/$25.00

- **Category 1: Methods based on scaling function expansion**
  The unknown solution is expanded in scaling function at some chosen level J and differential equation is solved using a Galerkin approach. This approach can't exploit wavelet compression hence methods in this category are not adaptive [15–18].

- **Category 2: Methods based on wavelet expansion**
  The differential equation is solved using a Galerkin approach as in the first category. However, the unknown solution is expressed in terms of wavelets rather than scaling function. Therefore, this approach can exploit wavelet compression; either to the solution [19], the differential operator [20], or both [21–23].

- **Category 3: Wavelets and finite differences**
  Here wavelets are used to derive adaptive finite difference methods. Instead of expanding the solution in terms of scaling function or wavelet expansion, the wavelet transform is used to determine where the finite difference grid must be refined or coarsened [5, 6, 24–26].

The detailed explanation of these wavelet based numerical methods will be given in subsequent sections.

II. SOME COMPUTATIONAL ASPECTS OF WAVELETS

A. Wavelets

A wavelet is a mathematical function used to divide a given function or continuous-time signal into different scale components. The word wavelet is due to Morlet and Grossmann in the early 1980s. They used the French word ondelette, meaning "small wave". Soon it was transferred to English by translating "onde" into "wave", giving "wavelet". The study of wavelets has attained the present growth due to mathematical analysis of wavelets by Stromberg [27], Grossmann and Morlet [28] and Meyer [29]. The concept of Multiresolution Analysis (MRA) was introduced by S. Mallat [30] and Y. Meyer [29]. The first orthonormal bases of compactly supported wavelets are introduced by I. Daubechies in 1988 [31]. A review of the basic properties of the wavelets and the decomposition and the reconstruction of functions in terms of the wavelet bases is given by Strang [32] and the details of the mathematical analysis of wavelets as approximation of functions in $L_2(\mathbb{R})$ are described in [33].

Given a basis $\{f_k\}_{k\in I}$ in a Hilbert space $\mathcal{H}$, every $f \in \mathcal{H}$ can be uniquely represented as

$$f = \sum_{k\in I} c_k(f) f_k. \tag{1}$$

A frame is also a set $\{f_k\}_{k\in I}$ in $\mathcal{H}$ which allows every f to be written like (1) but it may be linearly dependent (if frame is linearly independent set for $L_2(\mathbb{R})$ then frame gives Riesz basis for $L_2(\mathbb{R})$). Thus, one may get redundant representation. More precisely, a family $\{f_k\}_{k\in I}$ in a Hilbert space $\mathbb{H}$ is a frame for $\mathbb{H}$ if $\exists$ two constant $m > 0$, $M < \infty$ such that

$$m||f||^2 \le \sum_{k\in I} |<f, f_k>|^2 \le M||f||^2, \forall f \in \mathbb{H}. \tag{2}$$

Moreover, for every frame $\exists$ a dual frame $\{\tilde{f}_k\}_{k\in I}$ such that

$$f = \sum_{k\in I} <f, \tilde{f}_k> f_k, \forall f \in \mathbb{H}. \tag{3}$$

Therefore, bases are optimal for fast data processing, whereas the redundancy inherent in frames increase flexibility and robustness to the noise, but usually at the price of high computational cost.

Multiresolution analysis (MRA) [30] is the theory that was used by I. Daubechies to show that for any non negative integer n there exists an orthogonal wavelet with compact supports such that all the derivatives up to order n exist and characterized by the following axioms:

- $\mathcal{V}^j \subset \mathcal{V}^{j+1}$ (subspaces are nested),

- $f \in \mathcal{V}^j$ iff $f(2(.)) \in \mathcal{V}^{j+1}$ for all $j \in \mathbb{Z}$ (invariance to dialation),

- $\overline{\bigcup_{j\ge 0} \mathcal{V}^j} = L_2(\mathbb{R})$,

242

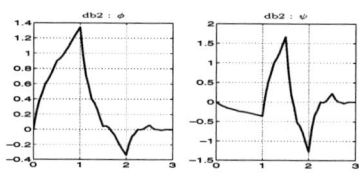

FIG. 1:

- $\{\phi(x - k)|k \in \mathbb{Z}\}$ is an orthonormal basis for $\mathcal{V}^0$ (invariance to translation).

We define $\mathcal{W}^j$ to be the orthogonal complement of $\mathcal{V}^j$ in $\mathcal{V}^{j+1}$, i.e. $\mathcal{V}^j \perp \mathcal{W}^j$ and

$$\mathcal{V}^{j+1} = \mathcal{V}^j + \mathcal{W}^j. \tag{4}$$

There exists a function, which is called a **scaling function** $\phi(x) \in \mathcal{V}^0$, such that the sequence $\phi_k^j(x) = 2^{j/2}\phi(2^j x - k)_{k \in \mathbb{Z}}$ is an orthonormal basis for $\mathcal{V}^j$ and similarly there exist a function $\psi(x) \in \mathcal{W}^0$ (which is called **mother wavelet**) such that $\psi_k^j(x) = 2^{j/2}\psi(2^j x - k)_{k \in \mathbb{Z}}$ is an orthonormal basis for $\mathcal{W}^j$. Since $\phi_0^0(x) = \phi(x) \in \mathcal{V}^0 \subset \mathcal{V}^1$, so

$$\phi(x) = \sum_{k=-\infty}^{\infty} h_k \phi_k^1(x). \tag{5}$$

Eq. (5) is called **dialation equation** (two scale relation for scaling function) and for Daubechies compactly supported scaling function only finitely many $h_k, k = 0, 1, \cdots D - 1$ will be nonzero. Where D is even positive integer called the **wavelet genus** and $h_0, h_1, \cdots, h_{D-1}$ are called **low pass filter coefficients**. Similarly, Daubechies compactly supported wavelet $\psi(x) \in \mathcal{W}^0 \subset \mathcal{V}^1$, therefore

$$\psi(x) = \sum_{k=0}^{D-1} g_k \phi_k^1(x). \tag{6}$$

Eq. (6) is called **wavelet equation** (two scale relation for wavelet function) and $g_0, g_1, \cdots, g_{D-1}$ are called **high pass filter coefficients**. These filter coefficients are connected by the relation $g_k = (-1)^k h_{D-1-k}, k = 0, 1, \cdots, D-1$.

The MATLAB function $[h_k, g_k] = $**wfilters('dbM','r')** computes these filter coefficients, where $M = \frac{D}{2}$, see [34] for details (e.g.: For Haar wavelet $M = 1, D = 2, h_k = [\frac{1}{\sqrt{2}}, \frac{1}{\sqrt{2}}]$ and $g_k = [\frac{1}{\sqrt{2}}, -\frac{1}{\sqrt{2}}]$).

One should notice that there is no closed form analytic (explicit) formula for Daubechies scaling function ($\phi(x)$) and wavelet function ($\psi(x)$) (except Haar scaling function ($\phi(x) = 1$ if $x \in [0,1]$, $\phi(x) = 0$ otherwise) and Haar wavelet function ($\psi(x) = 1$ if $x \in [0,.5]$, $\psi(x) = -1$ if $x \in [.5,1]$, $\psi(x) = 0$ otherwise)) and it's value can be computed only at dyadic points using the cascade algorithm [34, 35].

The MATLAB function
$[\phi_i, \psi_i, x_i] = $**wavefun('dbM',iter)** computes the value of $\phi(x)$ and $\psi(x)$ at the grid $x_i = [0, \frac{1}{2^{iter}}, \cdots, D - 1]$, see [35] for details. The functions $\phi(x)$ and $\psi(x)$ are plotted in Fig. 1 for 'db2' and iter=4.

In any wavelet bases or frames, a number of additional properties are desirable, such as smoothness, orthogonality, compact (local) support, Riesz stability, vanishing moments. We will justify few of them.

- **Smoothness**: Smoothness is needed to approximate smooth data.

- **Orthogonality**: For numerical computation orthogonality leads to fast algorithm. However, orthogonality is difficult to achieve (like there is no symmetric orthogonal wavelet ψ with compact support) otherwise leads to the loss of other useful properties. Therefore, in numerical applications orthogonality is relaxed to either semiorthogonality of biorthogonality. Important references for biorthogonal wavelets are [34, 36]. The orthogonal wavelet of Daubechies involve a scaling function $\phi(x)$ and $\psi(x)$ as discussed in Sec. II A. In case of biorthogonal

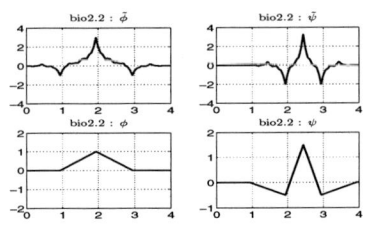

FIG. 2:

wavelet every multiresolution analysis is accompanied by dual multiresolution analysis consisting of nested space $\tilde{\mathcal{V}}^j$ with bases given by dual scaling function $\tilde{\phi}_k^j(x)$ which are biorthogonal to scaling function

$$< \phi_k^j, \tilde{\phi}_{k'}^j >= \delta_{k,k'}, \text{ for } k, k^{'} \in I.$$

Moreover the pair $(\psi_k^j, \tilde{\psi}_{k'}^{j'})$ satisfy the following biorthogonality property

$$< \psi_k^j, \tilde{\psi}_{k'}^{j'} >= \delta_{j,j'} \delta_{k,k'}, \text{ for } j, j^{'}, k, k^{'} \in I.$$

The MATLAB function
$[\tilde{\phi}_i, \tilde{\psi}_i, \phi_i, \psi_i, x_i]$=**wavefun('bior2.2',4)** computes the value of $\tilde{\phi}(x)$, $\tilde{\psi}(x)$, $\phi(x)$ and $\psi(x)$ at the grid $x_i = [0, \frac{1}{2^{iter}}, \cdots, D - 1]$, see [34] for details. The functions $\tilde{\phi}(x)$, $\tilde{\psi}(x)$, $\phi(x)$ and $\psi(x)$ are plotted in Fig. 2 for 'bio2.2' (both wavelet and dual wavelet have 2 vanishing moments) and iter=4.

If $\phi_k^j = \tilde{\phi}_k^j, \forall j, k \implies$ orthogonal scaling function.
If $\mathcal{V}^j = \tilde{\mathcal{V}}^j, \forall j \implies$ semiorthogonal scaling function.

- **Compact support**: It also leads to the fast algorithm for numerical computation.

- **Vanishing moment**: The wavelet is said to have $M (M \in \mathbb{N})$ vanishing moment if it verifies the following condition

$$\int_{\mathbb{R}} x^n \psi(x) = 0, \text{for}, n = 0, 1, \cdots, M - 1,$$

(equivalently scaling function can represents polynomials of degree up to $M - 1$ exactly). This property improves the efficiency of $\psi(x)$ at detecting singularities in the signal (therefore, wavelet bases are suitable for representing piecewise smooth function).

1. Univariate wavelet

A "Wavelet System" consists of the scaling function $\phi(x)$ and the wavelet function $\psi(x)$. In literature, several wavelets with different properties have been derived. Few examples of them are given below.

- Haar wavelet ([37]),

- Daubechies wavelets with different compact, supports ([31]),

- Coiflet (Beylkin *et al.* [20], Daubechies [34]),

- Block spline semi-orthogonal wavelets (Chui and Wang [38]),

- Battle-Lemarie's wavelets (Battle [39]),

- Biorthogonal wavelets of Cohen *et al.* ([36]),

- Shannon's wavelet and Meyer's wavelet ([29]).

The simplest way to obtain multivariate wavelets is to employ anisotropic or isotropic tensor products: (MRA-d) Here, the multivariate wavelets are defined by

$$\psi_l^j(x) := \psi_{l_1}^{j_1}(x_1). \cdots .\psi_{l_d}^{j_d}(x_d),$$

$$j := (j_1, \cdots, j_d) \quad x, l \text{ analogous.}$$

(MRA) Here, anisotropy is avoided by setting $j_1 = j_2 \cdots j_d = j$. The scaling functions are simply the tensor products of the univariate scaling functions. A two-dimensional MRA can be constructed from the following decomposition:

$$\mathcal{V}^{\mathbf{j}} = \mathcal{V}^j \otimes \mathcal{V}^j = (\mathcal{V}^{j-1} \oplus W^{j-1}) \otimes (V^{j-1} \oplus W^{j-1})$$

$$= (W^{j-1} \otimes W^{j-1}) \oplus (W^{j-1} \otimes \mathcal{V}^{j-1})$$

$$\oplus (\mathcal{V}^{j-1} \otimes W^{j-1}) \oplus (\mathcal{V}^{j-1} \otimes \mathcal{V}^{j-1})$$

$$= W^{\mathbf{j-1}} \oplus \mathcal{V}^{\mathbf{j-1}}.$$

Then we have $\mathcal{V}^{\mathbf{J}} = W^{\mathbf{J}-1} \oplus \cdots \oplus W^{\mathbf{0}} \oplus \mathcal{V}^{\mathbf{0}}$ and wavelet basis is given by

$$\{\psi_k^j \otimes \psi_l^j, \psi_k^j \otimes \phi_l^j, \phi_k^j \otimes \psi_l^j\}_{k,l \in \mathbb{Z}, 0 \leq j \leq J-1}$$

$$\cup \{\phi_k^0 \otimes \phi_l^0\}_{k,l \in \mathbb{Z}}.$$

B. Periodized Wavelets

Most of the wavelet algorithms can handle periodic boundary conditions easily. However, different possibilities of dealing general boundary conditions have been studied. The few of them are as follows.

- One approach is to use wavelets specified on an interval where wavelets are constructed satisfying certain boundary conditions. To achieve wavelet approximation on a bounded interval is to keep all Daubechies's wavelets whose supports are totally inside the interval, while modifying those wavelets intersecting the boundary by an orthonormalization [40–43] (semiorthogonalization in [44, 45]). The disadvantages of this approach are inconvenience of implementation and wavelet dependence on boundary conditions.

- The variational approach suggested by Glowinski et al. [1] is not applicable for some nonlinear problems, furthermore, it is impractical for higher dimensions.

- The use of antiderivatives of wavelets as trial functions in [46]. In this way singularities in the wavelets are smoothened and the boundary conditions can be treated more easily.

- A more satisfactory approach is to use **second generation wavelets** introduced in [47]. In Sec. II A wavelet function $\psi_k^j(x)$ are traditionally defined as the dyadic translates and dilates of one particular $L_2(\mathbb{R})$ function, the mother wavelet $\psi(x)$: $\psi_k^j(x) = \psi(2^j x - j)$. These wavelets are called **first generation wavelets**. Second generation wavelets are more general, where wavelets are not necessarily dilates and translates of single function but have all nice properties of first generation wavelets, which can be used for general boundary conditions as well complex geometry [6].

We will discuss wavelet based numerical methods for numerical example with periodic boundary conditions, therefore, we explain periodized wavelets in detail. As pointed out by Y. Meyer (1990) the complete toll box built in $L_2(\mathbb{R})$ can be used in the periodic case $L_2([0,p])$ by introducing a standard periodization technique. This technique consists at each scale in folding, around the integer values, the wavelet $\psi_k^j(x)$ and the scaling functions $\phi_k^j(x)$ centered in $[0,p]$. Let $\phi(x) \in L_2(\mathbb{R})$ and $\psi(x) \in L_2(\mathbb{R})$ be the scaling and wavelet function from a multiresolution analysis as defined in Sec. II A. For any $j, l \in \mathbb{Z}$ and $x \in \mathbb{R}$, we define the p-periodic scaling function

$$\tilde{\phi}_l^{j,p}(x) = \sum_{n=-\infty}^{\infty} \phi_l^j(x + pn) = 2^{j/2} \sum_{n=-\infty}^{\infty} \phi(2^j(x + pn) - l), \tag{7}$$

and the p-periodic wavelet

$$\tilde{\psi}_l^{j,p}(x) = \sum_{n=-\infty}^{\infty} \psi_l^j(x+pn) = 2^{j/2} \sum_{n=-\infty}^{\infty} \psi(2^j(x+pn)-l). \tag{8}$$

The p periodicity can be verified as follows

$$\tilde{\phi}_l^{j,p}(x+p) = \sum_{n=-\infty}^{\infty} \phi_l^j(x+pn+p) = \sum_{m=-\infty}^{\infty} \phi_{j,l}(x+pm)$$

$$= \tilde{\phi}_l^{j,p}(x),$$

and similarly $\tilde{\psi}_l^{j,p}(x+p) = \tilde{\psi}_l^{j,p}(x)$. Where $\tilde{\phi}_l^{j,p}(x)$ and $\tilde{\psi}_l^{j,p}(x)$ generates the spaces $\mathcal{V}^{j,p}$ and $\mathcal{W}^{j,p}$ respectively. For notational convenience $\tilde{\phi}_l^{j,1}(x) = \tilde{\phi}_l^j(x)$, $\tilde{\psi}_l^{j,1}(x) = \tilde{\psi}_l^j(x)$.

C. Projection onto space $\mathcal{V}^j$

Let $P_{\mathcal{V}^j}f$ be the projection of a function f on $\mathcal{V}^j$ and

$$P_{\mathcal{V}^j}f(x) = \sum_{k=-\infty}^{\infty} c_k^j \phi_k^j(x), \quad x \in \mathbb{R}. \tag{9}$$

Two natural ways of representing f on $\mathcal{V}^j$ (equivalently going to wavelet space from physical space) are orthogonal projection and interpolation.

1. Orthogonal projection

The orthogonality of the basis suggest the orthogonal projection. In this case, the expansion coefficients are defined as

$$c_k^j = \int_{-\infty}^{\infty} f(x)\phi_k^j(x)dx.$$

This integral can be approximated by a quadrature method.

2. Interpolation

Using interpolation is also a popular choice for projecting f on $\mathcal{V}^j$. To explain the idea consider the projection of f on periodic space $\mathcal{V}^{j,1}$ such that f coincide at node points of level j, where c_k^j are the expansion coefficients.

$$f(l/2^j) = \sum_{k=0}^{2^j-1} c_k^j \tilde{\phi}_k^j(l/2^j), \quad l = 0, \cdots, 2^j - 1.$$

This can be rewritten as

$$f(l/2^j) = \sum_{k=0}^{2^j-1} c_k^j \tilde{\phi}_{k-l}^j(l/2^j), \quad l = 0, \cdots, 2^j - 1.$$

Therefore calculating the coefficients c_k^j reduces to solving a matrix equation

$$F^j = T^j c^j,$$

where F^j is the vector of components $f_l^j = f(l/2^j)$ for $l = 0, \cdots, 2^j - 1$, c^j is the vector of coefficients c_k^j for $k = 0, \cdots, 2^j - 1$, T^j is the circulant matrix of size $N = 2^j$.
cost : If simple Gaussian elimination is used to solve this system, then the cost of finding the c_k^j is heavy: $O(n^3)$ operations, where n is the order of the matrix. Significantly better performance can be achieved by use of sparse matrix routines. However, because this system is circulant, using FFT, the solution can be found in $O(n\log_2 n)$ operations [48].

246

D. Wavelet transform (WT)

Wavelet transforms have advantages over traditional Fourier transforms for representing functions that have discontinuities and sharp peaks, and for accurately deconstructing and reconstructing finite, non-periodic and/or non-stationary signals. Wavelet transforms are classified into **discrete wavelet transform** (DWT) and **continuous wavelet transform** (CWT). The continuous wavelet transform $W_\psi f$ of $f \in L_2(\mathbb{R})$ with respect to ψ is defined as

$$(W_\psi)(b,a) = |a|^{-\frac{1}{2}} \int_{-\infty}^{\infty} f(x)\overline{\psi(\frac{x-b}{a})}.$$

In order to reconstruct f from $W_\psi f$, we need to know the constant

$$C_\psi = \int_{-\infty}^{\infty} \frac{|\widehat{\psi}(w)|^2}{|w|} dw < \infty.$$

The finiteness of this constant (**admissibility condition**) restrict the class of $L_2(\mathbb{R})$ functions that can be used as wavelets. Which implies

$$\int_{-\infty}^{\infty} \psi(x) = 0, \text{see [33] for details.}$$

With the constant C_ψ, we have the following reconstruction formula

$$f(x) = \frac{1}{C_\psi} \int_{\mathbb{R}^2} \int W_\psi(b,a)\overline{\psi(\frac{x-b}{a})} \frac{dadb}{a^2}, f \in L_2(\mathbb{R}).$$

Notice that the possibility of reconstruction is guaranteed by the admissibility condition. Now, in practice, numerical implementation requires the CWT to be discretized.

$$f(x) = \sum_{j,k \in \mathbb{Z}} (W_\psi f)(b_k^j, a^j)\tilde{\psi}_k^j(x), \tag{10}$$

where $a^j = \frac{1}{2^j}$ is called the binary dilation (or dyadic dilation) and $b_k^j = \frac{k}{2^j}$ is the binary or dyadic position. However, this procedure leads to the frames not to the bases. For bases, one needs another approach of DWT based on the notion of multiresolution analysis.

The orthogonal projection $P_{\mathcal{V}^j}$ is given by Eq. (9). Since space $\mathcal{V}^J$ is decomposed into wavelet space $\mathcal{V}^{J_0} + \mathcal{W}^{J_0} + \mathcal{W}^{J_0-1} + \cdots + \mathcal{W}^{J-1}$ using the relation (4), we obtain wavelet series

$$P_{\mathcal{V}^J} f(x) = \sum_{k=-\infty}^{\infty} c_k^{J_0} \phi_k^{J_0}(x) + \sum_{j=J_0}^{J-1} \sum_{k=-\infty}^{\infty} d_k^j \psi_{j,k}(x), \tag{11}$$

where J_0 (coarsest level of approximation) satisfy $0 \leq J_0 \leq J$. The $c_k^{J_0}$ (scaling coefficients) and d_k^j (wavelet coefficients) for $j = J_0, \cdots, J-1$ are given by

$$c_k^{J_0} = \int_{-\infty}^{\infty} f(x)\phi_k^{J_0}(x)dx, \tag{12}$$

$$d_k^j = \int_{-\infty}^{\infty} f(x)\psi_k^j(x)dx. \tag{13}$$

The orthonormality properties of the scaling and wavelet functions arising from a multiresolution of $L_2(\mathbb{R})$ lead to simple relations connecting the scaling coefficients and the wavelet coefficients of different levels. Mallat [49] exploited these relations to develop a fast algorithm which transforms the coefficients from one level of resolution j to the next coarse level $j-1$. This yields a fast and accurate algorithm denoted by pyramid algorithm. For Daubechies compactly supported wavelet system of genus D, using the relation (5)

$$\phi_l^{j-1} = 2^{\frac{(j-1)}{2}} \phi(2^{j-1}x - l) = 2^{\frac{j}{2}} \sum_{k=0}^{D-1} h_k \phi(2^j x - 2l - k)$$

$$= \sum_{k=0}^{D-1} h_k \phi_{2l+k}^j(x).$$

Similarly, $\psi_l^{j-1}(x) = \sum_{k=0}^{D-1} g_k \phi_{2l+k}^j(x)$ using (6). Using these results in the definitions of the scaling and wavelet coefficients

$$c_l^{j-1} = \sum_{k=0}^{D-1} h_k c_{2l+k}^j, \tag{14}$$

$$d_l^{j-1} = \sum_{k=0}^{D-1} g_k c_{2l+k}^j. \tag{15}$$

Applying the Eqs. (14) and (15) recursively for $j = J, J-1, \cdots, J_0+1$, starting with the initial sequence $c_l^J, l \in \mathbb{Z}$ gives the wavelet coefficients. Once the coefficients d_l^j are computed, they remain unaltered in the subsequent calculations. This gives a very efficient algorithm **fast wavelet transform (FWT**, decomposition procedure) for the computation of wavelet coefficients. In matrix vector product form Eqs. (14) and (15) are written as

$$\mathbf{d} = \mathcal{W}\mathbf{c},$$

where $\mathbf{c} = \mathbf{c}^{\mathbf{J}} = [c_0^J, c_1^J, \cdots, c_{2^J-1}^J]^T$ and $\mathbf{d} = [\mathbf{c}^{\mathbf{J_0}}, \mathbf{d}^{\mathbf{J_0}}, \mathbf{d}^{\mathbf{J_0}+1}, \cdots, \mathbf{d}^{\mathbf{J-1}}]^T, \mathbf{d}^{\mathbf{J}} = [d_0^J, d_1^J, \cdots, d_{2^J-1}^J]^T$. The **inverse fast wavelet transform (IFWT**, reconstruction procedure) can be obtained in a similar manner.

The MATLAB function $[\mathbf{d}] = \mathbf{fwt(x,D,\lambda)}$ gives the fast discrete periodized wavelet transform, where x is any vector of periodic function values, D ia wavelet genus and $\lambda = J - J_0$ is the depth of transform.

E.   Connection coefficient

Any numerical scheme for solving differential equations must adequately represent the derivatives and non-linearities of the unknown function. In the case of wavelet bases, these approximations give rise to certain L_2 inner products of the basis functions, their derivatives and their translates, called the connection coefficients. In Fourier-based methods, since the products of the basis elements are also basis elements, the procedure does not face any difficulty. The numerical approximation of the connection coefficients which appear with the wavelet bases is unstable since the integrands are highly oscillatory. Scaling functions and wavelets do not have explicit analytical expressions but are implicitly determined by the two scale relations (5) and (6), it is necessary to develop algorithms to compute several connection coefficients, which occur in the application of the wavelet-Galerkin to differential equations. Specific algorithms have been devised by Latto et al. [50]. In the most general case we allow ϕ_l to be differentiated which gives rise to the n-term connection coefficients:

$$\wedge(l_1, l_2, \cdots, l_n, d_1, d_2, \cdots, d_n) = \wedge_{l_1 l_2 \cdots l_n}^{d_1 d_2 \cdots d_n} = \int_{-\infty}^{\infty} \prod_{i=1}^{n} \phi_{l_i}^{d_i}(x).$$

We can alter a doubly subscripted connection coefficient in to a singly subscripted one, and a triply subscripted connection coefficient in to a doubly subscripted one. We therefore define the two and three term connection coefficients as

$$\wedge_l^{d_1 d_2} = \int_{-\infty}^{\infty} \phi^{d_1}(x)\phi_l^{d_2}(x)dx, \tag{16}$$

and

$$\wedge_{lm}^{d_1 d_2 d_3} = \int_{-\infty}^{\infty} \phi^{d_1}(x)\phi_l^{d_2}(x)\phi_m^{d_3}(x)dx, \tag{17}$$

where $d_i \geq 0$. The question of differentiability of ϕ and ψ is not fully understood (see [34] for details). Let space $C^\alpha(\mathbb{R})$ denotes the space of functions having continuous derivatives of order$\leq \alpha$ and $H^\beta(\mathbb{R}) = \{f \in L_2(\mathbb{R}) : f^d \in L_2(\mathbb{R}), |d| \leq \beta\}$. The regularity of Daubechies compactly supported scaling and wavelet functions is shown in Table I. We observe from Table I that same regularity could be achieved with less number of vanishing moments (e.g. for $D/M = 6/3$, $\phi, \psi \in C^1(\mathbb{R})$ and for $D/M = 12/6$ also $\phi, \psi \in C^1(\mathbb{R})$). So there are two different issues of maximum regularity and maximum number of vanishing moment (M) for $\psi(x)$ which should be chosen according to the application [34].

248

TABLE I: $\phi, \psi \in C^\alpha(\mathbb{R}), H^\beta(\mathbb{R})$.

D/M	2/1	4/2	6/3	8/4	10/5	12/6	14/7	16/8	18/9	20/10
α	-	0	1	1	1	1	2	2	2	2
β	0	0	1	1	2	2	2	2	3	3

Let $f \in \mathcal{V}^{j,1} \cap C^d(\mathbb{R})$ then

$$f^{(d)}(x) = \sum_{l=-\infty}^{\infty} c_l^j \tilde{\phi}_l^{j,(d)}(x), \quad x \in \mathbb{R}, \tag{18}$$

$f^{(d)}$ will in general not belong to $\mathcal{V}^{j,1}$ so we project $f^{(d)}$ back onto $\mathcal{V}^{j,1}$

$$P_{\mathcal{V}^{j,1}} f^{(d)}(x) = \sum_{k=-\infty}^{\infty} c_k^{j(d)} \tilde{\phi}_k^j, \quad x \in \mathbb{R}, \tag{19}$$

$$c_k^{j(d)} = \int_{-\infty}^{\infty} f^{(d)} \tilde{\phi}_k^j(x) dx. \tag{20}$$

Then substituting (18) into (20) systems of equation can be represented in matrix-vector form

$$c^{j(d)} = \mathcal{D}^{(d)} c^j. \tag{21}$$

We will refer to the matrix $\mathcal{D}^{(d)}$ as the differentiation matrix of order d. Derivation of Eq. (21) is given in [51] where

$$[\mathcal{D}^{(d)}]_{k, <n+k>_{2^j}} = 2^{jd} \Lambda_n^d, \quad k = 0, 1, \cdots, 2^j - 1,$$

$$n = 2 - D, 3 - D, \cdots, D - 2.$$

If the function to be differentiated is periodic with period p then we have

$$c^{j(d)} = D_1^{(d)} c^j,$$

where $\mathcal{D}_1^{(d)} = \frac{\mathcal{D}^{(d)}}{p^d}$ and $c^{j(d)} = [c_0^{j(d)}, c_1^{j(d)}, \cdots, c_{2^j-1}^{j(d)}]$.

Consider the $f \in C^M(\mathbb{R})$, then the approximation error will be

$$e^{j,1}(x) = f(x) - (P_{\mathcal{V}^{j,1}} f)(x) \tag{22}$$

and $\|e^{j,1}\|_\infty = O(2^{-jM})$ (this is exponential decay with respect to j. Furthermore, the greater the number of vanishing moments M, the faster the decay. In contrast, FDM and FEM yields convergence rates that are only algebraic in N (degree of freedom), typically $O(N^{-2})$ or $O(N^{-4})$), where $P_{\mathcal{V}^{j,1}} f(x)$ is orthogonal projection of f on $\mathcal{V}^{j,1}$ as defined in Sec. II C. It is comparable to accuracy of spectral method :for problems with smooth solutions convergence rates are $O(e^{-cN})$ or $O(e^{-c\sqrt{N}})$

Furthermore, the space $\mathcal{V}^{j,1}$ can exactly represent polynomials up to degree $M - 1$ (error term in Eq. (22) will be zero), however differentiation matrix is accurate of order $2M$. This doubling of the accuracy is also encountered in FEM and is known as **superconvergence** (which is lost in case of general boundary condition).

III. WAVELET BASED NUMERICAL METHODS

A. Wavelet Galerkin method

In Galerkin method the degrees of freedom are the expansion coefficients of a set of basis functions and these expansion coefficients are not in physical space (means in wavelet space). Moreover, in wavelet Galerkin methods the treatment of nonlinearities is complicated which can be handled with couple of techniques.

- Using the connection coefficients discussed in Sec. II E (expensive approach due to the summation over multiple indices).

- Using the quadrature formula [52] (loses its accuracy due to approximate calculation).

- Pseudo approach [23] (first map wavelet space to physical space, compute nonlinear term in physical space and then back to wavelet space, this approach is not very practical because it requires transformation between the physical space and wavelet space).

The derivatives can be obtained using the differential matrix discussed in Sec. II E.

B.   Wavelet collocation method

Collocation method involve numerical operators acting on point values (collocation points) in the physical space. Generally, wavelet collocation methods are created by choosing a wavelet and some kind of grid structure which will be computationally adapted. In effect, one obtains finite differences on nonuniform grid. The treatment of nonlinearities in wavelet collocation method is straightforward task due to collocation nature of algorithm. The derivatives can be obtained like interpolation method for projecting f on $\mathcal{V}^j$ discussed in Sec. II C 2.

Moreover, proofs are easier with Galerkin methods, whereas implementation is more practical with collocation methods.

C.   Numerical example

We illustrate various wavelet based numerical methods for heat equation in one–dimension with periodic boundary conditions.

$$u_t = \nu u_{xx} + f(x), t > 0$$
$$u(x,0) = h(x), \quad 0 \le x \le 1, \tag{23}$$

where ν is a positive constant, $f(x) = f(x+1)$ and $h(x) = h(x+1)$.

1.   Method based on scaling function expansion

Let us first leave the spatial variable x continuous and discretize only the time to obtain the Euler scheme:

$$u_t^n = \frac{u^{n+1} - u^n}{\delta t} = \nu u_{xx}^n + f(x). \tag{24}$$

Now to obtain solution $u_j \in \mathcal{V}^{j,1}$ we apply **wavelet-Galerkin method** to (24) with approximation of the form

$$u^j(x) = \sum_{k=0}^{2^j - 1} c_k^j \tilde{\phi}_k^j(x), \tag{25}$$

where c_k^j is the unknown coefficient of scaling function expansion of $u^j(x)$. The Galerkin discretization scheme gives

$$\left(c_u^{n+1} - c_u^n\right) = \nu \delta t \mathcal{D}^{(2)} c_u^n + \delta t c_f, \tag{26}$$

where c_u denote the vector of scaling function coefficients corresponding to u and c_f is given by

$$(c_f)_l^j = \int_0^1 f(x) \tilde{\phi}_l^j(x) dx,$$

and $\mathcal{D}^{(2)}$ is discussed in Sec. II E. Finally, we arrive at the linear algebraic system

$$\mathcal{A} c_u = \mathcal{F}, \tag{27}$$

which can be solved using any linear algebraic solver. Hence this comes under the category 1 discussed in Sec. I.

By rewriting Eq. (27) in the following form

$$\mathcal{W}\mathcal{A}\mathcal{W}^T \mathcal{W}c_u = \mathcal{W}\mathcal{F},$$ (28)

and substituting $d_u = \mathcal{W}c_u$ and $d_f = \mathcal{W}c_f$ in Eq. (28)

$$\mathcal{W}\mathcal{A}\mathcal{W}^T d_u = d_{\mathcal{F}}.$$ (29)

This comes under the category 2 discussed in Sec. I.

3. Method based on physical space representation

Multiplying Eq. (26) by T^j

$$T^j \left(c_u^{n+1} - c_u^n \right) = \nu \delta t T^j \mathcal{D}^{(2)} c_u^n + \delta t T^j c_f,$$ (30)

$$\left(u^{n+1} - u^n \right) = \nu \delta t T^j \mathcal{D}^{(2)} c_u^n + \delta t f.$$ (31)

By redefining $f = \{f(x_l)\}_{l=0}^{2^J-1}$, this can also seen as a **wavelet collocation method** for the solution of (23). This is also based on scaling function expansion and does not exploit any wavelet compression, hence it also comes under the category 1 discussed in Sec. I.

---

[1] R. Glowinski, W. Lawton, M. Ravachol, and E. Tenenbaum, in *Proceedings of the Ninth International Conference on Computing methods in applied sciences and Engineering* (SIAM, PA, 1990), pp. 55–120.
[2] A. Amaratunga, J. Wiliams, S. Qian, and J. Weiss, Int. J. Numer. Meth. Eng. **37**, 2703 (1994).
[3] M. Holmstrom and J. Walden, J. Sci. Comput. **13**, 19 (1998).
[4] M. Farge, K. Schneider, and N. K.-R. Kevlahan, Phys. Fluids **11**, 2187 (1999).
[5] O. V. Vasilyev and C. Bowman, J. Comput. Phys. **165**, 660 (2000).
[6] M. Mehra and N. K.-R. Kevlahan, J. Comput. Phys. **227**, 5610 (2008).
[7] K. Morton and D. Mayers, *Numerical Solution of Partial Differential Equations* (Cambridge University Press, 2005).
[8] A. Mitchell and D. Griffiths, *The Finite Difference Method in Partial Differential Equations* (Wiley, Chichester, 1985).
[9] G. Smith, *Numerical Solution of Partial Differential Equations: Finite Difference Methods* (Clarendon Press, Oxford, 1985).
[10] R. J. Leveque, *Finite difference methods for hyperbolic problems* (Cambridge University Press, 2002).
[11] J. N. Reddy, *Introduction to the Finite Element Method* (McGraw-Hill Science/Engineering/Math; 2 edition, 1993).
[12] C. Canuto, M. Y. Hussaini, A. Quarteroni, and T. A. Zang, *Spectral Method in Fluid Dynamics* (Springer Verleg, 1988).
[13] L. N. Trefethen, *Spectral Methods in MATLAB* (SIAM, Philadelphia, PA, 2000).
[14] R. O. J. W. H. L Resnikoff, *Wavelet Analysis: The Scalable Structure of Information* (Springer Verleg, 2002).
[15] A. Latto and E. Tenenbaum, C. R. Acad. Sci. Paris **311**, 903 (1990).
[16] L. Jameson, J. Sci. Comput. **8**, 267 (1993).
[17] B. V. R. Kumar and M. Mehra, International Journals of Computational Methods **2**, 75 (2005).
[18] B. V. R. Kumar and M. Mehra, BIT Numerical Mathematics **45**, 543 (2005).
[19] J. Liandrat and P. Tchamitchian, NASA Contactor Report 187480: ICASE report No. 90-83 (1990).
[20] G. Beylkin, R. Coifman, and V. Rokhlin, Comm. Pure Appl. Math. **XLIV**, 141 (1991).
[21] P. Charton and V. Perrier, Mathematical Aplicada e Computational **15**, 139 (1996).
[22] M. Mehra and B. V. R. Kumar, Communications in Numerical Methods in Engineering **21**, 313 (2005).
[23] B. V. R. Kumar and M. Mehra, International Journal of Wavelets, Multiresolution and Information Processing **3**, 1 (2005).
[24] L. Jameson, ICASE CR-191601 (1994).
[25] M. Holmstrom, SIAM J. Sci. Comput. **21**, 405 (1999).
[26] M. Mehra and N. K.-R. Kevlahan, SIAM J. on Sci. Comput. **30**, 3073 (2008).
[27] J. O. Stromberg, in *Proceedings of Harmonic Analysis* (Univ. of Chicago, 1981), pp. 475–494.
[28] A. Grossmann and J. Morlet, SIAM J. Math. Anal. **15**, 723 (1984).
[29] Y. Meyer, *Analysis at Urbana I: Analysis in Function Spaces* (Cambridge University Press, Cambridge, 1989).
[30] S. G. Mallat, Trans. Amer. Math. Soc. **315**, 69 (1989).
[31] I. Daubechies, Comm. Pure Appl. Math. **41**, 909 (1988).

[32] G. Strang, SIAM Rev. **31**, 614 (1989).
[33] C. K. Chui, *An introduction to wavelets* (Academic Press, Boston, MA, 1992).
[34] I. Daubechies, *Ten Lectures on Wavelets* (SIAM, Philadelphia, 1992).
[35] G. Strang and T. Nguyen, *Wavelets and Filter Banks* (Wellesley-Cambridge press, 1996).
[36] A. Cohen, I. Daubechies, and J. C. Feauveau, Comm. Pure Appl. Math. **45**, 485 (1992).
[37] A. Harr, Math. Ann. **69**, 331 (1910).
[38] C. K. Chui and J. Z. Wang, Trans. Amer. Math. Soc. **330**, 903 (1992).
[39] G. Battle, Comm. Math. Phys. **110**, 601 (1987).
[40] A. Cohen, I. Daubechies, and P. Vial, Appl. Comput. Harmonic Anal. **1**, 54 (1993).
[41] Y. Meyer, Seminaire Bourbaki **7**, 115 (1991).
[42] L. Andersson, N. Hall, B. Jawerth, and G. Peters, Academic Press, San Diego (1993).
[43] C. K. Chui and E. Quak, CAT Report 265, Department of Mathematics, Texax A & M University, College Station, TX (1992).
[44] W. Cai and J. Wang, SIAM J. Numer. Anal. **333**, 937 (1996).
[45] V. Kumar and M. Mehra, Internationa Journal of Wavelets Multiresolution and Information Processing **5**, 317 (2007).
[46] J. C. Xu and W. C. Shann, Numer. Math. **63**, 123 (1992).
[47] W. Sweldens, SIAM J. Math. Anal. **29**, 511 (1998).
[48] A. Garba, Tech. Rep., International Center for Theorytical Physics (1996).
[49] S. G. Mallat, IEEE Trans. on patt. Anal. Mach. Intell. **11**, 674 (1989).
[50] A. Latto, H. L. Resnikoff, and E. Tenenbaum, in *Proceedings of the French-USA Workshop on Wavelets and Turbulence* (Springer-Verlag, Princeton, New York, 1991).
[51] O. M. Nilsen, Ph.D. thesis, Technical University of Denamark, Lyngby (1998).
[52] W. Sweldens and R. Piessens, SIAM J. Numer. Anal. **31**, 1240 (1994).

New classes of Wavelets

P.Manchanda* and Meenakshi†

*Department of Mathematics, Guru Nanak Dev University, Amritsar, India
†Dev Samaj College for Women, Ferozepur, India

Abstract. Recently Manchanda, Meenakshi and Siddiqi [9, 10] have studied Haar-Vilenkin wavelet and a special type of non-uniform multiresolution analysis. Haar-Vilenkin wavelet is a generalization of Haar wavelet. Motivated by the paper of Gabardo and Nashed [5] we have introduced [10] a class of multiresolution analysis extending the concept of classical multiresolution analysis. We present here a resume of these results. We hope that applications of these concepts to some significant real world problems could be found.

Keywords: Haar type Vilenkin system, Haar-Vilenkin multiresolution analysis, Haar Vilenkin wavelet filter
PACS: 42A38, 42A55, 42C15, 42C40, 43A70

1. INTRODUCTION

The following system which is a generalization of Haar system is connected with the name of Vilenkin [11, 14]. Very often it is termed as a generalized Haar system or a Haar type Vilenkin system.
Let $m = (m_k, k \in \mathbf{N})$ be a sequence of natural numbers such that $m_k \geq 2$, $\mathbf{N}$ denote the set of non-negative integers. Let $M_0 = 1$ and $M_k = m_{k-1}M_{k-1}$, $k \in \mathbf{P}$.
Let $\mathbf{P}$ denote the set of positive integers and let $k \in \mathbf{P}$ can be written as

$$k = M_n + r(m_n - 1) + s - 1. \tag{1}$$

where $n \in \mathbf{N}$, $r = 0, 1, \ldots, M_n - 1$ and $s = 1, 2, \ldots, m_n - 1$. This expression is unique for each $k \in \mathbf{P}$. Let us write an arbitrary element $t \in [0, 1)$ in the form

$$t = \sum_{k=0}^{\infty} \frac{t_k}{M_{k+1}}, (0 \leq t_k < m_k). \tag{2}$$

It may be noted that there exists two such expressions (2), for so called m-adic rational numbers. In such cases we use the expression which contains only a finite number of terms different from zero.

Define the function system $(h_n, n \in \mathbf{N})$ by $h_0 = 1$ and

$$h_k(t) = \begin{cases} \sqrt{M_n}exp\frac{2\pi i s t_n}{m_n} & \frac{r}{M_n} \leq t < \frac{r+1}{M_n} \\ 0 & otherwise \end{cases}. \tag{3}$$

This system can be extended to $\mathbf{R}$ (the set of real numbers) by periodicity of period 1: $h_k(t+1) = h_k(t)$, $t \in [0, 1)$. It can be checked that $\{h_k(t)\}$ is a complete orthonormal system in $L^2(\mathbf{R})$. It is clear that

$$h_k(t) = \chi_{[\frac{r}{M_n}, \frac{r+1}{M_n}]}(t)\sqrt{M_n}exp\frac{2\pi ist_n}{m_n}.$$

The *Haar system* $H = (H_n, n \in \mathbf{N})$ is defined as follows:
$H_0 = 1$. For $n, r \in \mathbf{N}$ and $0 \le r < 2^n$ the function H_n is defined on $[0, 1)$ by

$$H_{2^n+r}(x) = \begin{cases} 2^{\frac{n}{2}} & x \in I(2r, n+1) \\ -2^{\frac{n}{2}} & x \in I(2r+1, n+1) \\ 0 & otherwise \end{cases}$$

where

$$\begin{aligned} I(2r, n+1) &= [2r2^{-(n+1)}, (2r+1)2^{-(n+1)}) \\ &= \left[\frac{2r}{2^{n+1}}, \frac{2r+1}{2^{n+1}} \right). \end{aligned}$$

It can be extended to $\mathbf{R}$ by the periodicity of period 1. Each Haar function is continous from the right and the Haar system H is orthonormal on $L^2(\mathbf{R})$ [11].

Very often, (see for example [16]), the function defined below is called the Haar wavelet(mother Haar wavelet)

$$H(t) = \begin{cases} 1 & t \in [0, \frac{1}{2}) \\ -1 & t \in [\frac{1}{2}, 1) \\ 0 & otherwise \end{cases}.$$

It can be expressed in the form

$$H(t) = \chi_{[0,\frac{1}{2})}(t) - \chi_{[\frac{1}{2},1)}(t).$$

By taking translations and dilation of $H(t)$ the system $\{H_{m,n}(t)\}$, where $H_{m,n} = 2^{m/2}H(2^m t - k)$ has been extensively studied. For example it has been proved that it is orthonormal basis in $L^2(\mathbf{R})$. Decomposition of a function $f \in L^p(\mathbf{R}), 1 < p < \infty$ with respect to the system has been studied and its convergence investigated. The family $\{H_{m,n}\}$ will generate a multiresolution analysis, for example let $S_n = span\{H_{j,k}\}$ and $L_n = \{$all functions in$L^2(\mathbf{R})$constant on all intervals$[k2^{-n}, (k+1)2^{-n}], k \in \mathbf{N}\}$. It can be proved that $L_n = S_n$ for all $n \in \mathbf{N}$. $\{L_n\}_{n=-\infty}^{\infty}$ form a multiresolution analysis. In this case the function

$$\chi_{[0,1)}(t) = \begin{cases} 1 & t \in [0, 1) \\ 0 & otherwise \end{cases}.$$

can be taken as a scaling function. Comparison of Fourier series of a function $f \in L^2(\mathbf{R})$ and its expansion with respect to the Haar system has been investigated. Behaviour of Haar coefficients are also studied, (for details see [15]) . For relevant literature of wavelet we refer to [3, 12, 15, 16].

Recently Manchanda, Meenakshi and Siddiqi[9, 10] have studied the concept of Haar-Vilenkin wavelet and a special class of Multiresolution analysis. Haar-Vilenkin wavelet is a generalization of Haar-wavelet while special class of nonuniform multiresolution analysis is a generalization of the classical multiresolution analysis.

In this paper we give a resume of results concerning orthogonality of Haar-Vilenkin wavelet, convergence of Haar-Vilenkin wavelet series, properties of Haar-Vilenkin wavelet coefficients, properties of multiresolution analysis where translation and dilation are respectively $\frac{b}{M_n}$, $b \in \mathbf{Z}$ (the set of integers) and m_n respectively. More precisely we describe properties of the projection operator associated with this class of multiresolution analysis and related filters as well as analogue of Jackson's and Berstein's inequalities for such a projection. Applications of these concepts to real world problems are envisaged.

2. HAAR-VILENKIN WAVELET

2.1. Haar-Vilenkin mother wavelet

The function $h_k(t)$ as defined in (3) can also be written as

$$
h_k(t) = \begin{cases}
\sqrt{M_n} & \frac{r}{M_n} \leq t < \frac{r}{M_n} + \frac{1}{M_{n+1}} \\
\sqrt{M_n} exp\frac{2\pi is}{m_n} & \frac{r}{M_n} + \frac{1}{M_{n+1}} \leq t < \frac{r}{M_n} + \frac{2}{M_{n+1}} \\
\sqrt{M_n} exp\frac{4\pi is}{m_n} & \frac{r}{M_n} + \frac{2}{M_{n+1}} \leq t < \frac{r}{M_n} + \frac{3}{M_{n+1}} \\
\cdots & \\
\sqrt{M_n} exp\frac{2\pi is(m_n-1)}{m_n} & \frac{r}{M_n} + \frac{m_n-1}{M_{n+1}} \leq t < \frac{r+1}{M_n}
\end{cases} \tag{4}
$$

It can be seen easily that $h_k(t) \in L^2[0,1)$ and $th_k(t) \in L^1[0,1)$ for $k \in \mathbf{P}$ and

$$
\begin{aligned}
\int_{-\infty}^{\infty} h_k(t)\,dt &= \int_{\frac{r}{M_n}}^{\frac{r}{M_n}+\frac{1}{M_{n+1}}} \sqrt{M_n}\,dt + \int_{\frac{r}{M_n}+\frac{1}{M_{n+1}}}^{\frac{r}{M_n}+\frac{2}{M_{n+1}}} \sqrt{M_n} exp\frac{2\pi is}{m_n}\,dt + \ldots \\
&\quad + \ldots + \int_{\frac{r}{M_n}+\frac{m_n-1}{M_{n+1}}}^{\frac{r+1}{M_n}} \sqrt{M_n} exp\frac{2\pi is(m_n-1)}{m_n}\,dt \\
&= \frac{\sqrt{M_n}}{M_{n+1}} \left[1 + exp\frac{2\pi is}{m_n} + exp\frac{4\pi is}{m_n} + \cdots + exp\frac{2\pi is(m_n-1)}{m_n} \right] \\
&= \frac{\sqrt{M_n}}{M_{n+1}} \left[\frac{1 - exp2\pi is}{1 - exp\frac{2\pi is}{m_n}} \right] \\
&= 0.
\end{aligned}
$$

Thus the function $h_k(t)$ is a *mother wavelet* for $k \in \mathbf{P}$ and for $t \in [0,1)$. The function $h_k(t)$ is called a **Haar Vilenkin Wavelet**.
Define

$$
\psi_{a,b}(t) = m_n^{a/2} h_k(m_n^a t - b) \tag{5}
$$

The collection $\{\psi_{a,b}(t)\}_{a,b \in \mathbf{Z}}$ is referred to as the Haar-Vilenkin system. $\psi_{a,b}(t)$ is supported on the interval $I_{a,b}$ where

$$I_{a,b} = \left[\frac{r}{m_n^a M_n} + \frac{b}{m_n^a}, \frac{r+1}{m_n^a M_n} + \frac{b}{m_n^a} \right), a,b \in \mathbf{Z}.$$

The system $\psi_{a,b}(t)$ can also be written as $\{ m_n^{\frac{a}{2}} h_k(m_n^a t - b) \} = D_{m_n^a} T_b h_k(t)$.

2.2. Haar-Vilenkin scaling function

For $k \in \mathbf{P}$ and $t \in [0,1)$ as defined in (1) and (2) the Haar -Vilenkin scaling function is defined as:

$$
\begin{aligned}
p_k(t) &= \sqrt{M_n} \chi_{\left[\frac{r}{M_n}, \frac{r+1}{M_n} \right)} \\
&= \begin{cases} \sqrt{M_n}, & \frac{r}{M_n} \le t < \frac{r+1}{M_n} \\ 0 & otherwise \end{cases}
\end{aligned}
\tag{6}
$$

Define

$$\phi_{a,b}(t) = m_n^{a/2} p_k(m_n^a t - b) \tag{7}$$

The collection $\{\phi_{a,b}(t)\}_{a,b \in \mathbf{Z}}$ is referred to as the system of Haar Vilenkin scaling functions. For a given $a \in \mathbf{Z}$, the collection $\{\phi_{a,b}(t)\}_{b \in \mathbf{Z}}$ is referred to as the system of scale a Haar-Vilenkin scaling functions.
$\phi_{a,b}(t)$ is supported on the interval $I_{a,b}$ where $a,b \in \mathbf{Z}$, where
$$I_{a,b} = \left[\frac{r}{m_n^a M_n} + \frac{b}{m_n^a}, \frac{r+1}{m_n^a M_n} + \frac{b}{m_n^a} \right)$$

For each $a,b \in \mathbf{Z}$

$$\int_{\mathbf{R}} \phi_{a,b}(t)\, dt = \int_{I_{a,b}} \phi_{a,b}(t)\, dt = m_n^{a/2} \sqrt{M_n} \frac{1}{m_n^a M_n} = m_n^{-a/2} M_n^{-1/2}.$$

and

$$\int_{\mathbf{R}} |\phi_{a,b}(t)|^2\, dt = \int_{I_{a,b}} |\phi_{a,b}(t)|^2\, dt = m_n^a M_n \frac{1}{m_n^a M_n} = 1.$$

Remark 2.1 *1. Haar system is a special case of Haar-Vilenkin system for $m_n = 2$ for all $n \in \mathbf{P}$.*
2. Given any $a \in \mathbf{Z}$, the collection of scale a Haar-Vilenkin scaling functions is an orthonormal system on $\mathbf{R}$.

Remark 2.2 *we have*

$$
\begin{aligned}
\phi_{a,b}(t) &= m_n^{-1/2} \phi_{a+1, \frac{r(m_n-1)}{M_n} + m_n b}(t) + m_n^{-1/2} \phi_{a+1, \frac{r(m_n-1)+1}{M_n} + m_n b}(t) + \cdots \\
&\quad \cdots + m_n^{-1/2} \phi_{a+1, \frac{(r+1)(m_n-1)}{M_n} + m_n b}(t).
\end{aligned}
\tag{8}
$$

and

$$\psi_{a,b}(t) \;=\; m_n^{-1/2}\phi_{a+1,\frac{r(m_n-1)}{M_n}+m_n b}(t) + exp\frac{2\pi is}{m_n}m_n^{-1/2}\phi_{a+1,\frac{r(m_n-1)+1}{M_n}+m_n b}(t) + \ldots$$

$$\ldots + exp\frac{2\pi is(m_n-1)}{m_n}m_n^{-1/2}\phi_{a+1,\frac{(r+1)(m_n-1)}{M_n}+m_n b}(t). \tag{9}$$

The equations (8)and (9) shows the relationship between Haar-Vilenkin scaling function and Haar-Vilenkin wavelet.

Remark 2.3 *For $k = 1$, $h_k(t)$ is the well known mother Haar wavelet.*

Theorem 2.1 *Given any $a \in \mathbf{Z}$, the collection $\{\phi_{a,\frac{b}{M_n}}\}_{b\in\mathbf{Z}}$ is an orthonormal system on* **R**.

Theorem 2.2 *The system $\{m_n^{\frac{a}{2}}h_k(m_n^a t - b)\} = \{\psi_{a,b}\}, a, b \in \mathbf{Z}$ is an orthonormal system in $L^2(\mathbf{R})$.*

The Approximation Operator in context of Haar-Vilenkin system:

Definition 2.1 *For each $a \in \mathbf{Z}$ define the approximation operator P_a on the functions $f(x) \in L^2(\mathbf{R})$ by*

$$P_a f(x) = \sum_b \left\langle f, \phi_{a,\frac{b}{M_n}}\right\rangle \phi_{a,\frac{b}{M_n}}(x)$$

Remark 2.4   *1. For each $a \in \mathbf{Z}$, define the approximation space V_a by*

$$V_a = span\left\{\phi_{a,\frac{b}{M_n}}\right\}_{b\in\mathbf{Z}}$$

Since $\left\{\phi_{a,\frac{b}{M_n}} : b \in \mathbf{Z}\right\}$ is an orthonormal system on **R**. *This implies that $P_a f(x)$ is a function in V_a best approximating $f(x)$ in L^2-sense.*
2.

$$\phi_{a,\frac{b}{M_n}}(x) = m_n^{a/2}\sqrt{M_n}\chi_{I_{a,\frac{b}{M_n}}}(x)$$

Thus

$$\left\langle f, \phi_{a,\frac{b}{M_n}}\right\rangle \phi_{a,\frac{b}{M_n}}(x) = m_n^a M_n \left(\int_{I_{a,\frac{b}{M_n}}} f(t)\,dt\right)\chi_{I_{a,\frac{b}{M_n}}}(x)$$

In other words, on the interval $I_{a,\frac{b}{M_n}}$, $P_a f(x)$ is the average value of $f(x)$ on $I_{a,\frac{b}{M_n}}$.
3. *The approximation operator P_a has the following properties:*
   (a) *for each $a \in \mathbf{Z}$, P_a is linear, given $f(x), g(x) \in L^2(\mathbf{R})$ and $\alpha, \beta \in \mathbf{C}$ (the set of complex numbers)*

$$P_a(\alpha f + \beta g)(x) = \alpha P_a(f)(x) + \beta P_a(g)(x)$$

(b) For each $a \in \mathbf{Z}$, P_a is idempotent, that is given $f(x) \in L^2(\mathbf{R})$

$$P_a(P_a f)(x) = P_a f(x)$$

(c) Given integers a, a' with $a \leq a'$ and $g(x) \in V_a$

$$P_{a'} g(x) = g(x)$$

(d) Given $a \in \mathbf{Z}$ and $f(x) \in L^2(\mathbf{R})$

$$\|P_a f\|_2 \leq \|f\|_2$$

3. CONVERGENCE OF HAAR-VILENKIN WAVELET SERIES

The system $\{\psi_{a,b}\}_{a,b \in \mathbf{Z}}$ is an orthonormal basis in $L^2(\mathbf{R})$. Thus for a function $f \in L^2(\mathbf{R})$ has a decomposition

$$f = \sum_{a \in \mathbf{Z}} \sum_{b \in \mathbf{Z}} \langle f, \psi_{a,b} \rangle \psi_{a,b}.$$

The number

$$C_{a,b} = \langle f, \psi_{a,b} \rangle \tag{10}$$

is called the a, b^{th} wavelet coefficient and

$$\sum_{a \in \mathbf{Z}} \sum_{b \in \mathbf{Z}} \langle f, \psi_{a,b} \rangle \psi_{a,b} \tag{11}$$

is called the wavelet series of $f \in L^2(\mathbf{R})$.
Following theorem has been proved, see for example [9]

Theorem 3.1 *If $f \in L^p(\mathbf{R})$ with $1 < p < \infty$ or f is continous function, then*

$$\lim_{p \to \infty} P_p(f) = f \tag{12}$$

where

$$P_p(f) = \sum_{a < p} \sum_{b \in \mathbf{Z}} \langle f, \psi_{a,b} \rangle \psi_{a,b}$$

i.e.

$$\lim_{p \to \infty} \|P_p(f) - f\|_p \to 0$$

Futhermore for each $p \in \mathbf{Z}$

$$\lim_{\mu \to \infty} P_a(f) + Q_a^\mu(f) = P_{a+1}(f) \tag{13}$$

where

$$Q_a^\mu(t) = \sum_{b \leq \mu} <f, \psi_{a,b}> \psi_{a,b}$$

It may be observed that the convergence part may be obtained from the fact that P_p are conditional expectations but we present here a direct proof.

4. BEHAVIOUR OF HAAR-VILENKIN COEFFICIENTS NEAR JUMP DISCONTINUITIES

Suppose that $f(x)$ is defined on interval $\left[\frac{r}{M_n}, \frac{r+1}{M_n}\right]$ with a jump discontinuity at $x_0 \in \left(\frac{r}{M_n}, \frac{r+1}{M_n}\right)$ and continous at all other points in $\left[\frac{r}{M_n}, \frac{r+1}{M_n}\right]$. We have to check whether Haar Vilenkin coefficients $< f, \psi_{a,b} >$ such that $x_0 \in I_{a,b}$ behave differently than do the Haar Vilenkin coefficients.

Let us assume that given function $f(x)$ is C^2 on the intervals $[0, x_0]$ and $[x_0, 1]$. This means that both $f'(x)$ and $f''(x)$ exist, are continous functions and hence are bounded on these intervals. Fix integers $a \geq 0$ and $0 \leq b \leq m_n^a - 1$ and let $x_{a,b}$ be the mid point of the interval $I_{a,b}$. i.e.

$$x_{a,b} = \frac{r+1/2}{m_n^a M_n} + \frac{b}{m_n^a}.$$

Case I If $x_0 \notin I_{a,b}$, then expanding $f(x)$ about $x_{a,b}$ by Taylor's formulae, it follows that for all $x \in I_{a,b}$

$$f(x) = f(x_{a,b}) + f'(x_{a,b})(x - x_{a,b}) + \frac{1}{2}f''(\xi_{a,b})(x - x_{a,b})^2$$

where $\xi_{a,b}$ is some point in $I_{a,b}$.
Since $\int_{I_{a,b}} \psi_{a,b}(x)\, dx = 0$,

$$
\begin{aligned}
\langle f, \psi_{a,b} \rangle &= \int_{I_{a,b}} f(x)\overline{\psi_{a,b}(x)}\, dx \\
&= f(x_{a,b}) \int_{I_{a,b}} \overline{\psi_{a,b}(x)}\, dx + f'(x_{a,b}) \int_{I_{a,b}} \overline{\psi_{a,b}(x)}(x - x_{a,b})\, dx \\
&\quad + \frac{1}{2} \int_{I_{a,b}} \overline{\psi_{a,b}(x)}(x - x_{a,b})^2 f''(\xi_{a,b})\, dx \\
&= \alpha_{a,b}(x) + \beta_{a,b}(x)
\end{aligned}
$$

where

$$\alpha_{a,b}(x) = f'(x_{a,b}) \int_{I_{a,b}} \overline{\psi_{a,b}(x)}(x - x_{a,b})\, dx$$

and

$$\beta_{a,b}(x) = \frac{1}{2} \int_{I_{a,b}} \overline{\psi_{a,b}(x)}(x - x_{a,b})^2 f''(\xi_{a,b})\, dx$$

Now

$$
\begin{aligned}
|\alpha_{a,b}(x)| &\leq |f'(x_{a,b})| \int_{I_{a,b}} |\overline{\psi_{a,b}(x)}| |(x - x_{a,b})|\, dx \\
&= |f'(x_{a,b})| m_n^{a/2} \sqrt{M_n} \int_{I_{a,b}} |(x - x_{a,b})|\, dx
\end{aligned}
$$

$$= |f'(x_{a,b})| m_n^{a/2} \sqrt{M_n} \left(\frac{1}{2m_n^{2a}M_n^2} \right)$$

$$= |f'(x_{a,b})| \frac{m_n^{-3a/2} M_n^{-3/2}}{4}.$$

and

$$
\begin{aligned}
|\beta_{a,b}(x)| &= \frac{1}{2} \left| \int_{I_{a,b}} \overline{\psi_{a,b}(x)}(x-x_{a,b})^2 f''(\xi_{a,b})\, dx \right| \\
&\leq \frac{1}{2} max_{x \in I_{a,b}} |f''(x)| \int_{I_{a,b}} |\overline{\psi_{a,b}(x)}|(x-x_{a,b})^2\, dx \\
&= \frac{1}{2} \sqrt{M_n} m_n^{a/2} max_{x \in I_{a,b}} |f''(x)| \int_{I_{a,b}} (x-x_{a,b})^2\, dx \\
&= \frac{1}{6} \sqrt{M_n} m_n^{a/2} max_{x \in I_{a,b}} |f''(x)| \left(\frac{1}{4m_n^{3a}M_n^3} \right) \\
&= \frac{1}{24} M_n^{-5/2} m_n^{-5a/2} max_{x \in I_{a,b}} |f''(x)|.
\end{aligned}
$$

If j is large, then $m_n^{-5a/2}$ will be very small compared with $m_n^{-3a/2}$, so we conclude that for the large values of j

$$|\langle f, \psi_{a,b} \rangle| \approx \frac{1}{4} m_n^{-3a/2} M_n^{-3/2} |f'(x_{a,b})|.$$

Case II: If $x_0 \in I_{a,b}$, then it is contained in one of the m_n subintervals of $I_{a,b}$. Assume that $x_0 \in \left[\frac{r}{m_n^a M_n} + \frac{b}{m_n^a}, \frac{r+1/m_n}{m_n^a M_n} + \frac{b}{m_n^a} \right]$.

Expanding $f(x)$ in a taylor series about x_0, we have

$$f(x) = f(x_0^-) + f'(\xi_-)(x-x_0), x \in \left[\frac{r}{m_n^a M_n} + \frac{b}{m_n^a}, x_0 \right), \xi_- \in [x, x_0]$$

and

$$f(x) = f(x_0^+) + f'(\xi_+)(x-x_0), x \in \left[x_0, \frac{r+1/m_n}{m_n^a M_n} + \frac{b}{m_n^a} \right), \xi_+ \in [x_0, x].$$

Thus

$$
\begin{aligned}
\langle f, \psi_{a,b} \rangle &= \int_{I_{a,b}} f(x) \overline{\psi_{a,b}(x)}\, dx \\
&= \int_A^{x_0} f(x_0^-) \overline{\psi_{a,b}(x)}\, dx + \int_{x_0}^{A+\frac{1}{m_n^a M_{n+1}}} f(x_0^+) \overline{\psi_{a,b}(x)}\, dx \\
&\quad + \int_{A+\frac{1}{m_n^a M_{n+1}}}^{A+\frac{m_n}{m_n^a M_{n+1}}} f(x_0^+) \overline{\psi_{a,b}(x)}\, dx + \varepsilon_{a,b}
\end{aligned}
$$

$$= m_n^{a/2}\sqrt{M_n}\left[f(x_0^-)(-A+x_0)+f(x_0^+)\left(A+\frac{1}{m_n^aM_{n+1}}-x_0\right)\right]$$

$$+f(x_0^+)\int_{A+\frac{1}{m_n^aM_{n+1}}}^{A+\frac{m_n}{m_n^aM_{n+1}}}\psi_{a,b}(x)\,dx+\varepsilon_{a,b}$$

$$= m_n^{a/2}\sqrt{M_n}\left[f(x_0^-)(-A+x_0)+f(x_0^+)\left(A+\frac{1}{m_n^aM_{n+1}}-x_0\right)\right]$$

$$-\sqrt{M_n}m_n^{a/2}f(x_0^+)\frac{1}{m_n^aM_{n+1}}+\varepsilon_{a,b}$$

$$= \sqrt{M_n}m_n^{a/2}(x_0-A)[f(x_0^-)-f(x_0^+)]+\varepsilon_{a,b}$$

where $A=\frac{r}{m_n^aM_n}+\frac{b}{m_n^a}$ and

$$\varepsilon_{a,b} = \int_A^{x_0}f(\xi_-)(x-x_0)\overline{\psi_{a,b}(x)}\,dx+\int_{x_0}^{A+\frac{1}{m_n^aM_{n+1}}}f(\xi_+)(x-x_0)\overline{\psi_{a,b}(x)}\,dx.$$

$$|\varepsilon_{a,b}| \le max_{t\in I_{a,b}\setminus\{x_0\}}|f'(t)|\int_{I_{a,b}}|x-x_0||\overline{\psi_{a,b}(x)}|\,dx$$

$$= max_{t\in I_{a,b}\setminus\{x_0\}}|f'(t)|\sqrt{M_n}m_n^{a/2}\int_{I_{a,b}}|x-x_0|\,dx$$

$$\le max_{t\in I_{a,b}\setminus\{x_0\}}|f'(t)|\frac{M_n^{-3/2}m_n^{-3a/2}}{4}.$$

If j is large, then $M_n^{-3/2}m_n^{-3a/2}$ will be very small compared with $M_n^{-1/2}m_n^{-a/2}$, so for large values of j

$$|<f,\psi_{a,b}>| \approx m_n^{a/2}\sqrt{M_n}\left|x_0-\frac{r}{m_n^aM_n}-\frac{b}{m_n^a}\right||f(x_0^-)-f(x_0^+)|.$$

The quantity $\left|x_0-\frac{r}{m_n^aM_n}-\frac{b}{m_n^a}\right|$ can be small if x_0 is close to $\frac{r}{m_n^aM_n}+\frac{b}{m_n^a}$ and can even be zero. We can expect that in middle of $\left[\frac{r}{m_n^aM_n}+\frac{b}{m_n^a},\frac{r+1/m_n}{m_n^aM_n}+\frac{b}{m_n^a}\right]$ so that $\left|x_0-\frac{r}{m_n^aM_n}-\frac{b}{m_n^a}\right|\approx\frac{1}{2m_n^aM_{n+1}}$.
Thus for the large values of j

$$|<f,\psi_{a,b}>| \approx m_n^{a/2}\sqrt{M_n}\frac{1}{2m_n^aM_{n+1}}|f(x_0^-)-f(x_0^+)|$$

$$= \frac{m_n^{-a/2}M_n^{1/2}}{2M_{n+1}}|f(x_0^-)-f(x_0^+)|.$$

Comparing the two cases, we see that the decay of $|<f,\psi_{a,b}>|$ for the large j is considerably slower if $x_0\in I_{a,b}$ than if $x_0\notin I_{a,b}$.

The large coefficient in the Haar-Vilenkin expansion of the coefficient f(x) that persist for all scales suggests the presence of jump discontinuity in the intervals $I_{a,b}$ corresponding to the large coefficient.

5. A SPECIAL NON-UNIFORM MULTIRESOLUTION ANALYSIS

The concept of multiresolution analysis is known as the heart of wavelet theory. A multiresolution on the set of real numbers $\mathbf{R}$, introduced by Mallat [8] is an increasing sequence of closed subspaces $\{V_j\}_{j \in \mathbf{Z}}$ of $L^2(\mathbf{R})$ such that $\cup_{j \in \mathbf{Z}} V_j = \{0\}, \cup_{j \in \mathbf{Z}} V_j$ is dense in $L^2(\mathbf{R})$ and which satisfies $f(x) \in V_j$ if and only if $f(2x) \in V_{j+1}$. Furthermore, there should exist an element $\phi \in V_0$ such that the collection of integer translates of ϕ, $\{\phi(x - n)\}_{n \in \mathbf{Z}}$ is a complete orthonormal system for V_0. A generalization of multiresolution can be made in higher dimensions by considering matrix dilations [8, 16]. Gabardo and Nashed [5] have studied the multiresolution where translation is taken for elements in $\wedge = \{0, \frac{r}{N}\} + 2\mathbf{Z}$, $N \geq 1$(an integer) and r is an odd integer with $1 \leq r \leq 2N - 1$ such that r and N are relatively prime and $\mathbf{Z}$ is the set of all integers. Farkov [4] has extended the notion of multiresolution analysis on locally compact abelian groups and has constructed orthogonal wavelets with compact support on locally compact abelian groups by the scaling function associated with this multiresolution analysis.

Definition 5.1 *For k as in (1),special nonuniform multiresolution analysis is a sequence $\{V_j\}_{j \in \mathbf{Z}}$ of closed subspaces of $L^2(\mathbf{R})$ such that*

1. *$V_j \subset V_{j+1}$ for all $j \in \mathbf{Z}$.*
2. *$span \cup_{j \in \mathbf{Z}} V_j = L^2(\mathbf{R})$.*
3. *$\cap_{j \in \mathbf{Z}} V_j = \{0\}$.*
4. *$f(x) \in V_j$ iff $f(m_n^{-j} x) \in V_0$ for all $j \in \mathbf{Z}$.*
5. *There exists a function $g_k(x)$ in $L^2(\mathbf{R})$, called the scaling function such that the collection $\{g_k(t - \frac{b}{M_n})\}_{b \in \mathbf{Z}}$ is an orthonormal system of translates and*

$$V_0 = \overline{span}\{T_{\frac{b}{M_n}} g_k(x)\}.$$

Remark 5.1 *A special nonuniform multiresolution analysis is defined by first identifying the space V_0, defining V_j by letting*

$$V_j = \{f(x) : f(x) = D_{m_n^j} g(x), g(x) \in V_0\}$$

so that the Definition 5.1(4) is satisfied and then proving that Definition 5.1(1), (2),(3)and(5) hold. V_0 can be defined by just identifying the function $g_k(x)$ such that $\{T_{\frac{b}{M_n}} g_k(x)\}_{b \in \mathbf{Z}}$ is an orthonormal system of translates and then defining

$$V_0 = \overline{span}\{T_{\frac{b}{M_n}} g_k(x)\}.$$

Definition 5.2 *For each $a, b \in \mathbf{Z}$ define $\gamma_{a,b}(x)$ by*

$$\gamma_{a,b}(x) = m_n^{a/2} g_k(m_n^a x - b) = D_{m_n^a} T_b g_k(x).$$

For each $a \in \mathbf{Z}$, define the approximation operator P_a on the functions $f(x) \in L^2(\mathbf{R})$ by

$$P_a f(x) = \sum_b < f, \gamma_{a,\frac{b}{M_n}} > \gamma_{a,\frac{b}{M_n}}. \tag{14}$$

Lemma 5.1 *For all continuous function $f(x)$ having compact support on* $\mathbf{R}$

1. $\lim_{a \to \infty} \|P_a f - f\|_2 = 0$, *and*
2. $\lim_{a \to -\infty} \|P_a f\|_2 = 0$.

Proof   1. Let $\varepsilon > 0$ be arbitrary. By Definition 5.1(2) there exists $A \in \mathbf{Z}$ and $g(x) \in V_A$ such that $\|f - g\|_2 < \varepsilon/2$.
By Definition 5.1(1), $g(x) \in V_a$ and $P_a g(x) = g(x) \forall a \geq A$.
Thus

$$
\begin{aligned}
\|f - P_a f\| &= \|f - g + P_a g - P_a f\|_2 \\
&\leq \|f - g\|_2 + \|P_a (f - g)\|_2 \\
&\leq 2\|f - g\|_2 < \varepsilon,
\end{aligned}
$$

where we have applied Minkowski's and Bessel's inequality. Since this holds for all $a \geq A$, thus the result is proved.

2. Suppose that $f(x)$ is supported in $[-\alpha, \alpha]$ and let $\varepsilon > 0$. By the orthonormality of $\{\gamma_{a,\frac{b}{M_n}}\}_{b \in \mathbf{Z}}$ and applying Cauchy-Schwarz and Minkowski inequalities:

$$
\begin{aligned}
\|P_a f\|_2^2 &= \left\| \sum_b < f, \gamma_{a,\frac{b}{M_n}} > \gamma_{a,\frac{b}{M_n}} \right\|_2^2 \\
&= \sum_b | < f, \gamma_{a,\frac{b}{M_n}} > |^2 \\
&= \sum_b \left| \int_{-\alpha}^{\alpha} f(x) m_n^{a/2} g_k(m_n^a x - \frac{b}{M_n}) dx \right|^2 \\
&\leq \sum_b \left(\int_{-\alpha}^{\alpha} |f(x)|^2 dx \right) m_n^a \int_{-\alpha}^{\alpha} \left| g_k(m_n^a x - \frac{b}{M_n}) \right|^2 dx \\
&= \|f\|_2^2 \sum_b \int_{-m_n^a \alpha - \frac{b}{M_n}}^{m_n^a \alpha - \frac{b}{M_n}} |g_k(x)|^2 dx.
\end{aligned}
$$

We need to show that

$$\lim_{a \to -\infty} \sum_b \int_{-m_n^a \alpha - \frac{b}{M_n}}^{m_n^a \alpha - \frac{b}{M_n}} |g_k(x)|^2 dx = 0.$$

To prove this, let $\varepsilon > 0$ and choose β so large that

$$\sum_{|\frac{b}{M_n}| \geq \beta} \int_{-\frac{1}{m_n} - \frac{b}{M_n}}^{\frac{1}{m_n} - \frac{b}{M_n}} |g_k(x)|^2 dx = \int_{|x| \geq \beta - \frac{1}{m_n}} |g_k(x)|^2 dx < \varepsilon.$$

263

Therefore, if $m_n^a \alpha < \frac{1}{m_n}$, then

$$\sum_{|\frac{b}{M_n}| \geq \beta} \int_{-m_n^a \alpha - \frac{b}{M_n}}^{m_n^a \alpha - \frac{b}{M_n}} |g_k(x)|^2 \, dx \leq \sum_{|\frac{b}{M_n}| \geq \beta} \int_{-\frac{1}{m_n} - \frac{b}{M_n}}^{\frac{1}{m_n} - \frac{b}{M_n}} |g_k(x)|^2 \, dx < \varepsilon.$$

Since for each $b \in \mathbf{Z}$

$$\lim_{a \to -\infty} \int_{-m_n^a \alpha - \frac{b}{M_n}}^{m_n^a \alpha - \frac{b}{M_n}} |g_k(x)|^2 \, dx = 0,$$

$$\begin{aligned}
\lim_{a \to -\infty} \|P_a f\|_2^2 &\leq \|f\|_2^2 \lim_{a \to -\infty} \sum_b \int_{-m_n^a \alpha - \frac{b}{M_n}}^{m_n^a \alpha - \frac{b}{M_n}} |g_k(x)|^2 \, dx \\
&= \|f\|_2^2 \lim_{a \to -\infty} \left(\sum_{|\frac{b}{M_n}| \leq \beta} \int_{-m_n^a \alpha - \frac{b}{M_n}}^{m_n^a \alpha - \frac{b}{M_n}} |g_k(x)|^2 \, dx + \sum_{|\frac{b}{M_n}| > \beta} \int_{-m_n^a \alpha - \frac{b}{M_n}}^{m_n^a \alpha - \frac{b}{M_n}} |g_k(x)|^2 \, dx \right) \\
&\leq \|f\|_2^2 \lim_{a \to -\infty} \left(\varepsilon + \sum_{|\frac{b}{M_n}| > \beta} \int_{-m_n^a \alpha - \frac{b}{M_n}}^{m_n^a \alpha - \frac{b}{M_n}} |g_k(x)|^2 \, dx \right) \\
&= \|f\|_2^2 \varepsilon. \quad \blacksquare
\end{aligned}$$

We now prove the orthogonality of the scaling system.

Theorem 5.1 *For each $a \in \mathbf{Z}$,* $\left\{ \gamma_{a, \frac{r(m_n-1)+l}{M_n}}(x) \right\}_{l \in \mathbf{Z}}$ *is an orthonormal basis for V_a.*

Proof Since $\gamma_{0, \frac{r(m_n-1)+l}{M_n}} \in V_0$ for all $l \in \mathbf{Z}$, definition of special nonuniform multiresolution analysis implies that

$$D_{m_n^a} \gamma_{0, \frac{r(m_n-1)+l}{M_n}} \in V_a \, \forall l \in \mathbf{Z}$$

Since $\left\{ \gamma_{0, \frac{r(m_n-1)+l}{M_n}} \right\}$ is an orthonormal sequence of translates, hence

$$\begin{aligned}
\left\langle \gamma_{0, \frac{r(m_n-1)+l}{M_n}}, \gamma_{0, \frac{r(m_n-1)+l'}{M_n}} \right\rangle &= \left\langle D_{m_n^a} \gamma_{0, \frac{r(m_n-1)+l}{M_n}}, D_{m_n^a} \gamma_{0, \frac{r(m_n-1)+l'}{M_n}} \right\rangle \\
&= \delta_{\frac{l-l'}{M_n}}.
\end{aligned}$$

Hence $\left\{ \gamma_{0, \frac{r(m_n-1)+l}{M_n}} \right\}_{l \in \mathbf{Z}}$ is an orthonormal system on $\mathbf{R}$.

Given $f(x) \in V_a$, $D_{m_n^a} f(x) \in V_0$, so by definition 5.1(5)

$$\begin{aligned}
D_{m_n^{-a}} f(x) &= \sum_l \left\langle D_{m_n^{-a}} f, \gamma_{0, \frac{r(m_n-1)+l}{M_n}} \right\rangle \gamma_{0, \frac{r(m_n-1)+l}{M_n}} \\
&= \sum_l \left\langle f, D_{m_n^a} \gamma_{0, \frac{r(m_n-1)+l}{M_n}} \right\rangle \gamma_{0, \frac{r(m_n-1)+l}{M_n}}.
\end{aligned}$$

Applying $D_{m_n^a}$ to the both sides of above equation, we get

$$
\begin{aligned}
f(x) &= D_{m_n^a} D_{m_n^{-a}} f(x) \\
&= D_{m_n^a} \sum_l \left\langle f, D_{m_n^a} \gamma_{0,\frac{r(m_n-1)+l}{M_n}} \right\rangle \gamma_{0,\frac{r(m_n-1)+l}{M_n}} \\
&= \sum_l \left\langle f, D_{m_n^a} \gamma_{0,\frac{r(m_n-1)+l}{M_n}} \right\rangle D_{m_n^a} \gamma_{0,\frac{r(m_n-1)+l}{M_n}}.
\end{aligned}
$$

This proves the theorem, as $\{\gamma_{0,\frac{r(m_n-1)+l}{M_n}}\}_{l \in \mathbf{Z}}$ is an orthonormal sequence and every element of V_a is its linear combination. $\blacksquare$

Examples of Special Non-uniform multiresolution analysis

Example 5.1 *Haar-Vilenkin MRA*
Let V_0 consist of all step functions $f(x)$ such that
(i)$f(x) \in L^2(\mathbf{R})$.
(ii) $f(x)$ is constant in the intervals $I_{0,\frac{b}{M_n}} \equiv \left[\frac{r+b}{M_n}, \frac{r+b+1}{M_n}\right)$ for all $a \in \mathbf{Z}$.
It can be verified that

$$
V_0 = \overline{span}\{p_k(x)\}
$$

where $p_k(x) = \sqrt{M_n} \chi_{[\frac{r}{M_n}, \frac{r+1}{M_n})}(x)$.

Example 5.2 *Let V_0 consists of all funtions $f(x) \in L^2$ and C^0 on $\mathbf{R}$ and linear on $I_{0,\frac{b}{M_n}}$ for $b \in \mathbf{Z}$ and V_j is defined by Definition 5.1(4). $\{V_j\}_{j \in \mathbf{Z}}$ is Special Non-uniform multiresolution analysis.*

6. PROPERTIES OF HAAR-VILENKIN FILTERS

Theorem 6.1 *Let g_k be a scaling function associated with a Special Non-uniform multiresolution analysis $\{V_a\}_{a \in \mathbf{Z}}$. Then there exists a sequence $\lambda_b \in l_2$ such that*

$$
g_k(x) = \sum_b \lambda_b \gamma_{1,\frac{r(m_n-1)+b}{M_n}}(x) \tag{15}
$$

is a function in $L^2(\mathbf{R})$. Moreover

$$
\hat{g}_k(t) = m_{g_k}(t/m_n)\hat{g}_k(t/m_n), \tag{16}
$$

where

$$
m_{g_k}(t) = \frac{1}{\sqrt{m_n}} \sum_b \lambda_b e^{-2\pi i\left(\frac{r(m_n-1)+b}{M_n}\right)t}. \tag{17}
$$

Proof Since $g_k \in V_0 \subset V_1$ and by Theorem 5.1 $\{\gamma_{1,\frac{r(m_n-1)+l}{M_n}}\}_{l \in \mathbf{Z}}$ is an orthonormal basis for V_1. Thus

$$
g_k(x) = \sum_l \left\langle g_k, \gamma_{1,\frac{r(m_n-1)+l}{M_n}} \right\rangle m_n^{1/2} g_k\left[m_n x - \frac{r(m_n-1)+l}{M_n}\right]
$$

Thus the result holds with

$$h_b = \left\langle g_k, \gamma_{1, \frac{r(m_n-1)+l}{M_n}} \right\rangle$$

which is in l_2 by Bessel's inequality.

By taking the Fourier transform on both sides of equation (15) we get (16). ∎

Definition 6.1 *Let g_k be a scaling function associated with a Special Non-uniform multiresolution analysis $\{V_a\}_{a \in \mathbf{Z}}$. The sequence $\{\lambda_b\}_{b \in \mathbf{Z}}$ satisfying equation (15) is called the scaling filter associated with $g_k(x)$.*
The function $m_{g_k}(t)$ as defined by (17) is called the auxillary function associated with $g_k(x)$ and $\alpha_b = \lambda_{1-r(m_n-1)-b} exp \frac{-2\pi i (r(m_n-1)+b)}{M_{n+1}}$ is called wavelet filter.
$w_k(x) = \sum_b \alpha_b \sqrt{m_n} g_k \left(m_n x - \frac{b}{M_n} \right)$ *is called the wavelet associated with wavelet filter.*

Theorem 6.2 *Let g_k be a scaling function of the Special Non-uniform multiresolution analysis $\{V_a\}$ belonging to $L^1(\mathbf{R}) \cap L^2(\mathbf{R})$ and let w_k be the associated wavelet and w_k also belongs to $L^1(\mathbf{R})$. Then*

1. $|\int_{\mathbf{R}} g_k(x) dx| = \frac{1}{\sqrt{M_n}}$.
2. $\int_{\mathbf{R}} w_k(x) dx = 0$.
3. $\hat{g}_k(b) = 0$ *for all integers $b \neq 0$.*

Proof 1. Let $f(x)$ be given such that $\|f\|_2 = 1$, $\hat{f}(t)$ is continuous and supported in $[-\alpha, \alpha], \alpha > 0$. It can be verified that

$$\hat{\gamma}_{a,b}(t) = m_n^{-a/2} \hat{g}_k(m_n^{-a} t) e^{-2\pi i b m_n^{-a} t}. \qquad (18)$$

By Parseval's formulae

$$\|P_a f\|_2^2 = \sum_b \left| < f, \gamma_{a, \frac{b}{M_n}} > \right|^2 = \sum_b \left| < \hat{f}, \hat{\gamma}_{a, \frac{b}{M_n}} > \right|^2$$

$$= \sum_b \left| \int_{\mathbf{R}} \hat{f}(t) m_n^{-a/2} \hat{g}_k(m_n^{-a} t) e^{-2\pi i \frac{b}{M_n} m_n^{-a} t} \right|^2.$$

Since $\{m_n^{-a/2} e^{-2\pi i b m_n^{-a} t}\}_{b \in \mathbf{Z}}$ is complete orthonormal system on $[-m_n^{a-1}, m_n^{a-1}]$, therefore as long as $m_n^{-a} > \alpha$, the above sum is the sum of squares of the Fourier coefficients of the period m_n^a extension to the function $\hat{f}(t) \hat{g}_k(m_n^{-a} t)$.
By the Plancherrel Theorem, we have

$$\|P_a f\|_2^2 = M_n \int_{-\alpha}^{\alpha} |\hat{f}(t)|^2 |\hat{g}_k(m_n^{-a} t)|^2 dt.$$

Since $g_k \in L^1(\mathbf{R})$, $\hat{g}_k(t)$ is continuous on $\mathbf{R}$ by the Riemann-Lebesgue theorem, it follows that $\lim_{a \to \infty} \hat{g}_k(m_n^{-a} t) = \hat{g}_k(0)$ uniformly on $[-\alpha, \alpha]$. By taking the limit under the integral sign, we conclude that

$$\lim_{a \to \infty} \|P_a f\|_2^2 = |\hat{g}_k(0)| \int_{-\alpha}^{\alpha} |\hat{f}(t)|^2 dt.$$

Since
$$\lim_{a \to \infty} \|P_a f\|_2 = \|f\|_2,$$

$$
\begin{aligned}
\|f\|_2^2 &= \lim_{a \to \infty} \|P_a f\|_2^2 \\
&= \lim_{a \to \infty} \int_{-\alpha}^{\alpha} |\hat{f}(t)|^2 |\hat{g}_k(m_n^{-a} t)|^2 \, dt \\
&= |\hat{g}_k(0)|^2 \int_{-\alpha}^{\alpha} |\hat{f}(t)|^2 \, dt \\
&= |\hat{g}_k(0)|^2 \|f\|_2^2.
\end{aligned}
$$

Thus $|\hat{g}_k(0)|^2 = 1$ and since $g_k(x) \in L^1$. Therefore

$$|\hat{g}_k(0)| = \left| \int_{\mathbf{R}} g_k(x) \, dx \right| = 1. \tag{19}$$

2. Since
$$\hat{g}_k(\xi) = m_{g_k}(\frac{\xi}{m_n}) \hat{g}_k(\frac{\xi}{m_n})$$

where $m_{g_k}(\xi)$ is designed by (16) and since $\hat{g}_k(0) \neq 0$ by (19), $m_{g_k}(0) = 1$ we have

$$w_k(x) = \sum_b \alpha_b \sqrt{m_n} g_k(m_n x - \frac{b}{M_n})$$

which gives us

$$\hat{w}_k(\xi) = m_{w_k}(\frac{\xi}{m_n}) \hat{g}_k(\frac{\xi}{m_n})$$

$$\text{or } \hat{w}_k(\xi) = e^{-2\pi i(\frac{\xi}{m_n} + \frac{1}{m_n})} \overline{m_{g_k}(\frac{\xi}{m_n} + \frac{1}{m_n})} \hat{g}_k(\frac{\xi}{m_n})$$

It can be checked that $\{T_b g_k(x)\}_{b \in \mathbf{Z}}$ is orthonormal. It implies that

$$|m_{g_k}(\xi)|^2 + |m_{g_k}(\xi + \frac{1}{2})|^2 = 1,$$

$m_{g_k}(\frac{1}{m_n}) = 0$ and hence $\hat{w}_k(0) = 0$. Therefore desired result holds in view that $w_k(x) \in L^1(\mathbf{R})$.
3. Proceeding on the lines of Lemma 7.4 [15] it can be proved that $\{T_b g_k(x)\}$ is an orthonormal system of translates iff

$$\sum_b |\hat{g}_k(\xi + b)|^2 = 1, \text{ for all } \xi \in \mathbf{R}.$$

Letting $\xi = 0$, this gives

$$\sum_b |\hat{g}_k(b)|^2 = 1.$$

By (19) $|\hat{g}_k(0)| = 1$ so that

$$\sum_{b \neq 0} |\hat{g}_k(b)| = 0$$

This implies that $\hat{g}_k(b) = 0$ for all integers $b \neq 0$. ∎

Theorem 6.3 *Let g_k be a scaling function of the Special Non-uniform multiresolution analysis $\{V_a\}$ and w_k be the associated wavelet. Then $\{\omega_{a,b}\}_{a,b \in \mathbf{Z}}$ is a wavelet orthonormal basis on $\mathbf{R}$ where $\omega_{a,b}(x) = m_n^{a/2} w_k(m_n^a x - b)$.*

We refer to [10] for its proof.

Theorem 6.4 *Let $\{V_a\}_{a \in \mathbf{Z}}$ be a Special Non-uniform multiresolution analysis and $\{\lambda_b\}$ and $\{\alpha_b\}$ be respectively the scaling and wavelet filter, then*

1. $\sum_b \lambda_b = \sqrt{m_n}$
2. $\sum_b \alpha_b = 0$
3. $\sum_b \lambda_b \lambda_{b-m_n l} = \sum_b \alpha_b \alpha_{b-m_n l} = \delta_l$
4. $\sum_b \alpha_b \lambda_{b-m_n l} = 0 \forall l \in \mathbf{Z}$

Proof 1. Since $\int_{\mathbf{R}} g_k(x) \neq 0$, so that

$$\begin{aligned}
\int_{\mathbf{R}} g_k(x)\,dx &= \int_{\mathbf{R}} \sum_b \lambda_b m_n^{1/2} g_k(m_n x - \frac{b}{M_n})\,dx \\
&= \sum_b \lambda_b m_n^{1/2} \int_{\mathbf{R}} g_k(m_n x - \frac{b}{M_n})\,dx \\
&= \sum_b \lambda_b m_n^{-1/2} \int_{\mathbf{R}} g_k(x)\,dx
\end{aligned}$$

This implies $\sum_b \lambda_b = \sqrt{m_n}$.

2. Since, we have $\int_{\mathbf{R}} w_k(x) = 0$

This implies $\quad \int_{\mathbf{R}} \sum_b \alpha_b m_n^{1/2} g_k(m_n x - \frac{b}{M_n})\,dx = 0$

or $\quad \sum_b \alpha_b m_n^{1/2} m_n^{-1} \int_{\mathbf{R}} g_k(x)\,dx = 0$

or $\quad \frac{1}{\sqrt{m_n}} \sum_b \alpha_b = 0$

or $\quad \sum_b \alpha_b = 0.$

3. Since $\{\gamma_{0,\frac{b}{M_n}}\}$ and $\{\gamma_{0,\frac{b'}{M_n}}\}$ are orthonormal systems on $\mathbf{R}$, we have

$$\delta_l = \int_{\mathbf{R}} g_k(x) g_k(x-l)\,dx$$

268

$$= \int_{\mathbf{R}} \sum_b \lambda_b \gamma_{1,\frac{b}{M_n}} \sum_{b'} \lambda_{b'} \gamma_{1,\frac{b'}{M_n}} (x-l)\, dx$$

$$= \sum_b \sum_{b'} \lambda_b \lambda_{b'-m_n l} \int_{\mathbf{R}} \gamma_{1,\frac{b}{M_n}} \gamma_{1,\frac{b'}{M_n}}\, dx$$

$$= \sum_b \lambda_b \lambda_{b-m_n l}.$$

This implies $\sum_b \lambda_b \lambda_{b-m_n l} = \delta_l$.

Similarly we can prove that $\sum_b \alpha_b \alpha_{b-m_n l} = \delta_l$ as $\{\omega_{0,b}(x)/b \in \mathbf{Z}\}$ is an orthonormal system on $\mathbf{R}$.

4. Since $<\omega_{0,b}, \gamma_{0,b'}>= 0$ for all $b, b' \in \mathbf{Z}$, the above argument yields that

$$\sum_b \alpha_b \lambda_{b-m_n l} = 0 \forall l \in \mathbf{Z}. \quad \blacksquare$$

7. JACKSON'S AND BERNSTEIN'S INEQUALITIES FOR THE SPECIAL NON-UNIFORM MULTIRESOLUTION ANALYSIS

With each Special Non-uniform multiresolution analysis, we can associate projections P_a defined by the equation

$$P_a f(x) = \int_{-\infty}^{\infty} f(t) m_n^a g_k(m_n^a t, m_n^a x)\, dt, \tag{20}$$

where

$$g_k(t,x) = \sum_{b \in \mathbf{Z}} g_k(t-b) g_k(x-b), \tag{21}$$

and $g_k(x)$ is a real scaling function satisfying the conditions

$$|g_k(x)| \le C(1+|x|)^{-\beta}, \beta > 3 \tag{22}$$

$$|g_k'(x)| \le C(1+|x|)^{-\beta}, \beta > 3 \tag{23}$$

$$\int_{-\infty}^{\infty} g_k(x)\, dx = 1. \tag{24}$$

It may be observed that (24) is automatically satisfied in view of (22) and Theorem 6.2(1). It follows from (22) that

$$|g_k(t,x)| \le C \sum_{b \in \mathbf{Z}} \frac{1}{(1+|t-b|)^{\beta}} \frac{1}{(1+|x-b|)^{\beta}}$$

This implies$|g_k(t,x)| \le C\dfrac{1}{(1+|t-x|)^{\beta-1}}$.

From (24), it follows that

$$\int_{-\infty}^{\infty} g_k(x,t)\, dt = 1. \tag{25}$$

Let $w_p(f,\delta)$ denote the p-modulus of continuity of a function f defined on $\mathbf{R}$, that is, for $1 \leq p \leq \infty$

$$w_p(f,\delta) = \sup_{0<|h|<\delta}\|f(x) - f(x-h)\|_p \tag{26}$$

defined for $\delta > 0$. The set of all functions having a p-modulus of continuity is denoted by $W_p(\mathbf{R})$.

Definition 7.1 (m_n-dilation operator) *Given an integer a, the m_n-dilation operator J_a acting on functions of $\mathbf{R}$ is defined by*

$$J_a f(x) = f(m_n^a x) \tag{27}$$

Remark 7.1 *Proceeding on the lines of $(6.2),(9.6),(9.7)$ [16] the following result will hold*
For an integer a and for the function f defined on $\mathbf{R}$

1. *$\|J_a f\|_p = m_n^{-a/p}\|f\|_p$*
2. *$w_p(T_h f;\delta) = w_p(f;\delta)$*
3. *$w_p(J_a f;\delta) = m_n^{-a/p} w_p(f;m_n^a \delta)$*

Theorem 7.1 *The projections P_a with $a \in \mathbf{Z}$ have the following properties*

1. *For the m_n-dilation J_r with $r \in \mathbf{Z}$ we have*

$$P_a J_r = J_r P_{a-r} \tag{28}$$

   in particular $P_a = J_a P_0 J_{-a}$.
2. *There exist constants $C \geq c > 0$ such that for each $a \in \mathbf{Z}$ and $1 \leq p \leq \infty$ and each sequence of scalars $\{d_b\}_{b\in\mathbf{Z}}$ we have*

$$\frac{C}{M_n}m_n^{a(\frac{1}{2}-\frac{1}{p})}(\sum_{b\in\mathbf{Z}}|d_b|^p)^{1/p} \leq \|\sum_{b\in\mathbf{Z}}d_b\gamma_{a,\frac{b}{M_n}}(x)\|_p \leq \frac{C}{M_n}m_n^{a(\frac{1}{2}-\frac{1}{p})}(\sum_{b\in\mathbf{Z}}|d_b|^p)^{1/p}. \tag{29}$$

Lemma 7.1 *Suppose $g_k(x)$ is a function on $\mathbf{R}$ such that $|g_k(x)| \leq C(1+|x|)^{-2}$. Then for any sequence of scalars $\{d_b\}_{b\in\mathbf{Z}}$ and any p, $1 \leq p \leq \infty$, we have*

$$\left\|\sum_{b\in\mathbf{Z}}d_b g_k\left(x - \frac{b}{M_n}\right)\right\|_p \leq C\|(d_b)\|_p. \tag{30}$$

We get the proof of Theorem 7.1 and Lemma 7.1 by proceeding on the lines of Proposition 8.1, Lemma 8.2 [16] .

Theorem 7.2 *(Analogue of Jackson's inequality)*
There exists a constant C such that for any $f \in W_p(\mathbf{R})$

$$\|f - P_a f\|_p \leq C w_p(f;m_n^{-a})\forall a \in \mathbf{Z}. \tag{31}$$

Theorem 7.3 *(Analogue of Bernstein's inequality)*
Under the assumptions about Special Non-uniform multiresolution analysis described by (22) and (23), for each p, $1 \leq p \leq \infty$, there exists a constant C such that for $f \in V_a$ we have

$$w_p(f;t) \leq C min(m_n^a t, 1) \|f\|_p. \tag{32}$$

We refer to [10] for proofs of these theorems.

We take this opportunity to thank Prof.A.H.Siddiqi who encouraged to write this paper.

REFERENCES

1. O.Christensen, *An introduction to Frames and Riesz Bases*, Birkhäuser, Boston- Basel- Berlin, 2003.
2. Z.Ciesielski, *Haar orthogonal Functions in Analysis and Probability*, Colloquia Societatis James Bolyai, 49, Alfred Haar Memorial Conference, Budapest, 1985, 25-27.
3. I.Daubechies, *Ten lectures on wavelets*, Society of Industrial and Applied Mathematics, Philadelphia, 1992.
4. Yu A.Farkov, *Orthogonal wavelet with compact support on locally compact abelian groups*, Izv. Ross. Akad. Nauk Ser. Mat. 69(3)(2005),193-220, English translation in Izv. Math. 69(3)(2005), 623-650.
5. J.P.Gabardo and M.Z.Nashed, *Nonuniform Multiresolution Analysis and Spectral Pairs*, Journal of Functional Analysis, 158(1998), 209-241.
6. V.Grozdanov and S.Stoilova: *Price and Haar type functions and uniform distribution of sequences*, Journal of inequalities in Pure and Applied Mathematics, 5(2)(2004), 1-17.
7. A.Haar, *Zur Theorie der orthogonalen funktionen systeme*, Math.An. 69(1910), 331-371.
8. S.Mallat, *Multiresolution approximation and wavelet orthogonal bases of $L_2(\mathbf{R})$*, Trans. Amar. Math. Soc. 315(1989), 69-87.
9. P.Manchanda, Meenakshi and A.H.Siddiqi *Haar-Vilenkin wavelet*, The Aligarh Bulletin of Maths 27(2008),59-73.
10. P.Manchanda, Meenakshi and A.H.Siddiqi, *A special type of non-uniform Multiresolution Analysis*, Preprint, December, 2008.
11. F.Schipp,W.R.Wade and P.Simon, *Walsh series: An Introduction to Dyadic Harmonic Analysis*, Adam Hilger, Bristol, 1990.
12. A.H.Siddiqi, *Applied Functional Analysis*, Marcel Dekker, New York, 2005.
13. P.L.Uljanov, *Haar series and related questions*, Colloqui Mathematica Societatis Janos Bolyai 49, Alfred Haar Memorial Conference, Budapest, Hungary, 1985, 57-96.
14. N.Ya.Vilenkin, *On the theory of lacunary orthogonal system with gaps*, Izo.Akad.Nauk.SSSR Ser. Math. 13(1949), 242-252.
15. D.Walnut, *An Introduction to Wavelet Analysis*, Birkhauser, Boston,2001.
16. P.Wojtaszczyk, *A Mathematical Introduction to Wavelets*, London Mathematical Society Student Texts 37, Cambridge University Press, 1997.

ESTIMATION OF HURST EXPONENT FOR THE FINANCIAL TIME SERIES

Kumar J*., Manchanda P.**

Department of Mathematic,
Guru Nanak Dev University, Amritsar 143001, India
*Email: *meenujkumar@rediffmail.com,** pmanch2k1@yahoo.co.in*

Abstract. Till recently statistical methods and Fourier analysis were employed to study fluctuations in stock markets in general and Indian stock market in particular. However current trend is to apply the concepts of wavelet methodology and Hurst exponent, see for example the work of Manchanda, J. Kumar and Siddiqi, Journal of the Frankline Institute 144 (2007), 613-636 and paper of Cajueiro and B. M. Tabak. Cajueiro and Tabak, Physica A, 2003, have checked the efficiency of emerging markets by computing Hurst component over a time window of 4 years of data. Our goal in the present paper is to understand the dynamics of the Indian stock market. We look for the persistency in the stock market through Hurst exponent and fractal dimension of time series data of BSE 100 and NIFTY 50.

Keywords: Time series; Hurst Exponent; Fractal dimension; Multifractal; Rescaled range

INTRODUCTION

Research on stock market volatility is central for the regulation of financial Institutions and for financial risk management. Its implications for economic, social and public welfare issues have a great significance.

In recent times, a lot of activity has been witnessed in the field of econophysics [2-4, 22].Various concepts of Mathematics and Physics have been applied to study financial time series both for short and long range studies. In the recent past some new mathematical methods such as wavelets, fractal and multifractal are being used for classification and understanding these systems[5-7] It has been found that financial time series (like stock indices, currency exchange rates etc.) have either fractal or multifractal features [8,9] . Another important feature that has been observed in recent times is that stock market indices can be represented by fractional Brownian motion instead of a classical Brownian motion. The similarity between the fluid turbulence and financial market is well known .The information transfer in financial markets is similar to energy flow in hydrodynamics [10].

Analyzing the time series statistically has been problem of considerable recent interest. The accuracy of measured financial risk models depends on the assumptions about the statistical properties of asset prices. The statistical properties of data during financial crisis are very different from those in normal markets. This will require considerable adjustment of the existent regulatory risk monitoring techniques which were developed for non extreme financial market behavior. With the surge of data outpouring from various fields such as biology, geophysics and finance (digitization of fingerprint data, seismic wave data etc.), it is becoming imperative to use proper Mathematical tools for classification and understanding these systems. For example, in the field of DNA sequence analysis, there exists immense mathematical literature [11-12]. More recently

CP1146, *Modelling of Engineering and Technological Problems*, edited by A. H. Siddiqi, A. K. Gupta, and M. Brokate
© 2009 American Institute of Physics 978-0-7354-0683-4/09/$25.00

(as of late 1995 and beyond), Wall Street analysts have started using calculus based methods for their own analysis. These advances have motivated certain groups of physicists and applied mathematicians to try some of their own classical methodologies to understand the dynamics of the stock market fluctuations. Besides the classical Black-Scholes formulation [13], a tremendous amount of work using wavelet applications [14] L'evy distributions, spectral analysis and, multifractal models [15] (to name a few) have been used as tools to understand the mathematical nature of the financial time series. The basic goal of all these methodologies is to understand the mathematical properties of "long term" memory versus "short term" memory (long range versus short range correlations) of the market.

The mathematics and the images derived from fractal geometry exploded the world in 1970s and 1980s.It is difficult to think of an area of science that has not been influenced by fractal geometry. Along with providing new insight in mathematics and science, fractal geometry helped us see the world around us. The Hurst exponent occurs in several areas of applied mathematics including fractals and chaos theory, long memory process and spectral analysis. Hurst exponent estimation has been applied in areas ranging from biophysics to computer networking [16-18]. The Hurst exponent is directly related to the fractal dimension which gives a measure of roughness of a surface.

Estimating the Hurst exponent for a data set provides a measure of whether the data is a pure random walk or has underlying trends. Another way to state this is that a random process with an underlying trend has some degree of autocorrelation. When the autocorrelation has a very long (or mathematically infinite) decay this kind of Gaussian process is sometimes referred to as a long memory process.

Standard statistical analysis begins by assuming that the system under study is primarily random, i.e., the underlying components has many parts and the interaction of those components is so complex that a deterministic explanation is not possible. In this set up the system is usually modeled by a random walk-type process, which implies that the events measured are independently and identically distributed (IID). In other words, the events must not influence one another and they must all be equally likely to occur. However there exists strong evidence that price processes of financial assets do not fall in this category.

In general, much research [1, 2, 4, 7, 13-15, 20] has been carried out to measure the fluctuations of the market indices. For example, a lot of work has been done in efforts to detect trends in S&P 500 (for various different time periods). There has also been some work done on currency exchange dynamics [19]. In most of these cases, the correlation content has been measured for long term periods. This paper will deal with the fluctuations in the market indices. Here we present study of financial data set of BSE 100 and Nifty 50 (The Indian stock exchange). In particular we examined the BSE 100 over 17.5 years, taken at one day interval and Nifty 50 over the 2 years, taken at every 5 minute interval. We estimated the Hurst Exponent using wavelets and Rescaled range (R/S) methods.

FINANCIAL MARKETS

'Finance' is one of the fastest developing areas of the modern corporate world. This, together with the sophistication of modern financial products, provides a rapidly growing impetus for the new mathematical models and modern mathematical methods; the area is an expanding source for novel and relevant real world mathematics. There are many kinds of financial market but the most important one is the stock markets which deal in shares.

The Stock Exchange, "share market" or a "bourse" is a mutual organization for traders or "stock brokers" who trade in different company securities and stocks. Companies or businesses have to be "listed" in the bourses in order for any trading or exchange in their "shares" or equities to be carried out. Stock markets are also the place for trading in unit trusts and bonds issued by the government.

New York, Chicago, Frankfurt, London and Bombay have well known share markets.
A stock index is a mathematical measurement of the performance of a company or a number of companies. It is a barometer of the Economy. At the Stock Exchange, share prices rise and fall depending, largely, on market forces. Share prices tend to rise or remain stable when companies and the economy in general show signs of stable growth. Therefore the movement of share prices can be an indicator of the general trend in the economy.

In India, we have two main stock exchanges, Bombay stock exchange (BSE) and National Stock Exchange (NSE). Both open and close at the same time, that is, these are synchronous. The BSE is the oldest stock exchange in Asia. It was started in 1875.The BSE index is a market capitalization weighted index of 30 stocks of sound Indian and multinational companies. The NSE was established to provide access to investors from all over India. The NSE started equity operations in November 1994 and operations in derivatives in   June 2000.The NSE also known as Nifty 50 is determined from 50 stocks of companies taken from 23 sectors of economy. In India equity trading is most active in the National Stock Exchange (NSE) and the Bombay Stock Exchange (BSE). Since the NSE's inception in 1994, it has caught up with the BSE in terms of capitalization but exceeded it in turnover. The BSE boasts of over 4,000 listed companies, surpassing stock exchanges in the U.S.

The sample employed in our study consists of daily closing of the indices of BSE 100 and Nifty 50 values over every five minutes return. For BSE100 we consider the daily closing from 06 January 1986 to 07 August 2003.This generates 4096 points for the said period. In case of Nifty50, the data length is from 29 August 2002 to 12 August 2004.we study this data over every 5 minutes return, generating in total data 32768 points.

FRACTALS AND MULTIFRACTALS

Fractals describe objects that are too irregular to fit into traditional geometric settings. Various phenomena show fractal features when plotted as functions of time. Some common examples include atmospheric pressure, water level of a reservoir and stock market index, usually when recorded over a fairly long time spans. Multifractals are the fractal objects which cannot be completely described using a single fractal dimension (monofractals). They have in fact infinite number of fractal dimensions associated with them. Signals that are singular at almost every point are multifractals and they appear in maintenance of economic records, physiological data including Electrocardiographs, electromagnetic fluctuations in galactic radiation noise, textures in images of natural terrain and variation in traffic flow etc.

HURST ANALYSIS

In the early 20th century, a celebrated British hydrologist H.E. Hurst worked on the Nile River Dam project. He had studied the eight and a half century long record that the Egyptians had kept the Nile River's overflow. This record did not appear to be a white noise. Larger than average overflows were more likely to be followed by still larger overflows. Abruptly the process would change a lower than average overflow, which was followed by other lower-than average overflows. This appears to be cycles and long-term memory in the data. However standard analysis revealed no statistically significant correlations between observations. As a result Hurst developed a method called the rescaled range (R/S) for distinguishing completely random time series from correlated time series [16].

DAILY RETURN

The return is the profit or loss in buying share of stock (or some other traded item), holding it for some period of time and then selling it. The most common way to calculate a return is the log return.

$$(ret)_{t-\Delta,t} = \log(P_t) - \log(P_{t-\Delta}) \tag{3.1}$$

This equation calculates log return for a share of stock that is purchased and held Δ days and then sold. Here P_t and $P_{t-\Delta}$ are the prices at time t (when we sell the stock and $t-\Delta$ (when we purchased the stock, Δ days ago).Taking the log of price removes most of the market trend from the return calculations.

Another way to calculate the return is to use the percentage return, which is as shown in Equation (3.2)

$$(ret)_{t-\Delta,t} = \frac{P_t - P_{t-\Delta}}{P_{t-\Delta}} \tag{3.2}$$

Equations (3.1) and (3.2) yield similar results.

An n-day return time series is created by dividing the given time series data into a set of blocks and calculating log return (Equation 3.1) for $P_{t-\Delta}$, the price at the start of the block and P_t, the price at the end of the block. We considered the 1-day return time series.

$$x_0 = \log p_1 - \log p_0$$
$$x_1 = \log p_2 - \log p_1$$
$$x_2 = \log p_3 - \log p_2 \qquad (3.3)$$
$$\cdots\cdots$$
$$x_n = \log p_t - \log p_{t-1}$$

EVALUATION OF HURST EXPONENT
RESCALED RANGE METHOD

Let $X(t)$ be the closing of the index on a time t and $r(t)$ be the logarithmic return denoted by $r(t) = \ln(X(t+1)/X(t))$. The R/S statistic is the range of partial sums of deviations of time series from its mean, rescaled by its standard deviation. So, consider a sample of continuously compounded asset returns $\{x_0, x_1, x_2, \dots, x_\tau\}$ and let $\overline{x}_\tau$ denote the sample mean $1/\tau \sum_\tau x_\tau$ where τ is the time span considered. Then is the R/S statistic is given by

$$\left(\frac{R}{S}\right)\tau = \frac{1}{S_\tau}\left[\max_{1<t<<\tau}\sum_{t=1}^{\tau}(x_t - \overline{x}_\tau) - \min_{1<t<<\tau}\sum_{t=1}^{\tau}(x_t - \overline{x}_\tau)\right], \qquad (3.4)$$

where S_τ is the usual standard deviation estimator

$$S_\tau = \sqrt{\frac{1}{\tau}\sum_{t=1}^{\tau}\left(x_t - \overline{x}_\tau\right)^2 \sum_{t=1}^{\tau}(x_t - \overline{x}_\tau)^2} \qquad (3.5)$$

Hurst found that the rescaled range, R/S, for many records in time is very well described by the following empirical relation:

$$\left(R/S\right)_\tau = \left(\tau/2\right)^H \qquad (3.6)$$

In practice for a given window length w, one subdivides the input series in a number of intervals of length w, measures R(w) and S(w) in each interval and calculates R/S(w) as the average ratio R(w)/S(w). This process is repeated for a number of window lengths, and the logarithms of R/S(w) are plotted versus the logarithms of w. The plot follows a straight line whose slope equals the Hurst exponent H. The fractal dimension of the trace is then calculated from the relationship between the Hurst exponent H and the fractal dimension:

D = 2-H

where D denotes the fractal dimension estimated from the rescaled range method.

The values of the Hurst exponent range between 0 and 1.A value of 0.5 indicates a true random walk (a Brownian time series).In a random walk there is no correlation between any element and future element. A Hurst exponent value H, $0.5 < H < 1$ indicates "persistent behavior" (e.g., a positive autocorrelation. If there is any increase from step ti-1 to ti , there will probably be an increase from time step ti to ti+1. The same is true for decrease, where a decrease will tend to follow a decrease. A Hurst exponent value $0 < H < 0.5$ will exist for a time series with "anti-persistent behavior" (or negative autocorrelation). Here an increase will tend to be followed by a decrease. Or decrease will be followed by an increase. This behavior is also sometimes is called "mean reversion".

BENOIT

Benoit [21] is software that enables to measure the fractal dimension and/or Hurst exponent of the data sets. It includes ruler, box, information, perimeter-area and mass for analysis of self similar patterns and R/S power spectral analysis, variogram roughness length and wavelets for analysis of self affine traces. It is particularly useful to find orders and patterns in seemingly chaotic data where traditional statistical approaches of analysis fail. It is widely used in diverse disciplines like physics, chemistry, biology, economics and medicine. We used wavelet method in Benoit to find Hurst exponent for the financial time series data of BSE 100 and Nifty50.

RESULTS

We perform the estimation of the Hurst exponent for time windows with 4096 data values (log returns) for the BSE. This data is processed for on one day return basis. NSE data for 2 years is processed on every 5 minute return basis thus taking 32768 data values in total. First R/S is calculated over the single time window over the full time range. Then the successive calculations were performed by subdividing the data in respective halves at each stage of calculation. We developed a C program to carry out the above calculations.

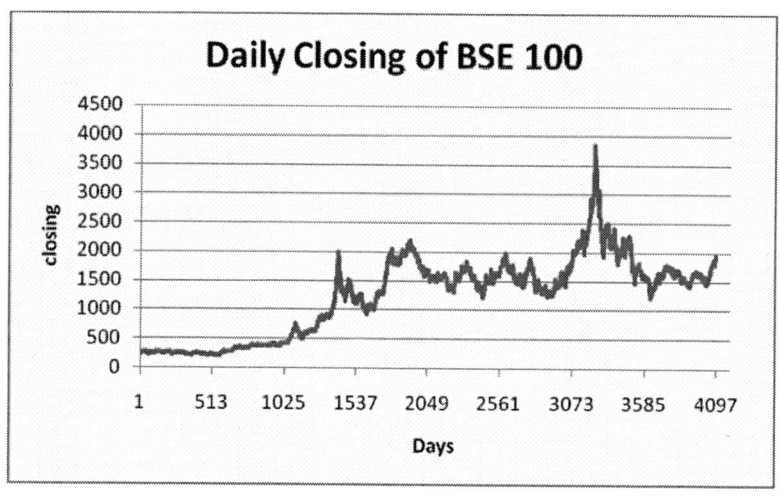

FIGURE 4.1: This Plot shows the daily closing of 4096 data points of BSE100

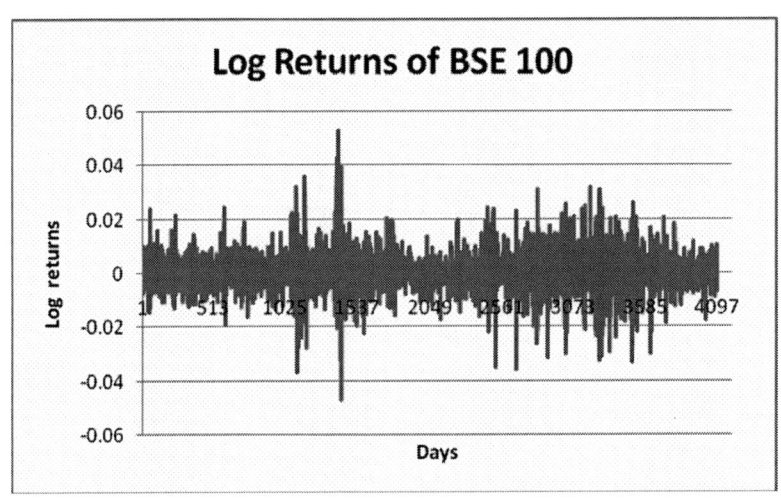

FIGURE 4.2: This Plot shows the log returns of 4096 data points of BSE100

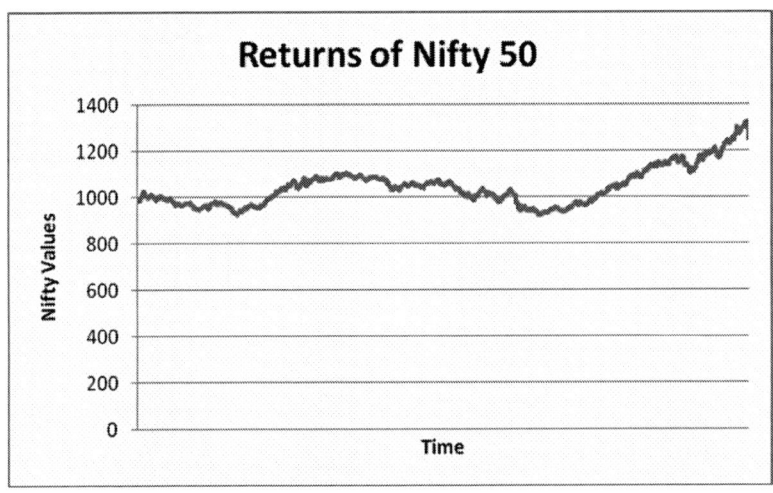

FIGURE 4.3: This plot shows the values of Nifty 50 for the first half data values (from 1 to 16384) over every five minutes return.

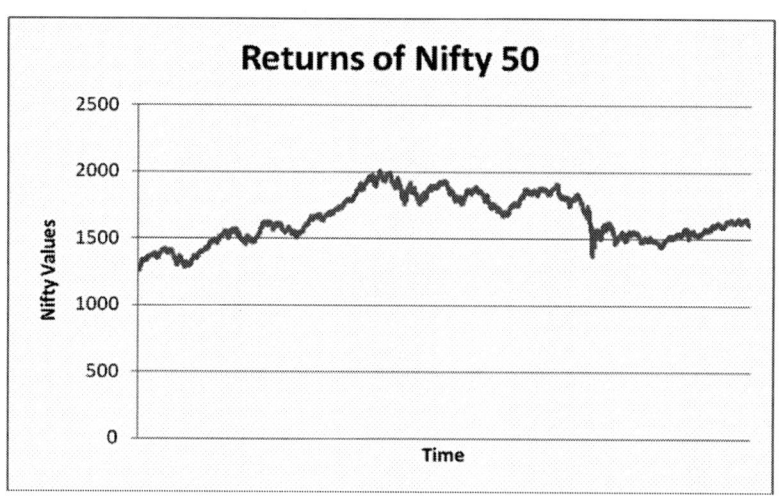

FIGURE 4.4: This plot shows the daily closing of second half data values (from 16385 to 32768) of Nifty 50

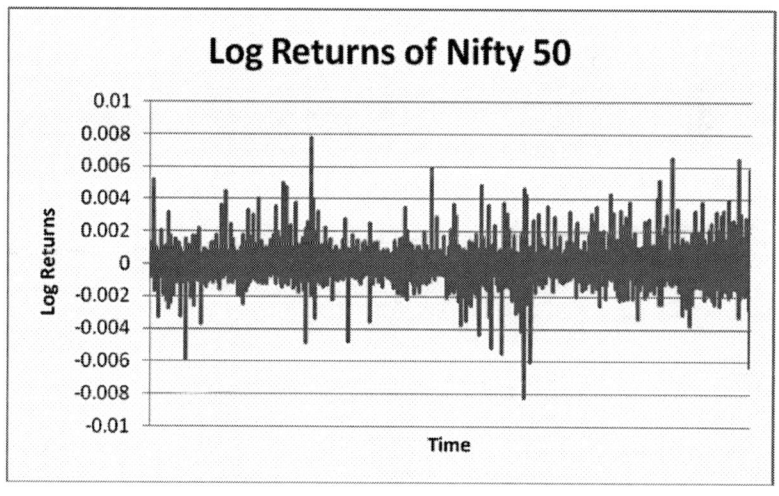

FIGURE 4.5: This plot shows the log of daily returns for first 16384 data points of Nifty 50

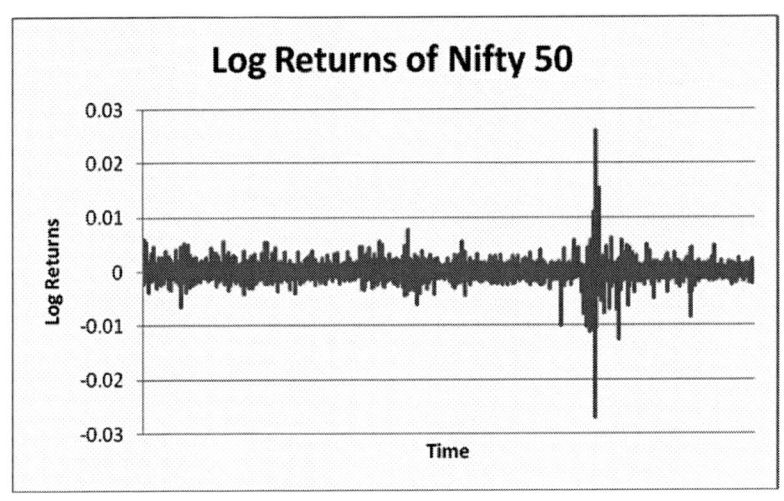

FIGURE 4.6: This plot shows the log returns for second half (from data values 16385 to 32768) data points of Nifty 50

TABLE

Table1. Results for BSE

Region Size	R/S Avg.	Log2(region size)	Log2(R/S Avg.)
4096	13.744919	12	3.780826
2048	11.671974	11	3.544977
1024	9.517603	10	3.250598
512	8.397802	9	3.070012
256	7.204024	8	2.848803
128	6.325675	7	2.661219
64	5.311925	6	2.409235
32	4.462337	5	2.157799
16	3.766132	4	1.913084
8	3.177631	3	1.667952

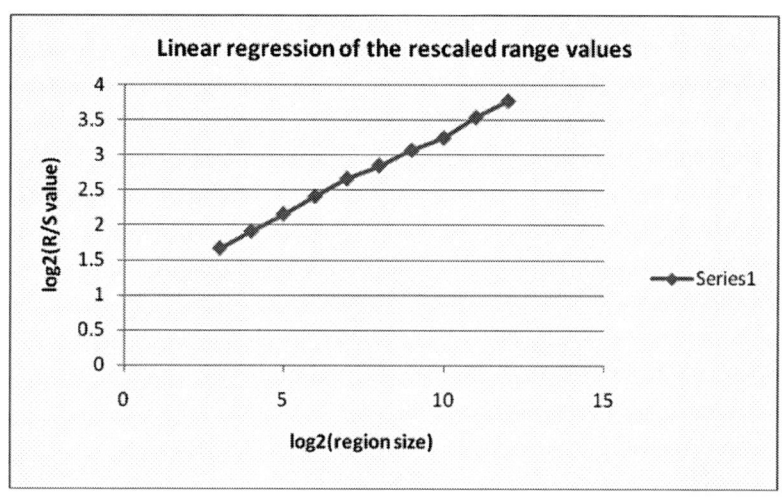

H.E.for BSE=0.230746

Table2. Results for NSE

Region Size	R/S Avg.	log2(Region Size)	log2(R/S Avg.)
32768	68.262917	15	6.093030
16384	42.396065	14	5.405858
8192	30.002571	13	4.907014
4096	23.591646	12	4.560204
2048	18.46965	11	4.207085
1024	14.741668	10	3.881828
512	12.012975	9	3.586522
256	9.805068	8	3.293528
128	7.866143	7	2.975656
64	6.231544	6	2.639590
32	4.720032	5	2.238797
16	3.802612	4	1.926991
8	3.15664	3	1.658390

HURST EXPONENT FOR NIFTY 50

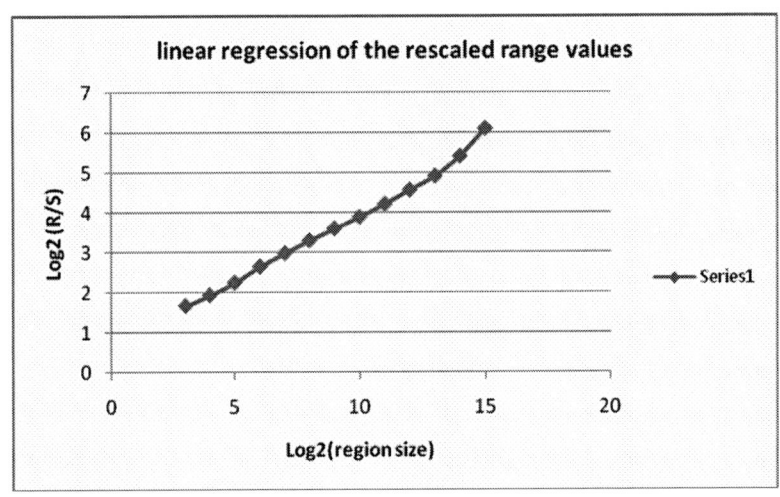

H.E.for NSE=0.348835

Table3. Rescaled Range Method

	BSE 100	Nifty 50
Hurst Exponent	0.231	0.349
Fractal Dimension	1.769	1.651

Table4. Benoit (Wavelet Method)

	BSE 100	Nifty 50
Hurst Exponent	0.634	0.566
Fractal Dimension	1.366	1.434

CONCLUSIONS

We compared the results calculated using rescaled range method with those obtained from Benoit.
Using rescaled Range (R/S) Method the values of the Hurst exponents obtained suggests an anti persistent behavior. But the graphs of two stock markets BSE 100 and Nifty 50

shows a persistent behavior of the data under consideration and this assertion is also supported by the values of Hurst exponents those obtained with Benoit

For the estimation of Hurst exponent, using Benoit is a better approach than Rescaled range (R/S) method.

ACKNOWLEDGEMENT

We thank Prof. Shafiqur Rehman for motivating discussions and suggestions.

REFRENCES:

1. Daniel O. Cajueiro, Benjamin M. Tabak: The Hurst exponent over time: Testing the assertion that emerging markets are becoming more efficient, Physa 7978 pp: 1-18.
2. Refal Weron, Beata Przybylowicz: Hurst Analysis of Electricity Price Dynamics, Physica A 283 (2000) 462-468.
3. R.N. Mantegna, H.E. Stanely, An Introduction to Econophysics: Correlations and Complexity in Finance, Cambridge University Press, Cambridge, 1999.
4. E.F.Fama, Efficient capital market: A review, J. Finance (1970) 383
5. A.Arneodo, B.Audit, N. Decoster, J.F. Muzy, C. Vaillant: wavelet based multifractal formalism: Applications to DNA sequences, satellite images of the cloud structure and stock market data, in; The science Disasters: A. Bunde, J. Kropp, H.J. Shennhuber Eds., Springer-Verlag, pp. 27-102, (2002).
6. Ashok Razdan: Wavelet correlation coefficient of 'strongly correlated' time series, Physica A 333 (2004) 335-342.
7. R.Gencay, F.Seluk, B. Whitcher: An Introduction to Wavelets and other filtering Methods in Finance and Economics, Academic Press, 2002.
8. A. Razdan, Pramana, J. Phys. 58 (2002) 537.
9. M. Ausloos, K. Ivanova, cond-mat/0108394, 24 August 2001.
10. N. Vandewalle, M. Ausloos, Physica A 246 (1997) 454.
11. S.Guharay, B.R. Hunt, J.A. Yorke, O.R. White, Physica D 146 2000 388.
12. P.Bernaola-Galvan, I. Grosse, P. Carpena, J.L. Oliver, R.Roman-Roldan, H.E. Stanley,Physical Review Letters 85 2000 1342.
13. F.Black, M. Scholes, Journal of Political Economy 81 1973 637.
14. J-F Muzy, D Sornette, J. Delour, A. Arneodo, Quantitative Finance 1 2001 131.
15. B.B. Mandelbrot, Fractals and scaling in Finance: Discontinuity, Concentration, Risk: Selecta vol. E, Springer-Verlag New York 1997.
16. H.E. Hurst, "Long-term storage capacity of reservoirs", Trans. Amer. Soc. Civ. Engrs., Vol.116, pp. 770-808, 1951.
17. J. Moody and L. Wu, "Long memory and Hurst Exponents of tick-by-tick interbank foreign exchange rates", Proceedings of Computational Intelligence in Financial Engineering, IEEE Press, Piscataway, NJ, pp. 26-30, 1995.
18. Rafal weron, Beata Przybylowicz:Hurst analysis of electricity price dynamics, Hugo Steinhaus center for stochastic methods, Physica A 283 (2000) 462-468.
19. N. Vandewalle, M. Ausloos, Eur. Phys. J.B 4 1998 257.
20. P.Manchanda, J.Kumar and A.H. Siddiqi, Mathematical Methods for Modelling Price Fluctuations of financial Time Series, J. of the Frankline Institute 144 (2007), 613-636.
21. Benoit T^m software developed by B.Mandelbrot.
22. Andrew W. Lo, Long-Term memory in stock market prices, Econometrica, Vol.59, No.5(1991), 1279-1313.

Wavelet based Simulation of Reservoir Flow

A. H. Siddiqi[a], A. K. Verma[b]

Noor - e - Zahra[c], Ashish Chandiok[d] and A. Hasan[e]

[a,b,d]BMAS Engineering College, NH2, Keetham, Agra- 282 007, UP, India.

[c,e]Hindustan College of Technology, Knowledge Park – III, Greater Noida, UP, India.

Abstract. Petroleum reservoirs consist of hydrocarbons and other chemicals trapped in the pores of a rock. The exploration and production of hydrocarbon reservoirs is still the most important technology to develop natural energy resources. Therefore, fluid flow simulators play a key role in order to help oil companies. In fact, simulation is the most important tool to model changes in a reservoir over the time. The main problem in petroleum reservoir simulation is to model the displacement of one fluid by another within a porous medium. A typical problem is characterized by the injection of a wetting fluid, for example water into the reservoir at a particular location displacing to the non wetting fluid, for example oil, which is extracted or produced at another location. Buckley – Leverett equation [1] models this process and its numerical simulation and visualization is of paramount importance. There are several numerical methods applied for numerical solution of partial differential equations modeling real world problems. We review in this paper the numerical solution of Buckley – Leverett equation for flat and non flat structures with special focus on wavelet method. We also indicate a few new avenues for further research.

Keywords: Wavelets, Buckley – Leverett equations, Numerical methods.

1. INTRODUCTION

The Buckley – Leverett equation was introduced by S. E. Buckley and M. C. Leverett in 1942, which is used for the oil recovery in oil industry. Buckley – Leverett equation can be written in the form:

$$\frac{\partial u}{\partial t} + \frac{\partial}{\partial x}\left(\frac{u^2}{u^2 + \frac{1}{4}(1 - u^2)}\right) = 0 \qquad (1.1)$$

It may be observed that,

$$\frac{\partial u}{\partial t} + \frac{\partial f(u)}{\partial x} = 0 \qquad (1.2)$$

where u is the conserved quantity and $f(u)$ is the flux, known as conservative law. Buckley – Leverett is its particular case and another special case, where $f(u) = \frac{1}{2}u^2$, it is a convex function. This special case is known as Burger's equation. Equation (1.2) models oil recovery, flow of gas, water, traffic flow etc.

CP1146, *Modelling of Engineering and Technological Problems*, edited by A. H. Siddiqi, A. K. Gupta, and M. Brokate
© 2009 American Institute of Physics 978-0-7354-0683-4/09/$25.00

The discovery and recovery of oil is highly complex process in which mathematical modeling and numerical simulation play a very significant role. Usually when an oil well is discovered, the pressure at early stage is naturally high because of that there may not be any problem in recovering the oil. The rate at which oil flows out of the well will naturally diminish with time. Well known techniques of keeping up oil flowing is to inject water to drive the oil towards the producing well.

An oil reservoir consists of layers of porous rocks sandwiched between impervious rocks. The oil reservoir is formed when oil is produced over geological time scales at great depths, migrating into reservoir which is filled with water. This movement cause's displacement to the water, gas may also be present depending on the pressure conditions. The proportion of oil, gas and water which are present are known as saturations. In equation, (1.1) Buckley – Leverett equation, there are two saturations, namely water and oil, say one of these saturation is u. In short Buckley – Leverett equation models a reservoir containing a mixture of water and oil, both incompressible fluids. In this model, diffusive effects such as capillary pressure of the physical mixing of fluids and gravitational forces are neglected, in the next section; we briefly outline the important features of this model.

During the last half of century, attempts have been made to develop techniques and methods to find appropriate solutions of these models and to examine whether the approximate solutions converge to exact solution or not. New endeavors made to find methods and techniques which require minimum time in evaluating and visualization of the solution of the model under consideration with tools constitute the subject of scientific computation. It is well known to mathematicians, scientists and engineers that Raleigh – Ritz, Galerkin, weighted residuals, collocation, least square, finite difference, multi grid, finite element, finite volume, particle, boundary element and wavelet methods are integral part of scientific computation. These methods have been used for computation and visualization of solution of the Buckley – Leverett equation in particular and other equations representing real world problems in general, see for example [1-13] and other references related to particle and wavelet methods in subsequent sections.

The main goal of this paper is to present an updated review of the literature related to Buckley – Leverett equation with special focus on wavelet method. In the next section, we present important features of this equation along with the five – spot problem [4]. Section 3 and 4 are devoted to wavelet methods and their application to Buckley – Leverett equation. An outline of a solution of B-L equation on a sphere by wavelet method is given in Section 5. In Section 6, certain recent work on wavelet based particle method and homogenization of B-L equation is indicated.

2. MODELLING OF RESERVOIR FLOW (BUCKLEY – LEVERETT EQUATION)

Each fluid (phase) is conserved and so their behavior is modeled by the mass balance equations. These equations are given below:

Mass balance equation of water:

$$\varphi(x)\frac{\partial u_w(t,x)}{\partial t} + \nabla a_w(t,x) = 0 \quad (2.1)$$

Mass balance equation of oil:

$$\varphi(x)\frac{\partial u_o(t,x)}{\partial t} + \nabla a_o(t,x) = 0 \quad (2.2)$$

Here, the scalar field $\varphi(x)$ denotes the porosity of the rock. The vector fields $a_w(t,x)$ and $a_o(t,x)$ are the phase velocities , $u_w(t,x)$ and $u_o(t,x)$ are the saturation of water and oil respectively. It may be observed that u_w and u_o are the fractions of the pore space, that are filled with water or oil, that is, $0 \leq u_{w,o} \leq 1$.

Equation (2.1) and (2.2) show that a change of mass for each phase in a given region of the reservoir is equal to the next flux of the phase across the boundary of that region. The equations describing reservoir fluid flow are given by Buckley – Leverett equation:

$$\frac{\partial u}{\partial t} + a.\nabla f(u) = 0 \qquad\qquad (2.3)$$

The incompressibility relation:
$$\nabla.a(t,x) = 0 \qquad\qquad (2.4)$$

And Darcy's law:
$$a(t,x) = -K(x)\,M(u)\,\nabla p(t,x), \qquad\qquad (2.5)$$

where $0 \leq u \leq 1$ is the water saturation, $p(t,x)$ is the reservoir pressure (It can be estimated from down – hole measurements in the field), $K(x)$ is the permeability tensor of the porous rock (it describes the ability of the rock to transmit the fluid. It can be determined by the laboratory measurements of case examples),

$$M = M_w + M_o.$$

M_w and M_o are termed as phase mobility's which are defined in term of viscosities of fluids:

$$M_w(u_w) = kr_w(u_w)/\mu_w \text{ and } M_o(u_o) = kr_o(u_o)/\mu_o,$$

where μ_w and μ_o denote the viscosity of two fluids (Viscosities are determined by laboratory measurements). M *is* called the total mobility. $f_w(u_w)$ is the flux tensor, given by the ratio

$$f_w(u_w) = \frac{M_w(u_w)}{M(u_w)}$$

between the phase mobility $M_w(u_w)$ and the total mobility M (In the field of reservoir simulation and engineering, the flux tensor $f_w(u_w)$ is usually called fractional flow of the water phase) $kr_w(u_w)$ and $kr_o(u_o)$ are the relative permeability of water and oil phase.

In reservoir modeling, the function $f: u \to f(u)$ is monotonically increasing and stratifies $0 \le f(u) \le 1$ for all $u \in [0,1]$. In core model [10], one takes

$kr_w(u) = u^2, kr_o(u_o) = (1-u)^2$ that yields

$$M(u) = [\frac{u^2}{\mu_w} + \frac{(1-u)^2}{\mu_o}]$$

for total mobility and so

$$f(u) = \left[\frac{u^2}{\left\{ u^2 + \frac{\mu_w}{\mu_o(1-u)^2} \right\}} \right].$$

This $f(.)$ is often called the Buckley – Leverett flux and it is used to model the displacement of oil by water.

Remarks 2.1

(i)   Let vector fields $a_w(t,x)$ and $a_o(t,x)$ be the phase velocities of water and oil and let $u_w(t,x)$ and $u_o(t,x)$ be saturations of water and oil respectively. u_w and u_o are fractions of the pore space, that are field with water or oil, that is $0 \le u_{w,o} \le 1$.

We have $u_w(t,x) + u_o(t,x) = 1$

$a_w(t,x) = -K(x)[\{\frac{kr_w(u_w)}{\mu_w}\} \nabla p(t,x)]$

$a_o(t,x) = -K(x)[\{\frac{kr_o(u_o)}{\mu_o}\} \nabla p(t,x)]$

(ii) There could be several special cases of Buckley – Leverett equation for example, we can assume $M(u) = 1$ in (2.5) and assume pressure constant in time, that is, $a(t,x) = a(x)$.

(iii) In general, the total velocity field will change during the simulation as the mobility $M(u)$ depends on the saturation u. substituting equation (2.5) in (2.4) yields a set of elliptic equations, which would have to be solved for the pressure, which in turn provides an updated total velocity field $a(t,x)$ through (2.5). Wavelet methods can be used to solve elliptic equations at different consideration providing solutions at different scales. However we focus our attention here to solve nonlinear conservation law modelled by equation (2.3).

The Five-spot problem: simulation of a model of five spot problem with one production well in the centre and four injection wells at corners has been studied by a variant of finite volume method called ADER schemes, sec [4]. Our group is studying this problem and also similar problem using wavelet methods on circular region.

3. WAVELET METHODS FOR EVOLUTION EQUATION

In this section we present outline of wavelet methods used to solve evaluation equation. These techniques could be adopted for special cases such as Buckley – Leverett equation.

Let us consider the following evolution partial differential equation,

$$(EPDE) \begin{cases} \dfrac{\partial u}{\partial t} + Au = 0, \\ \text{with the periodic boundary conditions,} \\ u(x+1,t) = u(x,t) \\ \text{and initial condition,} \\ u(x,0) = u_o(x), \end{cases} \quad (3.1)$$

where the operator A is linear or non linear of space variables.

Let V be a Hilbert space. Let u be a solution in V of the following problem:

$$(P) \begin{cases} \text{Find } u \in V \text{ so that for all } v \in V \\ \langle u_t, v \rangle + \langle Au, v \rangle = 0 \\ \langle u(.,0) - u_o, v \rangle = 0. \end{cases} \quad (3.2)$$

For the subspaces $\{V_J\}$ of the MRA, the problem (P) can be converted into the following approximate problem(P_j):

$$(P_J) \begin{cases} \text{Find } u_J \in V_J \text{ so that for all } v_J \in V_J \\ \langle u_{Jt}, v \rangle + \langle Au_J, v \rangle = 0 \\ \langle u_J(.,0) - P_J u_o, v \rangle = 0 \end{cases} \quad (3.3)$$

for some positive integer J.

Let consider the Burger's equation with the boundary conditions

$$(BE) \begin{cases} \dfrac{\partial u}{\partial t} + u\dfrac{\partial u}{\partial x} = \kappa \dfrac{\partial^2 u}{\partial x^2} \\ u(x+1,t) = u(x,t) \\ u(x,0) = u_0(x), \end{cases} \qquad (3.4),$$

where κ is a small positive real number. Let Δt be the time step and $u_n(x) = u(x, n\Delta t)$, where u is the solution of (BE).
The time discretization is given by

$$\frac{u_{n+1} - u_n}{\Delta t} + u_n \frac{\partial u_n}{\partial x} = \kappa \frac{\partial^2 u_{n+1}}{\partial x^2}, \qquad (3.5)$$

which can also be written as

$$\begin{cases} \left(I - \kappa\Delta t \dfrac{\partial^2}{\partial x^2}\right) u_{n+1} = u_n - \Delta t u_n \dfrac{\partial u_n}{\partial x}, \\ u_{n+1}(0) = u_{n+1}(1) \end{cases} \qquad (3.6)$$

where, I stands for identity operator. As indicated in [28], introduce a family of functions $\theta_{j,k}, 0 \le j \le J - 1, 0 \le k \le 2^j - 1$, as

$$\left(I - \kappa\Delta t \frac{\partial^2}{\partial x^2}\right) \theta_{j,k} = \psi_{j,k} \qquad (3.7)$$

and computethe wavelet coefficients of u_{n+1} by

$$\langle u_{n+1}, \psi_{j,k} \rangle = \langle u_n, \theta_{j,k} \rangle + \frac{\Delta t}{2} \langle u_n^2, \frac{\partial \theta_{j,k}}{\partial x} \rangle. \qquad (3.8)$$

The use of wavelets as basis functions for the discretization of PDEs has been great success. The properties of MRA seem to be a generalization of finite element methods with some characteristics of multi-grid methods. It is the localizing ability of wavelet expansions that gives rise to sparse operators and good numerical stability to the method.

Wavelet based numerical simulation method allows one to represent a function in terms of a set of basis functions called wavelets, which are localized both in position and scale. The wavelet bases combine the advantages of all classical methods. Let $f \in H^1(R)$ (Sobolev Space of order one) can be written as

$$Let \quad f(x) = \sum_{j=-\infty}^{\infty} \sum_{k=-\infty}^{\infty} c_{j,k} \psi_{j,k}(x) \qquad (3.9)$$

where, $c_{j,k} = \langle f, \psi_{j,k} \rangle$ and $\psi_{j,k}$ be a wavelet system, see [2, 3, 15, 18] for the wavelet systems.

$$f_J(x) = \sum_{j=0}^{J} \sum_{k \in Z} c_{j,k} \psi_{j,k}(x) \quad 1 < j \le J. \tag{3.10}$$

Let us write approximation of (3.9) for any intermediate resolution level j $(0 \le j \le J)$ as

$$f_j(x) = f_{j-1}(x) + \sum_{k \in Z} \psi_{j,k}(x), \quad 1 \le j \le J \tag{3.11}$$

$$f^o(x) = f_o(x) = \sum_{k \in Z} c_{o,k} \psi_{o,k}(x) \tag{3.12}$$

Let $\{x_{j,i} : i \in Z\}$ be a set of collocation points at the j level of resolution. Then

$$f_j(x_{l,i}) = f_{j-1}(x_{l,i}) + \sum_{k \in Z} c_{j,k} \psi_{j,k}(x_{l,i}) \tag{3.13}$$

$$1 \le j \le J, 0 \le l \le J, i \in Z$$

$$f_o(x_{l,i}) = \sum_{k \in Z_\Omega^o} c_{o,k} \psi_{o,k}(x_{l,i}), \quad 0 \le l \le J, \quad i \in Z_\Omega^l \tag{3.14}$$

$$A_{i,k}^{l,j} = \psi_{j,k}(x_{l,i}), \ 0 \le l, \ j \le J, \ i \in Z_\Omega^j, \quad k \in Z_\Omega^j \tag{3.15}$$

Let $R_{i,m}^{l,j} = \begin{cases} 1, for \ x_{l,i} = x_{j,m} \\ 0, \quad otherwise \end{cases} \tag{3.16}$

$$f_{l,i} = \sum_{m \in Z} R_{i,m}^{l,j} f_{j,m}, \ 0 \le l \le j \le J, \ i \in Z \tag{3.17}$$

where, $f_{j,i} = f_j(x_{j,i})$. Note that $f_j(x_{l,i}) = u_{l,i} \ for \ 0 \le l \le j \le J.$

$$\sum_{m \in Z} \Delta_{i,m}^{j,s} f_{s,m} = \sum_{k \in Z} A_{i,k}^{j,j} c_{j,k} \tag{3.18}$$

$$\text{where, } \Delta_{i,m}^{j,s} = R_{i,m}^{j,s} - \sum_{k \in Z} P_{i,p}^{j,j-1} R_{p,m}^{j-1,s}$$

$$f_j(z, I_i^l) = \sum_{m \in Z_\Omega^j} P_{i,m}^{l,i} f_{j,m} \quad 0 \le j \le l \le J, \quad i \in Z_\Omega^l \tag{3.19}$$

$$c_{j,k} = \sum_{m \in Z_\Omega^s} D^{j,s} f_{s,m} \quad 0 \le j \le s \le J, \qquad k \in Z_\Omega^j \quad (3.20)$$

where $D_{k,m}^{j,s} = \sum_{p \in Z_\Omega^j} (A^{j,j})_{k,m}^{-1} \Delta_{p,m}^{j,s}, \qquad 0 \le j \le s \le J, \ k \in Z_\Omega^j, \ m \in Z_\Omega^s \quad (3.21)$

The Prolongation operator $P_{i,m}^{l,j}$ is defined as

$$P_{i,m}^{l,j} =$$
$$\begin{cases} \sum_{p \in Z_\Omega^{j-1}} P_{i,p}^{l,j-1} R_{p,m}^{j-1,j} + \sum_{k \in Z_\Omega^j} A_{i,k}^{l,j} D_{k,m}^{j,j}, & 1 \le j \le l \le J, i \in Z_\Omega^l, m \in Z_\Omega^j \\ \sum_{k \in Z_\Omega^0} A_{k,m}^{l,0} D_{k,m}^{0,0}, & j = 0, \ 0 \le l \le J, i \in Z_\Omega^l, m \in Z_\Omega^0 \end{cases} \quad (3.22)$$

With derivative of the approximation function can be written as

$$f^{J(m)}(x) = \sum_{i \in Z_\Omega^j} f_i^{(m)}(x) f_{j,i} \quad (3.23)$$

where $f_i^{(m)}(x) = \sum_{j=0}^J \sum_{k \in Z_\Omega^j} \psi_k^{j(m)}(x) D_{k,m}^{j,J} \quad (3.24a)$

Let us consider the non linear parabolic partial differential equation
$$\frac{\partial u}{\partial t} = F(t, x, u, u_x, u_{xx}) for \ t > 0 \quad (3.24b)$$

$$u(x,0) = u_o(x) \quad (3.25)$$

where F is a non linear operator.

If the boundary conditions are inhomogeneous, the solution can always be written as a sum of a particular solution which satisfies the inhomogeneous boundary conditions and a complementary solution which satisfies homogeneous boundary conditions. Thus without loss of generality, we consider the problem with homogeneous boundary conditions. In the view of it, let us consider the boundary conditions

$$u(x_l, t) = u(x_r, t) = 0 \quad (3.26)$$

Following the classical collocation approach and evaluating f by (3.23) and (3.24), we obtain

$$u_i^{J(m)} = \sum_{k \in Z_\Omega^J} U_{i,k}^{(m)} u_{j,k}(t) \quad (3.27)$$

291

$$U_{i,k}^{(m)} = \sum_{j=0}^{J} \sum_{p \in z_\Omega^j} \psi_p^{j(m)}(x_{j,i}) D_{p,k}^{i,j} \quad (3.28)$$

Where, $i \in Z_\Omega^J$ and $D_{p,k}^{i,j}$ is given by (3.21). If we number the collocation points in such a way then the equations (3.24) and (3.25) reduce to a system of $2^{L+J} + N_l + N_r - 1$ non linear differential equations

$$\frac{d}{dt} u_{J,i}(t) = F\left(t, x_{J,i}, u_{i,J}(t), U_{ik}^{(1)} f_{J,k}(t), U_{ik}^{(2)} u_{j,k}(t)\right) \quad (3.29)$$

$$u_{J,i}(0) = u_o(x_{J,i}) \quad (3.30)$$

where repeated indices imply summation from $-2^{L+J-1} - N_l$ to $2^{L+J-1} + N_r$ and $i = -2^{L+J-1} - N_l + 1, \dots \dots \dots, 2^{L+J-1} + N_r - 1$ (3.31)

The boundary conditions (3.26) become

$$u_{j,-2^{L+J-1}-N_l}(t) = u_{j,2^{L+J-1}+N_r}(t) = 0 \quad (3.32)$$

after solving (3.29) and (3.30), we get the solution

$$u_J(x,t) = \sum_{i \in z_\Omega^J} I_i(x) u_i^J(t) \quad (3.33)$$

where, $I_i(x) = \sum_{j=0}^{J} \sum_{k \in z_\Omega^j} D_{k,i}^{i,J} \psi_{i,k}(x)$ (3.34)

There is an alternative integral approach in which the highest derivative appearing in the partial differential equation is approximated and then the approximation function is integrated while incorporating the boundary conditions using integration constant, see references [28-33].

4. WAVELET BASED SOLUTION OF BUCKLEY – LEVERETT EQUATION

In this section, we essentially describe the results in [26]. However, we indicate the problems on which we are engaged in.

Wavelets constitute conditional bases for a variety of function spaces, such as $L_2(R)$, Sobolev space $H^1(R)$ and Besov Space $B_2^\alpha (L_2(R))$; See for example [13, 14, and 15]. Thus, they can provide accurate approximations of functions in such spaces. With multi-resolution (MRA) properties, compact support and orthogonality, wavelets provide an attractive alternative as bases for numerical solution of differential equations. In contrast to the traditional trigonometric sine and cosine basis functions, which have infinite support and are not in $L_2(R)$, wavelet bases are $L_2(R)$, functions and may have compact support. Unlike trigonometric approximation with wavelet bases does not rely on cancellation. When an abrupt change, such as a shock wave or a spike, occurs in a function, only local coefficients in a wavelet approximation will be affected. Due to the MRA properties, wavelets are inherent multi-level bases. Mallat's decomposition and reconstruction algorithm provides a fast transition between approximations at different levels. We briefly outline the method of applying wavelet techniques for numerical simulation. For technical terms and concepts involved in this discussion we refer to [2, 3, 13, 15, 17, 18, 25, 28-32]. For orthogonal basis, the best approximation of an $L_2(R)$, function, $f(x)$, by a function $\sum_k c_{n,k}\varphi_{n,k}(x)$ in the subspace V_n of $L_2(R)$ (at resolution level n) is given by the orthogonal projection of f on V_n, as follows:

$$f(x) \cong \sum_k c_{n,k}\varphi_{n,k}(t) \qquad (4.1)$$

where, $c_{n,k}$ is given by the inner product of $f(x)$ and $\varphi_{n,k}(x)$,

$$c_{n,k} = \langle f, \varphi_{n,k} \rangle = \int_{-\infty}^{\infty} f(x)\varphi_{n,k}(t)dt \qquad (4.2)$$

The operator $P_n: f(t) \in L_2(R) \rightarrow P_n f(t) \in V_n$, is a linear orthogonal projection operator resulting in the best approximation of $f(x)$ in V_n.
Similarly, we can define an operator

$$Q_n: f(t) \in L_2(R) \rightarrow Q_n f(x) \in W_n$$

where W_n is a subspace of $L_2(R)$ is defined as

$$Q_n f(t) = \sum_k d_{n,k}\psi_{n,k}(t) \qquad (4.3)$$

where

$$d_{n,k} = \langle f, \psi_{n,k} \rangle = \int_{-\infty}^{\infty} f(t)\psi_{n,k}(t)dt \qquad (4.4)$$

Approximation of a function, $f(t)$, can be carried out at different resolution levels (corresponding to different n's) and the approximations in the subspace are recursive. Thus for a function, $f(t)$, we have at resolution level n,

$$P_n f(t) = P_{n-1} f(t) + Q_{n-1} f(t) \in W_{n-1} \qquad (4.5)$$

where, $P_{n-1}f(t) \in V_{n-1}$ and $Q_{n-1}f(t) \in W_{n-1}$

$P_{n-1}f(t)$ can be further decomposed and the process can be continued to a desired level.

Mallat's decomposition and reconstruction algorithm, see for example [3] for details, provides interesting relationship between $c_{n,k}, c_{n+1,k}, d_{n,k}$, namely-

$$c_{j,k} = \sum_{l \in Z} h_{l-2k} c_{j+1,l} \quad (4.6)$$

$$d_{j,k} = \sum_{l \in Z} (-1)^l h - l + 2k + 1 \, c_{j+1,l} \quad (4.7)$$

$$c_{j,k} = \sum_{l \in Z} h_{k-2l} \, c_{j-1,l} + (-1)^k h_{2l-k+1} d_{j-1,l} \quad (4.8)$$

where, h_k is a filter.

$\varphi_{n,k}(t) = 2^{\frac{n}{2}} \varphi(2^n t - k)$, where φ is called father wavelet or scaling function.

$\psi_{j,k}(t) = 2^{n/2} \, \psi(2^{n/2} t - k)$, where ψ is called mother wavelet.

φ and ψ are related through MRA. ψ is called orthonormal wavelet if $\{\psi_{n,k}(t)\}$ is orthonormal. Some well known wavelets are Haar, Daubechies, Coiflets, Shannon, Morlett and Meyer wavelets, see for example [2, 3, 14, 15, 16, 18, 27].

Wavelet – Galerkin Method

Let solution $u(x,t)$ of Buckley – Leverett equation be approximated by a scaling series $\sum_{k=1}^{m} c_{n,k}(t)\varphi_{n,k}(t)$ where coefficients $c_{n,k}(t)$ to be determined to satisfy B-L equation and m is the number of basis functions. Using Galerkin's method with $\varphi_{n,j}(t)$ as weighting functions, see the reference [2] for details, we get-

$$\sum \frac{d\, c_{n,k}(t)}{dt} \int_0^1 \varphi_{n,j}(x)\varphi_{n,k}(x)dx = -\varphi_{n,j}(x)f(x) + \int_0^1 \varphi_{n,j}(u)f(u)dx \quad (4.9)$$

where, $f(u) = u^2 / \{u^2 + \frac{1}{4}(1 - u^2)\}$

In weighted residuals method [2], we chose $R(x,t) = \frac{\partial u}{\partial t} + \frac{\partial f(u)}{\partial x}$ \hspace{1cm} (4.10)

where f is given above and u is saturation of water as in (1.1). By solving (4.9), we will get the approximate solution at the desired time.

Solution of B-L equation by Wavelet Collocation Method:

When using the collocation method [2], we minimize the residuals function by using the delta function at some points x_j that is-

$$\int_\Omega R(x,t)\delta(x_j)dx = 0, \qquad j = 1,2,3,\dots\dots\dots\dots,p \qquad (4.11)$$

which is equivalent to

$$R(x_j,t) = 0, \quad j = 1,2,3,\dots\dots\dots\dots,p;$$

where $x_j's$ are collocation points. The collocation points can be evenly distributed in the domain of interest. Starting from equation:

$$R_n(x,t) = \sum_k \frac{dc_{n,k}(t)}{dt}\,\varphi_{n,k}(x) + \frac{\partial f(u)}{\partial x} \qquad (4.12)$$

where f is as in (5.9), we use an upstream weighting scheme between collocation points for the nonlinear term $f(u)$. The initial condition is approximated by the collocation method, that is by solving the linear system,

$$\sum_k c_{n,k}(0)\,\varphi_{n,k}(x_j) = u(0,x_j) \qquad (4.13)$$

where, $x_j's$ are collocation points. For evenly distributed collocation points $x_j = \frac{(j-1)}{(n_p-1)}$, $j = 1,2,3,\dots\dots\dots\dots,n_p$, where n_p is the total number of coefficients.

It is pointed out in [26] that collocation method gives superior result than Galerkin method. All the collocation wavelet solutions are free from the oscillations observed in the Galerkin wavelet solutions. It is also observed that improvement is more pronounced at lower resolution levels and seems to have a minimal effect on the solution accuracy for $n=6$. Daubechies and Chui-Wang wavelets have been used for solution of B-L equation.

In our present work, we are looking simulation of B-L equation by using Coiflet and spline wavelets. Wavelet packets could also be tried. Barnika et al [30] could be very useful for numerical simulation of B-L equation by converting it into an elliptic equation, see Remark 2.1 (iii)

In [21] multiple-stabilized finite element method has been introduced for simulation of reservoir modeled by two dimensional and one dimensional B-L equation (General than (1.1), where 2 is replaced by index p, $p \neq 2$).

5. WAVELET BASED SOLUTION ON SPHERICAL REGION

To the best of our knowledge, no one has studied solution of B-L equation on spherical region and moreover using wavelet techniques. In [36] PDE on spherical region has been studied using wavelet methods. Similar approach may be adopted for B-L equation. We present here, notion of multi resolution analysis on sphere whose understanding is essential for wavelet based solution of B-L equation on spherical region. As in the flat case, wavelet on spherical structure can be obtained from a multi resolution analysis on sphere. A comprehensive study of this theme is given in reference [34].

Definition 5.1 Let $L_2(S^2)$ be the Hilbert space of all square integrable functions on the unit disk S^2. A nested sequence of subspaces V_j is called a multi resolution analysis (MRA) of

$$L_2(S^2), \text{ if } \{0\} \subseteq V_0 \subseteq \ldots \ldots \subseteq V_{j-1} \subseteq V_j \subseteq V_{j+1} \subseteq \ldots \ldots \ldots \ldots \subseteq L_2(S^2) \qquad (5.1)$$

$$\overline{\bigcup_{j \varepsilon Z} V_j} = L_2(S^2) \qquad (5.2)$$

holds and for each j scaling functions $\varphi_{j,k}, k \varepsilon K(j)$ exist, such that
$span\{\varphi_{j,k} / k \varepsilon K(j)\} = V_j$ (5.3)
is valid and index sets $K(j)$

$$K(j) \subseteq K(j-1) \qquad (5.4)$$

Remark 5.1 (a) In contrast to the classical case the scaling functions $\varphi_{j,k}$ do not have to be translated of one particular or mother function φ_j. relation (5.1) implies the existence of scaling coefficients $h_{j,k,l}$ such that

$$\varphi_{j,k} = \sum_{l \in K(j-1)} h_{j,k,l} \, \varphi_{j-1,l} \qquad (5.5)$$

holds.

Remark (b): Wavelets are base functions of the wavelet spaces W_j, which themselves are defined as the orthogonal complement of V_j in V_{j-1}

$$V_j \oplus W_j = V_{j-1} \qquad (5.6)$$

Definition 5.2 (Spherical wavelet base)
Consider the set of functions

$$W = \left\{ \frac{\psi_{j,m}}{j \varepsilon Z}, m \varepsilon M(j) \right\} \qquad (5.7)$$

where $M(j) \subseteq K(j-1)$ is an index set again. If

a) W is a base of $L_2(S^2)$, and
b) the set $W = \left\{ \frac{\psi_{j,m}}{j \varepsilon Z}, m \varepsilon M(j) \right\}$ is a base of W_j then the functions $\psi_{j,m}$ are called
   spherical wavelet base.

Remark 5.2

a) Since $W_j \subseteq V_{j-1}$, there have to exist coefficients $g_{j,m}$ with

$$\psi_{j,m} = \sum_{l \in K(j-1)} g_{j,m} \varphi_{j-1,l}, m \varepsilon M(j) \qquad (5.8)$$

b) It is difficult or even impossible to construct orthogonal wavelets on the sphere with a support smaller than the whole surface of the sphere. Therefore, one considers bi orthogonal wavelets on the sphere.

Construction of bi orthogonal wavelets on S^2: The dual scaling function $\tilde{\varphi}_l, m$ of $\varphi_{l,m}$ are defined by

$$\langle \varphi_{j,k}, \tilde{\varphi}_{j,l} \rangle = \delta k, l \qquad (5.9)$$

For a dual scaling function, the scaling equation

$$\tilde{\varphi}_{j,k} = \sum_{l \in K(j-1)} \tilde{h}_{j,k} \tilde{\varphi}_{j-1,l} \qquad (5.10)$$

holds.

In addition, for a given set of spherical wavelet $\psi_{j,k}$, there exist dual wavelets $\tilde{\psi}_{n,m}$ which are bi orthogonal to each other,

$$\langle \psi_{j,k}, \tilde{\psi}_{n,m} \rangle = \delta_{j,n} \delta_{k,n} \qquad (5.11)$$

This implies $\langle \tilde{\psi}_{j,k}, \varphi_{n,m} \rangle = \langle \psi_{j,k}, \tilde{\varphi}_{n,m} \rangle = 0$.
The essence of bi orthogonality is the use of dual bases for the decomposition and reconstruction of a signal f:

$$f = \sum_{j,m} \langle \tilde{\psi}_{j,m}, f \rangle \psi_{j,m} = \sum_{j,m} \langle \psi_{j,m}, f \rangle \tilde{\psi}_{j,m} \qquad (5.12)$$

Once the scaling coefficients $h_{j,k,l}$ and $g_{j,m,l}$ are given, an analog of Mallat's algorithm can be constructed.

Example (5.1) (Bi orthogonal wavelets)

Let $\varphi_{j,k}(x) = \chi_{[2^j k, 2^j k + 2^j]}(x) \qquad (5.13)$

where, χ_M is the characteristics function of M, are primal scaling function $\tilde{\varphi}_{j,l}$ are defined as

$$\tilde{\varphi}_{n,m}(x) = \frac{1}{2^j} \chi_{[2^j k, 2^j k + 2^j]}(x) \qquad (5.14)$$

Which leads to the orthogonality condition

$$\langle \varphi_{j,l}, \tilde{\varphi}_{j,m} \rangle = \delta_{l,m}$$

The primal wavelets are given by

$$\psi_{j,k}(x) = \varphi_{j-1,2k}(x) - \varphi_{j-1,2k+1} \qquad (5.15)$$

And the dual wavelets are given by

$$\tilde{\psi}_{j,k}(x) = \tilde{\varphi}_{j-1,2k}(x) - \tilde{\varphi}_{j-1,2k+1} \qquad (5.16)$$

6. WAVELET BASED PARTICLE METHOD

For introduction of the particle method, we refer to [2]. First author of this book, Prof. Helmut Neunzert, founder Director of the world wide famous institute "Frauenhofer Institute of Industrial and Business Mathematics" at Kaiserslautern had established a well known research group on particle methods in the Kaiserslautern University. Some former and present researchers of this group are making substantial contribution in this area, see [37]. Simulation of B-L equation is presented at the website cited at [37].

In [38], a novel multi resolution Lagrangian method with enhanced, wavelet based adaptivity has been introduced. The method is formulated for transport problem and combines the efficiency of wavelet collocation problems with inherent numerical stability of the particle methods. The efficiency and accuracy of this method has been tested on the certain real world models. We propose to test it on B-L equation. Wavelet based particle method for solution of B-L equation on spherical region is investigated. References [38-42] contain some recent and interesting results about B-L equation.

7. CONCLUDING REMARKS

A mathematical model of the oil recovery called Buckley – Leverett equation is described. Various fast numerical methods used to solve it have been introduced and explained in detail. Open problems are mentioned, which are under investigation, particularly, wavelet particle method and wavelet method for B-L equation on spherical regions.

ACKNOWLEDGMENTS

We take this opportunity to thank Chairman, Mr. P. K. Gupta, and Vice Chairman, Mr. Y. K. Gupta, and administrators of Sharda Group of Institutions (SGI) for providing facilities to write this research paper.

REFERENCES

1. S. E. Buckley and M. C. Leverett, Trans. AIME 146, 107 (1942).
2. H. Neunzert and A. H. Siddiqi, Topics in Industrial Mathematics case studies and Related Mathematical Methods (Kluwer Academic Publishers, 2000).
3. A.H. Siddiqi, Applied Functional Analysis-Numerical Methods and Image Processing (Marcal-Dekker, New York 2004).
4. A. Iske and T. Randen (eds.), Mathematical Methods and Modeling in Hydrocarbon Exploration and Production (Springer Schlumberger, 2006).
5. G. Chavent and J. Jaffre, Mathematical Models and Finite elements for Reservoir Simulation (North Holland, Amsterdam, 1986).
6. A. Quarteroni and A. Valli, Numerical Approximation of Partial Differential Equations (Springer Verlag, Heidelberg, 1994).
7. R. Ansorge, Mathematical Models of Fluid Dynamics, (Wiley-VCH GmbH and Co. KGaA, 2003.
8. E. B. Beckar, G. F. Grey and J. T. Oden, Finite Elements: An Introduction (Prentice-Hall, New Jersey, 1981).
9. M. F. Wheeler (ed.), Numerical Simulation in Oil Recovery (Springer, New York, 1988).
10. K. Aziz and A. Settari, Petroleum Reservoir Simulation (Sevier Science Publ., London, 1979).
11. C. Canuto, M. Y. Hussaini, A. Quartevoni, and T. A. Zang, Spectral Methods in Fluid Dynamics (Springer Verlag, 1988).
12. L. N. Trefethan, Spectral Methods in MATLAB (SIAM, Philadelphia, 2000).
13. R. Glowinski, W. Lawton, M. Ravachol and E. Tenenbaum, in proceedings of the Ninth International Conference on Computing Methods in Applied Sciences and Engineering (SIAM, 1990).
14. I. Daubechies, Comm. Pure Appl. Math: 41, 909 (1988).
15. I. Daubechies, Ten Lectures on Wavelets (SIAM, Philadelphia, 1992).
16. S. G. Mallat, IEEE Trans on Patt. Amal. Mach. Intell 11674 (1989).
17. A. Latto, H. L. Resnikoff, and E. Tenenbaum, proceedings of the French-USA Workshop on Wavelets and Turbulence (Springer-Verlug, Princeton, New York, 1991).
18. H. L. Resnikoff and R.O. Wells Jr. Wavelet Analysis, The scalable Structure of Information (Springer, New York, Berlin, 1998).
19. T. Schwartzkopff, C. D. Munz and E. F. Toro J. Sci. Comput. 17, 231 (2002).

20. V. A. Titarev and E. F. Toro, J. Sci. Comput, 17, 609 (2002).
21. R. Jaunes and T.W. Patzek, SPE 75231, SPE/ DOE Thirteenth Symposium on improved oil recovery, Tulso, Oklahoma, 13-17 April 2002.
22. F. Bressi and M. Fortin, Mixed and Hybrid Finite Element Methods, 15 (Springer-Verlag, New York, 1994).
23. R. E. Eqing and H. Wang, J. Comput. Appl. Math. 128, 423 (2001).
24. C. Baiocchi, F. Brezzi, and L. P. Franca, Virtual Bubbles comput. Methods Appl. Mech. Engrg. 105,125(1993).
25. L. A. Diaz and R. M. Blanco, Revista CIENCIAS Mathematics, 19, 27 (2001).
26. G. J. Moridis, M. Nikolaou and Y. You , Lawrence Berkley Laboratory University of California, LBL-36328, UC 400 conf. 950206-II, Oct. 1994.
27. Mani Mehra, Modeling of Engineering and Technical Problems, this volume....
28. K. M. Furati, P. Manchanda, M. K. Ahmad, A. H. Siddiqi, In Mathematical Models and Methods for Real World Systems (In Chapman * Hall / CRC, Taylor Frances Group, Bocaraton, London 2006).
29. G. Beylkin, R. Coifman, and V. Rokhlin Comm. Pure Appl. Maths. 44 (1993), 1471-183.
30. A. Barinka etal. ,SIAM J. Sci. Comput. 30 (2001) 910-939.
31. W. Dahmen, Acta Numer 6 155 (1997).
32. W. Dahmen, Comput. Appl. Maths. 128, 133 (2001).
33. A. H. Siddiqi, P. Manchanda, and M. Kocuara The 6th World Multi conference on Systematic Cybernetics and Informatics Editor N. Callaos, G. Whymark and W-Lesso, Vol. XIII, 2002, 141-146.
34. W. Keller, Wavelets in Geodesy and Geodynamics (Wather de Grecyter, Berlina, 2004).
35. R. Nguyen Vanyen, Marie Farge, Dmitry Kolomenskig Kai Schneider, Nick Kingsbury, Physica 237,215 (2008).
36. Mani Mehra, Nicholas K-R Kevlahan, Journal of Computation Physics, 227, 5610 (2008).
37. Benjamin Seibold, Particle law, Particle Methods, http://www.math.mit.edu/research/particleclaw/, http:/www.math.mit.edu/~seibold/research/meshfree
38. M. Bergdort and P. Koumoutsakos, Multiscale Modeling & Simulation 5,980 (2006).
39. Strinopoulas Uspscaling Immiscible Two-Phase Flows in ban Adaptive Frame, Ph.D. Thesis, California Institute of Technology, Pasadena, California, November 2005.
40. G. I. Barenblatt, T. W. Patzek, D. B. Sillin, The Mathematical Model of Non-Equilibrium Effects in Water –Oil Displacement, SPE 75169 (2002).
41. A. Westhead, Uscaling Two-Phase Flows in Porous Media, Caltech Thesis, 2005.
42. Rakib Ahmed, Numerical Schemes Applied to the Burgers and Buckley-Leverett Equations Department of Mathematics, University of Keading, M.Sc. Degree Thesis 2004.

SECTION D:
DYNAMICAL SYSTEMS

Global Orbit Patterns for Dynamical Systems On Finite Sets

René LOZI [†*] and Clarisse FIOL[†]

Laboratoire J.A. Dieudonné - UMR du CNRS n°6621
Université de Nice Sophia-Antipolis
Parc Valrose
06108 NICE CEDEX 2 FRANCE
†IUFM Célestin Freinet - Université de Nice Sophia-Antipolis
89, av. George V
06046 NICE CEDEX 1 FRANCE

Abstract. In this paper, the study of the global orbit pattern (gop) formed by all the periodic orbits of discrete dynamical systems on a finite set X allows us to describe precisely the behaviour of such systems. We can predict by means of closed formulas, the number of gop of the set of all the function from X to itself. We also explore, using the brute force of computers, some subsets of locally rigid functions on X, for which interesting patterns of periodic orbits are found.

Keywords: dynamical systems, chaotic analysis, combinatorial dynamics, global orbit pattern, locally rigid functions
PACS: 05.10-a, 05.45-a, 02.10.0x

INTRODUCTION

In some engineering applications such as chaotic encryption, chaotic maps have to exhibit required statistical and spectral properties close to those of random signals. There is a growing industrial interest to consider and study thoroughly the property of such map [10, 11, 12].

Very often, dynamical systems in several dimensions are obtained coupling 1-dimension ones and their properties are strongly linked [5].

Quasi-periodic or chaotic motion is frequently present in complicated dynamical systems whereas simple dynamical systems often involve only periodic motion. The most famous theorem in this field of research is the Sharkovskii's theorem, which addresses the existence of periodic orbits of continuous maps of the real line into itself. This theorem was once proved toward the year 1962 and published only two years after [4].

Mathematical results concerning periodic orbits are often obtained for functions on real intervals. However, most of the time, as the complex behaviour of chaotic dynamical systems is not explicitly tractable, mathematicians have recourse to computer simulations. The main question which arises then is: does these numerical computations are reliable ?

As an example we report the results of some computer experiments on the orbit structure of the discrete maps on a finite set which arise when the logistic map is iterated "naively" on the computer.

CP1146, *Modelling of Engineering and Technological Problems*, edited by A. H. Siddiqi, A. K. Gupta, and M. Brokate
© 2009 American Institute of Physics 978-0-7354-0683-4/09/$25.00

Due to the discrete nature of floating points used by computers, there is a huge gap between these results and the theoretical results obtained when this map is considered on a real interval. This gap can be narrowed in some sense (i.e. avoiding the collapse of periodic orbits) in higher dimensions when ultra weak coupling is used [6, 7].

Nowadays the claim is to understand precisely which periodic orbit can be observed numerically in such systems. In a first attempt we study in this paper the orbits generated by the iterations of a one-dimensional system on a finite set X_N with a cardinal N. The final goal of a good understanding of the actual behaviour of dynamical systems acting on floating numbers (i.e. the numbers used by computers) will be only reached after this first step will be achieved.

On finite set, only periodic orbits can exist. For a given function we can compute all the orbits, all together they form a global orbit pattern. We formalise such a gop as the ordered set of periods when the initial value thumbs the finite set in the increasing order. We are able to predict, using closed formulas, the number of gop for the set $\mathscr{F}_N$ of all the functions on X. We also explore by computer experiments special subsets of $\mathscr{F}_N$, such as sets of locally "rigid" functions which presents interesting patterns of gop.

This article is organized as follows : in the section "Computational divergences" we display some examples of such computational divergences for the logistic map in various ways of discretization. In the section "Pattern defined by all the orbits of a dynamical system" we introduce a new mathematical tool: the global orbit pattern, in order to describe more precisely the behaviour of dynamical systems on finite sets. In the section "Cardinal of classes" we give some closed formulas related to the cardinal of classes of gop of $\mathscr{F}_N$. In the section "Functions with local properties" we study the case of sets of functions with a kind of local "rigidity" versus their gop, in order to show the usefulness of these new tools.

COMPUTATIONAL DIVERGENCES

Discretized logistic map

As an example of collapsing effects which happen when using computers in numerical experiments, we presents the results of a sampling study in double precision of a discretization of the logistic map $f_4 : [0, 1] \to [0, 1]$ (see Fig. 1)

$$f_4(x) = 4x(1 - x) \tag{1}$$

and its associated dynamical system

$$x_{n+1} = 4x_n(1 - x_n) \tag{2}$$

which has excellent ergodic properties on the real interval.

There exists an unstable fixed point 0.

The set $\left\{ \frac{5-\sqrt{5}}{8}, \frac{5+\sqrt{5}}{8} \right\} = \{0.3454915, 0.9045084\}$ is the period-2 orbit.

In fact there exist an infinity of periodic orbits and an infinity of periods. This dynamical system possesses an invariant measure (see Fig. 2):

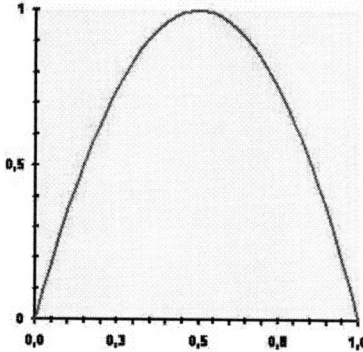

FIGURE 1. Graph of the map $f(x) = 4x(1-x)$ on $[0,1]$

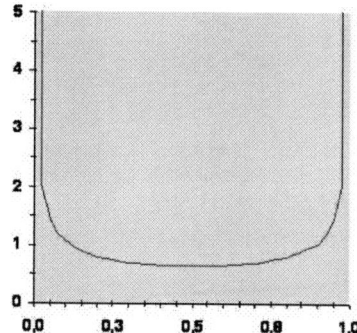

FIGURE 2. Invariant measure of the logistic map

$$P(x) = \frac{1}{\pi\sqrt{x(1-x)}} \tag{3}$$

However, in numerical computations using ordinary (IEEE-754) double precision numbers - so that the working interval contains of the order of 10^{16} representable points - out of 1,000 randomly chosen initial points (see Table 1),

- 596, i.e., the majority, converged to the fixed point corresponding to the unstable fixed point $\{0\}$ in equation 2,
- 404 converged to a cycle of period 15,784,521.

Thus, in this case at least, the very long-term behaviour of numerical orbits is, for a substantial fraction of initial points, in flagrant disagreement with the true behaviour of typical orbits of the original smooth logistic map.

In others numerical experiments we have performed, the computer working with fixed finite precision is able to represent finitely many points in the interval in question. It is probably good, for purposes of orientation, to think of the case where the representable points are uniformly spaced in the interval. The true logistic map is then *approximated* by a discretized map, sending the finite set of representable points in the interval to itself.

Describing the discretized mapping exactly is usually complicated, but it is *roughly* the mapping obtained by applying the exact smooth mapping to each of the discrete representable points and "rounding" the result to the nearest representable point. In our experiments uniformly spaced points in the interval with several order of discretization (ranging from 9 to 2,001 points) are involved. In each experiment the questions addressed are:

- how many periodic cycles are there and what are their periods ?
- how large are their respective basins of attraction, i.e. , for each periodic cycle, how many initial points give orbits with eventually land on the cycle in question ?

TABLE 1. Coexisting periodic orbits found using 1,000 random initial points for double precision numbers

Period	Orbit	Relative Basin size
1	{0} (unstable fixed point)	596 over 1,000
15,784,521	Scattered over the interval	404 over 1,000

TABLE 2. Coexisting periodic orbits for the discretization with regular meshes of $N = 9$, 10 and 11 points

N	Period	Orbit	Basin size
9	1	{0}	3 over 9
9	1	{6}	2 over 9
9	1	{3,7}	4 over 9
10	1	{0}	2 over 10
10	2	{3,8}	8 over 10
11	1	{0}	3 over 11
11	4	{3,8,6,9}	8 over 11

TABLE 3. Coexisting periodic orbits for the discretization with regular meshes of $N = 99$, 100 and 101 points

N	Period	Orbit	Basin size
99	1	{0}	3 over 99
99	10	{3,11,39,93,18,58,94,15,50,97}	96 over 99
100	1	{0}	2 over 100
100	1	{74}	2 over 100
100	6	{11,39,94,18,58,96}	72 over 100
100	7	{7,26,76,70,82,56,97}	24 over 100
101	1	{0}	3 over 101
101	1	{75}	2 over 101
101	1	{16,61,95}	96 over 101

On an another hand, for relatively coarse discretizations the orbit structure is determined completely, i.e., all the periodic cycles and the exact sizes of their basins of attraction are found. Some representative results are given in Tables 2 to 4. In theses tables, N

TABLE 4. Coexisting periodic orbits for the discretization with regular meshes of $N = 1,999; 2,000$ and $2,001$ points

N	Period	Orbit	Basin size
1,999	1	$\{0\}$	3 over 1,999
1,999	4	$\{554; 1,601; 1,272; 1,848\}$	990 over 1,999
1,999	8	$\{3; 11; 43; 168; 615; 1,702; 1,008; 1,997\}$	1,006 over 1,999
2,000	1	$\{0\}$	2 over 2,000
2,000	1	$\{1,499\}$	14 over 2,000
2,000	2	$\{691; 1,808\}$	138 over 2,000
2,000	3	$\{276; 1,221; 1,900\}$	6 over 2,000
2,000	8	$\{3; 11; 43; 168; 615; 1,703; 1,008; 1,998\}$	1,840 over 2,000
2,001	1	$\{0\}$	5 over 2,001
2,001	1	$\{1,500\}$	34 over 2,001
2,001	2	$\{691; 1,809\}$	92 over 2,001
2,001	8	$\{3; 11; 43; 168; 615; 1,703; 1,011; 1,999\}$	608 over 2,001
2,001	18	$\{35; 137; 510; 1,519; 1,461; 1,574; ...\}$	263 over 2,001
2,001	25	$\{27; 106; 401; 1,282; 1,840; 588; ...\}$	1,262 over 2,001

denotes the order of the discretization, i.e., the representable points are the numbers, $\frac{j}{N}$, with $0 \leq j < N$.

The Table 2 shows coexisting periodic orbits for the discretization with regular meshes of $N = 9$, 10 and 11 points. There are exactly 3, 2 and 2 cycles.

The Table 3 shows coexisting periodic orbits for the discretization with regular meshes of $N = 99$, 100 and 101 points. There are exactly 2, 4 and 3 cycles.

The Table 4 shows coexisting periodic orbits for the discretization with regular meshes of $N = 1,999$, $N = 2,000$ and $N = 2,001$ points.

It seems that the computation of numerical approximations of the periodic orbits leads to unpredictable results.

Statistical properties

Many others examples could be given, but those given may serve to illustrate the intriguing character of the results: the outcomes proves to be extremely sensitive to the details of the experiment, but the results all have a similar flavour : a relatively small number of cycles attract near all orbits, and the lengths of these significant cycles are much larger than one but much smaller than the number of representable points.

In [1], P. Diamond and A. Pokrovskii, suggest that statistical properties of the phenomenon of computational collapse of discretized chaotic mapping can be modelled by random mappings with an absorbing centre. The model gives results which are very much in line with computational experiments and there appears to be a type of universality summarised by an Arcsine law. The effects are discussed with special reference to the family of mappings

$$x_{n+1} = 1 - |1 - 2x_n|^\ell \quad 0 \leq x \leq 1 \quad 1 \leq \ell \leq 2 \tag{4}$$

Computer experiments show close agreement with prediction of the model.

However these results are of statistical nature, they do not give accurate information on the exact nature of the orbits (e.g. length of the shortest one, of the greater one, size of their basin of attraction ...). It is why we consider the problem of computational discrepancies in an original way in the next section.

PATTERN DEFINED BY ALL THE ORBITS OF A DYNAMICAL SYSTEM

In this section in order to describe more precisely which kind of behaviour occurs in discretized dynamical systems on finite sets we conceive a new mathematical tool: the global orbit pattern of a function that is the set of the periods of every different orbits of the dynamical system associated to the function when the initial points are took in increasing order.

General definitions

For every $x_0 \in X$, let $\{x_i\}$ be the sequence of the orbit of the dynamical system associated to the function f which maps X onto X defined by

$$x_{i+1} = f(x_i) \quad \text{for} \quad i \geq 0. \tag{5}$$

For convenience $\forall x_0 \in X$ we denote

$$f^0(x_0) = x_0 \tag{6}$$

and

$$\forall p \geq 1, \forall x_0 \in X, \quad f^p(x_0) = \underbrace{f \circ f \circ \ldots \circ f}_{p \text{ times}}(x_0). \tag{7}$$

Hence

$$x_i = f^i(x_0). \tag{8}$$

The orbit of x_0 under f is the set of points $\mathcal{O}(x_0, f) = \{f^i(x_0), i \geq 0\} = \{x_i, i \geq 0\}$.
The starting point x_0 for the orbit is called the initial value of the orbit.
A point x is a fixed point of the map f if $f(x) = x$.
A point x is a periodic point with period p if $f^p(x) = x$ and $f^k(x) \neq x$ for all k such that $0 \leq k < p$, p is called the order of x.
If x is periodic of order p, then the orbit of x under f is the finite set $\{x, f(x), f^2(x), \ldots, f^{p-1}(x)\}$. We will call this set the periodic orbit of order p or a p-cycle.
A fixed point is then a 1-cycle.
The point x is an eventually periodic point of f with order p if there exists $K > 0$ such that $\forall k \geq K \ f^{k+p}(x) = f^k(x)$.
$\forall x \in X$, we denote $\omega(x, f)$ the order of x under f or simply $\omega(x)$ when the map f involved is obvious.

A subset T of X is invariant under f if $f^{-1}(T) = T$. That is equivalent to say that T is invariant under f if and only if $f(T) \subset T$ and $f^{-1}(T) \subset T$.

Notation $\sharp X$ is the cardinal of the finite set X.

Map on finite set

Along this paper, N is a non-zero integer and $\sharp A$ stands for the cardinal of any finite set A. In this article, we consider X as an ordered finite set with N elements. We denote it X_N, it is isomorphic to the interval $[\![0, N-1]\!] \subset \mathbb{N}$. Then $\sharp X_N = N$. Let f be a map from X_N into X_N. We denote by $\mathscr{F}_N$ the set of the maps from X_N into X_N. Clearly, $\mathscr{F}_N$ is a finite set and $\sharp \mathscr{F}_N = N^N$ elements. For all $x \in X_N$, $\mathscr{O}(x, f)$ is necessarily a finite set with at most N elements. Indeed, let us consider the sequence $\{x, f(x), f^2(x), \ldots, f^{N-1}(x), f^N(x)\}$ of the first $N+1$ iterated points. Thanks to the Dirichlet's box principle, two elements are equals because X_N has exactly N different values. Thus, every initial value of X_N leads ultimately to a repeating cycle. More precisely, if x is a fixed point $\mathscr{O}(x, f)$ is the unique element x and if x is a periodic point with order p, $\mathscr{O}(x, f)$ has exactly p elements. In this case, the orbit of x under f is the set $\mathscr{O}(x, f) = \{x, f(x), f^2(x), \ldots, f^{p-1}(x)\}$. If x is an eventually periodic point with order p, there exists $K > 0$ such that $\forall k \geq K$ $f^{k+p}(x) = f^k(x)$. In this case, the orbit of x under f is the set $\mathscr{O}(x, f) = \{x, f(x), f^2(x), \ldots, f^K(x), f^{K+1}(x), \ldots, f^{K+p-1}(x)\}$.

Equivalence classes

Components

Let $f \in \mathscr{F}_N$. We consider on X_N the relation $\sim$ defined by : $\forall x, x' \in X_N$, $x \sim x' \Leftrightarrow \exists k \in \mathbb{N}$ such that $f^k(x) \in \mathscr{O}(x', f)$. The relation $\sim$ is an equivalence relation on X_N. $\mathscr{S}_N / \sim$ is the collection of the equivalence classes that we will call components of X_N under f which constitute a partition of X_N. The number of components are given in [3]. Asymptotic properties of the number of cycles and components are studied in [8]. For each component, we take as class representative element the least element of the component. The components will be written $T_N(x_0, f), \ldots, T_N(x_{p_{f,N}}, f)$ where x_i is the least element of $T_N(x_i, f)$.

By analogy with real dynamical systems, we can define attractive and repulsive components in discretized dynamical systems as follows.

Definition 1 A component is repulsive when it is a cycle. Otherwise, the component is attractive.

Remark In other words, a component is attractive when the component contains at least an eventually periodic element. The corresponding cycle is strictly contained in an attractive component.

Examples are given in Tables 5, 6 and 7.

For instance, in Table 6, the fonction f has $\{2,7\}$ as period-2 orbit and $\{1,2,7,9\}$ as component which is attractive because 1 and 2 are eventually periodic elements.

TABLE 5. Orbits and components of a function belonging to $\mathscr{F}_{11}$ with gop $[2,2,1,3]_{11}$.

Function			orbit/component/nature		
0	$\to$	6			
1	$\to$	3			
2	$\to$	2	period-2 orbit : $\{6,9\}$	$\{0,6,9\}$	attractive
3	$\to$	5			
4	$\to$	8	period-2 orbit : $\{5,10\}$	$\{1,3,5,10\}$	attractive
5	$\to$	10			
6	$\to$	9	fixed point : $\{2\}$	$\{2\}$	repulsive
7	$\to$	4			
8	$\to$	7	period-3 orbit : $\{4,8,7\}$	$\{4,8,7\}$	repulsive
9	$\to$	6			
10	$\to$	5			

TABLE 6. Orbits and components of a function belonging to $\mathscr{F}_{11}$ with gop $[2,2,1,3]_{11}$.

Function			orbit/component/nature		
0	$\to$	4			
1	$\to$	2			
2	$\to$	7	period-2 orbit : $\{4,8\}$	$\{0,4,8\}$	attractive
3	$\to$	3			
4	$\to$	8	period-2 orbit : $\{2,7\}$	$\{1,2,7,9\}$	attractive
5	$\to$	10			
6	$\to$	5	fixed point : $\{3\}$	$\{3\}$	repulsive
7	$\to$	2			
8	$\to$	4	period-3 orbit : $\{5,10,6\}$	$\{5,10,6\}$	repulsive
9	$\to$	1			
10	$\to$	6			

Order of elements

Here are some remarks on the order of elements of components.

Remark The order of every element of a component is the length of its inner cycle.

310

TABLE 7. Orbits and components of a function belonging to $\mathscr{F}_{11}$ with gop $[2,2,1,3]_{11}$.

Function			orbit/component/nature		
0	$\rightarrow$	9			
1	$\rightarrow$	6			
2	$\rightarrow$	4	period-2 orbit : $\{0,9\}$	$\{0,9\}$	repulsive
3	$\rightarrow$	7			
4	$\rightarrow$	10	period-2 orbit : $\{1,6\}$	$\{1,6\}$	repulsive
5	$\rightarrow$	3			
6	$\rightarrow$	1	fixed point : $\{10\}$	$\{2,4,8,10\}$	attractive
7	$\rightarrow$	5			
8	$\rightarrow$	2	period-3 orbit : $\{3,7,5\}$	$\{3,7,5\}$	repulsive
9	$\rightarrow$	0			
10	$\rightarrow$	10			

Definition 2   For all $x \in X_N$, there exists $i \in [\![0, p_{f,N}]\!]$ such that x belongs to the component $T_N(x_i, f)$. Then $\omega(x, f)$ is equal to the order $\omega(x_i, f)$.

Remark   For all $i \in [\![0, p_{f,N}]\!]$, $T_N(x_i, f)$ is an invariant subset of X_N under f.

In the example given in Table 5, the order of the element 0 is 2, the order of the element 1 is 2, the order of the element 4 is 3. The elements 1 and 3 have the same order.

Definition of global orbit pattern

For each $f \in \mathscr{F}_N$, we can determine the components of X_N under f. For each component, we determine the order of any element. Thus, for each $f \in \mathscr{F}_N$, we have a set of orders that we will denote $\Omega(f, N)$. Be given f, there exist $p_{f,N}$ components and $p_{f,N}$ representative elements such that $x_0 < x_1 < \ldots < x_{p_{f,N}}$.

For each $f \in \mathscr{F}_N$, the sequence $[\omega(x_0), \omega(x_1), \ldots, \omega(x_{p_{f,N}}); f]_{\mathscr{F}_N}$ with $x_0 < x_1 < \ldots < x_{p_{f,N}}$ will design the global orbit pattern of $f \in \mathscr{F}_N$.

We will write $gop(f) = [\omega(x_0), \omega(x_1), \ldots, \omega(x_{p_{f,N}}); f]_{\mathscr{F}_N}$.

When the reference to $f \in \mathscr{F}_N$ is obvious, we will write shortly $gop(f) = [\omega(x_0), \omega(x_1), \ldots, \omega(x_{p_{f,N}})]_N$ or $gop(f) = [\omega(x_0), \omega(x_1), \ldots, \omega(x_{p_{f,N}})]$.

For example, the same gop associated to the functions given in Tables 5, 6 and 7 is $[2,2,1,3]_{11}$.

Another example is given in Table 8. In that example, we have $\omega(0) = 2$, $\omega(3) = 1$, $\omega(4) = 4$.

TABLE 8. Orbits and components of a function belonging to $\mathscr{F}_8$ with gop $[2,1,4]_8$.

Function	orbit/component/nature		
$0 \rightarrow 1$			
$1 \rightarrow 0$	period-2 orbit : $\{0,1\}$	$\{0,1,2\}$	attractive
$2 \rightarrow 0$			
$3 \rightarrow 3$	fixed point : $\{3\}$	$\{3\}$	repulsive
$4 \rightarrow 5$			
$5 \rightarrow 6$	period-4 orbit : $\{4,5,6,7\}$	$\{4,5,6,7\}$	repulsive
$6 \rightarrow 7$			
$7 \rightarrow 4$			

Definition 3 The set of all the global orbit patterns of $\mathscr{F}_N$ is called $\mathscr{G}(\mathscr{F}_N)$.

For example, for $N = 5$, the set $\mathscr{G}(\mathscr{F}_5)$ is
$\{[1]; [1,1]; [1,1,1]; [1,2]; [1,1,1,1]; [1,1,2]; [1,3]; [1,1,1,1,1]; [1,1,1,2]; [1,1,2,1];$
$; [1,2,1,1]; [1,2,2]; [1,3,1]; [1,4]; ; [2,1]; [2,1,1]; [2,2]; [2,1,1,1]; [2,1,2]; [2,2,1]$
$; [3]; [3,1]; [3,1,1]; [3,2]; [4]; [4,1]; [5]\}.$

Class of gop

We give the following definitions :

Definition 4 Let be $A = [\omega_1, \ldots, \omega_p]_N$ a gop. Then the class of A, written $\overline{A}$, is the set of all the functions $f \in \mathscr{F}_N$ such that the global orbit pattern associated to f is A.

For example, for $N = 11$, the class of the gop $\overline{[2,2,1,3]}_{11}$ contains the following few of many functions defined in Tables 5, 6 and 7. The periodic orbit which are encountered have the same length nevertheless there are different.

Definition 5 Let be $A = [\omega_1, \ldots, \omega_p]_N$ a gop.
Then the modulus of A is $|A| = \sum_{k=1}^{p} \omega_k$.

Remark $\left|[\omega_1, \ldots, \omega_p]_N\right| \leq N.$

Notation $[\omega_{\tilde{k}}]_N$ means $[\underbrace{\omega, \ldots, \omega}_{k \text{ times}}]_N$ and $[\omega_{\tilde{k}}, v_{\tilde{m}}]_N$ means $[\underbrace{\omega, \ldots, \omega}_{k \text{ times}}, \underbrace{v, \ldots, v}_{m \text{ times}}]_N.$

Threshold functions

Ordering the discrete maps

Theorem 1 The sets $\mathscr{F}_N$ and $[\![1, N^N]\!]$ are isomorphic.

Proof We define the function ϕ from $\mathscr{F}_N$ to $[\![1, N^N]\!]$ by : for each $f \in \mathscr{F}_N$, $\phi(f)$ is the integer n such that $n = \sum_{k=0}^{N-1} f(k)N^{N-1-k} + 1$.

Then ϕ is well defined because $n \in [\![1, N^N]\!]$.

Let n be a given integer between 1 and N^N. We convert $n - 1$ in base N : there exists a unique N-tuple $(a_{n-1,0}; a_{n-1,1}; \ldots; a_{n-1,N-1}) \in [\![0, N - 1]\!]^N$ such that $\overline{n-1}^N = \sum_{i=0}^{N-1} a_{n-1,N-1-i} N^{N-i-1}$. We can thus define the map f_n with : $\forall i \in X_N$, $f_n(i) = a_{n-1,N-i-1}$. Then ϕ is one to one.

Remark This implies $\mathscr{F}_N$ is totally ordered.

Definition 6 Let $f \in \mathscr{F}_N$. Then

$$n = \sum_{k=0}^{N-1} f(k)N^{N-1-k} + 1 \tag{9}$$

is called the rank of f.

Threshold functions

Be given a global orbit pattern A, we are exploring the class $\overline{A}$.

Theorem 2 For every $A \in \mathscr{G}(\mathscr{F}_N)$, the class $\overline{A}$ has a unique function with minimal rank.

Definition 7 For every class $\overline{A} \in \mathscr{G}(\mathscr{F}_N)$, the function defined by the previous theorem will be called the threshold function for the class $\overline{A}$ and will be denoted by $Tr(A)$ or $Tr(\overline{A})$.

To prove the theorem, we need the following definition :

Definition 8 Let $f \in \mathscr{F}_N$ be a function. Let p a non zero integer smaller than N. Let be $x_1, \ldots, x_p$ p consecutive elements of X_N. Then $x_1, \ldots, x_p$ is a canonical p-cycle in relation to f if $\forall j \in [\![1, p-1]\!]$, $f(x_j) = x_{j+1}$ and $f(x_p) = x_1$.

Proof Let $[\omega_1,\ldots,\omega_p]$ be a global orbit pattern of $\mathscr{G}(\mathscr{F}_N)$. We construct a specific function f belonging to the class $\overline{[\omega_1,\ldots,\omega_p]}$ and we prove that the function so obtained is the smallest with respect to the order on $\mathscr{F}_N$. With the first ω_1 elements of $[\![0,N-1]\!]$, that is the set of integers $[\![0,\omega_1-1]\!]$, we construct the canonical ω_1-cycle : if $\omega_1 = 1$, we define $f(0) = 0$, else $f(0) = 1$, $f(1) = 2$, ..., $f(\omega_1-2) = \omega_1-1$, $f(\omega_1-1) = 0$. Then $\forall j \in [\![\omega_1-1,\omega_1+N-s-1]\!]$, we define $f(j) = 0$.
Then with the next ω_2 integers $[\![\omega_1+N-s,\omega_1+N-s+\omega_2-1]\!]$ we construct the canonical ω_2-cycle. We keep going on constructing for all $j \in [\![3,p]\!]$ the canonical ω_j-cycle.
In consequence, we have found a function f belonging to the class $\overline{[\omega_1,\ldots,\omega_p]}$.
Assume there exists a function $g \in \mathscr{F}_N$ belonging to the class of f such that $g < f$. Let $I = \{i \in [\![0,N-1]\!]$ such that $f(i) \neq 0\}$. As $g < f$, there exists $i_0 \in I$ such that $g(i_0) < f(i_0)$. There exists also j_0 such that $i_0 \in \omega_{j_0}$. If $f(i_0) = i_0$, then $\omega_{j_0} = 1$, $g(i_0) < i_0$ and then $g(i_0) \notin \omega_{j_0}$. Then the global orbit pattern of g doesn't contain anymore 1 as cycle. The global orbit pattern of g is different from the global orbit pattern of f. If $f(i_0) = i_0+1$, then $g(i_0) \leq i_0$. Either $g(i_0) = i_0$ and then the global orbit pattern of g is changed, or $g(i_0) < i_0$ and we are in the same situation as previously. Thus, in any case, the smallest function belonging to the class $\overline{[\omega_1,\ldots,\omega_p]}$ is the one constructed in the first part of the proof.

The proof of the theorem gives an algorithm of construction of the threshold function associated to a given gop.
The threshold function associated to the gop $[2_{\bar{2}},1,3]_{11}$ is explained in Table 9. Its rank is $n = 25,938,474,637$.

TABLE 9. Algorithm for the threshold function construction for the gop $[2_{\bar{2}},1,3]_{11}$.

First step		Second step		Third step		Fourth step		Fifth step	
Construction of the first canonical 2-cycle		Construction of the last canonical 3-cycle		Construction of the canonical 1-cycle		Construction of the canonical 2-cycle		Filling the remaining images with 0	
0	→ 1	0	→ 1	0	→ 1	0	→ 1	0	→ 1
1	→ 0	1	→ 0	1	→ 0	1	→ 0	1	→ 0
2	→	2	→	2	→	2	→	2	→ 0
3	→	3	→	3	→	3	→	3	→ 0
4	→	4	→	4	→	4	→	4	→ 0
5	→	5	→	5	→	5	→ 6	5	→ 6
6	→	6	→	6	→	6	→ 5	6	→ 5
7	→	7	→	7	→ 7	7	→ 7	7	→ 7
8	→	8	→ 9	8	→ 9	8	→ 9	8	→ 9
9	→	9	→ 10	9	→ 10	9	→ 10	9	→ 10
10	→	10	→ 8	10	→ 8	10	→ 8	10	→ 8

Theorem 3 There are exactly $2^N - 1$ different global orbit patterns in $\mathscr{F}_N$.

That is

$$\sharp \mathscr{G}(\mathscr{F}_N) = 2^N - 1. \tag{10}$$

For example, for $N = 4$, $\sharp \mathscr{G}(\mathscr{F}_4) = 2^4 - 1 = 15$.

Proof   Let p an integer between 1 and N. Consider the set $L(p,N)$ of p-tuples $(\alpha_1, \ldots, \alpha_p) \in (\mathbb{N}^*)^p$ such that $\alpha_1 + \ldots + \alpha_p \leq N$.
We write $L(N) = \{L(p,N), p = 1 \ldots N\}$. $L(N)$ and $\mathscr{G}(\mathscr{F}_N)$ have the same elements.
Then

$$\sharp \mathscr{G}(\mathscr{F}_N) = \sum_{p=1}^{p=N} \sharp L(p,N) = \sum_{p=1}^{p=N} \binom{N}{p} = 2^N - 1.$$

Ordering the global orbit patterns

We define an order relation on $\mathscr{G}(\mathscr{F}_N)$.

Proposition 1   Let A and B be two global orbit patterns of $\mathscr{G}(\mathscr{F}_N)$. We define the relation $\prec$ on the set $\mathscr{G}(\mathscr{F}_N)$ by

$$A \prec B \ \text{iff} \ Tr(A) < Tr(B)$$

Then the set $(\mathscr{G}(\mathscr{F}_N), \prec)$ is totally ordered.

Proof   As the order $\prec$ refers to the natural order of $\mathbb{N}$, the proof is obvious.

Let $r \geq 1$, $p \geq 1$ be two integers. Let $[\omega_1, \ldots, \omega_p]$ and $[\omega'_1, \ldots, \omega'_r]$ be two global orbit patterns of $\mathscr{G}(\mathscr{F}_N)$. For example, if $p < r$, in order to compare them, we admit that we can fill $[\omega_1, \ldots, \omega_p]$ with $r - p$ zeros and write $[\omega_1, \ldots, \omega_p] = [\omega_1, \ldots, \omega_p, 0, \ldots, 0]$.

Proposition 2   Let $r \geq 1$, $p \geq 1$ be two integers such that $p \leq r$. Let $A = [\omega_1, \ldots, \omega_p]$ and $B = [\omega'_1, \ldots, \omega'_r]$ be two global orbit patterns.

- If $r = p = 1$ and $\omega_1 < \omega'_1$ then $A \prec B$.
- If $r \geq 2$ then
  * If $\omega_1 < \omega'_1$ then $A \prec B$.
  * If $\omega_1 = \omega'_1$ then there exists $K \in [\![2;r]\!]$ such that $\omega_K \neq \omega'_K$ and $\forall i < K$, $\omega_i = \omega'_i$.
    - If $|A| < |B|$, then $A \prec B$.
    - If $|A| = |B|$, then if $\omega_K < \omega'_K$ then $A \prec B$.

For example, for $N = 5$, the global orbit patterns are in increasing order : $[1] \prec [1_{\tilde{2}}] \prec [1_{\tilde{3}}] \prec [1,2] \prec [1_{\tilde{4}}] \prec [1_{\tilde{2}},2] \prec [1,2,1] \prec [1,3] \prec [1_{\tilde{5}}] \prec [1_{\tilde{3}},2] \prec [1_{\tilde{2}},2,1] \prec [1_{\tilde{2}},3] \prec$

$[1,2,1_{\tilde{2}}] \prec [1,2_{\tilde{2}}] \prec [1,3,1] \prec [1,4] \prec [2] \prec [2,1] \prec [2,1_{\tilde{2}}] \prec [2_{\tilde{2}}] \prec [2,1_{\tilde{3}}] \prec [2,1,2] \prec$
$[2_{\tilde{2}},1] \prec [2,3] \prec [3] \prec [3,1] \prec [3,1_{\tilde{2}}] \prec [3,2] \prec [4] \prec [4,1] \prec [5].$

Algorithm for ordering the global orbit patterns : a pseudo-decimal order

The Table 10 gives a method for ordering the gop : indeed, we consider each gop as if each one represents a decimal number : we begin to order them in considering the first order ω_1. Considering two gops $A = [\omega_1,...,\omega_p]$ and $A' = [\omega_1',...,\omega_r']$, if $\omega_1 < \omega_1'$, then $A \prec A'$. For example, $[2,1,2] \prec [4,1]$. If $\omega_1 = \omega_1'$ and $|A| - \omega_1 < |A'| - \omega_1'$, then $A \prec A'$. For example to compare the gop $[1,2]$ and the gop $[1_{\tilde{2}}]$, we say that the first order ω_1 stands for the unit digit - which is $\omega_1 = 1$ here, then the decimal digits are respectively 0.2 and 0.111. We calculate for each of them the modulus-ω_1 : we find $\|[1,2]\| - 1 = 2$ and $\|[1_{\tilde{4}}]\| - 1 = 3$, thus $[1,2] \prec [1_{\tilde{4}}]$. Finally, if $\omega_1 = \omega_1'$ and $|A| - \omega_1 = |A'| - \omega_1'$, then also we use the order of the decimal part. For example, $[1_{\tilde{5}}] \prec [1,2,1_{\tilde{2}}]$ because $1.1111 < 1.211$. Applying this process, we have the sequence of the ordered gop for $N = 4$ given in the previous paragraph.

TABLE 10. Ordered gop for $N = 5$ with modulus and modulus-ω_1

Gop	Modulus	Modulus-ω_1	Gop	Modulus	Modulus-ω_1
$[1]$	1	0	$[2]$	2	0
$[1_{\tilde{2}}]$	2	1	$[2,1]$	3	1
$[1_{\tilde{3}}]$	3	2	$[2,1_{\tilde{2}}]$	4	2
$[1,2]$	3	2	$[2_{\tilde{2}}]$	4	2
$[1_{\tilde{4}}]$	4	3	$[2,1_{\tilde{3}}]$	5	3
$[1_{\tilde{2}},2]$	4	3	$[2,1,2]$	5	3
$[1,2,1]$	4	3	$[2_{\tilde{2}},1]$	5	3
$[1,3]$	4	3	$[2,3]$	5	3
$[1_{\tilde{5}}]$	5	4			
$[1_{\tilde{3}},2]$	5	4	$[3]$	3	0
$[1_{\tilde{2}},2,1]$	5	4	$[3,1]$	4	1
$[1_{\tilde{2}},3]$	5	4	$[3,1_{\tilde{2}}]$	5	2
$[1,2,1_{\tilde{2}}]$	5	4	$[3,2]$	5	2
$[1,2_{\tilde{2}}]$	5	4			
$[1,3,1]$	5	4	$[4]$	4	0
$[1,4]$	5	4	$[4,1]$	5	1
			$[5]$	5	1

For example, for $N = 5$, we construct one branch of a tree with $\omega_1 = 1$ (see Fig. 3) : each vertex is an ordered orbit, the modulus of the gop is written on the last edge. However, the sequence of ordered gop differs from the natural downward lecture of the tree and has to be done following the algorithm.

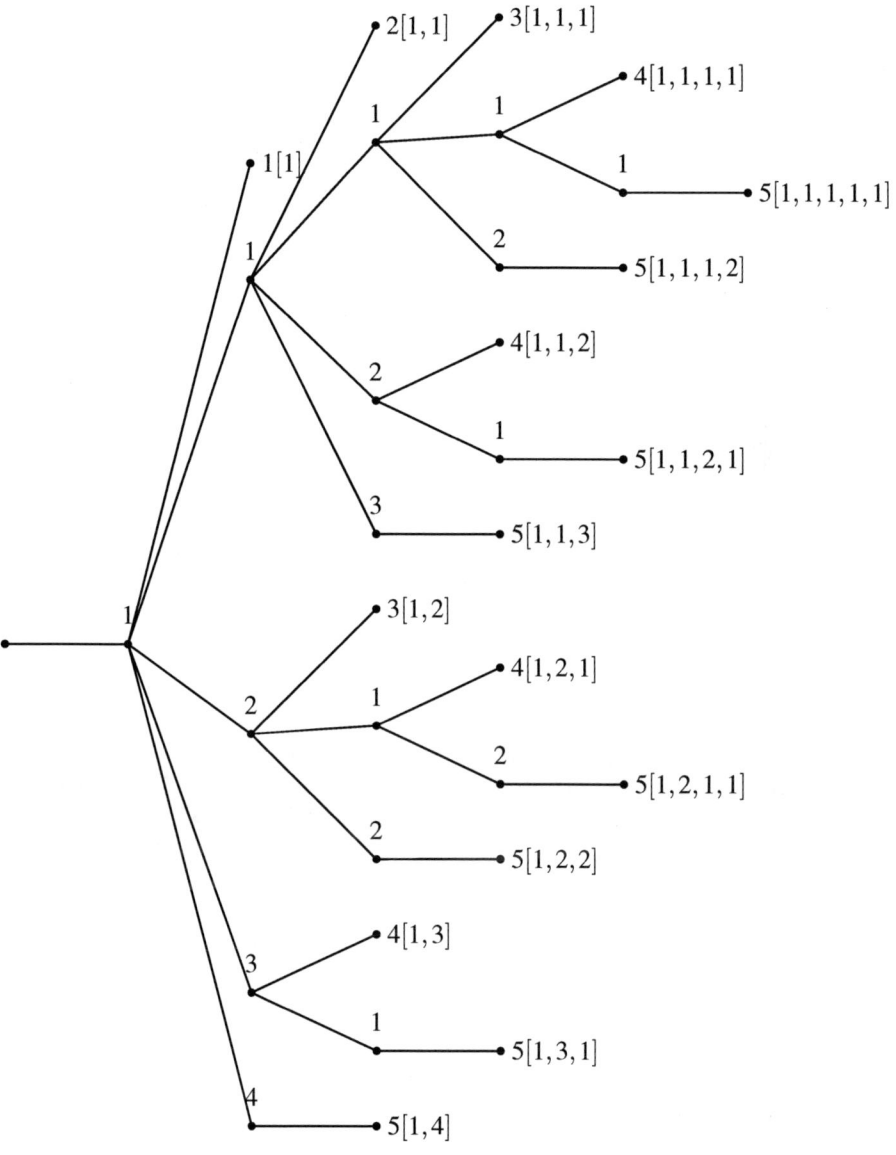

FIGURE 3. Branch of the tree for the construction of the gop with $\omega_1 = 1$ on $\mathcal{G}(\mathcal{F}_5)$

CARDINAL OF CLASSES

In this section we emphasize some closed formulas giving the cardinal of classes of gop. Recalling first the already known formula for the class $[1_{\overline{k}}]_N$ for which we give a detailed proof, we consider the case were the class possesses exactly one k-cycle, the case with only two cycles belonging to the class and finally the main general formula of any cycles with any length. We give rigorous proof of all. The general formula is very interesting in the sense that even using computer network it is impossible to check every function of $\mathscr{F}_N$ when N is larger than 100.

Discrete maps with 1-cycle only

The theorem 4 gives the number of discrete maps of $\mathscr{F}_N$ which have only fixed points and no cycles of length greater than one. This formula is explicit in [2] and [9]. A complete proof is given here in detail.

Theorem 4   Let k be an integer between 1 and N. The number of functions whose global orbit pattern is $[1_{\tilde{k}}]_N$ (i.e. belonging to the class $\overline{[1_{\tilde{k}}]}_N$) is $\begin{pmatrix} N-1 \\ N-k \end{pmatrix} N^{N-k}$.

That is

$$\sharp\overline{[1_{\tilde{k}}]}_N = \begin{pmatrix} N-1 \\ N-k \end{pmatrix} N^{N-k}. \tag{11}$$

Proof   Let k be a non-zero integer. Let f be a function of $\mathscr{F}_N$. There are $\begin{pmatrix} N \\ k \end{pmatrix}$ possibilities to choose k fixed points. There remain $N-k$ points. Let p be an integer between 1 and $N-k$. We assume that p points are directly connected to the k fixed points. For each of them, there are k manners to choose one fixed point. There are k^p ways to connect directly p points to k fixed points. There remains $N-k-p$ points that we must connect to the p points. There are $\sharp[1_{\tilde{p}}]_{N-k}$ functions. Finally, the number of functions with k fixed points is $\begin{pmatrix} N \\ k \end{pmatrix} \sum_{p=1}^{N-k} k^p \, \sharp\overline{[1_{\tilde{p}}]}_{N-k}$. We now prove recursively on N for every $0 \le k \le N$ that $\sharp\overline{[1_{\tilde{k}}]}_N = \begin{pmatrix} N-1 \\ N-k \end{pmatrix} N^{N-k}$. We have $\sharp\overline{[1]}_1 = 1$. The formula is true.

We suppose that $\forall k \le N \; \sharp\overline{[1_{\tilde{k}}]}_N = \begin{pmatrix} N-1 \\ N-k \end{pmatrix} N^{N-k}$.

Let X be a set with $N+1$ elements. We look for the functions of $\mathscr{F}_{N+1}$ which have k fixed points. Thanks to the previous reasoning, we have

$$\sharp\overline{[1_{\tilde{k}}]}_{N+1} = \begin{pmatrix} N+1 \\ k \end{pmatrix} \sum_{p=1}^{N+1-k} k^p \, \sharp\overline{[1_{\tilde{p}}]}_{N+1-k}.$$

$$\sharp\overline{[1_{\tilde{k}}]}_{N+1} = \begin{pmatrix} N+1 \\ k \end{pmatrix} \sum_{p=1}^{N+1-k} k^p \, \sharp\overline{[1_{\tilde{p}}]}_{N-(k-1)}.$$

We use the recursion assumption.

$$\sharp\overline{[1_{\tilde{k}}]}_{N+1} = \begin{pmatrix} N+1 \\ k \end{pmatrix} \sum_{p=1}^{N+1-k} k^p \begin{pmatrix} N-k \\ p-1 \end{pmatrix} (N-k+1)^{N-k+1-p}.$$

$$\#[\overline{1_{\tilde{k}}}]_{N+1} = k \binom{N+1}{k} \sum_{p=0}^{N-k} \binom{N-k}{p} k^p (N-k+1)^{N-k-p}.$$

$$\#[\overline{1_{\tilde{k}}}]_{N+1} = k \binom{N+1}{k} (N+1)^{N-k}.$$

$$\#[\overline{1_{\tilde{k}}}]_{N+1} = \binom{N}{k-1} (N+1)^{N-k+1}.$$

$$\#[\overline{1_{\tilde{k}}}]_{N+1} = \binom{N}{N-k+1} (N+1)^{N-k+1}. \quad \text{q.e.d.}$$

Discrete maps with k-cycle

We look now for the number of functions with exactly one k-cycle.

Theorem 5 Let k be an integer between 1 and N. The number of functions whose global orbit pattern is $[k]_N$ is $\#[\overline{1_{\tilde{k}}}]_N \times (k-1)!$.

i.e.

$$\#\overline{[k]}_N = \#[\overline{1_{\tilde{k}}}]_N \times (k-1)!. \tag{12}$$

Proof There are $\binom{N}{k}$ ways of choosing k elements among N. Then, there are $(k-1)!$ choices for the image of those k elements in order to constitute a k-cycle by f. We must now count the number of ways of connecting directly or not the remaining $N-k$ elements to the k-cycle. We established already this number which is equal to $\sum_{p=1}^{N-k} k^p \#[\overline{1_{\tilde{p}}}]_{N-k}$. Finally, we have $\#\overline{[k]}_N = (k-1)! \binom{N}{k} \sum_{p=1}^{N-k} k^p \#[\overline{1_{\tilde{p}}}]_{N-k}$. That is, $\#[k]_N = \#[\overline{1_{\tilde{k}}}]_N \times (k-1)!.$ q.e.d.

Discrete maps with only two cycles

We give the number of functions with only two cycles.

Theorem 6 Let $N \geq 2$. Let p and q be two non-zero integers such that $p + q \leq N$. Then,

$$\#\overline{[p,q]}_N = \#[\overline{1_{\widetilde{p+q}}}]_N \frac{(p+q-1)!}{q} = \frac{(N-1)! \, N^{N-(p+q)}}{(N-(p+q))! \, q}. \tag{13}$$

Proof We consider a function f which belongs to the class $[\overline{1_{\widetilde{p+q}}}]_N$. We search the number of functions constructed from f whose gop is $[p,q]_N$. From the p fixed points of

f, we construct a p-cycle. Thus, there are $\binom{p+q-1}{p-1}$ ways to choose $p-1$ integers among the $p+q-1$ fixed points. Counting the first given fixed point of f, we have p points which allow to construct $(p-1)!$ functions with a p-cycle. Then there remain q points which give $(q-1)!$ different functions with a q-cycle. Finally, the number of functions whose gop is $[p,q]_N$ is : $\binom{p+q-1}{p-1}(p-1)!(q-1)!$ that is the formula $\frac{(p+q-1)!}{q}$.

Remark   We notice that for all k non-zero integer such that $k \leq N-1$, $\sharp\overline{[k,1]}_N = \sharp\overline{[k+1]}_N$.

General case : discrete maps with cycles of any length

We introduce now the main theorem of the section which gives the number of gop of discrete maps thanks to a closed formula.

Given a global orbit pattern α, the next theorem gives a formula which gives the number of functions which belong to $\overline{\alpha}$.

Theorem 7   Let $p \geq 2$ be an integer. Let $[\omega_1,\ldots,\omega_p]_N$ be a gop of $\mathscr{G}(\mathscr{F}_N)$. Then,

$$\sharp\overline{[\omega_1,\ldots,\omega_p]}_N = \sharp\overline{[1_{\omega_1+\ldots+\omega_p}]}_N \frac{(\omega_1+\ldots+\omega_p-1)!}{\omega_p \times (\omega_{p-1}+\omega_p) \times \ldots \times (\omega_2+\ldots+\omega_p)} \tag{14}$$

$$\sharp\overline{[\omega_1,\ldots,\omega_p]}_N = \frac{(N-1)!\, N^{N-(\omega_1+\ldots+\omega_p)}}{(N-(\omega_1+\ldots+\omega_p))! \displaystyle\prod_{k=2}^{p} \left(\sum_{j=k}^{p} \omega_j \right)} \tag{15}$$

Proof   We consider a function f which belongs to $\overline{[1_{\omega_1+\ldots+\omega_p}]}_N$. We search the number of functions constructed from f whose gop is $[\omega_1,\ldots,\omega_p]_N$. From the ω_1 fixed points of f, we construct a ω_1-cycle. Thus, there are $\binom{\omega_1+\ldots+\omega_p-1}{\omega_1-1}$ ways to choose ω_1-1 integers among the $\omega_1+\ldots+\omega_p-1$ fixed points. Counting the first given fixed point of f, we have ω_1 points which allow to construct $(\omega_1-1)!$ functions with a ω_1-cycle. Then, the first fixed point of f which has not be chosen for the ω_1-cycle, will belong to the ω_2-cycle. Thus, there are $\binom{\omega_2+\ldots+\omega_p-1}{\omega_2-1}$ ways to choose ω_2-1 integers among the $\omega_2+\ldots+\omega_p-1$ fixed points. So we have ω_2 points which allow to construct $(\omega_2-1)!$ functions with a ω_2-cycle. We keep going on that way until there remain ω_p fixed points which allow to construct $(\omega_p-1)!$ functions with a ω_p-cycle. Finally, we have constructed :
$\binom{\omega_1+\ldots+\omega_p-1}{\omega_1-1}(\omega_1-1)!\binom{\omega_2+\ldots+\omega_p-1}{\omega_2-1}(\omega_2-1)! \times \ldots \times \binom{\omega_{p-1}+\omega_p-1}{\omega_{p-1}-1}(\omega_{p-1}-1)!(\omega_p-1)!$
functions. We simplify and obtain the formula.

Corollary 1 Let p be a non-zero integer. Let $[\omega_1,\ldots,\omega_p]_N$ be a gop of $\mathscr{G}(\mathscr{F}_N)$. We suppose that there exists j such that $\omega_j \geq 2$. Let h be an integer between 1 and $\omega_j - 1$. Then

$$\sharp\overline{[\omega_1,\ldots,\omega_j,\ldots,\omega_p]}_N = \sharp\overline{[\omega_1,\ldots,\omega_j-h,h,\omega_{j+1},\ldots,\omega_p]}_N \times (h+\omega_{j+1}+\ldots+\omega_p).$$
(16)

Proof $\sharp\overline{[\omega_1,\ldots,\omega_j-h,h,\omega_{j+1},\ldots,\omega_p]}_N \times (h+\omega_{j+1}+\ldots+\omega_p) = \sharp\overline{[1_{\widetilde{\omega_1+\ldots+\omega_p}}]}_N$

$\times \dfrac{(\omega_1+\ldots+\omega_p-1)!(h+\omega_{j+1}+\ldots+\omega_p)}{\omega_p(\omega_{p-1}+\omega_p)\ldots(\omega_{j+1}+\ldots+\omega_p)(h+\omega_{j+1}+\ldots+\omega_p)(\omega_j+\omega_{j+1}+\ldots+\omega_p)\times\ldots\times(\omega_2+\ldots+\omega_p)}.$

We simplify and we exactly obtain

$$\sharp\overline{[\omega_1,\ldots,\omega_j-h,h,\omega_{j+1},\ldots,\omega_p]}_N \times (h+\omega_{j+1}+\ldots+\omega_p) = \sharp\overline{[\omega_1,\ldots,\omega_j,\ldots,\omega_p]}_N.$$

Examples :

$\sharp\overline{[2_{\tilde{2}},1,3]}_{11} = 11,180,400.$

$\sharp\overline{[5,2,10,8,15,2,3]}_{50} = 29,775,702,147,667,389,218,762,343,520,975,006,$
$348,329,578,044,480,000,000,000,000,000.$

$\sharp\overline{[5,2,10,8,15,2,3]}_{50} \cong 2.98 \times 10^{63}$ among the 8.88×10^{84} functions of $\mathscr{F}_{50}.$

FUNCTIONS WITH LOCAL PROPERTIES

Locally rigid functions

Obviously it is not possible to transpose to the functions on finite sets the notions of continuity and derivability which play a dramatic role in mathematical analysis since several centuries. In fact the class $\mathscr{C}_0(I)$ of the continuous functions on the real interval I is a very small subset of the set $I^{\mathbb{R}}$ of all the functions on I. Hence by analogy to this fact and trying to mimic some others properties of continuous functions, we introduce some subsets of particular functions of $\mathscr{F}_N$, which have local properties such as locally bounded range in a sense we precise further. Limiting the range of the function in a neighbourhood of any point of the interval induces a kind of "rigidity" of the function, hence we call these functions locally rigid functions. In these subsets, the gop are found to be fully efficient in order to describe very precisely the dynamics of the orbits. We first consider the very simple subset $\mathscr{LR}_{1,N}$ of functions for which the difference between $f(p)$ and $f(p+1)$ is drastically bounded. In next subsection we consider more sophisticated subsets.

We consider the set :

$$\mathscr{L}\mathscr{R}_{1,N}=\{f\in\mathscr{F}_N \text{ such that } \forall p, 0 \le p \le N-2, |f(p)-f(p+1)| \le 1\}.$$

Orbits of $\mathscr{L}\mathscr{R}_{1,N}$

Theorem 8 If $f \in \mathscr{L}\mathscr{R}_{1,N}$ then f has only periodic orbits of order 1 or 2.

Proof We suppose that $f \in \mathscr{L}\mathscr{R}_{1,N}$ has a 3-cycle. We denote $(a; f(a); f^2(a))$ taking a the smallest value of the 3-cycle. If $a < f(a) < f^2(a)$ then there exist two non-zero integers e and e' such that $f(a) = a+e$ and $f^2(a) = f(a)+e'$. Thus, $f^2(a) - e' \le f^3(a) \le f^2(a)+e'$. That is $f(a) \le a \le f(a)+2e'$. And finally we have the relation $a+e \le a$ which is impossible.

If $a < f^2(a) < f(a)$ then there exist two non-zero integers e and e' such that $f^2(a) = a+e$ and $f(a) = f^2(a)+e'$. Thus, $f(a) - e \le f^3(a) \le f(a)+e$. That is $f(a) - e \le a \le f(a)+e$. But $f(a) - e = a+e'$. And finally we have the relation $a+e' \le a$ which is impossible.

We can prove in the same way that the function f can't have either 3-cycle or greater order cycle than 3.

Numerical results and conjectures

We have done numerical studies of the $\mathscr{G}(\mathscr{L}\mathscr{R}_{1,N})$ for $N = 1$ to 16, using the brute force of a desktop computer (i.e. checking every function belonging to these sets).

The Tables 11, 12, 13, 14, 15 and 16 show the sequences for $\mathscr{L}\mathscr{R}_{1,1}$ to $\mathscr{L}\mathscr{R}_{1,16}$.

In theses Tables we display in the first column all the gop of $\mathscr{G}(\mathscr{L}\mathscr{R}_{1,N})$ for every value of N. For a given N, there are two columns; the left one displays the cardinal of every existing class of gop (- stands for non existing gop). Instead the second shows more regularity, displaying on the row of the gop $[2_{\overline{k}}]$ the sum of the cardinals of all the classes of the gop of the form $[2,2,\ldots\ldots,\underbrace{1}_{i\text{th}},\ldots,2]$ which exist.

$$\underbrace{}_{k+1 \text{ orders}}$$

Then we are able to formulate some statements which have not yet been proved.

Statement 1

$$\sharp[1_{\overline{k}}]\mathscr{L}\mathscr{R}_{1,N} = \sharp[1_{\overline{k+1}}]\mathscr{L}\mathscr{R}_{1,N+1} \text{ for } k \le \frac{N+1}{2}. \tag{17}$$

322

TABLE 11. Numbering the locally rigid functions for $f \in \mathscr{LR}_{1,1}$, $f \in \mathscr{LR}_{1,2}$, $f \in \mathscr{LR}_{1,3}$, $f \in \mathscr{LR}_{1,4}$.

g.o.p.	N=1	N=1	N=2	N=2	N=3	N=3	N=4	N=4
Total number		1		4		17		68
$[1]$	1	+	2	+	7	+	26	+
$[1_{\tilde 2}]$	-	+	1	+	4	+	14	+
$[1_{\tilde 3}]$	-	+	-	+	1	+	4	+
$[1_{\tilde 4}]$	-	+	-	+	-	+	1	+
$[2]$	-	+	1	1	4	4	18	18
$[2,1]$	-	+	-	+	1	1	3	4
$[1,2]$	-	+	-	+	-	+	1	+
$[2_{\tilde 2}]$	-	+	-	+	-	+	1	1

TABLE 12. Numbering the locally rigid functions for $f \in \mathscr{LR}_{1,5}$, $f \in \mathscr{LR}_{1,6}$, $f \in \mathscr{LR}_{1,7}$.

g.o.p.	N=5	N=5	N=6	N=6	N=7	N=7
Total number		259		950		387
$[1]$	95	+	340	+	1,193	+
$[1_{\tilde 2}]$	50	+	174	+	600	+
$[1_{\tilde 3}]$	16	+	58	+	204	+
$[1_{\tilde 4}]$	4	+	16	+	60	+
$[1_{\tilde 5}]$	1	+	4	+	16	+
$[1_{\tilde 6}]$	-	+	1	+	4	+
$[1_{\tilde 7}]$	-	+	-	+	1	+
$[2]$	70	70	264	264	952	952
$[2,1]$	12	18	45	70	166	264
$[1,2]$	6	+	25	+	98	+
$[2_{\tilde 2}]$	4	4	18	18	70	70
$[2_{\tilde 2},1]$	1	1	4	4	17	18
$[1,2_{\tilde 2}]$	-	+	-	+	1	+
$[2,1,2]$	-	+	-	+	-	+
$[2_{\tilde 3}]$	-	+	1	1	4	4
$[2_{\tilde 3},1]$	-	+	-	+	1	1

Statement 2

$$\#[\overline{2_{\tilde k}}]\mathscr{LR}_{1,N} = \#[\overline{2_{\widetilde{k+1}}}]\mathscr{LR}_{1,N+2} \text{ for } k \leq \frac{N}{2}. \tag{18}$$

Statement 3

$$\#[\overline{2_{\tilde k}}]\mathscr{LR}_{1,N} = \#[\overline{2_{\tilde k},1}]\mathscr{LR}_{1,N+1} \text{ for } 2k \leq N \leq 3k-1. \tag{19}$$

TABLE 13. Numbering the locally rigid functions for $f \in \mathscr{LR}_{1,8}$, $f \in \mathscr{LR}_{1,9}$, $f \in \mathscr{LR}_{1,10}$.

g.o.p.	N=8	N=8	N=9	N=9	N=10	N=10
Total number		11,814		40,503		13,6946
$[1]$	4,116	+	14,001	+	47,064	+
$[1_{\tilde{2}}]$	2,038	+	6,852	+	22,806	+
$[1_{\tilde{3}}]$	700	+	2,366	+	7,896	+
$[1_{\tilde{4}}]$	214	+	742	+	2,520	+
$[1_{\tilde{5}}]$	60	+	216	+	754	+
$[1_{\tilde{6}}]$	16	+	60	+	216	+
$[1_{\tilde{7}}]$	4	+	16	+	60	+
$[1_{\tilde{8}}]$	1	+	4	+	16	+
$[1_{\tilde{9}}]$	-	+	1	+	4	+
$[1_{\widetilde{10}}]$	-	+	-	+	1	+
$[2]$	3,356	3,356	11,580	11,580	39,364	39,364
$[2,1]$	590	952	2,062	3,356	7,072	11,580
$[1,2]$	362	+	1,294	+	4,508	+
$[2_{\tilde{2}}]$	264	264	952	952	3,356	3,356
$[2_{\tilde{2}},1]$	62	70	222	264	770	952
$[1,2_{\tilde{2}}]$	6	+	28	+	113	+
$[2,1,2]$	2	+	14	+	69	+
$[2_{\tilde{3}}]$	18	18	70	70	264	264
$[2_{\tilde{3}},1]$	4	4	18	18	69	70
$[1,2_{\tilde{3}}]$	-	+	-	+	1	+
$[2,1,2_{\tilde{2}}]$	-	+	-	+	-	+
$[2_{\tilde{2}},1,2]$	-	+	-	+	-	+
$[2_{\tilde{4}}]$	1	1	4	4	18	18
$[2_{\tilde{4}},1]$	-	+	1	1	4	4
$[1,2_{\tilde{4}}]$	-	+	-	+	-	+
$[2,1,2_{\tilde{3}}]$	-	+	-	+	-	+
$[2_{\tilde{2}},1,2_{\tilde{2}}]$	-	+	-	+	-	+
$[2_{\tilde{3}},1,2]$	-	+	-	+	-	+
$[2_{\tilde{5}}]$	-	+	-	+	1	1

***Statement* 4**

$$\sharp \overline{[2_{\tilde{k}}]} \mathscr{LR}_{1,N} = \sum_{i=1}^{k+1} \sharp \underbrace{\overline{[2,2,\ldots\ldots,\underbrace{1}_{i\text{th}},\ldots,2]}}_{k+1 \text{ orders}} \mathscr{LR}_{1,N} \quad \text{for } 2k+1 \leq N$$

$$= \sum_{i=1}^{k+1} \sharp \overline{[2_{\widetilde{i-1}},1,2_{\widetilde{k-i+1}}]} \mathscr{LR}_{1,N} \quad \text{for } 2k+1 \leq N \qquad (20)$$

324

TABLE 14. Numbering the locally rigid functions for $f \in \mathscr{L}\mathscr{R}_{1,11}$, $f \in \mathscr{L}\mathscr{R}_{1,12}$, $f \in \mathscr{L}\mathscr{R}_{1,13}$.

g.o.p.	N=11	N=11	N=12	N=12	N=13	N=13
Total number		457,795		1,515,926		4,979,777
[1]	156,629	+	516,844	+	1,693,073	+
$[1_{\tilde{2}}]$	75,292	+	246,762	+	803,706	+
$[1_{\tilde{3}}]$	26,098	+	85,556	+	278,580	+
$[1_{\tilde{4}}]$	8,434	+	27,904	+	91,488	+
$[1_{\tilde{5}}]$	2,756	+	8,658	+	28,738	+
$[1_{\tilde{6}}]$	756	+	2,590	+	8,730	+
$[1_{\tilde{7}}]$	216	+	756	+	2,592	+
$[1_{\tilde{8}}]$	60	+	216	+	756	+
$[1_{\tilde{9}}]$	16	+	60	+	216	+
$[1_{\widetilde{10}}]$	4	+	16	+	60	+
$[1_{\widetilde{11}}]$	1	+	4	+	16	+
$[1_{\widetilde{12}}]$	-	+	1	+	4	+
$[1_{\widetilde{13}}]$	-	+	-	+	1	+
[2]	132,104	132,104	438,846	438,846	1,445,258	1,445,258
[2, 1]	23,941	39,364	80,108	132,104	265,548	438,846
[1, 2]	15,423	+	51,996	+	173,298	+
$[2_{\tilde{2}}]$	11,580	11,580	39,364	39,364	132,104	132,104
$[2_{\tilde{2}}, 1]$	2,634	3,356	8,883	11,580	29,659	39,364
$[1, 2_{\tilde{2}}]$	429	+	1,555	+	5,478	+
[2, 1, 2]	293	+	1,142	+	4,227	+
$[2_{\tilde{3}}]$	952	952	3,356	3,356	11,580	11,580
$[2_{\tilde{3}}, 1]$	255	264	899	952	3,098	3,356
$[1, 2_{\tilde{3}}]$	7	+	35	+	152	+
$[2, 1, 2_{\tilde{2}}]$	2	+	16	+	86	+
$[2_{\tilde{2}}, 1, 2]$	-	+	2	+	20	+
$[2_{\tilde{4}}]$	70	70	264	264	952	952
$[2_{\tilde{4}}, 1]$	18	18	70	70	263	264
$[1, 2_{\tilde{4}}]$	-	+	-	+	1	+
$[2, 1, 2_{\tilde{3}}]$	-	+	-	+	-	+
$[2_{\tilde{2}}, 1, 2_{\tilde{2}}]$	-	+	-	+	-	+
$[2_{\tilde{3}}, 1, 2]$	-	+	-	+	-	+
$[2_{\tilde{5}}]$	4	4	18	18	70	70
$[2_{\tilde{5}}, 1]$	1	1	4	4	18	18
$[1, 2_{\tilde{5}}]$	-	+	-	+	-	+
$[2_{\tilde{6}}]$	-	+	1	1	4	4
$[2_{\tilde{6}}, 1]$	-	+	-	+	1	1

Statement 5

$$\sharp[\overbrace{1}^{N-k+1}]\mathscr{L}\mathscr{R}_{1,N} = \begin{cases} 1 & \text{if } k=1 \\ 2 & \text{if } k=2 \\ \left(\frac{4}{27}\right)(k+1) \times 3^k & \text{for } 3 \leq k \leq \frac{N+1}{2} \end{cases} \tag{21}$$

TABLE 15. Numbering the locally rigid functions for $f \in \mathscr{LR}_{1,14}$, $f \subset \mathscr{LR}_{1,15}$.

g.o.p.	N=14	N=14	N=15	N=15
Total number		16,246,924		52,694,573
$[1]$	5,511,218	+	17,841,247	+
$[1_{\widetilde{2}}]$	2,603,258	+	8,391,360	+
$[1_{\widetilde{3}}]$	901,802	+	2,904,592	+
$[1_{\widetilde{4}}]$	297,728	+	962,888	+
$[1_{\widetilde{5}}]$	94,440	+	307,848	+
$[1_{\widetilde{6}}]$	29,050	+	95,676	+
$[1_{\widetilde{7}}]$	8,746	+	29,140	+
$[1_{\widetilde{8}}]$	2,592	+	8,748	+
$[1_{\widetilde{9}}]$	756	+	2,592	+
$[1_{\widetilde{10}}]$	216	+	756	+
$[1_{\widetilde{11}}]$	60	+	216	+
$[1_{\widetilde{12}}]$	16	+	60	+
$[1_{\widetilde{13}}]$	4	+	16	+
$[1_{\widetilde{14}}]$	1	+	4	+
$[1_{\widetilde{15}}]$	-	+	1	+
$[2]$	4,725,220	4,725,220	15,352,392	15,352,392
$[2,1]$	873,149	1,445,258	2,851,350	+
$[1,2]$	572,109	+	1,873,870	+
$[2_{\widetilde{2}}]$	438,846	438,846	1,445,258	1,445,258
$[2_{\widetilde{2}},1]$	98,135	132,104	322,310	438,846
$[1,2_{\widetilde{2}}]$	18,873	+	63,967	+
$[2,1,2]$	15,096	+	52,569	+
$[2_{\widetilde{3}}]$	39,364	39,364	132,104	132,104
$[2_{\widetilde{3}},1]$	10,460	11,580	34,845	39,364
$[1,2_{\widetilde{3}}]$	605	+	2,282	+
$[2,1,2_{\widetilde{2}}]$	389	+	1,596	+
$[2_{\widetilde{2}},1,2]$	126	+	641	+
$[2_{\widetilde{4}}]$	3,356	3,356	11,580	11,580
$[2_{\widetilde{4}},1]$	942	952	3,292	3,356
$[1,2_{\widetilde{4}}]$	8	+	44	+
$[2,1,2_{\widetilde{3}}]$	2	+	18	+
$[2_{\widetilde{2}},1,2_{\widetilde{2}}]$	-	+	2	+
$[2_{\widetilde{3}},1,2]$	-	+	-	+
$[2_{\widetilde{5}}]$	264	264	952	952
$[2_{\widetilde{5}},1]$	70	70	264	264
$[1,2_{\widetilde{5}}]$	-	+	-	+
$[2_{\widetilde{6}}]$	18	18	70	70
$[2_{\widetilde{6}},1]$	4	4	18	18
$[2_{\widetilde{7}}]$	1	1	4	4
$[2_{\widetilde{7}},1]$	-	+	1	1

Remark We call $u_k = \sharp[1_{\underbrace{\widetilde{N-k+1}}}]\mathscr{LR}_{1,N}$. For $k > 2$, then u_k is the sequence A120926 On-line Encyclopedia of integer Sequences : it is the number of sequences where 0 is isolated in ternary words of length N written with $\{0,1,2\}$.

326

These statements show that first the set $\mathscr{L}\mathscr{R}_{1,N}$ is an interesting set to be considered for dynamical systems and secondly the gop are fruitful in this study. However the set

$$\mathscr{L}\mathscr{R}_{2,N}=\{f\in\mathscr{F}_N \text{ such that } \forall p, 0\leq p\leq N-2, |f(p)-f(p+1)|\leq 2\}$$

is too much large to give comparable results. Then we introduce more sophisticated sets we call sets with locally bounded range which more or less correspond to an analogue of the discrete convolution product of the local variation of f with a compact support function $\overrightarrow{\alpha_t}$.

Orbits and patterns of locally rigid function sets

Consider now the set :

$$\mathscr{L}\mathscr{R}_{\overrightarrow{\alpha_t},q,N}=\{f\in\mathscr{F}_N \text{ such that } \forall p, 0\leq p\leq N-r-1, \sum_{r=1}^{r=t}\alpha_r|f(p)-f(p+r)|\leq q\}\cap\{f\in$$

$\mathscr{F}_N$ such that $\forall p, t\leq p\leq N-1, \sum_{r=1}^{r=t}\alpha_r|f(p)-f(p-r)|\leq q\}$ for the vector $\overrightarrow{\alpha_t}=(\alpha_1,\alpha_2,\ldots,\alpha_t)\in\mathbb{N}^t$, for $q\in\mathbb{N}$.

TABLE 17. Numerical study of the set $\mathscr{L}\mathscr{R}_{\overrightarrow{\alpha_t},q,N}$ for $N=10$, $t=5$, $\alpha_1=20$, $\alpha_2=9$, $\alpha_3=5$, $\alpha_4=2$ and $\alpha_5=1$, for $q=20,\ldots,142$

q	maximal period	modulus	gop number	functions number
20	1	1	1	10
26	2	2	3	82
44	2	3	6	21,764
49	3	3	7	48,112
50	3	3	7	53,210
56	3	4	9	208,692
59	4	4	15	330,800
63	4	5	19	626,890
66	4	10	37	952,228
67	4	10	46	1,064,316
72	5	10	50	1,630,018
74	6	10	60	1,816,826
76	6	10	61	2,152,450
77	6	10	88	2,416,368
78	6	10	91	2,762,434
79	6	10	97	3,188,080
80	6	10	99	3,735,666
84	6	10	100	5,876,324
85	6	10	103	6,473,288
87	6	10	105	7,851,728
88	7	10	121	8,644,178
89	8	10	129	9,521,920
91	8	10	136	11,414,556
92	8	10	165	12,454,440
94	8	10	175	14,756,058

Following next page

TABLE 17. (Next)

q	maximal period	modulus	gop number	functions number
95	8	10	177	16,077,780
96	8	10	184	17,208,654
97	8	10	185	18,369,854
98	8	10	188	19,585,746
100	8	10	192	22,083,852
101	8	10	199	23,584,452
102	8	10	204	25,513,892
103	8	10	244	27,912,772
104	8	10	304	30,560,238
105	9	10	333	33,516,466
106	9	10	380	36,682,960
107	9	10	424	40,004,280
108	10	10	491	43,685,352
109	10	10	517	47,655,856
110	10	10	529	51,785,410
111	10	10	562	55,907,120
112	10	10	583	60,341,276
113	10	10	612	64,930,790
114	10	10	647	69,766,178
115	10	10	706	74,989,752
116	10	10	747	80,087,120
117	10	10	791	85,570,272
118	10	10	820	91,206,218
119	10	10	836	97,040,288
120	10	10	852	103,121,916
121	10	10	872	109,650,464
122	10	10	896	116,345,296
123	10	10	919	123,241,156
124	10	10	924	130,360,938
125	10	10	928	137,636,628
126	10	10	930	145,536,068
127	10	10	932	154,370,862
128	10	10	938	164,145,928
129	10	10	960	174,942,026
130	10	10	986	186,438,038
131	10	10	1,006	198,594,118
132	10	10	1,013	211,550,402
133	10	10	1,015	225,324,700
134	10	10	1,021	239,976,118
135	10	10	1,022	255,106,866
137	10	10	1,023	286,726,234
142	10	10	1,023	374,355,356

The functions belonging to these sets show a kind of "rigidity": the less is q, the more "rigid" is the function, this "rigidity" being modulated by the vector $\overrightarrow{\alpha_t}$. Furthermore, the maximal length of a periodic orbit increases with q, and so the number of gop $\sharp \mathscr{G}(\mathscr{L}\mathscr{R}_{\overrightarrow{\alpha_t},q,N})$ and the maximal modulus of the gop.

Remark Using this generalized notation, one has : $\mathscr{LR}_{1,n} = \mathscr{LR}_{1,1,n}$ and $\mathscr{LR}_{2,n} = \mathscr{LR}_{1,2,n}$.

As an example, we explore numerically the case : $N = 10$, $t = 5$, $\alpha_1 = 20$, $\alpha_2 = 9$, $\alpha_3 = 5$, $\alpha_4 = 2$ and $\alpha_5 = 1$, for $q = 20,\ldots,142$. The results are displayed in Table 17. In this Table "modulus" means the maximal modulus of the gop belonging to this set for the corresponding value of q in the row, "gop number" stands for $\sharp\mathscr{G}(\mathscr{LR}_{\overrightarrow{\alpha_t},q,N})$ and "functions number" for $\sharp\mathscr{LR}_{\overrightarrow{\alpha_t},q,N}$. One can point out that for the particular function $\overrightarrow{\alpha_t}$ of the example; it is possible to find 10 intervals $I_1, I_2,\ldots,I_{10} \subset \mathbb{N}$ such that if $q \in I_r$ then there is no periodic orbit whose period is strictly greater than r, (e.g., $I_6 = [\![74,87]\!]$). Furthermore it is possible to split these intervals into subintervals $I_{r,s}$ in which $\sharp\mathscr{G}\left(\mathscr{LR}_{\overrightarrow{\alpha_t},q,N}\right)$ is constant when q thumbs $I_{r,s}$. This is not the case for $\sharp\mathscr{LR}_{\overrightarrow{\alpha_t},q,N}$.

CONCLUSION

A discrete dynamical system associated to a function on finite ordered set X can only exhibit periodic orbits. However the number of the periods and the length of each are not easily predictable. We formalise such a gop as the ordered set of periods when the initial value thumbs X in the increasing order. We can predict by means of closed formulas, the number of gop of the set of all the function from X to itself. We also explore, using the brute force of computers, some subsets of locally rigid functions on X, for which interesting patterns of periodic orbits are found. Further study is needed to understand the behaviour of dynamical systems associated to functions belonging to these sets.

TABLE 16. Numbering the locally rigid functions for $f \in \mathscr{L}\mathscr{R}_{1,16}$.

g.o.p.	N=16	N=16
Total number		170,028,792
[1]	57,477,542	+
$[1_{\tilde{2}}]$	26,932,398	+
$[1_{\tilde{3}}]$	9,314,088	+
$[1_{\tilde{4}}]$	3,097,650	+
$[1_{\tilde{5}}]$	996,764	+
$[1_{\tilde{6}}]$	312,456	+
$[1_{\tilde{7}}]$	96,096	+
$[1_{\tilde{8}}]$	29,158	+
$[1_{\tilde{9}}]$	8,748	+
$[1_{\widetilde{10}}]$	2,592	+
$[1_{\widetilde{11}}]$	756	+
$[1_{\widetilde{12}}]$	216	+
$[1_{\widetilde{13}}]$	60	+
$[1_{\widetilde{14}}]$	16	+
$[1_{\widetilde{15}}]$	4	+
$[1_{\widetilde{16}}]$	1	+
[2]	49,610,818	49,610,818
[2,1]	9,255,822	15,352,392
[1,2]	6,096,570	+
$[2_{\tilde{2}}]$	4,725,220	4,725,220
$[2_{\tilde{2}},1]$	1,051,686	1,445,258
$[1,2_{\tilde{2}}]$	213,975	+
[2,1,2]	179,597	+
$[2_{\tilde{3}}]$	438,846	438,846
$[2_{\tilde{3}},1]$	114,798	132,104
$[1,2_{\tilde{3}}]$	8,284	+
$[2,1,2_{\tilde{2}}]$	6,146	+
$[2_{\tilde{2}},1,2]$	2,876	+
$[2_{\tilde{4}}]$	39,364	39,364
$[2_{\tilde{4}},1]$	11,246	11,580
$[1,2_{\tilde{4}}]$	204	+
$[2,1,2_{\tilde{3}}]$	106	+
$[2_{\tilde{2}},1,2_{\tilde{2}}]$	22	+
$[2_{\tilde{3}},1,2]$	2	+
$[2_{\tilde{5}}]$	3,356	3,356
$[2_{\tilde{5}},1]$	951	952
$[1,2_{\tilde{5}}]$	1	+
$[2_{\tilde{6}}]$	264	264
$[2_{\tilde{6}},1]$	70	70
$[2_{\tilde{7}}]$	18	18
$[2_{\tilde{7}},1]$	4	4
$[2_{\tilde{8}}]$	1	1

REFERENCES

1. Diamond, P. and Pokrovskii, A., Statistical laws for computational collapse of discretized chaotic mappings. *International Journal of Bifurcation and Chaos*, 1996, Vol.6, 12A, pp. 2389-2399.
2. Donald E. Knuth, *The Art Of Computer Programming (Third Edition)*, Addison-Wesley, 2005, Vol. 2, pp.8-9.
3. Martin D. Kruskal, *The expected number of components under a random mapping function*, INST CNRS, June 1954, p.392.
4. A.N. Sharkovskii [1964], *Coexistence of cycles of continuous mapping of the line into itself.* *Ukrainian Math.*, 1995, International Journal of Bifurcation and Chaos, No.5, pp.1263-1273.
5. R. Lozi, *The importance of strange attractors for industrial mathematics*, Trends in Industrial and Applied Mathematics, 2002, Proceedings of the 1st International Conference on Industrial and Applied Mathematics of the Indian Subcontinent, A.H. Siddiqi and M. Kocvara (Eds) Kluwer Academic Publisher, pp.275-303.
6. R. Lozi, *Giga-periodic Orbits for Weakly Coupled tent and Logistic Discretized Maps*, Modern Mathematical Models, Methods and Algorithms for Real World Systems, 2007, A.H. Siddiqi, I.S. Duff and O. Christensen (Editors), Anamaya Publishers, New Delhi, India, pp.81-124.
7. R. Lozi, *New Enhanced Chaotic Number Generators*, Indian Journal of Industrial and Applied Mathematics, vol.1 No.1, pp.1-23.
8. Ljuben R. Mutafchiev, *Limit theorem concerning random mapping patterns*, Combinatorica, 1988, Vol. 8, pp.345-356.
9. Purdom, P.W, and Williams, J.H., *Cycle length in a random function*, Trans. of Amer. Math. Soc., Sept. 1968, Vol. 133, No.2, pp.547-551.
10. (2007) US Patent 7,170,997 - Method of generating pseudo-random numbers in an electronic device, and a method of encrypting and decrypting electronic data.
11. (2006) US Patent 6,999,445 - Multiple access communication system using chaotic signals and method for generating and extracting chaotic signal.
12. (2001) U.S. Pat. No. 5,048,086 assigned to Hughes Aircraft Company is related to an encryption system based on chaos theory. The system uses the logistic equation $x_{n+1} = \mu x_n(1 - x_n)$, which is a mapping exhibiting chaos for certain values of μ. In the computations, floating-point operations are used.

The Cauchy problem for BBGKY hierarchy of quantum kinetic equations with coulomb potential

Brokate M.[*] and Rasulova M.Yu.[†]

[*]*Munich Technical University,Boltzmannstr 3, 85748 Garching GERMANY,
email:brokate@ma.tum.de*
[†]*The Institute of Nuclear Physics, Ulughbek, Tashkent 100124 UZBEKISTAN,
email:rasulova@live.com*

Abstract. In the present work by a semigroup method the existence of a unique solution in terms of initial data of the BBGKY hierarchy of quantum kinetic equations with coulomb potential is proved.

Keywords: coulomb potential, quantum kinetic equations
PACS: 82C10

INTRODUCTION

In the last few year we can see an increasing interest in the Bogolyubov-Born-Green-Kirkwood-Yvon's (BBGKY) hierarchy. This interest is quite natural, since this hierarchy is relating the equation of Liouville, which is describing the evolution of a system interacting many particles with the Boltzmann and Vlasov equations. Last equations describe the evolution of one particle. It is known, that the later equations so far are fundamental equations, describing the evolution of particles in solids, semi-conductors, in gas and in plasma. Unlike Liouville's equation, the structure of the BBGKY's hierarchy permits the generalization of the physical results for one particle to system of many particles.

Since the time, when it was formulated in 1946, the BBGKY's hierarchy was the object of investigation for physicists as well as mathematicians [1]-[6],[9]-[12],[14]-[18],[20]-[24].

Well known, that charged particles interact via the coulomb potential. Until present, there is no solution of the BBGKY's hierarchy of quantum kinetic equations in the case when the particles interact via a coulomb potential. This is a important problem for many researchers. The present paper addresses the solution of this problem.

FORMULATION OF THE PROBLEM

We consider the hierarchy BBGKY of quantum kinetic equations, which describes the evolution of a system of identical particles with mass m and charge q interacting via a coulomb potential [8],[11] $\phi(x_i,x_j) = \frac{q^2}{|x_i-x_j|}$, which depends on the distance between

CP1146, *Modelling of Engineering and Technological Problems*, edited by A. H. Siddiqi, A. K. Gupta, and M. Brokate
© 2009 American Institute of Physics 978-0-7354-0683-4/09/$25.00

particles $|x_i - x_j| = ((x_i^1 - x_j^1)^2 + (x_i^2 - x_j^2)^2 + (x_i^3 - x_j^3)^2)^{1/2}$ and charges q. We assume that the charge is a real constant.

The BBGKY's hierarchy is given by [2],[3]:

$$i\frac{\partial \rho_s(t,x_1,...,x_s;x_1',...,x_s')}{\partial t} = [H_s,\rho_s](t,x_1,...,x_s;x_1',...,x_s') +$$

$$+\frac{N}{V}(1-\frac{s}{N})Tr_{x_{s+1}} \sum_{1\leq i\leq s} (\phi_{i,s+1}(|x_i - x_{s+1}|) - \phi_{i,s+1}(|x_i' - x_{s+1}|)) \times$$

$$\times \rho_{s+1}(t,x_1,...,x_s,x_{s+1};x_1',...,x_s',x_{s+1}), \tag{1}$$

with the initial condition

$$\rho_s(t,x_1,...,x_s;x_1',...,x_s')|_{t=0} = \rho_s(0,x_1,...,x_s;x_1',...,x_s'). \tag{2}$$

In the problem, given by equation (1),(2) the vector represented by x_i gives the position of i th particle in the 3-dimensional Euclidean space R^3, $x_i = (x_i^1,x_i^2,x_i^3)$, $i = 1,2,....,s$, and $x_i^\alpha, \alpha = 1,2,3$ are coordinates of a vector x_i. The length of the vector x_i is denoted by

$$|x_i| = ((x_i^1)^2 + (x_i^2)^2 + (x_i^3)^2)^{\frac{1}{2}}.$$

The reduced statistical operator of s particles is $\rho_s(x_1,...,x_s;x_1',...,x_s')$ related by positive symmetric density matrix of N particles by [2],[3]

$$\rho_s(x_1,...,x_s;x_1',...,x_s') =$$

$$= V^s Tr_{x_{s+1},......,x_N} D(x_1,...,x_s,x_{s+1},....x_N;x_1',...,x_s',x_{s+1},....,x_N),$$

$s \in N$, N is the number of particles, V the volume of the system of particles. The trace is defined in terms of the kernel $\rho(x,x')$ by the formula

$$Tr_x\rho = \int \rho(x,x)dx$$

In equation (1) $\hbar = 1$ is the Plank constant and $[,]$ denotes the Poisson bracket. The Hamiltonian of system is defined as:

$$H_s = \sum_{1\leq i\leq s} T_i + \sum_{1\leq i<j\leq s} \phi_{i,j},$$

where

$$T_i = -\frac{\hbar^2 \triangle_i}{2m},$$

$\triangle_i$ is the Laplacian

$$\triangle_i = \frac{\partial^2}{\partial(x_i^1)^2} + \frac{\partial^2}{\partial(x_i^2)^2} + \frac{\partial^2}{\partial(x_i^3)^2},$$

and

$$\phi_{i,j} = \frac{q^2}{|x_i - x_j|}.$$

333

The operator given by:

$$(\Phi \rho_s)(x_1,...,x_s;x_1',...,x_s') = \sum_{i \leq j} \phi_{i,j}(|x_i - x_j|)\rho_s(x_1,...,x_s;x_1',...,x_s')$$

is symmetric.

In the present work, the Cauchy problem (1),(2) is solved for a quantum system finite number paricles contained in the finite bounded region (vessel) with volume $V = |\Lambda|$.

A state of this system is described by a density matrix $\rho_s^{\Lambda}(t,x_1,...,x_s;x_1',...,x_s')$ that satisfies the Cauchy problem

$$i\frac{\partial \rho_s^{\Lambda}(t,x_1,...,x_s;x_1',...,x_s')}{\partial t} = [H_s^{\Lambda},\rho_s^{\Lambda}](t,x_1,...,x_s;x_1',...,x_s')+$$

$$+\frac{N}{V}(1-\frac{s}{N})Tr_{x_{s+1}}\sum_{1 \leq i \leq s}(\phi_{i,s+1}(|x_i - x_{s+1}|) - \phi_{i,s+1}(|x_i' - x_{s+1}|)) \times$$

$$\times \rho_{s+1}^{\Lambda}(t,x_1,...,x_s,x_{s+1};x_1',...,x_s',x_{s+1}), \qquad (3)$$

with the initial condition

$$\rho_s^{\Lambda}(t,x_1,...,x_s;x_1',...,x_s')|_{t=0} = \rho_s^{\Lambda}(0,x_1,...,x_s;x_1',...,x_s'). \qquad (4)$$

In (3) a Hamiltonian of a system is defined as

$$H_s^{\Lambda}(x_1,...,x_s) = \sum_{1 \leq i \leq s}(-\frac{1}{2m}\triangle_{x_i} + u^{\Lambda}(x_i)) + \sum_{1 \leq i < j \leq s}\phi_{i,j}(|x_i - x_j|),$$

where $u^{\Lambda}(x)$ is an external field which keeps the system in the region Λ ($u^{\Lambda}(x) = 0$ if $x \in \Lambda$ and $u^{\Lambda}(x) = +\infty$ if $x \notin \Lambda$).

The trace is defined in the region by the formula

$$Tr_x \rho = \int_{\Lambda} \rho(x,x)dx$$

Introducing the notation

$$\left(\mathscr{H}^{\Lambda}\rho^{\Lambda}\right)_s(t,x_1,...,x_s;x_1',...,x_s') = \left[H_s^{\Lambda},\rho_s^{\Lambda}(t,x_1,...,x_s;x_1',...,x_s')\right];$$

$$\left(\mathscr{D}_{x_{s+1}}^{\Lambda}\rho^{\Lambda}\right)_s(x_1,\cdots,x_s;x_1',\cdots,x_s') = \rho_{s+1}^{\Lambda}(x_1,\cdots x_s,x_{s+1};x_1',\cdots,x_s',x_{s+1});$$

$$(\mathscr{A}_{x_{s+1}}^{\Lambda}\rho^{\Lambda})_s(t,x_1,...,x_s;x_1',...,x_s') = \frac{N}{V}(1-\frac{s}{N})\sum_{1 \leq i \leq s}(\phi_{i,s+1}(|x_i - x_{s+1}|)-$$

$$-\phi_{i,s+1}(|x_i' - x_{s+1}|))\rho_s^{\Lambda}(t,x_1,...,x_s;x_1',...,x_s');$$

$$\rho^{\Lambda}(t) = \{\rho_1^{\Lambda}(t,x_1;x_1'),\cdots\cdots\cdots,\rho_s^{\Lambda}(t,x_1,\cdots,x_s:x_1',\cdots,x_s'),\cdots\}, \qquad s = 1,2,\cdots$$

we can cast (3), (4) in the form

$$i\frac{\partial}{\partial t}\rho_s^\Lambda(t,x_1,...,x_s;x_1',...,x_s') = \left(\mathcal{H}^\Lambda\rho^\Lambda\right)_s(t,x_1,...,x_s;x_1',...,x_s')+$$

$$+\int_\Lambda \left(\mathcal{A}_{x_{x+1}}^\Lambda \mathcal{D}_{x_{x+1}}^\Lambda \rho^\Lambda\right)_s(t,x_1,...,x_s,x_{s+1};x_1',...,x_s',x_{s+1})dx_{x+1}, \qquad (5)$$

$$\rho_s^\Lambda(t,x_1,...,x_s;x_1',...,x_s')|_{t=0} = \rho_s^\Lambda(0,x_1,...,x_s;x_1',...,x_s') \equiv \rho_s^\Lambda(x_1,...,x_s;x_1',...,x_s'). \qquad (6)$$

SOLUTION OF THE CAUCHY PROBLEM FOR BBGKY HIERARCHY OF QUANTUM KINETIC EQUATIONS WITH COULOMB POTENTIAL

To obtain the solution of the Cauchy problem defined by (3),(4) or by the reduced form in (5),(6), we use a semigroup method [7],[13],[16],[17].

Let $L_2^s(\Lambda)$ be the Hilbert space of functions $\psi_s^\Lambda(x_1,...,x_s)$, $x_i \in R^3(\Lambda)$, and B_s^Λ be the Banach space of positive-definite, self adjoint nuclear operators $\rho_s^\Lambda(x_1,...,x_s;x_1',...,x_s')$ on $L_2^s(\Lambda)$:

$$(\rho_s^\Lambda\psi_s^\Lambda)(x_1,...,x_s) = \int_\Lambda \rho_s^\Lambda(x_1,...,x_s;x_1',...,x_s')\,\psi_s^\Lambda(x_1',...,x_s')dx_1'...dx_s',$$

with norm

$$|\rho_s^\Lambda|_1 = \sup_{1\le i\le\infty} \sum |(\rho_s^\Lambda\psi_i^s,\varphi_i^s)|,$$

where the upper bound is taken over all orthonormalized systems of finite, twice differentiable functions with compact support $\{\psi_i^s\}$ and $\{\varphi_i^s\}$ in $L_2^s(\Lambda)$, $s \ge 1$ and $\|\rho_0^\Lambda\| = |\rho_0^\Lambda|$.

We'll suppose that the operators $\rho_s^\Lambda(t)$ and H_s^Λ act in the space $L_2^s(\Lambda)$ with zero boundary conditions.

Let B^Λ be the Banach space of sequences of nuclear operators

$$\rho^\Lambda = \{\rho_0^\Lambda,\rho_1^\Lambda(x_1;x_1'),...,\rho_s^\Lambda(x_1,...,x_s;x_1',...,x_s'),...\},$$

where ρ_0^Λ are complex numbers, $\rho_s^\Lambda \subset B_s^\Lambda$,

$$\rho_s^\Lambda(x_1,...,x_s;x_1',...,x_s') = 0, \qquad when \qquad s > s_0,$$

s_0 finite and norm is

$$|\rho^\Lambda|_1 = \sum_{s=0}^\infty |\rho_s^\Lambda|_1.$$

The coulomb potential $\phi_{i,j} = \frac{q^2}{|r_{i,j}|}$ can be represented as

$$\phi_{i,j} = \phi_{i,j}^1 + \phi_{i,j}^2,$$

335

where

$$\phi_{i,j}^1 = \frac{q^2}{|r_{i,j}|}\left(\frac{1}{1+|r_{i,j}|}\right) \subset L_2(R^3), \qquad \phi_{i,j}^2 = \frac{q^2}{1+|r_{i,j}|} \subset L_\infty(R^3),$$

$$r_{i,j} = ((x_i^1 - x_j^1)^2 + (x_i^2 - x_j^2)^2 + (x_i^3 - x_j^3)^2)^{1/2}.$$

Therefore the coulomb potential satisfies the conditions of Theorem X.15 [19] and the Hamiltonian with coulomb potential:

$$H_s^C(x_i, x_j) = -\sum_i^s \frac{1}{2}\triangle_{x_i} + \sum_{1 \leq i < j \leq s} \frac{q^2}{|x_i - x_j|}$$

is essentially a self-adjoint operator on the set $D(H_s^C)$ of finite, twice differentiable functions with compact support [8].

Let $\tilde{B}_s^\Lambda$ be a dense set of "good" elements of B_s^Λ of type $B_s^\Lambda \cap D(H_s^C) \otimes D(H_s^C)$, where $D(H_s^C)$ is the domain of the operator H_s^C [8] and $\otimes$ denote the algebraic tensor product.

Consider the operators

$$(\omega^\Lambda(t)\rho^\Lambda)_s(x_1, ..., x_s; x_1', ..., x_s') = (e^{-iH_s^\Lambda t}\rho^\Lambda e^{iH_s^\Lambda t})_s(x_1, ..., x_s; x_1', ..., x_s'),$$

on $\rho_s^\Lambda(x_1, ..., x_s; x_1', ..., x_s') \subset B_s^\Lambda$.

Theorem 1 *The operators $\omega^\Lambda(t)$ define a strongly continuous group of isometries on B^Λ whose generators coincides with $-i\mathcal{H}^\Lambda$ on $\tilde{B}^\Lambda$ everywhere dense in B^Λ.*

Proof: The prove is summarized in the following four stage.

1. The operator $\omega^\Lambda(t)$ is an isometry in the nuclear norm on B^Λ:

$$|\omega^\Lambda(t)\rho^\Lambda|_1 = |exp(-iH^\Lambda t)\rho^\Lambda exp(iH^\Lambda t)|_1 =$$

$$= sup \sum_{1 \leq i \leq \infty} |(e^{-iH^\Lambda t}\rho^\Lambda e^{iH^\Lambda t}\varphi_i^\Lambda, \psi_i^\Lambda)| =$$

$$= sup \sum_{1 \leq i \leq \infty} |\rho^\Lambda e^{iH^\Lambda t}\varphi_i, e^{iH^\Lambda t}\psi_i| = |\rho^\Lambda|_1,$$

where the upper bound is taken over all orthonormalized systems of finite, twice differentiable functions with compact support $\{\psi_i^s\}$ and $\{\varphi_i^s\}$ in $L_2^s(\Lambda)$.

2. Operator $\omega^\Lambda(t)$ is strongly continuous on t in the nuclear norm on B^Λ:
The strong continuity of $\omega^\Lambda(t)$ on B^Λ follows from the relations

$$|exp(-iH^\Lambda t)\rho^\Lambda exp(iH^\Lambda t) - \rho^\Lambda|_1 = |exp(-iH^\Lambda t)\rho^\Lambda exp(iH^\Lambda t) - \rho^\Lambda +$$

$$+exp(-iH^\Lambda t)\rho_n^\Lambda exp(iH^\Lambda t) - \rho_n^\Lambda - exp(-iH^\Lambda t)\rho_n^\Lambda exp(iH^\Lambda t) + \rho_n^\Lambda|_1 \leq$$

$$\leq (|exp(-iH^\Lambda t)(\rho^\Lambda - \rho_n^\Lambda)exp(iH^\Lambda t)|_1 +$$

$$+|\rho^\Lambda - \rho_n^\Lambda|_1 + |exp(-iH^\Lambda t)\rho_n^\Lambda exp(iH^\Lambda t) - \rho_n^\Lambda|_1) =$$

$$= 2|\rho^\Lambda - \rho_n^\Lambda|_1 + |exp(-iH^\Lambda t)\rho_n^\Lambda exp(iH^\Lambda t) - \rho_n^\Lambda|_1. \tag{7}$$

The term $2|\rho^\Lambda - \rho_n^\Lambda|_1$ in (7) can be made as small as desired because the ρ_n^Λ is dense in the space of nuclear operators [16],[17]. Therefore

$$|exp(-iH^\Lambda t)\rho^\Lambda exp(iH^\Lambda t) - \rho^\Lambda|_1 \le |exp(-iH^\Lambda t)\rho_n^\Lambda exp(iH^\Lambda t) - \rho_n^\Lambda|_1. \quad (8)$$

As follows from (8)

$$Lim_{t\to 0}|\omega^\Lambda(t)\rho^\Lambda - \rho^\Lambda|_1 = Lim_{t\to 0}|exp(-iH^\Lambda t)\rho^\Lambda exp(iH^\Lambda t) - \rho^\Lambda|_1 \le$$

$$Lim_{t\to 0}|exp(-iH^\Lambda t)\rho_n^\Lambda exp(iH^\Lambda t) - \rho_n^\Lambda|_1 \le$$

$$\le Lim_{t\to 0}|exp(-iH^\Lambda t)\rho_n^\Lambda (exp(iH^\Lambda t) - I)|_1 +$$

$$+ Lim_{t\to 0}|(exp(-iH^\Lambda t) - I)\rho_n^\Lambda|_1 \le Lim_{t\to 0}(\sum_{1\le i\le n} \lambda_i \|\psi_i\| \|(exp(iH^\Lambda t) - I)\varphi_i\| +$$

$$+ \sum_{1\le i\le n} \lambda_i \|(exp(-iH^\Lambda t) - I)\psi_i\| \|\varphi_i\|) = 0. \quad (9)$$

This follows from the strong continuity of the group $exp(\mp iH^\Lambda t)$, according to which

$$Lim_{t\to 0}\|(exp(iH^\Lambda t) - I)\varphi_i\| = 0,$$

$$Lim_{t\to 0}\|(exp(-iH^\Lambda t) - I)\psi_i\| = 0.$$

In (9) we used

$$|\rho_n^\Lambda|_1 \le \sum_{1\le i\le n} \lambda_i \|\psi_i\| \|\varphi_i\|,$$

$$|exp(-iH^\Lambda t)\rho_n^\Lambda (exp(iH^\Lambda t) - I)|_1 \le (\sum_{1\le i\le n} \lambda_i \|\psi_i\| \|(exp(iH^\Lambda t) - I)\varphi_i\|,$$

$$|(exp(-iH^\Lambda t) - I)\rho_n^\Lambda(t_0)|_1 \le \sum_{1\le i\le n} \lambda_i \|(exp(-iH^\Lambda t) - I)\psi_i\| \|\varphi_i\|),$$

where $\{\varphi_i\}$ and $\{\psi_i\}$ from $L_2^s(\Lambda)$ and $i = 1, 2, ...$, are systems of finite, twice differentiable functions with compact support.

3. The operator $\omega^\Lambda(t)$ satisfies the group property:

$$\omega^\Lambda(t_1)\omega^\Lambda(t_2)\rho^\Lambda = \omega^\Lambda(t_1)e^{-iH^\Lambda t_2}\rho^\Lambda e^{iH^\Lambda t_2} =$$

$$= e^{iH^\Lambda t_1}e^{iH^\Lambda t_2}\rho^\Lambda e^{iH^\Lambda t_2}e^{iH^\Lambda t_1} = e^{-iH^\Lambda(t_1+t_2)}\rho^\Lambda e^{iH^\Lambda(t_1+t_2)} = \omega^\Lambda(t_1 + t_2)\rho^\Lambda.$$

Analogously:

$$\omega^\Lambda(t_2)\omega^\Lambda(t_1)\rho^\Lambda = \omega^\Lambda(t_2 + t_1)\rho^\Lambda.$$

4. The generator of the group $\omega^\Lambda(t)$ is defined on B^Λ coincides with $-i\mathscr{H}$ on $\tilde{B}^\Lambda$:

$$lim_{t\to 0}|\frac{\omega^\Lambda(t)\rho^\Lambda - \rho^\Lambda}{t}|_1 = lim_{t\to 0}sup \sum_{1\le i\le \infty} |(e^{-iH^\Lambda t}\rho^\Lambda \frac{e^{iH^\Lambda t} - I}{t}\varphi_i +$$

$$+ \frac{e^{iH^\Lambda t} - I}{t}\rho^\Lambda e^{iH^\Lambda t}\varphi_i, \psi_i)| = lim_{t\to 0}sup \sum_{1\le i\le \infty} |((\rho^\Lambda iH^\Lambda - iH^\Lambda \rho^\Lambda)\varphi_i, \psi_i)| =$$

337

$$= |-i[H^\Lambda, \rho^\Lambda]|_1,$$

where the upper bound is taken over all orthonormalized systems of finite, twice differentiable functions with compact support $\{\psi_i^s\}$ and $\{\varphi_i^s\}$ in $L_2^s(\Lambda)$.

We introduce the operator $\Omega(\Lambda)$ on the space B^Λ by:

$$(\Omega(\Lambda)\rho^\Lambda)_s(x_1, ..., x_s; x_1', ..., x_s') =$$

$$= \frac{N}{V}\left(1 - \frac{s}{N}\right) \int_\Lambda \sum_i \rho_{s+1}^\Lambda(x_1, ..., x_s, x_{s+1}; x_1', ..., x_s', x_{s+1}) \times$$

$$\times g_i^1(x_{s+1})\tilde{g}_i^1(x_{s+1})dx_{s+1},$$

where $g_i^1(x_{s+1})$ is a complete orthonormal system of vectors in the one-particle space $L_2(\Lambda)$.

We introduce the operator $U^\Lambda(t)$ on B_s^Λ by the formula

$$\rho_s^\Lambda(t, x_1, ..., x_s; x_1', ..., x_s') = (U^\Lambda(t)\rho^\Lambda)_s(x_1, ..., x_s; x_1', ..., x_s') =$$

$$= (e^{\Omega(\Lambda)}e^{-iH^\Lambda t}e^{-\Omega(\Lambda)}\rho^\Lambda e^{iH^\Lambda t})_s(x_1, ..., x_s; x_1', ..., x_s'). \qquad (10)$$

The structure of the group (10) first time for quantum system of particles was derived for potential, which satisfied the Kato's criterion. The history of derivation of this formula quoted in [16].

Theorem 2. *The operator $U^\Lambda(t)$ generates a strongly continuous group of bounded operators on B^Λ, whose generators coincide with the operator $\mathscr{H} + Tr_x\mathscr{A}_x\mathscr{D}_x$ on $\tilde{B}^\Lambda$ everywhere dense in B^Λ.*

Proof: The proof is summarized in the following four stage:

1. Let us show that the operator $U^\Lambda(t)$ is bounded on B^Λ. We begin by evaluating the operator Ω^Λ.

Let $|\frac{N}{V}(1 - \frac{s}{N})|$ satisfy condition

$$\left|\frac{N}{V}\left(1 - \frac{s}{N}\right)\right| = 1. \qquad (11)$$

In this case [16],[17]

$$|\Omega(\Lambda)|_1 = \frac{1}{|\rho^\Lambda|_1} \sum_{s=1}^\infty |(\Omega(\Lambda)\rho^\Lambda)_s|_1 = \frac{1}{|\rho^\Lambda|_1} \sum_{s=1}^\infty sup \sum_i |(\psi_i^s, (\Omega(\Lambda)\rho^\Lambda)_s \varphi_i^s)|_1 =$$

$$= \frac{1}{|\rho^\Lambda|_1} \sum_{s=1}^\infty sup \sum_i \left|\frac{N}{V}\left(1 - \frac{s}{N}\right)\right| |(\psi_i^{s+1}, \rho_{s+1}^\Lambda \varphi_i^{s+1})|_1 \leq$$

$$\leq \frac{1}{|\rho^\Lambda|_1} \sum_{s=1}^\infty sup \sum_i |(\psi_i^s, \rho_s^\Lambda \varphi_i^s)|_1 = \frac{1}{|\rho^\Lambda|_1} \sum_{s=1}^\infty |\rho_s^\Lambda|_1 = \frac{|\rho^\Lambda|_1}{|\rho^\Lambda|_1},$$

where the upper bound is taken over all orthonormal system of vectors $\{\psi_i^s\}$ and $\{\varphi_i^s\}$ in the s-particle space $L_2^s(\Lambda)$ and $\psi_i^{s+1} = g_i\psi_i^s$, $\varphi_i^{s+1} = g_i\varphi_i^s$.

If we take into account condition (11) and positivity of $\rho_s^\Lambda(t)$, we get

$$|\Omega(\Lambda)|_1 = \frac{|\rho^\Lambda|_1}{|\rho^\Lambda|_1} = 1.$$

From the boundedness of the operator $\Omega(\Lambda)$, it follows that $e^{\Omega(\Lambda)}$ is bounded $|e^{\pm\Omega(\Lambda)}|_1 \leq e^1$.

The operator $U^\Lambda(t)$, as a product of the bounded operators of $e^{\pm\Omega(\Lambda)}$ and the unitary operators $e^{\mp iH_s^\Lambda t}$, is bounded and satisfies the estimate $U^\Lambda(t) \leq e^2$ on B^Λ.

2. Strong continuity of the operator $U^\Lambda(t)$ on B^Λ follows from boundedness of the operator $e^{\pm\Omega(\Lambda)}$ and the strong continuity of the operator $\omega^\Lambda(t)$ on B^Λ [16].

$$Lim_{t\to 0}|e^{\Omega(\Lambda)}exp(-iH^\Lambda t)e^{-\Omega(\Lambda)}\rho^\Lambda(t_0)exp(iH^\Lambda t) - \rho^\Lambda(t_0)|_1 = 0.$$

Proof is analogously to (7)-(9).

3. The operator $U^\Lambda(t)$ satisfies the group property on B^Λ:

$$U^\Lambda(t_1)U^\Lambda(t_2)\rho^\Lambda = U^\Lambda(t_1)e^{\Omega(\Lambda)}e^{-iH^\Lambda t_2}(e^{-\Omega(\Lambda)}\rho^\Lambda)e^{iH^\Lambda t_2} =$$

$$= e^{iH^\Lambda t_1}e^{\Omega(\Lambda)}e^{iH^\Lambda t_2}e^{-\Omega(\Lambda)}\rho^\Lambda e^{iH^\Lambda t_2}e^{iH^\Lambda t_1} =$$

$$= e^{\Omega(\Lambda)}(e^{-iH^\Lambda(t_1+t_2)}(e^{-\Omega(\Lambda)}\rho^\Lambda)e^{iH^\Lambda(t_1+t_2)}) = U^\Lambda(t_1+t_2)\rho^\Lambda.$$

Analogously:

$$U^\Lambda(t_2)U^\Lambda(t_1)\rho^\Lambda = U^\Lambda(t_1+t_2)\rho^\Lambda.$$

4. The generator of the operator $U^\Lambda(t)$ is defined on B^Λ consides with $-i(\mathcal{H}^\Lambda + Tr_x\mathcal{A}_x^\Lambda\mathcal{D}_x^\Lambda)$ on $\check{B}^\Lambda$:

The infinitesimal generator of the group $U^\Lambda(t)$ is defined on the set of finite sequences of nuclear operators

$$\rho^\Lambda = \{\rho_0^\Lambda, \rho_1^\Lambda(x_1,;x_1'),...,\rho_s^\Lambda(x_1,...,x_s;x_1',...,x_s'),...\},$$

$$\rho_s^\Lambda(x_1,...,x_s;x_1',...,x_s') = 0, \qquad s > s_0,$$

with the property: the commutator $[H_s^\Lambda, \rho_s^\Lambda]$ belongs to B_s^Λ together with ρ_s^Λ. This set is everywhere dence in B^Λ and belongs to $D(-i(\mathcal{H}^\Lambda + Tr_x\mathcal{A}_x^\Lambda\mathcal{D}_x^\Lambda))$

$$lim_{t\to 0}|\frac{U^\Lambda(t)\rho^\Lambda - \rho^\Lambda}{t}|_1 = lim_{t\to 0}|\frac{1}{t}(\omega^\Lambda(t)\rho^\Lambda - \rho^\Lambda + \Omega(\Lambda)\omega^\Lambda(t)\rho^\Lambda -$$

$$-\omega^\Lambda(t)\Omega(\Lambda)\rho^\Lambda + \sum_{n=2}^{\infty}\sum_{k=1}^{n}\frac{(-1)^k}{k!(n-k)!}(\Omega^{n-k}(\Lambda)\omega^\Lambda(t)\Omega^k(\Lambda)\rho^\Lambda)|_1 =$$

$$= lim_{t\to 0}|(\frac{1}{t}(\omega^\Lambda(t)\rho^\Lambda - \rho^\Lambda + \Omega(\Lambda)(\omega^\Lambda(t)\rho^\Lambda - \rho^\Lambda) - (\omega^\Lambda(t)\rho^\Lambda - \rho^\Lambda)\times$$

$$\times\Omega(\Lambda)\rho^\Lambda + \sum_{n=2}^{\infty}\sum_{k=1}^{n}\frac{(-1)^k}{k!(n-k)!}(\Omega^{n-k}(\Lambda)(\omega^\Lambda(t)\rho^\Lambda - \rho^\Lambda)\Omega^k(\Lambda)\rho^\Lambda)|_1 =$$

339

$$= |-i(\mathscr{H}^{\Lambda} + Tr_x \mathscr{A}_x^{\Lambda} \mathscr{D}_x^{\Lambda})\rho^{\Lambda} + \sum_{n=2}^{\infty} \sum_{k=1}^{n} \frac{(-1)^k}{k!(n-k)!} (\Omega^{n-k}(\Lambda) \mathscr{H}^{\Lambda} \times$$

$$\times \Omega^k(\Lambda)\rho^{\Lambda})|_1 = \sum_{s=1}^{s=s_0} |(-i(\mathscr{H}^{\Lambda} + Tr_{x_{s+1}} \mathscr{A}_x^{\Lambda} \mathscr{D}_x^{\Lambda})\rho^{\Lambda})_s +$$

$$+ \sum_{n=2}^{\infty} (\frac{N}{V}(1 - \frac{s}{N}))^n \sum_{k=1}^{n} \frac{(-1)^k}{k!(n-k)!} \Omega^{n-k}(\Lambda)[H_s^{\Lambda} + H_{n-k}^{\Lambda} +$$

$$+ H_{s,n-k}^{\Lambda}, \Omega^k(\Lambda)\rho_{s+n}^{\Lambda}]|_1 = \sum_{s=1}^{s=s_0} |(-i(\mathscr{H}^{\Lambda} + Tr_{x_{s+1}} \mathscr{A}_x^{\Lambda} \mathscr{D}_x^{\Lambda})\rho^{\Lambda})_s +$$

$$+ \sum_{n=2}^{\infty} (\frac{N}{V}(1 - \frac{s}{N}))^n \sum_{k=1}^{n} \frac{(-1)^k}{k!(n-k)!} \Omega^n(\Lambda)[H_{s,n-k}^{\Lambda}, \rho_{s+n}^{\Lambda}]|_1 =$$

$$= \sum_{s=1}^{s=s_0} |(-i(\mathscr{H}^{\Lambda} + Tr_{x_{s+1}} \mathscr{A}_x^{\Lambda} \mathscr{D}_x^{\Lambda})\rho^{\Lambda})_s + \sum_{n=2}^{\infty} (\frac{N}{V}(1 - \frac{s}{N}))^n \times$$

$$\times \sum_{k=1}^{n} \frac{(-1)^k(n-k)}{k!(n-k)!} \Omega^n(\Lambda)[H_{s,1}^{\Lambda}, \rho_{s+n}^{\Lambda}]|_1 =$$

$$= \sum_{s=1}^{s=s_0} |(-i(\mathscr{H}^{\Lambda} + Tr_{x_{s+1}} \mathscr{A}_x^{\Lambda} \mathscr{D}_x^{\Lambda})\rho^{\Lambda})_s|_1 = |(-i(\mathscr{H}^{\Lambda} + Tr_x \mathscr{A}_x^{\Lambda} \mathscr{D}_x^{\Lambda})\rho^{\Lambda})_s|_1. \quad (12)$$

In (12) we used following identities:

$$\sum_{k=0}^{n} \frac{(-1)^k}{k!(n-k)!} \Omega^{n-k}(\Lambda)[H_s, \Omega^k(\Lambda)\rho_{s+n}^{\Lambda}] =$$

$$= \sum_{k=0}^{n} \frac{(-1)^k}{k!(n-k)!} \Omega^n(\Lambda)[H_s, \rho_{s+n}^{\Lambda}] = 0, \quad \sum_{k=0}^{n} \frac{(-1)^k}{k!(n-k)!} = 0,$$

and

$$\Omega^{n-k}(\Lambda)[H_{s,n-k}^{\Lambda}, \Omega^k(\Lambda)\rho_{s+n}^{\Lambda}] = 0,$$

since the operators H_{n-k}^{Λ} and $\Omega^k(\Lambda)$ under the sign of the trace commute.
Here

$$\sum_{k=0}^{n} \frac{(-1)^k}{k!(n-k)!} \Omega^{n-k}(\Lambda) \left[H_{s,n-k}^{\Lambda}, \Omega^k(\Lambda)\rho_{s+n}^{\Lambda} \right] =$$

$$= \sum_{k=0}^{n} \frac{(-1)^k}{k!(n-k)!} \Omega^n(\Lambda) \left[H_{s,n-k}^{\Lambda}, \rho_{s+n}^{\Lambda} \right]$$

and from identity of particles:

$$\sum_{k=0}^{n} \frac{(-1)^k(n-k)}{k!(n-k)!} \Omega^n(\Lambda) \left[H_{1,n-k}^{\Lambda}, \rho_{s+n} \right] = 0.$$

340

So:

$$\lim_{t \to 0} |(\frac{U^\Lambda(t)\rho^\Lambda - \rho^\Lambda}{t})_s(x_1,...,x_s;x_1',...,x_s') - (i[H_s^\Lambda, \rho_s^\Lambda](x_1,...,x_s;x_1',...,x_s')+$$

$$+\frac{N}{V}(1 - \frac{s}{N})Tr_{x_{s+1}} \sum_{1 \le i \le s} (\phi_{i,s+1}(|x_i - x_{s+1}|) - \phi_{i,s+1}(|x_i' - x_{s+1}|)) \times$$

$$\times \rho_{s+1}^\Lambda(x_1,...,x_s,x_{s+1};x_1',...,x_s',x_{s+1}))|_1 = 0.$$

This implies that the infinitesimal operator of the group $U^\Lambda(t)$ on B_s^Λ concides with the operator

$$-i([H_s^\Lambda,] + \frac{N}{V}(1 - \frac{s}{N})Tr_{x_{s+1}} \sum_{1 \le i \le s} (\phi_{i,s+1}(|x_i - x_{s+1}|) - \phi_{i,s+1}(|x_i' - x_{s+1}|))) \quad (13)$$

on the right-hand side of the BBGKY hierarchy of quantum kinetic equations on $\tilde{B}_s^\Lambda$.

According to [8] and Theorem 2 of Chapter XIX of reference [7], since $U^\Lambda(t)$ is a strongly continuous semigroup on B^Λ with generator (13) on $\tilde{B}_s^\Lambda$ which is dense in B_s^Λ, the abstract Cauchy problem (3),(4) associated with operator (13) has the unique solution

$$\rho_s^\Lambda(t,x_1,...,x_s;x_1',...,x_s')) = (U^\Lambda(t)\rho^\Lambda)_s(x_1,...,x_s;x_1',...,x_s') =$$

$$= (e^{\Omega(\Lambda)}e^{-iH^\Lambda t}e^{-\Omega(\Lambda)}\rho^\Lambda e^{iH^\Lambda t})_s(x_1,...,x_s;x_1',...,x_s') \quad (14)$$

for each $\rho_s^\Lambda(x_1,...,x_s;x_1',...,x_s') \subset \tilde{B}_s^\Lambda$. For the initial data ρ_s^Λ belonging to a certain subset of B_s^Λ (to the domain of definition of $D(-i(\mathcal{H} + Tr_{x_{s+1}}\mathcal{A}_x\mathcal{D}_x)_s)$ of the operator $-i(\mathcal{H} + Tr_{x_{s+1}}\mathcal{A}_x\mathcal{D}_x)_s$, which is everewhere dense in B_s^Λ, (14) is strong solution of Cauchy problem (3),(4).

The proof is finished.

SUMMARY

In this paper we have proved the existance of a unique solution for BBGKY's hierarchy of quantum kinetic equations with coulomb potential.

ACKNOWLEDGEMENTS

Authors gratefully acknowledge Prof. H.Spohn for discussion. Rasulova M.Yu. also thanks Munich Technical University for invitation and hospitality and Deutscher Akademicher Austausch Dienst (DAAD) for the financial support, the Academy of Sciences of Republic of Uzbekistan, for grant FA F1-F067.

341

REFERENCES

1. O.Benedetto, F.Castella, R.Esposito and M.Pulvirenti, *Comm.in Math.Phys.* **277**,1-44 (2008).
2. N.N.Bogolyubov, *Problems of a dynamical theory in statistical physics*, Moscow 1946; in: *Studies in Statistical Mechanics*, **1** (J. de Boer and G. E. Uhlenbeck, editors), North-Holland, Amsterdam, 1962; N.N.Bogolyubov,*Lectures on Quantum Statistics*, London, 1970; *Selected Works*, [in Russian], **2** Naukova Dumka, Kiev, 1970.
3. N.N.Bogolyubov, N.N.(Jn.)Bogolyubov,*Introduction to Quantum Statistical Mechanics*, Nauka, Moscow, 1984.
4. M.Born, H.S.Green,*A General Kinetic Theory of Fluids*, Cambridje University Press, Cambridje, 1949.
5. C.Cercignani,*Theory and application of the Boltzmann equation*, Scottish Academic Press, Edinburg and London, 1975.
6. E.G.D.Cohen, *The generalization of the Boltzmann Equation to Higher densities* in: Statistical Mechanics at the turn of the decade, Cohen,E.G.D., Marcel Dekker, INC., New York, 1971.
7. I.Gohberg, S.Goldberg and M.A.Kaashoek, *Classes of Linear Operators*, **1**, Birkhäuser Verlag, Basel-Boston-Berlin, 1990.
8. T.Kato,*Perturbation theory for linear operators*, Springer-Verlag, Berlin-Heidelberg-New York, 1966.
9. F.King, *BBGKY hierarchy for positive potentials*, in: PhD.Thesis, Dep.of Mathematics, University of California at Berkeley, 1975.
10. J.G.Kirkwood, *J.Chem.Phys.*, **14**,180 (1946); *J.Chem.Phys.*, **15**, 72 (1947).
11. E.Lieb, *The stability of Matter: from atoms to star*, in: Bulletin (New Series)of the American Mathematical Society, **22**, N.1, 1990.
12. O.Lanford III, *The evolution of large classical systems. Dynamical systems, theory and applications* Springer, Berlin, 1975.
13. A.Pazy,*Semigroups of Linear Operators and Applications to Partial Differential Equations*, Springer-Verlag, New York-Berlin, Heidelberg-London-Paris-Tokio-Hong Kong-Barcelona-Budapest, 1983.
14. D.Ya.Petrina, *Theoretical and Mathematical Physics*, **13**, 391-404 (1972).
15. D.Ya.Petrina, A.K.Vidibida, *Trudi MI ANUSSR*, **136**, 370-378 (1975).
16. D.Ya. Petrina,*Mathematical Foundation of Quantum Statistical Mechanics, Continuous Systems*, Kluwer Acadec Publishers, Dordrecht-Boston-London, 1995.
17. M.Yu.Rasulova, Cauchy problem for the Bogolyubov kinetic Equations. Quantum Case. in: Preprint ITP-44R, Kiev (1975); *Docl.Acad.Nauk Uzbek SSR*, **2**, 248-254 (1976).
18. M.Yu.Rasulova, *Theoretical and Mathematical Physics*, **42**, 124-132 (1980).
19. M.Reed, B.Saymon,*Methods of modern mathematical physics*, **2**, Academic Press, New York.-San Francisco-London, 1975.
20. Ya.Sinai, Yu.M.Suhov, *Theoretical and Mathematical Physics*, **19**, 344-363(1974).
21. H.Spohn, *Rev. Mod. Phys.* **52**, 569-615 (1980).
22. H.Spohn,*On the Integrated Form of the BBGKY Hierarchy for Hard Spheres, Large Scale Dynamics of Interacting Particles*, Springer-Verlag, Heydelberg, 1991; math-ph\0605068v1 25 May (2006).
23. Yu.M.Suhov, *Theoretical and Mathematical Physics*, **55** 78-87 (1983).
24. J.Yvon,*La Theorie Statistique des Fluides*, Actualites Scientifiques et Industrielles, Herrman, Paris, 1935.

Semilinear Elliptic Problems, Mountain Pass Index and Break of Symmetry

P.N. Srikanth

Tata Institute of Fundamental Research, Centre for Applicable Mathematics
Post Bag 6503, Chikkabommasandra
Sharadanagar
GKVK Post
Bangalore -560 065, India

Abstract. The paper highlights how the topological information of the Mountain Pass Solutions can be used to obtain interesting break of symmetry results.

DISCUSSIONS AND RESULTS

A typical description of break of symmetry is: A system that is symmetric with respect to some symmetry group goes into a vacuum state that is not symmetric. Such a statement is a typical way a physicist would explain break of symmetry. Let us understand this from a mathematician's point of view.

Suppose we consider a parameter dependent problem:

$$(P_\lambda) \qquad \begin{array}{ll} -\Delta u = f(x,u,\lambda) & \text{in } \Omega \\ u = 0 & \text{on } \partial\Omega \end{array}$$

where $\Omega = \left\{ x \in \mathbb{R}^n : |x| < 1 \right\}$, $f : \Omega \times \mathbb{R} \times \mathbb{R}$ is a smooth function where $f(\mathfrak{X}, .\,, .) =$

$f((\mathfrak{X}), .\,,)$, then a typical question concerning break of symmetry is: Does there exist non-radial solutions to (P_λ). If we try to understand this in the context of what is stated at the beginning then the Break of Symmetry or existence of Non-Radial solution would mean there exists a branch of solutions of (P_λ) say (u_λ^R, λ) where u_λ^R's are radial and at some critical $(u_{\lambda_0}^R, \lambda_0)$ emanates an other branch of solutions (u_λ, λ) where u_λ's are no more radial.

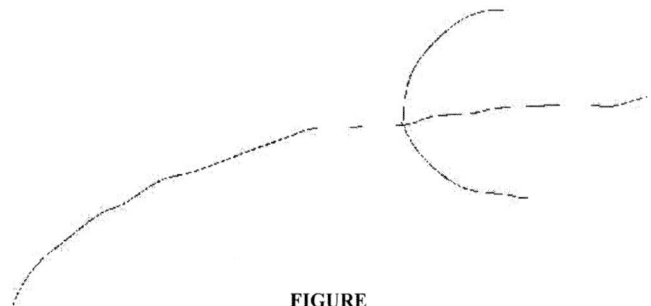

FIGURE

CP1146, *Modelling of Engineering and Technological Problems,* edited by A. H. Siddiqi, A. K. Gupta, and M. Brokate
© 2009 American Institute of Physics 978-0-7354-0683-4/09/$25.00

Though there are many examples, the difficulty in proving them through rigorous mathematical means arise due to proving that certain special degeneracy occurs along the Branch. In view of these difficulties a new approach is suggested which exploits the topological information about Mountain Pass solutions (Mountain Pass Lemma, by Ambrosetti-Rabinowitz [1] is a typical tool for existence of solutions for problems of type P_λ):

Results of N. Ghoussoub and collaborators [2] show that under suitable hypothesis the Mountain Pas Lemma yields a solution which has Morse Index less than or equal to one. It is this information on Morse Index that can be exploited to prove certain break of symmetry results. See [3] for details.

REFERENCES.

1. Ambrosetti and P.H. Rabinowitz, Dual variational methods in critical point theory and applications, J. Functional Analysis,14, 349-381 (1973).
2. G. Fang and N. Ghoussoub, Morse-type information on Palais-Smale sequences obtained byMin-Max principles, Comm. Pure Appl. Math. 47 (1994) 1593-1653.
3. D.G. de Figueredo, P.N. Srikanth and S. Santra, Non-Radially Symmetric solutions for a Superlinear Ambrosetti-Prodi Type problems in a Ball, Comm. in contemporary Mathematics, 7 (2005) 849-866.

AN INTEGRAL EQUATION APPROACH TO ORIENTATIONAL PHASE TRANSITIONS IN QUADRUPOLAR GAY-BERNE FLUID USING DENSITY-FUNCTIONAL THEORY

R. C. Singh

Department of Physics,
Hindustan Institute of Technology,
32, 34 Knowledge Park-3,
Greater Noida – 201306, India
e-mail:rcsingh_physics@yahoo.com

Abstract. The effects of quadrupole moments on the phase behaviour of isotropic-nematic transition are studied by using density functional theory for a system of molecules which interact via the Gay-Berne pair potential. The pair correlation functions of isotropic phase, which enter in the theory as input information, are found from the Percus-Yevick integral equation theory. The method used involves an expansion of angle-dependent functions appearing in the integral equations in terms of spherical harmonics and the harmonic coefficients are obtained by an iterative algorithm. All the terms of harmonic coefficients which involve l indices up to less than or equal to six have been considered. The dependence of the accuracy of the results on the number of terms taken in the basis set is explored for both fluids at different densities, temperatures and quadrupole moments. The results have been compared with the available computer simulation results.

1. Introduction

The ability of highly anisotropic molecules to form liquid crystal phases, which possess intermediate order between the fully disordered isotropic liquid and the fully ordered isotropic liquid and the fully ordered crystalline solid is well established [1-2]. Liquid crystals are characterized by long range orientational order. This order originates from the anisotropic nature of the intermolecular interactions. Detailed information about these complex interactions is not directly accessible from experiments. Such knowledge is, however, essential for the understanding of physical and chemical properties of liquid crystalline systems. Therefore, there is a considerable interest in developing and improving theoretical models for inter- as well as for intra-molecular interactions which can be employed in the analysis of experimental data. A number of computationally efficient pair interaction models have been proposed over the past decades for the study of liquid crystalline mesogens. In particular, fluids composed of elongated (prolate) molecules constitute an important class of liquid crystals with relevant industrial and biological applications. Therefore, different models have been used to explore their behavior such as the hard ellipsoids of revolution, hard spherocylinders [3], cut sphere, the Kihara core model [4] and the Gay-Berne [5]. All these are single site models and refer to rigid molecules of cylindrical symmetry. Even for these simple models calculating the phase diagram is difficult.

In recent years, the most studied potential describing interactions between anisotropic particles has been the Gay-Berne (GB) potential. The GB pair potential, that is well established to serve as a model potential for systems of thermotropic liquid

CP1146, *Modelling of Engineering and Technological Problems*, edited by A. H. Siddiqi, A. K. Gupta, and M. Brokate
© 2009 American Institute of Physics 978-0-7354-0683-4/09/$25.00

crystals, has been employed. In the GB pair potential, ith ellipsoidal molecule is represented by the position r_i of its centre of mass and a unit vector $\hat{\mathbf{e}}_i$ along the symmetry axis of the ith molecule. The interaction between two ellipsoidal molecules i and j is given by

$$U^{GB}\left(\hat{\mathbf{e}}_i,\hat{\mathbf{e}}_j,\hat{\mathbf{r}}_{ij}\right)=4\varepsilon\left(\hat{\mathbf{e}}_i,\hat{\mathbf{e}}_j,\hat{\mathbf{r}}_{ij}\right)\left(R_{ij}^{-12}-R_{ij}^{-6}\right),\tag{1}$$

where

$$R_{ij}=\frac{r_{ij}-\sigma\left(\hat{\mathbf{e}}_i,\hat{\mathbf{e}}_j,\hat{\mathbf{r}}_{ij}\right)+\sigma_0}{\sigma_0}.\tag{2}$$

Here the scaling parameter σ_0 defines the thickness or equivalently, the separation between the two ellipsoids in a side-by-side configuration, r_{ij} is the distance between the centers of mass of the molecules i and j, and $\hat{\mathbf{r}}_{ij}=\mathbf{r}_{ij}/\left|\mathbf{r}_{ij}\right|$ is a unit vector along the intermolecular separation vector $\mathbf{r}_{ij}=\mathbf{r}_i-\mathbf{r}_j$. $\sigma\left(\hat{\mathbf{e}}_i,\hat{\mathbf{e}}_j,\hat{\mathbf{r}}_{ij}\right)$ is the distance (for given molecular orientation) at which the intermolecular potential vanishes and is given by

$$\sigma_{ij}\left(\hat{\mathbf{e}}_i,\hat{\mathbf{e}}_j,\hat{\mathbf{r}}_{ij}\right)=$$

$$\sigma_0\left[1-\chi\left\{\frac{\left(\hat{\mathbf{e}}_i\cdot\hat{\mathbf{r}}_{ij}\right)^2+\left(\hat{\mathbf{e}}_j\cdot\hat{\mathbf{r}}_{ij}\right)^2-2\chi\left(\hat{\mathbf{e}}_i\cdot\hat{\mathbf{r}}_{ij}\right)\left(\hat{\mathbf{e}}_j\cdot\hat{\mathbf{r}}_{ij}\right)\left(\hat{\mathbf{e}}_i\cdot\hat{\mathbf{e}}_j\right)}{1-\chi^2\left(\hat{\mathbf{e}}_i\cdot\hat{\mathbf{e}}_j\right)^2}\right\}\right]^{-\frac{1}{2}}\tag{3}$$

The parameter defining the molecular anisotropy χ is given by

$$\chi=\frac{x_0^2-1}{x_0^2+1},$$

where x_0 is the ratio σ_e/σ_s of the contact distances σ_e and σ_s for the molecules in an end-to-end arrangement and a side-by-side arrangement, respectively. It follows then that for a sphere χ vanishes while for an infinitely long rod it is unity and for infinitely thin disc it is minus one. The orientation dependence of the potential well depth is given by a product of two functions

$$\varepsilon\left(\hat{\mathbf{e}}_i,\hat{\mathbf{e}}_j,\hat{\mathbf{r}}_{ij}\right)=\varepsilon_0\left[\varepsilon_1\left(\hat{\mathbf{e}}_i,\hat{\mathbf{e}}_j\right)\right]^{\nu}\left[\varepsilon_2\left(\hat{\mathbf{e}}_i,\hat{\mathbf{e}}_j,\hat{\mathbf{r}}_{ij}\right)\right]^{\mu},\tag{4}$$

where the exponents υ and μ are the adjustable parameters. The first of these functions

$$\varepsilon_1\left(\hat{\mathbf{e}}_i,\hat{\mathbf{e}}_j\right)=\left[1-\chi^2\left(\hat{\mathbf{e}}_i.\hat{\mathbf{e}}_j\right)^2\right]^{-\frac{1}{2}}, \tag{5}$$

favours the parallel alignment of the particle and so aids liquid crystal formations. The second function has a form analogous to $\sigma\left(\hat{\mathbf{e}}_i,\hat{\mathbf{e}}_j,\hat{\mathbf{r}}_{ij}\right)$, i.e.

$$\varepsilon_2\left(\hat{\mathbf{e}}_i,\hat{\mathbf{e}}_j,\hat{\mathbf{r}}_{ij}\right)=\left[1-\chi'\left\{\frac{\left(\hat{\mathbf{e}}_i.\hat{\mathbf{r}}_{ij}\right)^2+\left(\hat{\mathbf{e}}_j.\hat{\mathbf{r}}_{ij}\right)^2-2\chi'\left(\hat{\mathbf{e}}_i.\hat{\mathbf{r}}_{ij}\right)\left(\hat{\mathbf{e}}_j.\hat{\mathbf{r}}_{ij}\right)\left(\hat{\mathbf{e}}_i.\hat{\mathbf{e}}_j\right)}{1-\chi'^2\left(\hat{\mathbf{e}}_i.\hat{\mathbf{e}}_j\right)^2}\right\}\right], \tag{6}$$

and

$$\chi'=\frac{k'^{\frac{1}{\mu}}-1}{k'^{\frac{1}{\mu}}+1}.$$

Here, k' provides a measure of the anisotropy in the well depth and for rod like molecules is the ratio $\varepsilon_s/\varepsilon_e$, where ε_s and ε_e are the well depths for a pair of ellipsoids aligned in a side-by-side (s) and end-to-end (e) configuration, respectively. Here ε_0 is the depth of the minimum of the pair potential between two ellipsoids aligned in cross configuration $\left(\hat{\mathbf{e}}_i.\hat{\mathbf{e}}_j=\hat{\mathbf{r}}_{ij}.\hat{\mathbf{e}}_i=\hat{\mathbf{r}}_{ij}.\hat{\mathbf{e}}_j=0\right)$.

The GB model contains four parameters (x_0,k,μ,υ) that determine the anisotropy in the repulsive and attractive forces in addition to two parameters (σ_0,ε_0) that scale the distance and energy, respectively. Though x_0 measures the anisotropy of the repulsive core, it also determine the difference in the depth of the attractive well between the side-by-side and the cross configurations. Both x_0 and k' play important role in stabilizing the liquid crystalline phases. The exact role of the other two parameters μ and υ are not very obvious; though they appear to effect the anisotropic attractive forces in a subtle way. Varying these parameters give rise to an infinite number of Gay-Berne potentials. These have shown to give rise to stable nematic and smectic phases. The computer simulation studies show that this potential has been widely studied for a number of parameterizations [6-16] and can be regarded as one of the most important anisotropic potentials in use at present. Some theoretical attempts have also been made to calculate the GB phase diagram using density functional theory (DFT), perturbation method and virial approximations [17-21].

The work presented here concentrates on the study of effect of quadrupole moments to locate the isotropic-nematic (I-N) transition for (reduced) temperatures $T^*\le1.25$ from density functional theory of freezing calculations. This allows us to analyse the dependence of the coexistence properties upon changes in quadrupole moment and temperature. The intermolecular potential studied can be written as the sum of the anisotropic GB potential and the contribution of an embedded quadrupole along the molecular axis of prolate ellipsoid,

$$u(i,j) = U\left(\hat{\mathbf{e}}_i, \hat{\mathbf{e}}_j, \hat{\mathbf{r}}_{ij}\right) = U^{GB}\left(\hat{\mathbf{e}}_i, \hat{\mathbf{e}}_j, \hat{\mathbf{r}}_{ij}\right) + U^{QQ}\left(\hat{\mathbf{e}}_i, \hat{\mathbf{e}}_j, \hat{\mathbf{r}}_{ij}\right). \tag{7}$$

The first term in equation (7) represents the GB potential which is expressed by equation (1) and the second term is the potential due to the electrostatic interactions such as the quadrupolar-quadrupolar interaction, U^{QQ} [22],

$$U^{QQ}\left(\hat{\mathbf{e}}_i, \hat{\mathbf{e}}_j, \hat{\mathbf{r}}_{ij}\right) = \frac{3}{4} \frac{Q^2}{r_{ij}^5} \left[1 - 5\left(c_1^2 + c_2^2 + 3c_1^2 c_2^2\right) + 2\left(s_1 s_2 c' - 4c_1 c_2\right)^2 \right], \tag{8}$$

where $c_i = \cos\theta_i$, $s_i = \sin\theta_i$, $c' = \cos(\varphi_i - \varphi_j)$ and Q is the permanent quadrupole moment.

The paper is organized as follows: in section 2, the Percus-Yevick integral equation theory for the calculation of the pair correlation functions of the isotropic phase has been discussed briefly. In section 3 the density functional formalism has been discussed which is used to locate the freezing parameters for the I-N transition. The paper ends with a discussion given in section 4.

2. Isotropic Phase: Pair-Correlation Functions

The values of pair correlation functions (PCFs) as a function of intermolecular separation and orientations at a given temperature and pressure are found either by computer simulations or by solving the Ornstein-Zernike (OZ) equation

$$\gamma(1,2) = h(1,2) - c(1,2) = \rho_f \int c(1,3)\left[\gamma(2,3) + c(2,3)\right] dx_3, \tag{9}$$

where ρ_f is the number density of the fluids and $i = x_i$ denotes both the location r_i of the center of the ith molecule and its relative orientation Ω_i, described by the Euler angles θ, φ, and ψ with suitable closure relations such as the Percus-Yevick (PY) integral equation, hypernetted chain (HNC) equation, mean spherical approximation (MSA), etc. Approximations are introduced in the theory through these closure relations. In equation (9), $h(1,2) = g(1,2) - 1$ and $c(1,2)$ are, respectively, the total and direct correlation functions (DCFs).

The PY closure relations are written in various equivalent forms. The form adopted here is [23]

$$c(1,2) = f(1,2)\left[1 + \gamma(1,2)\right], \tag{10}$$

where

$$f(1,2) = \exp\left[-\beta u(1,2)\right] - 1,$$

and $\beta = (k_B T)^{-1}$. Here $u(1,2)$ is a pair potential energy of interactions. Since for the isotropic liquid DCF is an invariant pair wise function, it has an expansion in body fixed (BF) frame in terms of basic set of rotational invariants, as

$$c(1,2) = c(\mathbf{r}_{12}, \mathbf{\Omega}_1, \mathbf{\Omega}_2) = \sum_{l_1 l_2 m} c_{l_1 l_2 m}(r_{12}) Y_{l_1 m}(\mathbf{\Omega}_1) Y_{l_2 \underline{m}}(\mathbf{\Omega}_2), \tag{11}$$

Where $\underline{m} = -m$. The coefficients $c_{l_1 l_2 m}(r_{12})$ are defined as

$$c_{l_1 l_2 m}(r_{12}) = \int c(\mathbf{r}_{12}, \mathbf{\Omega}_1, \mathbf{\Omega}_2) Y_{l_1 m}^*(\mathbf{\Omega}_1) Y_{l_2 \underline{m}}^*(\mathbf{\Omega}_2) d\mathbf{\Omega}_1 d\mathbf{\Omega}_2. \tag{12}$$

Expanding all the angle dependent functions in BF frame, the OZ equation reduces to a set of algebraic equation in Fourier space.

$$h_{l_1 l_2 m}(k) = c_{l_1 l_2 m}(k) + (-1)^m \frac{\rho_f}{4\pi} \sum_{l_3} c_{l_1 l_3 m}(k) h_{l_3 l_2 m}(k), \tag{13}$$

where the summation is over allowed values of l_3. The PY closure is expanded in spherical harmonics in the body or space fixed (SF) frame. The expansion coefficients of the DCF in SF frame are related to BF coefficients by

$$C_{l_1 l_2 l}(r_{12}) = \sum_m \left[\frac{4\pi}{2l+1}\right]^{\frac{1}{2}} C_g(l_1 l_2 l; m\underline{m}0) C_{l_1 l_2 m}(r_{12}), \tag{14}$$

where $C_g(l_1 l_2 l; m\underline{m}0)$ are the Clebsch-Gordan coefficients. The PCFs are then found by solving these equations self-consistently [23].

3. Density-Functional Theory of Freezing

The density-functional theory (DFT) has been clearly discussed several times in literature [24-27] and the essentials of the approach are well known. With more and more reliable approximations available, the DFT has been applied to different problems of increasing complexity [28]. The grand thermodynamic potential has the general form

$$-W = \beta A - \beta \mu_c \int d\mathbf{x} \rho(\mathbf{x}), \tag{15}$$

where A is the Helmholtz free energy, μ_c the chemical potential and $\rho(x)$ is a singlet distribution function. It is convenient to subtract the isotropic fluid thermodynamic potential from W and write it as [27]

$$\Delta W = W - W_f = \Delta W_1 + \Delta W_2, \tag{16}$$

with

$$\frac{\Delta W_1}{N} = \frac{1}{\rho_f V} \int d\mathbf{r}\, d\mathbf{\Omega} \left[\rho(\mathbf{r},\mathbf{\Omega}) \ln \left\{ \frac{\rho(\mathbf{r},\mathbf{\Omega})}{\rho_f} \right\} - \Delta\rho(\mathbf{r},\mathbf{\Omega}) \right], \tag{17}$$

and

$$\frac{\Delta W_2}{N} = -\frac{1}{2\rho_f} \int d\mathbf{r}_{12}\, d\mathbf{\Omega}_1\, d\mathbf{\Omega}_2 \, \Delta\rho(\mathbf{r}_1,\mathbf{\Omega}_1) c(\mathbf{r}_{12},\mathbf{\Omega}_1,\mathbf{\Omega}_2) \Delta\rho(\mathbf{r}_2,\mathbf{\Omega}_2). \tag{18}$$

Here $\Delta\rho(\mathbf{r},\mathbf{\Omega}) = \rho(\mathbf{r},\mathbf{\Omega}) - \rho_f$, where ρ_f is the density of the coexisting liquid.

The order parameter equation is obtained by minimizing ΔW with respect to the arbitrary variation in the ordered phase density subject to a constraint that corresponds to some specific feature of the ordered phase. This leads to

$$\ln \frac{\rho(\mathbf{r}_1,\mathbf{\Omega}_1)}{\rho_f} = \lambda_L + \int d\mathbf{r}_2 d\mathbf{\Omega}_2 \, c(r_{12},\mathbf{\Omega}_1,\mathbf{\Omega}_2;\rho_f) \Delta\rho(\mathbf{r}_2,\mathbf{\Omega}_2) \tag{19}$$

where λ_L is Lagrange multiplier which appears in the equation because of constraint imposed on the minimization.

Equation (19) is solved by expanding the singlet distribution $\rho(\mathbf{r},\mathbf{\Omega})$ in terms of the order parameters that characterize the ordered structures using the Fourier series and Wigner rotation matrices. Thus,

$$\rho(\mathbf{r},\mathbf{\Omega}) = \rho_0 \sum_q \sum_{lmn} Q_{lmn}(G_q) \exp(i\mathbf{G}_q.\mathbf{r}) D_{mn}^l(\mathbf{\Omega}), \tag{20}$$

where the expansion coefficients

$$Q_{lmn}(G_q) = \frac{2l+1}{N} \int d\mathbf{r} \int d\mathbf{\Omega}\, \rho(\mathbf{r},\mathbf{\Omega}) \exp(-i\mathbf{G}_q.\mathbf{r}) D_{mn}^{*l}(\mathbf{\Omega}) \tag{21}$$

are the order parameters which measure the nature and strength of the ordering, $\mathbf{G}_q$ the reciprocal lattice vectors, ρ_0 the mean number density of the ordered phase, and $D_{mn}^{*l}(\mathbf{\Omega})$ the generalized spherical harmonics or Wigner rotation matrices . Note that for a uniaxial system consisting of cylindrically symmetric molecules $m = n = 0$ and, therefore, one has

$$\rho(\mathbf{r},\mathbf{\Omega}) = \rho_0 \sum_l \sum_q Q_{lq} \exp(i\mathbf{G}_q.\mathbf{r}) P_l(\cos\theta) \tag{22}$$

and

$$Q_{lq} = \frac{2l+1}{N} \int d\mathbf{r} \int d\Omega \, \rho(\mathbf{r},\Omega) \, \exp(-i\mathbf{G_q}.\mathbf{r}) P_l(\cos\theta) \qquad (23)$$

where $P_l(\cos\theta)$ is the Legendre polynomial of degree l and θ is the angle between the cylindrical axis of a molecule and the director.

In the present calculation, two orientational order parameters have been considered,

$$\overline{P}_l = \frac{Q_{l0}}{2l+1} = <P_l(\cos\theta)>, \qquad (24)$$

with $l = 2$ and 4, one order parameter corresponding to the positional order along the z-axis,

$$\overline{\mu} = Q_{00}(G_z) = \left\langle \cos\left(\frac{2\pi}{d}z\right) \right\rangle \qquad (25)$$

(d, being the layer spacing) and one mixed order parameter that measures the coupling between the positional and orientational ordering and is defined as,

$$\tau = \frac{1}{5} Q_{20}(G_z) = \left\langle \cos\left(\frac{2\pi}{d}z\right) P_l(\cos\theta) \right\rangle. \qquad (26)$$

The angular brackets in the above equations indicate the ensemble average.
The following order parameter equations are obtained by using equations (20)-(23):

$$\overline{P}_l = \frac{1}{2d} \int_0^d dz_1 \int_0^\pi \sin\theta_1 d\theta_1 \, P_l(\cos\theta_1) \exp[sum], \qquad (27)$$

$$\overline{\mu} = \frac{1}{2d} \int_0^d dz_1 \int_0^\pi \sin\theta_1 d\theta_1 \cos\left(\frac{2\pi z_1}{d}\right) \exp[sum], \qquad (28)$$

$$\tau = \frac{1}{2d} \int_0^d dz_1 \int_0^\pi \sin\theta_1 d\theta_1 \, P_2(\cos\theta_1) \cos\left(\frac{2\pi z_1}{d}\right) \exp[sum], \qquad (29)$$

and the change in density at the transition is found from the relation

$$1 + \Delta\rho^* = \frac{1}{2d} \int_0^d dz_1 \int_0^\pi \sin\theta_1 d\theta_1 \exp[sum]. \qquad (30)$$

Here

$$sum = \Delta\rho^* \hat{c}_{00}^0 + 2\bar{\mu}\cos\left(\frac{2\pi z_1}{d}\right)\hat{c}_{00}^1(\theta_1) + \bar{P}_2\,\hat{c}_{20}^0(\theta_1) + \bar{P}_4\,\hat{c}_{40}^0(\theta_1)$$

$$+2\tau\cos\left(\frac{2\pi z_1}{d}\right)\hat{c}_{20}^1(\theta_1) \tag{31}$$

and

$$\hat{c}_{L0}^q(\theta_1) = \left(\frac{2l+1}{4\pi}\right)^{\frac{1}{2}} \rho_f \sum_{l_1 l} i^l (2l_1+1)^{\frac{1}{2}} (2l+1)^{\frac{1}{2}} P_{l_1}(\cos\theta_1) Cg(l_1 Ll;000)$$

$$\times \int_0^\infty c_{l_1 Ll}(r_{12}) j_l(G_q r_{12}) r_{12}^2 \, dr_{12}, \tag{32}$$

where $Cg(l_1 Ll;000)$ are Clebsch-Gordon coefficients and $G_q = 2\pi/d$.

In the isotropic phase all the four order parameters become zero. In the nematic phase the orientational order parameters $\bar{P}_2$ and $\bar{P}_4$ become nonzero but the other two parameters $\bar{\mu}$ and τ remain zero. This is because the nematic phase has no long range positional order. In the smectic a phase all the four order parameters are nonzero showing that the system has both the long range orientational and positional order along one direction. In order to evaluate the transition parameters such as order parameters, change in density etc., Equations (24)-(30) were solved self-consistently using the values of harmonics of DCFs $c_{l_1 l_2 l}(r)$ evaluated at given temperature and density. The calculation was done with inter layer spacing $d = x_0$ for the smectic A phase. These solutions are substituted in equations (16)-(18) to find the grand thermodynamic potential difference between ordered and isotropic phases, i.e.

$$-\frac{\Delta W}{N} = -\Delta\rho^* + \frac{1}{2}\Delta\rho^*(2+\Delta\rho^*)\hat{c}_{00}^0 + \frac{1}{2}(\bar{P}_2^2\,\hat{c}_{22}^0 + \bar{P}_4^2\,\hat{c}_{44}^0) + \bar{\mu}^2\,\hat{c}_{00}^1$$

$$+2\bar{\mu}\tau\hat{c}_{20}^1 + \tau^2\hat{c}_{22}^1 \tag{33}$$

where

$$\hat{c}_{LL'}^q = (2L+1)^{\frac{1}{2}}(2L'+1)^{\frac{1}{2}}\rho_f \sum_l i^l \left(\frac{2l+1}{4\pi}\right)^{\frac{1}{2}} C_g(LL'l;000)$$

$$\times \int_0^\infty c_{LL'l}(r_{12}) j_l(G_q r_{12}) r_{12}^2 \, dr_{12} \tag{34}$$

352

At a given temperature and density a phase with lowest grand potential is taken as the stable phase. Phase coexistence occurs at the values of ρ_f that makes $-\Delta W / N = 0$ for the ordered and the liquid phases.

4. Results and Discussions

The quadrupolar GB potential model considered here is characterized by anisotropy parameters $x_0 = 3.0$, $k' = 5$, $\mu = 2$, $\upsilon = 1$ and the reduced quadrupole moment $Q^* = Q/\left(4\pi\varepsilon_0\sigma_0^5\right)^{1/2}$. The molecular packing fraction and reduced temperature are given respectively as $\eta = (\pi/6)\rho_f\sigma_0^3$ and $T^* = k_B T/\varepsilon_0$. The PCFs needed as input information in the DFT are calculated using the PY integral equation theory for quadrupolar GB fluid for a wide range of reduced temperatures and densities at $Q^{*2} = 0.0$, 1.0 and 1.5. In Table -1, the I-N coexistence parameters of the quadrupolar GB fluid found using the DFT of freezing are compared with the simulation results of Miguel et.al. [7] for GB fluid. It can be seen that the coexistence densities $\left(\rho_f^*, \rho_n^*\right)$ and the change in density at transitions $\Delta\rho^* \left(= \dfrac{\rho_n - \rho_f}{\rho_f}\right)$ are found to increase as T^* is increased. The transition parameters found from DFT are higher than those of the simulation results. One of the possible reasons for this is the inaccuracy in the values of the PCFs at higher temperatures. This is due to the fact that the PY theory underestimates the angular correlations.

The effect of quadrupole moments on the I-N transition of GB fluid can also be seen from Table -1. It has been observed that the coexistence densities decrease as the strength of the quadrupole moments increase. Clearly, as Q^{*2} increases the phase stability decreases towards lower fluid densities and the isotropic to nematic transition takes place at lower densities [29]. The DFT predicts that the order parameter $\overline{P}_2$ and $\overline{P}_4$ decrease as the transition temperature is increased. This nature reveals quite well the behavior of the nematic order as predicted, e.g. by Maier-Saupe theory [30]. As the Q^{*2} increases the values of $\overline{P}_2$ and $\overline{P}_4$ also increase but decrease as temperature is increased. The general feature of $\overline{P}_2$ and $\overline{P}_4$ is in agreement with the experiment. The order parameter approaches unity in a phase that is completely orientationally ordered and approaches zero in the isotropic phase.

The Gay-Berne potential model has been used to study the effect of quadrupole moments and temperature on the isotropic-nematic transition with the original set of parameters $\mu = 2$ and $\upsilon = 1$, and anisotropy parameters $x_0 = 3.0$ and $k' = 5$. The model includes (anisotropic) attractive interactions, and therefore allows for a systematic study of the effect of varying temperature and quadrupole moments on liquid crystal properties. The PCFs of the isotropic fluid are calculated using the PY integral equation theory and have been used in the DFT to locate the I-N transitions. The freezing parameters obtained from DFT have been compared with those of the computer simulations and found that DFT reproduces all the features of the phase diagrams which are in good qualitative agreement. To have a better quantitative agreement there is a need to evaluate the isotropic pair correlations more accurately than those given in the Percus-Yevick theory. Also the use of correlations evaluated directly in the nematic phase may improve the results. The DFT calculations of quadrupolar GB fluids have shown that the introduction of the quadrupolar interaction can have a profound influence on the phase behavior.

5. Acknowledgements

The work was supported by the Department of Science and Technology (India) through a project grant. The support facility provided by the Director and Management of Hindustan Institute of Technology, Greater Noida is acknowledged.

Table

TABLE 1. Isotropic-nematic transition parameters for the quadrupolar GB fluid using the harmonics of the DCFs from the PY theory at $x_0 = 3.0$, $k' = 5$.

T^*	Q^*	Theory	ρ_f^*	ρ_n^*	$\Delta\rho^*$	$\overline{P}_2$	$\overline{P}_4$
1.25	0.0	MC	0.3152	0.3219	0.021	0.55	
		DFT	0.3786	0.3820	0.009	0.683	0.381
	1.0	DFT	0.3549	0.3596	0.013	0.705	0.409
	1.5	DFT	0.3102	0.3199	0.031	0.749	0.457
1.20	0.0	MC	0.3147	0.3213	0.021	0.52	
		DFT	0.3703	0.3741	0.010	0.680	0.377
	1.0	DFT	0.3478	0.3529	0.015	0.706	0.408
	1.5	DFT	0.3028	0.3136	0.036	0.756	0.463
1.15	0.0	MC	0.3129	0.3192	0.020	0.56	
		DFT	0.3617	0.3658	0.011	0.679	0.374
	1.0	DFT	0.3403	0.3459	0.016	0.707	0.409
	1.5	DFT	0.2948	0.3070	0.041	0.764	0.472
1.10	0.0	MC	0.3097	0.3158	0.020	0.55	
		DFT	0.3527	0.3571	0.012	0.679	0.373
	1.0	DFT	0.3324	0.3386	0.019	0.709	0.410
	1.5	DFT	0.2863	0.3004	0.049	0.776	0.483
1.05	0.0	MC	0.3082	0.3144	0.020	0.57	
		DFT	0.3431	0.3481	0.015	0.680	0.373
	1.0	DFT	0.3239	0.3309	0.022	0.713	0.412
	1.5	DFT	0.2771	0.2936	0.059	0.793	0.498
1.00	0.0	MC	0.3076	0.3128	0.017	0.52	
		DFT	0.3329	0.3386	0.017	0.684	0.376
	1.0	DFT	0.3149	0.3228	0.025	0.718	0.417
	1.5	DFT	0.2671	0.2869	0.074	0.815	0.520
0.95	0.0	MC	0.3045	0.3116	0.023	0.54	
		DFT	0.3219	0.3286	0.021	0.690	0.382
	1.0	DFT	0.3052	0.3144	0.030	0.726	0.423
	1.5	DFT	0.2560	0.2812	0.098	0.849	0.553
0.90	0.0	MC	0.3015	0.3069	0.018	0.49	
		DFT	0.3099	0.3179	0.026	0.701	0.392
	1.0	DFT	0.2946	0.3056	0.037	0.737	0.434
	1.5	DFT	0.2431	0.2792	0.148	0.916	0.617
0.85	0.0	MC	0.3013	0.3079	0.022	0.53	
		DFT	0.2969	0.3068	0.033	0.717	0.408
	1.0	DFT	0.2830	0.2964	0.047	0.755	0.450
	1.5	DFT	-	-	-	-	-

0.80	0.0	MC	0.2987	0.3023	0.012	0.50	
		DFT	0.2821	0.2952	0.046	0.743	0.434
	1.0	DFT	0.2699	0.2872	0.064	0.782	0.476
	1.5	DFT	-	-	-	-	-
0.75	0.0	MC	0.2943	0.3009	0.022	0.49	
		DFT	0.2649	0.2836	0.070	0.788	0.479
	1.0	DFT	0.2551	0.2794	0.095	0.829	0.522
	1.5	DFT	-	-	-	-	-

References

1. S. Chandrasekhar, *Liquid Crystals*, 2nd ed. (Cambridge University Press, Cambridge, England, 1992).
2. P. M. Chaikin and T. C. Lubensky, *Principles of Condensed Matter Physics*, (Cambridge University Press, Cambridge, England, 1997).
3. D. Frenkel, 1987, *J. Phys. Chem.*, **91**, 4912; 1988, **92**, 3280.
4. J. G. Gay and B. J. Berne, 1981, *J. Chem. Phys.*, **74**, 3316.
5. T. Kihara, *Intermolecular Forces*, (Wiley, New York, 1976).
6. J. G. Gay and B. J. Berne, 1981, *J. Chem. Phys.*, **74**, 3316.
7. G. R. Luckhurst and P. S. J. Simmonds 1993, *Mol. Phys.*, **80**, 233; M. A. Bates and G. R. Luckhurst 1999, *J. Chem. Phys.*, **110**, 7087.
8. E. de Miguel, L. F. Rull, M. K. Chalam, K. E. Gubbins and F. V. Swol, 1991, *Mol. Phys.*, **72**, 593; E. de Miguel, L. F. Rull, M. K. Chalam, K. E. Gubbins, 1991, *Mol. Phys.*, **74**, 405; E. de Miguel, E. Martin del Rio, J. T. Brown and M. P. Allen, 1996, *J. Chem.*, *Phys.* **105**, 4234; J. T. Brown and M. P. Allen, E. Martin del Rio and E. de Miguel, 1998, *Phys. Rev. E* **57**, 6685; E. de Miguel, 2002, Mol. Phys., **100**, 2449; E. de Miguel and E. Martin del Rio, 2003, *J. Chem. Phys.*, **118**, 1852.
9. M. P. Allen, J. T. Brown and M. A. Warren, 1996, *J. Phys.: Condens. Matter*, **8**, 9433.
10. D. J. Adams, G. R. Luckhurst and R. W. Phippen, 1987, *Mol. Phys.*, **61**, 1575; G. R. Luckhurst, R. A. Stephens and R. W. Phippen, 1990, *Liq. Cryst.*, **8**, 451.
11. G. R. Luckhurst, 1993, *Ber. Bunsenge. Phys. Chem.*, **97**, 1.
12. R. Berardi, A. P. J. Emerson and C. Zannoni, 1993, *J. Chem. Soc. Faraday Trans.*, **89**, 4069.
13. J. W. Emsley, G. R. Luckhurst, W. E. Palke and D. J. Tildesley, 1992, *Mol. Phys.*, **11**, 519.
14. M. A. Bates and G. R. Luckhurst, 1999, *J. Chem. Phys.*, **110**, 7087.
15. E. de Miguel and C. Vega, 2002, *J. Chem. Phys.*, **117**, 6313.
16. E. de Miguel, 2002, *Mol. Phys.*, **100**, 2449.
17. E. de Miguel, E. Martin del Rio, and F. J. Bias, 2004, *J. Chem. Phys.*, **21**, 11183.
18. E. Velasco, A. M. Somoza and L. J. Mederos, 1995, *Chem. Phys.*, **102**, 8107; E. Velasco and L. J. Mederos, 1998, *J. Chem. Phys.*, **109**, 2361.
19. V. V. Ginzburg, M. A. Glaser and N. A. Clark, 1996, *Liq. Cryst.*, **21**, 265; 1997, *Liq. Cryst.*, **23**, 227.
20. R. C. Singh and J. Ram, 2003, *Physica A*, **326**, 13; R. C. Singh, 2006 *Mol. Cryst. Liq. Cryst.*, **457**, 67; R. C. Singh, 2007, *J. Phys.: Condens. Matter*, **19**, 376101.
21. W. L. Wagner and L. Bennett, 1998, *Mol. Phys.*, **94**, 571.
22. P. Mishra, J. Ram and Y. Singh, 2004, *J. Phys.: Condens. Matter*, **16**, 1695; P. Mishra, J. Ram, 2005, *Eur. Phys. J. E*, **17**, 345.
23. C. G. Gray & K.E. Gubbins, *Theory of Molecular Fluids*, Vol. I (Oxford: Clarendon, 1984).
24. J. Ram, R.C. Singh, and Y. Singh, 1994, *Phys. Rev. E*, **49**, 5117; R.C. Singh, J. Ram and Y. Singh, 1996, *Phys. Rev. E*, **54**, 977; 2002, *Phys. Rev. E*, **65**, 031711.
25. R. Evans., 1979, *Adv. Phys.*, **28**, 143.
26. J. S. Rowlinson, and B. Widom, *Molecular Theory of Capillarity*, Chaps. 4 and 7, Oxford University Press: Oxford, 1982.

27. D. W. Oxtoby, 1990, *Nature*, **347**, 725; J. P. Hansen, D. Levesque and J. Zinh-Justin, In: *Liquids, Freezing, and the Glass Transition* (Elsevier; New York, 1991); and references therein.
28. Y. Singh, 1991, *Phys. Rep.* **207**, 351; and reference therein.
29. S. Zhou, 2008, *Phys. Rev. E*, **77**, 041110; 2001, *J. Chem. Phys.* **115**, 2212; 2006, *J. Chem. Phys.* **124**, 144501; S. Zhou and E. Ruckenstein, 2000, *J. Chem. Phys.* 112, 8079.
30. R. C. Singh, Braj Mohan Singh, J. Ram, 2009, *J. Phys.: Condens. Matter* **21**, 115101.
31. G. R. Luckhurst and G. W. Gray, *The Molecular Physics of Liquid Crystals* (New York: Academics,1981)

Equilibrium Points In The Photogravitational Non-Planar Restricted Three Body Problem

Manoj Kumar Singh

Department of Mathematics,

R.B.S. College, Dhanuki - 844 101

Bihar

Abstract. In photogrvitational restricted three body problem there exist equilibrium points which, in addition to the five coplanar such points of the classical problem, lie out of the orbital plane. Radzievskii found two equilibrium points on OXZ plane symmetrical with respect to the orbital plane. We have found equilibrium points which lie out of the orbital plane, when both primaries are radiating.

Keywords: Equilibrium points / Photogravitational / Non-Planar / RTBP

PACS: 0.1.30. Cc

1. INTRODUCTION

In the photogravitational restricted three body problem there exist equilibrium points which, in addition to the five coplanar such points of the classical problem, lie out of the orbital plane.

The existence of these points was first pointed out by Radzievskii [1] who primarily studied the cases of sun-planet-particle. He found two equilibrium points, L_6 and L_7 on the OXZ plane symmetrical with respect to the orbital plane. Chernikov [2] added some notes in the case where the Poynting-Robertson effect is considered.

The stability of these points was first studied in the solar problem by Perezhogin[3], [4] in the whole range of existence when the smaller body is considered non-luminous.

In this paper we find the existence of out of plane equilibrium points. We suppose that both primaries are radiating. These points lie on the OXZ plane and appear in pairs with members symmetrical with respect to the OX-axis.

CP1146, *Modelling of Engineering and Technological Problems*, edited by A. H. Siddiqi, A. K. Gupta, and M. Brokate
© 2009 American Institute of Physics 978-0-7354-0683-4/09/$25.00

2. Out Of Plane Equilibrium Points

In a rotating, barycentric, dimensionless coordinate system with the two radiating bodies on the OX-axis, the equations of motion for the photogravitational restricted three-body problem take the form

$$\ddot{X} - 2\dot{Y} = X - \frac{Q_1(X+\mu)}{r_1^3} - \frac{Q_2(X+\mu-1)}{r_2^3}$$

$$\ddot{Y} + 2\dot{X} = Y\left(1 - \frac{Q_1}{r_1^3} - \frac{Q_2}{r_2^3}\right)$$

$$\ddot{Z} = Z\left(-\frac{Q_1}{r_1^3} - \frac{Q_2}{r_2^3}\right)$$

where

$$Q_1 = q_1(1-\mu), \quad Q_2 = q_2\mu$$

$$r_1^2 = (X+\mu)^2 + Y^2 + Z^2, \quad r_2^2 = (X+\mu-1)^2 + Y^2 + Z^2$$

and $q_1, q_2 \ (q_1 \leq 1, q_2 \leq 1)$ the radiation pressure constants corresponding to the bodies with masses

$1-\mu$ and $\mu \ \left(\mu \leq \dfrac{1}{2}\right)$.

We suppose $\overline{R}_o = (X_0, Y_0, Z_0) \ Z_0 \neq 0$ the position of out of plane equilibrium points. For the existence and position of equilibrium points, the following equations must be satisfied.

$$X_0 - \frac{Q_1(X_0+\mu)}{r_{10}^3} - \frac{Q_2(X_0+\mu-1)}{r_{20}^3} = 0 \tag{2}$$

$$\frac{Q_1}{r_{10}^3} + \frac{Q_2}{r_{20}^3} = 0, \quad Y_0 = 0$$

where

$$r_{10} = r_1\left(\overline{R}_0\right), \quad r_{20} = r_2\left(\overline{R}_0\right)$$

$$r_{10}^2 = (X_0+\mu)^2 + Z_0^2$$

$$r_{20}^2 = (X_0+\mu-1)^2 + Z_0^2$$

From equations (2), we have that

$$X_0 = \frac{Q_1}{r_{10}^3} = -\frac{Q_2}{r_{20}^3} \tag{3}$$

or $\quad X_0^{5/3} - 2(1-2\mu)X_0^{2/3} - \left(Q_1^{2/3} - Q_2^{2/3}\right) = 0 \tag{4}$

These points on the OXZ plane appear in pairs with members symmetrical with respect to the OX-axis under the necessary condition that $Q_1 Q_2 < 0$

We define the auxiliary parameter

$$Q = \left(\frac{Q_1}{Q_2}\right)^{1/3} = \left(\frac{q_1(1-\mu)}{q_2\mu}\right)^{1/3} < 0$$

We suppose Q = -1

Then for $q_2 < -(1-2\mu)/16\mu$ and $\mu \neq \frac{1}{2}$ there exist a pair of equilibrium points named L_6, L_7

with coordinates $(X_0, Y_0, Z_0) = \left(\frac{1}{2} - \mu, 0, \pm \left[\left(\frac{2q_2\mu}{1-2\mu}\right)^{2/3} - \frac{1}{4}\right]^{1/2}\right)$

Thus, we conclude that out of plane equilibrium points are affected by radiation factor of smaller primary.

3.Stability Of Equilibrium Points.

The linearized system of equations about the out of plane equilibrium points take the form

$$F(D)\begin{bmatrix} X \\ Y \\ Z \end{bmatrix} = \overline{O} \tag{6}$$

where

$$F(D) = \begin{bmatrix} D^2 - A_1 & -2D & -A_2 \\ 2D & D^2 - 1 & 0 \\ -A_2 & 0 & D^2 + A_1 - 1 \end{bmatrix} \tag{7}$$

$$A_1 = 1 + 3\left[\frac{Q_1}{r_{10}^5}(X_0 + \mu)^2 + \frac{Q_2}{r_{20}^5}(X_0 + \mu - 1)\right]$$

$$A_2 = 3Z_0 \left[\frac{Q_1}{r_{10}^5} (X_0 + \mu) + \frac{Q_2}{r_{20}^5} (X_0 + \mu - 1) \right]$$

and D stands for differentiation with respect to the time. The characteristics equation of (6) is

$$\lambda^6 + 2\lambda^4 + b\lambda^2 + c = 0 \tag{8}$$

where

$$b = 1 + 9 \frac{X_0 Z_0^2}{r_{10}^2 r_{20}^2} \left[5X_0 - (1 - 2\mu) \right]$$

$$c = \frac{3X_0 Z_0^2}{r_{10}^2 r_{20}^2} \left[5X_0 - (1 - 2\mu) \right] \tag{9}$$

As the roots λ_i of 8 exist in pairs of opposite numbers, it is clear that the differential system (6) has bounded solutions only in the case that $\lambda_j \in I$, $(j = 1, \ldots \ldots 6)$ Putting $\lambda^2 = \rho$, we get the equation.

$$\rho^3 + 2\rho^2 + b\rho + c = 0 \ldots \ldots \tag{10}$$

Stability holds for $\rho_i < 0, i = 1, 2, 3$ which is equivalent to the relations

$$\Delta < 0, \ b > 0, \ c > 0 \tag{11}$$

where

$$\Delta = \left(\frac{3b - 8}{9} \right)^3 + \left(\frac{18b - 27c - 16}{54} \right)^2$$

or equivalently

$$0 < b \le \frac{4}{3} \text{ and } c_m (b) < c < c_M (b) \tag{12}$$

where

$$c_m (b) = \begin{cases} 0, & for \ b \le 1 \\ \dfrac{2}{27} \left[9b - 8 - (4 - 3b)^{3/2} \right], & for \ b > 1 \end{cases} \tag{13}$$

and

$$c_M (b) = \frac{2}{27} \left[9b - 8 + (4 - 3b)^{3/2} \right] \tag{14}$$

For L_6 and L_7 equilibrium points we have $\left(\mu = 0.5\right)$

$$b = 1 + 9 \frac{X_0^2 \, Z_0^2}{r_{10}^2 \, r_{20}^2} > 1 \qquad (15)$$

$$c = 15 \frac{X_0^2 \, Z_0^2}{r_{10}^2 \, r_{20}^2} = b - 1 > 0$$

while from (12) we have

$$b < b^* = \frac{-45 + \sqrt{5145}}{24} \qquad (16)$$

Combining relations (15), (16) and the equations which determine Xo and Zo, we get the condition for stability for $\mu = 0.5$

$$\left[q_1^{2/3} - q_2^{2/3} \right]^{8/5} - 2\left(q_1^{2/3} + q_2^{2/3} \right) + \left[q_1^{2/3} - q_2^{2/3} \right]^{2/5} > 0 \qquad (17)$$

REFERENCES

1. V.V.Radzievskii, 1950, Astron. Zh. 27, 1950, pp, 250.

2. Yu. A. Chernikov, Soviet Astron A.J. 14, No. 1. 1970.

3. A.A. Perezhogin, Astron J. Letters 2, 1976 pp.448.

4. A.A Perezhogin, Academy of Science U.S.S.R. 20, Moscow, 1982.

5. ORagos,. and C. Zagouras, , Earth Moon and Planets, Vol. 41, No. 3 . June 1988,pp. 257-278.

6. J.F.L., Simmons, A.J.C. McDonald, and J.C Brown, Celest. Mech. 35 , 1985, pp.145.

7. C.Zagouras, and M.Kalogeropoulou, Astron and Astrophysics, Supl. 32, 1977, pp.307.

SECTION E:
NUMERICAL METHODS

Algorithm for a New System of Completely Generalized Multi-valued Variational Inclusions

ABUL HASAN SIDDIQI[†] AND SYED SHAKAIB IRFAN [‡]

[†]B.M.A.S. Engineering College
Agra 282007, U.P., India
[‡]College of Engineering, Qassim University
P. O. Box 6677, Buraidah 51452, Al-Qassim
Kingdom of Saudi Arabia
E-mail: siddiqi.abulhasan@gmail.com; shakaib11@rediffmail.com

Abstract

In this work, we introduce and study a new system of completely generalized multi-valued variational inclusions in uniformly smooth Banach spaces. By using the resolvent operator technique associated with m-accretive mappings, we propose an iterative algorithm for computing the approximate solutions of system of completely generalized multi-valued variational inclusions. We prove that approximate solutions obtained by the proposed algorithm converge to the exact solutions of system of completely generalized multi-valued variational inclusions. The results in this paper unify, extend and improve some known results from the literature.

Key words: System of completely generalized multi-valued variational inclusions, resolvent operator, retraction, m-accretive mappings, strongly accretive mappings, uniformly smooth Banach space, iterative algorithm, convergence.
2000 Mathematics Subject Classification: 47H19, 49J40.

1.INTRODUCTION

Variational inclusion problems are among the most interesting and intensively studied classes of mathematical problems and have wide applications in the fields of optimization and control, transport network modelling, economics, operation research, equlibrium and engineering sciences, etc; see for example [1, 2, 3, 6, 7, 10, 13, 14, 15, 16, 19, 21].

In 2001, Verma [22] introduced and investigated a class of system of variational inequalities and develop some iterative algorithms for approximating the solutions

CP1146, *Modelling of Engineering and Technological Problems,* edited by A. H. Siddiqi, A. K. Gupta, and M. Brokate
© 2009 American Institute of Physics 978-0-7354-0683-4/09/$25.00

of systems of variational inequalities. Wang [24], Huang and Fang [12], Fang et al. [9], Agarwal et al. [1], Verma [23] and Ansari and Yao [6] considered a system of variational inequalities. Ansari et al. [5] introduced and studied a system of vector variational inequalities using fixed point theorem. Kim and Kim [17] introduced a new system of generalized nonlinear quasi-variational inequalities and obtained some existence and uniqueness results on solutions for this system of generalized nonlinear quasi-variational inequalities in Hilbert spaces. Cho et al. [8] introduced and studied a new system of nonlinear variational inequalities in Hilbert spaces. They proved some existence and uniqueness theorems for solutions for the system of nonlinear variational inequalities. Very recently Ahmad and Usman [3] introduced a system of generalized variational inclusions with H-accretive operators defined in uniformly smooth Banach spaces.

Inspired and motivated by the above results, we aim in this paper to introduce a new mathematical model, which is called a system of completely generalized multi-valued variational inclusions in uniformly smooth Banach spaces. By using the resolvent operator technique associated with m-accretive mappings, we propose an iterative algorithm for computing the approximate solutions of system of completely generalized multi-valued variational inclusions. We prove that approximate solutions obtained by the proposed algorithm converge to the exact solutions of system of completely generalized multi-valued variational inclusions. The results in this paper unify, extend and improve some known results from the literature.

2. Preliminaries

We suppose that $\mathcal{B}$ is a real Banach space with dual spaces, norm and the generalized dual pair denoted by $\mathcal{B}^*, \|.\|$ and $\langle .,. \rangle$, respectively, $2^{\mathcal{B}}$ is the family of all nonempty subsets of $\mathcal{B}, CB(\mathcal{B})$ is the family of all nonempty closed and bounded subsets of $\mathcal{B}$. $\mathcal{H}(.,.)$ is the Hausdorff metric on $CB(\mathcal{B})$ defined by

$$\mathcal{H}(A, B) = \max \left\{ \sup_{x \in A} d(x, B), \sup_{y \in B} d(A, y) \right\},$$

where $d(x, B) = \inf_{y \in B} d(x, y)$ and $d(A, y) = \inf_{x \in A} d(x, y)$
and $J : \mathcal{B} \to \mathcal{B}^*$ is the *normalized duality mapping* defined by

$$\|Jx\|_* = \|x\| \text{ and } \langle x, Jx \rangle = \|x\|^2, \ \forall \ x \in \mathcal{B}.$$

The *uniform convexity* of the Banach space $\mathcal{B}$ means that for any given $\epsilon > 0$, there exists $\delta > 0$ such that $\forall \ x, y \in \mathcal{B}$, $\|x\| \leq 1$, $\|y\| \leq 1$, $\|x - y\| = \epsilon$ ensure the following inequality

$$\|x + y\| \leq 2(1 - \delta).$$

The function

$$\delta_{\mathcal{B}}(\epsilon) = \inf\left\{1 - \frac{\|x + y\|}{2} : \|x\| = 1,\ \|y\| = 1,\ \|x - y\| = \epsilon\right\}$$

is called the *modulus of the convexity* of the space $\mathcal{B}$.

The *uniform smoothness* of the space $\mathcal{B}$ means that for any given $\epsilon > 0$, there exists $\delta > 0$ such that

$$\frac{1}{2}(\|x + y\| + \|x - y\|) - 1 \le \epsilon\|y\|$$

holds. The *modulus of smoothness* of the space $\mathcal{B}$ is the function $\rho_{\mathcal{B}} : [0, \infty) \to [0, \infty)$ defined by

$$\rho_{\mathcal{B}}(t) = \sup\left\{\frac{1}{2}(\|x + y\| + \|x - y\|) - 1 : \|x\| = 1,\ \|y\| = t\right\}.$$

A Banach space $\mathcal{B}$ is called *uniformly convex* if

$$\delta_{\mathcal{B}}(\epsilon) > 0; \quad \forall\, \epsilon > 0.$$

$\mathcal{B}$ is called *uniformly smooth* if

$$\lim_{t\to 0}\frac{\rho_{\mathcal{B}}(t)}{t} = 0.$$

The following proposition will be used in the proof of our main result.

PROPOSITION 2.1 [4]. Let $\mathcal{B}$ be a uniformly smooth Banach space and J be a normalized duality mapping from $\mathcal{B}$ to $\mathcal{B}^*$. Then $\forall\, x, y \in \mathcal{B}$, we have

(i) $\|x + y\|^2 \le \|x\|^2 + 2\langle y, J(x + y)\rangle$;

(ii) $\langle x - y, J(x) - J(y)\rangle \le 2C^2\rho_{\mathcal{B}}(4\|x - y\|/C)$, where $C = \sqrt{(\|x\|^2 + \|y\|^2)/2}$.

DEFINITION 2.1. A mapping $M : \mathcal{B} \to \mathcal{B}^*$ is said to be

(i) *strongly accretive* if there exists a constant $\gamma > 0$ such that

$$\langle M(x) - M(y), J(x - y)\rangle \ge \gamma\|x - y\|^2, \ \forall\, x, y \in \mathcal{B};$$

(ii) *m-accretive* if M is accretive and $(I + \rho M)(\mathcal{B}) = \mathcal{B}$, $\forall \rho > 0$, where I is the identity mapping;

(iii) *Lipschitz continuous* if there exists a constant $\lambda_M > 0$ such that

$$\|M(x) - M(y)\| \le \lambda_M\|x - y\|, \ \forall\, x, y \in \mathcal{B}.$$

367

DEFINITION 2.2. A multivalued mapping $G : \mathcal{B} \to CB(\mathcal{B})$ is said to be $\mathcal{H}$−Lipschitz continuous if there exists a constant $\lambda_{\mathcal{H}_G} > 0$ such that

$$\mathcal{H}(G(x), G(y)) \le \lambda_{\mathcal{H}_G} \|x - y\|; \ \forall \ x, y \in \mathcal{B}.$$

DEFINITION 2.3. Let $N : \mathcal{B} \to 2^{\mathcal{B}}$ be a m-accretive mapping. For any $\gamma > 0$ the mapping $J_N^\gamma : \mathcal{B} \to \mathcal{B}$ associated with N defined by

$$J_N^\gamma(x) = (I + \gamma N)^{-1}(x); \ \forall \ x \in \mathcal{B}$$

is called the *resolvent operator*.

REMARK 2.1. It is well known that J_N^λ is a single-valued and non-expansive mapping.

DEFINITION 2.4. The resolvent operator $J_N^\lambda : \mathcal{B} \to \mathcal{B}$ is said to be *retraction* if

$$\left\{ J_N^\lambda(x) \right\}^2 = J_N^\lambda(x) \ \forall \ x \in \mathcal{B}.$$

3. A SYSTEM OF COMPLETELY GENERALIZED MULTI-VALUED VARIATIONAL INCLUSIONS AND AN ITERATIVE ALGORITHM

Let $\mathcal{B}$ be a real Banach space. Let $G, F : \mathcal{B} \to CB(\mathcal{B})$ are set-valued mappings and $f, g, p, q : \mathcal{B} \to \mathcal{B}$; $S, T : \mathcal{B} \times \mathcal{B} \to \mathcal{B}$ are single-valued mappings. Let $M, N : \mathcal{B} \times \mathcal{B} \to 2^{\mathcal{B}}$ be multi-valued mappings. We consider the following *system of completely generalized multi-valued variational inclusions:*

$$(3.1) \quad \begin{cases} \text{Find } x, y \in \mathcal{B}, \ u \in G(x), \ v \in F(y) \text{ such that} \\ 0 \in f(x) - g(y) + S(y - p(y), v) + M(f(x), x) \\ 0 \in g(y) - f(x) + T(u, x - q(x)) + N(g(y), y) \end{cases}$$

SPECIAL CASES:

(i) If $f(x) = g(y)$, then problem (3.1) reduces to the following problem:

$$(3.2) \quad \begin{cases} \text{Find } x, y \in \mathcal{B}, \ u \in G(x), \ v \in F(y) \text{ such that} \\ 0 \in S(y - p(y), v) + M(f(x), x) \\ 0 \in T(u, x - q(x)) + N(g(y), y) \end{cases}$$

Problem (3.2) is considered by Ahmad et al. [3] in Banach spaces with H-accretive operators.

(ii) If $f(x) = g(y)$, $p(y) = 0 = q(x)$, $M(f(x), x) = M(f(x))$ and $N(g(y), y) = N(g(y))$, then problem (3.1) reduces to the following problem:

$$(3.3) \quad \begin{cases} \text{Find } x, y \in \mathcal{B}, \ u \in G(x), \ v \in F(y) \text{ such that} \\ 0 \in S(y, v) + M(f(x)) \\ 0 \in T(u, x) + N(g(y)) \end{cases}$$

Problem (3.3) is considered by Lan et al. [18] in Hilbert spaces.

(iii) If $f(x) = g(y)$, $p(y) = 0 = q(x)$, $M(f(x), x) = M(x)$ and $N(g(y), y) = N(y)$ then problem (3.1) reduces to the following problem:

$$(3.4) \quad \begin{cases} \text{Find } x, y \in \mathcal{B}, \ u \in G(x), \ v \in F(y) \text{ such that} \\ 0 \in S(y, v) + M(x) \\ 0 \in T(u, x) + N(y) \end{cases}$$

Problem (3.4) is considered by Huang et al. [11] in Hilbert spaces.

(iv) If $p(x) = 0 = q(x)$, f and g are identity mappings, $S(y, v) = A(y) + B(y)$ and $T(u, x) = C(x) + D(x)$ where A, B, C and D are single-valued mappings, then the problem (3.1) reduces to the following problem:

$$(3.5) \quad \begin{cases} \text{Find } x, y \in \mathcal{B} \text{ such that} \\ 0 \in x - y + (A(y) + B(y)) + M(x) \\ 0 \in y - x + (C(x) + D(x)) + N(y) \end{cases}$$

Problem (3.5) is considered by Agarwal et al. [1] in Hilbert spaces.

LEMMA 3.1. $x, y \in \mathcal{B}$, $u \in G(x)$ and $v \in F(y)$ is the solution of the system of completely generalized multi-valued variational inclusion (3.1) if and only if it satisfies

$$f(x) = J_M^\rho[g(y) - \rho S(y - p(y), v)]$$

$$g(y) = J_N^\gamma[f(x) - \gamma T(u, x - q(x))]$$

where $\rho > 0$ and $\gamma > 0$ are constants.

Based on Lemma 3.1, we construct the following iterative algorithm for solving the system of completely generalized multi-valued variational inclusion (3.1).

ALGORITHM 3.1. For any $x_0, y_0 \in \mathcal{B}$ we choose $u_0 \in G(x_0)$, $v_0 \in F(y_0)$ and compute $\{x_n\}$, $\{y_n\}$, $\{u_n\}$ and $\{v_n\}$ by iterative schemes

$$x_{n+1} = x_n - f(x_n) + J_M^\rho[g(y_n) - \rho S(y_n - p(y_n), v)]; \ \rho > 0$$

$$y_{n+1} = y_n - g(x_n) + J_N^\gamma[f(x_n) - \gamma T(u, x_n - q(x_n))]; \ \gamma > 0$$

where

$$f(x_n) = J_M^\rho[g(y_n) - \rho S(y_n - p(y_n), v]$$

and

$$g(y_n) = J_N^\gamma[f(x_n) - \gamma T(u, x_n - q(x_n)]$$

and choose $u_{n+1} \in G(x_n)$ and $v_{n+1} \in F(y_{n+1})$ such that

$$\|u_n - u_{n+1}\| \leq \mathcal{H}(G(x_n), G(x_{n+1}));$$

$$\|v_n - v_{n+1}\| \leq \mathcal{H}(F(y_n), F(y_{n+1}))$$

where $\rho > 0$ and $\gamma > 0$ are constants and $n = 0, 1, 2, 3, \dots..$

4. Existence of solutions and convergence of an iterative algorithm

THEOREM 4.1. Let $\mathcal{B}$ be a real uniformly smooth Banach space with module of smoothness $\rho_{\mathcal{B}}(t) \le Dt^2$ for some $D > 0$. Let $f, g, p, q : \mathcal{B} \to \mathcal{B}$ be strongly accretive mappings with constants α, β, δ and ψ respectively; and Lipschitz continuous with constants $\lambda_f, \lambda_g, \lambda_p$ and λ_q respectively. Suppose that $S, T : \mathcal{B} \times \mathcal{B} \to \mathcal{B}$ are both Lipschitz continuous mappings in the first argument with constants $\lambda_{S_1}, \lambda_{T_1}$, respectively; and in the second argument with constants $\lambda_{S_2}, \lambda_{T_2}$ respectively. Let $G, F : \mathcal{B} \to CB(\mathcal{B})$ be $\mathcal{H}$-Lipschitz continuous mappings with constants $\lambda_{\mathcal{H}_G}$ and $\lambda_{\mathcal{H}_F}$, respectively. Suppose that $M, N : \mathcal{B} \to 2^{\mathcal{B}}$ be m-accretive mappings such that the resolvent operators associated with M and N are retractions. If there exist constants $\rho > 0$ and $\gamma > 0$ such that

$$0 < \sqrt{1 - 2\alpha + 64D\lambda_f^2} + \frac{\phi_2 \phi_3}{\beta} < 1 \tag{4.1}$$

where

$$\phi_2 = 1 + \sqrt{1 - 2\beta + 64D\lambda_g^2} + 1 + \rho(\lambda_{S_1}\sqrt{1 - 2\delta + 64D\lambda_p^2} + \lambda_{S_2}\lambda_{\mathcal{H}_F})$$

and

$$\phi_3 = \lambda_f + \gamma(\lambda_{T_2}\sqrt{1 - 2\psi + 64D\lambda_q^2} + \lambda_{T_1}\lambda_{\mathcal{H}_G})$$

PROOF. From Algrithm 3.1 and using the non-expansiveness of the resolvent operator, we have

$$
\begin{aligned}
\|x_{n+1} - x_n\| &= \|x_n - f(x_n) + J_M^\rho[g(y_n) - \rho S(y_n - p(y_n), v_n)] - [x_{n-1} - f(x_{n-1}) \\
&\quad + J_M^\rho[g(y_{n-1}) - \rho S(y_{n-1} - p(y_{n-1}), v_{n-1})]]\| \\
&\le \|x_n - x_{n-1} - (f(x_n) - f(x_{n-1}))\| + \|g(y_n) - g(y_{n-1}) \\
&\quad - \rho[S(y_n - p(y_n), v_n) - S(y_{n-1} - p(y_{n-1}), v_{n-1})]\| \\
&\le \|x_n - x_{n-1} - (f(x_n) - f(x_{n-1}))\| + \|y_n - y_{n-1} - (g(y_n) - g(y_{n-1}))\| \\
&\quad + \|y_n - y_{n-1}\| + \rho\|S(y_n - p(y_n), v_n) - S(y_{n-1} - p(y_{n-1}), v_{n-1})\| \quad (4.2)
\end{aligned}
$$

By proposition (2.1), we have

$$\|x_n - x_{n-1} - (f(x_n) - f(x_{n-1}))\|^2$$

$$
\begin{aligned}
&\le \|x_n - x_{n-1}\|^2 + 2\langle -(f(x_n) - f(x_{n-1})), J(x_n - x_{n-1} - (f(x_n) - f(x_{n-1})))\rangle \\
&\le \|x_n - x_{n-1}\|^2 + 2\langle -(f(x_n) - f(x_{n-1})), J(x_n - x_{n-1})\rangle \\
&\quad + 2\langle -(f(x_n) - f(x_{n-1})), J(x_n - x_{n-1} - (f(x_n) - f(x_{n-1}))) - J(x_n - x_{n-1})\rangle \\
&\le \|x_n - x_{n-1}\|^2 - 2\alpha\|x_n - x_{n-1}\|^2 + 4C^2 \rho_{\mathcal{B}}\left(\frac{4\|f(x_n) - f(x_{n-1})\|}{C}\right) \\
&\le \|x_n - x_{n-1}\|^2 - 2\alpha\|x_n - x_{n-1}\|^2 + 64D\|f(x_n) - f(x_{n-1})\|^2 \\
&\le \|x_n - x_{n-1}\|^2 - 2\alpha\|x_n - x_{n-1}\|^2 + 64D\lambda_f^2\|x_n - x_{n-1}\|^2 \\
&\le (1 - 2\alpha + 64D\lambda_f^2)\|x_n - x_{n-1}\|^2. \tag{4.3}
\end{aligned}
$$

370

Similarly, we have

$$\|y_n - y_{n-1} - (g(y_n) - g(y_{n-1}))\|^2 \le (1 - 2\beta + 64D\lambda_g^2)\|y_n - y_{n-1}\|^2 \tag{4.4}$$

Now as S is Lipschitz continuous in the first and second argument with constant λ_{S_1} and λ_{S_2} and by using the same arguments as for (4.3), we have

$$\|S(y_n - p(y_n), v_n) - S(y_{n-1} - p(y_{n-1}), v_{n-1})\|$$

$$\begin{aligned}
&\le \|S(y_n - p(y_n), v_n) - S(y_{n-1} - p(y_{n-1}), v_n) \\
&\quad + S(y_{n-1} - p(y_{n-1}), v_n) - S(y_{n-1} - p(y_{n-1}), v_{n-1})\| \\
&\le \|S(y_n - p(y_n), v_n) - S(y_{n-1} - p(y_{n-1}), v_n)\| \\
&\quad + \|S(y_{n-1} - p(y_{n-1}), v_n) - S(y_{n-1} - p(y_{n-1}), v_{n-1})\| \\
&\le \lambda_{S_1}\|y_n - p(y_n) - (y_{n-1} - p(y_{n-1}))\| + \lambda_{S_2}\|v_n - v_{n-1}\| \\
&\le \lambda_{S_1}\sqrt{1 - 2\delta + 64D\lambda_p^2}\|y_n - y_{n-1}\| + \lambda_{S_2}\mathcal{H}(F(y_n), F(y_{n-1})) \\
&\le (\lambda_{S_1}\sqrt{1 - 2\delta + 64D\lambda_p^2} + \lambda_{S_2}\lambda_{\mathcal{H}_F})\|y_n - y_{n-1}\|.
\end{aligned} \tag{4.5}$$

By using (4.3)-(4.5), we have

$$\begin{aligned}
\|x_{n+1} - x_n\| &\le \sqrt{1 - 2\alpha + 64D\lambda_f^2}\|x_n - x_{n-1}\| + \sqrt{1 - 2\beta + 64D\lambda_g^2}\|y_n - y_{n-1}\| \\
&\quad + \|y_n - y_{n-1}\| + \rho(\lambda_{S_1}\sqrt{1 - 2\delta + 64D\lambda_p^2} + \lambda_{S_2}\lambda_{\mathcal{H}_F})\|y_n - y_{n-1}\|.
\end{aligned} \tag{4.6}$$

Since g is strongly accretive with constant β and J is normalized duality mapping, we have

$$\begin{aligned}
\|g(y_n) - g(y_{n-1})\|\|y_n - y_{n-1}\| &= \|g(y_n) - g(y_{n-1})\|\|J(y_n - y_{n-1})\| \\
&\ge \langle g(y_n) - g(y_{n-1}), J(y_n - y_{n-1})\rangle \\
&\ge \beta\|y_n - y_{n-1}\|^2 \\
\|y_n - y_{n-1}\| &\le \frac{1}{\beta}\|g(y_n) - g(y_{n-1})\| \\
&\le \frac{1}{\beta}\|J_N^\gamma[f(x_n) - \gamma T(u_n, x_n - q(x_n))] \\
&\quad - J_N^\gamma[f(x_{n-1}) - \gamma T(u_{n-1}, x_{n-1} - q(x_{n-1}))]\| \\
&\le \frac{1}{\beta}\{\|f(x_n - f(x_{n-1})\| + \gamma\|T(u_n, x_n - q(x_n)) \\
&\quad - T(u_{n-1}, x_{n-1} - q(x_{n-1}))\|\}.
\end{aligned} \tag{4.7}$$

As f is Lipschitz continuous and T is Lipschitz continuous in both the arguments and H is $\mathcal{H}$-Lipschitz continuous, we have

$$\|f(x_n) - f(x_{n-1})\| \le \lambda_f\|x_n - x_{n-1}\| \tag{4.8}$$

$$\|T(u_n, x_n - q(x_n)) - T(u_{n-1}, x_{n-1} - q(x_{n-1}))\|$$

$$\leq (\lambda_{T_2}\sqrt{1 - 2\psi + 64D\lambda_q^2} + \lambda_{T_1}\lambda_{\mathcal{H}_G})\|x_n - x_{n-1}\|. \tag{4.9}$$

Using (4.8) and (4.9), (4.7) becomes

$$\|y_n - y_{n-1}\| \leq \frac{1}{\beta}\{\lambda_f\|x_n - x_{n-1}\| + \gamma(\lambda_{T_2}\sqrt{1 - 2\psi + 64D\lambda_q^2} + \lambda_{T_1}\lambda_{\mathcal{H}_G})\}\|x_n - x_{n-1}\|$$

$$\leq \frac{1}{\beta}(\lambda_f + \gamma(\lambda_{T_2}\sqrt{1 - 2\psi + 64D\lambda_q^2} + \lambda_{T_1}\lambda_{\mathcal{H}_G})\|x_n - x_{n-1}\|. \tag{4.10}$$

Now using (4.10), (4.6) becomes

$$\|x_{n+1} - x_n\| \leq (\phi_1 + \frac{1}{\beta}\phi_2\phi_3)\|x_n - x_{n-1}\|,$$

where

$$\phi_1 = \sqrt{1 - 2\alpha + 64D\lambda_f^2}$$

$$\phi_2 = \sqrt{1 - 2\beta + 64D\lambda_g^2} + 1 + \rho(\lambda_{S_1}\sqrt{1 - 2\delta + 64D\lambda_p^2} + \lambda_{S_2}\lambda_{\mathcal{H}_F})$$

and

$$\phi_3 = \lambda_f + \gamma(\lambda_{T_2}\sqrt{1 - 2\psi + 64D\lambda_q^2} + \lambda_{T_1}\lambda_{\mathcal{H}_G}).$$

Hence, we have

$$\|x_{n+1} - x_n\| \leq \phi\|x_n - x_{n-1}\|, \tag{4.11}$$

where

$$\phi = \phi_1 + \frac{1}{\beta}\phi_2\phi_3.$$

By (4.1), $0 < \phi < 1$ and so (4.11) implies that $\{x_n\}$ and $\{y_n\}$ are Cauchy sequences. Thus there exists $x, y \in \mathcal{B}$ such that $x_n \to x$ and $y_n \to y$ as $n \to \infty$.

Now we prove that $u_n \to u \in G(x)$ and $v_n \to v \in F(y)$. In fact, it follows from the $\mathcal{H}$-Lipschitz continuity of G, F and by Algorithm 3.1, that

$$\|u_{n+1} - u_n\| \leq \mathcal{H}(G(x_{n+1}), G(x_n)) \leq \lambda_{\mathcal{H}_G}\|x_{n+1} - x_n\|;$$

$$\|v_{n+1} - v_n\| \leq \mathcal{H}(F(y_{n+1}), F(y_n)) \leq \lambda_{\mathcal{H}_F}\|y_{n+1} - y_n\|.$$

Which implies that $\{u_n\}$ and $\{v_n\}$ are also cauchy sequence $\mathcal{B}$. Thus there exists $u, v \in \mathcal{B}$ such that $u_n \to u$ and $v_n \to v$ as $n \to \infty$. Further

$$\begin{aligned} d(u, G(x)) &\leq \|u - u_n\| + d(u_n, G(x)) \\ &\leq \|u - u_n\| + \mathcal{H}(G(x_n), G(x)) \\ &\leq \|u - u_n\| + \lambda_{\mathcal{H}_G}\|x_n - x\| \to 0, \text{ as } n \to \infty, \end{aligned}$$

which implies that $d(u, G(x)) = 0$. Since $G(x)) \in CB(\mathcal{B})$, it follows that $u \in G(x)$. Similarly, we can show that $v \in F(y)$. By continuity of $f, g, p, q, S, T, G, F, J_M^\rho, J_N^\gamma$ and Algorithm 3.1, we have

$$f(x) = J_M^\rho [g(y) - \rho S(y - p(y), v]$$

$$g(y) = J_N^\gamma [f(x) - \gamma T(u, x - q(x)].$$

By Lemma (3.1) (x, y, u, v) is a solution of problem (3.1). This completes the proof. □

References

[1] R. P. Agarwal, N. J. Huang, M. Y. Tan, Sensitivity analysis for a new system of generalized nonlinear mixed quasi-variational inclusions, *Appl. Math. Lett.*, **17(3)**, 345-352, (2004).

[2] R. Ahmad, Q. H. Ansari, S. S. Irfan, Generalized Variational Inclusions and Generalized Resolvent Equations in Banach Spaces, *Comput.Math. Appl.*, **(49)**, 1825-1835, (2005).

[3] R. Ahmad, F. Usman, System of generalized variational inclusions with *H*-accretive operators in uniformly smooth Banach spaces, *J.Comput. Appl. Math.*, doi: 10.1016/j.cam.2008.12.008, (2008).

[4] Y. Albar, J. C. Yao, Algorithm for generalized multi-valued co-variational inequalities in Banach spaces, *Funct. Diff. Equ.*, **7**, 5-13, (2000).

[5] Q. H. Ansari, S. Schaible, J. C. Yao, System of vector equilibrium problems and its applications, *J. Optim. Theory Appl.*, **107**, 547-557, (2000).

[6] Q. H. Ansari, J. C. Yao, A fixed point theorem and its applications to a system of variational inequalities, *Bull. Austral. Math. Soc.*, **59**, 433-442, (1999).

[7] V. A. Barbu, *Nonlinear Semigroups and Differential Equations in Banach Spaces*, Noordhoff, Leydon, (1996).

[8] Y. J. Cho, Y. P. Fang, N. J. Huang, Algorithms for system of nonlinear variational inequalities, *J. Korean Math. Soc.*, **41**, 489-499, (2004).

[9] Y. P. Fang, N. J. Huang, H. B. Thompson, A new system of variational inclusions with (H, η)-monotone operators in Hilbert spaces, *Comput. Math. Appl.*, **49(2-3)**, 365-374, (2005).

[10] P. T. Harker, J. S. Pang, Finite-dimensional variational inequality and nonlinear complementarity problems: A survey of theory, algorithms and applications, *Math. Program.*, **48**, 161-220, (1990).

[11] N. J. Huang, Y. P. Fang, Fixed point theorem and a new system of multivalued generalized order complementarity problems, *Positivity*, **7**, 257-265, (2003).

[12] N. J. Huang, Y. P. Fang, A new class of general variational inclusions involving maximal η-monotone mappings, *Publ. Math. Debrecen,* **62,** 83-98, (2003).

[13] G. Isac *Complementarity problems,* Springer-Verlag, Berlin, (1992).

[14] G. Isac *Topological methods in complementarity theory,* Kluwer Academic Publishers, Dordrecht, Boston, London, (2000).

[15] S. S. Irfan, R. Ahmad Generalized Multivalued Vector Variational-like Inequalities, *J. Global Optimization,* doi: 10.1007/s10898-009-9404-4 (2009).

[16] K. R. Kazmi, M. I. Bhat, Iterative algorithm for a system of nonlinear variational-like inclusions, *Comput. Math. Appl.,* **48,** 1929-1935, (2004).

[17] J. K. Kim, D. S. Kim, A new system of generalized nonlinear mixed variational inequalities in Hilbert spaces, *J. Convex Anal.,* **11(1),** 235-243, (2004).

[18] H. Y. Lan, J. H. Kim, Y. J. Cho, On a new system of nonlinear A-monotone multivalued variational inclusions, *J. Math. Anal. Appl.,* **327,** 481-493, (2007).

[19] P. Marcotte, D. L. Zhu, Weak sharp solutions and the finite convergence of algorihtms for solving variational inequalities, *SIAM J. Optim.,* **9,** 179-189, (1999).

[20] S. B. Nadler, Jr., Multivalued contraction mappings, *Pacific J. Math.* **30,** 475-488, (1969).

[21] A. H. Siddiqi, Q. H. Ansari, M. F. Khan, Variational-like inequalities for multivalued maps, *Indian J. Pure appl. Math.* **30(2),** 161-166, (1999).

[22] R. U. Verma, Projection methods, algorithms and a new system of nonlinear variational inequalities, *Comput. Math. Appl.,* **41,** 1025-1031, (2001).

[23] R. U. Verma, Generalized system for relaxed cocoercive variational inequalities and projection methods, *J. Optim. Theory Appl.,* **121(1),** 203-210, (2004).

[24] Z. Wang, C. Wu, A system of nonlinear variational inclusions with (A, η)-monotone mappings, *J. Inequal. Appl.,* doi: 10.1155/2008/681734, (2008).

Finite Element Approach for the Study of Thermoregulation in Human Head Exposed to Cold Environment

M. A. Khanday [*] and V. P. Saxena [†]

*Department of Mathematics
University of Kashmir, Srinagar-190006, India
khanday@gmail.com
†Sagar Institute of Research Technology and Science,
Ayodhya Bypass Road Bhopal-462041, India
Saxena_vp@hotmail.com

Abstract. The temperature of outer parts of human head exposed to cold environment shows large variations. In this paper a theoretical model has been envisaged for the comprehensive analysis of thermoregulation in human head which is taken as a divided heterogeneous medium surrounded by natural tissue layers. The model incorporates biochemical reactions concerning heat generation, blood circulation and other biophysical activities. The model obtained in terms of partial differential equations has been treated with the help of finite element method. This results in the estimation of temperature distribution under the influence of (i) atmospheric conditions (ii) cerebral blood circulation with fluctuating flow in scalp. This study leads to the estimation of risk factor analysis in cold environment.

Keywords: Human Head, Thermoregulation, Bio-heat Equation and Finite Element Method.
PACS: AIP Classification: 44.05. +e

INTRODUCTION

Heat plays a significant role in the functioning of biological systems. For human subjects the body core temperature is within narrow bounds about 37 ^{0}C. Changes

CP1146, *Modelling of Engineering and Technological Problems*, edited by A. H. Siddiqi, A. K. Gupta, and M. Brokate
© 2009 American Institute of Physics 978-0-7354-0683-4/09/$25.00

in thermoregulation have noteworthy consequences for the activities of individual cells and the body as a whole. Heat transfer within the body takes place by means of conduction and convection. Convective transfer by blood plays a key role. Densities of blood vessels are very high and hence it is hard to compute a detailed temperature distribution in even a small part of the body while accounting for all of the blood vessels individually [2]. Major mechanisms responsible for body temperature regulation in mammals are well known in [11]. In this study, the human head is assumed to be exposed to cold environment and the resulting temperature changes at the different parts of head are estimated.

Blood flow serves various roles in brain functioning. It delivers nutrients, removes waste products and supports temperature regulation. Blood flow acts as a heat exchanger with tubes (blood vessels) penetrating into all brain structures, equilibrating head and body temperature. As a result, the temperature has very little dependence on brain metabolism and is primarily defined by the temperature of incoming arterial blood. A recently proposed theoretical model [10] attributed this phenomenon to the temperature shielding effect of blood flow, which efficiently replenishes tissue heat diffusing down a temperature gradient.

MATHEMATICAL FORMULATION

In 1948, Pennes [6] devised the model determining the temperature distribution in human body tissue by the following standard bio-heat equation:

$$\rho c \frac{\partial T}{\partial t} = \nabla.(K\nabla T) - \rho_b c_b m(T - T_A) + S \qquad \ldots (1)$$

where

K - thermal conductivity of the tissue

ρ - density of the tissue

c - specific heat of the tissue
c_b - specific heat of the blood
m - blood mass flow rate
ρ_b - volumetric blood perfusion of the tissue
T_A - temperature of the arterial blood flow
S - rate of metabolic heat generation

The bio-heat equation (1) is simplified to study heat exchange which takes place in the capillaries and tissues. Blood reaches the capillaries with the temperature of the major supply artery. While during return veins heat transfer with the tissue is negligible. The bio-heat equation has established itself as the most used continuum description of tissue heat transfer. The heat transfer between tissue and blood does take place in vessels of all sizes. Much effort is put in the development of accurate models to take those effects into account and they are recently reviewed by Van Leeuwen et al [11]. Here we will focus on the more comprehensive approach using the bio-heat equation together with the variational finite element method for the solution.

MATERIALS AND METHODS

Consider the head as a domain consisting of central tissue (brain) with three overlaying layers of cerebrospinal fluid (CSF), skull and scalp. The temperature distribution within these regions $T^{(j)}(x)$ ($j = 0,1,2,3$ corresponds to the layers respectively) can be found as a solution of a set of static bio-heat equations. The equation (1) can be re-defined for each of the layer by taking into account the biophysical aspects of the system:

$$\frac{\partial}{\partial x}\left(K_0 \frac{\partial T^{(0)}}{\partial x}\right) - \rho_b c_b m_0 \left(T^{(0)} - T_{a_0}\right) + S_0 = 0 \qquad \ldots (2)$$

$$K_1 \frac{\partial^2 T^{(1)}}{\partial x^2} = 0 = K_2 \frac{\partial^2 T^{(2)}}{\partial x^2} \qquad \ldots (3)$$

$$\frac{\partial}{\partial x}\left(K_3 \frac{\partial T^{(3)}}{\partial x}\right) - \rho_b c_b m_3 \left(T^{(3)} - T_{a_3}\right) + S_3 = 0 \qquad \ldots (4)$$

where $K_j, m_j, T_{aj}, S_j, (j = 0,1,2,3)$ are thermal conductivity, blood mass flow rate, arterial blood temperature and metabolic heat production rate in the corresponding regions respectively; ρ_b and c_b are the density and specific heat of blood. The temperature distribution near the outer surface of head can be treated as a one dimensional problem with unequal element size $d_1, (d_2 - d_1), (d_3 - d_2), (d_4 - d_3)$ corresponding to brain, CSF, skull and scalp respectively. The boundary and interface conditions at the internal interfaces brain/CSF $(x = d_1)$, CSF/skull $(x = d_2)$, skull/scalp $(x = d_3)$ and scalp/air $(x = d_4)$ are given below:

$$T_1(d_1) = T_2(d_1), \quad K_0 \dot{T}_1(d_1) = K_1 \dot{T}_2(d_1) \qquad \ldots (5)$$

$$T_2(d_1 + d_2) = T_3(d_1 + d_2), \; K_1 \dot{T}_2(d_1 + d_2) = K_2 \dot{T}_3(d_1 + d_2) \qquad \ldots (6)$$

$$T_3(d_1 + d_2 + d_3) = T_4(d_1 + d_2 + d_3), \; K_2 \dot{T}_3(d_1 + d_2 + d_3) = K_3 \dot{T}_4(d_1 + d_2 + d_3) \qquad \ldots (7)$$

$$-K_3 \dot{T}_4(d_1 + d_2 + d_3 + d_4) = h\left(T_4(d_1 + d_2 + d_3 + d_4) - T_a\right) + LE \qquad \ldots (8)$$

where $\dot{T} = dT / dx$

These boundary and interface conditions given in equations (5) to (8) reflect the fact that no heat dissipation takes place on the interfaces and hence temperature and heat flux are continuous at the interfaces. The boundary condition between scalp and atmosphere takes into account heat exchange with the environment. Here, T_1, T_2, T_3, T_4 denote the nodal temperatures at different interfaces, T_a is an ambient temperature, h is the heat transfer coefficient which includes heat transfer due to air convection and radiation. The latent heat and evaporation rate term L and E in the last boundary condition lead only to a renormalization of the ambient temperature. The temperature at the deep brain T_0 is taken as constant due to thermal stability in the human body.

FINITE ELEMENT FORMULATION

The variational integral

$$I = \int F(T,T',x)dx \qquad \qquad \dots (9)$$

in optimum form is equivalent to the Euler- Lagrange differential equation [4])

$$\frac{\partial F}{\partial x} - \frac{d}{dx}\left(\frac{\partial F}{\partial T'}\right) = 0; \quad where \quad T' = \frac{\partial T}{\partial x} \qquad \dots (10)$$

On comparing equations (2) to (4) with Euler-Lagrange equation (11), we arrive at the variational integral

$$I_i = \frac{1}{2}\int_{d_i}^{d_{i+1}} [K_j(\partial T^{(i)}/\partial x)^2 + \rho_b c_b m_i(T^{(i)})^2 - 2S_i T^{(i)}]dx$$

$$+ \delta_i \left(\frac{1}{2}h(T - T_a)^2 + LET_4\right) \qquad \dots (11)$$

where

$$i = 0,1,2,3 \quad and \quad \delta_i = \begin{cases} 1 & for \ i = 3 \\ 0 & elsewhere \end{cases}$$

This formulation is equivalent to equation (9) for optimum value of I [9]).

SOLUTION OF THE MODEL

The solution of the model was carried out by approximating the temperature distribution at different nodal points of the regions mentioned above. Due to the minute thickness of the sub-regions and for the simplicity of calculations, the linear shape functions approximate the temperature profiles. The linear variation between successive layers has been refined by using Lagrange's interpolation method. Since the blood mass flow rate and metabolic heat generation depend on density of blood vessels at different depths. It is reasonable to consider the terms m and S negligible in CSF and skull. The layer wise description of the parameters is described in the following equations (12) to (16):

$$T^{(j)} = \frac{\left(d_{j+1}T_j - d_j T_{j+1}\right)}{\left(d_{j+1} - d_j\right)} - \frac{T_{j+1} - T_j}{\left(d_{j+1} - d_j\right)} x \; ; \qquad \ldots (12)$$

$$for \ j = 0,1,2,3 \ and \ d_0 = 0$$

Accordingly, the variational integrals corresponding to each of the regions are obtained as:

$$I_0 = \frac{1}{2} \int_0^{d_1} \left\{ K_0 \left(\frac{T_1 - T_0}{d_1} \right)^2 + \rho_b c_b m_0 \left(T_0 + \frac{(T_1 - T_0)}{d_1} x - T_{a0} \right)^2 \right.$$
$$\left. - 2S_0 \left(T_0 + \frac{(T_1 - T_0)}{d_1} x \right) \right\} dx \qquad \ldots (13)$$

$$I_1 = \frac{1}{2} \int_{d_1}^{d_2} K_1 \left(\frac{T_2 - T_1}{d_2 - d_1} \right)^2 dx \qquad \ldots (14)$$

$$I_2 = \frac{1}{2} \int_{d_2}^{d_3} K_2 \left(\frac{T_3 - T_2}{d_3 - d_2} \right)^2 dx \qquad \ldots (15)$$

$$I_3 = \frac{1}{2} \int_{d_3}^{d_4} \left\{ K_3 \left(\frac{T_4 - T_3}{d_4 - d_3} \right)^2 + \rho_b c_b m_3 \left(\frac{d_4 T_3 - d_3 T_4}{d_4 - d_3} + \frac{(T_4 - T_3)}{d_4 - d_3} x - T_{a3} \right)^2 \right.$$
$$\left. - 2S_3 \left(\frac{d_4 T_3 - d_3 T_4}{d_4 - d_3} + \frac{(T_4 - T_3)}{d_4 - d_3} \right) \right\} dx + \frac{1}{2} \left[h(T_4 - T_a)^2 + 2LET_4 \right] \quad \ldots (16)$$

Evaluating the integrals in equation (13) to (16), using

380

$$I = \sum_{i=0}^{3} I_i \qquad \qquad \cdots (17)$$

and differentiating eqn. (17) successively with respect to nodal points T_1, T_2, T_3, T_4, we arrive at the system of equations

$$A_{11}T_1 + A_{12}T_2 = B_1 \qquad \cdots (18)$$

$$A_{21}T_1 + A_{22}T_2 + A_{23}T_3 = B_2 \qquad \cdots (19)$$

$$A_{32}T_2 + A_{33}T_3 + A_{34}T_4 = B_3 \qquad \cdots (20)$$

$$A_{43}T_3 + A_{44}T_4 = B_4 \qquad \cdots (21)$$

where the coefficients of the equations (18) to (21) are given in the Appendix.

On solving equations (18) to (21) for the nodal temperatures T_1, T_2, T_3, T_4 and using Lagrange's interpolation method. We obtain temperature T between different head regions as an approximate solution of equation (1).

The following numerical values of physiological parameters were used [1] to draw the temperature profiles given in Figures (i), (ii) and (iii).

TABLE 1. Numerical values of parameters

Thickness of layers (cm, d_1-d_4)	0.25, 0.40, 0.55, 0.70
Arterial temperature (^{0}C, T_{a0}, T_{a3})	37, 37
Thermal conductivity (W/cm ^{0}C, K_{0-3})	[4.98, 3.4 3.02, 2.96]x 10^{-3}
Specific heat of blood (J/g ^{0}C, c_b)	3.8
Heat transfer coeff. (W/cm^2 -^{0}C, h)	4 x 10^{-4}
Blood flow rate (lit /kg.sec, m_0, m_3)	[5.24, 11.47]x 10^{-3}
Rate of metabolic heat generation (cal/cm^3-min, S_0, S_3)	2.3, 1.4
Latent heat (cal/g, L)	579 cal/g
Evaporation rate (g/cm^2-min, E)	0.12 x 10^{-3}

DISCUSSION AND CONCLUSION

The standard bioheat equation determining the temperature in the living tissue of a human body has been modeled by using one of the emerging technique know as FEM. In this study, the domain is being discretized into four subdomains depending upon on the physiology and other allied properties of the regions. The variational integrals corresponding to each of the regions were calculated by virtue of which the temperature at different interfaces has been approximated. The linear shape function assumed to be the solution of one dimensional steady state temperature distribution were refined by using Lagrange's interpolation method for the curvilinear temperature profiles of different regions. The analysis predicts change in the temperature as a function of major internal and external parameters namely temperature of arterial blood, blood flow rate, ambient temperature and heat exchange with the environment. The model can be used for predicting head temperature response to extreme cold conditions. It will be helpful for estimating the extent of possible changes in brain temperature during selective head cooling as practice during surgery. The estimation of temperature profiles using finite element technique seems to be more realistic. The study confirms the fact that the brain endures the thermal fluctuations in the cold environment and keeps itself intact to stabilize the tissue temperature up to large scale. Even the low ambient temperature leads to changes only in the vicinity of the head surface. Due to cold exposure the variation in brain and scalp blood flow has been taken into account and the corresponding changes in temperature profiles are given in figures.

Figures

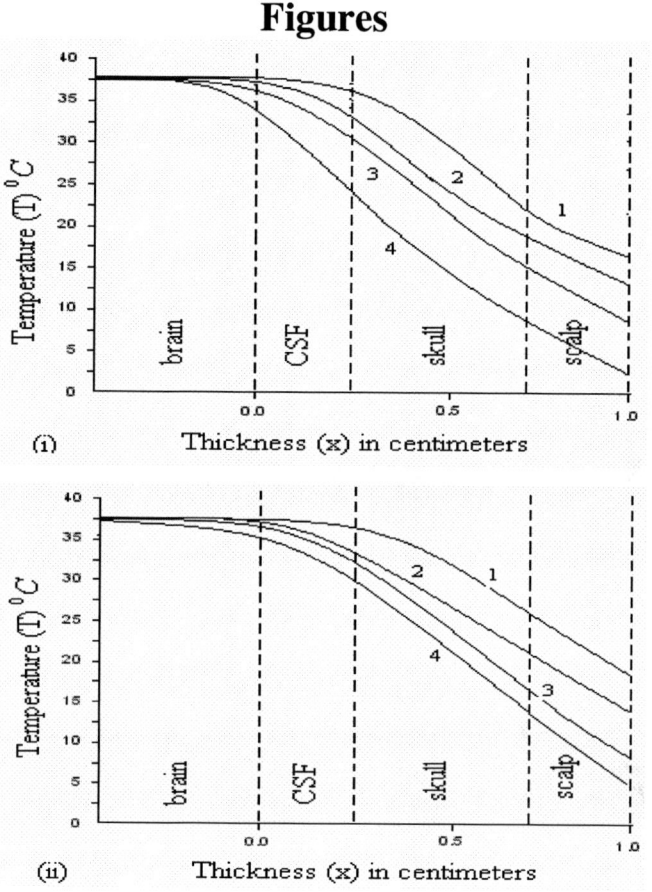

(i)

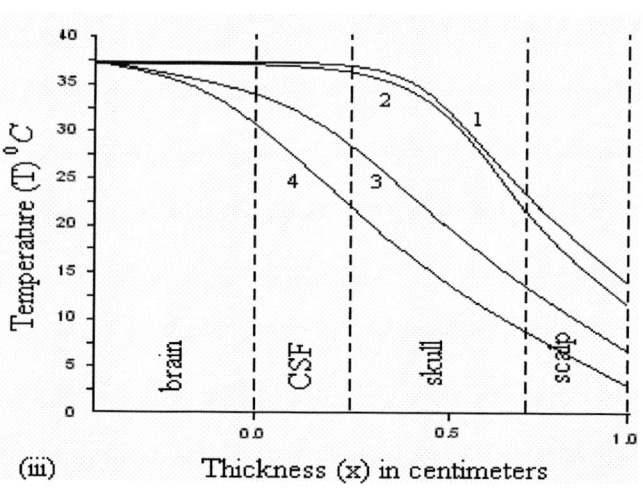

(ii)

(iii)

383

FIGURE (i), (ii) and (iii). Temperature distribution in human head surface (x = 0 on horizontal axis corresponding to the brain surface) at different values of (i) effective ambient temperature Ta, (ii) blood flow in the brain, m_0 (iii) blood flow in the scalp, m_3.

Lines 1-4 in Figure (i) corresponds to $T_a = 5, 0, -5$ and $-10[^0C]$ respectively, the lines in Figure (ii) corresponds to the blood flow in the brain $m_0 = 25.2, 12.4, 8.1$ and 0.8 x $[10^{-4} s^{-1}]$. Also, the lines 1-4 in (iii) correspond to the scalp blood flow $m_3 = 15, 2.5, 0.32$ and $0.14[\times 10^{-4} s^{-1}]$.

REFERENCES

1. K. N. Chao, J. C. Eisely and W. J. Yang, *Heat and Water Migration in Regional Skin and Subcutaneous tissues*, Bio-Medical Sym. ASME, 1973, pp. 69-72.
2. A. C. Guyton, *Text Book of Medical Physiology*, 10th edition, W.B. Saunders, 2000, pp. 904-984.
3. T. Hodgson, *The Development of Instrumentation for Direct Measurement of Heat Loss from a Man in Normal Working Mode*, Ph.D. Thesis south Africa, 1974, CSIR.
4. G. E. Myer, *Analytic Methods in Conduction Heat Transfer*, McGraw Hill Com, 1971, pp. 320-428.
5. D. A. Nelson and S. A. Nunneley, *Brain Temperature and Limits on Transcranial Cooling in Humans: Quantitative Modeling Results.* Eur J Appl Phys **78**, 1998, pp. 353-359.
6. H. H. Pennes, *Analysis of Tissue and Arterial Blood Temperatures in the Resting Human Forearm.* J. Appl. Physiol **1**, 1948, pp. 93-122.
7. T. C. Ruch and H. D. Patton, *Physiology and Biophysics*, W.B. Saunders, **2**, 1976, pp. 325-415.
8. V. P. Saxena, *Temperature Distribution in Human Skin and Subdermal Tissues,* Journal of Theo. Biol. **102**, 1983, pp. 277-286.
9. V. P. Saxena, *Variational Finite element approach to Heat Distribution Problems in Skin and Subdermal Tissues. Numerical Methods in Thermal Problems.* Edited by Lewis and Morgan, Pineridge Press, U.K., 1979, pp. 1067-1076.
10. A. L. Sukstanskii and D. A. Yablonsiy, *An Analytical Model of Temperature Regulation in Human Head*, J. Therm Biol, **29**, 2004, pp. 583-587.

11. H. Swan, *Thermoregulation and Bioenergetics*. New York: Elsevier, 1974.

12. G. M. J. Van Leeuwen, J. W. Hand et al, *Numerical Modeling of Temperature Distributions within the Neonatal Head, Pediatr. Res*, **48**, 2000, pp. 351-356.

APPENDIX

$$K_0/d_1=g_1, \quad K_1/(d_2-d_1)=g_2, \quad K_2/(d_3-d_2)=g_3, \quad K_3/(d_4-d_3)=g_4,$$

$$\rho_b c_b m_0 = \alpha_0, \quad \rho_b c_b m_3 = \alpha_1, \quad d_1 T_{a0}\alpha_0/2=h_1, \quad d_1 S_0/2=h_2,$$

$$d_1\alpha_0/3=h_3, \quad d_4^2\alpha_1 = h_4, \quad d_4(d_3+d_4)/(d_4-d_3)=c_3,$$

$$d_3 d_4\alpha_1 = h_5, \quad (d_3^2+d_4^2+d_3 d_4)\alpha_1/3(d_4-d_3)=b_3,$$

$$(d_3+d_4)^2\alpha_1 T_{a3}/2(d_4-d_3)=c_4, \quad d_4(d_4-d_3)\alpha_1 T_{a3}=h_6,$$

$$S_3 d_4 = h_7, \quad (d_3+d_4)S_3/2=c_5, \quad A_{11}=g_1+h_3+g_2,$$

$$A_{12}=-g_2, \quad A_{22}=g_2+g_3, \quad A_{21}=-g_2,$$

$$B_1=(-g_1+h_1-h_3)T_0+h_2, \quad A_{23}=-g_3, B_2=0,$$

$$A_{32}=-g_3, \quad A_{33}=g_3+g_4+h_4+b_3+c_3, \quad B_3=h_6+h_7-c_5,$$

$$A_{34}=-g_4-h_5-b_3-c_4, \quad A_{43}=-(g_4+h_5+b_3+c_4),$$

$$A_{44}=g_4+h_8+b_3+c_6+h, \quad B_4=c_5-h_9-h_{10}+hT_a-LE$$

$$d_3^2\alpha_1/(d_4-d_3)=h_8, \quad (d_3+d_4)\alpha_1 T_{a3}/(d_4-d_3)=c_6,$$

$$d_3\alpha_1 T_{a3}=h_9, \quad d_3 S_3=h_{10}$$

Numerical Solution of BVPs by OCFE using a Hermite Basis

Vijay Kukreja[*], Ajay Mittal[†], Nabendra Parumasur[**] and Pravin Singh[**]

*Department of Mathematics, SLIET, Longowal, 148106 (Punjab) INDIA
†Department of Mathematics, BGIET, Sangrur, 148101 (Punjab) INDIA
**School of Mathematical Sciences, University of KwaZulu-Natal, Durban, 4000, SOUTH AFRICA

Abstract. The numerical solution of problems occurring in the simulation process of washing of packed bed of porous particles via the method of orthogonal collocation on finite elements (OCFE) is considered. Essentially, OCFE combines the classical orthogonal collocation method (OCM) and finite element method (FEM). Hermite basis is used in place of Lagrange polynomial basis for the computation.

Keywords: Orthogonal collocation on finite elements; OCFE; Hermite basis
PACS: 02.60.Cb

INTRODUCTION

Two-point boundary value problems have been used for decades to describe mass transport processes of solid and semi solid particles in chemical and process industries. For different boundary and initial conditions these models were solved analytically [1]-[6] and numerically by methods, such as finite-difference method [7]-[8], quasi-linearization method [9], orthogonal collocation method [10]-[14], fitted mesh collocation method [15], orthogonal collocation on finite elements [16]-[20] and Galerkin / Petrov Galerkin method [21]-[24]. The analytic solution is complicated therefore one has to take the recourse of approximation techniques. In finite-difference method, however, the solution of this system is very unstable and requires strict selection of step size. Nevertheless, the accuracy of numerical solution is not so high. For large values of the parameters, the nature of the equations becomes stiff; as a result the oscillations increase. Therefore, in such a situation the orthogonal collocation method does not give good results even for large number of collocation points [25]. The discretization of even a few PDEs by the MOL can lead to an extremely large system of ODEs, the numerical solution of which may have severe cost and storage implications [26]. To overcome this situation, the method of orthogonal collocation on finite elements (OCFE) has been used [19]-[20], in which orthogonal collocation method is associated with the finite element method. In OCFE, the domain $0 \leq x \leq 1$, is divided into small sub domains of finite length Δx, called elements. Then the orthogonal collocation is applied within each element. In this process, it is mandatory that the trial function and its first derivative should be continuous at the nodal points or at the boundaries of the elements. The trial function is usually expressed in terms of Lagrangian interpolation polynomials.

In the present paper the technique of OCFE is used with a cubic Hermite basis instead of Lagrangian basis. With the Hermite basis, the coefficients of the basis functions are easily chosen so that the solution and its first derivative are automatically continuous at the boundary of the elements. This results in a significant saving in computational effort as compared to the Lagrange basis case where the continuity equations have to be solved.

MATHEMATICAL MODEL

An extended mathematical model is for the study of washing of pulp fibers (wood, non wood and synthetic) in packed bed is available elsewhere [19]. This model assumes that the packed bed is macroscopically uniform and system is isothermal. Fibers are of uniform cylindrical size and porous in nature. The movement of the solute within the fiber pores is described by Fick's law of diffusion. The flow of non-reactive species through the packed bed of fibers is described by the dispersed plug flow model in terms of Peclet number (Pe) and Biot number (Bi). These numbers depend on axial dispersion coefficient (D_L) and cake thickness (L). The intrapore solute concentration (q) and the concentration of solute adsorbed on the fiber surface (n) are interrelated by Langmuir adsorption isotherm whereas the intrapore solute concentration (q) and the bulk fluid concentration (c) are related by linear relation. The flow of fluid

CP1146, *Modelling of Engineering and Technological Problems*, edited by A. H. Siddiqi, A. K. Gupta, and M. Brokate
© 2009 American Institute of Physics 978-0-7354-0683-4/09/$25.00

through the thin film wall around the fibrous material is controlled by the film resistance mass transfer coefficient (k_f). As the molecular diffusion coefficient is very small, the effect of all mechanisms contributing towards the breakthrough time distribution is lumped together into a single axial dispersion coefficient. The mathematical equations for particle diffusion and bulk fluid in dimensionless form can be written as:

$$\frac{\partial^2 Q}{\partial \eta^2} + \frac{1}{\eta}\frac{\partial Q}{\partial \eta} = \frac{\partial Q}{\partial \tau} + N_1 \frac{(1-\beta)}{\beta}\frac{\partial N}{\partial \tau},$$ (1)

$$\frac{\partial Q}{\partial \eta} = 0 \qquad \text{at } \eta = 0,$$ (2)

$$\frac{\partial Q}{\partial \eta} = -Bi(Q|_{\eta=1} - C) \qquad \text{at } \eta = 1,$$ (3)

$$\frac{\partial N}{\partial \tau} = \frac{R^2 k_1}{D_F}\left(C_1 Q(1-N) - \frac{N}{k}\right),$$ (4)

$$\frac{\partial C}{\partial \tau} = \frac{\psi}{Pe}\frac{\partial^2 C}{\partial \xi^2} - \psi\frac{\partial C}{\partial \xi} - \theta Bi\left(C - Q|_{\eta=1}\right),$$ (5)

$$C - \frac{1}{Pe}\frac{\partial C}{\partial \xi} = 0, \qquad \text{at } \xi = 0,$$ (6)

$$\frac{\partial C}{\partial \xi} = 0, \qquad \text{at } \xi = 1,$$ (7)

$$C = Q = N = 1 \qquad \text{at } \tau = 0.$$ (8)

OCFE USING HERMITE BASIS

In the method of orthogonal collocation on finite elements (OCFE) the radial domain $0 \le \eta \le 1$ is divided into w small subdomains called elements by placing the dividing points at η_l, $l = 1, 2, ..., w+1$, with $\eta_1 = 0$ and $\eta_{w+1} = 1$. Each subdomain is mapped to [0,1] by using the transformation

$$u = \frac{\eta - \eta_l}{\eta_{l+1} - \eta_l} = \frac{\eta - \eta_l}{h_l}, \qquad l = 1, 2, ..., w,$$ (9)

where $h_l = \eta_{l+1} - \eta_l$ is the length of the l_{th} element in the radial direction. Within each element the zeros $u_2 = 1/2(1 - 1/\sqrt{3})$ and $u_3 = 1/2(1 + 1/\sqrt{3})$ of the shifted Legendre polynomial $P_2(x) = (2x-1)^2 - 1/3$ are used as collocation points. The end points of the elements are denoted by $u_1 = 0$ and $u_4 = 1$.

Similarly the axial domain $0 \le \xi \le 1$ is divided into n_e small subdomains by placing the dividing points at ξ_m, $m = 1, 2, ..., n_e + 1$ with $\xi_1 = 0$ and $\xi_{n_e+1} = 1$. Each subdomain is mapped to [0,1] by using the transformation

$$v = \frac{\xi - \xi_m}{\xi_{m+1} - \xi_m} = \frac{\xi - \xi_m}{h_m}, \qquad m = 1, 2, ..., n_e,$$ (10)

where h_m is the length of the m_{th} element in the axial direction. Within each element the zeros $v_2 = 1/2(1 - 1/\sqrt{2})$ and $v_3 = 1/2(1 + 1/\sqrt{2})$ of the shifted Chebyshev polynomial $T_2(x) = 2(2x-1)^2 - 1$ are used as collocation points. The end points of the elements are denoted by $v_1 = 0$ and $v_4 = 1$.

The Hermite interpolant basis functions in 2-D are defined by

$$H_i^j(u,v) = H_i(u)H_j(v), \qquad i, j = 1, 2, 3, 4,$$ (11)

where $H_i(u)$ and $H_j(v)$ are the Hermite basis functions [27] in 1-D at position m along the axial direction and position l along the radial direction respectively.

Let $\hat{Q}_l^m(\eta, \xi, \tau)$ denote the solution $Q(u, v, \tau)$ in this element then $\hat{Q}_l^m(\eta, \xi, \tau)$ can be approximated and interpolated according to

$$\hat{Q}_l^m(\eta, \xi, \tau) \simeq Q_l^m(u, v, \tau) = \sum_{i,j=1}^{4} q_{i+2(l-1)}^{j+2(m-1)}(\tau) H_i^j(u, v).$$ (12)

TABLE 1. Parameters and Values

$N_1 = N_i/C_1$	$6.3361 \times 10^{-2}/C_1$
β	0.88
$B = (1-\beta)/\beta$	0.13636
$Bi = k_f \beta R/KD_F$	7.2861
$P_1 = k_1 R^2/D_f$	1.5×10^{-3}
k^*	0.062208
$C_1 = C_i/C_F$	8.3782/68.760
ψ	0.024475/0.0273
Pe	18.990
θ	2(1-0.968)/0.968

The above choices of coefficients ensure the continuity of the solution and its first derivatives at the boundaries of the elements, that is,

$$Q_l^m(1, v, \tau) = Q_{l+1}^m(0, v, \tau), \qquad \frac{\partial Q_l^m}{\partial \eta}(1, v, \tau) = \frac{\partial Q_{l+1}^m}{\partial \eta}(0, v, \tau),$$

$$Q_l^m(u, 1, \tau) = Q_l^{m+1}(u, 0, \tau), \qquad \frac{\partial Q_l^m}{\partial \xi}(u, 1, \tau) = \frac{\partial Q_l^m}{\partial \xi}(u, 0, \tau).$$

Similarly N and C can be approximated along with their respective boundary conditions.

The collocation points are chosen in the interior of the elements and the residuals are satisfied there. The boundary conditions are satisfied at the boundary points. Evaluating the residual at the collocation points from the partial differential equations (1), (4) and (5), following equation corresponding to (1) is obtained:

$$\sum_{i,j=1}^{4} \frac{dq_{i+2(l-1)}^{j+2(m-1)}(\tau)}{d\tau} H_i^j(u_r, v_s) = \frac{1}{h_l^2} \sum_{i,j=1}^{4} q_{i+2(l-1)}^{j+2(m-1)}(\tau) \frac{\partial^2 H_i^j}{\partial u^2}(u_r, v_s)$$

$$+ \frac{1}{(u_r h_l + \eta_l) h_l} \sum_{i,j=1}^{4} q_{i+2(l-1)}^{j+2(m-1)}(\tau) \frac{\partial H_i^j}{\partial u}(u_r, v_s)$$

$$- N_1 \frac{1-\beta}{\beta} \sum_{i,j=1}^{4} \frac{dn_{i+2(l-1)}^{j+2(m-1)}(\tau)}{d\tau} H_i^j(u_r, v_s).$$

(13)

Here, $l = 1, 2, ..., w$, $m = 1, 2, ..., n_e$ and the indices r and s are chosen appropriately for each residual equation to include all the collocation points appropriate to that equation. The discretized equations for N and C can be obtained from equations (4) and (5). Similarly, the boundary equations (2), (3), (6), (7) and initial conditions (8) are evaluated at the boundary and the initial points respectively.

The above discretization leads to a system of $(8w + 10)(n_e + 1)$ differential algebraic equations (DAEs) and a same number of unknowns. The system of DAEs can be solved using MATLAB software.

To check the applicability of the model for an industrial system, the actual industrial data of the fourth stage of a rotary vacuum washer reported by [19] is used. The corresponding parameters are given in Table 1. Figure 1 illustrates the exit solute concentration profile for $w = 5$, $n_e = 5$.

CONCLUSION

The numerical solution of boundary value problems involving two spatial variables by OCFE is easily and conveniently handled using a Hermite basis. In this case, a DAE system of reduced size is obtained and this is an advantage when compared to the Lagrange basis case [20].

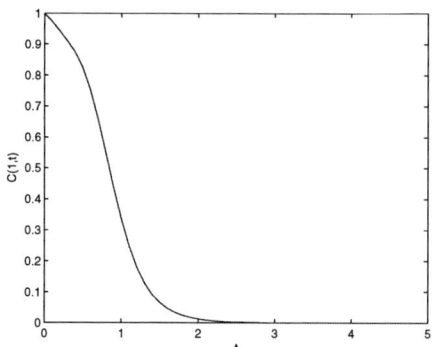

FIGURE 1. Exit solute concentration

REFERENCES

1. H. Brenner, *Chem. Eng. Sci.*, **17**, 229-243 (1962).
2. G.L. Pellett, *Tappi J.*, **49(2)**, 75-82 (1966).
3. Y. Zheng and T. Gu, *Chem. Eng. Sci.*, **51**, 3773-3779 (1996).
4. A.K. Ray and V.K. Kukreja, "Solving Pulp Washing Problems Through Mathematical Models", in *Fundamentals and Numerical Modeling of Unit Operations in the Forest Products Industries*, edited by B.B. Brogdon, American Institute of Chemical Engineers, **96(324)**, 42-47 (2000).
5. H.T. Liao and C.Y. Shiau, *AIChE J.*, **46(6)**, 1168-1176 (2000).
6. A. Rasmuson and I. Neretnieks, *AIChE J.*, **26(4)**, 686-690 (1980).
7. L.T. Fan and R.C. Baille, *Chem. Eng. Sci.*, **13**, 63-68 (1960).
8. L.M. Sun and F. Meunier, *AIChE J.*, **37(2)**, 244-254 (1991).
9. E.S. Lee, *AIChE J.*, **14**, 490-496 (1968).
10. L.T. Fan, G.K.C. Chen and L.E. Erickson, *Chem. Eng. Sci.*, **26**, 379-387 (1971).
11. M.M. Hassen and S.A. Beg, *Chem. Eng. J.*, **36**, B15-B27 (1987).
12. G. Juncu, S. Bildea and O. Floarea, *Chem. Eng. Sci.*, **49(1)**, 123-130 (1994).
13. N.S. Raghvan and D.M. Ruthven, *AIChE J.*, **29(6)**, 922-925 (1983).
14. L.E. Grahs, *Svensk Papperstidning*, **78(12)**, 446-450 (1975).
15. F. Liu and S.K. Bhatia, *Comp. Chem. Eng.*, **23**, 933-943 (1999).
16. G.F. Carey and B.A. Finlayson, *Chem. Eng. Sci.*, **30**, 587-596 (1975).
17. Z. Ma and G. Guiochon, *Comp. Chem. Eng.*, **15(6)**, 415-426 (1991).
18. W.E. Jacobsen and Yi. Liu, *Comp. Chem. Eng.*, **28**, 161-169 (2004).
19. S. Arora, S.S. Dhaliwal and V.K. Kukreja, *Comp. Chem. Eng.*, **30(6-7)**, 1054-1060 (2006).
20. S. Arora, S.S. Dhaliwal and V.K. Kukreja, *App. Math. Comp.*, **183**, 1170-1180 (2006).
21. S.L. Grotch, *AIChE J.*, **15**, 463-465 (1969).
22. S.E. Onah, *App. Math. Comp.*, **127**, 207-213 (2002).
23. M. Al-Jabari, A.R.P. Van Heiningen and T.G.M. Van De Ven, *J. Pulp Paper Sci.*, **20(9)**, J249-J253 (1994).
24. F. Liu and S.K. Bhatia, *Chem. Eng. Sci.*, **56**, 3727-3735 (2001).
25. F. Shirashi, *Chem. Eng. J.*, **83**, 175-183 (2001).
26. W.E. Schiesser, *The Numerical Method of Lines*, Academic Press, San Diego, New York (1991).
27. B.A. Finlayson, *Nonlinear Analysis in Chemical Engineering*, McGraw-Hill Inc., USA (1980).

Analysis of Parametric Effects on Efficiency of the Brown Stock Washer in Paper Industry Using MATLAB

Deepak Kumar, Vivek Kumar and V. P. Singh

Department of Paper Technology, Indian Institute of Technology Roorkee, Saharanpur Campus, Saharanpur -247001, India
E-mail: dkr2009@gmail.com, vivekfpt@iitr.ernet.in, singhvp3@gmail.com

Abstract. In the present paper, the effects of cake thickness and time on the efficiency of brown stock washer of the paper mill are studied by using mathematical model of pulp washing for the species of sodium and lignin ions. The mechanism of the diffusion- dispersion washing of the bed of the pulp fibers is mathematically modeled by the basic material balance and adsorption isotherm is used to describe the equilibrium between the concentration of the solute in the liquor and concentration of the solute on the fibers. To study the parametric effect, numerical solutions of the axial domain of the system governed by partial differential equations (transport and isotherm equations) for different boundary conditions are obtained by the "pdepe" solver in MATLAB source code. The effects of both the parameters are shown by three dimensional graphical representation as well as concentration profiles.

Keywords: MATLAB "pdepe" solver; Peclet number; Adsorption isotherm; diffusion-dispersion, Parameters, Brown stock washer.

1. INTRODUCTION

Pulp washing is a key unit operation in pulp and paper mills which significantly affect the economy as well as the environment. The objective of the pulp washing is to remove the soluble impurities such as organic solids (lignin), fines, inorganic solids (Na, Mg, Ca and K ions) and bleach byproducts after each processing step. Removable of soluble impurities reduces the consumption of chemicals in subsequence sections, which in turn produce less pollutant. Ideally pulp should be washed using less amount of wash water and with maximum removal of black liquor solids. The mathematical models describing the pulp washing process are fairly well established in terms of transport equations by some earlier researchers as Lapidus and Amundson (1952), Kuo (1960), Brenner (1962), Pellett (1966), Grahs (1975), Perron & Lebeau (1977) and Sherman (1964). These transport equations together with the corresponding equations of isotherms and various boundary conditions provide the mathematical model of pulp washing, which are extremely intricate in nature and solution of these models is highly complex. It is important to mention that the problem with its simplified version has been solved analytically by Brenner (1962), Kukreja (1996) using Laplace transform and numerically using Orthogonal Collocation by Grahs

CP1146, *Modelling of Engineering and Technological Problems*, edited by A. H. Siddiqi, A. K. Gupta, and M. Brokate
© 2009 American Institute of Physics 978-0-7354-0683-4/09/$25.00

(1975) & Arora (2006). Kumar (2002) attempted to solve the washing model using Finite difference method. All these methods are very complex and time consuming. Application of such solution techniques in control systems is very difficult due to the more processing time and involvement of high mathematical skills at operator level. Recently system of such equations has been successfully solved with easy approach, "pdepe" solver in MATLAB source code by Singh et. al. (2008) which is more convenient and consumes less time than the other techniques mentioned above.

In the present paper the effect of two main parameters cake thickness and time on exit solute concentration for both species sodium (Na^+) and lignin ions in brown stock washing operation of the paper mill is investigated. The mathematical model combines the transport equation based on diffusion–dispersion during flow through multiparticle system given by Sherman (1964) and Pellett (1966) and adsorption isotherm to describe the equilibrium between the concentration of the solute in the liquor and fibers given by Lapidus and Amundson (1952). The "pdepe" solver is applied on the parallel lines as adopted by Singh et al. (2008) on axial domain of the system governed by partial differential equations.

2. Description of Mathematical Model

	Nomenclature
c	: Concentration of the solute in the liquor, kg/m3
c_i	: Concentration of solute inside the vat, kg/m3
c_s	: Concentration of solute in the wash liquor, kg/m3
D_L	: Longitudinal dispersion coefficient, m2/s
D_v	: Molecular diffusion coefficient, m2/s
k_1, k_2	: Mass transfer coefficients, 1/s
L	: Cake thickness, m
N	: Concentration of solute on fibers, kg/m3
T	: Time, s
C	: Dimensionless Concentration of solute in the liquor
N	: Dimensionless Concentration of solute in the fiber
Z	: Dimensionless Distance
T	: Dimensionless Time
U	: Liquor speed in cake pores, m/s
z	: Variable cake thickness, m
Δz	: Small increment in cake thickness, m
ε	: Porosity of the bed

Using simple material balance for setting up a differential equation, consider a thin slice of a filter cake (pulp mat) as shown in FIGURE 1, through which filtrate or wash water flows.

Material balance across the simple shell given in FIGURE 1, in the z direction can be written as:

$$(uc\varepsilon_t A')_{z,t} - (uc\varepsilon_t A')_{z+\Delta z,t} = [\frac{\partial}{\partial t}\{(c\,\varepsilon_t A'\Delta z + n(1-\varepsilon_t)A'\Delta z\}]_{\overline{z},t} \qquad (1)$$

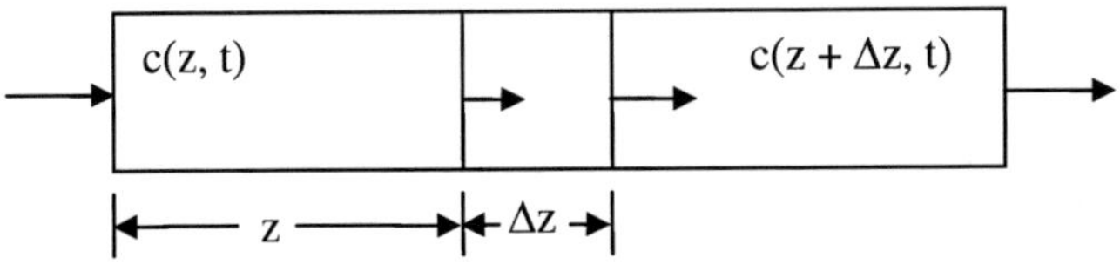

FIGURE 1. A Simple Shell balance

where $z \prec \overline{z} \prec z + \Delta z$. By taking ε_t and A' as constant and taking the limit as $\Delta z \rightarrow 0$, one can obtain the following expression,

$$-\varepsilon_t c\left(\frac{\partial u}{\partial z}\right) = \varepsilon_t u\left(\frac{\partial c}{\partial z}\right) + \varepsilon_t\left(\frac{\partial c}{\partial t}\right) + (1-\varepsilon_t)\left(\frac{\partial n}{\partial t}\right) \qquad (2)$$

2.1 Transport Equation

Using simple material balance for the mat of pulp fibers, assumed to be stationary packed bed of homogeneous symmetrical cylindrical fibers with constant area A_c and the total porosity ε_t, the mathematical equation is obtained given by Eq. 2. After that Using Fick's second law of diffusion and neglecting the molecular diffusion coefficient D_v [According as Sherman (1964)], the transport equation can be obtained as:

$$D_L\left(\frac{\partial^2 c}{\partial z^2}\right) = u\left(\frac{\partial c}{\partial z}\right) + \left(\frac{\partial c}{\partial t}\right) + \frac{(1-\varepsilon_t)}{\varepsilon_t}\left(\frac{\partial n}{\partial t}\right) \qquad (3)$$

Where u is the velocity of the liquor in the mat, c is the concentration in the liquid phase and n is the concentration in fiber.

Writing $(1-\varepsilon_t)/\varepsilon_t$ as μ for convenience, the Eq. 3 may be written as

$$D_L\left(\frac{\partial^2 c}{\partial z^2}\right) = u\left(\frac{\partial c}{\partial z}\right) + \left(\frac{\partial c}{\partial t}\right) + \mu\left(\frac{\partial n}{\partial t}\right) \qquad (4)$$

This is a non homogeneous, non linear, first degree, second order, parabolic partial differential equation. Here u, ε_t and D_L are functions of z while c and n are functions of z and t.

2.2. Adsorption Isotherm

For the present study, the adsorption isotherm has been proposed to describe the equilibrium between the concentration of the solute in the liquor and fiber. Unfortunately the adsorption desorption isotherm dynamics not clearly known for Na^+ and lignin ions. Various adsorption-desorption equilibrium equations are available in the literature. In the present study, adsorption isotherm given by Lapidus and Amundson (1952) is used by taking the rate of adsorption is finite and initial adsorbate concentration to be zero. For ready reference the equations are given as

$$\frac{\partial n}{\partial t} = k_1 c - k_2 n \qquad (5)$$

2.3. Initial and Boundary conditions

According to Brenner (1952), initial condition is $c\,(z,\,t) = n\,(z,\,t) = c_i$ for $0 < t <$ L/u, where L/u corresponds to displacement time and boundary condition at the inlet of the bed is

$$uc - D_L \frac{\partial c}{\partial z} = uc_s, \text{ at } z = 0 \text{ and } t > 0 \qquad (6)$$

Perron and Lebeau (1977) give the boundary condition at the inlet of the bed
$c = c_s$, at $z = 0$ and $t > 0$ (7)
Boundary condition at bed exit is same for both the above cases of inlet boundary conditions i.e. $(\partial c / \partial z) = 0$, at $z = L$ and $t > 0$ (8)
The conditions given in Eq. 6, Eq. 7 & Eq. 8 have been used in the present investigation.

2.4. Conversion into dimensionless form

Before obtaining the solution, pulp washing model with adsorption isotherm is converted into dimensionless form by using certain dimensionless parameters like Peclet number (or Bodenstein number), dimensionless time, dimensionless cake thickness and dimensionless concentration given below:
Peclet Number, $Pe = uL/D_L$
Dimensionless cake thickness, $Z = z/L$
Dimensionless concentrations of solute in liquor, $C = (c - c_s)/(c_i - c_s)$,
Dimensionless time, $T = ut/L$
Dimensionless concentrations of solute in fiber, $N = (n - c_s)/(c_i - c_s)$.
And some constant parameters are as follows, $K = k_1/k_2$, $G = k_2 L/u$, and $H = (K-1)\,c_s/(c_i - c_s)$

2.5. Existing mathematical models for washing zone used in present investigation (Dimensionless Form)

Transport equation in dimensionless form is obtained and given by

$$\left(\frac{\partial^2 C}{\partial Z^2}\right) = Pe\left(\frac{\partial C}{\partial Z} + \frac{\partial C}{\partial T} + \mu\frac{\partial N}{\partial T}\right) \qquad (9)$$

Similarly adsorption isotherm in dimensionless form is reduces as

$$\frac{\partial N}{\partial T} = G(H + KC - N) \qquad (10)$$

Various boundary conditions as given by Eq. 6, Eq. 7 & Eq. 8 in dimensionless form are given below.

$$\frac{\partial C}{\partial Z} = PeC \quad \text{for } (Z = 0, T > 0)$$

(11)

$$C = 0 \quad \text{for } (Z = 0, T > 0)$$

(12)

$$\frac{\partial C}{\partial Z} = 0 \quad \text{at } (Z=1)$$

(13)

The mathematical model of pulp washing given by transport Eq. 9, combined with the adsorption isotherm given by Eq.10 is to be solved subject to two different inlet boundary conditions respectively. Thus two cases arise for both inlet boundary conditions given by Eq.11 & Eq. 12 together with the bed exit boundary condition given by Eq. 13. For convenience both cases are representing by two different sets of models i.e. Model 1 for boundary conditions given by Eq.11 & Eq. 13 and Model 2 for Eq.12 & Eq. 13 respectively.

3. Results and Discussion

For control purpose the transient behavior of the solute concentration in the black liquor is of more interest rather than solute concentration in fiber and so the value of $\partial N/\partial T$ from adsorption isotherm substituted in the transport equation and then the solution is obtained. The effects of both the parameters are shown by 3-D graphical representations by the FIGURES 1(a), 2(a), 3(a) & 4(a). To study the effect of dimensionless cake thickness, the behavior of solute concentration is checked by C vs T profiles at various values of dimensionless cake thickness for Z=0.0 to Z=1.0 and given in the FIGURES 1(c), 2(c), 3(c) & 4(c). The behavior is also checked at dimensionless time T=1.0 by C vs Z profiles depict by the FIGURES 1(b), 2(b), 3(b) & 4(b) for both sodium (Na^+) and lignin ions respectively.

3.1 Effect of dimensionless cake thickness on dimensionless solute concentration for Na^+

In this investigation the entire range of dimensionless cake thickness is divided into 21 equal parts and then the influence of Z on C is estimated. FIGURE 1(a) depicts the 3-D graph of the behavior of the dimensionless concentration with respect to both dimensionless cake thickness Z as well as dimensionless time T for Model 1. 3-D behavior of the solute concentration shows the excellent agreement with the results of earlier investigators Grahs (1974) and Kumar (2002). FIGURE 1(b) shows the C vs Z profile for Model 1 at T=1.0. It is clear from the FIGURE 1(b) that dimensionless solute concentration increases with the increase in the bed depth Z (dimensionless cake thickness). It is observed that with change in the cake thickness, significant changes occur in the concentration profiles. These results give excellent agreement with the results of Gren and Strom (1985), Trinh et al. (1989) & Kumar (2002). In similar manner the effects for Model 2 are shown by concentration profiles in the FIGURES 2(a), 2(b) for sodium ion (Na^+).

3.2 Effect of dimensionless time on dimensionless solute concentration for Na^+

FIGURES 1(c) & 2(c) show C vs T profiles for Na^+ at various bed depths for both the models respectively. The various values of bed depths are obtained by dividing the range 0.0 to 1.0 of dimensionless cake thickness into 21 equal parts. For both the models FIGURES 1(c) & 2(c) respectively show the variabilities in dimensionless solute concentration for sodium species at the top four layers become insignificant whereas for the bottommost layer C shows a typical breakthrough curve. Lowering the bed depth (from top to bottom) the shape of the breakthrough curves change to a non linear curve which is practically asymptotic in nature. It is further evident that the deeper bed gives the higher washing efficiency.

It is also important to noted that for Model 2, the curves at Z=0.0 shows a different characteristic at time T=0.0 to T=0.05 this is due to using different boundary condition in the Model 2.

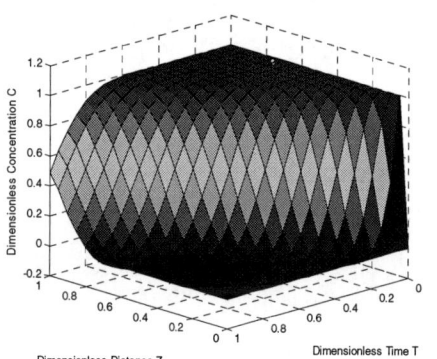

FIGURE 1(a). 3-D Behavior of C vs Z &T for Na^+ for Model 1

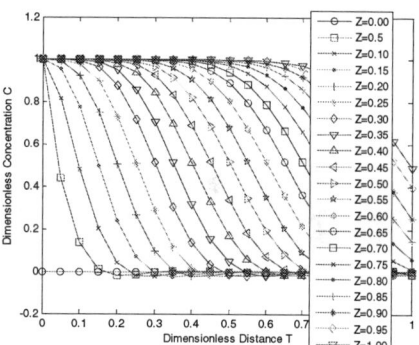

FIGURE 1(c). C vs T for Na^+ for Z =0 to Z=1.0 using 21 mess points for Model 1

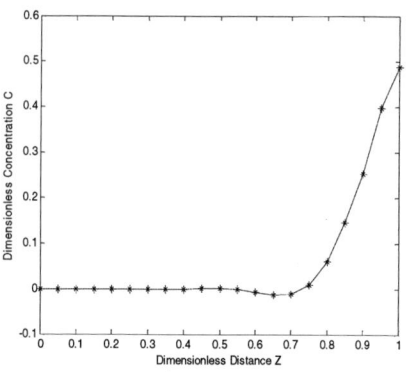

FIGURE 1(b). C vs Z for Na^+ at T =1 for Model 1

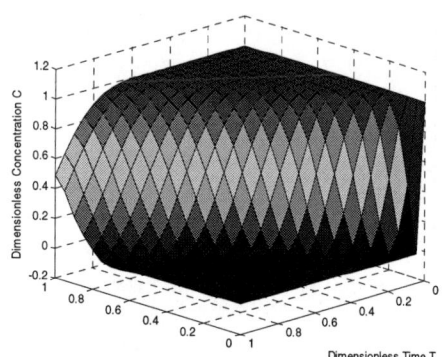

FIGURE 2(a). 3-D Behavior of C vs Z &T for Na^+ for Model 2

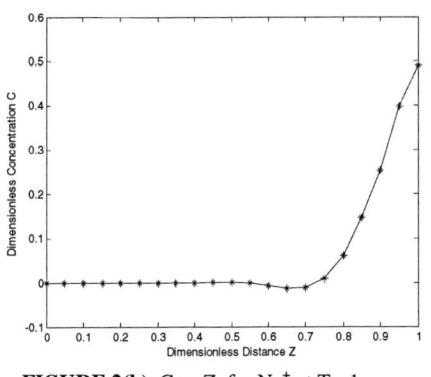

FIGURE 2(b). C vs Z for Na$^+$ at T =1
for Model 2

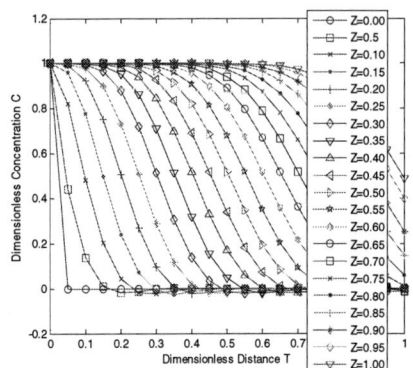

FIGURE 2(c). C vs T for Na$^+$ for Z =0 to
Z=1.0 using 21 mess points for Model 2

3.3 Effect of dimensionless cake thickness on dimensionless solute concentration for Lignin ion

In a similar manner as done for Na$^+$ in section 3.1, the effects of dimensionless cake thickness for lignin ion is also studied for both the models. FIGURE 3(a) depicts the 3-D graph of the behavior of the dimensionless concentration with respect to both dimensionless distance Z as well as dimensionless time T and also FIGURE 3(b) shows the C vs Z profile for lignin ion at T=1.0 for Model 1. The profiles have the same trend as that for Na$^+$. This is also verified from the fact that the dimensionless concentration for lignin also increases as we move towards the bottom layer, starting from the top layer where the dimensionless concentration of lignin is zero. Similarly, the effects of dimensionless cake thickness on Model 2 are shown by the FIGURES 4(a) & 4(b) for the lignin ion. The results are in good agreement with the results of Kumar (2002).

3.4 Effect of dimensionless time on dimensionless solute concentration for lignin ion

FIGUREs 3(c) & 4(c) show C vs T profiles for lignin ion at various bed depths for both models respectively. The various values of bed depths are obtained as above for both the models. Like sodium, lignin species also show the variations in dimensionless solute concentration at the top four layers become insignificant whereas for the bottommost layer (at the bed exit point) shows a typical breakthrough curve. Lowering the bed depth (from top to bottom), the shape of the breakthrough curves analogous to Na$^+$ again change to a non linear asymptotic curve for lignin. The profiles of the curves display the similar characteristic as shown by Kumar (2002).

Similar as for Na$^+$, it is also important to note that for Model 2, the curves at Z=0.0 shows a different characteristic at time T=0.0 to T=0.05 this is due to the fact of using different boundary condition for the Model 2. It may be concluded that boundary conditions influence the shape of profiles.

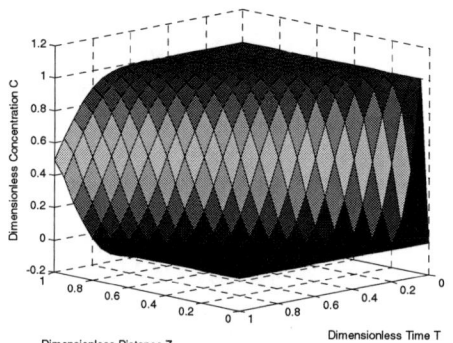

FIGURE 3(a). 3-D Behavior of C vs Z &T for Lignin for Model 1

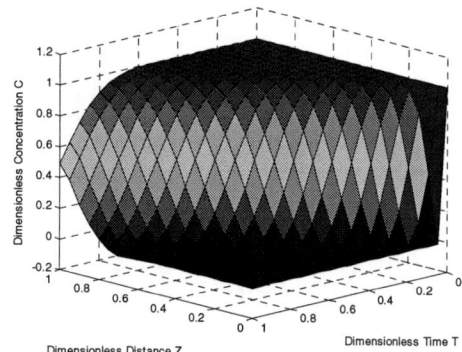

FIGURE 4(a). 3-D Behavior of C vs Z &T for Lignin for Model 2

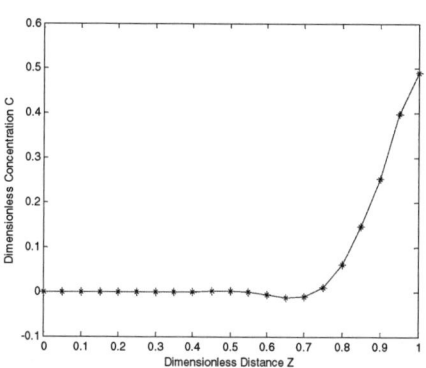

FIGURE 3(b). C vs Z for Lignin at T =1 for Model 1

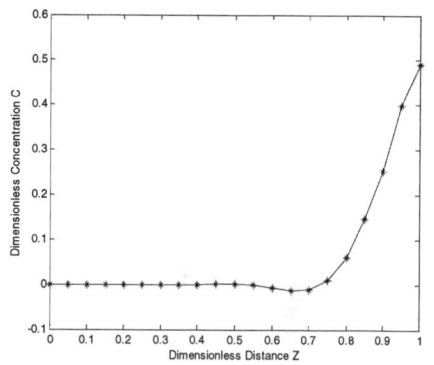

FIGURE 4(b). C vs Z for Lignin at T =1 for Model 2

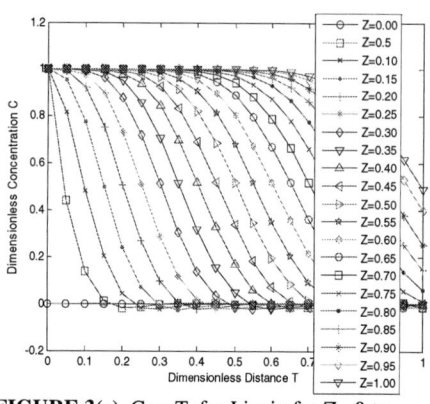

FIGURE 3(c). C vs T for Lignin for Z =0 to Z=1.0 using 21 mess points for Model 1

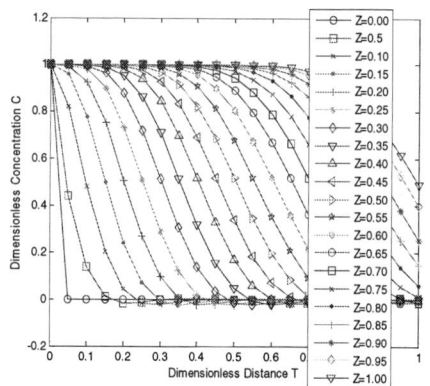

FIGURE 4(c). C vs T for Lignin for Z =0 to Z=1.0 using 21 mess points for Model 2

397

4. Conclusion

Based on the results and discussion the following conclusion may be drawn:
1. FIGURES 1(a), 2(a), 3(a) & 4(a) depict 3-D behavior of the dimensionless concentration with respect to both dimensionless cake thickness Z as well as dimensionless time T. 3-D behavior of the solute concentration gives the excellent agreement with the results of earlier investigators.
2. FIGURES 1(b), 2(b), 3(b) & 4(b) show that the C vs Z profile for sodium and Lignin ions at T=1.0. It is clear from the FIGURES that dimensionless solute concentration increases with the increase in the bed depth Z. It is further evident that the deeper bed gives the higher washing efficiency.
3. FIGURES 1(c), 2(c), 3(c) & 4(c) show that C vs T profiles for sodium and lignin ions at various bed depths for both the models respectively. FIGURES show that the variabilities in dimensionless solute concentration at the top four layers become insignificant whereas for the bottommost layer C depicts a typical breakthrough curve for both the models. Lowering the bed depth (from top to bottom), the shape of the breakthrough curves change to a non linear curve which is practically asymptotic in nature.
4. For both the species of sodium and lignin ions, it is important to mention that the curves at Z=0.0 shows a different characteristic at time T=0.0 to T=0.05 for Model 2 which is contrary to Model 1. This is due to the use of different inlet boundary conditions. So in this case, it may be concluded that the boundary conditions influence the shape of profiles significantly.

REFERENCES

1. Arora. S., Dhaliwal. S.S. and Kukreja. V.K. "Modeling of the displacement washing of the pulp fiber bed", Indian Journal of Chemical Technology, 13(9), 1-6 (2006).
2. Brenner. H. "The diffusion model of longitudinal mixing in beds of finite length", Numerical values, Chemical Engineering Science, 17, 229-243 (1962).
3. Grahs. L.E. "Washing of cellulose fibers, analysis of displacement washing operation", PhD. Thesis. Chalmers University of Technology, Goteborg, Sweden (1974).
4. Gren. U.B. and Strom. K.H.U. "Displacement washing of packed beds of cellulose fibres", Pulp and Paper Canada, 86(9), T261-T264 (1985).
5. Kuo. M.T. "Filter cake washing performance", AICHE J., 6(4), 566-568 (1960).
6. Kuo. M.T. and Barret E.C. "Continuous Filter cake washing performance", AICHE J., 16(4), 633-638 (1970).
7. Kukreja. V.K. "Modeling of washing of brown stock on rotary vacuum washer", PhD. Thesis. University of Roorkee, Roorkee, India (1996).
8. Kumar. Mukesh. "Mathematical modeling of pulp washing systems and solutions", PhD. Thesis. Indian Institute of Technology Roorkee, Roorkee, India (2002).
9. Lapidus. L. & Amundson. N.R. "Mathematics of Adsorption in beds, part-vi: The effect of longitudinal diffusion in ion exchange and chromatographic columns", Journal of Physical Chemistry, 56, 984-988 (1952).
10. Neretnieks. I. "Mathematical model for continuous counter current adsorption", Svensk Paperstidning, 77(11), 407-411 (1974).

11. Potucek. F. "Washing of pulp fibre bed. Collect", Czeck. Chem. Commun., 62, 626-644 (1997).
12. Potucek. F. and Pulcer. M. "Displacement of black liquor from pulp bed. 31st International Conference of the Slovak Society of Chemical Engineering, Tatranske Matliare, 58(6), 337-381 (2004).
13. Pellett. G. L. "Longitudinal dispersion, intra particle diffusion and liquid-phase mass transfer during flow through multi particle systems", Tappi J., 49(2), 75-82 (1966).
14. Perron. M. and Lebeau. B. "A mathematical model of pulp washing on rotary drums", Pulp and paper Canada, 78(3), TR1-TR5 (1977).
15. Sherman. W.R. "The movement of a soluble material during the washing of a bed of packed solids", AICHE J., 10(6), 855-860 (1964).
16. Trinh. D.T., Poirier. N.A., Crotogino. R.H. and Douglas. W.J.M. "Displacement washing of wood Pulps-An experimental study", Journal of Pulp and paper science, 15(1), J28-J35 (1989).
17. Singh. V.P., Kumar. Vivek and Kumar. Deepak. "Numerical Solution of diffusion model of Brown Stock washing beds of finite length Using MATLAB", IEEE Proceeding of '2nd UKSim European Symposium on Computer Modeling and Simulation' pp 295-300 (2008).

An Investigation Into Effect of Electromagnetic Fields On Separation Tendency In Generalised Coutte Flow

B.S.Mudagi

Department of Mathematics, Nowrosjee Wadia College,Pune-411001.(India).
Email:basaweshwarmudagi@ yahoo.co.in

Abstract. One of the basic problem, the plane Coutte flow, has been a source of many research workers in dealing with the interplay of various fluid forces and their interaction with the electromagnetic forces .The combined effect of magnetic and electric fields on separation tendency in generalized Coutte flow has been studied. The velocity distribution has been numerically computed for various values of magnetic parameter, electric parameter and pressure gradient parameter. It is observed that for electric parameter $k \neq 0$ (open circuited) the tendency of separation does not exist.

Keywords: Open Circuit, back flow, Hartmann number.
PACS: 0.1.30.Cc

1. INTRODUCTION

One of the basic problem, the plane Coutte flow, has been a source for investigation of many research workers in dealing with the interplay of various fluid forces and their interaction with the electromagnetic forces. In view of the applications of the Coutte flow results to modern technology; it has attracted the attention of several research workers. Lenhert [1] investigated the behavior of an electrically conducting fluid in a magnetic field. The plane Couette flow of a conducting gas in presence of transverse magnetic field has been studied by Leadon [2]. Morgan [3] analysed the Couette flow of a compressible viscous heat conducting perfect gas. Bleviss [4] studied magnetogas dynamics of hypersonic Couette flow. Yen and Chang [5] discussed hydromagnetic Channel flow under time- dependent pressure. Agarwal [6] studied generalized incompressible Couette flow in MHD. Rammurthy [7] investigated the generalized Couette flow between two porous plates with suction at the stationary plate and injection at the other plate. Rathi [8] discussed the same problem without neglecting the induced magnetic field. Suttan and Shermann [9] studied the joint influence of electric and magnetic fields on plane Couette flow. .Soundalgekar [10] made an investigation of combined influence of electric and magnetic field on generalized MHD Couette flow. He also studied generalized Couette flow with heat transfer. Verma and Mathur [11]) discussed hydromagnetic flow between two parallel plates, one in uniform motion and other at rest with uniform suction at stationary plate. Maiti and Seth [12] made an investigation into MHD Couette flow and heat transfer in a rotating system. Choudhary et. al .[13] studied MHD flow with heat transfer

CP1146, *Modelling of Engineering and Technical Problems*, edited by A. H. Siddiqi, A. K. Gupta, and M. Brokate
© 2009 American Institute of Physics 978-0-7354-0683-4/09/$25.00

Hazem Ali Attia [14] studied MHD Couette flow with temperature dependent viscosity and the ion slip. Jain N.C. and Gupta P. [15] discussed three dimensional free convection Couette flow with transpiration cooling.

In the present analysis an attempt has been made to observe the joint effect of electric and magnetic fields on the tendency of separation (back – flow) in the generalized Couette flow of viscous incompressible and electrically conducting fluid. The velocity distribution has been numerically computed for various values of magnetic parameter (M), electric parameter (K) and pressure gradient parameter (N). The numerical and graphical analysis led to several significant results.

2. Formulation Of The Problem And Governing Equations

We suppose that the steady laminar flow of viscous incompressible electrically conducting fluid is between two non-conducting parallel flat plates. The plates are infinite in extent in both x-direction and y-direction, the x-axis is chosen along lower stationary plate in the direction of flow and z axis is chosen normal to the direction of flow.

We suppose that the steady laminar flow of viscous incompressible electrically conducting fluid is between two non-conducting parallel flat plates. The plates are infinite in extent in both x-direction and y-direction, the x-axis is chosen along lower stationary plate in the direction of flow and z axis is chosen normal to the direction of flow.

Under these assumptions the governing equation of motion for the steady flow of a viscous incompressible and electrically conducting fluid is

$$\mu \frac{d^2 u}{dz^2} + \sigma (E_y - u B_0) B_0 = \frac{\partial p}{\partial x} \tag{1}$$

where $\overline{E} = (0, E_y, 0)$ is the electric field, σ the electric conductivity of the fluid,

$\overline{B} = (B_x, 0, B_0)$ the magnetic induction vector, $\overline{q} = (u, 0,0)$ the velocity, p the pressure, ρ the density, μ is the coefficient of viscosity.

The boundary conditions are

z = d,   u = U

z = 0,   u = 0   (2)

where d is the separation distance between the two parallel plates.

3. Solution For Velocity Distribution $\overline{u}$

In order to reduce equation (1) in dimensionless form, the following non-dimensional quantities are introduced.

$$\overline{u} = \frac{u}{U}$$

$$\overline{x} = \frac{x}{d}$$

$$\overline{z} = \frac{z}{d}$$

$$\overline{p} = \frac{p}{\rho U^2}$$

$$K = \frac{E_y}{UB_0}$$

$$M^2 = \frac{\sigma B_0^2 d^2}{\mu}$$

$$\bar{u} = \frac{u}{U}$$

$$\bar{x} = \frac{x}{d}$$

$$\bar{z} = \frac{z}{d}$$

$$\bar{p} = \frac{p}{\rho U^2}$$

$$K = \frac{E_y}{UB_0}$$

$$R_e = \frac{\rho U d}{\mu} = \frac{U d}{v}$$

$$N = R_e \overline{p_x}$$

$$\overline{p}_x = \frac{\partial \overline{p}}{\partial \overline{x}}$$

These substitutions reduce equation (1) to

$$\frac{d^2\overline{u}}{d\overline{z}^2} - M^2\overline{u} = N - M^2K \tag{3}$$

where M is the Hartmann number, K is the electric parameter and N is the pressure gradient parameter respectively.

The boundary conditions (2) become

$$\overline{z} = 1 \ ; \ \overline{u} = 1 \tag{4}$$

$$\overline{z} = 0 \ ; \ \overline{u} = 0$$

Solving equation (3) we get

$$\overline{u} = c_1 \cosh M\overline{z} + c_2 \sinh M\overline{z} - \frac{\left(N - M^2K\right)}{M^2} \tag{5}$$

where c_1 and c_2 are constants.

After further calculation the solution of equation (5) under the boundary conditions (4) give velocity distribution $\overline{u}$ as

$$\overline{u} = \frac{M^2 \sinh M\overline{z} + \left(N - M^2K\right)\left[\sinh M\overline{z} + \sinh M\left(1 - \overline{z}\right) - \sinh M\right]}{M^2 \sinh M} \tag{6}$$

The velocity profile $\overline{u}$ has been numerically computed for various values of M, N and K.

The results of calculations are entered in the following table.

TABLE 1

Values of $\overline{u}$ for different Parameters

K	M	N	Z→ 0.0	0.1	0.2	0.3	0.4	0.5	0.6	0.7	0.8	0.9	1
0	2	2	0	-0.0111	-0.0026	0.0258	0.0754	0.1481	0.2467	0.3754	0.5391	0.7446	1
0	4	1	0	-0.0463	0.00012	0.0150	0.0425	0.0870	0.1558	0.2599	0.4163	0.6504	1
0	6	1	0	-0.0928	-0.0117	-0.0819	0.0259	0.0246	0.0661	0.1467	0.282	0.5364	1
0	4	2	0	-0.0243	-0.0323	-0.0252	-0.020	0.0411	0.1112	0.2599	0.3839	0.6307	1
0	6	2	0	-0.0217	-0.0309	-0.0309	-0.022	-0.004	0.0416	0.1197	0.2628	0.5239	1
0	6	-2	0	0.02805	0.04589	0.0601	0.0761	0.0997	0.1397	0.2108	0.3396	0.5736	1
0	3	-2	0	0.08157	0.1504	0.2127	0.2742	0.3403	0.4172	0.5117	0.6325	0.7905	1
0.1	2	-1	0	0.10216	0.19437	0.2803	0.3635	0.4472	0.5348	0.6298	0.7361	0.8579	1
0.2	2	-2	0	0.1488	0.27548	0.3851	0.4821	0.5704	0.6534	0.7346	0.8172	0.9045	1
0.2	2	2	0	0.01553	0.04373	0.0857	0.1432	0.2184	0.3145	0.4352	0.5855	0.7712	1
0.2	4	2	0	0.03867	0.07145	0.1036	0.1405	0.188	0.2538	0.3485	0.4876	0.6937	1
0.4	4	1	0	0.12133	0.20761	0.2728	0.3276	0.3807	0.4408	0.5177	0.6238	0.7763	1
0	3	-1.5	0	0.06818	0.12869	0.1851	0.2433	0.3084	0.3863	0.4841	0.6108	0.7777	1
0	6	-1.5	0	0.02182	0.03629	0.0488	0.0639	0.0872	0.1274	0.1994	0.3299	0.5675	1

405

4. Numerical Discussion And Conclusions

When the electric parameter K≠0 (open circuited) the velocity profile $\bar{u}$ is positive over the entire range for all values of magnetic parameter M and pressure gradient N, so the tendency of separation does not exist.

1. For a pressure decreasing (N < 0) in the direction of motion, $\bar{u}$ is positive over the whole width of the channel for all values of M or K. Therefore the back flow does not occur.

2. For a pressure increasing (N >0) in the direction of motion the back flow occurs near the stationary plate in case of K=0 (Short Circuited) even in the presence of transverse magnetic field. Also, tendency of separation increases with increase in M in the case N > 0, K = 0

3. The Separation tendency in the generalized Couette flow with increasing pressure gradient can be completely prevented by the joint application of magnetic field and electric fields.

Expression (6) for velocity profile can be utilized for obtaining shearing stress at the lower plate and velocity gradient at the upper plate.

ACKNOWLEDGMENTS

I am grateful to Dr. N.P. Patil, Department of Mathematics K.T.H.M. College, Gangapur Road, Nashik Maharashtra (India) for giving suggestions for improvements.

406

REFERENCES

1. B.U. Lehnert Ark. Fys., 5 1952, pp. 60-90 .

2. B.M Leadon , Convair Scientific Research Laboratory R.N1957, pp. 113 .

3. A.J.:Morgan , A Jaz. 24, 1957, pp. 315-316.

4. Z. O. Bleviss , J. Aero Space Sc. 25,1958, pp. 601-615.

5. J.T. Yen . and. S.C Chang , Phys. Fluids. 1961, 4 pp. 135-136.

6. M.L. Agarawal and. R.C. Ram, Proc. Nat. Acad. Sci India 42 A, 1962.

7. Rammurthy , J. Aero. Space. Sc. 29, 1962, N .I.

8. Rathi R.K.: ZAAM 43, 1963, pp. 370-374.

9. G.W Suttan. and Shermann , Engineering Magnetohydrodynamic, Mc Graw Hill book. Co. Inc. NewYork 1965..

10. V.M. :Soundalgekar ,India Acad. Sci, 64, 1966, pp. 304-314 .

11. P.D. Verma and A.K.: Mathur , Proc. Inst .Sci .India, Vol.1. .35A ,1969, pp.507-517.

12. M.K.Maiti . and G.S Seth , Int. J of Pure and App. Maths Vol. 13 , 1982, pp.981.

13. R.C Choudhary , A.K. Mathur . and. H.S. Kalsi. , Acta. Cinecia .Indica Vol. XVIII, N4, 2002, pp. 541 .

14. Hazem Ali Attia , Tamkang Journal of Science and Engineering, Vol.8. No.1, 2006, pp. 11-16 ..

15. N. C. Jain and P. J.Gupta , Zhejiang Univ SCIENCEA Vol. 7.No.3, 2006, pp..340-346 .

SECTION F:
TYPICAL METHODS OF ENGINEERING
AND TECHNOLOGICAL PROBLEMS

Algebraic computations of some Tensors in geometry and general Relativity

Kamran Asghar[a], Ashfaque H. Bokhari and F. D. Zaman[b]

[a] *Department of Mathematics, Quaid-i-Azam University, Islamabad, Pakistan*

[b] *Department of Mathematics and Statistics, King Fahd University of Petroleum and Minerals, Dhahran 31261, Saudi Arabia*

Abstract. To be able to algebraically compute tensors that arise in geometry and general relativity an algebraic computational package is written. This package computes Christoffel symbols, Riemann, Ricci, Einstein and Weyl tensors and some curvature invariants.

Keywords: Algebraic computations, Relativity and Geometry.
PACS: 02.70.Wz, 04.20.-q

INTRODUCTION

In geometry and general relativity [2, 3] studies of manifolds are based on tensor differential calculus. Whereas these calculations are of immense importance in understanding physics of manifolds, these computations themselves are purely mechanical and involve lengthy calculations. With a point in mind that time should be saved to perform such mechanical calculations, we have developed a package to perform such calculations algebraically using Mathematica. This package is capable of not only computing most tensors arising in geometry and general relativity, the calculations performed are error free and executed immediately. An-other interesting feature of this package is that it is more user friendly as compared with other such packages available on commercial basis [1]. We hope that this package can be of great use to relativists and is obtainable from authors for academic purposes free of cost. The paper is prepared by keeping in mind that it explains the use of package in as simple a form as possible. The plan of the paper is as follows: In the next section we briefly give definitions of the tensors which the package computes. In section 3 we illustrate use of the package with the help of an example of Schwarzschild geometry [3]. The complete package is given at the end of the paper in appendices.

DEFINITIONS OF TENSORS COMPUTED BY PACKAGE

In this section we briefly introduce tensors the package can compute.

The metric tensor

Given an arbitrary coordinate system a quadratic function, ds^2, of the coordinates differentials dx^a is defined by

$$ds^2 = g_{ij}(x^k)dx^a dx^b \qquad \forall i,j,k = 1,\cdots,n \qquad (1)$$

CP1146, *Modelling of Engineering and Technological Problems*, edited by A. H. Siddiqi, A. K. Gupta, and M. Brokate
© 2009 American Institute of Physics 978-0-7354-0683-4/09/$25.00

The g_{ij} in above equation forms components of a symmetric tensor, called metric tensor. The inverse of g_{ij} is denoted by g^{ij} and it exists if the determinant of g_{ij} is non-zero and satisfies $g_{ij}g^{jk} = \delta_i^k$.

The Christoffel symbol

In a spacetime with metric g_{ij}, there is a relationship between Christoffel symbol Γ^i_{jk} and the metric. This relationship is obtained by requiring that the spacetime metric is covariantly constant. Mathematically this condition is translated by the relation [3]

$$g_{ij;k} = 0, \tag{2}$$

where a semi colon represents covariant derivative which is defined, for a tensor of rank (p+q), by

$$T^{\overset{p}{\overline{i\cdots j}}}{}_{\underset{q}{\underline{k\cdots l};m}} = T^{i\cdots j}{}_{k\cdots l,m} + \Gamma^i_{i_1 m} T^{i_1\cdots j}{}_{k\cdots l} + \cdots \Gamma^j_{j_1 m} T^{i\cdots j_1}{}_{k\cdots l}$$
$$- \Gamma^{k_1}_{km} T^{i\cdots j}{}_{k_1\cdots l} - \Gamma^{l_1}_{lm} T^{i\cdots j}{}_{k\cdots l_1} \tag{3}$$

In the light of above definition equation (2) takes the form,

$$g_{ij,k} - \Gamma^a_{ik} g_{aj} - \Gamma^a_{jk} g_{ia} = 0 \tag{4}$$

Cyclically permuting indices in (4) and using resulting expressions one gets,

$$\Gamma^p_{ik} = \frac{1}{2} g^{pj} \left(g_{ij,k} + g_{jk,i} - g_{ki,j} \right) \tag{5}$$

Equation (5) defines Christoffel symbol.

The Riemann tensor

Riemann tensor is a tensor which defines curvature of a spacetime and is defined by taking difference of the action of two successive covariant derivative operations on a vector performed in one order and then the order reversed. For a contravariant vector, ξ^i, this difference becomes [3],

$$\xi^i_{;j;k} - \xi^i_{;k;j} = \left(\Gamma^i_{km,j} - \Gamma^i_{jm,k} + \Gamma^i_{nj}\Gamma^n_{km} - \Gamma^i_{nk}\Gamma^n_{jm} \right)\xi^m. \tag{6}$$

Representing $R^i_{mjk} = \Gamma^i_{km,j} - \Gamma^i_{jm,k} + \Gamma^i_{nj}\Gamma^n_{km} - \Gamma^i_{nk}\Gamma^n_{jm}$, equation (6) takes the simple form

$$\xi^i_{;j;k} - \xi^i_{;k;j} = R^i_{mjk}\xi^m. \tag{7}$$

The R^i_{mjk} defined above is called the Riemann tensor. In n-dimensional spacetime this tensor has a total of n^4 components. To find independent components of the Riemann tensor, one can write it in a fully covariant form $R_{ijkm}=g_{ip}R^p_{jkm}$. This form of the Riemann tensor satisfies certain symmetry and cyclic properties in its indices given respectively by $R_{ijkm}=-R_{ijmk}=-R_{jikm}=R_{kmij}$ and $R_{ijkm}+R_{ikmj}+R_{imjk}=0$. Using these properties, it is easy to see that the Riemann tensor possesses a total of $n^2(n^2-1)/12$ components in n-dimensional space.

The Ricci tensor

The Ricci tensor relates matter with geometry [3]. It is obtained from trace of the Riemann tensor and given mathematically by

$$R_{ij} = R^k_{ikj} \qquad (8)$$

The Ricci tensor is a symmetric tensor and in n-dimensional space has $n(n+1)/2$ independent components.

The Ricci scalar

The trace of the Ricci tensor defines the Ricci scalar given by,

$$R = R^i_{i} \qquad (9)$$

Scalar invariants

In Relativity one encounters two types of singularities, namely, coordinate and essential singularities [3]. To determine which singularity is coordinate or essential, scalar invariants are used. These invariants are respectively given by

$$R^1 = R^{ij}_{kl} R^{kl}_{ij} \qquad (10)$$

and

$$R^2 = R^{ij}_{kl} R^{kl}_{mn} R^{mn}_{ij} \qquad (11)$$

If these invariants are not divergent on a value on which the spacetime metric is, the value gives coordinates singularity, otherwise it is an essential singularity.

The Weyl tensor

The curvature tensor can be split into its trace free part, Ricci tensor and Ricci scalar by [3]

$$C_{ijkm} = R_{ijkm} - \frac{1}{n-2}(g_{ik}R_{mj} + g_{im}E_{kj} - g_{jk}R_{mi} - g_{jm}R_{ki}) +$$
$$\frac{1}{(n-1)(n-2)}(g_{ik}g_{mj} - g_{im}g_{kj})R \qquad (12)$$

The C_{ijkm} above is called the Weyl tensor.

The Einstein tensor

The Einstein tensor is an important constituent of the Einstein equations [3],

$$R_{ij} - \frac{1}{2} g_{ij} R = \kappa T_{ij} \, , \tag{13}$$

where κ is coupling between geometry and matter and T_{ij} is the energy-momentum tensor [3]. The left hand side of Einstein equations defines the Einstein tensor,

$$E_{ij} = R_{ij} - \frac{1}{2} g_{ij} R \tag{14}$$

THE PACKAGE[1]

In this section we give steps that are needed to use the package to compute the tensors discussed in the previous section.

Step 1: Loading the Package

To load the package first requires it to be copied into the Mathematica directory and then open a new note book and enter "<<load.m". Pressing insert button or "Shift + Enter" keys activates the package and following screen appears:

```
In[1]:=  << load.m


                    Algebraic Computations of Some

                 Tensors in Geometry & Relativity

                                by

                 KAMRAN ASGHAR, ASHFAQUE H.BOKHARI

                          F. D. ZAMAN

                 Department of Mathematics and Statistics

              King Fahd University of Petroleum and Minerals

                          Dhahran 31261

                          SAUDI ARABIA
```

Step 2: Entering the metric

Since all the tensorial computations for our purpose metric dependent, we first show how a metric is entered in the package. To illustrate how a metric is entered we choose Schwarzschild metric [3],

$$ds^2 = (1\text{-}2m/r)\, dt^2 - (1\text{-}2m/r)^{-1} dr^2 - r^2 d\theta^2 - r^2 sin^2\theta d\varphi^2. \tag{10}$$

---

[1] The package is available on a CD that can be obtained from the authors or can be fed in to Mathematica in the form as given in Appendices A - E

Note that this metric is written in '4' spacetime dimensions using t, r, ϑ and φ coordinates. To enter it in the package, we enter this metric as follows:

Metric[4,{t, r, theta, phi},{{(1-2*m/r),0,0, 0},{0,-(1-2m/r)^-1,0,0},{0,0,-r^2,0}

$$\{0, 0, 0,-r^2*\sin[theta]^2\}\}]$$ (11)

Pressing the "Shift+Enter" keys activates the input given by Eq. (10). After loading the metric the package displays a message "you have entered a 4 x 4 metric" confirming that the metric has been successfully loaded. The following screen appears in Mathematica note book:

$$\text{In[2]:= } \mathbf{Metric}\left[4, \{t, r, th, ph\},\right.$$

$$\left\{\left\{\left(1-2\frac{m}{r}\right), 0, 0, 0\right\}, \left\{0, -\left(1-2\frac{m}{r}\right)^{-1}, 0, 0\right\},\right.$$

$$\left.\left\{0, 0, -r^2, 0\right\}, \left\{0, 0, 0, -(r*\mathrm{Sin}[th])^2\right\}\right\}\right]$$

You have entered a 4 x 4 metric

At this stage we can in principle compute components of all tensors defined above in the Schwarzschild geometry. However, we limit ourselves to only exhibit procedure by computing two Christoffel symbols (Γ^1_{12}, Γ_{112}), three Riemann (R^1_{212}, R^{12}_{12}, R_{1212}) and Ricci (R^{11}, $R^1{}_1$, R_{11}) tensor components, three Weyl tensor components ($W^1{}_{212}$, W^{12}_{12}, W_{1212}), three Einstein tensor components (E^{11}, $E^1{}_1$, E_{11}) as well as the scalar invariants. The detail of all the routines is given in Appendix A to G.

3a. Computation of Christoffel symbols ($\Gamma^1{}_{12,}\Gamma_{112}$)

In[2]:=Christoffelsymbol[-1,1,2]
Out[2]=   m/(-2mr+r²)
In[3]=Christoffelsymbol[1,1,2]
Out[3]=   m/r²

The input and output of these two calculations as shown on the Mathematica note book are respectively given by[2],

In[3]:= **Christoffelsymbol[-1, 1, 2]**

Hang on! The Calculations are underway!!!

$$C^1_{12}$$

$$\text{Out[3]= } \frac{m}{-2\,m\,r + r^2}$$

and

---

[2] In Mathematica computations we have used $\Gamma^i{}_{jk} = C^i{}_{jk}$

In[4]:= **Christoffelsymbol[1, 1, 2]**

Hang on! The Calculations are underway!!!

$$\text{Out[4]}= \frac{C_{112}}{r^2}$$

3b. Computation of the Riemann tensor components ($R^1{}_{212}, R^{12}{}_{12}, R_{1212}$)

In[4]:=   Riemanntensor[-1,2,1,2]
Out[4]= $2m/r^2(-2m+r)$
In[5]:=Riemanntensor[-1,-2,1,2]
Out[5]= $-2m/r^3$
In[6]:=   Riemanntensor[1,2,1,2]
Out[6]= $2m/r^3$

The input and output of these two calculations as shown on the Mathematica note book are respectively given by,

In[5]:= **Riemanntensor[-1, 2, 1, 2]**

Hang on! The Calculations are underway!!!

$R^1{}_{212}$

$$\text{Out[5]}= \frac{2m}{r^2(-2m+r)}$$

In[6]:= **Riemanntensor[-1, -2, 1, 2]**

Hang on! The Calculations are underway!!!

$R^{12}{}_{12}$

$$\text{Out[6]}= -\frac{2m}{r^3}$$

and

In[7]:= **Riemanntensor[1, 2, 1, 2]**

Hang on! The Calculations are underway!!!

R_{1212}

$$\text{Out[7]}= \frac{2m}{r^3}$$

3c. Computation of the Ricci tensor components (**R11, R1 1, R11**)

In[7]:=Riccitensor[-1,-1]
Out[7]=   0

416

In[8]:=Riccitensor[-1,1]
Out[8]=   0
In[9]:=Riccitensor[1,1]
Out[9]=   0

The input and output of these two calculations as shown on the Mathematica note book are respectively given by,

In[8]:= **Riccitensor[-1, -1]**

Hang on! The Calculations are underway!!!
11
R

Out[8]= 0

In[9]:= **Riccitensor[-1, 1]**

Hang on! The Calculations are underway!!!
 1
R
  1

Out[9]= 0

and

In[10]:= **Riccitensor[1, 1]**

Hang on! The Calculations are underway!!!

R
11

Out[10]= 0

3d. Computation of Ricci scalar and scalar invariants(R, R_1, R_2)

In[10]:=Ricciscalar
Out[10]=   0
In[11]:=Scalarinvariant[1]
Out[11]= $48m^2/r^6$
In[12]:=Scalarinvariant[2]
Out[12]= $-96m^3/r^9$

The input and output of these two calculations as shown on the Mathematica note book are respectively given by,

417

In[11]:= **Ricciscalar**

Hang on! The Calculations are underway!!!

R

Out[11]= 0

In[13]:= **Scalarinvariant[1]**

Hang on! The Calculations are underway!!!

$$\overset{1}{R}$$

Out[13]= $\dfrac{48\ m^2}{r^6}$

and

In[14]:= **Scalarinvariant[2]**

Hang on! The Calculations are underway!!!

$$\overset{2}{R}$$

Out[14]= $-\dfrac{96\ m^3}{r^9}$

3e. Computation of Weyl tensor components ($W^1{}_{212}$, $W^{12}{}_{12}$, W_{1212})

In[13]:=Weyltensor[-1,2,1,2]
Out[13]= $2m/r^2(-2m+r)$
In[14]:=Weyltensor[-1,-2,1,2]
Out[14]= $-2m/r^3$
In[15]:=Weyltensor[1,2,1,2]
Out[15]= $2m/r^3$

The input and output of these two calculations as shown on the Mathematica note book are given respectively by,

In[15]:= **Weyltensor[-1, 2, 1, 2]**

Hang on! The Calculations are underway!!!

$$\overset{1}{W}{}_{212}$$

Out[15]= $\dfrac{2\ m}{r^2\ (-2\ m + r)}$

418

In[16]:= **Weyltensor[-1, -2, 1, 2]**

Hang on! The Calculations are underway!!!

12

W

   12

Out[16]= $-\dfrac{2\,m}{r^3}$

and

In[17]:= **Weyltensor[1, 2, 1, 2]**

Hang on! The Calculations are underway!!!

W

   1212

Out[17]= $\dfrac{2\,m}{r^3}$

3f. Computation of Einstein tensor components $(E^{11}, E^1{}_1, E_{11})$

```
In[16]:=Einsteintensor[-1,-1]
Out[16]=      0
In[17]:=Einsteintensor[-1,1]
Out[17]=      0
In[18]:=Einsteintensor[1,1]
Out[18]=      0
```

The input and output of these two calculations as shown on the Mathematica note book are given respectively by,

In[18]:= **Einsteintensor[-1, -1]**

Hang on! The Caculations are Underway!!!

   11

E

Out[18]= 0

In[19]:= **Einsteintensor[-1, 1]**

Hang on! The Caculations are Underway!!!

   1

E

   1

Out[19]= 0

and

419

```
In[20]:= Einsteintensor[1, 1]
```

Hang on! The Caculations are Underway!!!

E
  11

```
Out[20]= 0
```

Apart from computing tensorial components individually, the package also has the capability to compute all the non-zero components of tensors using one command. For example *AllChristoffelsymbol[1,1,1]* computes all the non-zero Christoffel symbols Γ_{ijk} and *AllChristoffelsymbol[-1,1,1]* computes all the non-zero Christoffel symbol $\Gamma^i_{\ jk}$ .Similarly, the *AllRiemanntensor*, *AllRiccitensor*, *AllWeyltensor* and *AllEinsteintensor* with relevant arguments in any combination of 1 or −1 can be computed in one step.

ACKNOWLEDGMENTS

Authors are thankful to Professors A K Gupta and A H Siddiqi for their invitation to participate and present this work at the International Conference on Modeling of Engineering and Technological Problems, and 9th Biennial Conference of Indian Society of Industrial and Applied Mathematics.

REFERENCES

1. R. J., Kamin S. N. and Wellin P. R, *Introduction to Programming with Mathematica*, Verlog: Springer, (1993)
2. M. P. Docormo, Differential geometry of curves and surfaces, *Prentice Hall,* (1976).
3. C. W. Misner, K. S. Thorne, J. A. Wheeler Gravitation, W. H. Freeman, , (1973)

Package to Calculate the Christoffel Symbols for a Given Metric Tensor

```
Metric[dim_,coordinatelist_,componentlist_]:=Block[{kint,kkd,kd,kn,ke,ki,kj,kk,kl},
Clear[dimension,metriccoordinate,metriclist,metriclistup,chris,chrisder,cur];
dimension=dim;
metriccoordinate=coordinatelist;
metriclist=componentlist;
metriclistup=Inverse[metriclist];
chris=Array[b,{dimension,dimension,dimension}];
chrisder=Array[c,{dimension,dimension,dimension,dimension}];
cur=Array[d,{dimension,dimension,dimension,dimension}];
Do[Do[Do[chris[[ki,kj,kk]]=Sum[1/2*(metriclistup[[ki,kd]]*
(metricderivative[kj,kd,kk]+metricderivative[kk,kd,kj]-metricderivative[kj,kk,kd])),
{kd,1,dimension}],{kk,1,dimension}],{kj,1,dimension}],{ki,1,dimension}];
Do[Do[Do[Do[chrisder[[ki,kj,kk,kl]]=D[chris[[ki,kj,kk]],metriccoordinate[[kl]]]],
{kl,1,dimension}],{kk,1,dimension}],{kj,1,dimension}],{ki,1,dimension}];
Do[Do[Do[Do[cur[[ki,kj,kk,kl]]=kjk,{kl,1,dimension}],
{kk,1,dimension}],{kj,1,dimension}],{ki,1,dimension}];
Do[Do[Do[Do[If[cur[[ki,kj,kk,kl]]===kjk,cur[[ki,kj,kk,kl]]=chrisder[[ki,kj,kl,kk]]-
chrisder[[ki,kj,kk,kl]]+Sum[chris[[ki,kf,kk]]*chris[[kf,kj,kl]],{kf,1,dimension}]-
Sum[chris[[ki,kf,kl]]*chris[[kf,kj,kk]],{kf,1,dimension}];
cur[[ki,kj,kl,kk]]=-cur[[ki,kj,kk,kl]],
Continue[]],
{kl,1,dimension}],{kk,1,dimension}],{kj,1,dimension}],{ki,1,dimension}];
Print["You have entered a ",dimension," x ",dimension," metric"]]
metricinverse[a_,b_]:= metriclistup[[a,b]];
metricinversederivative[a_,b_,wrt_]:= D[metriclist[[a,b]],metriccoordinate[[wrt]]];
metricderivative[a_,b_,wrt_]:=Block[{kd},
kkd=Position[metriclist[[a,b]],metriccoordinate[[wrt]],Infinity];
If[Length[kkd]===0,0,
D[metriclist[[a,b]],metriccoordinate[[wrt]]]]]]
chrisruletwo[a_,b_,c_]:=Block[{tmp,kd},tmp = 0;
tmp = Sum[metriclistup[[kd,Abs[b]]]*chris[[Abs[a],Abs[kd],c]],{kd,1,dimension}];
Simplify[tmp]]chrisrulethree[a_,b_,c_]:=Block[{tmp,kd},tmp = 0;
tmp = Sum[metriclistup[[kd,Abs[c]]]*chrisruletwo[Abs[a],Abs[b],Abs[kd]],
{kd,1,dimension}];
Simplify[tmp]]chrisrule1[a_,b_,c_]:=Block[{tmp,kd},tmp = 0;
tmp = Sum[metriclist[[kd,Abs[a]]]*chris[[Abs[kd],b,c]],{kd,1,dimension}];
Simplify[tmp]]chrisrule2[a_,b_,c_]:=Block[{tmp,kd},tmp = 0;
tmp = Sum[metriclistup[[kd,Abs[c]]]*chris[[Abs[a],b,Abs[kd]]],{kd,1,dimension}];
Simplify[tmp]]chrisrule3[a_,b_,c_]:=Block[{tmp,kd},tmp = 0;
tmp = Sum[metriclistup[[kd,Abs[b]]]*chrisrule1[Abs[a],Abs[kd],Abs[c]],
{kd,1,dimension}];
Simplify[tmp]]chrisrule4[a_,b_,c_]:=Block[{tmp,kd},tmp = 0;
tmp = Sum[metriclistup[[kd,Abs[c]]]*chrisrule3[Abs[a],Abs[b],Abs[kd]],
{kd,1,dimension}];
Simplify[tmp]]chrisrule5[a_,b_,c_]:=Block[{tmp,kd},tmp = 0;
tmp = Sum[metriclistup[[kd,Abs[c]]]*chrisrule1[Abs[a],Abs[b],Abs[kd]],
{kd,1,dimension}];
Simplify[tmp]]Christoffelsymbol[a_,b_,c_]:=Block[{tmp,kd},tmp = 0;
Print[crform[a,b,c]];
If[(Sign[a]    ===    -1  &&   Sign[b]   ===   1   &&   Sign[c]   ===
1),Simplify[chris[[Abs[a],b,c]]],
```

```
If[(Sign[a] === 1 && Sign[b] === 1 && Sign[c] === 1),
tmp = Sum[metriclist[[kd,Abs[a]]]*chris[[Abs[kd],b,c]],{kd,1,dimension}];
Simplify[tmp],
If[(Sign[a] === -1 && Sign[b] === -1 && Sign[c] === 1),chrisruletwo[a,b,c],
If[(Sign[a] === -1 && Sign[b] === -1 && Sign[c] === -1),chrisrulethree[a,b,c],
If[(Sign[a] === -1 && Sign[b] === 1 && Sign[c] === -1),chrisrule2[a,b,c],
If[(Sign[a] === 1 && Sign[b] === -1 && Sign[c] === 1),chrisrule3[a,b,c],
If[(Sign[a] === 1 && Sign[b] === -1 && Sign[c] === -1),chrisrule4[a,b,c],
If[(Sign[a] === 1 && Sign[b] === 1 && Sign[c] === -1),chrisrule5[a,b,c],
Print["Wrong Input"]]]]]]]]]
AllChristoffelsymbol[a_,b_,c_]:=Block[{tmp,ki,kj,kk,kd,de},
Print["Hang on Calculations are underway"];
de = 0;
If[(a===-1 && b===1 && c===1),Do[Do[Do[tmp = Simplify[chris[[kk,kj,ki]]];
If[tmp === 0,
Continue[],
de = de+1;
Print[de," : ","c[",-kk,",",kj,",",ki,"] = ",tmp];
Print[" "]],{ki,kj,dimension}],{kj,1,dimension}],{kk,1,dimension}],
If[(a===1 && b===1 && c===1),
Do[Do[Do[tmp = Sum[metriclist[[kd,ki]]*chris[[kd,kj,kk]],{kd,1,dimension}];
If[tmp === 0,
Continue[],
de = de+1;
Print[de," : ","c[",ki,",",kj,",",kk,"] = ",Simplify[tmp]];
Print[" "]],{ki,1,dimension}],{kj,1,dimension}],{kk,1,dimension}],
If[(a===-1 && b===-1 && c===-1),Do[Do[Do[tmp = chrisrulethree[ki,kj,kk];
If[tmp === 0,
Continue[],
de = de+1;
Print[de," : ","c[",-ki,",",-kj,",",-kk,"] = ",Simplify[tmp]];
Print[" "]],{ki,1,dimension}],{kj,1,dimension}],{kk,1,dimension}],
If[(a===-1 && b===-1 && c===1),Do[Do[Do[tmp = chrisruletwo[ki,kj,kk];
If[tmp === 0,
Continue[],
de = de+1;
Print[de," : ","c[",-ki,",",-kj,",",kk,"] = ",Simplify[tmp]];
Print[" "]],{ki,1,dimension}],{kj,1,dimension}],{kk,1,dimension}],
If[(a===-1 && b===1 && c===-1),Do[Do[Do[tmp = chrisrule2[ki,kj,kk];
If[tmp === 0,
Continue[],
de = de+1;
Print[de," : ","c[",-ki,",",kj,",",-kk,"] = ",Simplify[tmp]];
Print[" "]],{ki,1,dimension}],{kj,1,dimension}],{kk,1,dimension}],
If[(a===1 && b===-1 && c===-1),Do[Do[Do[tmp = chrisrule4[ki,kj,kk];
If[tmp === 0,
Continue[],
de = de+1;
Print[de," : ","c[",ki,",",-kj,",",-kk,"] = ",Simplify[tmp]];
Print[" "]],{ki,1,dimension}],{kj,1,dimension}],{kk,1,dimension}],
If[(a===1 && b===-1 && c===1),Do[Do[Do[tmp = chrisrule3[ki,kj,kk];
If[tmp === 0,
Continue[],
de = de+1;
Print[de," : ","c[",ki,",",-kj,",",kk,"] = ",Simplify[tmp]];
```

```
Print["   "]],{ki,1,dimension}],{kj,1,dimension}],{kk,1,dimension}],
If[(a===1 && b===1 && c===-1),Do[Do[Do[tmp = chrisrule5[ki,kj,kk];
If[tmp === 0,
Continue[],
de = de+1;
Print[de," : ","c[",ki,",",kj,",",-kk,"] = ",Simplify[tmp]];
Print["   "]],{ki,1,dimension}],{kj,1,dimension}],{kk,1,dimension}]]]]]]];
If[de===0,Print["No non zero ChristoffelSymbol"]]]
christoffelsymbolderivative[a_,b_,c_,wrt_]:=Simplify[chrisder[[a,b,c,wrt]]]
Ricciscalar[a_] := Scalarinvariant[a]
```

Appendix B: Package for All Riemann Tensor Components

```
AllRiemanntensor[a_,b_,c_,d_]:= If[( Abs[a] =!= 1 || Abs[a] =!= 1 ||Abs[c] =!= 1 ||
Abs[d] =!= 1 ),Print["Enter only +ve or -ve 1 only for all four arguments"],
If[(Sign[a]===-1 && Sign[b]===1 && Sign[c]===1 && Sign[d]===1),
(* 1. R[-a,b,c,d] *)Block[{tmp,cnt,ki,kj,kk,kl},cnt = 0;
Do[Do[Do[Do[If[tmp = Simplify[cur[[ki,kj,kk,kl]]];
tmp === 0,
Continue[],
cnt+=1;
Print[cnt,": ",R[-ki,kj,kk,kl]," = ",tmp]],{kl,kk,dimension}],{kk,1,dimension}],
{kj,1,dimension}],{ki,1,dimension}];
If[cnt === 0,Print["No non zero R[-a,b,c,d] for this metric"]]],
If[(Sign[a]===-1 && Sign[b]===-1 && Sign[c]===1 && Sign[d]===1),
(* 2. R[-a,-b,c,d] *)Block[{tmp,cnt,ki,kj,kk,kl},cnt = 0;
Do[Do[Do[Do[If[(kl === kk)||(kj === ki)||(kl <= kk)||(kj <= ki),
Continue[],
tmp = Simplify[temp[ki,kj,kk,kl]];
If[tmp === 0,
Continue[],
cnt+=1;
Print[cnt,": ",R[-ki,-kj,kk,kl]," = ",tmp]]],{kl,kj,dimension}],{kk,ki,dimension}],
{kj,1,dimension}],{ki,1,dimension}];
If[cnt === 0,Print["No non zero R[-a,-b,c,d] for this metric"]]],
If[(Sign[a]===1 && Sign[b]===1 && Sign[c]===1 && Sign[d]===1),
(* 3. R[a,b,c,d] *)Block[{tmp,cnt,ki,kj,kk,kl},cnt = 0;
Do[Do[Do[Do[If[(kl === kk)||(kj === ki)||(kl <= kk)||(kj <= ki),
Continue[],
tmp = Simplify[curlow[ki,kj,kk,kl]];
If[tmp === 0,
Continue[],
cnt+=1;
Print[cnt,": ",R[ki,kj,kk,kl]," = ",tmp]]],{kl,kj,dimension}],{kk,ki,dimension}],
{kj,1,dimension}],{ki,1,dimension}];
If[cnt === 0,Print["No non zero R[a,b,c,d] for this metric"]]],
If[(Sign[a]===-1 && Sign[b]===-1 && Sign[c]===-1 && Sign[d]===1),
(* 4.R[-a,-b,-c,d] *)Block[{tmp,cnt,ki,kj,kk,kl},cnt = 0;
Do[Do[Do[Do[If[tmp = Simplify[Riem4[ki,kj,kk,kl]];
tmp === 0,
Continue[],
cnt+=1;
Print[cnt,": ",R[-ki,-kj,-kk,kl]," = ",tmp]],{kl,1,dimension}],{kk,1,dimension}],
{kj,1,dimension}],{ki,1,dimension}];
If[cnt === 0,Print["No non zero R[-a,-b,-c,d] for this metric"]]],
```

```
If[(Sign[a]===-1 && Sign[b]===-1 && Sign[c]===-1 && Sign[d]===-1),
(*5. R[-a,-b,-c,-d] *)Block[{tmp,cnt,ki,kj,kk,kl},cnt = 0;
Do[Do[Do[Do[If[(kl === kk)|| (kj === ki)||(kl <= kk)||(kj <= ki),
Continue[],
tmp = Simplify[Riem5[ki,kj,kk,kl]];
If[tmp === 0,
Continue[],
cnt+=1;
Print[cnt,": ",R[-ki,-kj,-kk,-kl]," = ",tmp]]],{kl,kj,dimension}],{kk,ki,dimension}],
{kj,1,dimension}],{ki,1,dimension}];
If[cnt === 0,Print["No non zero R[-a,-b,-c,-d] for this metric"]]],
If[(Sign[a]===-1 && Sign[b]===1 && Sign[c]===-1 && Sign[d]===-1),
(* 6. R[-a,b,-c,-d] *)Block[{tmp,cnt,ki,kj,kk,kl},cnt = 0;
Do[Do[Do[Do[If[tmp = Simplify[Riem6[ki,kj,kk,kl]];
tmp === 0,
Continue[],
cnt+=1;
Print[cnt,": ",R[-ki,kj,-kk,-kl]," = ",tmp]],{kl,1,dimension}],{kk,1,dimension}],
{kj,1,dimension}],{ki,1,dimension}];
If[cnt === 0,Print["No non zero R[-a,b,-c,-d] for this metric"]]],
If[(Sign[a]===1 && Sign[b]===-1 && Sign[c]===-1 && Sign[d]===-1),
(* 7. R[a,-b,-c,-d] *)Block[{tmp,cnt,ki,kj,kk,kl},cnt = 0;
Do[Do[Do[Do[If[tmp = Simplify[Riem7[ki,kj,kk,kl]];
tmp === 0,
Continue[],
cnt+=1;
Print[cnt,": ",R[ki,-kj,-kk,-kl]," = ",tmp]],{kl,1,dimension}],{kk,1,dimension}],
{kj,1,dimension}],{ki,1,dimension}];
If[cnt === 0,Print["No non zero R[a,-b,-c,-d] for this metric"]]],
If[(Sign[a]===1 && Sign[b]===1 && Sign[c]===-1 && Sign[d]===-1),
(*8. R[a,b,-c,-d] *)Block[{tmp,cnt,ki,kj,kk,kl},cnt = 0;
Do[Do[Do[Do[If[(kl === kk)|| (kj === ki)||(kl <= kk)||(kj <= ki),
Continue[],
tmp = Simplify[Riem8[ki,kj,kk,kl]];
If[tmp === 0,
Continue[],
cnt+=1;
Print[cnt,": ",R[ki,kj,-kk,-kl]," = ",tmp]]],{kl,kj,dimension}],{kk,ki,dimension}],
{kj,1,dimension}],{ki,1,dimension}];
If[cnt === 0,Print["No non zero R[a,b,-c,-d] for this metric"]]],
If[(Sign[a]===-1 && Sign[b]===-1 && Sign[c]===1 && Sign[d]===-1),
(* 9. R[-a,-b,c,-d] *)Block[{tmp,cnt,ki,kj,kk,kl},cnt = 0;
Do[Do[Do[Do[If[tmp = Simplify[Riem9[ki,kj,kk,kl]];
tmp === 0,
Continue[],
cnt+=1;
Print[cnt,": ",R[-ki,-kj,kk,-kl]," = ",tmp]],{kl,1,dimension}],{kk,1,dimension}],
{kj,1,dimension}],{ki,1,dimension}];
If[cnt === 0,Print["No non zero R[-a,-b,c,-d] for this metric"]]],
If[(Sign[a]===-1 && Sign[b]===1 && Sign[c]===1 && Sign[d]===-1),
(* 10. R[-a,b,c,-d] *)Block[{tmp,cnt,ki,kj,kk,kl},cnt = 0;
Do[Do[Do[Do[If[tmp = Simplify[Riem10[ki,kj,kk,kl]];
tmp === 0,
Continue[],
cnt+=1;
```

```
Print[cnt,": ",R[-ki,kj,kk,-kl]," = ",tmp]],{kl,1,dimension}],{kk,1,dimension}],
{kj,1,dimension}],{ki,1,dimension}];
If[cnt === 0,Print["No non zero R[-a,b,c,-d] for this metric"]]],
If[(Sign[a]===1 && Sign[b]===-1 && Sign[c]===1 && Sign[d]===1),
(* 11. R[a,-b,c,d] *)Block[{tmp,cnt,ki,kj,kk,kl},cnt = 0;
Do[Do[Do[Do[If[tmp = Simplify[Riem11[ki,kj,kk,kl]];
tmp === 0,
Continue[],
cnt+=1;
Print[cnt,": ",R[ki,-kj,kk,kl]," = ",tmp]],{kl,1,dimension}],{kk,1,dimension}],
{kj,1,dimension}],{ki,1,dimension}];
If[cnt === 0,Print["No non zero R[a,-b,c,d] for this metric"]]],
If[(Sign[a]===1 && Sign[b]===1 && Sign[c]===1 && Sign[d]===-1),
(* 12. R[a,b,c,-d] *)Block[{tmp,cnt,ki,kj,kk,kl},cnt = 0;
Do[Do[Do[Do[If[tmp = Simplify[Riem12[ki,kj,kk,kl]];
tmp === 0,
Continue[],
cnt+=1;
Print[cnt,": ",R[ki,kj,kk,-kl]," = ",tmp]],{kl,1,dimension}],{kk,1,dimension}],
{kj,1,dimension}],{ki,1,dimension}];
If[cnt === 0,Print["No non zero R[a,b,c,-d] for this metric"]]],
If[(Sign[a]===-1 && Sign[b]===1 && Sign[c]===-1 && Sign[d]===1),
(* 13. R[a,b,-c,d] *)Block[{tmp,cnt,ki,kj,kk,kl},cnt = 0;
Do[Do[Do[Do[If[tmp = Simplify[Riem13[ki,kj,kk,kl]];
tmp === 0,
Continue[],
cnt+=1;
Print[cnt,": ",R[ki,kj,-kk,kl]," = ",tmp]],{kl,1,dimension}],{kk,1,dimension}],
{kj,1,dimension}],{ki,1,dimension}];
If[cnt === 0,Print["No non zero R[a,b,-c,d] for this metric"]]],
If[(Sign[a]===1 && Sign[b]===-1 && Sign[c]===-1 && Sign[d]===1),
(* 14. R[a,-b,-c,d] *)Block[{tmp,cnt,ki,kj,kk,kl},cnt = 0;
Do[Do[Do[Do[If[tmp = Simplify[Riem14[ki,kj,kk,kl]];
tmp === 0,
Continue[],
cnt+=1;
Print[cnt,": ",R[ki,-kj,-kk,kl]," = ",tmp]],{kl,1,dimension}],{kk,1,dimension}],
{kj,1,dimension}],{ki,1,dimension}];
If[cnt === 0,Print["No non zero R[a,-b,-c,d] for this metric"]]],
If[(Sign[a]===1 && Sign[b]===1 && Sign[c]===-1 && Sign[d]===1),
(* 15. R[a,-b,c,-d] *)Block[{tmp,cnt,ki,kj,kk,kl},cnt = 0;
Do[Do[Do[Do[If[tmp = Simplify[Riem15[ki,kj,kk,kl]];
tmp === 0,
Continue[],
cnt+=1;
Print[cnt,": ",R[ki,-kj,kk,-kl]," = ",tmp]],{kl,1,dimension}],{kk,1,dimension}],
{kj,1,dimension}],{ki,1,dimension}];
If[cnt === 0,Print["No non zero R[a,-b,c,-d] for this metric"]]],
If[(Sign[a]===1 && Sign[b]===-1 && Sign[c]===1 && Sign[d]===1),
(* 16. R[-a,b,-c,d] *)Block[{tmp,cnt,ki,kj,kk,kl},cnt = 0;
Do[Do[Do[Do[If[tmp = Simplify[Riem16[ki,kj,kk,kl]];
tmp === 0,
Continue[],
cnt+=1;
Print[cnt,": ",R[-ki,kj,-kk,kl]," = ",tmp]],{kl,1,dimension}],{kk,1,dimension}],
```

{kj,1,dimension}],{ki,1,dimension}];
If[cnt === 0,Print["No non zero R[-a,b,-c,d] for this metric"]]]]]]]]]]]]]]]]]]]]]]

Appendix C: Package for Ricci Tensor Components

```
Ricci1[a_,b_] := Block[{kd,tmp,sum}, (*R[a,b]*)
sum = 0;
tmp = 0;(*Print[rform[a,b]];(* R[a,b] *)*)
Do[tmp = cur[[kd,a,kd,b]];
sum = sum + tmp,{kd,1,dimension}];Simplify[sum]]
Ricci2[a_,b_] := Block[{kd,tmp,sum}, (*R[-a,-b]*)
sum = 0 ;
tmp = 0;(*Print[rform[a,b]];(* R[a,b] *)*)
Do[tmp = metriclistup[[kd,Abs[b]]] * Ricci3[-a,kd];
sum = sum + tmp,{kd,1,dimension}];Simplify[sum]]
Ricci3[a_,b_] := Block[{kd,tmp,sum}, (*R[-a,b]*)
sum = 0 ;
tmp = 0;(*Print[rform[a,b]];(* R[a,b] *)*)
Do[tmp = metriclistup[[Abs[a],kd]] * Ricci1[kd,b];
sum = sum + tmp,{kd,1,dimension}];Simplify[sum]]
Ricci4[a_,b_] := Block[{kd,tmp,sum}, (*R[a,-b]*)
sum = 0 ;
tmp = 0;(*Print[rform[a,b]];(* R[a,b] *)*)
Do[tmp = metriclistup[[kd,Abs[b]]] * Ricci1[a,kd];
sum = sum + tmp,{kd,1,dimension}];Simplify[sum]]
Riccitensor[a_,b_]:= Block[{kd,tmp,sum},
If[(Sign[a]===1 && Sign[b]=== 1),Print[ricciform[a,b]];
Ricci1[Abs[a],Abs[b]],If[(Sign[a]=== -1 && Sign[b]=== -1),
Print[ricciform[a,b]];Ricci2[Abs[a],Abs[b]],
If[(Sign[a]=== -1 && Sign[b]=== 1),Print[ricciform[a,b]];
Ricci3[Abs[a],Abs[b]],If[(Sign[a]=== 1 && Sign[b]=== -1),
Print[ricciform[a,b]];Ricci4[Abs[a],Abs[b]]]]]]]]
```

Appendix D: Package for All Ricci Tensor Components

```
AllRiccitensor[a_,b_] := Block[{tmp,cnt,ki,kj},
cnt = 0;Print["Hang on! The Calculations are Underway !!!"];
Do[Do[If[Sign[a] === -1 && Sign[b] === -1,
tmp = Simplify[Ricci2[-ki,-kj]],If[Sign[a] === -1 && Sign[b] === 1,
tmp = Simplify[Ricci3[-ki,kj]],If[Sign[a] === 1 && Sign[b] === -1,
tmp = Simplify[Ricci4[ki,-kj]],If[Sign[a] === 1 && Sign[b] === 1,
tmp = Simplify[Ricci1[ki,kj]]]]]];
If[tmp === 0,Continue[],cnt+=1;Print[cnt,": ",
R[Sign[a]*ki,Sign[b]*kj]," = ",tmp]],{kj,1,dimension}],{ki,1,dimension}];
If[cnt === 0,Print[" "];Print["No non zero R[ ",a,",",b," ] for this metric"]]]
```

Appendix E: Package for All Weyl Tensor Components

```
AllWeyltensor[a_,b_,c_,d_]:=If[( Abs[a] =!= 1 || Abs[a] =!= 1 || Abs[c] =!= 1 ||
```

426

```
Abs[d] =!= 1 ),Print["Enter only +ve or -ve 1 only for all four arguments"],
If[(Sign[a]===-1 && Sign[b]===1 && Sign[c]===1 && Sign[d]===1),
(* 1. W[-a,b,c,d] *)Block[{tmp,cnt,ki,kj,kk,kl},cnt = 0;
Do[Do[Do[If[tmp = Simplify[weyleup[Abs[a],b,c,d]];
tmp === 0,
Continue[],
cnt+=1;
Print[cnt,": ",W[-ki,kj,kk,kl]," =   ",tmp]],{kl,kk,dimension}],{kk,1,dimension}],
{kj,1,dimension}],{ki,1,dimension}];
If[cnt === 0,Print["No non zero W[-a,b,c,d] for this metric"]]],
If[(Sign[a]===-1 && Sign[b]===-1 && Sign[c]===1 && Sign[d]===1),
(* 3. W[-a,-b,c,d] *)Block[{tmp,cnt,ki,kj,kk,kl},cnt = 0;
Do[Do[Do[Do[If[(kl === kk)|| (kj === ki)||(kl <= kk)||(kj <= ki),
Continue[],
tmp = Simplify[Weyl3];
If[tmp === 0,
Continue[],
cnt+=1;
Print[cnt,": ",W[-ki,-kj,kk,kl]," = ",tmp]]],{kl,kj,dimension}],{kk,ki,dimension}],
{kj,1,dimension}],{ki,1,dimension}];
If[cnt === 0,Print["No non zero W[-a,c,d] for this metric"]]],
If[(Sign[a]===1 && Sign[b]===1 && Sign[c]===1 && Sign[d]===1),
(* 2. W[a,b,c,d] *)Block[{tmp,cnt,ki,kj,kk,kl},cnt = 0;
Do[Do[Do[Do[If[(kl === kk)|| (kj === ki)||(kl <= kk)||(kj <= ki),
Continue[],
tmp = Simplify[weylten[ki,kj,kk,kl]];
If[tmp === 0,
Continue[],
cnt+=1;
Print[cnt,": ",W[ki,kj,kk,kl]," = ",tmp]]],{kl,kj,dimension}],{kk,ki,dimension}],
{kj,1,dimension}],{ki,1,dimension}];
If[cnt === 0,Print["No non zero W[a,b,c,d] for this metric"]]],
If[(Sign[a]===-1 && Sign[b]===-1 && Sign[c]===-1 && Sign[d]===1),
(* 4. W[-a,-b,-c,d] *)Block[{tmp,cnt,ki,kj,kk,kl},cnt = 0;
Do[Do[Do[If[tmp = Simplify[Weyl4[ki,kj,kk,kl]];
tmp === 0,
Continue[],
cnt+=1;
Print[cnt,": ",W[-ki,-kj,-kk,kl]," = ",tmp]],{kl,1,dimension}],{kk,1,dimension}],
{kj,1,dimension}],{ki,1,dimension}];
If[cnt === 0,Print["No non zero W[-a,-b,-c,d] for this metric"]]],
If[(Sign[a]===-1 && Sign[b]===-1 && Sign[c]===-1 && Sign[d]===-1),
(*5. W[-a,-b,-c,-d] *)Block[{tmp,cnt,ki,kj,kk,kl},cnt = 0;
Do[Do[Do[Do[If[(kl === kk)|| (kj === ki)||(kl <= kk)||(kj <= ki),
Continue[],
tmp = Simplify[Weyl5[ki,kj,kk,kl]];
If[tmp === 0,
Continue[],
cnt+=1;
Print[cnt,": ",W[-ki,-kj,-kk,-kl]," = ",tmp]]],{kl,kj,dimension}],{kk,ki,dimension}],
{kj,1,dimension}],{ki,1,dimension}];
If[cnt === 0,Print["No non zero W[-a,-b,-c,-d] for this metric"]]],
If[(Sign[a]===-1 && Sign[b]===1 && Sign[c]===-1 && Sign[d]===-1),
(* 6. W[-a,b,-c,-d] *)Block[{tmp,cnt,ki,kj,kk,kl},cnt = 0;
Do[Do[Do[Do[If[tmp = Simplify[Weyl6[ki,kj,kk,kl]];
```

```
tmp === 0,
Continue[],
cnt+=1;Print[cnt,":                    ",W[-ki,kj,-kk,-kl],"                        =
",tmp]],{kl,1,dimension}],{kk,1,dimension}],
{kj,1,dimension}],{ki,1,dimension}];
If[cnt === 0,Print["No non zero W[-a,b,-c,-d] for this metric"]]],
If[(Sign[a]===1 && Sign[b]===-1 &&Sign[c]===-1 && Sign[d]===-1),
(* 7. W[a,-b,-c,-d] *)Block[{tmp,cnt,ki,kj,kk,kl},cnt = 0;
Do[Do[Do[Do[If[tmp = Simplify[Weyl7[ki,kj,kk,kl]];
tmp === 0,
Continue[],
cnt+=1;Print[cnt,":                    ",W[ki,-kj,-kk,-kl],"                        =
",tmp]],{kl,1,dimension}],{kk,1,dimension}],
{kj,1,dimension}],{ki,1,dimension}];
If[cnt === 0,Print["No non zero W[a,-b,-c,-d] for this metric"]]],
If[(Sign[a]===1 && Sign[b]===1 && Sign[c]===-1 && Sign[d]===-1),
(*8. W[a,b,-c,-d] *)Block[{tmp,cnt,ki,kj,kk,kl},cnt = 0;
Do[Do[Do[Do[If[(kl === kk)|| (kj === ki)||(kl <= kk)||(kj <= ki),
Continue[],
tmp = Simplify[Weyl8[ki,kj,kk,kl]];
If[tmp === 0,
Continue[],
cnt+=1;
Print[cnt,": ",W[ki,kj,-kk,-kl]," = ",tmp]]],{kl,kj,dimension}],{kk,ki,dimension}],
{kj,1,dimension}],{ki,1,dimension}];
If[cnt === 0,Print["No non zero W[a,b,-c,-d] for this metric"]]],
If[(Sign[a]===-1 && Sign[b]===-1 && Sign[c]===1 && Sign[d]===-1),
(* 9. W[-a,-b,c,-d] *)Block[{tmp,cnt,ki,kj,kk,kl},cnt = 0;
Do[Do[Do[Do[If[tmp = Simplify[Weyl9[ki,kj,kk,kl]];
tmp === 0,
Continue[],
cnt+=1;Print[cnt,":                    ",W[-ki,-kj,kk,-kl],"                        =
",tmp]],{kl,1,dimension}],{kk,1,dimension}],
{kj,1,dimension}],{ki,1,dimension}];
If[cnt === 0,Print["No non zero W[-a,-b,c,-d] for this metric"]]],
If[(Sign[a]===-1 && Sign[b]===1 && Sign[c]===1 && Sign[d]===-1),
(* 10. W[-a,b,c,-d] *)Block[{tmp,cnt,ki,kj,kk,kl},cnt = 0;
Do[Do[Do[Do[If[tmp = Simplify[Weyl10[ki,kj,kk,kl]];
tmp === 0,
Continue[],
cnt+=1;Print[cnt,":                    ",W[-ki,kj,kk,-kl],"                        =
",tmp]],{kl,1,dimension}],{kk,1,dimension}],
{kj,1,dimension}],{ki,1,dimension}];
If[cnt === 0,Print["No non zero W[-a,b,c,-d] for this metric"]]],
If[(Sign[a]===1 && Sign[b]===-1 && Sign[c]===1 && Sign[d]===1),
(* 11. W[a,-b,c,d] *)Block[{tmp,cnt,ki,kj,kk,kl},cnt = 0;
Do[Do[Do[Do[If[tmp = Simplify[Weyl11[ki,kj,kk,kl]];
tmp === 0,
Continue[],
cnt+=1;Print[cnt,":                    ",W[ki,-kj,kk,kl],"                        =
",tmp]],{kl,1,dimension}],{kk,1,dimension}],
{kj,1,dimension}],{ki,1,dimension}];
If[cnt === 0,Print["No non zero W[a,-b,c,d] for this metric"]]],
If[(Sign[a]===1 && Sign[b]===1 && Sign[c]===1 && Sign[d]===-1),
(* 12. W[a,b,c,-d] *)Block[{tmp,cnt,ki,kj,kk,kl},cnt = 0;
```

428

```
Do[Do[Do[Do[If[tmp = Simplify[Weyl12[ki,kj,kk,kl]]];
tmp === 0,
Continue[],
cnt+=1;
Print[cnt,": ",W[ki,kj,kk,-kl]," = ",tmp]],{kl,1,dimension}],{kk,1,dimension}],
{kj,1,dimension}],{ki,1,dimension}];
If[cnt === 0,Print["No non zero W[a,b,c,-d] for this metric"]]],
If[(Sign[a]===-1 && Sign[b]===1 && Sign[c]===-1 && Sign[d]===1),
(* 13. W[a,b,-c,d] *)Block[{tmp,cnt,ki,kj,kk,kl},cnt = 0;
Do[Do[Do[Do[If[tmp = Simplify[Weyl13[ki,kj,kk,kl]]];
tmp === 0,
Continue[],
cnt+=1;Print[cnt,":                    ",W[ki,kj,-kk,kl],"                    =
",tmp]],{kl,1,dimension}],{kk,1,dimension}],
{kj,1,dimension}],{ki,1,dimension}];
If[cnt === 0,Print["No non zero W[a,b,-c,d] for this metric"]]],
If[(Sign[a]===1 && Sign[b]===-1 && Sign[c]===-1 && Sign[d]===1),
 (* 14. W[a,-b,-c,d] *)Block[{tmp,cnt,ki,kj,kk,kl},cnt = 0;
Do[Do[Do[Do[If[tmp = Simplify[Weyl14[ki,kj,kk,kl]]];
tmp === 0,
Continue[],
cnt+=1;
Print[cnt,": ",W[ki,-kj,-kk,kl]," = ",tmp]],{kl,1,dimension}],{kk,1,dimension}],
{kj,1,dimension}],{ki,1,dimension}];
If[cnt === 0,Print["No non zero W[a,-b,-c,d] for this metric"]]],
If[(Sign[a]===1 && Sign[b]===1 && Sign[c]===-1 && Sign[d]===1),
(* 15. W[a,-b,c,-d] *)Block[{tmp,cnt,ki,kj,kk,kl},cnt = 0;
Do[Do[Do[Do[If[tmp = Simplify[Weyl15[ki,kj,kk,kl]]];
tmp === 0,
Continue[],
cnt+=1;Print[cnt,":                    ",W[ki,-kj,kk,-kl],"                    =
",tmp]],{kl,1,dimension}],{kk,1,dimension}],
{kj,1,dimension}],{ki,1,dimension}];
If[cnt === 0,Print["No non zero W[a,-b,c,-d] for this metric"]]],
If[(Sign[a]===1 && Sign[b]===-1 && Sign[c]===1 && Sign[d]===1),
 (* 16. W[-a,b,-c,d] *)Block[{tmp,cnt,ki,kj,kk,kl}, cnt = 0;
Do[Do[Do[Do[If[tmp = Simplify[Weyl16[ki,kj,kk,kl]]];
tmp === 0,
Continue[],
cnt+=1;Print[cnt,":                    ",W[-ki,kj,-kk,kl],"                    =
",tmp]],{kl,1,dimension}],{kk,1,dimension}],
{kj,1,dimension}],{ki,1,dimension}];
If[cnt === 0,Print["No non zero W[-a,b,-c,d] for this metric"]]]]]]]]]]]]]]]]]]]
```

Appendix F: Package for All Einstein Tensor Components

```
Ein1[a_,b_] := Block[{kd,tmp,sum}, (* Einstein[a,b]*)
Simplify[Ricci1[a,b] - (1/2)*Ricciscalar1*metriclist[[a,b]]]]
Ein2[a_,b_] := Block[{kd,tmp,sum}, (* Einstein[-a,-b]*)
sum = 0 ;
tmp = 0;
Do[tmp = metriclistup[[kd,Abs[b]]] * Ein3[a,kd];
sum = sum + tmp,{kd,1,dimension}];
Simplify[sum]]Ein3[a_,b_] := Block[{kd,tmp,sum}, (* Einstein[-a,b]*)sum = 0 ;
```

429

```
tmp = 0;
Do[tmp = metriclistup[[Abs[a],kd]] * Ein1[kd,b];
sum = sum + tmp,{kd,1,dimension}];
Simplify[sum]]Ein4[a_,b_] := Block[{kd,tmp,sum}, (* Einstein[a,-b]*)sum = 0 ;
tmp = 0;
Do[tmp = metriclistup[[kd,Abs[b]]] * Ein1[a,kd];
sum = sum + tmp,{kd,1,dimension}];
Simplify[sum]]Einsteintensor[a_,b_]:= Block[{kd,tmp,sum},
If[(Sign[a]===1  &&  Sign[b]=== 1),Print["Hang on! The Caculations are
Underway!!!"];
Print[ColumnForm[{StringJoin[ToString[""],ToString[""]],ToString["E"],
StringJoin[ToString[" "],ToString[Abs[a]],ToString[Abs[b]]]}]];
Ein1[a,b],If[(Sign[a]=== -1 && Sign[b]=== -1),
Print["Hang on! The Caculations are Underway!!!"];
Print[ColumnForm[{StringJoin[ToString["],ToString[Abs[a]],
ToString[Abs[b]]],ToString["E"]}]];
Ein2[a,b],If[(Sign[a]=== -1 && Sign[b]=== 1),
Print["Hang on! The Caculations are Underway!!!"];
Print[ColumnForm[{StringJoin[ToString[" "],ToString[Abs[a]]],ToString["E"],
StringJoin[ToString[" "],ToString[Abs[b]]]}]];
Ein3[a,b],If[(Sign[a]=== 1 && Sign[b]=== -1),
Print["Hang on! The Caculations are Underway!!!"];
Print[ColumnForm[{StringJoin[ToString[" "],ToString[Abs[b]]],ToString["E"],
StringJoin[ToString[" "],ToString[Abs[a]]]}]];
Ein4[a,b]]]]] (* To Find all non zero E[a,b] together*)
AllEinsteintensor[a_,b_] := Block[{tmp,cnt,ki,kj},cnt = 0;
Print["Hang on! The Calculations are Underway !!!"];
Do[Do[If[Sign[a] === -1 && Sign[b] === -1,tmp = Simplify[Ein2[-ki,-kj]],
If[Sign[a] === -1 && Sign[b] === 1,tmp = Simplify[Ein3[-ki,kj]],
If[Sign[a] === 1 && Sign[b] === -1,tmp = Simplify[Ein4[ki,-kj]],
If[Sign[a] === 1 && Sign[b] === 1,tmp = Simplify[Ein1[ki,kj]]]]]];
If[tmp === 0,
Continue[],
cnt+=1;
Print[cnt,":                        ",E[Sign[a]*ki,Sign[b]*kj]," =
",tmp]],{kj,1,dimension}],{ki,1,dimension}];
If[cnt === 0,Print[" "];
Print["No non zero R[ ",a,",",b," ] for this metric"]]]
```

Appendix G: Package for Three Scalar Invariants

```
Ricciscalar := Block[{tmp,kf,f1,f2,f3,f4,f5,f6},
Print["Hang on! The Calculations are underway!!!"];
Print["R"];
Simplify[Sum[Sum[metriclistup[[f1,f2]]*Ricci1[f1,f2],{f2,1,dimension}],
{f1,1,dimension}]]]
Ricciscalar1                                                          :=
Block[{tmp,kf,f1,f2,f3,f4,f5,f6},Simplify[Sum[Sum[metriclistup[[f1,f2]]*Ricci1[f1,f2
],
{f2,1,dimension}],{f1,1,dimension}]]]
Scalarinvariant[a_]:=Block[{tmp,kf,f1,f2,f3,f4,f5,f6},
If[a===1,Print["Hang on! The Calculations are underway!!!"];
Print[ColumnForm[{ToString[" 1"],ToString["R"]}]];
Simplify[Sum[Sum[Sum[Sum[temp[f1,f2,f3,f4]*temp[f3,f4,f1,f2],
{f4,1,dimension}],{f3,1,dimension}],{f2,1,dimension}],{f1,1,dimension}]],
```

```
If[a===2,Print["Hang on! The Calculations are underway!!!"];
Print[ColumnForm[{ToString[" 2"],ToString["R"]}]];
Simplify[Sum[Sum[Sum[Sum[Sum[Sum[temp[f1,f2,f3,f4]*temp[f3,f4,f5,f6]*
temp[f5,f6,f1,f2],{f6,1,dimension}],{f5,1,dimension}],{f4,1,dimension}],
{f3,1,dimension}],{f2,1,dimension}],{f1,1,dimension}]],
Print["This will find only R1,R2"]]]]
```

Appendix H: Package for Format Definitions

```
Format[rform[a_:ka,          b_:kb,          c_:kc,          d_:kd,e_:ke]]          :=
   Block[{R},If[e===1,R="W",R="R"];
If[(Sign[a]===-1 && Sign[b]===1 && Sign[c]===1 && Sign[d]===1),
ColumnForm[{StringJoin["Hang on! The Calculations are underway!!! \n",
ToString[Abs[a]]], StringJoin[ToString[R]," "], StringJoin[" ", ToString[b],
ToString[c], ToString[d]]}], If[(Sign[a]===-1 && Sign[b]===-1 && Sign[c]===1
&& Sign[d]===1),  ColumnForm[{StringJoin["Hang on! The Calculations are
underway!!! \n", ToString[Abs[a]], ToString[Abs[b]]], StringJoin[ToString[R], " "],
StringJoin[" ",  ToString[c], ToString[d]]}], If[(Sign[a]===1 && Sign[b]===1 &&
Sign[c]===1 && Sign[d]===1),
ColumnForm[{StringJoin["Hang on! The Calculations are underway!!! \n",
ToString[R], " "], StringJoin[" ", ToString[a],ToString[b],  ToString[c],
ToString[d]]}], If[(Sign[a]===-1 && Sign[b]===-1 && (*start*) Sign[c]===-1 &&
Sign[d]===1),   (*R#4*) ColumnForm[{StringJoin["Hang on! The Calculations are
underway!!! \n", " ",StringJoin[ToString[Abs[a]], ToString[Abs[b]],
ToString[Abs[c]]]], StringJoin[ToString[R]," "],  StringJoin[" ",ToString[Abs[d]]]}],
If[(Sign[a]===-1 && Sign[b]===-1 && Sign[c]===-1 && Sign[d]===-1), (*R#5*)
ColumnForm[{StringJoin ["Hang on! The Calculations are underway!!! \n", "
",StringJoin[ToString[Abs[a]], ToString[Abs[b]], ToString[Abs[c]],
ToString[Abs[d]]]], StringJoin[ToString[R]," "], StringJoin[" "]}],
If[(Sign[a]===-1 && Sign[b]===1 && (*R#6*)Sign[c]===-1 && Sign[d]===-1),
ColumnForm[{StringJoin["Hang on! The Calculationsare underway!!! \n", "",
StringJoin[ToString[Abs[a]]," ", ToString[Abs[c]], ToString[Abs[d]]]],
StringJoin[ToString[R]," "], StringJoin[" "," ", ToString[Abs[b]], " "," "]}],
If[(Sign[a]===1 && Sign[b]===-1 && (*R#7*)Sign[c]===-1 && Sign[d]===-1),
ColumnForm[{StringJoin["Hang on! The Calculations are underway!!! \n", " ",
StringJoin[" ",ToString[Abs[b]],ToString[Abs[c]], ToString[Abs[d]]]],
StringJoin[ToString[R]," "],  StringJoin[" ",ToString[Abs[a]]," ", " "," "]}],
If[(Sign[a]===1 && Sign[b]===1 && (*R#8*) Sign[c]===-1 && Sign[d]===-1),
ColumnForm[{StringJoin["Hang on! The Calculations are underway!!! \n", " ",
StringJoin[" "," ",ToString[Abs[c]],ToString[Abs[d]]]], StringJoin[ToString[R]," "],
StringJoin[" ",ToString[Abs[a]],ToString[Abs[b]], " "," "]}],
If[(Sign[a]===-1 && Sign[b]===-1 && (*R#9*) Sign[c]===1 && Sign[d]===-1),
ColumnForm[{StringJoin["Hang on! The Calculations are underway!!! \n", " ",
StringJoin[ToString[Abs[a]],ToString[Abs[b]]," ",ToString[Abs[d]]]],
StringJoin[ToString[R]," "], StringJoin[" ", " "," "," ",ToString[Abs[c]]," "]}],
If[(Sign[a]===-1 && Sign[b]===1 && (*R#10*)Sign[c]===1 && Sign[d]===-1),
ColumnForm[{StringJoin["Hang on! The Calculations are underway!!! \n", " ",
StringJoin[ToString[Abs[a]]," "," ",ToString[Abs[d]]]],StringJoin[ToString[R]," "],
StringJoin[" ", " "," ",ToString[Abs[b]],ToString[Abs[c]]," "]}],
If[(Sign[a]===1 && Sign[b]===-1 && (*R#11*)Sign[c]===1 && Sign[d]===1),
ColumnForm[{StringJoin["Hang on! The Calculations are underway!!! \n", " "," ",
StringJoin[" ",ToString[Abs[b]]," "," "]],StringJoin[ToString[R]," "],StringJoin[" ",
ToString[Abs[a]]," ",ToString[Abs[c]],ToString[Abs[d]]]}],
If[(Sign[a]===1 && Sign[b]===1 && (*R#12*)Sign[c]===1 && Sign[d]===-1),
```

```
ColumnForm[{StringJoin["Hang on! The Calculations are underway!!! \n", " ",
StringJoin[" "," "," ",ToString[Abs[d]]]],StringJoin[ToString[R]," "],StringJoin[" ",
ToString[Abs[a]],ToString[Abs[b]],ToString[Abs[c]]," "]}],
If[(Sign[a]===-1 && Sign[b]===1 && (*R#13*)Sign[c]===-1 && Sign[d]===1),
ColumnForm[{StringJoin["Hang on! The Calculations are underway!!! \n", " ",
StringJoin[ ToString[Abs[a]]," ",ToString[Abs[c]]," "]],StringJoin[ToString[R]," "],
StringJoin[" "," ",ToString[Abs[b]]," ",ToString[Abs[d]]]]}],
If[(Sign[a]===1 && Sign[b]===-1 && (*R#14*)Sign[c]===-1 && Sign[d]===1),
ColumnForm[{StringJoin["Hang on!The Calculations are underway!!! \n", " ",
StringJoin[" ", ToString[Abs[b]],ToString[Abs[c]]," "]],StringJoin[ToString[R]," "],
StringJoin[" ", ToString[Abs[a]]," "," ",ToString[Abs[d]]]]}],
If[(Sign[a]===1 && Sign[b]===1 && (*R#15*)Sign[c]===-1 && Sign[d]===1),
ColumnForm[{StringJoin["Hang on! The Calculationsare underway!!! \n", " ",
StringJoin[" "," ",ToString[Abs[c]]," "]], StringJoin[ToString[R]," "],
StringJoin[" ",ToString[Abs[a]],ToString[Abs[b]]," ", ToString[Abs[d]]]]}],
If[(Sign[a]===1 && Sign[b]===-1 && (*R#16*)Sign[c]===1 && Sign[d]===1),
ColumnForm[{StringJoin["Hang on! The Calculations are underway!!! \n", " ",
StringJoin[" ",ToString[Abs[b]]," "," "]],StringJoin[ToString[R]," "],
StringJoin[" ",ToString[Abs[a]]," ",ToString[Abs[c]],ToString[Abs[d]]]]}],
If[(a=!=ka && b=!=kb && c===kc && d===kd), ColumnForm[{StringJoin["Hang
on! The Calculations are underway!!! \n",ToString[R], " "],
StringJoin[" ", ToString[a], ToString[b]]}], If[(a=!=ka && b===kb && c===kc &&
d===kd), ColumnForm[{StringJoin["Hang on! The Calculations are underway!!! \n",
ToString[R], " "], StringJoin[" ", ToString[a]]}]]]]]]]]]]]]]]]]]]]
Format[crform[a_:ka, b_:kb, c_:kc]] := Block[{C}, (*[-a,b,c]*)
If[(Sign[a]===-1 && Sign[b]===1 && Sign[c]===1 ),
ColumnForm[{StringJoin["Hang on! The Calculations are underway!!! \n",
StringJoin[ToString[" "],ToString[Abs[a]]]],
StringJoin[ToString ["C"]," "], StringJoin[ToString[" "], ToString[b], ToString[c]]}],
(*[a,b,c]*) If[(Sign[a]=== 1 && Sign[b]===1 && Sign[c]===1 ),
ColumnForm[{StringJoin["Hang on! The Calculations are underway!!! \n",
StringJoin[ToString[" "],ToString[""]]], StringJoin[ToString["C"]], StringJoin[" ",
ToString[Abs[a]],ToString[b], ToString[c]]}],(*[-a,-b,c]*)
If[(Sign[a]===-1 && Sign[b]=== -1 && Sign[c]===1 ),
ColumnForm[{StringJoin["Hang on! The Calculations are underway!!! \n",
StringJoin[ToString[" "],ToString[Abs[a]],ToString[Abs[b]]]],
StringJoin[ToString["C"]," "], StringJoin[" ", ToString[" "], ToString[c]]}],(*[-a,-b,-
c]*)
If[(Sign[a]===-1 && Sign[b]=== -1 && Sign[c]=== -1 ),
ColumnForm[{StringJoin["Hang on! The Calculations are underway!!! \n",
StringJoin[ToString[" "],ToString[Abs[a]],ToString[Abs[b]],ToString[Abs[c]]]],
StringJoin[ToString["C"]," "], StringJoin[" ", ToString[" "], ToString[""]]}],(*[-a,b,-
  c]*)
If[(Sign[a]===-1 && Sign[b]=== 1 && Sign[c]=== -1 ),
ColumnForm[{StringJoin["Hang on! The Calculations are underway!!! \n",
StringJoin[ToString[" "],ToString[Abs[a]], ToString[" "],ToString[Abs[c]]]],
StringJoin[ToString["C"],""],StringJoin[ToString[" "], ToString[Abs[b]],
ToString[""]]}],
(*[a,-b,c]*)If[(Sign[a]=== 1 && Sign[b]=== -1 && Sign[c]=== 1 ),
ColumnForm[{StringJoin["Hang on! The Calculations are underway!!! \n",
StringJoin[ToString[" "],ToString[Abs[b]], ToString[" "],ToString[" "]]],
StringJoin[ToString["C"]," "], StringJoin[" ", ToString[Abs[a]], ToString[" "],
ToString[Abs[c]]]}], (*[a,b,-c]*) If[(Sign[a]=== 1 && Sign[b]=== 1 && Sign[c]===
-1), ColumnForm[{StringJoin["Hang on! The Calculationsare underway!!! \n",
StringJoin[ToString[" "],ToString[" "],ToString[" "],ToString[Abs[c]]]],
```

```
StringJoin[ToString["C"]," "], StringJoin[" ", ToString[Abs[a]], ToString[Abs[b]]]}],
(*[a,-b,-c]*) If[(Sign[a]=== 1 && Sign[b]=== -1 && Sign[c]=== -1 ),
ColumnForm[{StringJoin["Hang on! The Calculations are underway!!! \n",
StringJoin[ToString[" "],ToString[" "], ToString[Abs[b]],ToString[Abs[c]]]]],
StringJoin[ToString["C"]," "], StringJoin[" ", ToString[Abs[a]], ToString["
   "]]}]]]]]]]]]]
Format[ricciform[a_:ka, b_:kb]] := Block[{R}, If[(Sign[a]===-1 && Sign[b]===1),
ColumnForm[{StringJoin["Hang on! The Calculations are underway!!! \n",
StringJoin[ToString[" "],ToString[Abs[a]]]], StringJoin[ToString["R"]," "],
StringJoin[" ", ToString[b]]}], If[(Sign[a]=== -1 && Sign[b]=== -1),
ColumnForm[{StringJoin["Hang on! The Calculations are underway!!! \n",
StringJoin[ToString[" "],ToString[Abs[a]],ToString[Abs[b]]]]],
StringJoin[ToString["R"]], StringJoin[" ", ToString[""],ToString[""], ToString[""]]}],
If[(Sign[a]=== 1 && Sign[b]=== 1),ColumnForm[{StringJoin["Hang on! The
Calculationsare underway!!! \n", StringJoin[ToString["
"],ToString[""],ToString[""]]]],
StringJoin[ToString["R"]], StringJoin[" ", ToString[Abs[a]],ToString[Abs[b]],
ToString[""]]}]],If[(Sign[a]===1 && Sign[b]===-1),ColumnForm[{StringJoin["Hang
on! The Calculations are underway!!! \n", StringJoin[ToString["
"],ToString[Abs[b]]]]],
StringJoin[ToString["R"]," "], StringJoin[" ", ToString[Abs[a]]]}]]]]]]]
```

Max-Packet Generation Process For Removing Skew In Network Multimedia Communication- A Mathematical Model

S.Kandar[a], K.Roy[b], M.Barman[c], C.T.Bhunia[d]

[a,b,c]*Haldia Institute of Technology*
P.O.-HIT(W.B, India), PIN-721657
E-mail: shyamalenduk@yahoo.com
[d]*SMIEEE, Senior Associate ICTP, Italy, Director, BITM, Bolpur(W.B)*
E-mail:ctbhunia@vsnl.com

Abstract. Multimedia data are sensed by human. These types of data are error tolerable to some extent but delay intolerable. To provide multimedia services with a guaranteed QoS is a research challenge. The two important parameters that degrade QoS are jittering & Skew. Researchers studied different methods for reducing jittering & skew. Earlier investigations attempted to use buffer management and introducing variable delay in delivering buffer to meet the challenges of jittering and skew. In the current work, we investigate Max-packet (Large packet made of data of different services) to reduce the effect of skew.

Keywords: Skew, Max-Packet, Combined Packet, Only Video packet.

PACS: 89.20.Ff Computer science and technology

1. INTRODUCTION

In order to achieve some guaranteed QoS, two issues that are paramount importance particularly for continuous bit rate (CBR) services are: jitter and skew [1-4]. Jitter is caused due to the variable delay that occurs during transmission through network between the packets of a particular service, say only for audio or only for video. Jitter can be removed by providing a compensation buffer at the receiving side.

Skew refers to the variable delay between the two (or more) corresponding packets of two (or more services) during transportation in the network.

At the transmitter side say for two media components Audio & Video, Pi is the packet for Video and pi is the packet for Audio. At the time of transmission at the transmitter side say T is instant of transmission of K th sample of media1(Video). So in the transmitter side T will also be the instant of transmission of K th sample of media2(Audio). At the receiver side say at T/ instant of time K th sample of media1 (Video) is received. Say at T// instant of time the K th sample of media2 (Audio) is received. Skew will occur when T/ ≠ T// [Fig: 1].

If skew occurs, at the receiver side there will be a mismatch between audio & video. This will affect the user in realizing the multimedia data. This type of mismatch occurs due to different types of delay in the network. In order to remove skew and jittering, different researchers [7-8] investigated different techniques. From implementation point of view, these techniques are hard to provide guaranteed QoS. Basically they are supposed to provide a pre defined perceived Quality of Service.

CP1146, *Modelling of Engineering and Technological Problems,* edited by A. H. Siddiqi, A. K. Gupta, and M. Brokate
© 2009 American Institute of Physics 978-0-7354-0683-4/09/$25.00

Transmission Side

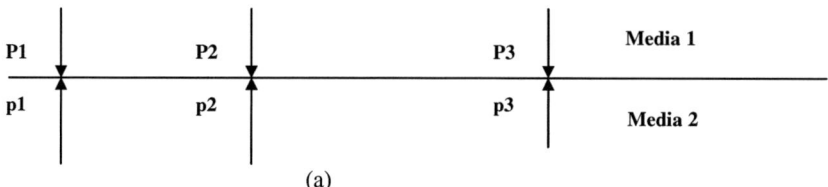

(a)

Receiving Side

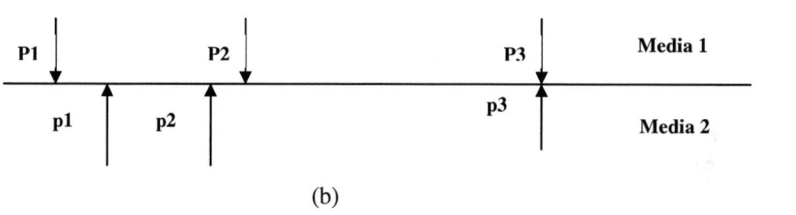

(b)

FIGURE 1. Occurrence of Skew

2. DISCUSSION ABOUT THE PROPOSED MAX-PACKET

Skew occurs due to the variable delay between the two (or more) corresponding packets of two (or more services) during transportation in the network. To remove skew we propose the concept of Max packet. Our goal is to provide guaranteed QoS. If we group the samples of two or mode media components generated with in a same time instant in a single packet then at the receiver side the variable delay between two or more corresponding packets of two or more services will not appear during transportation in the network. If we consider two media components as voice & video then the corresponding data of two Medias generated within a same time instant will be in the Max-Packet. The concept of Max-packet was studied elsewhere [5]

If a multimedia data consists of only audio & video then with in a same time instant it is seen that size of video data is much more than the size of audio data. For video the least frequency of a video signal is 33.4 MHz, where as for audio the maximum audible audio frequency is 20 KHz. For some instant of time the size of data for both of the components can vary. So the Max-Packet will be of variable length. In a Max-Packet there are more than one media components [For our case it is two, Audio & Video], so there must be a some extra bits in between two media data in the Max-Packet to differentiate between the two media components at the receiver side. As in the Max-Packet data of more than one media exist there must be some bits to indicate each type of media data within the packet.

In multimedia there are 6 types of multimedia data [Text, image, Graphics, Audio, Video & Animation], but those can be grouped into 4types [Text, Image, Audio, Video]. So two bits are sufficient to identify the media components [00, 01, 10, 11]. Yet we shall take 3 bits means 8 combinations. Say for 000 for Audio, 001 for Video, 011 for Text, 111 for Image. This bit pattern will be known to both sender & receiver side.

Flag(1 byte)	Address(1 to several byte) [Depends on the number of nodes]	Control bit (1 or two byte)	Voice (Variable Length)	Stuffed bit	Video (variable length)	Frame check sequence(2 to 4 byte)	Flag(1 byte)

FIGURE 2. Variable Length MAX-Packet

3. MAX-PACKET GENERATION PROCESS

On a laboratory scale Audio signal of frequency 20 KHz and video signal of 40 MHz are taken. These signals are sampled separately just at Nyquist rate (40 KHz for Audio and 80MHz for video) followed by conventional process of PCM (Pulse code modulation).Then the samples are needed to be quantized. In order to maintain the synchronization between audio and video signals the quantization level (Q_l) must be kept same for both of the signals. After quantization, the signals will be digitized. These digitized values will be put in packets. To remove skew the digitized values of the samples of audio & video with in a given time instant will be put in a single packet. For developing the packet, one media will be taken as Master and another will be taken as Slave. In our work Audio is taken as Master and Video is taken as Slave. In our research there is a frequency difference of 2×10^3 times to video signal than that of audio signal. That means if two audio samples are taken at time instant t_1 & t_2 , then for these two audio samples there will be $(t_2 - t_1) \times (2 * 10^3)$ number of video samples provided $t_2 > t_1$. If a combined packet is produced by taking the audio and video samples between t_1 & t_2, the packet size will be huge. By using the large packet the benefit of packet switching will not be obtained.

This can be handled in another way. One combined packets will be made from the quantized values, obtained from the samples at time instant t_1 ; and in time instant t_2 another combined packet will be made. Between t_1 & t_2 there will be (n-1) number of quantized slave sample, i.e. only video samples. Where n= C_f **video**/C_f **audio**, C_f stands for carrier frequency. m number of video samples will be grouped to form a Only Video packet. For (n-1) video samples number of packets will be

$$\lceil (n-1)/m \rceil \qquad\qquad (1)$$

For each packet [Combined or only video] there will be a header of 1 byte i.e. 8 bit. For combined packet the header is 00000000 & for only video packet the header is 11111111. Between two combined packets the only video packets will be numbered to do synchronization at the receiver side. There will no packet containing only audio data.

4. BLOCK DIAGRAM FOR MAX-PACKET GENERATION

From the discussion in the previous topic it is clear that the total process is divided into several stages.
a) Sampling of Audio signal
b) Sampling of Video signal
c) Combined packet Formation
d) Only video packet Formation

a) Sampling of Audio signal

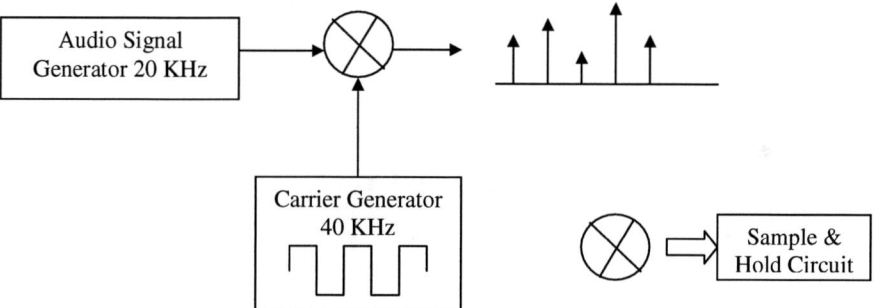

FIGURE 3. Sampling process of Audio signal

b) Sampling of Video signal

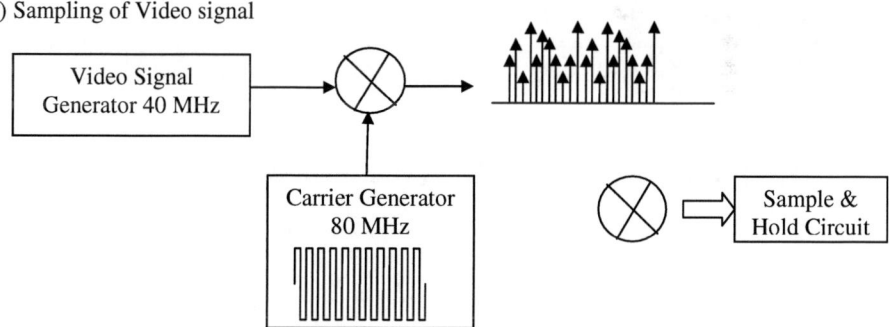

FIGURE 4. Sampling process of Video signal

c) Combined packet Formation

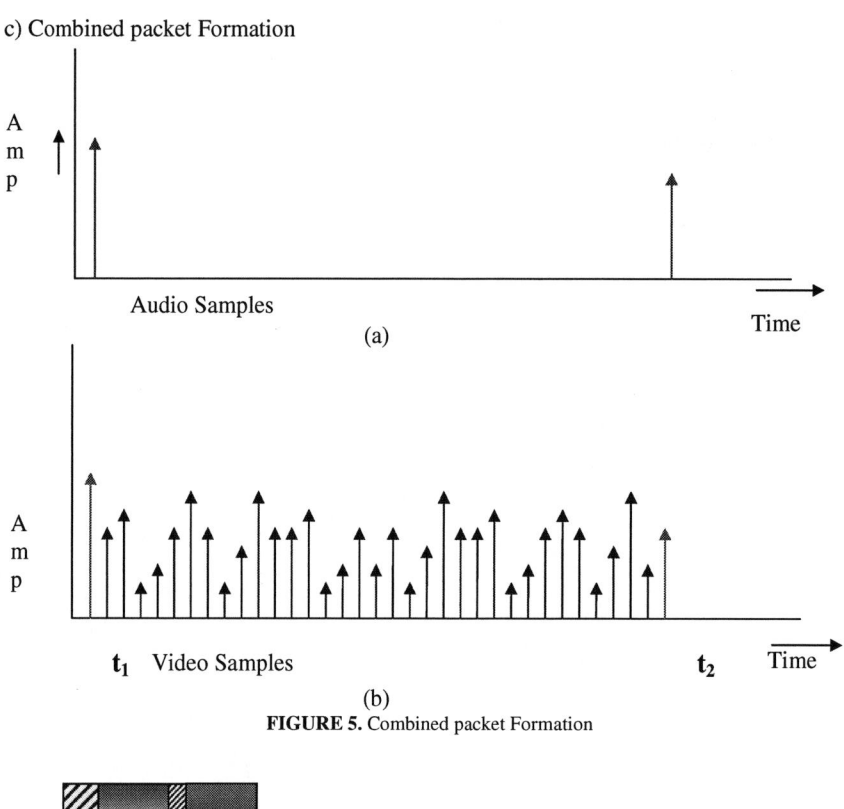

Audio Samples

(a)

Time

t_1 Video Samples

t_2 Time

(b)

FIGURE 5. Combined packet Formation

Header   Video Data   Separator Bit   Audio Data

FIGURE 6. Combined Packet containing Video & Audio data

d) Only Video Packet Formation

Header   m numbers of Video Samples

FIGURE 7. Only Video Packet

438

(n-1)/m number of Video Packets

Time

FIGURE 8. Packet Transmission from Source to Destination

5. CALCULATION OF OVERHEAD BITS

Overhead bits means how many number extra bits are sent with the original data at the time of packets transmission from sender to receiver. Overhead bits are added in the form of header of the packet or any stuffed bits. This calculation is necessary because number of overhead bit increases means size of the data increases means the transmission time from source to destination increases.

Let the multimedia data is transmitted within a time period T.

After a certain time period audio and video data are grouped and forms a combined packet. Let such type of first packet is formed at the time instant t_1 and the second such type of packet is formed at the time instant t_2.

So total number of combined packets within the time period T is $T/(t_2-t_1)$.

For each such type of combined packet the overhead bits are in the formed of Headed (h) and Separator bit (s).

So number of overhead bits for combined packets are

$$[T/(t_2-t_1)]*(h+s)] \tag{2}$$

Within the time period $(t_2 - t_1)$ there are

$\lceil (n-1)/m \rceil$ number of Only Video packets. Where n= quantization level and m= number of video samples in a single Only Video packet.

Total number of Only Video Packets within the time period T is

$$[T/(t_2-t_1)] * \lceil (n-1)/m \rceil \tag{3}$$

For each Only Video packet overhead bit is header(h). So total number of overhead bits for Only video packets is

$$h* [\{T/(t_2-t_1)\} * \lceil (n-1)/m \rceil] \tag{4}$$

So, during transmission of the total multimedia data the overhead bits are

$$[\{T/(t_2-t_1)\} * (h+s)] + h* [\{T/(t_2-t_1)\} * \lceil (n-1)/m \rceil]$$

$$= [T/(t_2-t_1)] * [h+s+h* \lceil (n-1)/m \rceil] \tag{5}$$

We like to make it future work a comparative gain of QoS with respect to overhead bits. The overhead bits provide a trade off with expected QoS.

6. CONCLUSION

We have proposed a concept of Max packet, which is proposed to eliminate effect of skew. By this technique skew will be totally removed but due to network delay there may occur jittering of video packets. We believe if the technique is implemented with other techniques namely accelerating & deaccelerating or buffer management technique of removing skew & jitter; we may derive it as a sound technique which will successfully remove skew & jittering in network multimedia communication and will provide Quality multimedia service. The proposed technique is under practical implementation & testing.

REFERENCES

1. C T bhunia, IT Network & Internet, New age International Publishers, New Delhi, 2005
2. Ralf Stteinmetz and Klara Nahrstedt, Multimedia: Computing, Communications and Applications, Pearson Education Asia, 1995
3. Jerry D Gibson, Multimedia Communications, Academic Press, 2001
4. John F Koegel Buford, Multimedia Systems, Addison Wesley, 2000
5. C.T. Bhunia, S.Kandar, M.Barman, Technique in Removing Skew and Analysis, National Conference RAFTICT, Seemanta Engineering College, Orissa ,2008, pp. 184 –189
6. Yan Xu et al, Calculation and Analysis of Compensation Buffer Size in Multimedia Systems, IEEE Communication Letters, Vol 5, No 8, 2001, pp. 355 –357
7. R Steinmetz, Human Perception of jitter and media synchronization, IEEE J Select Areas Commn, Vol 14, 1996, pp 61-72
8. C T Bhunia, Multimedia Services with Guaranteed QoS over Packet Networks/ Internet, J IE(I), Vol 87, Nov'2006, pp 38-42
9. C T Bhunia, Multilevel ATM --- a proposition, Proc National Conf on CCIS, Goa'2004, Vol II, pp 209-11

An Interior Point Method for Semidefinite Programming based on New Kernel Functions

M. Reza Peyghami[*][†]

[*] *Math Department, K.N. Toosi University of Technology, Tehran, Iran*
[†] *Institute for Studies in Theoretical Physics and Mathematics (IPM), Tehran, Iran*

Abstract

In this paper we consider generic primal-dual methods for solving semidefinite programming problems based on a new class of kernel function defined on the positive definite cone. By using some appealing and mild conditions of the new class, we prove by simple analysis that the new class based large update primal-dual interior point methods enjoy an $O\left(\sqrt{n}\log n\log\frac{n}{\varepsilon}\right)$ iteration bound to solve semidefinite programming problems with special choice of parameters.

Key words: Semidefinite programming, primal-dual interior-point method, large-update method, polynomial complexity.
MSC2000 Subject Classification: 90C22, 90C31

1 Introduction

Every one knows that Linear Optimization (LO) revitalized as an active area of research after the seminal paper of Karmarkar [7]. Lately the IPMs have shown their powers in solving LO and large classes of other optimization problems [15, 21, 23]. They are also the powerful tools to solve some widely used mathematical problems such as SDO, second order conic optimization and linear complementarity problems.

SDO problems are convex optimization problems over the intersection of an affine set and the cone of positive semidefinite matrices. Most of polynomial-time interior point methods designed for LO have been successfully extended to SDO [14, 19, 24]. Many researchers have studied SDO and obtained plentiful and beautiful results [18, 20].

Recently, Peng et al. [14] introduced a new variant of feasible IPMs based on Self-Regular (SR) proximity functions and also extended to SDO. They provided so far the best worst case theoretical complexity, namely $O\left(\sqrt{n}\log n\log\frac{n}{\varepsilon}\right)$, for large update feasible IPMs. Amini and Peyghami [1, 2] designed IPMs for LO based on a new class of kernel functions and they get $O\left(q\sqrt{n}\left(\log n\right)^{1+1/q}\log\frac{n}{\varepsilon}\right)$ iteration complexity for large update primal-dual IPMs which is $O\left(\sqrt{n}\left(\log n\right)^2\log\frac{n}{\varepsilon}\right)$.

Motivated by their work, and by introducing a new class of non SR kernel functions, we provide a simple analysis for the complexity of primal-dual IPMs for SDO, and prove that the complexity analysis of the new class for SDO matches the currently best known iteration bound for LO, with the special choices of the parameters of the new class.

The outline of the paper is as follows. In Section 2 we first describe special matrix functions used in later sections. Then we briefly recall the basic concepts of interior-point methods for SDO, such as central path, NT-search directions, etc.. In Section 3, we describe a primal-dual interior-point algorithm based on the proximity functions induced by the new class of kernel functions for SDO. In Section 4, we

CP1146, *Modelling of Engineering and Technological Problems*, edited by A. H. Siddiqi, A. K. Gupta, and M. Brokate
© 2009 American Institute of Physics 978-0-7354-0683-4/09/$25.00

present the properties of the kernel functions. We analyze the algorithm and derive the aforementioned complexity bound for large-update methods in Section 5. Finally, some concluding remarks follow in Section 6.

The following notations are used throughout the paper: The nonnegative and positive orthants are denoted by $\mathbb{R}^n_+$ and $\mathbb{R}^n_{++}$, respectively. $\mathcal{S}^n$ ($\mathcal{S}^n_+, \mathcal{S}^n_{++}$) denotes the cone of symmetric (positive semidefinite, positive) $n \times n$ matrices, respectively. We use the classical Löwner partial order "$\succeq$" for symmetric matrices means. The matrix inner product is defined by $A \bullet B = \mathrm{Tr}(A^T B)$. For any $Q \in \mathcal{S}^n_{++}$, the expression $Q^{\frac{1}{2}}$ denotes the symmetric square root of Q. For any $V \in \mathcal{S}^n_{++}$, the vector of eigenvalues of V arranged in non-increasing order is denoted by $\lambda(V)$, that is, $\lambda_1(V) \geq \lambda_2(V) \geq \ldots \geq \lambda_n(V)$. For any matrix M, we denote by $\sigma_1(M) \geq \sigma_2(M) \geq \ldots \geq \sigma_n(M)$ the singular values of M. Especially, if M is symmetric, then one has $\sigma_i(M) = |\lambda_i(M)|$, $i = 1, 2, \ldots, n$. The matrix E denotes $n \times n$ identity matrix.

2 Preliminaries

In this paper, we consider the SDO problem in the following standard form:

$$(P) \qquad \min \quad C \bullet X$$
$$s.t. \quad A_i \bullet X = b_i \qquad i = 1, \ldots, m$$
$$X \succeq 0$$

and its dual problem

$$(D) \qquad \max \quad b^T y$$
$$s.t. \quad \sum_{i=1}^{m} y_i A_i + S = C$$
$$S \succeq 0$$

where C and A_i ($1 \leq i \leq m$) are symmetric $n \times n$ matrices, and b, $y \in \mathbb{R}^m$. Furthermore, "$X \succeq 0$" (or $X \succ 0$) means that X is a symmetric positive semidefinite (or positive definite). The matrices A_i are assumed to be linearly independent. We also assume that (P) and (D) satisfy the Interior Point Condition (IPC), i.e., there exists $X \in \mathcal{P}$, $S \in \mathcal{D}$ with $X \succ 0$ and $S \succ 0$, where $\mathcal{P}$ and $\mathcal{D}$ denote the feasible set of problems (P) and (D), respectively. This can be achieved via the so-called self-dual embedding technique introduced by E. de Klerk et al. in [20]. Note that, as pointed out by various authors [8, 17, 20], for SDO it is impossible in general case to build up the duality theory as strongly as LO case.

We need to use some special matrix functions. Here, we list some known facts from linear algebra [19].

Theorem 2.1. (Spectral theorem for symmetric matrices [20]) *The real $n \times n$ matrix A is symmetric if and only if there exists an orthogonal basis with respect to which A is real and diagonal, i.e., if and only if there exists a matrix $U \in \mathbb{R}^{n \times n}$ such that $U^T U = E$ and $U^T AU = \Lambda$ where Λ is a diagonal matrix*

Note that the columns of the matrix U in Theorem 2.1 are the eigenvectors of the matrix A. Now, we are ready to define a matrix function.

Definition 2.2. Let V be any symmetric $n \times n$ matrix with eigen-decomposition

$$V = Q_V^T \mathrm{diag}\,(\lambda_1(V), \lambda_2(V), \ldots, \lambda_n(V))\, Q_V \tag{1}$$

where Q_V is any orthonormal matrix that diagonalizes V. The matrix function $\psi(V) : \mathcal{S}^n_{++} \to \mathcal{S}^n$ is defined by

$$\psi(V) \;=\; Q_V^T \mathrm{diag}\left(\psi\left(\lambda_1(V)\right), \psi\left(\lambda_2(V)\right), \ldots, \psi\left(\lambda_n(V)\right)\right) Q_V. \tag{2}$$

Moreover, the real valued matrix function (*proximity function* induced by a matrix function) $\Psi(V) : \mathcal{S}^n_{++} \to \mathbb{R}_+$ as follows:

$$\Psi(V) \;:=\; \mathrm{Tr}(\psi(V)) = \sum_{i=1}^{n} \psi(\lambda_i(V)) \tag{3}$$

where $\psi(V)$ is given by (2).

Suppose that $\psi(t)$ is twice differentiable for $t > 0$. The derivatives $\psi'(V)$ and $\psi''(V)$ are well-defined and obtain by replacing $\psi(\lambda_i(V))$ in (2) by $\psi'(\lambda_i(V))$ and $\psi''(\lambda_i(V))$, respectively, for each $1 \leq i \leq n$. In what follows we provide some concepts related to matrix functions. The results below can be found in [6, 9].

Definition 2.3. A matrix $M(t)$ is said to be a matrix of functions if each entry of $M(t)$ is a function of t, i.e., $M(t) = [Mij(t)]$.

If $M, N \in \mathcal{S}^n_+$, then

$$|\mathrm{Tr}(MN)| \leq |\lambda_1(M)| \sum_{i=1}^{n} |\lambda_i(N)|. \tag{4}$$

Furthermore, if $M_1 \preceq M_2$ and $N \succeq 0$, then

$$\mathrm{Tr}(M_1 N) \leq \mathrm{Tr}(M_2 N). \tag{5}$$

The usual concepts of continuity, differentiability and integrability can be naturally extended to matrices of functions, by interpreting them entry-wise. Let $M(t)$ and $N(t)$ be two matrices of functions. Then, it can easily be understood that

$$\frac{d}{dt}\mathrm{Tr}(M(t)) \;=\; \mathrm{Tr}\left(\frac{d}{dt}M(t)\right) = \mathrm{Tr}(M'(t)) \tag{6}$$

$$\frac{d}{dt}\mathrm{Tr}(\psi(M(t))) \;=\; \mathrm{Tr}\left(\psi'(M(t))\right) M'(t) \tag{7}$$

$$\frac{d}{dt}(M(t)N(t)) \;=\; \left(\frac{d}{dt}M(t)\right) N(t) + M(t)\left(\frac{d}{dt}N(t)\right) = M'(t)N(t) + M(t)N'(t). \tag{8}$$

In the rest, when we use the function $\psi(.)$ and its derivatives $\psi'(.)$ and $\psi''(.)$, these denote matrix functions if the argument is a matrix and a univariate function if the argument is in $\mathbb{R}_+$.

2.1   Generic IPM for SDO

The concept of the central path has been extended from LO to SDO [11]. If IPC holds, then the central path for SDO is defined by the solution sets $\{(X(\mu), y(\mu), S(\mu) \;:\; \mu > 0)\}$ of the system

$$
\begin{aligned}
A_i \bullet X &= b_i, \quad 1 \leq i \leq m & X \succ 0 \\
\sum_{i=1}^{m} y_i A_i + S &= C & S \succ 0 \\
XS &= \mu E
\end{aligned}
\tag{9}
$$

where $\mu > 0$. Under IPC assumption, the system (9) has a unique solution, for $\mu > 0$. We call $X(\mu)$ the μ-center of (P) and $(y(\mu), S(\mu))$ the μ-center of (D). The set of μ-centers (with μ running through positive real numbers) is called the central path of (P) and (D). If $\mu \to 0$ then the limit of the central path exists and since the limit points satisfy the complementarity condition, the limit yields optimal solutions for (P) and (D). IPMs follow the central path approximately. A direct application of Newton's method to (9) produces the following equations for the search direction $(\Delta X, \Delta y, \Delta S)$

$$
\begin{aligned}
A_i \bullet \Delta X &= 0, \qquad 1 \le i \le m \\
\sum_{i=1}^{m} \Delta y_i A_i + \Delta S &= 0 \\
X \Delta S + \Delta X \, S &= \mu E - XS
\end{aligned}
\tag{10}
$$

It can be showed that this system has a unique solution, see [20]. A crucial observation is that ΔX is not necessarily symmetric. In this paper, we consider the symmetrization scheme from which the NT direction [13] is derived. Let

$$
P := X^{\frac{1}{2}} \left(X^{\frac{1}{2}} S X^{\frac{1}{2}} \right)^{-\frac{1}{2}} X^{\frac{1}{2}} = S^{-\frac{1}{2}} \left(S^{\frac{1}{2}} X S^{\frac{1}{2}} \right)^{\frac{1}{2}} S^{-\frac{1}{2}},
$$

and $D = P^{\frac{1}{2}}$. The symmetric and positive definite matrix D can be used to scale the matrices X and S and to define the symmetric and positive definite matrix V as follows:

$$
V := \frac{1}{\sqrt{\mu}} D^{-1} X D^{-1} = \frac{1}{\sqrt{\mu}} DSD.
\tag{11}
$$

Thus, we have

$$
V^2 := \frac{1}{\mu} D^{-1} XSD.
\tag{12}
$$

Let us further define

$$
\begin{aligned}
\bar{A}_i &:= DA_iD, \qquad i = 1, 2, \dots, m \\
D_X &:= \frac{1}{\sqrt{\mu}} D^{-1} \Delta X D^{-1}, \qquad D_S := \frac{1}{\sqrt{\mu}} D \Delta SD.
\end{aligned}
\tag{13}
$$

Then it follows from (10) that the (scaled) NT search direction $(\Delta X, \Delta y, \Delta S)$ are defined by the system

$$
\begin{aligned}
\bar{A}_i \bullet D_X &= 0, \qquad 1 \le i \le m \\
\sum_{i=1}^{m} \Delta y_i \bar{A}_i + D_S &= 0 \\
D_X + D_S &= V^{-1} - V
\end{aligned}
\tag{14}
$$

Now, following [13], we turn to the new approach of this paper. An important observation is that with defining the univariate kernel function $\psi(t)$ as

$$
\begin{aligned}
\psi &: \quad \mathbb{R}_{++} \to \mathbb{R}_+ \\
\psi(t) &= \frac{t^2 - 1}{2} - \log t,
\end{aligned}
$$

the right hand side of the third equation in (14) equals to $-\psi'(V)$. Assume that $\psi(t)$ is a strictly convex function on $\mathbb{R}_{++}$ with minimizer at $t = 1$, and $\psi(1) = 0$. We call the univariate function $\psi(t)$

a *kernel function*. Given the kernel function $\psi(t)$ and the associated matrix functions $\psi(V)$ and $\psi'(V)$ as defined in Definition 2.2, we replace the right-hand side of the third equation in (14) by $-\psi'(V)$. Thus, the new search direction D_X and D_S are obtained by solving the following system.

$$\bar{A}_i \bullet D_X = 0, \qquad 1 \le i \le m$$

$$\sum_{i=1}^{m} \Delta y_i \bar{A}_i + D_S = 0 \tag{15}$$

$$D_X + D_S = -\psi'(V)$$

This system has a unique solution D_X, D_S and Δy [19], which can be used to compute ΔX and ΔS via (13). Note that D_X and D_S are orthogonal due to orthogonality of ΔX and ΔS. Moreover,

$$D_X = D_S = 0_{n \times n} \Leftrightarrow \psi'(V) = 0_{n \times n} \Leftrightarrow V = E \Leftrightarrow \Psi(V) = 0,$$

i.e., if and only if $XS = \mu E$, which implies that $X = X(\mu)$ and $S = S(\mu)$. Otherwise $\Psi(V) > 0$, hence, if $(X, y, S) \ne (X(\mu), y(\mu), S(\mu))$, then $(\Delta X, \Delta y, \Delta S) \ne 0$. By taking a step along the search direction, with the step size α by some line search rules, one constructs a new triple (X_+, y_+, S_+) according to

$$X_+ = X + \alpha \Delta X, \qquad y_+ = y + \alpha \Delta y, \qquad S_+ = S + \alpha \Delta S. \tag{16}$$

Now, we are in place to describe the generic form of primal-dual IPM for SDO [19].

Algorithm 1: Generic Primal-Dual IPM for SDO

Input:
    A proximity function $\Psi(v)$;
    a threshold parameter $\tau > 0$;
    an accuracy parameter $\varepsilon > 0$;
    a fixed barrier update parameter $0 < \theta < 1$;
    a strictly feasible pair (X^0, S^0) and $\mu^0 = 1$ such that $\Psi(X^0, S^0, \mu^0) \le \tau$;
begin
    $X := X^0$; $S := S^0$; $\mu := \mu^0$;
    while $n\mu > \varepsilon$ **do**
        $\mu := (1 - \theta)\mu$;
        while $\Psi(X, S, \mu) > \tau$ **do**
            solve system (15) and use (13) to obtain $(\Delta X, \Delta y, \Delta S)$;
            determine a step size α;
            update $X_+ = X + \alpha \Delta X$, $y_+ = y + \alpha \Delta y$, $S_+ = S + \alpha \Delta S$;
        end do
    end do
end

As we have mentioned before, the closeness of (X, y, S) to $(X(\mu), y(\mu), S(\mu))$ is measured by the value of proximity function $\Psi(V)$, with τ as a threshold value. If $\Psi(V) \le \tau$ then we start a new *outer iteration* by performing a μ-update, otherwise we enter an *inner iteration* by computing the search directions at the current iterates with respect to the current value of μ and apply (16) to get new iterates. This procedure is repeated until we find iterates that are in the neighborhood of $(X(\mu), y(\mu), S(\mu))$. Then, μ is again reduced by the factor $1 - \theta$ and we apply Newton's method targeting at the new μ-update, and so on. This process is repeated until μ small enough, say until $n\mu < \varepsilon$.

The parameters τ, θ and the step size α should be chosen in such a way that the algorithm is optimized in the sense that the number of iterations required by the algorithm is as small as possible.

3 New class of kernel (proximity) functions

In this paper, we consider a new class of kernel functions defined on $\mathbb{R}_{++}$ as

$$\psi_p(t) = \frac{t^2 - 1}{2} - \int_1^t e^{p(x^{-1}-1)} dx, \tag{17}$$

where $p \geq 1$ are the parameters of the new class. It is easily seen that the kernel function $\psi_p(t)$ does not belong to the family of self-regular kernel functions. In the analysis of Algorithm 1, we need first three derivatives of $\psi_p(t)$ with respect to t. For ease of reference, we give them here. We have,

$$\psi_p'(t) = t - e^{p(t^{-1}-1)} \tag{18}$$

$$\psi_p''(t) = 1 + \frac{p}{t^2} e^{p(t^{-1}-1)} \tag{19}$$

$$\psi_p'''(t) = -\left(2 + \frac{p}{t}\right)\frac{p}{t^3} e^{p(t^{-1}-1)} \tag{20}$$

From (19) we obtain that $\psi_p''(t) \geq 1$, for $t > 0$, therefore $\psi_p(t)$ is strongly convex over $\mathbb{R}_{++}$. We also have $\psi_p(1) = \psi_p'(1) = 0$. Thus, $\psi_p(t)$ is indeed a kernel function.

Due to conditions $\psi_p(1) = \psi_p'(1) = 0$, we can completely describe $\psi_p(t)$ by its second derivative as follows:

$$\psi_p(t) = \int_1^t \int_1^\xi \psi_p''(\zeta) d\zeta d\xi \tag{21}$$

In the analysis of the algorithm, we also use the norm-based proximity measure $\delta(V)$ defined by

$$\delta(V) := \frac{1}{2}\|\psi_p'(V)\| = \frac{1}{2}\sqrt{\sum_{i=1}^n \psi_p'(\lambda_i(V))^2} = \frac{1}{2}\|D_X + D_S\|. \tag{22}$$

Note that $\psi_p'(E) = \psi_p(E) = 0_{n \times n}$. We also have $\Psi_p(V)$ is strictly convex with respect to $V \succ 0$ and vanishes at its global minimal point $V = E$, i.e.,

$$\Psi_p(V) = 0 \Leftrightarrow \delta(V) = 0 \Leftrightarrow V = E$$

From aforementioned properties, we have the following results.

Lemma 3.1. $\psi_p''(t)$ is monotonically decreasing for all $t > 0$ and $p \geq 1$.

   Proof. The lemma is a direct consequence of (20). □

Lemma 3.2. $\psi_p(t)$ defined as (17) is exponentially convex. i.e. for all $t_1, t_2 > 0$, we have

$$\psi_p(\sqrt{t_1 t_2}) \leq \frac{1}{2}(\psi_p(t_1) + \psi_p(t_2)).$$

446

Proof. Recall from the Lemma 1.6 in [3], $\psi_p(t)$ is exponentially convex if and only if $\psi'_p(t) + t\psi''_p(t) \geq 0$, for all $t > 0$. Since $\psi''_p(t) \geq 1$ and $t = 1$ is the minimal point, thus $\psi'_p(t) + t\psi''_p(t) \geq 0$, for all $t \geq 1$. Using (18) and (19), for all $t \in (0,1)$, we have

$$\psi'_p(t) + t\psi''_p(t) = 2t + e^{p(t^{-1}-1)}\left(\frac{p}{t} - 1\right) \geq 0,$$

where the last inequality obtains from the fact that $\frac{p}{t} \geq 1$, for all $t \in (0,1)$ and $p \geq 1$. $\square$

As a consequence of Lemma 3.2, we have the following theorem.

Theorem 3.3. *(Proposition 5.2.6 in [12]) Suppose that matrices V_1 and V_2 are symmetric positive definite, then*

$$\Psi_p\left(\left[V_1^{\frac{1}{2}} V_2 V_1^{\frac{1}{2}}\right]^{\frac{1}{2}}\right) \leq \frac{1}{2}(\Psi_p(V_1) + \Psi_p(V_2)).$$

The super convexity property of $\psi_p(t)$ (i.e. $\psi''_p(t) \geq 1$ for $t > 0$) leads us to have more powerful tools, indicated in the following lemma, which will be useful in the complexity analysis of algorithm.

Lemma 3.4. *For the kernel function $\psi_p(t)$ defined as (17), we have:*

1. *for all $t > 0$, $\frac{1}{2}(t-1)^2 \leq \psi_p(t) \leq \frac{1}{2}\psi'_p(t)^2$.*

2. *for any $V \succ 0$, $\Psi_p(V) \leq 2\delta(V)^2$.*

3. *for any $V \succ 0$, $\|\lambda(V)\| \leq \sqrt{n} + \sqrt{2\Psi_p(V)}$.*

Proof. One can find a similar proof of the lemma in [3]. $\square$

Note that at the start of any outer iteration of the Algorithm 1, just before the update of μ with the factor $1 - \theta$, we have $\Psi_p(V) \leq \tau$. Thus, during the subsequent inner iterations, $\Psi_p(V)$ decreases until it passes the threshold τ again. Hence, during the process of the Algorithm 1 the largest values of $\Psi_p(V)$ occur just after the updates of μ. Now, we investigate the growth behavior of $\Psi_p(V)$ after μ-update.

Lemma 3.5. *Let $\beta \geq 1$, then*

$$\psi_p(\beta t) \leq \psi_p(t) + \frac{1}{2}(\beta^2 - 1)t^2.$$

Proof. The proof is similar to that in [3]. $\square$

Using (12), we have $V_+ = \frac{V}{\sqrt{1-\theta}}$. Thus, we have:

Corollary 3.6. *Let $0 < \theta < 1$ and $V_+ = \frac{V}{\sqrt{1-\theta}}$, then*

$$\Psi_p(V_+) \leq \Psi_p(V) + \frac{\theta}{2(1-\theta)}\left(2\Psi_p(V) + 2\sqrt{2n\Psi_p(V)} + n\right).$$

Proof. Using Lemma 3.4 and Lemma 3.5 by taking $\beta = \frac{1}{\sqrt{1-\theta}}$, the proof is straightforward. $\square$

447

4 Evaluation of step size

In this section we evaluate the largest possible feasible step size α to have decrease in the proximity function. In each inner iteration the search direction ΔX, Δy and ΔS are obtained by solving the system (15) and using (13). After a damped step, with step size α, by using (11) and (13) we may write

$$
\begin{aligned}
X_+ &= X + \alpha\Delta X = X + \alpha\sqrt{\mu}DD_XD = \sqrt{\mu}D(V + \alpha D_X)D \\
S_+ &= S + \alpha\Delta S = S + \alpha\sqrt{\mu}D^{-1}D_SD^{-1} = \sqrt{\mu}D^{-1}(V + \alpha D_S)D^{-1}.
\end{aligned}
$$

Thus, after a damped step, from (12) we obtain

$$
V_+ = \frac{1}{\sqrt{\mu}}(D^{-1}X_+S_+D)^{\frac{1}{2}}. \tag{23}
$$

It is trivial that V_+^2 is unitarily similar to the matrix $X_+^{\frac{1}{2}}S_+X_+^{\frac{1}{2}}$ and thus to $(V+\alpha D_X)^{\frac{1}{2}}(V+\alpha D_S)(V+\alpha D_X)^{\frac{1}{2}}$. This further implies that the eigenvalues of the matrix V_+ are precisely the same as those of

$$
\bar{V}_+ = \left((V + \alpha D_X)^{\frac{1}{2}}(V + \alpha D_S)(V + \alpha D_X)^{\frac{1}{2}}\right)^{\frac{1}{2}} \tag{24}
$$

From the definition of $\Psi_p(V)$, we obtain $\Psi_p(V_+) = \Psi_p(\bar{V}_+)$. Hence, using Theorem 3.3, we have

$$
\Psi_p(V_+) = \Psi_p(\bar{V}_+) \leq \frac{1}{2}\Psi_p(V + \alpha D_X) + \frac{1}{2}\Psi_p(V + \alpha D_S).
$$

Defining

$$
f(\alpha) = \Psi_p(V_+) - \Psi_p(V) = \Psi_p(\bar{V}_+) - \Psi_p(V),
$$

and

$$
f_1(\alpha) = \frac{1}{2}\Psi_p(V + \alpha D_S) + \frac{1}{2}\Psi_p(V + \alpha D_S) - \Psi_p(V),
$$

we have $f(\alpha) \leq f_1(\alpha)$. Obviously,

$$
f(0) = f_1(0) = 0.
$$

By using (6)-(8), we obtain

$$
f_1'(\alpha) = \frac{1}{2}\text{Tr}\left((\psi_p'(V + \alpha D_X)D_X + \psi_p'(V + \alpha D_S)D_S)\right) \tag{25}
$$

$$
f_1''(\alpha) = \frac{1}{2}\text{Tr}\left((\psi_p''(V + \alpha D_X)D_X^2 + \psi_p''(V + \alpha D_S)D_S^2)\right). \tag{26}
$$

Now, the third inequality of system (14) gives

$$
f_1'(0) = \frac{1}{2}\text{Tr}\left(\psi_p'(V)(D_X + D_S)\right) = \frac{1}{2}\text{Tr}\left(-\psi_p'(V)^2\right) = -2\delta(V)^2. \tag{27}
$$

The following lemma is a slight modification of Weyl Theorem [22].

Lemma 4.1. *Let A, $A + B \in \mathbb{S}_+^n$. One has*

$$
\lambda_i(A + B) \geq \lambda_n(A) - |\lambda_1(B)|, \qquad i = 1, 2, \ldots, n. \tag{28}
$$

To estimate an upper bound for the decrease of the proximity function during an iteration, we also need the following technical lemma which is one of the main results of this section. For notational convenience, we define $\delta := \delta(V)$.

448

Lemma 4.2. *We have*

$$f_1''(\alpha) \leq 2\delta^2 \psi_p''(\lambda_n(V) - 2\alpha\delta) \tag{29}$$

Proof. Using (22), we have $\|D_X + D_S\|^2 = \|D_X\|^2 + \|D_S\|^2 = 4\delta^2$, which implies that $|\lambda_1(D_X)| \leq 2\delta$ and $|\lambda_1(D_S)| \leq 2\delta$. Using Lemma 4.1 and the fact that $V + \alpha D_X \succeq 0$, we have

$$\lambda_i(V + \alpha D_X) \geq \lambda_n(V) - \alpha|\lambda_1(D_X)| \geq \lambda_n(V) - 2\alpha\delta, \qquad i = 1, 2, \ldots, n.$$

Since ψ_p'' is monotonically decreasing in $t > 0$, we get

$$\psi_p''(\lambda_i(V + \alpha D_X)) \leq \psi_p''(\lambda_n(V) - 2\alpha\delta), \qquad i = 1, 2, \ldots, n,$$

which concludes that

$$\text{diag}\left(\psi_p''(\lambda_1(V + \alpha D_X)), \ldots, \psi_p''(\lambda_n(V + \alpha D_X))\right) \leq \text{diag}\left(\psi_p''(\lambda_n(V) - 2\alpha\delta), \ldots, \psi_p''(\lambda_n(V) - 2\alpha\delta)\right).$$

Therefore, we have

$$\psi_p''(V + \alpha D_X) \leq \psi_p''(\lambda_n(V) - 2\alpha\delta)E.$$

Since $D_X^2 \in \mathbb{S}_+^n$, by using (4) and (5), we can get

$$\text{Tr}(\psi_p''(V + \alpha D_X)D_X^2) \leq \text{Tr}(\psi_p''(\lambda_n(V) - 2\alpha\delta)E D_X^2) \leq \psi_p''(\lambda_n(V) - 2\alpha\delta)\sum_{i=1}^{n}\lambda_i(D_X^2).$$

Similarly,

$$\text{Tr}(\psi_p''(V + \alpha D_S)D_S^2) \leq \text{Tr}(\psi_p''(\lambda_n(V) - 2\alpha\delta)E D_S^2) \leq \psi_p''(\lambda_n(V) - 2\alpha\delta)\sum_{i=1}^{n}\lambda_i(D_S^2).$$

Using (26), from the aforementioned relations, we have

$$f_1''(\alpha) \leq \frac{1}{2}\psi_p''(\lambda_n(V) - 2\alpha\delta)\sum_{i=1}^{n}\left(\lambda_i(D_X^2) + \lambda_i(D_S^2)\right) = 2\delta^2\psi_p''(\lambda_n(V) - 2\alpha\delta),$$

which completes the lemma. □

A suitable stepsize should be chosen such that X_+ and S_+ are feasible and $f(\alpha)$ decreases sufficiently. Since the procedure of selecting the largest possible stepsize is almost a "word-by-word" extension of the LO case in [4], for the proof of the following lemmas we refer to [4].

Lemma 4.3. (Lemma 3.2 in [4]) *We have $f_1'(\alpha) \leq 0$ if α satisfies the following inequality*

$$\psi_p'(\lambda_n(V)) - \psi_p'(\lambda_n(V) - 2\alpha\delta) \leq 2\delta \tag{30}$$

We note that for all $t > 0$, $\psi_p''(t) \geq 1$, so the function $-\frac{1}{2}\psi_p'(t)$ has the inverse function. Suppose that $\rho : [0, \infty) \to (0, 1]$ is the inverse function of the restriction of $-\frac{1}{2}\psi_p'(t)$ to the interval $(0, 1]$. We want to solve inequality (30) for the largest possible α. The following lemma derives the largest possible value of α satisfying (30).

Lemma 4.4. (Lemma 3.3 in [4]) *Let $\rho : [0, \infty) \to (0, 1]$ be the inverse function of the restriction of $-\frac{1}{2}\psi_p'(t)$ to the $t \leq 1$. Then, the largest possible value of the step size of α satisfying (30) is given by*

$$\bar{\alpha} = \frac{1}{2\delta}(\rho(\delta) - \rho(2\delta)). \tag{31}$$

Lemma 4.4 implies that

$$\bar{\alpha} = \frac{1}{2\delta}\left(\rho(\delta) - \rho(2\delta)\right) = \frac{1}{2\delta}\int_{2\delta}^{\delta}\rho'(\sigma)d\sigma \tag{32}$$

Using the definition of ρ, we have $-\psi_p'(\rho(\delta)) = 2\delta$. Taking the derivative with respect to δ, we obtain

$$\rho'(\delta) = -\frac{2}{\psi_p''(\rho(\delta))} < 0.$$

Thus, from (32), we can get the following upper and lower bounds for $\bar{\alpha}$:

$$\frac{1}{\psi_p''\left(\rho(2\delta)\right)} \leq \bar{\alpha} \leq \frac{1}{\psi_p''(\rho(\delta))}.$$

In the sequel, we use the following $\tilde{\alpha}$ as our default value for the step size,

$$\tilde{\alpha} = \frac{1}{\psi_p''\left(\rho(2\delta)\right)}. \tag{33}$$

Note that $\tilde{\alpha} \leq \bar{\alpha}$.

5 Complexity of the Algorithm 1

In this section, we first obtain an estimate for the value of $f(\tilde{\alpha})$, then, by using some technical lemmas, we conclude the iteration complexity.

To get an estimate value for $f(\tilde{\alpha})$, we need the following technical lemma where its elementary proof can be found in [14].

Lemma 5.1. *Let $h(t)$ be a twice differentiable convex function with $h(0) = 0$ and $h'(0) < 0$ such that its (global) minimum is attained at $t^* > 0$. If $h''(t)$ is increasing for $t \in [0, t^*]$, then*

$$h(t) < \frac{t\, h'(0)}{2}, \qquad t \in [0, t^*]. \tag{34}$$

Now, we can present an upper bound for $f(\tilde{\alpha})$ as follows:

Lemma 5.2. *If the step size α is such that $\alpha \leq \tilde{\alpha}$, then*

$$f(\alpha) \leq -\alpha\delta^2. \tag{35}$$

Proof. Let the univariate function $h(\alpha)$ be such that

$$h(0) = f_1(0) = 0, \quad h'(0) = f_1'(0) = -2\delta^2,$$

and

$$h''(\alpha) = 2\delta^2 \psi_p''\left(\lambda_n(V) - 2\alpha\delta\right).$$

Using Lemma 4.2, we have $f_1''(\alpha) \leq h''(\alpha)$ and as a consequence $f_1'(\alpha) \leq h'(\alpha)$ and $f_1(\alpha) \leq h(\alpha)$. We may write

$$
\begin{aligned}
h'(\alpha) &= h'(0) + \int_0^{\alpha} h''(\xi)d\xi \\
&= -2\delta^2 + 2\delta^2 \int_0^{\alpha} \psi_p''\left(\lambda_n(V) - 2\xi\delta\right)d\xi \\
&= -2\delta^2 - \delta\left(\psi_p'\left(\lambda_n(V) - 2\alpha\delta\right) - \psi_p'(\lambda_n(V))\right)
\end{aligned}
$$

450

Therefore, from Lemma 4.3, we can get $h'(\alpha) \leq 0$ for $\alpha \leq \bar{\alpha}$. On the other hand, since $\psi_p''(\alpha)$ is decreasing, $h''(\alpha)$ is increasing with respect to α. Hence, by applying Lemma 5.1, we obtain

$$f_1(\alpha) \leq h(\alpha) \leq \frac{1}{2}\alpha h'(0) = -\alpha\delta^2$$

This completes the proof, since $f(\alpha) \leq f_1(\alpha)$. $\square$

Corollary 5.3. *Let ρ be the inverse function of the restriction of $-\frac{1}{2}\psi_p'(t)$ to the interval $(0,1]$ and $\tilde{\alpha}$ be defined as in (33). Then*

$$f(\alpha) \leq -\frac{\delta^2}{\psi_p''(\rho(2\delta))} \tag{36}$$

Proof. The corollary follows immediately if we apply Lemma 5.2 to the default step size $\tilde{\alpha}$ defined as in (33). $\square$

Now, we apply the so far obtained results to our proximity function. To this end, we need to compute $\rho(2\delta)$. Letting $s = \rho(2\delta)$, we have $\psi_p'(s) = -4\delta$, from the definition of ρ. So, from (18) ,(19) and (33), we may write

$$\tilde{\alpha} = \frac{1}{\psi_p''(s)} = \frac{1}{1 + \frac{p}{s^2}e^{p(s^{-1}-1)}}, \tag{37}$$

and

$$e^{p(s^{-1}-1)} = s + 4\delta \leq 1 + 4\delta.$$

where the last inequality obtains from $s \leq 1$. Thus,

$$s^{-1} = 1 + \frac{1}{p}\log(s + 4\delta) \leq 1 + \frac{1}{p}\log(1 + 4\delta).$$

Using this inequality, we have

$$\frac{1}{s^2} \leq \left(1 + \frac{1}{p}\log(1 + 4\delta)\right)^2.$$

Therefore, (37) can be written as:

$$\tilde{\alpha} \geq \frac{1}{1 + p\left(1 + \frac{1}{p}\log(1 + 4\delta)\right)^2 (1 + 4\delta)}$$

$$\geq \frac{1}{2p\left(1 + \frac{1}{p}\log(1 + 4\delta)\right)^2 (1 + 4\delta)}.$$

Thus, using Lemma 5.2, one has

$$f(\tilde{\alpha}) \leq -\tilde{\alpha}\delta^2 \leq -\frac{\delta^2}{2p\left(1 + \frac{1}{p}\log(1 + 4\delta)\right)^2 (1 + 4\delta)} \tag{38}$$

451

Since the right hand side of expression in (38) is monotonically decreasing in δ, by using Lemma 3.4, we obtain

$$
\begin{aligned}
f(\tilde{\alpha}) &\leq -\frac{\Psi_p(V)}{4p\left(1 + \frac{1}{p}\log\left(1 + 2\sqrt{2\Psi_p(V)}\right)\right)^2\left(1 + 2\sqrt{2\Psi_p(V)}\right)} \\
&= \Theta\left(-\frac{\sqrt{\Psi_p(V)}}{p\left(1 + \frac{1}{p}\log\Psi_p(V)\right)^2}\right),
\end{aligned}
\tag{39}
$$

where the last equality obtains from the fact that during an inner iteration in Algorithm 1, we have $\Psi_p(V) > \tau$, and since we are working on large update methods in large neighborhood of the central path, we assume that $\tau = O(n)$.

Now, we are going to get iteration complexity for Algorithm 1 using the proximity function induced by kernel function (17). We know, from Corollary 3.6, that after the update of μ to $(1 - \theta)\mu$, we have,

$$
\Psi_p(V_+) \leq \Psi_p(V) + \frac{\theta}{2(1 - \theta)}\left(2\Psi_p(V) + 2\sqrt{2n\Psi_p(V)} + n\right)
$$

At the start of an outer iteration we have $\Psi_p(V) \leq \tau$. For the purpose of analyzing the large-update method, we have $\theta = \Theta(1)$. We need to count how many inner iterations are required to return to the situation where $\Psi_p(V) \leq \tau$ after μ-update. We denote the value of $\Psi_p(V)$ after the μ-update as Ψ_0, and the subsequent values are denoted as Ψ_j, $j = 1, \ldots, L$, with L denoting the total number of inner iterations in an outer iteration. By Corollary 3.6,

$$
\Psi_0 \leq \tau + \frac{\theta}{2(1 - \theta)}\left(2\tau + 2\sqrt{2n\tau} + n\right) = O(n).
\tag{40}
$$

The decrease on each inner iteration is given by (39), i.e.,

$$
\Psi_{j+1} \leq \Psi_j - \kappa\frac{\sqrt{\Psi_j}}{p\left(1 + \frac{1}{p}\log\Psi_j\right)^2}, \qquad j = 0, \ldots, L - 1.
\tag{41}
$$

where κ is some positive constant. Let

$$
\Delta\Psi_j = \frac{\sqrt{\Psi_j}}{p\left(1 + \frac{1}{p}\log\Psi_j\right)^2}
$$

Thus, from (41), we obtain

$$
\Psi_{j+1} \leq \Psi_j - \kappa\Delta\Psi_j, \qquad j = 0, \ldots, L - 1,
\tag{42}
$$

We need the following technical result to get the iteration bound. A proof of the following lemma can be found in [13].

Lemma 5.4. *For $\alpha \in [0, 1]$ and for $t \geq -1$, we have*

$$
(1 + t)^\alpha \leq 1 + \alpha t
\tag{43}
$$

We use the following lemma to get the number of inner iterations and therefore the total number of iterations for Algorithm 1.

Lemma 5.5. *Considering (42), we have*

$$L \leq \frac{2\Psi_0}{\kappa\Delta\Psi_0}$$

Proof. We may write $\Delta\Psi_j$ as

$$\Delta\Psi_j = f(\Psi_j)\sqrt{\Psi_j} \qquad (44)$$

where $f(\Psi_j) = \frac{1}{p\left(1+\frac{1}{p}\log\Psi_j\right)^2}$ is monotonically decreasing. Substitution on (42) gives

$$\Psi_{j+1} \leq \Psi_j - \kappa f(\Psi_j)\sqrt{\Psi_j}, \qquad j = 0,\dots,L-1.$$

From this inequality, we derive

$$
\begin{aligned}
0 \leq \sqrt{\Psi_{j+1}} &\leq \left(\Psi_j - \kappa f(\Psi_j)\sqrt{\Psi_j}\right)^{\frac{1}{2}} \\
&= \sqrt{\Psi_j}\left(1 - \kappa f(\Psi_j)\Psi_j^{-\frac{1}{2}}\right)^{\frac{1}{2}} \\
&\leq \sqrt{\Psi_j}\left(1 - \frac{1}{2}\kappa f(\Psi_j)\Psi_j^{-\frac{1}{2}}\right) \\
&= \sqrt{\Psi_j} - \frac{1}{2}\kappa f(\Psi_j)
\end{aligned}
$$

where the last inequality obtains from Lemma 5.4. Since $f(\Psi)$ is monotonically decreasing, we have

$$0 \leq \sqrt{\Psi_j} - \frac{1}{2}\kappa f(\Psi_j) \leq \sqrt{\Psi_j} - \frac{1}{2}\kappa f(\Psi_0), \qquad j = 0,\dots,L-1.$$

This implies

$$\sqrt{\Psi_j} \leq \sqrt{\Psi_0} - \frac{1}{2}\kappa j f(\Psi_0), \qquad j = 0,\dots,L-1.$$

Taking $j = L$, we obtain

$$0 \leq \sqrt{\Psi_0} - \frac{1}{2}\kappa L f(\Psi_0),$$

which implies

$$L \leq \frac{\sqrt{2\Psi_0}}{\kappa f(\Psi_0)} = \frac{2\Psi_0}{\kappa\Delta\Psi_0}.$$

The equality is due to (44). This proves the lemma. □

Using (40), we have $\Psi_0 = O(n)$. Thus, from Lemma 5.5, we obtain that the number of inner iterations is bounded above as

$$L \leq \left\lfloor \frac{2\Psi_0}{\kappa\Delta\Psi_0} \right\rfloor = \left\lfloor O\left(p\sqrt{n}\left(1 + \frac{1}{p}\log n\right)^2\right) \right\rfloor.$$

The iteration complexity of the algorithm is obtained by multiplying this number by the number of outer iteration, which is bounded above by $O\left(\log\frac{n}{\varepsilon}\right)$, see [15]. Omitting the integer brackets, which does not change the order, the iteration complexity is given by

$$O\left(p\sqrt{n}\left(1 + \frac{1}{p}\log n\right)^2 \log\frac{n}{\varepsilon}\right).$$

In special case, by choosing $p = O(\log n)$, we get the so far best known LO case complexity, i.e., $O\left(\sqrt{n}\log n \log \frac{n}{\varepsilon}\right)$, for large update primal dual interior point methods for SDO also.

Acknowledgement This research was in part supported by the grant from Institute for Studies in Theoretical Physics and Mathematics (IPM grant No. 86900017). The author also would like to thank the Research Council of K.N. Toosi University of Technology.

References

[1] K. Amini and M.R. Peyghami. An Interior-Point Algorithm for Linear Optimization Based on a New Kernel Function. *Southeast Asian Bulletin of Mathematics*, 29:651-667, 2005.

[2] K. Amini and M.R. Peyghami. An Interior-Point Algorithm for Linear Programming Based on a Class of Kernel Functions. *Bulletin of Australian Mathematical Society*, 71:139-153, 2005.

[3] Y.Q. Bai, M. El ghami and C. Roos. A new efficient large-update primal-dual interior point method based on a finite barrier. *SIAM Journal on Optimization*, 13(3)(electronic):766-782, 2003.

[4] Y.Q. Bai and C. Roos. A primal-dual interior-point method based on a new kernel function with linear growth rate. *Proceedings of Industrial Optimization Symposium and Optimization Day*, Australia, November 2002.

[5] R. Bellman and K. Fan. On systems of linear inequalities in Hermitian matrix variables. In V.L. Klee, Ed., *Convexity*, 7:1-11, *Proceeding of Symposia in Pure Mathematics*, Amer. Math. Soc. Providence, RI, 1963.

[6] R.A. Horn and R. Johnson Charles. *Topics in Matrix Analysis*. Cambridge University Press, 1991.

[7] N. K. Karmarkar. A new polynomial-time algorithm for linear programming. *Combinatorica*, 4:373–395, 1984.

[8] E. de Klerk. *Interior Point Methods for Semidefinite Programming*. Ph.D. Thesis, Faculty of ITS/TWI, Delft University of Technology, The Netherlands, 1997.

[9] H. Lütkepohl. *Handbook of matrices*. John Wiley & Sons, 1996.

[10] N. Megiddo. Pathways to the optimal set in linear programming. In N. Megiddo Ed., *Progress in Mathematical Programming: Interior Point and Related Methods*, Springer Verlag, New York, 131158, 1989. Identical version in : *Proceedings of the 6th Mathematical Programming Symposium of Japan*, Nagoya, Japan, 135, 1986.

[11] Y.E. Nestrov and A.S. Nemirovskii. Interior point polynomial algorithms in convex programming. *SIAM Studies in Applied Mathematics*, Vol. 13, SIAM, Philadelphia, USA, 1994.

[12] J. Peng. *New design and Analysis of Interior Point Methods*. Ph.D. Thesis, Universal Press, 2001.

[13] J. Peng, C. Roos and T. Terlaky. A new class of polynomial primal-dual methods for linear and semidefinite optimization. *European Journal of Operations Research*, 143(2):234-256, 2002.

[14] J. Peng, C. Roos and T. Terlaky. Self-regular functions and new search directions for linear and semidefinite optimization. *Mathematical Programming*, 93:129-171, 2002.

[15] C. Roos, T. Terlaky and J-P. Vial. *Theory and Algorithms for Linear Optimization: An Interior Point Approach*. Wiley-Interscience Series in Discrete Mathematics and Optimization. John Wiley, 1997.

[16] G. Sonnevend. An analytic center for polyhedrons and new classes of global algorithms for linear (smooth, convex) programming. In A. Prékopa, J. Szelezsán, and B. Strazicky Eds. *System Modelling and Optimization : Proceedings of the 12th IFIP-Conference held in Budapest,* Hungary, September, 1985, volume 84 of Lecture Notes in Control and Information Sciences, Springer Verlag, Berlin, WestGermany, 866876, 1986.

[17] J.F. Strum. *Theory and Algorithms of Semidefinite Programming.* In H. Frenk, C. Roos, T. Terlaky and S. Zhang, Eds., *High erformance Optimization,* pages 1-194, Kluwer Academic Publishers, Boston, 1999.

[18] L. Vandenberghe and S. Boyd. Semidefinite Programming. *SIAM Review,* 38(1):49-95, 1996.

[19] G.Q. Wang, Y.Q. Bai and C. Roos. Primal-dual interior point algorithm for semidefinite optimization based on a simple kernel function. *Journal of Mathematical Modelling and Algorithms,* 4:409-433, 2005.

[20] H. Wolkowicz, R. Saigal and L. Vandenberghe. *Handbook of Semidefinite Programming, Theory, Algorithms and Applications.* Kluwer Academic Publishers, 2000.

[21] S.J. Wright. *Primal-Dual Interior-Point Methods.* SIAM Publications, 1997.

[22] H. Wolkowicz, R. Saigal and L. Vanderberghe. *Handbook of Semidefinite Programming, Theory, Algorithms, and Applications.* Kluwer Academic Publishers, 2000.

[23] Y. Ye. *Interior-Point Algorithms: Theory and Analysis.* Wiley-Interscience Series in Discrete Mathematics and Optimization. John Wiley, 1997.

[24] Y. Zhang. On extending some primal-dual interior point algorithms from linear programming to semidefinite programming. *SIAM Journal on Optimization,* 8:365-386, 1998.

Reza Peyghami
DEPARTMENT OF SCIENCES, K.N. TOOSI UNIVERSITY OF TECHNOLOGY, IRAN
INSTITUTE FOR STUDIES IN THEORETICAL PHYSICS AND MATHEMATICS (IPM), IRAN
Email: peyghami@optlab.mcmaster.ca and peyghami@kntu.ac.ir

Solutions of Generalized Photogravitational Elliptic Restricted Three Body Problem

Dr. Sanjay Kumar [a] and Professor B. Ishwar [b]

[a] Department of Mathematics, R.D.S. College, Muzaffarpur
B.R.A. Bihar University, Muzaffarpur – 842001
[b] P.I. DST Project, University Department of Mathematics,
B.R.A. Bihar University, Muzaffarpur - 842001,
Email : - ishwar_bhola@hotmail.com

Abstract. We attempted to find the solutions of generalized photogravitational elliptic restricted three body problem. The problem is generalized in the sense that both primaries are supposed to be an oblate spheroid. By photogravitational we mean that both primaries are radiating as well. We have found the particular solutions. They depend upon radiation and oblateness of primaries. Classical results may be verified from this result.

Keywords: Solutions/ Generalized/ Photogravitational/ ERTBP.
PACS: 95.10.Ce

1. Introduction

Radzievskii [1] formulated the Photogravitational restricted three body problem. This arises from the classical problem when one of the masses is an intense emitter of radiation. Arnold [2] studied the stability of positions of equilibrium of a Hamiltanian system in the general elliptic case. Choudhary [3] studied the periodic orbits of the third kind and stability of the generating solution in the elliptical restricted three-body problem. Bhatanagar [4] examined periodic orbits of collision in the elliptic restricted three-body problem. Sharma and Subbarao [5] studied the restricted three-body problem when the primaries are oblate spheroids. Sharma [6] investigated the linear stability of triangular libration points when the more massive primary is a source of radiation and oblate spheroid as well. He also examined the linear stability of libration points of the photogravitational restricted three-body problem when the smaller primary is an oblate spheroid. Beauge [7] give a note on a global expansion of the disturbing function in the planar elliptic restricted three-body problem. Khasan [8] studied librational solutions to the photogravitational restricted three- body problem by considering both primaries as radiating. He also examined the stability of collinear and triangular points. Khasan [9] also studied three dimensional periodic solutions to the photogravitational Hill problem. He investigated restricted photogravitational elliptic three body problem. Kumar and Ishwar [10] established the equations of motion of the generalized photogravitational elliptic restricted three body problem.

In this paper we have obtained the location of triangular equilibrium points. It is observed that these points form nearly equilateral triangles with the primaries. The

CP1146, *Modelling of Engineering and Technological Problems*, edited by A. H. Siddiqi, A. K. Gupta, and M. Brokate
© 2009 American Institute of Physics 978-0-7354-0683-4/09/$25.00

oblateness and radiating properties of the primaries affect significantly the location of these points.

2. Equations of motion

The equations of motion of the third body of mass m_3 are obtained as in Kumar and Ishwar [10]

$$\xi''-2\eta'=\frac{\partial\Omega^*}{\partial\xi}$$

$$\eta''-2\xi''=\frac{\partial\Omega^*}{\partial\eta} \qquad (1)$$

$$\xi''=\frac{\partial\Omega^*}{\partial\xi}$$

where

$$\Omega^*=\frac{1}{(1-e^2)^{1/2}}\left\{\frac{\xi^2+\eta^2}{2}+\frac{1}{\eta^2}\left(\frac{(1-\mu)q_1}{r_1}+\frac{\mu q_2}{r_2}+\frac{(1-\mu)A_1q_1}{2r_1^3}+\frac{\mu A_2q_2}{2r_2^3}\right)\right\} \qquad (2)$$

Here A_1 and A_2 are the factors characterising the forces due to oblateness of the primaries and are given by

$$A_1=\frac{AF^2-AQ^2}{5R^2},\quad A_2=\frac{AE^2-AP^2}{5R^2}$$

where AE, AF and AP, AQ are the dimensional equatorial and polar radii of the primaries.

The force of radiation is given by

$$F=F_g-F_p=F_g\left(1-\frac{F_p}{F_g}\right)=q_iF_g$$

where $\quad F_g \quad$ = The Gravitational attraction force

$\quad\quad\quad F_p \quad$ = The radiation pressure.

$\quad\quad\quad q_i \quad$ = The mass reduction factor.

Here we have taken $m_1=1-\mu$, $m_2=\mu$ such that $\mu\le\frac{1}{2}$.

3. Solution of the equations of motions

Multiplying the three equations of (1) by $2\xi'$, $2\eta'$ and $2\zeta'$ and integrating after adding, we get

$$\xi'^2+\eta'^2+\zeta'^2=2\Omega^*+C=f(\xi,\eta,\zeta)$$

where

$$f(\xi,\eta,\zeta) = \frac{1}{(1-e^2)^{\frac{1}{2}}}\left[\xi^2+\eta^2+\frac{1}{n^2}\left\{\frac{2(1-\mu)q_1}{r_1}+\frac{2\mu q_2}{r_2}+\frac{(1-\mu)q_1 A_1}{r_1^3}+\frac{\mu q_2 A_2}{r_2^3}\right\}\right] \quad (3)$$

Singular point will occur if

$$\frac{\partial f}{\partial \xi}=\frac{\partial f}{\partial \eta}=\frac{\partial f}{\partial \xi}=0 \tag{4}$$

which gives

$$\xi-\frac{1}{n^2}\left\{\frac{(1-\mu)q_1(\xi-\xi_1)}{r_1^3}+\frac{\mu q_2(\xi-\xi_2)}{r_2^3}+\frac{3(1-\mu)q_1 A_1(\xi-\xi_1)}{2r_1^5}+\frac{3\mu q_2 A_2(\xi-\xi_2)}{2r_2^5}\right\}=0$$

$$\tag{5}$$

$$\eta\left[1-\frac{1}{n^2}\left\{\frac{(1-\mu)q_1}{r_1^3}+\frac{\mu q_2}{r_2^3}+\frac{3(1-\mu)q_1 A_1}{2r_1^5}+\frac{3\mu q_2 A_2}{2r_2^5}\right\}\right]=0 \tag{6}$$

$$\zeta\left\{\frac{(1-\mu)q_1}{r_1^3}+\frac{\mu q_2}{r_2^3}+\frac{3(1-\mu)q_1 A_1}{2r_1^5}+\frac{3\mu q_2 A_2}{2r_2^5}\right\}=0 \tag{7}$$

From (7), $\zeta=0$ gives planar solutions. For the planar solutions away from x-axis $\eta\neq0$. Hence from (6).

$$1-\frac{1}{n^2}\left\{\frac{(1-\mu)q_1}{r_1^3}+\frac{\mu q_2}{r_2^3}+\frac{3(1-\mu)q_1 A_1}{2r_1^5}+\frac{3\mu q_2 A_2}{2r_2^5}\right\} \tag{8}$$

Multiplying (8) by $\xi-\xi_1$ and $\xi-\xi_2$ respectively and subtracting from (5), we get

$$\xi_1+\frac{1}{n^2}\left\{\frac{\mu q_2(\xi_2-\xi_1)}{r_2^3}+\frac{3\mu q_2 A_2(\xi_2-\xi_1)}{2r_2^5}\right\}=0 \tag{9}$$

and

$$\xi_2-\frac{1}{n^2}\left\{\frac{(1-\mu)q_1(\xi_2-\xi_1)}{r_1^3}+\frac{3(1-\mu)q_1 A_1}{2r_1^5}(\xi_2-\xi_1)\right\}=0 \tag{10}$$

It is obvious that the distance between the two primaries i.e. $\xi_2-\xi_1=1$ and from the definition of the centre of mass

$(1-\mu)\xi_1+\mu\xi_2=0$

i.e. $\xi_1+\mu=0$

i.e. $\xi_1=-\mu$, Hence $\xi_2=1-\mu$

Substituting these values in (9), we get

458

$$\frac{q_2}{n^2 r_2^3} + \frac{3q_2 A_2}{2n^2 r_2^5} = 1 \quad \mu \neq 0$$

i.e.
$$n^2 = \frac{q_2}{r_2^3} + \frac{3q_2 A_2}{2r_2^5} \tag{11}$$

Similarly from (10), we get

$$n^2 = \frac{q_1}{r_1^3} + \frac{3q_1 A_1}{2r_1^5} \tag{12}$$

When the oblateness and the radiating effect of the primaries are neglected i.e. when $A_i = 0$ and $q_1 = 1 = q_2$

$$n^2 = \frac{1}{r_1^3}$$

i.e $r_1 = \left(\dfrac{1}{n}\right)^{\frac{2}{3}}$ and $r_2 = \left(\dfrac{1}{n}\right)^{\frac{2}{3}}$

which is the case of classical elliptical restricted three body problem.

Taking $r_1 = \delta_1 + \left(\dfrac{1}{n}\right)^{\frac{2}{3}}$ and $r_2 = \delta_2 + \left(\dfrac{1}{n}\right)^{\frac{2}{3}}$ $\tag{13}$

where δ_1 and δ_2 are infinitesimally small quantities. Substituting these values in (11) and (12) and neglecting the powers of δ_1 and δ_2, we get

$$r_1 = \left(\frac{1}{n}\right)^{\frac{2}{3}} + \frac{n^2 - \dfrac{q_1}{n^2} - \dfrac{3q_1 A_1}{2n^{10/3}}}{\dfrac{3q_1}{n^{4/3}} + \dfrac{15q_1 A_1}{2n^{8/3}}}.$$

Similarly

$$r_2 = \left(\frac{1}{n}\right)^{\frac{2}{3}} + \frac{n^2 - \dfrac{q_2}{n^2} - \dfrac{3q_2 A_2}{2n^{10/3}}}{\dfrac{3q_2}{n^{4/3}} + \dfrac{15q_2 A_2}{2n^{8/3}}}.$$

From the relation obtained for mean motion we have

$$n^2 = \frac{\{1 + \frac{3}{2}(A_1 + A_2)\}(1 + e^2)^{\frac{1}{2}}}{a(1 - e^2)}$$

When $A_1 = 0 = A_2$, we have

$$n^2 = \frac{1}{a}\left(1 + \frac{3}{2}e^2\right)$$

Therefore

$$\left(\frac{1}{n}\right)^{\frac{2}{3}} = a^{\frac{1}{3}}\left(1 - \frac{1}{2}e^2\right).$$

Substituting this value in (13), we get

$$r_1 = \frac{q_1 - 1 - \frac{3}{2}e^2 + \frac{3}{2}q_1e^2 + \frac{3}{2}q_1A_1a^{\frac{-2}{3}} + \frac{15}{4}q_1A_1a^{\frac{-2}{3}}e^2}{3q_1\left(a^{\frac{-1}{3}} + 2a^{\frac{-1}{3}}e^2 + \frac{5}{2}A_1a^{-2} + \frac{15}{2}A_1a^{-2}e^2\right)}$$

and

$$r_2 = \frac{q_2 - 1 - \frac{3}{2}e^2 + \frac{3}{2}q_2e^2 + \frac{3}{2}q_2A_2a^{\frac{-2}{3}} + \frac{15}{4}q_2A_2a^{\frac{-2}{3}}e^2}{3q_2\left(a^{\frac{-1}{3}} + 2a^{\frac{-1}{3}}e^2 + \frac{5}{2}A_2a^{-2} + \frac{15}{2}A_2a^{-2}e^2\right)} \qquad (14)$$

Thus we find that the triangular equilibrium points do not form equilateral triangle as in classical case. Also the oblateness and radiation of the primaries affect significantly the position of the equilibrium points. The exact location of the triangular equilibrium points are obtained from the relations

$$r_1^2 = (\xi - \xi_1)^2 + \eta^2 = (\xi + \mu)^2 + \eta^2$$

and

$$r_2^2 = (\xi - \xi_2)^2 + \eta^2 = (\xi - 1 + \mu)^2 + \eta^2$$

Therefore $\xi = \frac{1}{2} - \mu + \frac{r_1^2 + r_2^2}{2}$

and $\quad \xi = \frac{1}{2} - \mu + \frac{r_1^2 + r_2^2}{2} \qquad (15)$

where r_1 and r_2 are given in (14).

Substituting the values of r_1 and r_2 from (14) in (15), we get

$$\xi = \frac{1}{2} - \mu + \left(\frac{A_1 - A_2}{2n^{\frac{8}{3}}}\right) - \frac{5}{4n^{\frac{4}{3}}}\left(\frac{A_1^2}{q_1^{\frac{2}{3}}} - \frac{A_2^2}{q_2^{\frac{2}{3}}}\right) \qquad (16)$$

460

$$\eta = \pm \left[\frac{q_1^{\frac{2}{3}}}{n^{\frac{8}{4}}} - \frac{1}{4} + \left(\frac{A_1 + A_2}{2n^{\frac{8}{3}}} \right) - \frac{5}{4n^{\frac{4}{3}}} \left(\frac{A_1^2}{q_1^{\frac{2}{3}}} + \frac{A_2^2}{q_2^{\frac{2}{3}}} \right) \right]^{\frac{1}{2}} \tag{17}$$

ACKNOWLEDGMENTS

We are thankful to Dr. Manaziruddin of Dr. J.M. College, Muzaffarpur for his valuable suggestions.

REFERENCES

1. V.V. Radzievskii, The restricted problem of three bodies taking account of light pressure, *Astron, Journal 27*, 1950, pp.250 – 256.
2. V.I. Arnold, On the stability of positions of equilibrium of a Hamiltonian system of ordinary differential equations in the general elliptic case, *Soviet Math.*, 1961, pp. 2,247.
3. R.K. Chaudhary, Existence of periodic orbits of the third kind in the elliptical restricted three body problem and stability of the generating solution. *Bull. Ins. Theoret. Astronomy 10*, 1966, pp.523-536.
4. K.B. Bhatnagar, Periodic orbits of collision in the plane elliptic restricted problem of three bodies, *Proc. Natn. Inst. Sci. India, 35A*, 1969, pp.829-44.
5. R. K. Sharma and P.V. Subbarao, Stationary solutions and their characteristics exponents in the restricted three body problem when the more massive primary is an oblate spheroid. *Celest. Mech. 13*, 1975, pp.137-149.
6. R.K. Sharma, W. Fricke and G. Teleki (eds), Sun and Planetary system, *by D. Reidel Publishing company*, 1982, pp.435-436.
7. C. Beauge, *Cales Mech. and Dyn. Astron. Vol 64, No. 4.*, 1996,
8. S.N. Khasan, Librational solutions to the photogravitational restricted three body problem, *Cosmic Research Vol. 34, No.2*, 1996, pp.146-151.
9. S.N. Khasan , *Cosmic Research, Vol. 34, No. 5*, 1996 a, pp.504-507.
10. Sanjay Kumar and B. Ishwar, Equations of motion in the generalized photogravitational elliptic restricted three body problem. *Review Bull. Cal. Math. Soc., 12 (1 and 2)*, 2004, pp.115-118.

The Economic Optimization of Pulp and Paper Making Processes Using Computational Intelligence

Millie Pant, Radha Thangaraj and V. P. Singh

Department of Paper Technology,
Indian Institute of Technology Roorkee,
Saharanpur Campus, Saharanpur – 247 001,
Uttar Pradesh.

Abstract. In this paper we present an application of two Computational Intelligence Algorithms, namely Particle Swarm Optimization (PSO) Algorithm and Differential Evolution (DE) for finding and optimal solution to two optimization problems that occur in a paper industry. The first problem deals with the economic optimization of a hypothetical but realistic Kraft pulping process and in the second problem we have considered the optimization of Boiler load allocation problem. Both the problems form an integral part of paper making process. The simulation results show the efficiency and time effectiveness of DE and PSO.

Keywords: Particle Swarm Optimization, Differential Evolution, Pulp and Paper, constraint optimization

INTRODUCTION

Paper industry accounts for nearly 3.5% of world's industrial production and 2% of world trade. Current annual consumption of paper is of the order of 270 million tones. This industry is 10th major section in India, which has some major sections like pulping and recovery cycle, stock preparation and machine operation etc. The procedure of paper making is a complex integration of many sub-processes, many of which may be modeled as optimization problems.

Various algorithms that have been developed for solving optimization problems may be broadly divided into two categories: Deterministic Techniques (DT) and Stochastic Techniques (ST). One major difference between Deterministic and Stochastic Techniques is that ST does not depend on the mathematical nature of the objective function and the solution space unlike DT.

For this reason, ST have become very popular for solving complex optimization problems which are otherwise difficult to solve by the classical optimization techniques [1], [2].

Computational Intelligence Algorithms (CIA) form a part of ST and are generally based on social or biological metaphors.

Some well known CIA include Evolutionary and Genetic Algorithms [3] – [6], PSO [7], [8], DE [9], [10] etc. In this paper we have analyzed the performance of PSO and DE (which are relatively new CIA) on two common optimization problems arising in a

CP1146, *Modelling of Engineering and Technological Problems*, edited by A. H. Siddiqi, A. K. Gupta, and M. Brokate

Pulp and Paper industry. The following sections briefly describe the PSO and DE algorithms and give the mathematical models of the Kraft Pulping system and boil load allocation problem with their simulation results.

PARTICLE SWARM OPTIMIZATION

Particle swarm optimization technique is a population based stochastic search technique first suggested by Kennedy and Eberhart in 1995. The mechanism of PSO is inspired from the complex social behavior shown by the natural species. For a D-dimensional search space the position of the ith particle is represented as $X_i = (x_{i1}, x_{i2}, ..., x_{id}, ..., x_{iD})$. Each particle maintains a memory of its previous best position $P_i = (p_{i1}, p_{i2}, ..., p_{id}, ..., p_{iD})$ and a velocity $V_i = (v_{i1}, v_{i2}, ..., v_{id}, ..., v_{iD})$ along each dimension. At each iteration, the P vector of the particle with best fitness in the local neighborhood, designated g, and the P vector of the current particle are combined to adjust the velocity along each dimension and a new position of the particle is determined using that velocity. The two basic equations which govern the working of PSO are that of velocity vector and position vector are given by:

$$v_{id} = \omega v_{id} + c_1 r_1 (p_{id} - x_{id}) + c_2 r_2 (p_{gd} - x_{id}) \tag{1}$$

$$x_{id} = x_{id} + v_{id} \tag{2}$$

The first part of equation (1) represents the inertia of the previous velocity, the second part is tells us about the personal thinking of the particle and the third part represents the cooperation among particles and is therefore named as the social component. Acceleration constants c_1, c_2 and inertia weight ω are predefined by the user and r_1, r_2 are the uniformly generated random numbers in the range of [0, 1].

DIFFERENTIAL EVOLUTION

Differential Evolution is a simple powerful evolutionary algorithm for global optimization proposed by Storn and Price in 1995 [9]. It is a population based algorithm like genetic algorithms using the similar operator; crossover, mutation and selection. The main difference in constructing better solutions is that genetic algorithms rely on crossover while DE relies on mutation operator [11]. DE works as follows: First, all individuals are initialized with uniformly distributed random numbers and evaluated using the fitness function provided. Then the following will be executed until maximum number of generation has been reached or an optimum solution is found.

For a D-dimensional search space, each target vector $x_{i,g}$, a mutant vector is generated by

$$v_{i,g+1} = x_{r_1,g} + F * (x_{r_2,g} - x_{r_3,g}) \tag{3}$$

where $r_1, r_2, r_3 \in \{1, 2, ..., NP\}$ are randomly chosen integers, must be different from each other and also different from the running index i. F (>0) is a scaling factor which controls the amplification of the differential evolution $(x_{r_2,g} - x_{r_3,g})$. In order to increase the diversity of the perturbed parameter vectors, crossover is introduced [10]. The parent vector is mixed with the mutated vector to produce a trial vector $u_{ji,g+1}$,

$$u_{ji,g+1} = \begin{cases} v_{ji,g+1} & if \quad (rand_j \leq CR) \quad or \quad (j = j_{rand}) \\ x_{ji,g} & if \quad (rand_j > CR) \quad and \quad (j \neq j_{rand}) \end{cases}$$

where j = 1, 2,......, D; $rand_j \in [0,1]$; CR is the crossover constant takes values in the range [0, 1] and $j_{rand} \in (1,2,.....,D)$ is the randomly chosen index.

Selection is the step to choose the vector between the target vector and the trial vector with the aim of creating an individual for the next generation.

PENALTY METHOD FOR CONSTRAINED OPTIMIZATION

Many real-world optimization problems are solved subject to sets of constraints. The search space in COPs consists of two kinds of solutions: feasible and infeasible. Feasible points satisfy all the constraints, while infeasible points violate at least one of them. Therefore the final solution of an optimization problem must satisfy all constraints.

In this paper, the two algorithms PSO and DE handle the constraints using the concept of penalty functions. In the penalty function approach, the constrained problem is transformed into an unconstrained optimization problem by penalizing the constraints and building a single objective function, which is minimized using an unconstrained optimization algorithm. That is,

$$F(x) = f(x) + \lambda \, p(x) \tag{4}$$

Where $\quad p(x_i,t) = \sum_{m=1}^{n_g+n_h} \lambda_m(t) p_m(x_i)$ (5)

$$p_m(x_i) = \max\{0, g_m(x_i)^{\alpha}\}$$
$$\text{if} \quad m \in [1,......,n_g] \, (\text{inequality}) \tag{6}$$

$$p_m(x_i) = |h_m(x_i)|^{\alpha}$$
$$\text{if} \quad m \in [n_g+1,......,n_g+n_h] \, (\text{equality}) \tag{7}$$

with α a positive constant, representing the power of the penalty. The inequality constraints are considered as $g(x)$ and $h(x)$ represents the equality constraints. n_g and n_h denotes the number inequality and equality constraints respectively. λ is the constraint penalty coefficient.

ECONOMIC OPTIMIZATION OF A KRAFT PULPING FOR PULP AND PAPER INDUSTRY

Kraft process is a dominant chemical pulping process which uses NaOH and Na_2S as puling chemicals. The Kraft pulping [12] and recovery cycle operations comprise a reasonably isolated, yet complex, subsystem of an integrated papermaking process which requires sophisticated mathematical techniques for optimization. Figure 1 [12] shows the various subsystems involved in the process of paper making. It is evident from the figure that Pulping and recovery cycle operations form the center of integrated paper making process. The given hypothetical system consists of the interrelated digester and recovery cycle operations for a Kraft mill producing, under certain realistic conditions, a fixed daily amount of unbleached spruce pulp. Also there are certain typical revenues and variable costs, and a number of realistic constraints. For more details please refer to [12].

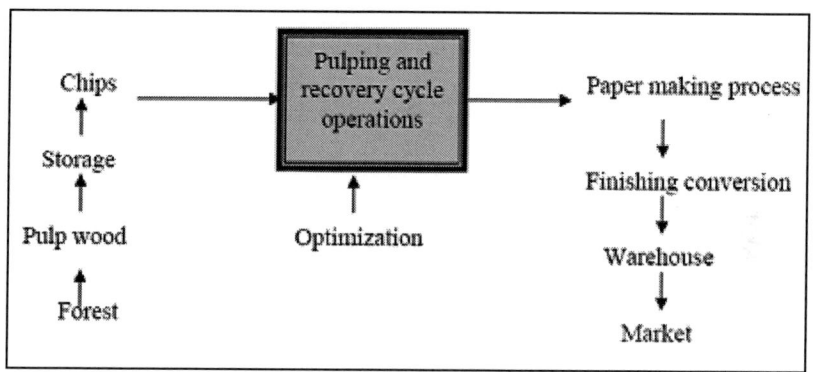

FIGURE 1 Integrated paper making process

The mathematical model of a hypothetical Kraft Pulping system is given as [13]:
Minimize
$$f(x) = 0.00011y_{14} + 0.1365 + 0.00002358y_{13}$$
$$+ 0.000001502y_{16} + 0.0321y_{12} + 0.004324y_5$$
$$+ 0.0001\frac{c_{15}}{c_{16}} + 37.48\frac{y_2}{c_{12}} - 0.0000005843y_{17}$$

Subject to:

$$g_1 = \frac{0.28}{0.72}y_5 - y_4 \le 0, \qquad g_2 = x_3 - 1.5x_2 \le 0, \qquad g_3 = 3496\frac{y_2}{c_{12}} - 21 \le 0,$$

$$g_4 = 110.6 + y_1 - \frac{62{,}212}{c_{17}} \le 0, \quad g_5 = y_1 - 405.23 \le 0, \qquad g_6 = 213.1 - y_1 \le 0$$

$$g_7 = y_2 - 1053.6667 \le 0, \qquad g_8 = 17.505 - y_2 \le 0, \qquad g_9 = y_3 - 35.03 \le 0,$$

$$g_{10} = 11.275 - y_3 \le 0, \qquad g_{11} = y_4 - 665.585 \le 0, \qquad g_{12} = 214.228 - y_4 \le 0,$$

$$g_{13} = y_5 - 584.463 \le 0, \qquad g_{14} = 7.458 - y_5 \le 0, \qquad g_{15} = y_6 - 265.916 \le 0,$$

$$g_{16} = 0.961 - y_6 \le 0, \qquad g_{17} = y_7 - 7.046 \le 0, \qquad g_{18} = 1.612 - y_7 \le 0,$$

$$g_{19} = y_8 - 0.222 \le 0, \qquad g_{20} = 0.146 - y_8 \le 0, \qquad g_{21} = y_9 - 273.366 \le 0,$$

$$g_{22} = 107.99 - y_9 \le 0, \qquad g_{23} = y_{10} - 1286.105 \le 0, \qquad g_{24} = 922.693 - y_{10} \le 0,$$

$$g_{25} = y_{11} - 1444.046 \le 0, \qquad g_{26} = 926.832 - y_{11} \le 0, \qquad g_{27} = y_{12} - 537.141 \le 0,$$

$$g_{28} = 18.766 - y_{12} \le 0, \qquad g_{29} = y_{13} - 3247.039 \le 0, \qquad g_{30} = 1072.163 - y_{13} \le 0,$$

$$g_{31} = y_{14} - 26844.086 \le 0, \qquad g_{32} = 8961.448 - y_{14} \le 0, \qquad g_{33} = y_{15} - 0.386 \le 0,$$

$$g_{34} = 0.063 - y_{15} \le 0, \qquad g_{35} = y_{16} - 140{,}000 \le 0, \qquad g_{36} = 71{,}084.33 - y_{16} \le 0,$$

$$g_{37} = y_{17} - 12{,}146{,}108 \le 0, \qquad g_{38} = 2{,}802{,}713 - y_{17} \le 0,$$

$$704.4148 \le x_1 \le 906.3855, \ 68.6 \le x_2 \le 288.88, \ 0 \le x_3 \le 134.75, \ 193 \le x_4 \le 287.0966,$$

$$25 \le x_5 \le 84.1988.$$

Calculations:

$$y_1 = x_2 + x_3 + 41.6, \qquad c_1 = 0.024x_4 - 4.62, \qquad y_2 = \frac{12.5}{c_1} + 12,$$

$$c_2 = 0.0003535x_1^2 + 0.5311x_1 + 0.08705y_2x_1, \qquad c_3 = 0.052x_1 + 78 + 0.002377y_2x_1,$$

$$y_3 = \frac{c_2}{c_3}, \qquad y_4 = 19y_3,$$

$$c_4 = 0.04782(x_1 - y_3) + \frac{0.1956(x_1 - y_3)^2}{x_2} + 0.6376y_4 + 1.594y_3,$$

$$c_5 = 100x_2, \qquad c_6 = x_1 - y_3 - y_4, \qquad c_7 = 0.95 - \frac{c_4}{c_5}, \qquad y_5 = c_6c_7,$$

$$y_6 = x_1 - y_5 - y_4 - y_3, \quad c_8 = (y_5 + y_4)0.995, \quad y_7 = \frac{c_8}{y_1}, \quad y_8 = \frac{c_8}{3798},$$

$$c_9 = y_7 - \frac{0.0663y_7}{y_8} - 0.3153, \quad y_9 = \frac{96.82}{c_9} + 0.321y_1,$$

$$y_{10} = 1.29y_5 + 1.258y_4 + 2.29y_3 + 1.71y_6, \quad y_{11} = 1.71x_1 - 0.452y_4 + 0.58y_3$$

$$c_{10} = \frac{12.3}{752.3}, \quad c_{11} = (1.75y_2)(0.995x_1), \quad c_{12} = 0.995y_{10} + 1998, \quad y_{12} = c_{10}x_1 + \frac{c_{11}}{c_{12}}$$

$$y_{13} = c_{12} - 1.75y_2, \quad y_{14} = 3623 + 64.4x_2 + 58.4x_3 + \frac{146312}{y_9 + x_5}$$

$$c_{13} = 0.995y_{10} + 60.8x_2 + 48x_4 - 0.1121y_{14} - 5095, \quad y_{15} = \frac{y_{13}}{c_{13}}$$

$$y_{16} = 148000 - 331000y_{15} + 40y_{13} - 61y_{15}y_{13}, \quad c_{14} = 2324y_{10} - 28740000y_2$$

$$y_{17} = 14{,}130{,}000 - 1328y_{10} - 531y_{11} + \frac{c_{14}}{c_{12}}, \quad c_{15} = \frac{y_{13}}{y_{15}} - \frac{y_{13}}{0.52}$$

$$c_{16} = 1.104 - 0.72y_{15}, \quad c_{17} = y_9 + x_5.$$

There are five independently adjustable variables, and they are defined as follows:

x_1 = total load of inorganic chemical as Na_2O in white liquor before losses

x_2 = volume of white liquor to digesters

x_3 = volume of black liquor to digesters

x_4 = rate of fresh wash water to washers

x_5 = time between cooks at each digester

BOILER LOAD ALLOCATION

Optimum boiler load allocation is another problem that occurs frequently in a paper making process. Consider the plant in Figure 2 [14] in which B4, B5 are base-loaded by bark and liquor, respectively. Since the boilers 1 – 3 are operating with high-cost fuel (gas and oil), their load will be optimized. It is not unusual to have these three boilers operating with bark and oil or gas. The cost curves can still be obtained without any undue difficulty.

The mathematical model of this problem is given below:

Minimize $C = \sum_{i=1}^{3} c_i$

Subject to:

$D = \sum_{i=1}^{3} m_i = 450$

$100 \leq m_i \leq 200, \ i = 1,2,3.$

Where,

C – Total cost of steam per hour

c_i – cost of steam of each boiler per hour

D – Steam demand in kg/hour

m_i – Individual boiler load, kg/hour

The values of c_1, c_2, c_3 are taken from [14] as

$c_1 = -0.0654 * 10^{-3} * m_1^2 + 2.453m_1 + 7.22$

$c_2 = 0.9513 * 10^{-3} * m_2^2 + 2.304m_2 - 0.6$

$c_3 = 0.8235 * 10^{-3} * m_3^2 + 2.404m_3 - 13.6$

The load values and demand are in thousands.

EXPERIMENTAL SETTINGS AND SIMULATION RESULTS

Experimental Settings

The initial population for both the algorithms is taken as 30. As mentioned previously, PSO and DE are ST; more than one execution of the algorithms is needed to reach to a conclusion. A total of 30 runs were performed and best result throughout the run was recorded. Maximum number of iterations allowed was set to 500.

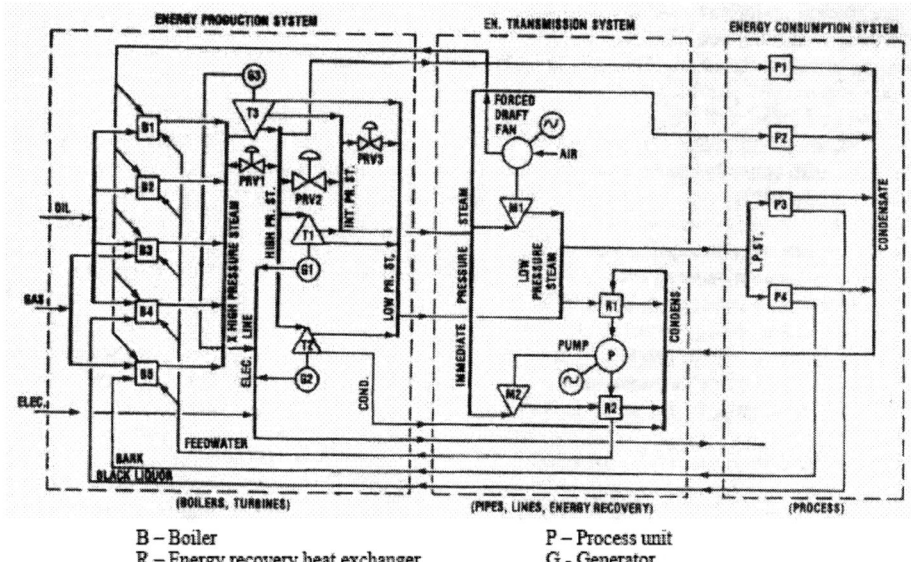

B – Boiler
R – Energy recovery heat exchanger
T – Electric power generating turbine
M - Mechanical power generating turbine

P – Process unit
G - Generator
PRV – Pressure release valve

FIGURE 2 Energy management system model for paper and pulp plant

Simulation Results

Table 1 shows the experimental results of Kraft pulping system. Figure 3 gives the convergence curves of PSO and DE for the Kraft pulping system. Table 2 shows the experimental results of boiler load allocation problem. For the present study, the algorithms are coded in Turbo C++ and executed on a P IV computer.

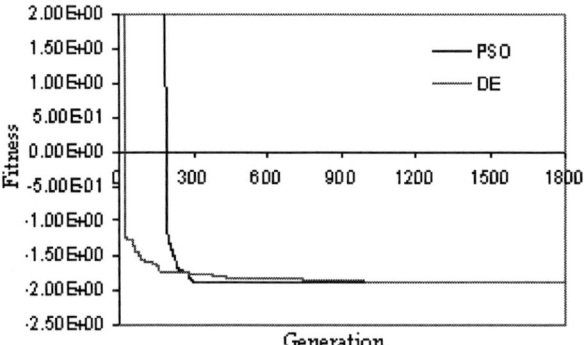

FIGURE 3 Convergence curves of PSO and DE for the Kraft Pulping System

TABLE 1 Simulation results of Kraft Pulping System

Item	PSO	DE	Results in [13]
x1	705.170955	705.180325	705.06
X2	68.6	68.6	68.6
x3	102.899995	102.899995	102.900
x4	282.324854	282.324033	282.341
x5	37.583506	37.571403	35.627
f (x)	-1.905168	-1.905134	-1.90500
Run time (sec)	0.63	1.23	---
Generation	504	1763	---

TABLE 2 Simulation results of Boiler Load Allocation

		[14]	PSO	DE
Boiler load assingments	#1	200	180.5	200
	#2	144.5	152.3	115.5
	#3	105.5	117.2	134.5
Demand, ton/h		450	450	450
Stream Cost, $/h		2247.39	2146.15	2207.67
Savings, $/h		0	101.24	39.72
Savings, %		0	4.7173	1.7992
Savings, $/yr		0	809920	317760
CPU time (sec)		----	0.1	0.32

DISCUSSION AND CONCLUSION

This paper presents an industrial application of two popular CIA namely PSO and DE, by taking two classical examples of optimization problems occurring a Pulp and Paper Industry. The mathematical model of the optimization of Kraft pulping system is a Linear Programming Problem model consisting of five unknown variables and thirty eight constraints. The second example deals with the optimum boiler load allocation problem and is nonlinear in nature.

Numerical results show the competence of PSO and DE in solving linear as well as non linear optimization problems. Another advantage of these algorithms is that they were able to solve the algorithms in very small time. For optimization of Kraft pulping system, PSO took only 0.63 seconds and 504 generations to solve the problem, where as DE took 1.23 seconds and 1763 generations to do the same and in case of Boiler load allocation problem, PSO took 0.1 seconds and DE took 0.32 seconds. Thus it may be concluded that PSO and DE can be used for solving large scale industrial optimization problems, linear or non linear in nature. However, we would like to add that since the nature of the optimization problems taken in this study is quite simple we have considered the basic versions of PSO and DE.

For more complex problems, the advanced versions of PSO and DE may be used. In future we shall be taking up more complex optimization problems that occur in Paper industry. Also we will modify the algorithms for solving integer programming problems as well.

APPENDIX

Comparison of PSO and DE on some selected benchmark problems for global Optimization

Problem 1: Sphere function
$f_1(x) = x_1^2 + x_2^2$, $-5.12 \leq x_i \leq 5.12$.
The optimum value is $f_1(x^*) = 0$ at $x^* = (0,0)$

Problem 2: Rastringin function
$f_2(x) = 20 + x_1^2 + x_2^2 - 10(\cos(2\pi x_1) + \cos(2\pi x_2))$, $-5.12 \leq x_i \leq 5.12$.
The optimum value is $f_1(x^*) = 0$ at $x^* = (0,0)$

Problem 3: Griewank function
$f_3(x) = \dfrac{1}{4000} \sum_{i=1}^{2} x_i^2 - \prod_{i=1}^{2} \cos(\dfrac{x_i}{\sqrt{i}}) + 1$,
$-600 \leq x_i \leq 600$.
The optimum value is $f_1(x^*) = 0$ at $x^* = (0,0)$.

Problem 4: Rosenbrock function
$f_4(x) = 100(x_2 - x_1^2)^2 + (x_1 - 1)^2$, $-30 \leq x_i \leq 30$.
The optimum value is $f_1(x^*) = 0$ at $x^* = (1,1)$
Table 3 shows the results above benchmark problems using PSO and DE.

TABLE 3 Numerical results of benchmark problems

Function	PSO	DE	True solution
F1	1.667377e-06	1.387779e-17	0.0
F2	1.181916e-10	1.309548e-23	0.0
F3	3.007688e-05	3.359894e-12	0.0
F4	5.634183e-05	2.108665e-09	0.0

REFERENCES

1. Pinter J.D., "Global Optimization in Action. Continuous and Lipschitz Optimization: Algorithms, Implementation and Applications", Kluwer, Dordrecht, 1996.
2. A. Torn and Zilinskas, "Global Optimization", Springer Verlag, Berlin, 1989.
3. I. Rechenberg, "Evolution Strategy: Optimization of Technical systems by means of biological evolution", Fromman-Holzboog, 1973.
4. L. J. Fogel, A. J. Owens, and M. J. Walsh, "Artificial intelligence through a simulation of evolution", In M. Maxfield, A. Callahan and L. J. Fogel, editors, Biophysics and Cybernetic systems. Proc. of the 2nd Cybernetic Sciences Symposium, pp. 131 – 155, Spartan Books, 1965.
5. Holland J. H., "Adaptation in Natural and Artificial Systems: An Introductory Analysis with Applications to Biology, Control, and Artificial Intelligence," Ann Arbor, MI: University of Michigan Press.
6. Goldberg D., "Genetic Algorithms in Search Optimization and Machine Learning", Addison Wesley Publishing Company, Reading, Massachutes.
7. Kennedy J. and Eberhart R., "Particle Swarm Optimization," IEEE International Conference on Neural Networks (Perth, Australia), IEEE Service Center, Piscataway, NJ, 1995, pg. IV: 1942-1948.
8. Kennedy J., "The Particle Swarm: Social Adaptation of Knowledge," IEEE International Conference on Evolutionary Computation (Indianapolis, Indiana), IEEE Service Center, Piscataway, NJ, pg.303-308, 1997.
9. R. Storn and K. Price, "Differential Evolution – a simple and efficient adaptive scheme for global optimization over continuous spaces", Technical Report, International Computer Science Institute, Berkley, 1995.
10. R. Storn and K. Price, "Differential Evolution – a simple and efficient Heuristic for global optimization over continuous spaces", Journal Global Optimization. 11, 1997, pp. 341 – 359.
11. D. Karaboga and S. Okdem, "A simple and Global Optimization Algorithm for Engineering Problems: Differential Evolution Algorithm", Turk J. Elec. Engin. 12(1), 2004, pp. 53 – 60.
12. C. W. Carroll, "An Operation Research Approach to the Economic Optimization of a Kraft Pulping Process", Ph. D. dissertation, The Institute of Paper Chemistry, Appleton, Wisconsin, 1959.
13. D. M. Himmelblau, "Applied Non-Linear programming", New York: McGraw-Hill, 1972.
14. A. Kaya and M. A. Keyes, "Energy Management Technology in Pulp, Paper, and Allied Industries", Automatica, Vol. 19 (2), 1983, pp.111 – 130.

Effect of Gravity Modulation on the Onset of Convection in a Horizontal Anisotropic Porous Layer

T. Sivakumar and S. Saravanan*

Department of Mathematics, Bharathiar University, Coimbatore 641 046, Tamil Nadu, INDIA,
*Email: sshravan@lycos.com

Abstract. A linear stability theory is used to investigate the effect of gravity modulation on the onset of convection in a homogeneous anisotropic porous layer heated from above. The Brinkman model with anisotropic permeability is considered in the current analysis. The Oberbeck - Boussinesq approximation is employed and the corresponding governing equations are reduced to Mathieu's equation. Floquet theory is used to find the critical Rayleigh number and the corresponding wave number as a function of modulation amplitude and frequency. It is found that the onset of convection can be advanced or delayed by adjusting amplitude and frequency depending upon the Darcy number.

Keywords: Porous medium, Brinkman model, Stability, Gravity modulation
PACS: 44.30.+v, 47.56.+r

INTRODUCTION

Natural convection in fluid saturated porous media confined between parallel surfaces of different temperatures is of engineering importance. The applications include exothermic reactions in packed bed reactors, cooling of electronic equipment, nuclear waste disposals and water movements in geothermal reservoirs. The stability of convection in porous media has been investigated extensively by several researchers and are well documented by Nield and Bejan (2000).

Most of the previous studies have been concerned with homogeneous isotropic porous media. But in many practical applications the porous matrix is anisotropic in nature which may be due to the preferential orientation or asymmetric geometry of the minute porous structures. Castinel and Combarnous (1974) were the first to conduct an experimental as well as theoretical investigation of the Rayleigh-Bénard convection in a porous medium with anisotropic permeability. Epherre (1977) then studied the effect of thermal anisotropy on the onset of motion. Govender (2006) investigated natural convection in an anisotropic porous layer subjected to centrifugal body force. He found that convection is stabilized when the anisotropy ratio, which is a function of the thermal and mechanical anisotropic parameters, is increased in magnitude. The same problem with a magnetically sensitive fluid has been studied by Saravanan (2009). It was found that the magnetic field has a destabilizing effect and can be suitably adjusted depending on the anisotropy parameters to enhance convection. Recently, Malashetty and Mahentesh Swamy (2007) have investigated the effect of rotation on the onset of convection in

CP1146, *Modelling of Engineering and Technological Problems*, edited by A. H. Siddiqi, A. K. Gupta, and M. Brokate
© 2009 American Institute of Physics 978-0-7354-0683-4/09/$25.00

a horizontal anisotropic porous medium and they found that the effect of mechanical anisotropy is to allow the onset of oscillatory convection instead of stationary.

Studies on time dependent gravity are of great interest in crystal growth, petroleum production and other areas. Gresho and Sani (1970) first studied the effect of sinusoidal gravity modulation in a differentially heated fluid layer with small amplitude approximation. They found that the system may be stabilized in the same manner as an inverted pendulum is stabilized by vertical oscillations. Much work has been done for pure fluids in this area and studies of modulation in porous media is quite recent. Zen'kovskaya and Rogovenko (1999) have investigated this problem by the use of averaging method subject to high frequency oscillations in an arbitrary direction. It was found that horizontal oscillation has a destabilizing effect in the case of zero and micro gravities. More recently, Natalia (2008) has dealt with a fluid saturated Darcy porous medium subjected to vertical oscillations of arbitrary amplitude and frequency.

The purpose of the present paper is to study the effect of gravity modulation of arbitrary amplitude and frequency on the onset of convection in a fluid saturated porous medium with anisotropic permeability using the Brinkman flow model.

MATHEMATICAL FORMULATION

We consider a fluid saturated anisotropic porous medium, confined between two horizontal surfaces $z = 0$ and $z = h$ of infinite extent and maintained at temperatures T_1 and T_2 respectively $(T_2 > T_1)$. We assume that a time dependent gravity field is acing on it and the permeability of the porous matrix is anisotropic. The equations governing the above system under the assumption of the Oberbeck-Boussinesq approximation are

$$\frac{1}{\varphi}\frac{\partial \mathbf{v}}{\partial t} + \frac{1}{\varphi^2}(\mathbf{v} \cdot \nabla)\mathbf{v} = -\frac{1}{\rho}\nabla p - \frac{v}{\overline{\overline{K}}}\mathbf{v} + v\nabla^2\mathbf{v} + \beta T g(t)\widehat{\mathbf{k}} \tag{1}$$

$$\varkappa\frac{\partial T}{\partial t} + (\mathbf{v} \cdot \nabla)T = \chi\nabla^2 T \tag{2}$$

$$\nabla \cdot \mathbf{v} = 0 \tag{3}$$

where $\mathbf{v} = (v_1, v_2, v_3)$ is the velocity, p the pressure, T the temperature, φ the porosity, ρ the density, v the kinematic viscosity, $\overline{\overline{K}} = K_x(\mathbf{ii} + \mathbf{jj}) + K_z\mathbf{kk}$ the permeability tensor, β the thermal expansion coefficient, $\widehat{\mathbf{k}}$ the unit vector directed vertically upward, $\varkappa = (\rho c_p)_m/(\rho c_p)_f$ the heat capacity ratio of the porous medium and fluid and χ the thermal diffusivity of the porous medium. The time dependent gravitational field is taken to be $g(t) = g_0 + \frac{A}{\varphi}\Omega^2 f''(\tau)$, where g_0 is a reference acceleration level, A the amplitude, Ω the modulation frequency and $f(\tau)$ the 2π-periodic function with zero 2π-average.

We study the stability of the following quiescent basic state $\mathbf{v}^0 = 0$, $T^0 = T_1 - Cz$ where $C = (T_1 - T_2)/h$, using the method of small perturbations. The non-dimensional governing equations are

473

$$c\frac{\partial \mathbf{u}}{\partial t} = -\nabla q - \mathbf{u_a} + Da\nabla^2\mathbf{u} + Gr(1 + \eta f''(\tau))\theta\widehat{\mathbf{k}} \tag{4}$$

$$\varkappa\frac{\partial \theta}{\partial t} - u_3 = \frac{1}{Pr}\nabla^2\theta \tag{5}$$

$$\nabla\cdot\mathbf{u} = 0 \tag{6}$$

where $\mathbf{u_a} = (u_1/Kr, u_2/Kr, u_3)$ is the modified velocity vector, $Kr = K_x/K_z$ the anisotropic parameter, $c = K_z/\varphi h^2$ the ratio of dimensionless permeability to the porosity, $Da = K_z/h^2$ the Darcy number, $Gr = \beta Ch^2 g_0 K_z/\nu^2$ the filtration Grashof number, $Pr = \nu/\chi$ the Prandtl number, $\eta = A\Omega^2/\varphi g_0$ the amplitude and $\omega = \Omega h^2/\nu$ the frequency of modulation. Eliminating the pressure and performing normal mode expansion in the form $(u_3, \theta) = \left(\widetilde{u}_3, \widetilde{\theta}\right)e^{i(\alpha_1 x + \alpha_2 y)}$, we get

$$c\frac{\partial}{\partial t}\left(\frac{1}{Pr}D^4\widetilde{\theta} - \frac{2\alpha^2}{Pr}D^2\widetilde{\theta} - \varkappa\frac{\partial}{\partial t}D^2\widetilde{\theta} + \frac{1}{Pr}\alpha^4\widetilde{\theta} + \varkappa\alpha^2\frac{\partial\widetilde{\theta}}{\partial t}\right) + \frac{1}{Kr}\left(\frac{1}{Pr}D^4\widetilde{\theta}\right.$$

$$-\frac{\alpha^2}{Pr}D^2\widetilde{\theta} - \varkappa\frac{\partial}{\partial t}D^2\widetilde{\theta}\right) + \frac{1}{Pr}\alpha^4\widetilde{\theta} + \varkappa\alpha^2\frac{\partial\widetilde{\theta}}{\partial t} - \frac{\alpha^2}{Pr}D^2\widetilde{\theta} = \alpha^2 Gr\left(1 + \eta f''(\tau)\right)\widetilde{\theta} \tag{7}$$

$$+Da\left(\frac{1}{Pr}D^6\widetilde{\theta} - \frac{3\alpha^2}{Pr}D^4\widetilde{\theta} - \varkappa\frac{\partial\widetilde{\theta}}{\partial t}D^4\widetilde{\theta}\frac{3\alpha^4}{Pr}D^2 + 2\varkappa\alpha^2\frac{\partial\widetilde{\theta}}{\partial t}D^2 - \varkappa\alpha^4\frac{\partial\widetilde{\theta}}{\partial t} - \frac{\alpha^6}{Pr}\widetilde{\theta}\right)$$

where $D \equiv \partial/\partial z$ and $\alpha^2 = \alpha_1^2 + \alpha_2^2$ is overall horizontal wave number. The appropriate boundary conditions we consider are

$$\widetilde{\theta} = \frac{\partial^2\widetilde{\theta}}{\partial z^2} = 0 \quad \text{at} \quad z = 0 \text{ and } z = 1 \tag{8}$$

From Eq.(8) the z-variable is separated by taking $\widetilde{\theta}(z, t) = \sin(\pi\ell z)\Theta(t), \ell = 1, 2, 3, \ldots$ and the resulting equation takes the form

$$\Theta''(\widetilde{t}) + \left(\frac{m_1^2}{P} + P\frac{m_2^2}{m_1^2} + Da\,m_1^2 P\right)\Theta'(\widetilde{t})$$

$$+\left(Da\,m_1^4 + m_2^2 - \frac{\alpha^2}{m_1^2}Ra(1 - \eta\cos(\widetilde{\omega}\widetilde{t}))\right)\Theta(\widetilde{t}) = 0 \tag{9}$$

where $\widetilde{t} = \dfrac{t}{\sqrt{Pr\varkappa c}}$, $P = \sqrt{\dfrac{Pr\varkappa}{c}}$, $\widetilde{\omega} = \omega\sqrt{Pr\varkappa c}$, $m_1^2 = (\pi\ell)^2 + \alpha^2$, $m_2^2 = (\pi\ell)^2 + \dfrac{\alpha^2}{Kr}$, $Ra = Gr\cdot Pr$ is the Rayleigh number which is negative for stable equilibrium and $f(\tau) = \cos\tau$. To reduce this equation to the canonical form of Mathieu's equation, we use the change of variables $\Theta(t) = e^{-\lambda t}F(t)$ and $2\tau = \omega t$ to obtain

$$\ddot{F} + [A - 2B\cos(2\tau)]F = 0 \tag{10}$$

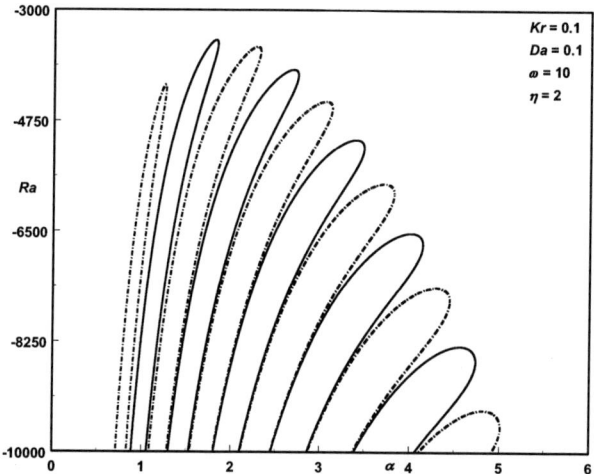

FIGURE 1. Synchronous (dash dot line) and subharmonic (solid line) branches of the marginal curve

where A and B are given by the expressions

$$A = \frac{4}{\omega^2}\left(Dam_1^4 + m_2^2 - \frac{\alpha^2}{m_1^2}Ra - \lambda^2\right), \ B = -\frac{4\alpha^2 Ra\eta}{2\omega^2 m^2} \text{ and } 2\lambda = \frac{m_1^2}{P} + P\frac{m_2^2}{m_1^2} + Dam_1^2 P$$

Following the Floquet theory, the general solution of Eq.(10) is of the form

$$F(\tau) = e^{\mu\tau}G(\tau) = e^{\mu\tau}\sum_{n=-\infty}^{+\infty} a_n\, e^{in\omega\tau} \tag{11}$$

where μ is the Floquet exponent and $G(\tau)$ is a periodic function with period π or 2π. Then the marginal stability condition corresponding to the periodic solutions is given by $\mu\omega/2 = \lambda$. Here the Floquet exponent $\mu = 0$ corresponds to synchronous mode with period π and $\mu = i\omega/2$ corresponds to subharmonic mode with period 2π. Substituting the Fourier decomposition into Mathieu's equation yields a Hill determinant. The resulting system is then solved numerically to obtain an eigenvalue Ra depending on α and other parameters. Further details of the procedure adopted can be found in Aniss et al. (2000).

RESULTS AND DISCUSSION

The effect of gravity modulation on the onset of convection in a fluid saturated anisotropic porous medium is investigated. Negative Rayleigh numbers, corresponding to heating of the layer from above, are considered. We present results for arbitrary values of amplitude η and frequency ω. We fixed $Pr = 1$, $c = 1$, $\varkappa = 1$ and $\ell = 1$. We note that the equation (9) reduces to that of Natalia (2007) for $Da = 0$ and $Kr = 1$. The

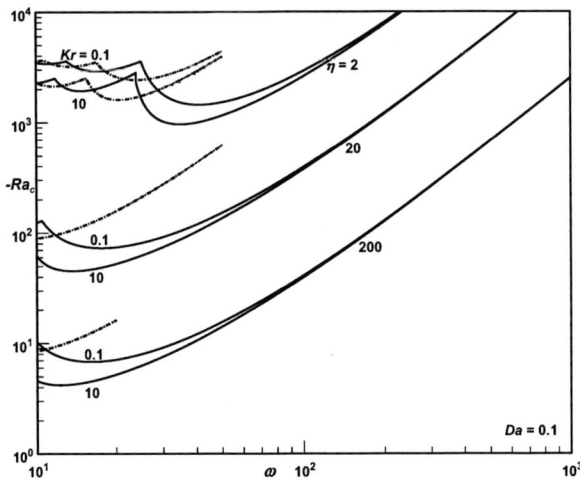

FIGURE 2. $-Ra_c$ against ω for $Da = 0.1$, $Kr = 0.1, 10$ and different amplitudes

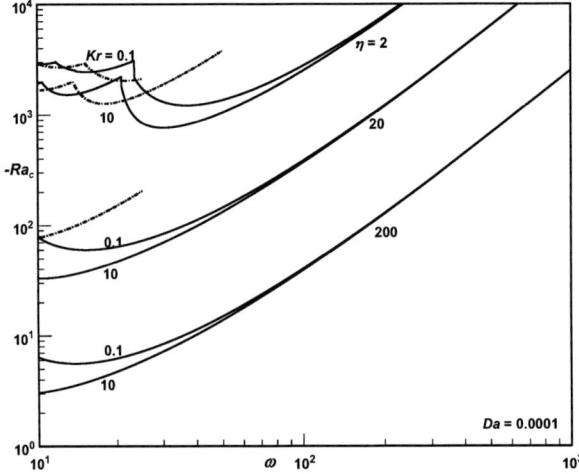

FIGURE 3. $-Ra_c$ against ω for $Da = 0.0001$, $Kr = 0.1, 10$ and different amplitudes

marginal curve corresponding to the solutions of Mathieu's equation for $Kr = Da = 0.1$ is shown in Fig.1. It consists of a group of loop shaped branches that indicates the appearance of instability in a stable setup. The instability region comprises of alternate regions of synchronous (dash dot lines) and subharmonic (solid lines) modes. Each marginal curve has a unique minimal Rayleigh number at which the unstable region terminates and the global minimum of these Rayleigh numbers is referred to as the critical Rayleigh number Ra_c representing the onset condition. The wavenumber corresponding

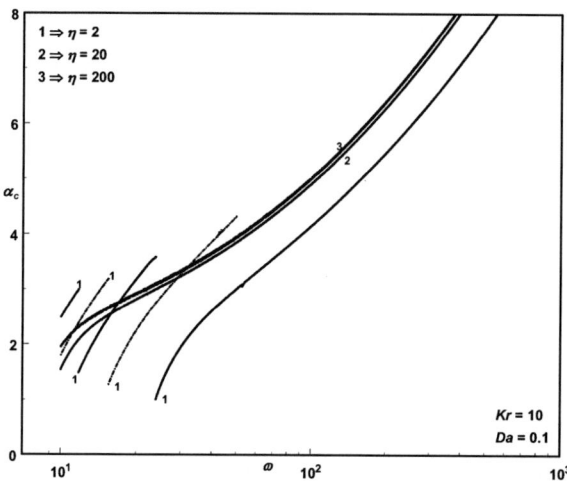

FIGURE 4. $-\alpha_c$ against ω for $Da = 0.1$, $Kr = 0.1$ and different amplitudes

to Ra_c is the critical wave number α_c. In Fig.1, Ra_c occurs at the second loop which is subharmonic.

The influence of Da and Kr on $-Ra_c$ and α_c for different values of η are plotted in Figs.2-4. From Figs.2 & 3, we notice that destabilization of the system becomes very strong as η increases to higher values. For high-ω limit, $|Ra_c| \to \infty$ as expected, i.e., modulation has no effect on high frequency range and it corresponds to the unmodulated problem. It is clear that low amplitude critical curves are somewhat more complicated as synchronous and subharmonic curves often intersect at low frequencies. The corresponding α_c plotted in Fig.4 shows an increasing trend against ω with several jumps in the low frequency range. In other words a new thin loop shaped marginal curve emerges and develops in the low wavenumber region and becomes dominant at frequencies in which the jump occurs. From Figs.2 & 3 it is clear that subharmonic mode is the dominant one at high amplitudes. We also observe that an increasing Kr always destabilizes the system and is significant at low frequencies. Moreover $|Ra_c|$ and α_c increase slightly when Da is increased. We also observe that the parameters Da and Kr have no much influence at high frequencies.

CONCLUSION

The effect of gravity modulation on the onset of convection in a horizontal anisotropic porous layer is studied using linear stability analysis. Anisotropic permeability and the Brinkman flow model are used to describe the system. It is found that Da stabilizes the system under consideration whereas Kr destabilizes it. The study also reveals that increasing amplitude destabilizes convection and increases the critical wave number.

ACKNOWLEDGMENTS

The authors thank UGC, India for its support through DRS Special Assistance Programme in Fluid Dynamics.

REFERENCES

1. D. A. Nield and A. Bejan, *Convection in Porous Media*, Second edn., Springer, New Yark, 1999.
2. G. Castinel, M. Combarnous, *Critere d'apparition de la convection naturelle dans une couche poreuse anisotrope horizontal*, C. R. Acad. Sci. B 278 (1974) 701–704.
3. J. F. Epherre, *Criterion for the appearance of natural convection in an anisotropic porous layer*, Int. Chem. Engg. 17 (1977) 615–616.
4. S. Govender, *On the effect of anisotropy on the stability of convection in rotating porous media*, Trans. Porous Media, 64 (2006) 413–422.
5. S. Saravanan, *Centrifugal acceleration induced convection in a magnetic fluid saturated anisotropic rotating porous medium*, Trans. Porous Media, 77 (2009) 79–86.
6. M. S. Malashetty and S. Mahantesh, *The effect of rotation on the onset of convection in a horizontal anisotropic porous layer*, Int. J. Thermal Sci., 46 (2007) 1023–1032.
7. P. M. Gresho and R. L. Sani, *The effects of gravity modulation on the stability of a heated fluid layer*, J. Fluid Mech., 40 (1970) 783–806.
8. S. M. Zen'kovskaya and T. N. Rogovenko, *Filtration convection in a high-frequency vibration field*, J. Appl. Mech. Tech. Phys., 40 (1999) 379–385.
9. S. Natalia, *Effect of vertical modulation on the onset of filtration convection*, J. Math. Fluid Mech., 10 (2008) 488–502.
10. S. Aniss, M. Souhar and M. Belhaq, *Asymptotic study of the convective parametric instability in Hele-Shaw cell*, Phys. Fluids, 12 (2000) 262–268.

TRANSIENT BEHAVIOUR OF BATCH ARRIVAL QUEUE WITH N-POLICY AND SINGLE VACATION (M^X/G/1/N-POLICY)

Anjana Solanki

Bundelkhand Institute of Engineering & Technology
Jhansi (INDIA)

Abstract. In this paper M^X/G/1 queuing system with N-policy and single vacation is considered. As soon as the system becomes empty, the server leaves the system for a vacation of random length V. When he returns from the vacation, if the system size is greater then or equal to predetermined value N (threshold), he begins to serve the customers. If not, the server waits in the system until the system size reaches or exceeds N. Here the time dependent system size distribution is obtained.

Keyword: Dormant period , Vacation.

INTRODUCTION

The present work is the modification of batch arrival queue with N-policy and single vacation, studied by Lee, Yoon and Chae [9]. The transient state behavior of the batch arrival queue with N-policy and single vacation is discussed. An expression for the time dependent system size distribution is obtained.

Vacation queues have attracted many attentions from numerous researchers since Levy and Yechiali [1]. Fuhrman and Cooper [2] provide the well known "decomposition property" for vacation queues. For applications of vacation models to polling system, see Takagi [3]. For comprehensive survey on vacation queues, see Doshi [4] and Takagi [5]. Hofri [6] studied the toe N-policy queus attend by a single server Kella [7] studied the M^X/G/1 queue with N policy and vacations. Batch arrival queue with threshold and with/without multiple vacations were first studied by Lee and Srinivasan [8].

Consider the manufacturing system in which the production does not start until some specified number of raw materials, say N, is accumulated during an idle period. We assume that the operator of the machine perform some extra operations (for example, machine repair, preventive maintenance, etc.) when there are no raw materials to process. To be more realistic, we also assume that the raw material arrives in batches. This production system can be modeled by an M^X/G/1 queue with N policy and single vacation. Here customer arrive according to the compound Poisson process with random arrive size X. As soon as the system becomes empty the server leaves for a vacation of random length V (vacation period). When he returns from the vacation and the system size is greater than or equal to a predetermined value N (threshold), the server begins to serve the customers until there is no customer to serve (busy period). If he find fewer customer than N, he waits in the system until the system size reaches or exceed N (dormant period). Thus, in this system, a vacation period, a dormant period (the length of which is zero if the returning server finds N or more customers) and a busy period constitute a cycle. This system is depicted in Fig.1.

CP1146, *Modelling of Engineering and Technological Problems*, edited by A. H. Siddiqi, A. K. Gupta, and M. Brokate

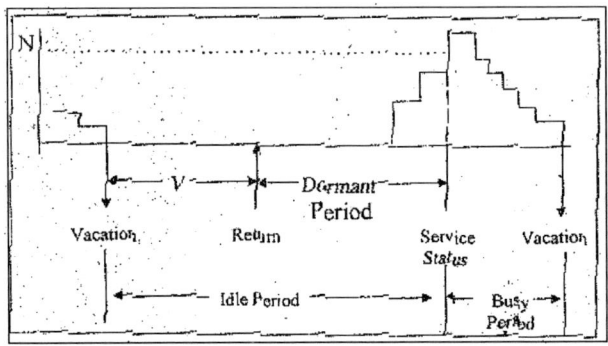

Fig.1. The queuing system (M^x/G/1/N- policy) and single vacation.

THE SYSTEM SIZE DISTRIBUTION

In this section, the system equations and the probability generating function (PGF) of the system size distribution are derived. The following notations and probabilities are defined:

N	Threshold
λ	Group arrival rate
X	Arrival size random variable
g_K	Pr $(X = k)$
α_K	Probability that k customers arrive during a vacation
X(z)	Probability generating function of X
S	Service time random variable
V	Vacation time random variable
s (x)	Probability density function of S
v (x)	Probability density function of V
$S^0(t)$	Remaining service time of the customer in service at time t
$V^0(t)$	Remaining vacation time of the server in vacation at time t
$S^*(\theta)$	Laplace Stieltjes Transform (LST) of S
$V^*(\theta)$	Laplace Stieltjes Transform (LST) of V
N(t)	System size at time t

$$Y(t) \begin{cases} 0, \text{ if server is on vacation} \\ 1, \text{ if server is in dormancy} \\ 2, \text{ if server is busy} \end{cases}$$

$R_n(t) = Pr \, [N(t)=n, Y(t) = 1], \qquad n = 0,1,2,\ldots\ldots\ldots,N-1$

480

$P_n(x,t) \, dt = P \, [N(t)=n, \, x \le S^0(t) \le x + dt, \, Y(t) =2], \quad n = 1,2,3\ldots\ldots\ldots$

$Q_n(x,t) \, dt = P \, [N(t)=n, \, x \le V^0(t) \le x + dt, \, Y(t) =0], \quad n = 0,1,2\ldots\ldots\ldots$

Forming Kolmogorov differential difference equations :

$$\frac{d}{dt} R_0(t) = - \lambda R_0(t) + Q_0(0,t) \tag{2.1}$$

$$\frac{d}{dt} Rn(t) = - \lambda R_n(t) + \lambda \sum_{k=1}^{n} R_{n-k}(t) g_k + Q_n(0,t) \quad , n = 1,2,3,\ldots\ldots\ldots N-1 \tag{2.2}$$

$$- \frac{\partial}{\partial x} P_1(x,t) = - \lambda P_1(x,t) + P_2(0,t) s(x) \tag{2.3}$$

$$- \frac{\partial}{\partial x} P_m(x,t) = - \lambda P_m(x,t) + \lambda \sum_{k=1}^{m-1} P_{m-k}(x,t) \, g_k + P_{m+k}(0,t) \, s(x)$$
$$\{m = 2,3,\ldots\ldots\ldots N-1\} \tag{2.4}$$

$$- \frac{\partial}{\partial x} P_n(x,t) = - \lambda P_n(x,t) + P_{n+1}(0,t) s(x) + \lambda \sum_{k-1}^{n-1} P_{n-k}(x,t) \, g_k +$$
$$\lambda s(x) \sum_{k=0}^{N-1} R_k(t) g_{n-k} + Q_n(0,t) s(x) \quad (n \ge N) \tag{2.5}$$

$$- \frac{\partial}{\partial x} Q_0(x,t) = - \lambda Q_n(x,t) + P_1(0,t) v(x) \tag{2.6}$$

$$- \frac{\partial}{\partial x} Q_m(x,t) = - \lambda Q_m(x,t) + \lambda \sum_{k=1}^{m} Q_{m-k}(x,t) \, g_k, \, (m \ge 1) \tag{2.7}$$

taking Laplace Stieltjes Transform (LST) on both side of the equations from (2.3) to (2.7) we have

$$\theta P_1^*(\theta,t) - P_1(0,t) = \lambda P_1^*(\theta,t) - P_2(0,t) S^*(\theta) \tag{2.8}$$

$$\theta P_m^*(\theta,t) - P_m(0,t) = \lambda P_m^*(\theta,t) - P_{m+1}(0,t) S^*(\theta) - \lambda \sum_{k=1}^{m-1} P_{m-k}^*(\theta,t) \, g_k$$
$$m = 2,3,\ldots N-1 \tag{2.9}$$

$$\theta P_n^*(\theta,t) - P_n(0,t) = \lambda P_n^*(\theta,t) - P_{n+1}(0,t) S^*(\theta) - \lambda \sum_{k-1}^{n-1} P_{n-k}^*(\theta,t) \, g_k -$$
$$\lambda S^*(\theta) \sum_{k=0}^{N-1} R_k^*(t) g_{n-k} + Q_n(0,t) S^*(\theta), \, n \ge N \tag{2.10}$$

$$\theta Q_0^* \ (\theta, t) \ - Q_0 \ (0, t) = \lambda \ Q_0^* \ (\theta, t) \ - P_1 \ (0, t) \ V^* \ (\theta) \tag{2.11}$$

$$\theta \ Q_m^* (\theta, t) - Q_m \ (0, t) = \lambda Q_m^* \ (\theta, t) \ - \lambda \sum_{k=1}^{m} Q_{m-k} \ (\theta, t) \ g_k, \ m \geq 1 \tag{2.12}$$

The following probability generating functions are considered:

$$P^* \ (z, \theta, t) = \sum_{n=1}^{\infty} \ P_n^* \ (\theta, t) \ z^n$$

$$P \ (z, 0, t) = \sum_{n=1}^{\infty} \ P_n \ (0, t) \ z^n$$

$$Q^* \ (z, \theta, t) = \sum_{n=0}^{\infty} \ Q_n^* \ (\theta, t) \ z^n$$

$$Q \ (z, 0, t) = \sum_{n=0}^{\infty} \ Q_n^* \ (0, t) \ z^n$$

Queue size distribution due to idle period
Multiplying both sides of (2.12) by z^n and taking summation from n = 1 to n = ∞. for n $\geq$ 1, following equation is obtained

$$\theta \sum_{n=1}^{\infty} Q_n^* \ (\theta, t) \ z^n - \sum_{n=1}^{\infty} \ Q_n \ (0, t) \ z^n = \lambda \sum_{n=1}^{\infty} \ Q_n^* \ (\theta, t) \ z^n - \lambda \sum_{n=1}^{\infty} \left(\sum_{k=1}^{n} Q_{n-k}^* \ (\theta, t) \ g_k \right) z^n$$

now using equation (2.11) and 2.12, in above equation following equation is obtained :

$$\theta Q^* \ (z, \theta, t) - Q_n \ (z, 0, t) = \lambda Q^* \ (z, \theta, t) - P_1 \ (0, t) \ V^* \ (\theta) - \lambda \sum_{n=1}^{\infty} \left(\sum_{k=1}^{n} Q_{n-k}^* \ (\theta, t) \ g_k \right) z^n \tag{2.13}$$

After solving the term $\displaystyle \sum_{n=1}^{\infty} \left(\sum_{k=1}^{\infty} Q_{n-k}^* \ (\theta, t) \ g_k \right) z^n$, its value is obtained as:

$$\sum_{n=1}^{\infty} \left(\sum_{k=1}^{\infty} Q_{n-k}^* \ (\theta, k) \ g_k \right) z^n = Q^* \ (z, \theta, t) \ x \ (z) - g_0 Q^* \ (z, \theta, t)$$

$$[g_0 = P \ (X = 0) = 0]$$

Substituting the value of the above term equation (2.13) becomes:
$$[\theta - \lambda \ X \ (z)] \ Q^* \ (z, \theta, t) = Q \ (z, 0, t) - P_1 \ (0, t) \ V^* \ (\theta) \tag{2.14}$$

Letting $\theta = \lambda - \lambda \ X \ (z)$, following expression is obtained:
$$Q \ (z, 0, t) \ = P_1 \ (0, t) \ V^* \ [\lambda - \lambda \ X \ (z)] \tag{2.15}$$
Thus equation (2.14) becomes
$$Q^* \ (z, \theta, t) \ = \frac{P_1 \ (0, t) \{ V^* [\lambda - \lambda \ X \ (z)] - V^* [\theta] }{\theta - \lambda + \lambda \ X \ (z)} \tag{2.16}$$

482

Queue size distribution due to busy period

Now using equations (2.8), (2.9) and (2.10) and multiplying equation (2.10) by z^n and taking summation from $n = N$ to $n = \infty$

$$\theta \sum_{n=N}^{\infty} P_n^*(\theta,t) z^n - \sum_{n=N}^{\infty} P_n(0,t) z^n = \lambda \sum_{n=N}^{\infty} P_n^*(\theta,t) z^n - \sum_{n=N}^{\infty} P_n^*(0,t) S^*(\theta) z^n -$$

$$\lambda \sum_{n=N}^{\infty} \left(\sum_{k=1}^{n-1} P_{n-k}^*(\theta,t) g_k \right) z^n - \lambda \sum_{n=N}^{\infty} S^*(\theta) \left(\sum_{k=0}^{N-1} R_k(t) g_{n-k} \right) z^n$$

$$- \sum_{n=N}^{\infty} Q_n(0,t) S^*(\theta) z^n \qquad (2.17)$$

Multiplying equation (2.9) by z^n and taking summation from $n = 2$ to $n = N-1$

$$\theta \sum_{n=2}^{N-1} P_n^*(\theta,t) z^n - \sum_{n=2}^{N-1} P_n(0,t) z^n = \lambda \sum_{n=2}^{N-1} P_n^*(\theta,t) z^n - \frac{S^*(\theta)}{z} \sum_{n=2}^{N-1} P_{n+1}^*(0,t) z^{n+1}$$

$$- \lambda \sum_{n=2}^{N-1} \left(\sum_{k=1}^{n-1} P_{n-k}^*(\theta,t) g_k \right) z^n \qquad (2.18)$$

from equation (2.8) we get

$$- \lambda P_1^*(\theta,t) z + P_2(0,t) S^*(\theta) z = - \theta P_1^*(\theta,t) z + P_1(0,t) z \qquad (2.19)$$

from equation (2.18)

$$\left[\sum_{n=1}^{N-1} \theta P_n^*(\theta,t) z^n - \theta P_1^*(\theta,t) z \right] - \left[\sum_{n=1}^{N-1} P_n(0,t) z^n - P_1(0,t) z \right]$$

$$= \lambda \left[\sum_{n=1}^{N-1} P_n^*(\theta,t) z^n - P_1^*(\theta,t) z \right] - \frac{S^*(\theta)}{z} \left[\sum_{n=1}^{N-1} P_{n+1}(0,t) z^{n+1} - P_2(0,t) z^2 \right]$$

$$- \lambda \sum_{n=2}^{N-1} \left(\sum_{k=1}^{n-1} P_{n-k}^*(\theta,t) g_k \right) z^n$$

using equation (2.19) we have

$$\sum_{n=1}^{N-1} \theta P_n^*(\theta,t) z^n - \sum_{n=1}^{N-1} P_n(0,t) z^n = \lambda \sum_{n=1}^{N-1} P_n^*(\theta,t) z^n - \frac{S^*(\theta)}{z} \sum_{n=1}^{N-1} P_{n+1}(0,t) z^{n+1}$$

$$- \lambda \sum_{n=2}^{N-1} \left(\sum_{k=1}^{n-1} P_{n-k}^*(\theta,t) g_k \right) z^n \qquad (2.20)$$

by using equation (2.17) in (2.19), we have

$$\sum_{n=1}^{\infty} \theta P_n^*(\theta,t) z^n - \sum_{n=1}^{\infty} P_n(0,t) z^n = \lambda \sum_{n=1}^{\infty} P_n^*(\theta,t) z^n - \frac{S^*(\theta)}{z}$$

$$- \lambda S^*(\theta) \left[\sum_{n=1}^{\infty} z^n \sum_{k=0}^{N-1} R_k(t) g_{n-k} - \sum_{n=1}^{N-1} z^n \sum_{k=0}^{N-1} R_k(t) g_{n-k} \right]$$

$$- S^*(\theta) \left[\sum_{n=1}^{\infty} Q_n(0,t) z^n - \sum_{n=0}^{N-1} Q_n(0,t) z^n \right] - \lambda \sum_{n=2}^{\infty} \left(\sum_{k=1}^{n-1} P_{n-k}^*(\theta,t) g_k \right) z^n$$

Also by changing the order of summation

$$\sum_{n=2}^{\infty} \left(\sum_{k=1}^{n-1} P_{n-k}^{*}(\theta,t)\, g_{k} \right) z^{n} = \sum_{k=1}^{\infty} \left(\sum_{n-k-1}^{\infty} P_{n-k}^{*}(\theta,t)\, z^{n-k} \right) g_{k}\, z^{k}$$

$$\Rightarrow \sum_{n=1}^{\infty} \theta P_{n}^{*}(\theta,t)\, z^{n} - \sum_{n=1}^{\infty} P_{n}(0,t)\, z^{n}$$

$$= -\lambda \sum_{n-1}^{\infty} P_{n}^{*}(\theta,t)\, z^{n} - \frac{S^{*}(\theta)}{z} \left[\sum_{n+k-1}^{\infty} P_{n+1}^{*}(0,t)\, z^{n+1} - P_{1}(0,t)\, z \right] -$$

$$\lambda S^{*}(\theta) \sum_{n-N}^{\infty} z^{n} \sum_{k=0}^{N-1} R_{k}(t)\, g_{n-k} - S^{*}(\theta) \left[\sum_{n=0}^{\infty} Q_{n}(0,t)\, z^{n} - \sum_{n=0}^{N-1} Q_{n}(0,t)\, z^{n} \right] -$$

$$\lambda \sum_{k=1}^{\infty} \left(\sum_{n=k-1}^{\infty} P_{n-k}^{*}(\theta,t)\, z^{n-k} \right) g_{k}\, z^{k}$$

$$\Rightarrow \theta P^{*}(z,\theta,t) - \lambda P^{*}(z,\theta,t) + \lambda X(z)\, P^{*}(z,\theta,t)$$

$$= P(z,0,t) - S^{*}(\theta) \left[\frac{P(z,0,t)}{z} - P_{1}(0,t) + Q(z,0,t) - \sum_{n=0}^{N-1} Q_{n}(0,t)\, z^{n} \right.$$

$$\left. + \lambda \sum_{n=N}^{\infty} z^{n} \sum_{k=0}^{N-1} R_{k}(t)\, g_{u-k} \right]$$

$$[\theta - \lambda + \lambda X(z)]\, P^{*}(z,\theta,t) = P(z,0,t) - S^{*}(\theta) \left[\frac{P(z,0,t)}{z} - P_{1}(0,t) + Q(z,0,t) - \right.$$

$$\left. \sum_{n=0}^{N-1} Q_{n}(0,t)\, z^{n} + \lambda \sum_{n=N}^{\infty} z^{n} \sum_{k=0}^{N-1} R_{k}(t)\, g_{n-k} \right] \qquad (2.21)$$

Define $R(z,t) = \sum_{n=0}^{N-1} R_{n}(t)\, z^{n}$

Multiplying z^{n} in equation (2.2) and taking summation from $n = 1$ to $n=N-1$

$$\sum_{n=1}^{N-1} \left\{ \frac{d}{dt} R_{n}(t) \right\} z^{n} = -\sum_{n=1}^{N-1} R_{n}(t)\, z^{n} + \sum_{n=1}^{N-1} Q_{n}(0,t)\, z^{n} + \lambda \sum_{n=1}^{N-1} \left[\sum_{k=1}^{n} R_{n-k}(t)\, g_{k} \right] z^{n}$$

$$\Rightarrow \left[\sum_{n=0}^{N-1} \left\{ \frac{d}{dt} R_{n}(t) \right\} z^{n} - \frac{d}{dt} R_{0}(t) \right]$$

$$= -\lambda \left[\sum_{n=0}^{N-1} R_{n}(t)\, z^{n} - R_{0}(t) \right] + \left[\sum_{n=0}^{N-1} Q_{n}(0,t)\, z^{n} - Q_{n}(0,t) \right] + \lambda \sum_{n=1}^{N-1} \left[\sum_{k=1}^{n} R_{n-k}(t)\, g_{k} \right] z^{n}$$

$$(2.22)$$

Let us define $\frac{d}{dt} R_{0}(t) = W_{n}(t)$ and the probability generating

function $W(z,t) = \sum_{n=0}^{N-1} W_{n}(t)\, z^{n}$, from equation (2.22), following equation is obtained

$$W(z,t) = -\lambda \sum_{n=0}^{N-1} R_{n}(t)\, z^{n} + \sum_{n=0}^{N-1} Q_{n}(0,t)\, z^{n} + \lambda \sum_{n=1}^{N-1} \left[\sum_{k=1}^{n} R_{n-k}(t)\, g_{k} \right] z^{n} \quad (2.23)$$

solving $\displaystyle\sum_{n=1}^{N-1} z^n \sum_{k=1}^{n} R_{n-k}(t)\, g_k$

$$\sum_{n=1}^{N-1} z^n \sum_{k=1}^{n} R_{n-k}(t)\, g_k = z\,[R_0(t)g_1] + z^2\,[R_1(t)g_1 + R_0(t)g_2] + z^3\,[R_2(t)g_1 +$$

$$R_1(t)g_2 + R_0(t)g_3] + \ldots\ldots\ldots\ldots$$

$$= \sum_{n=0}^{N-1} z^n \sum_{k=0}^{n-1} R_1(t)\, g_{n-k}$$

From equation (2.23) we have

$$W(z,t) = -\lambda \sum_{n=0}^{N-1} R_n(t)\, z^n + \sum_{n=0}^{N-1} Qn(0,t\, z^n + \lambda \sum_{n=0}^{N-1} z^n \sum_{k=0}^{n-1} R_k(t) g_{n-k})$$

$$\Rightarrow \sum_{n=0}^{N-1} Q_n(0,t)\, z^n = \lambda \sum_{n=0}^{N-1} R_n(t)\, z^n - \lambda \sum_{n=0}^{N-1} z^n \sum_{k=0}^{n-1} R_k(t)\, g_{n-k} + W(z,t) \quad (2.24)$$

thus using equation (2.24) we have

$$\sum_{n=0}^{N-1} Q_n(0,t)\, z^n - \lambda \sum_{n-N}^{\infty} z^n \sum_{k=0}^{N-1} R_k(t)\, g_{n-k} = -\lambda\,[X(z)-1]\, R(z,t) + W(z,t) \quad (2.25)$$

Thus using equation (2.15) and (2.25) in equation (2.21) we have

$$[\theta - \lambda + \lambda X(z)]\, P^*(z,\theta,t) = P(z,0,t) - S^*(\theta)\left[\frac{P(z,0,t)}{z} + P_1(0,t)\left\{V^*(\lambda\right.\right.$$

$$\left. - \lambda X(z)) - 1\right\} + \lambda R(z,t)(X(z)-1) - W(z,t)\Big] \quad (2.26)$$

letter $\theta - \lambda + \lambda X(z) = 0$ we have

$\theta = \lambda - \lambda X(z)$

$$0 = P(z,0,t) - S^*(\theta)\left[\frac{P(z,0,t)}{z} + P_1(0,t)\{V^*(\lambda - \lambda X(z)) - 1\} + \right.$$

$$\lambda R(z,t)(X(z)-1) - W(z,t)\,]$$

$$P(z,0,t)\left[1 - \frac{S^*(\lambda - \lambda X(z))}{z}\right]$$

$$= S^*(\lambda - \lambda X(z))\,[P_1(0,t)\{V^*(\lambda - \lambda X(z)) - 1\} + \lambda R(z,t)(x(z)-1) - W(z,t)]$$

$$P(z,0,t) = \frac{zS^*(\lambda - \lambda X(z))[P_1(0,t)\{V^*(\lambda - \lambda X(z)) - 1\} + \lambda R(z,t)(X(z)-1}{z - S^*(\lambda - \lambda X(z))}$$
$$\frac{-W(z,t)}{}$$

$$(2.27)$$

Thus we have

$$P^*(z,\theta,t) = \frac{zS^*[(\lambda - \lambda X(z)) - S^*(\theta)][P_1(0,t)\{V^*(\lambda - \lambda X(z)) - 1\} + \lambda R(z,t)(X(z) - 1}{z - S^*(\lambda - \lambda X(z))][\theta - \lambda + \lambda X(z)]} - W(z,t)$$

(2.28)

Finally the probability generating function of the system size distribution in transient state becomes

$$P(z,t) = P^*(z,0,t) + Q^*(z,0,t) + R(z,t)$$

by $S^*(0) = \int_0^\infty s\,dt = 1$ and

$$V^*(0) = \int_0^\infty v\,dt = 1$$

$$P(z,t) = \frac{z[S^*(\lambda - \lambda X(z)) - 1][P_1(0,t)\{V^*(\lambda - \lambda X(z)) - 1\} + \lambda R(z,t)(X(z) - 1 - W(z,t)}{\lambda[z - S^*(\lambda - \lambda X(z))][X(z) - 1]}$$

$$\frac{P_1(0,t)\{V^*(\lambda - \lambda X(z)) - 1\}}{\lambda[X(z) - 1]} + R(z,t)$$

After simplifying we get

$$P(z,t) = \frac{(z-1)S^*(\lambda - \lambda X(z))P_1(0,t)\{V^*(\lambda - \lambda X(z) - 1\} + (z-1)S^*(\lambda - \lambda X(z))\lambda R(z,t)(X(z) - 1) - zS^*(\lambda - \lambda X(z))W(z,t) + zW(z,t)}{\lambda[z - S^*(\lambda - \lambda X(z))][X(z) - 1]}$$

$\therefore$ the PGF of the system size distribution is

$$P(z,t) = \frac{(z-1)S^*(\lambda - \lambda X(z))}{z - S^*(\lambda - \lambda X(z))}\left[\frac{\{1 - V^*(\lambda - \lambda X(z))\}}{\lambda - \lambda X(z)} P_1(0,t) + R(z,t)\right] +$$

$$\frac{zW(z,t)[1 - S^*(\lambda - \lambda X(z))]}{\lambda[z - S^*(\lambda - \lambda X(z))][X(z) - 1]}$$

CONCLUSION

It is concluded that the probability generating function of the system size distribution depends upon the system size of the ordinary $M^X/G/1/N$- policy without vacation and also on the number of customer arriving in the system when the server is idle.

REFERENCES

1. Y. Levy and Y. Yechiali, "Utilization of idle time in an M/G/1 queuing sytems", *Mgmt. Sci.* 22, 202-211 (1975)
2. S.W. Fuhrmann and R.B. Cooper, "Stochastic decomposition in the M/G/1 queue with generalized vacations", *Ops. Res.* 22, 1117-1129 (1985).
3. H. Takagi, "Analysis of polling system", MIT Press, Boston, MA (1986).
4. B.T. Doshi, "Queuing systems with vacations a survey", *Queuing System.* 1, 29-66 (1986)
5. H. Takagi, "Queuing Analysis : a foundation of performance evaluation", Vol. 1, *Vacation and priority systems*, part 1 North Holland, Amsterdam (1991).
6. M. Hofri, "Queuing system with a procrastinating server". *Performance 86 and ACMSIGMETRICS 1980, Perform. Eval.*, Rev. 14 (1), 254-253 (1986).
7. O. Kella, "The threshold policy in the M/G/1 queue with server vacations". *Naval, Res. Logistics,* 36, 111-123 (1989).
8. H.S. Lee and M. M. Srinivasan, "Control Policies for the $M^X/G/1$ queuing system", *Mgmt. Sci.* 35 (6), 708-721 (1989).
9. S.S. Lee and S.H. Yoon, "Batch arrival queue with N-policy and single vacation", *Computer Ops. Res.* Vol. 22, No. 2, 173-189 (1995).
10. M.L. Chaudhary and J.G.C. Templeton, "A First Course in Bulk Queues" Wiley, New York (1983).

Fast Arithmetic Using Signed Digit Numbers and Ternary Logic

Rakesh Kumar Saxena[a], Neelam Sharma[b] and A. K. Wadhwani[c]

[a]*Asst. Prof., Sachdeva Institute of Technology, Mathura, U.P., India.,*
e-mail: saxenark06@rediffmail.com
[b]*Professor, Institute of Engg. & Technology, Alwar, Rajasthan, India., e-mail:*
neelam_sr@yahoo.com
[c]*Professor, M. I. T. S. Gwalior, M.P., India., e-mail: wadhwani_arun@rediffmail.com*

Abstract. Redundant Binary Signed Digit Number System may not be convenient for manual computations but may be useful in designing high-speed arithmetic machines. This number system is gaining popularity in computationally intensive environments particularly due to possessing of the carry-free addition / subtraction properties. This property has enabled arithmetic operations such as addition, multiplication, division, square root, etc., to be performed much faster than with conventional binary number systems. In RBSD number system carry propagation chains are eliminated which reduces the computational time substantially, thus enhancing the speed of the machine. The credit of RBSD number system goes to Robertson, who proposed it in 1959 and Avizienis in 1961.

In this paper, some of the recent contributions in the area of design of redundant arithmetic based addition and multiplication algorithms and architectures are briefly discussed. Also use of parallel implementation for architectures is discussed so that the enhancement in speed through the use of redundant arithmetic is possible. Also, in this paper, RBSD adder is designed. After calculation and comparison it is concluded that efficiency of RBSD adder is much better than the other adders. An addition of two's complement circuit will make an RBSD subtractor. These Adders / Subtractors can further be used as building blocks for fast multiplication, division and square root operation.

Keywords. Signed-Digit Numbers, Carry Free Addition, Carry Chain, Redundant Number Representation, Fast Computing.

1. INTRODUCTION

Digital computer arithmetic operations play a very important role in many applications where speed is essential e.g. in digital signal processing, communications, cryptography, etc. VLSI circuit implementation of sophisticated arithmetic operators, which were considered prohibitively complex to implement in the past, can be easily put on a tiny wafer of silicon. Such capabilities have enabled microprocessor designers to provide on-chip floating point arithmetic units and avoid off-chip floating point units provided as co-processors.

Redundant number systems [1] [2] are positional number systems similar to the conventional decimal system, but allow more values for a digit than the base. This will result in more than one representation for a number and hence its name as redundant.

CP1146, *Modelling of Engineering and Technological Problems*, edited by A. H. Siddiqi, A. K. Gupta, and M. Brokate
© 2009 American Institute of Physics 978-0-7354-0683-4/09/$25.00

Using this number system the arithmetic operators designed achieve considerable speed enhancement compared with operators designed using conventional number system. Speed of an arithmetic operator is directly related to the architecture chosen. Architectures and implementation styles are usually decided by the number system employed to represent numbers. Architectures are concerned with the direct hardware configuration of logic gates for the physical realization of the arithmetic algorithm. Implementation style dictates the manner in which communication between different basic units of the architecture and the communication with the external world has to be achieved.

For example, in conventional binary number systems, many implementation styles are possible such as bit-parallel (all bits at a time), bit-serial (one bit at a time), digit-serial (multiple bits at a time), word-parallel (multiple words at a time) etc. In redundant arithmetic architectures, two implementation styles are prominent. They are digit-parallel and digit-serial. Digit-parallel is similar to bit-parallel (for binary number system) implementation. Digit-serial arithmetic implementation is also referred to as the on-line arithmetic implementation [3]. An interesting feature of this type of implementation is that both least-significant digit first and the most-significant digit first realizations can be easily achieved with low computational latencies.

This paper is divided into the following sections. Section 2 gives a brief overview of redundant arithmetic number system. Section 3 discusses some of the recently proposed architectures for arithmetic operations addition and subtraction which employ redundant number systems. Section 4 gives idea to design RBSD adder cell. In section 5, some of the recently proposed architectures for multiplication operation are discussed.

2. SIGNED DIGIT NUMBER SYSTEM

Signed digit representation limits carry propagation to one position to the left during the operation of addition and subtraction in digital computers. Thereby carry-propagation chains are eliminated. Redundancy in the number representation allows a method of fast addition and subtraction in which each sum digit is a function only of the digits in two adjacent digit positions.

In SD representation with radix r a digit can have more than r values. SD (Signed Digit) number can assume $2\alpha+1$ values [2][11].

$$\Sigma_r = \{-\alpha,....-1,0,1,......... \alpha\} \tag{1}$$

α must be within the following region

$$(r-1)/2 <= \alpha <= r-1 \tag{2}$$

$$\text{Where } \alpha >= 1 \ \& \ r >= 2$$

A SD number is represented by n+m+1 digit

$$Z_i = (i=-n,.........-1,0,1,.........m)$$

has the algebraic value

$$Z = \sum_{i=-n}^{m} Z_i * r^{-1} \tag{3}$$

r is radix and a is +ve integer

Z = 0 has a unique representation
For every value of Z within a specified range there exist transformation between conventional and SD representation.

Changing of a Radix r number to its equivalent SD form & back to Conventional Radix:

Let Conventional number $X = (X_{n-1},.....,X_1,X_0)_r$
and equivalent SD number $Y = (Y_{n-1},.....,Y_1,Y_0)_r$
For every conventional digit x_i we generate interim digit d_i

$$d_i = x_i - r*b_{i+1} \tag{4}$$

where the borrow digit

$$b_{i+1} = \begin{cases} 0 \text{ if } x_i < \alpha \\ 1 \text{ if } x_i >= \alpha \end{cases} \tag{5}$$

And the SD no. can be calculated

$$y_i = d_i + b_i \tag{6}$$

Let $X=(1100)_2$ with $r = 2$, $n = 4$ and $\alpha = 1$ for a given SD set (-1,0,1), the SD number Y is $(10\ \overline{1}00)$ as obtained in the table1.

TABLE 1. Binary to RBSD Conversion

Digit Position	4	3	2	1	0
x_i		1	1	0	0
d_i		$\overline{1}$	$\overline{1}$	0	0
b_i	1	1	0	0	
y_i	1	0	$\overline{1}$	0	0

Conversion back to conventional radix:

$Y_i* = Y_i^+ + Y_i^-$
$Y_i^* = (1\ 0\ \overline{1}\ 0\ 0)_{rbsd}$
$Y_i^+ = (1\ 0\ 0\ 0\ 0)$ [only positive ones are replaced by 1]
$Y_i^- = -(0\ 0\ 1\ 0\ 0)_2$ [only negative ones are replaced by 1]
$X_i = (0\ 1\ 1\ 0\ 0)_2$ Subtract to obtain X_i
X_i= back to conventional binary

3. ARCHITECTURE: ADDER AND SUBTRACTOR

Some recent papers have proposed methods of designing fast arithmetic units using signed-digit carry free adders [16,19]. In these designs binary numbers are used to represent signed-digit numbers thus requiring more than 1-bit for each digit of the signed-digit number. This results in increased circuit complexity in terms of hardware requirement as well as the number of interconnections.

Avizienis[2] proposed that to allow carry-free addition / subtraction, for a radix β number system with digits {- a , . . . , a }, a should satisfy [(β+1)/2] $\leq$ a. This is true for any radix greater than 2. But for radix 2, the carry and sum digits had to be selected in a particular way only [12].

3.1 Ripple Carry Adder and Carry Save Adder

The most important feature of redundant arithmetic number systems is carry-free additions/subtractions. In adders carry propagation is the main cause of delay as sum of two numbers is not generated in an adder until carry has propagated from the least significant bit position to the most significant bit position. An example for this is a W-b ripple carry adder, which has a delay of W.δfa (δfa is the delay of l-bit full adder). And hence, eliminating carry propagation can result in fast adder and subtractor designs.

The carry-free addition and subtraction [11][12] can also be carried out using carry-save type of addition of conventional binary numbers. A W-digit carry-save adder has a delay of an l-bit full adder cell and is independent of the word-length, W. An important feature of carry-save architectures is that they allow carry-free multi-operand addition through the use of Wallace trees [20]. The number of levels required for adding k operands is approximately log (k/2) / log (3/2) [12]. But the major disadvantage of Wallace-trees is that they require complex interconnection, when implemented in VLSI.

3.2 Signed Digit Adder using Ternary Logic

In 1990, T. N. Rajashekhara and I-Shi Eric Chen [17] present an adder design using 3-valued called ternary logic which results in reduced circuit complexity and interconnections hence making it suitable for VLSI implementation. They used signed-digit numbers with radix 2 and digit set {-1,0,1} and for implementing the RBSD adder structure, MOS/CMOS ternary logic design used. It was suggested here that addition of two RBSD numbers can be carried out in three stages in parallel without the need for carry propagation. The choice of ternary logic and RBSD number system complemented each other well because one ternary bit could support one RBSD digit. This provides an advantage over using binary logic where more than one bit would be needed to support one RBSD digit.

The RBSD, ternary adder circuit is designed using Positive Ternary Inverter (PTI), Negative Ternary Inverter (NTI), and Ternary T Gate. PTI and NTI functions are defined as follows:

$$\overline{x} = \begin{cases} i & \text{if } x = 0 \\ -x & \text{if } x \neq 0 \end{cases}$$

Where i = 1 for PTI and -1 for NTI and x can take 3 values -1(0V), 0(2.5V), and 1 (5V). PTI (NTI) is implemented as a CMOS inverter shown in fig.1.

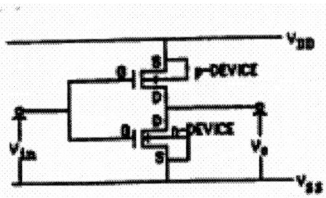

The T gate is defined as follows:

$T(y1, y2, y3: x) = y_i$;

$i = 1$ if $x = -1$

$i = 2$ if $x = 0$

$i = 3$ if $x = 1$

Fig. 2 shows the block diagram and truth table of T gate.

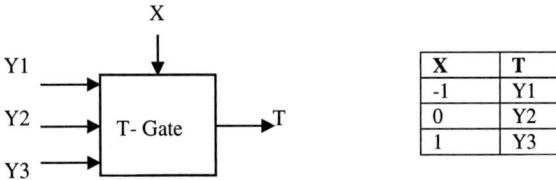

X	T
-1	Y1
0	Y2
1	Y3

FIGURE 2. Block Diagram & Truth Table for T Gate

In this design, RBSD number system offered faster add times because of carry free addition. Ternary logic offered reduced circuit complexity in terms of both transistor count and interconnections.

3.3 Adders based on Hybrid numbers

A variation of the carry-save type of addition that employs signed binary digits belonging to the set (- 1 , 0 , +1) and conventional binary bits (0,+1) has been used by H. R. Srinivas and Parhi in 1992 [10] and called hybrid number system representation. The adders used here are similar to the 1-bit full adders both in area and time complexity. Hybrid number based architectures that employ a representation formed by mixing of redundant and conventional representations is possible [10] [15]. In [10], the architectures presented accept inputs and generate outputs in binary form whereas internally radix 2 redundant arithmetic is used. In [15] an idea of allowing both positive and negative digits only at certain digit positions instead of all the digits has been presented. Such a technique limits the length of the carry propagation chain to the region between the signed digit positions. Radix 2 redundant arithmetic has also been used to design a fast adder for addition of 2 two's-complement binary numbers with carry-propagation [9].

4. RBSD ADDER STRUCTURE

RBSD adder structure with a digit set (-1,0,1) is discussed below. The addition operation is described by the equation (7), (8) and (9) called "two transfer addition technique". It is computed in three stages.

$$x_i + y_i = 2*t'_{i+1} + w'_i \qquad (7)$$

$$w'_i + t'_i = 2*t''_{i+1} + w''_i \qquad (8)$$

$$S_i = w''_i + t''_i \qquad (9)$$

Where x_i and $y_i \in (\overline{1},0,1)$, t'' and t' are intermediate carriers (transfer digits of ith stage), w' and w'' are the intermediate sums in the ith stage and S_i is the final sum.

The sum of the operands is realized in three steps:

Step I: Transfer digit $|t'_{i+1}| = 1$, whenever $|x_i + y_i| >= 1$ and w'_i is calculated using relation (7).

Step II: Transfer digit $|t''_{i+1}| = 1$ only if $|t' + w'| >= 2$ and w''_i is calculated using relation (8).

Step III: Sum digit Si is obtained using relation (9).

Note: Under the conditions of step (1) and (2) w" and t" cannot be both +1 or –1 thus giving carry free addition. RBSD adder structure based on equations (7), (8) and (9) is shown in fig. 3. Chosen the RBSD encoded digit set (1,0), (0,0) and (0,1) to represent (-1), (0), and (1) respectively.

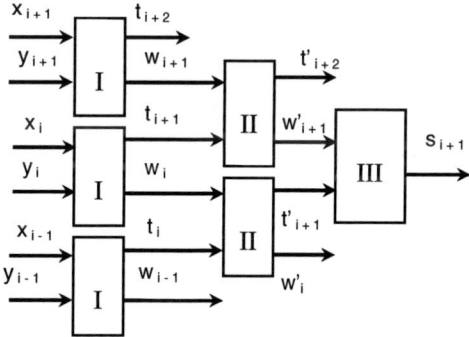

FIGURE 3. Parallel Adder / Subtractor in Signed Digit arithmetic

Using this structure, parallel addition of two numbers can be performed in a constant time that is independent of word length of the operands. The gate count required to realize the circuit is proportional to word length n and can be reduced further.

4.1 RBSD Adder Cell

In our paper we implemented the logic diagram of RBSD adder cell for two 4 bit RBSD numbers.

The following Boolean expressions (10) – (14) are used for RBSD addition.

493

$$d_i = m_i \oplus \overline{x_i^+}\ \overline{x_i^-} \oplus \overline{y_i^+}\ \overline{y_i^-} \tag{10}$$

$$m_{i+1} = \overline{x_i^+}\ \overline{y_i^+} \tag{11}$$

$$b_{i+1} = \overline{m_i}\ x_i^+\ \overline{x_i^-} + x_i^+ y_i^+ + \overline{y_i^+}\ \overline{y_i^-}\ m_i + x_i^+\ \overline{x_i^-}\ \overline{y_i^+}\ \overline{y_i^-} \tag{12}$$

$$s_i^+ = d_i b_i \tag{13}$$

$$s_i^- = d_i \overline{b_i} \tag{14}$$

where m_i, b_i and d_i are binary variables with digit set $\{0,1\}$ and x_i^*, y_i^* and s_i^* are RBSD variables with digit set $\{-1,0,1\}$ represented by $(0,1),(0,0),(1,0)$ respectively. The binary variable representation corresponding to x_i^*, y_i^* and s_i^* are $x_i^+ x_i^-$, $y_i^+ y_i^-$ and $s_i^+ s_i^-$ respectively.

Following example shows the RBSD addition.

```
  0 -1  0  1  1 -1  0  1   xᵢ (-43)
 -1  0 -1  1 -1  0  1  0   yᵢ (-150)
 ─────────────────────────
  0  0  0  1  0  0  1  1  1   mᵢ
  0  1  1  0  0  0  0  0  0   dᵢ
  1  1  0  0  0  0  0  0  1   bᵢ
 ─────────────────────────
 -1  0  1  0  0  0  0  0  1   sᵢ* (-193)
```

Fig. 4 shows the logic diagram and block diagram of the RBSD adder cell that is implemented using equation (10)-(14).

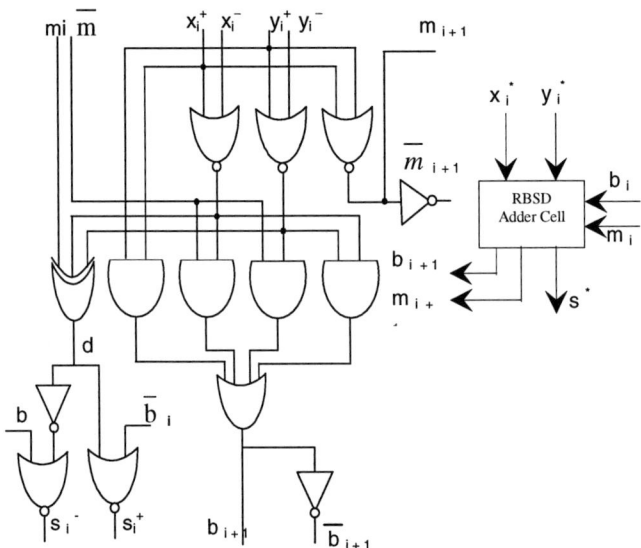

FIGURE 4. Logic diagram and Block diagram of RBSD adder cell

4.2 VHDL Implementation: Algorithm for Design of RBSD Adder

The RBSD adder shown in figure 4 is implemented using VHDL with the help of following steps

Step1. Make a package to convert conventional binary number to SD form using the eequations (4), (5) & (6). Apply the SD form to the RBSD adder and obtain the result in SD form of both binary numbers to be added.

Step 2: Write entities and architectures for various gates used in the RBSD adder cell.

Step 3: Input the $x_1^+y_1^+$, $x_2^+y_2^+$, $x_3^+y_3^+$, $x_4^+y_4^+$ bits to generate m_2, m_3, m_4, m_5, $\overline{m_2}$, $\overline{m_3}$, $\overline{m_4}$ & $\overline{m_5}$ bits as $m_{i+1} = x_i^+ \, y_i^+$ and where $X_i = x_4x_3x_2x_1$ and $Y_i = y_4y_3y_2y_1$

Step 4: Now input $x_i^+x_i^-$ and $y_i^+y_i^-$ where $i=1,2,3,4..$ and obtain the outputs of the NOR gates of second level to generate signals b'_i and a'_i.

Step 5: In the third and fourth level feed the outputs obtained from step 1 and 2 as follows: (we obtain b_{i+1} and $\overline{b}_{i+1}$ bits as outputs).

Step 6: In the fourth step we obtain the d output as shown in the figure 4.

Step 7: Now to obtain the sum, NOR gates are used and S_i^+ S_i^- are the desired outputs as in the chosen digit sets

Step 8: S_i^+ S_i^- generated in RBSD form can be converted to binary to check the result.

Hence the four sum bits are obtained and carry after the MSB bit is also recorded to obtain the final sum. This adder algorithm can be expanded to obtain n bit adder structure.

Here Algorithm for VHDL implementation of a single cell is described in step 1 to 7. The mode of programming followed should be structural. This structural module can be extended to make an eight bit adder structure. Note that the type of gates used are two input NOR, two input NOT, two input AND, four input OR, three input EX-OR. Programs of each of these gates may be written and used in the structural module of the eight bit adder program.

4.3 Result Analysis

A nxn RBSD adder was simulated and synthesized using VHDL. The software used for simulation was FPGA Advantage and that for synthesis was Xylinx Project Navigator. The timing report gave the results shown in the table 2.

Table 2. Timing Result

Logic Circuit	No. of Bits	Time Delay (ns)
	4x4	16.51
RBSD Adder	8x8	16.51
	16x16	16.51

Theoretically a RBSD Adder takes addition time equal to time taken for adding two bits since there are no carry propagation chains. The delay time of RBSD adder is constant for addition of numbers of any bit length as shown by result of analysis. With these results, it can be clearly observed that the speed of RBSD machines is much more than the conventional devices used for fast computation.

The algorithms report of the adder structure shows that the time to obtain output is greatly reduced for RBSD adder as compared to eight bit conventional adder structures of CLA, CSA & ripple adder.

5. RBSD MULTIPLIER

Multiplication plays an important role in various digital systems such as image processing, computer graphics, process controllers, signal processors etc. The demand placed by these applications on the speed of multiplication is motivating researchers to look for alternative approaches for designing high speed multipliers. Using non-conventional number system such as signed-digit numbers for designing fast arithmetic units is particularly gaining much attention in recent years.

Signed-digit number system offers the possibility of carry free addition by taking advantage of the redundancy associated with this representation. Multiplication is repetitive addition of partial products. An addition of two's complement circuit will make an RBSD subtractor. These adders / subtractor can be used as building blocks in designing a fast multiplication and division circuit.

Fast multiplication schemes use carry-free adders for addition of partial products to generate the product of two numbers. For example, the cellular array multiplier like the Baugh- Wooley [4] multiplier employs carry-free adders in each and every row of the array to add all the partial products. The final product is in redundant form which is converted to binary form using a fast adder or a converter. The latency of such multiplier arrays is of the order $O(W)$, where W is the word-length of the operands being multiplied.

Several papers have appeared in recent years exploring the use of signed digit numbers for designing the fast arithmetic units [17]. Takagi et. al. [19] have presented a high speed multiplier design using redundant binary signed-digit numbers. In this paper, the partial products are added pair-wise which reduces the number of the additions in proportion to the logarithm of the word length.

Rajashekhara and Kal [16] have presented an implementation of the same idea with reduced logic complexity making it suitable for VLSI implementation. In both these designs RBSD numbers is represented using binary logic which requires two bits per digit of RBSD number. Though these designs are able to achieve higher speeds, the circuit complexity and the number of interconnections can be significantly reduced if a proper choice of logic implementation compatible with the number system is made. Ternary or 3-valued logic blends itself with RBSD number system since each digit in the digit set (-1,0,1) of RBSD requires only one ternary bit as in the design of adder[5].

Booth recoding technique [1] is also utilized to reduce the number of partial products by a factor of 2. Two partial products are added in constant time, since the partial products are represented in RBSD number system. Modified Booth Encoding (MBE) techniques are used to reduce the number of partial products by a factor of two. The partial products are added pair-wise i.e. by using Wallace trees which again reduce the number of additions. The additions are carried out by carry free RBSD adders which reduce the delay time immensely. A single unit of RBSD adder is made up of logic gates using CMOS technology. Hence it can be clearly observed that the speed of RBSD machines is much more than the conventional devices used for fast computation. The radix 4 modified Booth multiplier [1], which recodes the multiplier bits using minimally redundant radix 4 digits (-2, -1 , 0 , +1, +2), also uses carry-save adders to add all the W/2 partial products. These array type multipliers actually employ redundant arithmetic (in the form of carry-save arithmetic) although this has not been indicated explicitly. The latency of the Booth multiplier arrays is of the order O(W/2). The Wallace-tree method of adding the partial products employed with modified Booth Encoding results in computational latencies of O(1og W/2).

6. CONCLUSION

This paper briefly discussed some of the recent work on redundant arithmetic based addition and multiplication operations. The interested reader is referred to reference [8] for further reading on redundant arithmetic techniques. The attractiveness of the redundant signed-digit number system lies in their "carry free" addition property. This feature makes them very useful in digit serial implementation of division and square root operations where the computation starts from most significant digit first. Note that carry free property leads to low latency or high speed. High speed adders leading to fast multiplication and thereafter fast division are a step up in digital technology.

REFERENCES

1. A. D. Booth, "A Signed Binary Multiplication Technique," &, J . mech. appl. math:4, pp. 236-240, Oxford University Press, 1951.
2. Avizienis, A.," Signed digit number representation for fast parallel arithmetic", *IRE Trans. Electron Computers.,* vol EC-10, pp.389-400, Sept.1961.
3. Besli, N. and Deshmukh, R.G.," A novel Redundant Binary Signed–Digit (RBSD) Booth's Encoding"., *Proceedings IEEE Southeastcon* 2002.
4. C. R. Baugh and B. A. Wooley, "A Two's- Complement Parallel Array Multiplication Algorithm," *IEEE Transactions on Computers,* vol. C-22, pp. 1045-1047, December 1973.
5. Chen, I Shi E. and Rajashekhara, T.N., "A fast multiplier Design Using Singed Digit Numbers and 3-Valved Logic*", Proc IEEE Binngamton* New York 1991.
6. Chow C. Y. and Robertson J. E. (1978) "Logical Design of a Redundant Binary Adder", *Proceedings of 4th Symposium on Computer Arithmetic,* pp. 109-115.
7. Cotolora, S. and Vassiliadis, S. "Signed Digit Addition and Related Operations with Threshold Logic" *IEEE Transaction on Computers,* vol. 49, No. 3, March 2000.
8. E. E. Swartzlander JR., ed., Computer Arithmetic, vol. 1 and 2. Los Alamitos, California: IEEE Computer Society Press, 1990.
9. H. R. Srinivas and K. K. Parhi, "A Fast VLSI Adder Architecture," *IEEE Journal of Solid State Circuits,* vol. 27, pp. 761-767, May 1992.
10. H. R. Srinivas and K. K. Parhi, "High-speed VLSI Arithmetic Processors Using Hybrid Number Systems," *Journal of VLSI Signal Processing,* vol. 4, pp. 177-198, April 1992.
11. Hwang, K., "Computer Arithmetic/ Principles, Architecture and Design", New York : Wiley 1979
12. I. Koren, Computer Arithmetic Algorithms. NJ 07632: Prentice Hall, Englewood Cliffs, 1993.
13. M. D. Ercegovac and T. lang, "Fast Multiplication Without Carry-Propagate Addition," *IEEE Transactions on Computers,* vol. 39, no. 11, pp. 1385-1390, November 1990.
14. M. J. Irwin and R. M. Owens, "Digit-Pipelined Arithmetic as Illustrated By the Paste-UP System: A Tutorial," IEEE Computer, pp. 61-73, April 1987.
15. Phatak, D.S. and Koren, I., "Hybrid Signed-Digit number systems: A unified Framework for Redundant number representations with Bounded Carry Propogation Chains", *IEEE Transactions on Computers,* vol. 43, No. 8 pp 880-891, August 1994.
16. Rajashekhar, T.N.and Kal, O., "Fast Multiplier Design using Redundant Signed- Digit Numbers", *International Journal of Electronics* vol .69, No. 3, pp 359-368, 1990.
17. Rajashekhar,T.N. and I-Shi Eric Chen," A Fast Adder Design Using Signed-Digit Numbers and Ternary Logic.", *Proceedings IEEE Southern Tier Technical Conference,* pp.187-194, Binngamton, New York, April 1990.
18. S. Kuninobu, T. Nishiyama, H. Edamatsu, T. Tanaguchi, and N. Takagi, "Design of High Speed MOS Multiplier and Divider Using Redundant Binary Representation," *Proceedings. of 8th Symposium on Computer Arithmetic,* (Como, Italy), pp. 80-86, 1987.

19. Takagi, N. et. al., "High Speed VLSI Multiplication Algorithm with A Redundant Binary Addition Tree," *IEEE Trans. Comput.*, Vol. C-34, No. 9, pp. 789-796, Sept. 1985.
20. Wallace, (1964) 'A Suggestion for a Fast Multiplier', *IEEE Transactions Electronics Computer,* Vol. EC-13, pp. 14-17.

Self-Similar Solution of Self-Gravitating, Magneto-Gas Dynamic Spherical Shock Wave Propagating in a Rotating Medium with Radiation Heat Flux

Pankaj Sharma[a] and Vivek Kumar Sharma[b]

[a] Department of Applied Mathematics, Jawaharlal Institute of Technology, Borawan,
Khargone - 451228, INDIA , Email: pankaj_sharma130969@rediffmail.com
[b] Department of Applied Mathematics and Computational Science,
Shri G.S. Institute of Technology and Science, 23, Park Road, Indore -452003, INDIA
Email: vkmps_07@ yahoo.co.in

Abstract. Self-similar solutions for the flow behind a self-gravitating, Magneto- gas dynamic, spherical shock wave are obtained. The ambient medium is a non uniformly rotating gas of infinite electrical conductivity with uniform azimuthal magnetic field and radiation heat flux. The medium is a perfect grey gas in local thermodynamic equilibrium .The total energy of the expanding shock wave is supposed to remain constant .The effects of variations of the shock Mach number , the Alfven Mach number , the radiation parameter and the gravitation on flow field behind the shock wave are shown graphically and discussed.

Keywords: Shock waves, Gravitation, Self-similar solution, rotating medium, radiation heat flux.
PACS: 47.40. –X , 47.40. Nm , 52. 35.Tc.

INTRODUCTION

The astrophysical phenomena like supernova explosion, sudden expansion of corona and solar wind have been the rich field of investigation. Interplanetary disturbances in atmosphere with rotation and radiation induce motion with supersonic speed which ultimately generates the shock wave. Also, the problem of propagation of shock wave is of great interest in exploring the effects of explosion in stars and in the atmosphere of earth. Similarity methods of Sedov [9] are used by most of the authors to solve the gas dynamical model of explosion problem.

The adiabatic motion of non-rotating gas model of stars have been studied by Sedov[9], Zel'dovich and Raizer [12] , Lee and Chen[5] and Summers [10]. Ojha et al. [6] obtained the similarity solutions for the flow behind spherical shock wave propagating in a non-uniform rotating interplanetary atmosphere with increasing energy. Ganguly and Jana [2] studied the theoretical model of propagation of shock wave in self-gravitating atmosphere with radiation heat flux in presence of a

CP1146, *Modelling of Engineering and Technological Problems*, edited by A. H. Siddiqi, A. K. Gupta, and M. Brokate
© 2009 American Institute of Physics 978-0-7354-0683-4/09/$25.00

magnetic field. They, also considered medium behind the shock to be rotating, but neglected the rotation of the undisturbed medium.

The influence of radiation on shock wave and on the flow field behind the shock front is important in the consideration of many magneto-hydrodynamic processes under astrophysical circumstances. Similarity models for classical blast wave problem have been extended taking radiation in to account by Elliot [1], Wang[11], Helliwell[4], Nicastro[7], Ghoniem et al. [3]. Ojha and Tiwari[8] have studied the cylindrical magneto-hydrodynamic model of shock with radiation heat flux. Sharan and Patel [13] studied the propagation of a spherical shock wave in a rotating medium with radiation, in presence of a uniform azimuthal magnetic field, but neglected the effects of self-gravitation.

In the present problem, the method of self-similarity is used to discuss the propagation of a spherical shock wave in a rotating medium with radiation, in presence of a self-gravitating and uniform azimuthal magnetic field. The angular velocity of the ambient medium is taken to vary inversely as the distance in order to similarity solution to exist. The problem is simplified under the assumption that the medium is a non-viscous ideal gas of infinite electrical conductivity, and radiation pressure and radiation energy are absent. The atmosphere is assumed to be grey and opaque and the shock wave is to be isothermal. Similarity solutions are obtained for the flow field behind the shock wave, and the effects of the variation of shock-Mach number, the Alfven-Mach number and a parameter of radiation are investigated.

BASIC EQUATIONS AND BOUNDARY CONDITIONS

The fundamental equations governing the motion of the ideal fluid in a self-gravitating, radiative magneto- hydrodynamic rotating medium in spherical symmetry are given by (Jana and Ganguly[14])

$$\frac{\partial \rho}{\partial t} + \rho \frac{\partial u}{\partial r} + u \frac{\partial \rho}{\partial r} + \frac{2\rho u}{r} = 0, \tag{1}$$

$$\frac{\partial u}{\partial t} + u \frac{\partial u}{\partial r} + \frac{1}{\rho} \frac{\partial p}{\partial r} + \frac{B}{\mu \rho r} \frac{\partial}{\partial r}(Br) + \frac{Gm}{r^2} - \frac{v^2}{r} = 0, \tag{2}$$

$$\frac{\partial B}{\partial t} + u \frac{\partial B}{\partial r} + B \frac{\partial u}{\partial r} + \frac{Bu}{r} = 0, \tag{3}$$

$$\frac{\partial e}{\partial t} + u \frac{\partial e}{\partial r} + p \frac{\partial}{\partial t}\left(\frac{1}{\rho}\right) + pu \frac{\partial}{\partial r}\left(\frac{1}{\rho}\right) + \frac{1}{\rho r^2} \frac{\partial}{\partial r}(Fr^2) = 0, \tag{4}$$

where $\quad e = \dfrac{p}{\rho(\gamma-1)}$,

$$\frac{\partial v}{\partial t} + u \frac{\partial v}{\partial r} + \frac{uv}{r} = 0, \tag{5}$$

$$\frac{\partial m}{\partial r} = 4\pi \rho r^2. \tag{6}$$

500

where u, v , ρ, p ,B, F, r, t , μ and γ are the radial and azimuthal components of the fluid velocity ,the density, the pressure , the azimuthal magnetic field , the radiation flux, the radial distance ,time , magnetic permeability and adiabatic exponent respectively.

Assuming local thermodynamic equilibrium and taking Rosseland's differential approximation , we have

$$F = -\frac{1}{3}c\mu^* \frac{\partial}{\partial r}\left(aT^4\right) \quad , \tag{7}$$

where c and μ^* are velocity of light and mean free path of radiation which is function of density and temperature . If we follow Wang[11] , we have

$$\mu^* = \mu_0 \rho^{\alpha'} T^{\beta'} \quad , \tag{8}$$

where μ_0 ,α',β' are constants. Also,

$$v = Ar \quad , \tag{9}$$

where A is the angular velocity of the medium at radial distance r from the point of symmetry. The shock jump conditions may be written as

$$\frac{\rho_0}{\rho_s} = 1 - \frac{u_s}{V} \quad , \tag{10}$$

$$p_s - p_0 + \frac{1}{2\mu}\left(B_s^2 - B_0^2\right) = \rho_0 u_s V_s \quad , \tag{11}$$

$$\frac{B_0}{B_s} = 1 - \frac{u_s}{V} \quad , \tag{12}$$

$$\left[V^2 - \frac{1}{2}(\gamma+1)Vu_s - a_0^2\right](V - u_s)^2$$

$$= b_0^2\left[V^2 - \frac{1}{2}(\gamma+2)Vu_s + \frac{1}{2}\gamma u_s^2\right] + \frac{F_s(\gamma-1)(V - u_s)V}{u_s \rho_s} \quad , \tag{13}$$

$$v_0 = v_s \quad , \tag{14}$$

$$m_0 = m_s \quad , \tag{15}$$

where subscript 0 and s denotes quantities just ahead and just behind the shock respectively. The quantity V is the shock speed , and a_0, b_0 are the sound and Alfven speeds in the gas at rest , given by

$$a_0^2 = \frac{\gamma p_0}{\rho_0} \quad , \qquad b_0^2 = \frac{B_0^2}{\mu \rho_0} \quad , \tag{16}$$

The equation (13) is derived by combining the energy equation with the continuity , momentum and induction equations across the shock . For an isothermal shock ,

501

$$\frac{p_s}{\rho_s} = \frac{a_0^2}{\gamma} \; , \tag{17}$$

The field variables just ahead the shock are ,

$$u = u_0 = 0 \; , \quad \rho = \rho_0 \quad , p = p_0 \quad , \quad B = B_0 \quad , \text{ and } A = \frac{A_0}{r_s} \; ,$$

where ρ_0, p_0, B_0 and A_0 are constants and r_s is the shock radius.
We have ,

$$v_0 = A \, r_s = A_0 \; ,$$

and from the momentum equation (2)

$$v_0^2 = A_0^2 = \frac{B_0^2}{\mu \rho_0} = b_0^2 \; .$$

SIMILARITY SOLUTIONS

From general dimensional considerations the form of the solutions of the equation (1)-(6) are

$$\rho = \rho_0 \sigma(\eta), \tag{18}$$

$$u = \frac{r}{t} U(\eta), \tag{19}$$

$$p = \rho_0 \left(\frac{r}{t}\right)^2 P(\eta), \tag{20}$$

$$B = (\mu \rho_0)^{1/2} \left(\frac{r}{t}\right) \beta(\eta), \tag{21}$$

$$F = \rho_0 \left(\frac{r}{t}\right)^3 f(\eta), \tag{22}$$

$$v = \left(\frac{r}{t}\right) K(\eta) \; , \tag{23}$$

$$m = \frac{4\pi \rho_0 r^3}{3} S(\eta) \; , \tag{24}$$

where $\sigma, U, P, \beta, f, K, S$ are non-dimensional functions of the non-dimensional variable

$$\eta = \frac{r}{Vt} \; , \tag{25}$$

The lines of constant η are straight lines in the (r,t) plane passing through the origin. In particular , the shock front and the inner contact front are two such lines;

these fronts , therefore , expand with constant radial velocity . It is convenient to introduce as independent variables instead of η and t , the stream function ψ defined by

$$\frac{\partial \psi}{\partial r} = \frac{\rho}{\rho_0} r^2 \quad , \tag{26}$$

$$\frac{\partial \psi}{\partial t} = -\frac{\rho}{\rho_0} u r^2 , \tag{27}$$

and the variable ϕ as

$$\phi = \frac{3\psi}{r^3} \quad , \tag{28}$$

for spherically symmetric flow. Physically , ϕ is the ratio of mass between the inner contact surface and the radius r to the mass initially in the same volume . The value of ϕ at the shock is , therefore $\phi_s = 1$, while at contact front $\phi_c = 0$.

In terms of the variable (ψ, ϕ), the form of the solutions are ,

$$\rho = \rho_0 \sigma(\phi), \tag{29}$$

$$u = V U(\phi), \tag{30}$$

$$p = \rho_0 V^2 P(\phi) \quad , \tag{31}$$

$$B = (\mu \rho_0)^{1/2} V \beta(\phi) \quad , \tag{32}$$

$$F = \rho_0 V^3 f(\phi) , \tag{33}$$

$$v = V K(\phi) \quad , \tag{34}$$

$$m = \frac{4\pi \rho_0 r^3}{3} S(\phi) . \tag{35}$$

where the quantities $\sigma, U, P, \beta, f, K$ and S are non-dimensional function of non-dimensional variable ϕ. The Jacobean of transformations defined by equations (26),(27) and (28) found to be

$$J = \frac{\partial(\psi, \phi)}{\partial(r,t)} = -3\sigma U V \phi \left(\frac{3\psi}{\phi} \right)^{1/3}$$

Since $J \neq 0$ (except at certain point), the transformation defines one to one correspondence between (r,t) and (ψ, ϕ). The differential equation

$$\frac{\partial t}{\partial \psi} = -\frac{1}{J} \frac{\partial \phi}{\partial r} = \frac{1}{\sigma U V} \frac{(\sigma - \phi)}{\sqrt[3]{9\phi\psi^2}},$$

is integrated to yield

$$\psi = \frac{1}{3} \phi \left(\frac{\sigma U V t}{\sigma - \phi} \right)^3 . \tag{36}$$

Finally it is possible to express the variable η as function of ϕ by combining equation (25) and (26) as

$$\eta = \frac{\sigma U}{(\sigma - \phi)} \tag{37}$$

The total energy E carried by shock wave is constant if

$$E = \int_0^{r_s} \left[\frac{1}{2} \rho (u^2 + v^2) + \frac{p}{(\gamma - 1)} + \frac{B^2}{2\mu} - \frac{Gm\rho}{r} \right] 4\pi r^2 dr$$

$$= \frac{4}{3} \pi \left[\frac{1}{2} \sigma (U^2 + K^2) + \frac{P}{(\gamma - 1)} + \frac{\beta^2}{2} + \frac{2\sigma}{M^2\gamma} \right] K_1$$

where $\quad K_1 = \rho_0 V^2 r^3 \quad$ is constant.

SOLUTION OF THE PROBLEM

In terms of the non-dimensional form , the fundamental equation (1)-(6) with (7) and (8) take the form

$$2\sigma U + 3\sigma (\sigma - \phi) \frac{dU}{d\phi} - 3U\phi \frac{d\sigma}{d\phi} = 0 \qquad , \tag{38}$$

$$\phi U \frac{dU}{d\phi} + \frac{(\phi - \sigma)}{\sigma} \left[\frac{dP}{d\phi} + \beta \frac{d\beta}{d\phi} \right] - \frac{\beta^2}{3\sigma} + \frac{2S}{3M^2\gamma} + \frac{K^2}{3} = 0 \qquad , \tag{39}$$

$$\beta U + 3(\sigma - \phi)\beta \frac{dU}{d\phi} - 3\phi U \frac{d\beta}{d\phi} = 0 \qquad , \tag{40}$$

$$\frac{3\phi U}{(\gamma - 1)} \left[\frac{\gamma P}{\sigma} \frac{d\sigma}{d\phi} - \frac{dP}{d\phi} \right] + 2f + 3(\sigma - \phi) \frac{df}{d\phi} = 0 \qquad , \tag{41}$$

$$\frac{dK}{d\phi} = \frac{K}{3\phi} \qquad , \tag{42}$$

$$\frac{dS}{d\phi} = \frac{(\sigma - S)}{(\sigma - \phi)} \tag{43}$$

and

$$f = N\sigma^{\alpha' - \beta' - 4} P^{\beta' + 3} (\sigma - \phi) \left[\frac{dP}{d\phi} - \frac{P}{\sigma} \frac{d\sigma}{d\phi} \right] \qquad ,$$

where $\quad N = -\dfrac{4ac\mu_0}{\Gamma^{\beta' + 4} r \rho_0^{1 - \alpha'} V^{-3\beta' - 9}} \qquad . \tag{44}$

In terms of the non-dimensional variables , the shock relations become

$$\frac{1}{\sigma_s} = 1 - U_s \quad , \qquad (45)$$

$$P_s = \frac{1}{\gamma M^2} - \frac{1}{2M_A^2}\left(\sigma_s^2 - 1\right) + U_s \qquad , \qquad (46)$$

$$\frac{1}{\beta_s} = \left(1 - U_s\right)M_A \quad , \qquad (47)$$

$$\left(1 - \frac{1}{2}(\gamma+1)U_s - \frac{1}{M^2}\right)\left(1 - U_s\right) = \frac{1}{M_A^2}\left(1 - \frac{1}{2}\gamma U_s\right) + \frac{f_s(\gamma-1)}{U_s \sigma_s} \quad , \qquad (48)$$

$$K_s = \frac{1}{M_A} \quad . \qquad (49)$$

where $M = \dfrac{V}{a_0}$ and $M_A = \dfrac{V}{b_0}$ are shock Mach number and the Alfven Mach number . From equation (2.17) , we have

$$\frac{P_s}{\sigma_s} = \frac{1}{\gamma M^2} \quad , \qquad (50)$$

The set of differential equations (38)-(44) reduce to the form

$$U' = \frac{\left[\dfrac{3f}{N} U\phi\sigma^{-\alpha'+\beta'+4} P^{-\beta'-3} - K^2\sigma\phi U + \beta^2\sigma U - \dfrac{2S\sigma\phi U}{M^2\gamma} + 2PU(\sigma-\phi)\right]}{3\left(\phi^2 U^2\sigma - \beta^2(\sigma-\phi)^2 - P(\sigma-\phi)^2\right)} \quad , \qquad (51)$$

$$\sigma' = \frac{2\sigma U + 3\sigma(\sigma-\phi)U'}{3U\phi} \quad , \qquad (52)$$

$$\beta' = \frac{\beta U + 3(\sigma-\phi)\beta U'}{3U\phi} \quad , \qquad (53)$$

$$P' = \frac{\left[K^2\sigma\phi U + 3U'\left(\phi^2 U^2\sigma - \beta^2(\sigma-\phi)^2\right) - \beta^2\sigma U\right]}{3(\sigma-\phi)\phi U} + \frac{2S\sigma}{3M^2\gamma(\sigma-\phi)} \quad , \qquad (54)$$

$$K' = \frac{K}{3\phi} \quad , \qquad (55)$$

$$f = \frac{\left[\begin{array}{l} K^2\sigma\phi U + 3U'\left(\phi^2 U^2\sigma - \beta^2(\sigma-\phi)^2 - \gamma P(\sigma-\phi)^2\right) - \beta^2\sigma U \\ -2\gamma PU(\sigma-\phi) - 2f(\sigma-\phi)(\gamma-1) \end{array}\right]}{3(\sigma-\phi)^2(\gamma-1)}$$
$$+ \frac{2S\sigma\phi U}{3M^2(\gamma-1)\gamma(\sigma-\phi)^2} \quad , \qquad (56)$$

$$S' = \frac{(\sigma-S)}{(\sigma-\phi)} \quad , \qquad (57)$$

The boundary conditions now become

$$U(1) = 1 - \frac{1}{2}\left[\left(\frac{1}{\gamma M^2} + \frac{1}{2M_A^2}\right) + \left(\left(\frac{1}{\gamma M^2} + \frac{1}{2M_A^2}\right)^2 + \frac{2}{M_A^2}\right)^{1/2}\right] \quad , \qquad (58)$$

$$\sigma(1) = \frac{1}{1-U(1)} \quad , \qquad (59)$$

$$\beta(1) = \frac{1}{[1-U(1)]M_A} \quad , \qquad (60)$$

$$P(1) = \frac{1}{[1-U(1)]\gamma M^2} \quad , \qquad (61)$$

$$K(1) = \frac{1}{M_A} \quad , \qquad (62)$$

$$f(1) = \frac{U(1)}{(\gamma-1)}\left[1 - \frac{1}{M^2} - \frac{1}{2}(\gamma+1)U(1)\right] - \frac{U(1)\left[1 - \frac{1}{2}\gamma U(1)\right]}{(\gamma-1)M_A^2(1-U(1))} \quad , \qquad (63)$$

$$S(1) = 1 \quad . \qquad (64)$$

RESULTS AND DISCUSSION

The equations (51) to (57) under boundary conditions (58) to (64) are numerically integrated by Runge-Kutta method. The values of constant parameters are taken as $\gamma = 1.4$, $M^{-2} = 0.01, 0.04$, $M_A^{-2} = 0.01, 0.0133, 0.04$, and $N = 0.1$. The solutions are shown in Fig.1 to Fig.7.

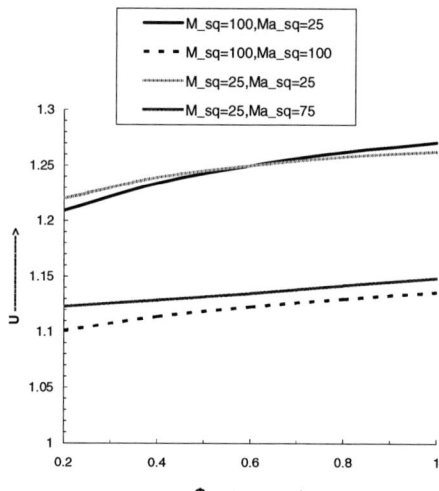

FIGURE 1. Variation of Non-dimensional radial velocity U with non-dimensional variable Φ in the flow field behind the shock front

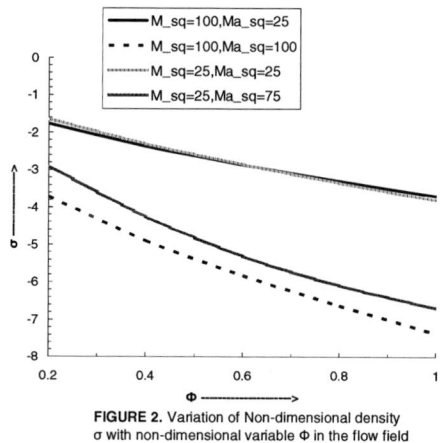

FIGURE 2. Variation of Non-dimensional density σ with non-dimensional variable Φ in the flow field behind the shock front

The Fig.1 shows that the radial velocity U decreases from the shock front to inner expanding surface. Fig.2 shows that the density increases.

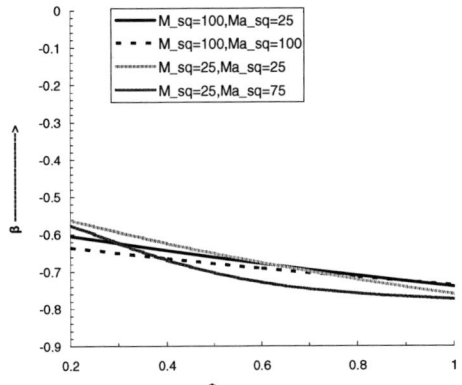

FIGURE 3. Variation of Non-dimensional azimuthal magnetic field β with non-dimensional variable Φ in the flow field behind the shock front

Fig.3 and Fig.4 show that azimuthal magnetic field increases whereas the pressure P increases slowly as we move inwards from the shock front . Fig.5 shows that radiation heat flux f increases from the shock front. Fig.6 shows that mass S decreases from shock front to inner surface , and is independent of shock Mach number. Fig.7 shows that azimuthal velocity K decreases from the shock front.

The effects of increasing M^{-2} i.e. decreasing M^2 are (from Fig.1-Fig.7)

(i) to increase the radial velocity U , density , azimuthal magnetic field.

(ii) to decrease pressure P , radiation heat flux f ,

(iii) the mass S and azimuthal fluid velocity K are unaffected by an increase of M^{-2} .

The effects of increasing azimuthal magnetic field (i.e. increasing M_A^{-2} i.e. decreasing M_A^2) are (from Fig.1- Fig.7)

(i) to increase the radial velocity U , density , azimuthal magnetic field, pressure P, azimuthal magnetic field K.

(ii) to decrease radiation heat flux f .

It should be noted that due to inclusion of gravitation in our problem, we observe a difference in variation of flow variables from that of Sharan and Patel [13], who did not include gravitation in their study.

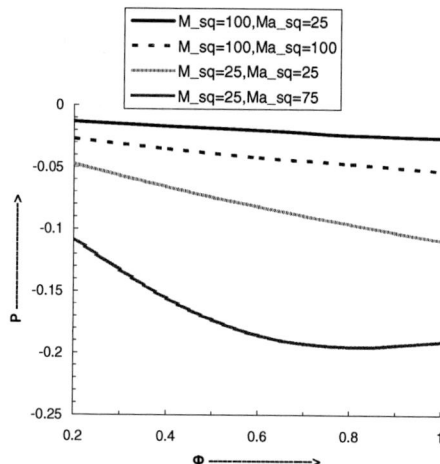

FIGURE 4. Variation of Non-dimensional pressure P with non-dimensional variable Φ in the flow field behind the shock front

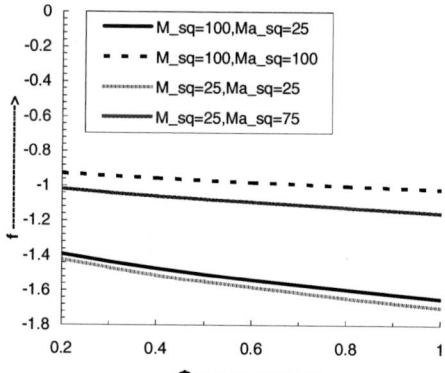

FIGURE 5. Variation of Non-dimensional radiation heat flux f with non-dimensional variable Φ in the flow fieldbehind the shock front

509

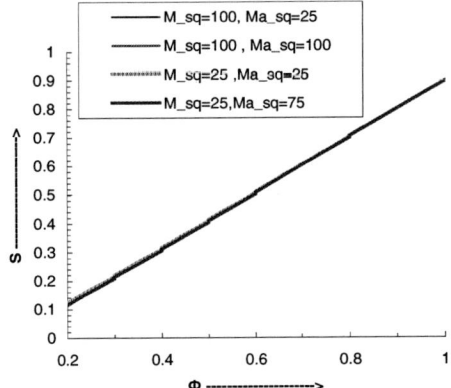

FIGURE 6. Variation of Non-dimensional mass S with non-dimensional Φ in the flow field behind the shock front

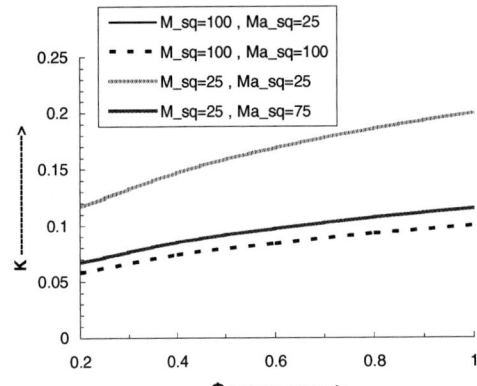

FIGURE 7. Variation of Non-dimensional azimuthal velocity K with non-dimensional variable Φ in the flow field behind the shock front

REFERENCES

1. L.A. Elliot, *Similarity methods in radiation hydrodynamics*, Proc.Royal Soc., London **258A** ,1960 , pp. 287-301.
2. A. Ganguly, and M. Jana, *Propagation of shock wave in self-gravitating, radiative, magneto-hydrodynamic non-uniform rotating atmosphere*, Bull.

Cal. Math. Soc. **90**, 1998, pp. 77-82.

3. A.F. Ghoneim , *Effects of internal heat transfer on the structure of self similar blast waves* , J. Fluid Mech. **117** ,1982 , pp. 473-491.

4. J.B. Helliwell , *Self-similar piston problem with radiative heat transfer* , J. Fluid Mech. **37**, 1969, pp. 497-512.

5. T.S. Lee and T. Chen , *Hydrodynamic interplanetary shock wave* , Planet. Space Sci. **16** ,1968 , pp. 1483-1502.

6. S.N. Ojha , Onkar Nath and H.S. Takhar , *A study of stellar point explosion in a radiative, magneto-hydrodynamic medium* , Astrophys. Space Sci. **183,** 1991 , pp. 135.

7. J.R. Nicastro, *Similarity analysis of radiative gas dynamics with spherical Symmetry,* Phys. Fluids **13** ,1970 , pp. 2000-2006.

8. S.N. Ojha and V.D. Tiwari , *Cylindrical magneto-hydrodynamic model of shock waves with radiation heat flux* , Astrophys. Space Sci. **162** ,1989, pp. 57-69.

9. L.I. Sedov , *Similarity and dimensional methods in Mechanics* , Academic Press , New York , 1959.

10. D. Summers, *An idealized model of a magneto -hydrodynamic spherical blast wave applied to a flare produced shock in the solar wind* , Astron. Astrophys. **45**, 1975, pp. 151-158.

11. K.C. Wang , *The piston problem with thermal radiation* , J. Fluid Mech. **20,** 1964, pp. 447-455.

12. Ya. B. Zel'dovich and W.P. Raizer , *Physics of shock waves and high temperature hydrodynamic phenomena* , Academic Press , New York ,1966.

13. V.Sharan and A. Patel, *Self-similar solution of magneto-gas-dynamic spherical shock propagation in a rotating medium with radiation heat flux,* South East Asian J. Math. and Math. Sc. **5** ,2007 , pp. 33-48.

14. M. Jana and A. Ganguly, *A comparative study between two models of propagation of spherical and cylindrical shock waves with varying energy in self- gravitating , magneto-radiative non-uniform atmosphere,* Astrophys. Space Sci., **275,** 2001, pp. 285-297.

511

Affiliation of Contributors

S.No.	Name & Address	Title
1.	Prof. Roddam Narasimha, FRS Chairman Of The Engineering Mechanics Unit, Jncasr And Director, National Institute Of Advanced Studies, Bangalore. rroddam@caos.iisc.ernet.in, roddam@cao.iisc.ernet.in, roddam@jncasr.ac.in	DR ZAKIR HUSSAIN AWARD LECTURE
2.	Dr. H.P Dikshit Director General School Of Good Governance And Policy Analysis (An Autonomous Institution Set Up By The State Govt. Of Madhaya Pradesh) C-403, 4th Floor, Narmada Bhawan 59 Area Hills, Bhopal 462001 India hpdsushasan@gmail.com	DR ZAKIR HUSSAIN AWARD LECTURE
3.	Prof. Arvind Gupta, Director ICIAM-2011 School Of Computing Science, 8888 University Drive, Simon Fraser University, Burnaby BC, V5a 1S6, Canada arvind@mitacs.ca Dr. Alireza Hadj Khodabakhshi School Of Computing Science, 8888 University Drive, Simon Fraser University, Burnaby BC, V5a 1S6, Canada Dr. Ján Ma˘nuch School Of Computing Science, 8888 University Drive, Simon Fraser University, Burnaby BC, V5a 1S6, Canada Dr. Arash Rafiey School Of Computing Science, 8888 University Drive, Simon Fraser University, Burnaby BC, V5a 1S6, Canada Dr. Ladislav Stacho Department Of Mathematics, 8888 University Drive, Simon Fraser University, Burnaby BC, V5a 1S6, Canada	PROTEIN DESIGNS IN HP MODELS
4.	Prof. Hans Mittelmann Department Of Mathematics, Arizona State University, Tempe, Az 85287-1804, USA MITTELMANN@asu.edu Dr. Jiming Peng Department Of Industrial And Enterprise System Engineering, University Of Illinois At Urbana-Champaign. Urbana, Il, 61801 Dr. Xiaolin Wu Department Of Electrical & Computer Engineering, Mcmaster University, Ontario, Canada	GRAPH MODELING FOR QUADRATIC ASSIGNMENT PROBLEMS ASSOCIATED WITH THE HYPERCUBE
5.	Prof. Kazuhiro Aoyama Deptt Of Systems Innovation, School Of Engineering, The University Of Tokyo, Japan. aoyama@sys.t.u-tokyo.ac.jp Prof. Tsuyoshi Koga Deptt Of Systems Innovation, School Of Engineering, The University Of Tokyo, Japan	INTEGRATE SYSTEM MODELING FOR DESIGN AND PRODUCTION PLANNING OF HIGH QUALITY PRODUCTS CONSIDERING FAILURE DATA
6.	Prof. P Achutan Department of Mathematics, Amrita Vishwa Vidyapeetham, Coimbatore, India 641 105. Dept. of Mathematics, Indian Institute of Technology, Madras, India 600 036. Prof. Narayanan Kutty Dept. of Physics, Amrita Vishwa Vidyapeetham, Coimbatore, India 641 105. k_narayanankutty@ettimadai.amrita.edu	ON MATHEMATICAL MODELING OF QUANTUM SYSTEMS
7.	Prof. Bhawani Shankar Chowdhry • Regular Associate, The Abdus Salam Ictp, Trieste, Italy • Mehran University Of Engineering & Technology, Jamshoro, Pakistan • School Of Electronics And Computer Science, University Of Southampton, UK bsc_itman@yahoo.com	VISUALIZATION AND ANALYSIS OF WIRELESS SENSOR NETWORK DATA FOR SMART CIVIL STRUCTURE APPLICATIONS BASED ON SPATIAL

	Dr. Neil M White School Of Electronics And Computer Science, University Of Southampton, UK	CORRELATION TECHNIQUE
	Dr. Jai Kumar Jeswani Mehran University Of Engineering & Technology, Jamshoro, Pakistan	
	Dr. Khalil Dayo Mehran University Of Engineering & Technology, Jamshoro, Pakistan	
	Manorma Rathi Mehran University Of Engineering & Technology, Jamshoro, Pakistan	
8.	Prof. Salim A. Messaoudi Department Of Mathematics And Statistics King Fahd University Of Petroleum And Minerals Dhahran 31261, Saudi Arabia. E-Mail: messaoud@kfupm.edu.sa	A STABILITY RESULT IN A WEAKLY DAMPED NONLINEAR TIMOSHENKO SYSTEM
	Dr. Muhammad I. Mustafa Department Of Mathematics And Statistics King Fahd University Of Petroleum And Minerals Dhahran 31261, Saudi Arabia. E-Mail: Mohmim@Kfupm.Edu.Sa	
9.	Dr.(Mrs) N A Sontakke Scientist 'D' Indian Institute Of Tropical Meteorology Dr. Homi Bhabha Road, Ncl Post Pashan, Pune - 411008 (India) sontakke@tropmet.res.in	RAINFALL SPATIAL VARIABILITY AND ITS IMPACT ON IMPORTANT ENVIRONMENT PROCESSES IN INDIA
	Dr. Nityanand Singh Scientist -F & Head Climatology And Hydrometeorology Division Indian Institute Of Tropical Meteorology (Ministry Of Earth Sciences, Govt. Of India) Dr. Homi Bhabha Road, Ncl Post, Pashan Pune - 411008 (India) nsingh@tropmet.res.in	
10.	Dr. Shelly Arora Department of Mathematics, Punjabi University, Patiala shellyarora_25@yahoo.co.in	MODELLING OF DISPLACEMENT WASHING OF PULP BED USING ORTHOGONAL COLLOCATION ON FINITE ELEMENTS
	Dr. František Potůček Department of Wood Pulp and Paper, University of Pardubice, Czech Republic	
	Dr. S.S. Dhaliwal Department of Mathematics, SLIET, Longowal	
	Dr. V. K. Kukreja Department of Mathematics, SLIET, Longowal	
11.	Dr. Millie Pant, Department Of Paper Technology, Indian Institute Of Technology Roorkee, Saharanpur Campus, Saharanpur – 247001, India. millifpt@iitr.ernet.in	A NEW DIFFERENTIAL EVOLUTION ALGORITHM AND ITS APPLICATION TO REAL LIFE PROBLEMS
	Mr. Musrrat Ali Department Of Paper Technology, Indian Institute Of Technology Roorkee, Saharanpur Campus, Saharanpur – 247001, India.	
	Prof. V. P. Singh Department Of Paper Technology, Indian Institute Of Technology Roorkee, Saharanpur Campus, Saharanpur – 247001, India.	
12.	Prof. Dr. Hans Georg Feichtinger Group Leader Of Nuhag Numerical Harmonic Analysis Group Institute Of Mathematics, University Of Vienna Office: Alserbachstrasse 23, A-1090 Wien, Austria Mail To: Nordbergstrasse 15, A-1090 Wien, Austria Email: Hans.Feichtinger@Univie.Ac.At	BANACH GELFAND TRIPLES FOR APPLICATIONS IN PHYSICS AND ENGINEERING
13.	Dr. M. Melek Vice Chairman Beykoz Logistic School Of Higher Education, 34805, Beykoz, Istanbul, Turkey	WAVELET ANALYSIS OF OIL PRICES, USD VARIATIONS AND IMPACT ON LOGISTICS
	Dr. A. Tokgozlu Suleyman Demirel University, Faculty Of Science, Isparta, Turkey	
	Prof. Z. Aslan Beykoz Logistic School Of Higher Education, 34805, Beykoz, Istanbul, Turkey zaferaslan@ beykoz.edu.tr	

14.	Dr. Mani Mehra Department Of Mathematics, Indian Institute Of Technology, New Delhi, India url: http://Web.iitd.ac.in/~mmehra mmehra@maths.iitd.ac.in	WAVELETS AND DIFFERENTIAL EQUATIONS– A SHORT REVIEW
15.	Prof P Manchanda Deptt Of Mathematics, GNDU, Amritsar, Punjab pmanch2k1@yahoo.co.in Miss Meenakshi. †*Dev Samaj College for Women, Ferozepur, India* meenakshi_wavelets@yahoo.com	New Classes of Wavelets
16.	Dr. J. Kumar Department Of Mathematics, Guru Nanak Dev University, Amritsar 143001, India Meenujkumar@Rediffmail.Com Prof. P. Manchanda Department Of Mathematics, Guru Nanak Dev University, Amritsar 143001, India Pmanch2k1@Yahoo.Co.In	ESTIMATION OF HURST EXPONENT FOR THE FINANCIAL TIME SERIES
17.	Prof. A H Siddiqi BMAS Engg College, SGI, Agra Mr A K Verma BMAS Engg College, SGI, Agra Mrs Noor-E-Zahra Hindustan Institute Of Technology (HIT), SGI 32, 34 Knowledge Park-3, Greater Noida – 201306, India, noor_zahra_india@yahoo.co.in Mr. Ashish Chandiok BMAS Engg College, SGI Agra Mr Ashique Hussai Hindustan Institute Of Technology (HIT), SGI 32, 34 Knowledge Park-3, Greater Noida – 201306, India,	WAVELET BASED SIMULATION OF RESERVOIR FLOW

18.	Prof. René Lozi
	• Laboratoire J.A. Dieudonné - Umr Du CNRS No6621
	   Université De Nice Sophia-Antipolis
	   Parc Valrose
	   06108 Nice Cedex 2 France
	• Iufm Célestin Freinet - Université De Nice Sophia-Antipolis
	   89, Av. George V
	   06046 Nice Cedex 1 France R.LOZI@unice.fr
	Dr. Clarisse Fiol
	Iufm Célestin Freinet - Université De Nice Sophia-Antipolis
	89, Av. George V
	06046 Nice Cedex 1 France
		GLOBAL ORBIT PATTERNS
		FOR DYNAMICAL SYSTEMS
		ON
		FINITE SETS

19.	Prof. Dr. M.Brokate Munich Technical University,Boltzmannstr 3, 85748 Garching Germany, Email:brokate@ma.tum.de Prof. Dr. M.Yu Rasulova The Institute Of Nuclear Physics, Ulughbek, Tashkent 100124 ,Uzbekistan, Email:rasulova@live.com	THE CAUCHY PROBLEM FOR BBGKY HIERARCHY OF QUANTUM KINETIC EQUATIONS WITH COULOMB POTENTIAL
20.	Prof. P.N. Srikanth, Dean Tata Institute Of Fundamental Research, Centre For Applicable Mathematics Post Bag 6503, Chikkabommasandra Sharadanagar Gkvk Post Bangalore -560 065, India srikanth@math.tifrbng.res.in	SEMILINEAR ELLIPTIC PROBLEMS, MOUNTAIN PASS INDEX AND BREAK OF SYMMETRY
21.	Prof. R. C. Singh, Dean Academics Department Of Physics, Hindustan Institute Of Technology, 32, 34 Knowledge Park-3, Greater Noida – 201306, India rcsingh_physics@yahoo.com	AN INTEGRAL EQUATION APPROACH TO ORIENTATIONAL PHASE TRANSITIONS IN QUADRUPOLAR GAY-BERNE FLUID USING DENSITY- FUNCTIONAL THEORY
22.	Dr. Manoj Kumar Singh Department Of Mathematics, R.B.S. College, Dhanuki - 844 101 Bihar	EQUILIBRIUM POINTS IN THE PHOTOGRAVITATIONAL NON-PLANAR RESTRICTED THREE BODY PROBLEM

23.	Prof. Dr. Abul Hasan Siddiqi B.M.A.S. Engineering College Agra 282007, U.P., India E-Mail: siddiqi.abulhasan@gmail.Com; Dr. Syed Shakaib Irfan College Of Engineering, Qassim University P. O. Box 6677, Buraidah 51452, Al-Qassim Kingdom Of Saudi Arabia shakaib11@rediffmail.Com	ALGORITHM FOR A NEW SYSTEM OF COMPLETELY GENERALIZED MULTI-VALUED VARIATIONAL INCLUSIONS
24.	Dr. M. A. Khanday School Of Mathematics And Allied Sciences, Jiwaji University Gwalior, Madhya Pradesh-474011, India khanday@gmail.com Prof. Dr. V. P. Saxena Sagar Institute Of Research Technology And Science, Ayodhya Bypass Road Bhopal, M.P-462041, India	FINITE ELEMENT APPROACH FOR THE STUDY OF THERMOREGULATION IN HUMAN HEAD EXPOSED TO COLD ENVIRONMENT
25.	Dr. Vijay Kukreja Department Of Mathematics, Sliet, Longowal, 148106 (Punjab) India vkkukreja@gmail.com Dr. Ajay Mittal Department Of Mathematics, Bgiet, Sangrur, 148101 (Punjab) India Prof. Nabendra Parumasur School Of Mathematical Sciences, University Of Kwazulu-Natal, Durban, 4000, South Africa Prof. Pravin Singh School Of Mathematical Sciences, University Of Kwazulu-Natal, Durban, 4000, South Africa singhp@ukzn.ac.za	NUMERICAL SOLUTION OF BVPS BY OCFE USING A HERMITE BASIS
26.	Mr. Deepak Kumar Department Of Paper Technology, Indian Institute Of Technology Roorkee, Saharanpur Campus, Saharanpur -247001, India dkr2009@gmail.Com Mr. Vivek Kumar Department Of Paper Technology, Indian Institute Of Technology Roorkee, Saharanpur Campus, Saharanpur -247001, India vivekfpt@iitr.ernet.in Prof. V. P. Singh Department Of Paper Technology, Indian Institute Of Technology Roorkee, Saharanpur Campus, Saharanpur -247001, India singhvp3@gmail.com	ANALYSIS OF PARAMETRIC EFFECTS ON EFFICIENCY OF THE BROWN STOCK WASHER IN PAPER INDUSTRY USING MATLAB
27.	Dr. B.S.Mudagi Department Of Mathematics, Nowrosjee Wadia College,Pune-411001.(India). Email:basaweshwarmudagi@ yahoo.co.in	AN INVESTIGATION INTO EFFECT OF ELECTROMAGNETIC FIELDS ON SEPARATION TENDENCY IN GENERALISED COUTTE FLOW
28.	Dr. Kamran Asghar Department Of Mathematics, Quaid-I-Azam University, Islamabad, Pakistan Prof. Ashfaque H. Bokhari Department Of Mathematics And Statistics, King Fahd University Of Petroleum And Minerals, Dhahran 31261, Saudi Arabia abokhari@kfupm.edu.sa Prof. F. D. Zaman Department Of Mathematics And Statistics, King Fahd University Of Petroleum And Minerals, Dhahran 31261, Saudi Arabia	ALGEBRAIC COMPUTATIONS OF SOME TENSORS IN GEOMETRY AND GENERAL RELATIVITY
29.	Dr. S.Kandar Haldia Institute Of Technology P.O.-Hit(W.B, India), Pin-721657 E-Mail: shyamalenduk@yahoo.com Dr K.Roy Haldia Institute Of Technology P.O.-Hit(W.B, India), Pin-721657 Dr. M.Barman Haldia Institute Of Technology P.O.-Hit(W.B, India), Pin-721657	MAX-PACKET GENERATION PROCESS FOR REMOVING SKEW IN NETWORK MULTIMEDIA COMMUNICATION- A MATHEMATICAL MODEL

	Prof. C.T.Bhunia SMIEE, Senior Associate ICTP, Italy, Director, BITM, Bolpur(W.B) E-Mail:ctbhunia@vsnl.com	
30.	Prof. M. Reza Peygham Math Department, K.N. Toosi University Of Technology, Tehran, Iran Institute For Studies In Theoretical Physics And Mathematics (IPM), Tehran, Iran rpeyghami@gmail.com	AN INTERIOR POINT METHOD FOR SEMIDEFINITE PROGRAMMING BASED ON NEW KERNEL FUNCTIONS
31.	Dr. Sanjay Kumar Department Of Mathematics, R.D.S. College, Muzaffarpur B.R.A. Bihar University, Muzaffarpur – 842001	SOLUTIONS OF GENERALISED PHOTOGRAVITATIONAL
	Professor B. Ishwar P.I. DST Project University Department Of Mathematics, B.R.A. Bihar University, Muzaffarpur - 842001 Email : - ishwar_bhola@hotmail.com	ELLIPTIC RESTRICTED THREE BODY PROBLEM
32.	Dr. Millie Pant, Department Of Paper Technology, Indian Institute Of Technology Roorkee, Saharanpur Campus, Saharanpur – 247 001, Uttar Pradesh.	THE ECONOMIC OPTIMIZATION OF PULP AND PAPER
	Dr. Radha Thangaraj Department Of Paper Technology, Indian Institute Of Technology Roorkee, Saharanpur Campus, Saharanpur – 247 001, Uttar Pradesh.	MAKING PROCESSES USING COMPUTATIONAL INTELLIGENCE
	Prof. V. P. Singh Department of Paper Technology, Indian Institute Of Technology Roorkee, Saharanpur Campus, Saharanpur – 247 001, Uttar Pradesh. singhvp3@gmail.com	
33.	Dr. T. Sivakumar Department Of Mathematics, Bharathiar University, Coimbatore 641 046, Tamil Nadu, India,	EFFECT OF GRAVITY MODULATION ON THE ONSET OF CONVECTION IN A HORIZONTAL
	Dr. S. Saravanan Department Of Mathematics, Bharathiar University, Coimbatore 641 046, Tamil Nadu, India, Email: sshravan@lycos.com	ANISOTROPIC POROUS LAYER
34.	Dr. Mrs. Anjana Solanki Bundelkhand Institute of Engineering & Technology Jhansi (INDIA) anjanabiet_2007@yahoo.co.in	TRANSIENT BEHAVIOUR OF BATCH ARRIVAL QUEUE WITH N-POLICY AND SINGLE VACATION (MX/G/1/N-POLICY)
35.	Mr Rakesh Kumar Saxena Sachdeva Institute Of Technology, Mathura, U.P., India., E-Mail: saxenark06@rediffmail.com	
	Dr. Neelam Sharma Institute Of Engg. & Technology, Alwar, Rajasthan, India., E-Mail: neelam_sr@yahoo.com	FAST ARITHMETIC USING SIGNED DIGIT NUMBERS AND TERNARY LOGIC
	Prof. A. K. Wadhwani M. I. T. S. Gwalior, M.P., India., E-Mail: Wadhwani_Arun@Rediffmail.Com	
36.	Dr. Pankaj Sharma Department Of Applied Mathematics, Jawaharlal Institute Of Technology, Borawan, Khargone - 451228, India , Email: pankaj_sharma130969@rediffmail.com	SELF-SIMILAR SOLUTION OF SELF-GRAVITATING, MAGNETO-GAS DYNAMIC SPHERICAL SHOCK WAVE
	Dr. Vivek Kumar Sharma Department Of Applied Mathematics And Computational Science, Shri G.S. Institute Of Technology And Science, 23, Park Road, Indore -452003, India Email: vkmps_07@ yahoo.co.in	PROPAGATING IN A ROTATING MEDIUM WITH RADIATION HEAT FLUX

HE DR. A.R. KIDWAI GOVERNOR OF HARYANA WAS THE CHIEF GUEST & PROF. K.R. SREENIVASAN, DIRECTOR ICTP WAS THE GUEST OF HONOR ON THIS OCCASION

Patron	Mr. P.K. Gupta, Chairman SGI	
	Mr. Y.K.Gupta, Vice Chairman, SGI	
Steering Committee	Prof. M.S. Teotia, Prof. A H Siddiqi, Dr. A.K. Gupta,	
	Prof. N.K. Gupta, Prof. P Manchanda	

LIST OF INVITED SPEAKERS & PARTICIPANTS

S.N.	Name	University/Address	Email/contact no.	Remark
1	Prof. Roddam Narasimha, FRS (Indian)	Chairman Of The Engineering Mechanics Unit, JNCASR And Director, National Institute Of Advanced Studies, Bangalore.	roddam@caos.iisc.ernet.in roddam@cao.iisc.ernet.in roddam@jncasr.ac.in	
2	Dr. H.P Dikshit (Indian)	Director General School of Good Governance and Policy Analysis (An autonomous Institution set up by the State Govt. of Madhaya Pradesh) C-403, 4th Floor, Naemada Bhawan 59 Area Hills, Bhopal 462001 India	Tele. 0755-2570217 Fax: 0755-2570218 Email: hpdsushasan@gmail.com	
3	Prof K.R Sreenivasan (US and Indian)	Director The Abdus Salam International Centre for Theoretical Physics Strada Costiera II 1-34100 Trieste, Italy	krs@ictp.it	
4	Prof Rolf Jeltsch (Switzerland)	President ICIAM Seminar for Applied Mathematics ETH Zurich, Switzerland	jeltsch@math.ethz.ch Mb: +41794566649	
5	Prof Ian H Sloan, FAA (Australian)	Former President ICIAM School of Mathematics and Statistics University of New South Wales Sydney NSW2052, Australia	i.sloan@unsw.edu.au	
6	Prof Arvind Gupta (Canada)	Scientific Director MITACS Inc TASC2, Room 9904 Simon Frazer University 8888 University drive Burnaby, B.C Canada v5A156	arvind@mitacs.ca	
7	Professor Ms Marie Farge (French)	Director CNNR LMD-IPSL-CNRS, Ecole Normale Superieure, 24, rue Lhomond, 75231 Peris Cedex 5	farge@lmd.ens.fr Tel: 33-(0)1-44-32-22-35 Fax: 33-(0)1-43-36-83-92	
8	Prof. Dr. Martin Brokate (Germany)	Zentrum Mathematik/M6, TU Muenchen, D-85747 Garching bei Muenchen, Germany Phone: ++ 49 - (0)89 - 289 - 16806, Fax - 16809 http://www-m6.ma.tum.de/	brokate@ma.tum.de Tel ++49-89-289-16806	
9	Prof Hans G Feichtinger (Austria)	NUHAG (Numerical Harmonic Analysis Faculty of Mathematics Group) University of Vienna Nardberg strasse 15 A-1090 Vienna Austria	Hans.feichtinger@univie.ac.at	
10	Prof. R Lozi (France)	Laboratiire J.A Dieudonne UMR du CNRS No 6621 University of Nice-Sophiya – Antipolis Parc Valrose, 06108 Nice Cedex 02, Evan	rfozi@unice.fr	
11	Prof. Hans D Mittelmann (USA Germany)	Department of Mathematics and Stastics, Arizona State University, empe, Arizona 85287, USA	mittelmann@asu.edu	
12	Prof. Adimurti (Indian)	TIFR Centre, P.O.Box No 1234 IISC Campus, Banglore-560012	aditi@math.tifrbng.res.in Tele No:- 91-80-23600062, 23600138	

13	Prof. S. Kesavan (Indian)	Institute Mathematical Science Deputy Director Chennai Mathematical Institute C.I.T. Campus, Taramani, Chennai-600113, India	kesh@imsc.ves.in.Kesh@cmi.ac.in Telephone No. 91-0-44022543209,0-44-32983441/2	
14	Prof. P. N. Srikanth (Indian)	Director Dean TIFR Centre, P.O.Box No 1234 IISC Campus, Banglore-560012	srikanth@math.tifrbng.ves.in	
15	Prof Kazuhiro Aoyama (Japan)	Professor 7-3-1 Hongo, Bunkyo-ku TOKYO, JAPAN 113-8656 Department of Systems Innovation, School of Engineering The University of Tokyo	aoyama@sys.t.u-tokyo.ac.jp office: +81-3-5841-6504 / fax:+81-3-3815-8364 mobile: +91-3-90-4622-9182 email: aoyama@sys.t.u-tokyo.ac.jp	
16	Prof Tom Lyche (Norway)	Centre of Mathematics for Applications (CMA), Department of informatics, University of Oslo, P.O. box 1053 Blindern No- 0316 Oslo, Norway	tom@ifi.uio.no	
17	Prof Michael Floater (Norway)	Centre of Mathematics for Applications (CMA) Department of Informatics, University of Oslo, P.O. box 1053 Blindern No- 0316 Oslo,Norway	michaelf@ifi.uio.no mobile number is 00 47 95 04 58 54	
18	Prof Hiromasa SUZUKI (Japan)	RCAST, the University of Tokyo, Japan Address: 4-6-1, Komaba,Meguro, Tokyo 153-8904, Japan	suzuki@den.rcast.u-tokyo.ac.jp	
19	Prof. N. RUDRAIAH (Indian)	Honorary Professor UGC-Centre for Advanced Studies in Fluid Mechanics Department of Mathematics Bangalore University Bangalore - 560 001.	rudraiahn@hotmail.com Phone: 91-080-22220483/ 22219714 (O) 91-080-26764356(R) 9448389763 (M) Fax: 91-080-22219714 91-080-26764356	
20	Prof. Dr. Osman N. Ucan, (Turkey)	Istanbul University, Faculty of Engineering Dept. Electrical & Electronics Engineering 34320 Avcilar, Istanbul, TURKEY Gönderen: uosman@istanbul.edu.tr	uosman@istanbul.edu.tr Telephone +90 542 3229115	
21	Prof. S.P. Singh (Canada/Indian)	Former Chairman, Department of Math & Stat, Memorial University,Canada	ssingh55@rogers.com	
22	Prof. Dinesh Singh (Indian)	Director South Campus Delhi University, Delhi	dineshsingh1@gmail.com	
23	Prof. Sanjeev Agarwal (Indian)	Delhi University, Delhi		
24	Prof. Karmeshu	School of Computer and System science JNU New Delhi	karmeshu@mail.jnu.ac.in	
25	Prof. M.Z. Khan (Indian)	Dean Faculty of Science AMU Aligarh	Phone: 91-9412596639	
26	Prof.Hiroshi Yamaguchi (Japan)	Department of Mechanical Engineering Doshisha University, Kyoto 610-0321, Japan Professor Director of Energy Conversion Research Centre	hyamaguc@mail.doshisha.ac.jp Tel. (0) 774-65-6462 Fax. (0) 774-65-6831	
27	Prof. Sunder Lal (Indian)	Pro-Vice Chancellor Dr. B.R. Ambedkar University, Agra	sunder_lal2@rediffmail.com	
28	Prof. Mukhayo Rasulova (Uzbekistan)	Institute of Nuclear Physics Academy of Sciences Republic of Uzbekistan, Ulughbek, Tashkent 702132 UZBEKISTAN	rasulova@live.com Tel.998-712-606753 (off.) 998-712-403591 (home)	

29	Prof. Salim A. Messaoudi (Algeria)	King Fahd University of Petroleum and Minerals Department of Mathematics and Statistics Dhahran 31261, Saudi Arabia.	messaoud@kfupmedu.sa	
30	Prof. P.Singh(Pravin) (South Africa)	University Of Kwazulu-Natal, Durban, 4000, south africa Maths Dept. PBX 54001, Durban 4000 South Africa tel 0312607686	singhp@ukzn.ac.za	
31	Prof. Nasser-eddine Tatar (Algeria)	King Fahd of Petroleum and Minerals Department of Mathematics and Statistics Dhahran, 31261 Saudi Arabia	tatarn@kfupm.edu.sa	
32	Prof. G.C. Sharma (Indian)	Exprovice Chancellor, Dr. B.R. Ambedkar University, Agra	9897417784	
33	Dr. Nityanand Singh (Indian)	Indian Institute of Tropical Meteorology, Dr. Homi Bhabha road,NCL post Pashan, Pune 411008, India, 9850978744	nsingh@tropmet.res.in	
34	Dr. (Mrs.) N.A. Sontakke, Dr. (Mrs.) Neelima A. Sontakke(Indian)	Indian Institute of Tropical Meteorology, Dr.Homi Bhabha Road, Pune-411008 INDIA	sontakke@tropmet.res.in Ph. +91 20 25893600 (O) Fax +91 20 25893825 (O) Ph +91 20 22952876 Mob. 9823813744	
35	Prof. B.Ishwar (Indian)	B.R.A. Bihar University, Muzaffarpur	ishwar_bhola@hotmail.com	
36	Prof. P Kandaswamy (Indian)	UGC-DRS Center for Fluid Dynamics, Department of Mathematics, Bharathiar University, Coimbatore - 641 046, INDIA	pgkswamy@yahoo.co.in	
37	Dr.L.P. Rai (Indian)	Scientist, NISTADS (CSIR), K. S. Krishnan Marg, New Delhi-110 012, INDIA	lp_rai@yahoo.com, lprai@nistads.res.in Fax:(+91-11) 25846640 Ph:(+91-11) 25843093, M: 9868413148	
38	Prof. V. P. Singh, (vedpal) (Indian)	Professor of Mathematics Indian Institute of Technology Roorkee Saharanpur Campus, Saharanpur-247001	singhvp3@gmail.com	
39	Dr. Sigrid Bettina Heineken (Austria)	Mathematics University of Vienna Address: Nordbergstr. 15, UZA4, Alserbachstr.23/4/7 Address for correspondence: Augasse 11/2/29B, 1090 Wien, Austria	sigrid.heineken@gmail.com Tel: +431427750735/ Mobile;+436767461802	
40	Ms. Yeliz Karaca (Turkey)	Istanbul Commerce University, Faculty of Science and Letters 34672, Istanbul, Turkey	ykaraca@iticu.edu.tr	
41	Ms. ROBABEH SAHANDI TOROGH, (Indian)	Department of Mathematics, Islamic Azad University of Varamin Branch, Iran	sahandi−1352@yahoo.com	
42	Prof. H.G.Sharma (Indian)	Professor of Mathematics Indian Institute of Technology Roorkee Roorkee-247667, INDIA	hgsfdfma@iitr.ernet.in Tel: +91-1332-285133(O) +91-1332-270368®	
43	Prof. DS Hooda (Indian)	Head of mathematics department, Jaypee institute of engineering&Technology, A.B.road, Raghogarh (M.P) Dist. Guna-473226 (India)	ds_hooda@rediffmail.com Tel:(o) 07544-267310-14Ext. 104, (M) 94251-31126	
44	Prof. V.K. Parashar (Indian)	Deptt. Of Applied Sc. University Polytechnic AMU	09897217321, parasharanshul@hotmail.com	
45	Dr. S.C. Gupta (Indian)	Head of Mathematics Department, Women college of AMU	9837124298	
46	Dr. V.K.Kukreja (Indian)	Department of Mathematics, SLIET, Longowal, Punjab	vkkukreja@gmail.com	
47	Dr. Millie Pant, (Indian)	Asst. Professor Department of Paper Technology, Indian Institute of Technology Roorkee,	millifpt@iitr.ernet.in, t.radha@ieee.org, singhfpt@iitr.erenet.in	

		Saharanpur Campus – 247001.		
48	Dr. Shelly Arora (Indian)	Department of Mathematics, Punjabi University, Patiala	shellyarora_25@yahoo.co.in Tele: 97791-85794	
49	Dr Sanjeev Kumar	Department of Mathematics, Dr B R A University, Agra	sanjeevibs@yahoo.co.in	
50	Dr. Md. Kalimuddin Ahmad (Indian)	DAAD Fellow (Germany) 96-98, Reader (Assoc. Prof.) Deptt. of Mathematics Aligarh Muslim Univ., Aligarh U.P. - 202 002, India, 9837000608 (Mobile): 0571-2704736 (Res.)	ahmadkalimuddin@yahoo.co.in	
51	Dr. Shyamalendu Kandar (Indian)	Haldia Institute of Technology P.O.-Hit, Pin-721657	shyamalenduk@yahoo.com 09433343304	
52	Prof. Khalil Ahmed (Indian)	Professor Department of Mathematics Jamia Millia Islamia New Delhi - 25	khalil_ahmad49@yahoo.com Ph: +91 11 26981717 extn. 3357 011-65684624 (Resi.)	
53	Dr. Jamal Hussain (Indian)	Reader & Head, Deptt. of Mathematics and CS Mizoram Univ., Tanhril - 796 009 Aizawl, Mizoram, INDIA	Ph. No. (+91) 389 2330873/2330874 (O) Fax No. (+91) 389 2330873 Mobile: (+91) 9436352389	
54	Dr. Meenu Sharma (Indian)	Sr. Lecturer,Deptt of Maths DAV College,Amritsar House no. 06, Streetn no.02, Pawan Ngr., Batala Rd. Amritsar-143001 Punjab	meenusharma03@yahoo.co.in 01832273248, 09872207005	
55	Ms. Meenakshi (Indian)	Dev Samaj College for Women Ferozepur, India, H.No. 145, Bazar no.2, Ferozepur Coult. Punjab	meenakshi_wavelets@yahoo. com, Phone: 09815277516	
56	Prof H H Khan	Chairman, Department of Mathematics, Aligarh Muslim University, Aligarh, India	huzoorkhan@yahoo.com	
57	Prof R C Singh	Dean Academic, Department of Physics, Hindustan Institute of Technology, 32, 34 Knowledge Park-3, Greater Noida – 201306, India	rcsingh_physics@yahoo.com	
58	Mr. Shaban Ahmad Siddiqui (Indian)	*Shaban Ahmad Siddiqui,Senior Lecturer,Azad Institute of Engg. & Tech.,Lucknow.	ahmadshaban@rediffmail.com	
59	Mr. Imran Khan (Indian)	A.P. in Deptt.of EE Azad Institute of Tech. Lucknow L-57/B, Fateh Ali Ka Talab, Jail Road, Lucknow-226005	pe.imran@gmail.com 0522-2463150, 09305361685	
60	Naveen Kumar Saxena, (Indian)	Microwave Lab, Department of Physics, Agra College Agra, PIN 282002 (U.P) India.	Nav3091@rediffmail.com	
61	Dr. D.L. Suthar (Indian)	Alwar Institute of Engg. & Tech. ,Alwar, Staff Qtr. Block A-201, North Ext. MIA, Alwar Mob. No. 9887887624, 9413854611	dd_suthar@yahoo.co.in	
62	Mr. Jagdish Chand Bansal (Indian)	Department of Mathematics Indian Institute of Technology Roorkee Roorkee – 247667 India	jcbansal@gmail.com	
63	Mr. Ranjeet Kumar Singh (Indian)	Department of Mathematics, SLIET, Longowal, Punjab	ranjeetsingh_don@yahoo.co.in	
64	Mr. Hasin Alam (Indian)	Asstt.Professor Depts.of ECE Integral University Lucknow Mobile - 09919030975 H.N.4, Gaurabagh Kursi Rd. Lucknow	alam_hasin@yahoo.com alamhasin@gmail.com	
65	Rajshree mishra (Indian)	Teacher Research Fellow Jiwaji University Gwalior(M.P.) INDIA TRF , Jiwaji University,Gwalior(M.P.) INDIA	rajshreemishraa@gmail.com	
66	Ms. Shweta Upadhyaya (Indian)	Department of Mathematics, Institute of Basic Science, Dr. B.R. Ambedkar University, Agra – 282002 (India)		

67	Ms. Khushbu Jain (Indian)	Department of Mathematics, Institute of Basic Science, Dr. B.R. Ambedkar University, Agra – 282002 (India)		
68	Dr. Sanjay Kumar (Indian)	Department of Mathematics, R.D.S. Colleges, Muzaffarpur-842001, Teacher Qtr. 15, RDS Campus	ishwar_bhola@hotmail.com 09234370450	
69	Dr. Gunjeshwar Shukla (Indian)	Reader Department of Mathematics UNPG College Padrauna, KushiNagar		
70	Dr. Manoj Kumar Singh (Indian)	Department of Mathematics, R.B.S. College, Dhanukhi-844101, Bihar	ishwar_bhola@hotmail.com 09835265936	
71	Mr. Surendra Kumar Sharma (Indian)	Department of Mathematics, SV College Aligarh		
72	Rekha Chaturvedi (Indian)	Department of Mathematics, MG College Jabalpur	rekha_mgmm@rediffmail.com	
73	Mr. Rakesh Kumar Saxena (Indian),	Asst. Prof. (ECE) Sachdeva Institute of Technology, Mathura, U.P., India.	saxenark06@rediffmail.com	
74	Dr. Pankaj Madan (Indian)	Lecturer, G.N. Khalsa College, Abohar Resi. 3,4 Sidhu Nagri, Abohar, Near Dr. Kartar	pankaj_madaan24@yahoo.com 09915836388	
75	Ms. Anupama (Indian)	Dev Samaj Colelge for Women, Ferozepur City House no. 145, Bazar No.2, Ferozepur Cantt, Punjab	anupma2512@yahoo.co.in988 8260369	
76	Mr. Mukhtar Ahmad Khanday (Indian)	SIRT Bhopal		
77	Pankaj Sharma (Indian)	Deptt. Of Applied Mathematics Jawaharlal Institute of Technology, Borawan, Khargon-451228, India	pankaj_sharma130969@rediffmail.com pankaj_13sep@yahoo.com	
78	Mr. Javed Iqbal (Indian)	Islamic University of Science and Technology, Aventipura Kashmir		
79	Dr. S. C. Shiralashetti, (Siddhu) (Indian)	Asst. Professor, Dept of Maths SDM College of Engineering and Technology, Dharwad. Karnataka, India.	naregal.sharada@gmail.com 9986323159, 9448923152	
80	Mr. Afroz (Indian)	Department of Mathematics Jamia Milia Islamia Okhla New Delhi		
81	Mr. Kashif Khan (Indian)	Department of Mathematics Jamia Milia Islamia Okhla New Delhi		
82	Mrs. Rashmi Priyadarshini (Indian)	Department of Electronis and Communication Engg., Indira Gandhi Institute of Technology Guru Gobind singh indraprastha University, Delhi.	priyadarshini.rashmi@rediffmail.com	
83	Dr. Vikas Kumar Rai (Indian)	Assistant Professor (E&C), Sachdeva Institute of Technology, Farah, Mathura HIG - A/1429, Sector-I, LDA Colony, Kanpur Road, Scheme, Lucknow - 226012	hellovikasrai@yahoo.com 9219615682 9451897898	
84	Mr. Shobhit Saraswat (Indian)	Sr. Lecturer Sachdeva Institute of Technology, Farah Mathura 8/153/2B-1, New Lawyers Colony, Tila Patpari, Agra		
85	Mr. Mohit Gaharwar, (Indian)	Assistant Professor (E&C), Sachdeva Institute of Technology, Farah, Mathura 17, Moti Bagh, Dayal Bagh Agra	mohitgaharwar@rediffmail.com	
86	Mr. Somnath Jha (Indian)	Centre for Atmospheric Sciences, Indian Institute of Technology, Delhi, New Delhi, 110016, India Resi: C/o Dr. Ramesh Raghava, CAS IIT, Delhi Hauz Khas, New Delhi-110016	somnath.jha@gmail.com Tel: 011-26591316 Mob : +91-9999277244	
87	Dr. Pranjali Arondekar (Indian)	Medi Caps Institute of Technology & Management Indore 104, Indrapuri Colony, Indore	parondekar@gmail.com 09826533329	

523

88	Ms. Pooja Khandekar(Indian)	GMU Sch. No. 54m Vijay Nagar, Indore		
89	Mr. A.M. Kulkarni (Indian)	Director Concentric Circles F-101, Koyna Appt. Pt. Malvlya Nagar, Khamla Nagarpur, 440025	kulkarniabhay51@yahoo.com Phone:9270049479	
90	Mrs. Sharayu Abhay Kulkarni (Indian)	F-101, Koyna Appt. Pt. Malviya Nagar, Khamla Nagarpur, 440025	Ph : 09373108435 09730578985 kulkarni.sharayu@yahoo.co.in	
91	Prof. Madhu Jain, (Indian)	Department of Mathematics IIT Roorkee	madhujain@sancharnet.in,	
92	Dr. Dhanapal Basti , (D. P. Basti) (Indian)	Department of Mathematics, S. D. M. College of Engineering & Technology, Dharwad – 580002, INDIA	dpbasti2002@yahoo.com	
93	Mr. Yogesh Gupta (Indian)	Assistant Professor Department of Mathematics, Motilal Nehru National Institute of Technology, Allahabad-211004(U.P)India	manoj@mnnit.ac.in, yogesh_uni@rediffmail.com	
94	Mr. Pankaj Kumar Srivastava (Indian)	Assistant Professor Department of Mathematics Motilal Nehru National Institute of Technology, Allahabad-211004(U.P) India	dr_kumar2@rediffmail.com dr_mksaini@yahoo.com (0532)2271906(R)2271255(O) Mobile:09451369162	
95	DR. Mrs. Sharada C. Venkateswarlu (Indian)	Selection Grade Lecturer Yeshwant Rao Charan College of Engg. (YCCE) Hingna Road,Wanadongri, Nagpur-441110 F-2, Shakti Apts, 19-A, Hill Road, Gandhi Nagar, Nagpur-440010	scv_nagpur@hotmail.com, principal @ ycce.edu	
96	Mr. Manish Kumar, (Indian)	Ramswaroop Memorial College of Engineering & Management, Lucknow,	mani_kr_2000@yahoo.com	
97	Mr. Vineet Kumar Sharma (Indian)	SV College Aligarh	9259255214	
98	Dr. Mrs. Anjana Solanki(Indian)	Bundelkhand Institute of Engineering & Technology Jhansi (INDIA)	anjanabiet_2007@yahoo.co.in	
99	Dr. Hitesh Kumar (Indian)	Sr. Lecturer, JIET School of Engg. & Tech. for Girls, Jodhpur, 12/6, Chapasni Housing Board, Jodhpur, (Rajasthan) India- 342008	hiteshrsharma@yahoo.com 09214657038	
100	Dr. Vivek Kumar Sharma (Indian)	Deptt. Of Applied Mathematics and Computational Sci. Shri G.S. Institute of Technology and Sci. 23, Park Road, Indore-452003 India 29, State Bank Cly., Jaipur House Agra	vkmps_07@yahoo.co.in 09926612116, 0562-2813283	
101	Mr. Pankaj Sharma (Indian)	Deptt. Of Applied Mathematics Jawaharlal Institute of Technology, Borawan, Khargon-451228, India	pankaj_sharma130969@rediff mail.com pankaj_13sep@yahoo.com	
102	Ms. Seema Agarwal (Indian)	Department of Mathematics, Institute of Basic Science, Dr B. R. Ambedkar University, Agra-282002 (India)	seema_ibs@indiatimes.com	
103	Ms. Neetu Singh (Indian)	World College of Technology and Management, Gurgaon	jadounneetu@yahoo.co.in	
104	Ms. Ragini Mittal	St. Johns' College, Agra (India)	m_ragini234@yahoo.co.in	
105	Mr. Ram Krishna Kumar (Indian)	Research scholar MNIT Allahabad	karamkrishna81@yahoo.com	
106	Dr. S. saravanan (shanmugam saravanan) (Indian)	UGC-DRS Center for Fluid Dynamics, Department of Mathematics, Bharathiar University, Coimbatore - 641 046, INDIA	sshravan@lycos.com +91-(0)422-2428414 (Off)	
107	Prof. M Eswaramurthy (Indian)	Deparment of Mathematcis, KONGU Engineering College, ERODE-638052		
108	Mr. Chetan Kotwal (Indian)	Research ScholarDept of Elect. Engg.IIT, RoorkeeUttarakhand 247667	apc67dec@iitr.ernet.in	

524

109	Dr. AK Abdul Hakeem (Indian)	Department of Mathematics, Sri Ramakrishna Mission Vidyalaya College of Arts & Science, Coimbatore- 641 020, Tamilnadu, INDIA	pqkswamy@yahoo.co.in	
110	Dr. Romain Nguyen van yen (France/Vietnam)	Romain Nguyen van yen LMD - ENS - PARIS	rnguyen@lmd.ens.fr	
111	Prof. K. Narayanankutty (Narayanankutty Karuppath) (Indian)	Narayanan kutty Karuppath (Asst.Professor), Department of Sciences(Physics), Amritha Vishwa Vidyapeetham(Deemed University), Ettimadai,Coimbatore, Tamil Nadu, India 641 105 Ph:(Office) 0422-2656422 Extn 274 Mobile: 09344875991	k_narayanankutty@ettimadai. amrita.edu	
112	Dr.Manoj Kumar (Indian)	Assistant Professor Department of Mathematics Motilal Nehru National Institute of Technology, Allahabad-211004(U.P) India Residence:C-6 Staff Colony MNNIT Allahabad (U.P)211004 INDIA	dr_kumar2@rediffmail.com dr_mksaini@yahoo.com (0532)2271906(R)2271255(O) Mobile:09451369162	
113	Mr. Anshuman Singh (Indian)	Department of Electronics Engineering, Sachdeva Institute of Technology, Mathura (U.P.), India	chatanshuman@rediffmail.co m,	
114	Mr. Rajan Singh (Indian)	Department of Electronics Engineering, Sachdeva Institute of Technology, Mathura (U.P.), India	rajan.1114@gmail.com	
115	Mr. Manish Sejwal (Indian)	Anamaya Publisher New Delhi		
116	Dr. Rashid Ali (Indian)	Assistant Professor, Integral University, Kursi road, Bas-Ha, Lucknow, H.No. 4/131 H, Street no. Tayyb Colony, Aligarh	rashid_ali10rediffmail.com, rashid_ali10@yahoo.co.in #+91-9450458907	
117	Mrs. Reena Thakur (Indian)	Deptt. Of computer Sc. & IT Anand Engineering College, Agra C/o V. Dayal, 18 Syndicate Bank Colony, Agra	rina151174@rediffmail.com 09410664044	
118	Mr. Vinay Kumar (Indian)	Deptt. Of computer Sc. & IT Anand Engineering College, Agra	vksingh100@rediffmail.com	
119	Mrs. NOOR E ZAHRA (Indian)	B74,Pocket-4,kendriya Vihar Sector 82 Noida 201301	noor zahara india@yahoo.com 09871744973	
120	Dr. R.P. Singh (Indian)	Principal, SSGB College of Engg. & Technology, Near ZTC, Bhusawal Dist. Jalgaon, (Maharastra)	rps125@rediffmail.com 09823092665, 02582-225649®	
121	Prof. Firdous Ahmad Shah (Indian)	Department of Applied Mathematics, BGSB University, Rajouri -185131, Jammu, INDIA.	ahasan@kfupm.edu.sa fashah jmi@yahoo.co.in, fashah79@gmail.com	
122	Dr. K.Sandeep, (Indian)	Department of Mechanical Engineering, Institute of Technology, Banaras Hindu University,Varanasi-221005	dr.sandeep.kumar@gmail.com	
123	Ms. Vandana Srivastava (Indian)	Department of Mathematics & , Jhansi Bundelkhand Unvierstiy,		
124	Mr. Jitender Kumar (Indian)	Department of Mathematics, Gurunanak Dev University , Delhi		
125	Ms R. Lallawmp (Indian)	Research Scholar, Deptt. of Mathematics and CS Mizoram Univ. Tanhril - 796 009, Aizawl, Mizoram, INDIA C-45, Chanmani West Aizawal - 796007	0389-2347113 (R) 0946193523 (M)	
126	Ms. D. Zadeng (Denghningliani Zadeng) (Indian)	Research Scholar, Deptt. of Mathematics and CS Mizoram Univ. A-77, Mizo Arsi Press, Upper Repulic, Aizawal Mizoram - 796001	zadeng08@rediffmail.com Phone : 0389-23134423 09436159596	
127	Mr. Sohrabali	Department of Mathematics Jamia Millia		

525

		(Indian)	Islamic N.D.		
128	Mr. Santosh Mathur (Indian)	Department of electrical engineering Faculty of Engineering and Technology Agra College, Agra	santosh_mathur2003@yahoo.com	
129	Mr. V. Prem Prakash (Indian)	5, Ashoka Enclave, Dayalbagh Agra	vpremprakash@acm.org	
130	Mr. Yajuvindra Kumar, (Indian)	Department of Mathematics, Indian Institute of Technology Roorkee, Roorkee – 247 667, C/o Prof. Roshan Lal, Dept. of Maths, IIT Roorkee (U.K.)	yaju_saini@yahoo.com	
131	Dr. B.S MUDAGI (Indian)	Nowrosjee Wadia College, Pune (India) S.No. 37-37/1 Sasson Road, Shardaram Park, C-12 Pune 9373991536	basaweshwarmudagi@yahoo.co.in	
132	Sh. Vikas Bhardwaj (Indian)	Scientist-C, ADRDE Agra		
133	Sh Vipin Kumar, (Indian)	Scientist-B, ADRDE Agra		
134	Sh Ajeet Kunar, (Indian)	Scientist-C, ADRDE Agra		
135	Sh Ajeet Gaur, (Indian)	Scientist-B, ADRDE Agra		
136	Sh Ravi Krishna, (Indian)	Scientist-B, ADRDE Agra		
137	Sh Anurag Yadav, (Indian)	Scientist- B, ADRDE Agra		
138	Sh. Manish Bhatnagar, (Indian)	Scientist- D, ADRDE Agra		
139	Dr. ChandraShekhar Salimath (Indian)	Prof. HOD Amruta Institute of Engineering & Management Science Banglore AIEMS Near Kirloskar Toyota off Mysore Rd. Bidadi Banglore Karnataka	salimathcs@yahoo.com	
140	Mr. Ajay Mittal (Indian)	BGIET		
141	Dr. Mamta Rani Singh (Indian)	Associate Prof & HeadDepartment of Comp. Applications Galgotias College of Engg. & Tech. Greater Noida	mamtarsingh@rediffmail.com Mobile-9910077916	
142	Mr. Saurabh Goel (Indian)	Galgotias Institute of Mgmt &Tech., Greater Noida,	Email: saurabh2me@yahoo.com	
143	Mr. Avinash Pokhriyal, (Indian)	Faculty of Management & Computer Applications, R.B.S. College, Agra	bmastp@gmail.com	
144	Ms. Kumud Saxena (Indian)	Institute of Information & Computer Sciences, Khandari, Agra	pokhariyal@hotmail.com	
145	Ms. Pinky Chauhan (Indian)	Institute of Information & Computer Sciences, Khandari, Agra		
146	Mr. Deepak Kumar, (Indian)	Research Scholar IIT Roorkee, Department of Paper Technology, Saharanpur Campus Saharanpur-247001, India	dkr80dpt@iitr.ernet.in, Mob No. +919927090165	
147	Mr. Musrrat Ali and (Indian)	IIT Roorkee, Roorkee-247667	musrrat.iitr@gmail.com	
148	Mr. Sunil Kumar, (Indian)	IIT Roorkee, DPT, Saharanpur Campus, Saharanpur, India,	singhvp3@gmail.com , Fax: +91-132-2714011.	
149	Dr. Shiv Kumar Tomar (Indian)	PhD Research scholar(QIP) Indian Institute of Technology, Roorkee, INDIA.	shivktomar@gmail.com	
150	Prof.Hradyesh Kumar Mishra (Indian)	Department of Mathematics SPMIT Kaushambi (U.P) India	hkm1975@yahoo.co.in	
151	Mrs. Arpita Johri (Indian)	B.M.A.S. Engineering College, Agra Anand Engineering College, Agra	Mob: 09219513240 Mob: 09259025860 E-mail: mudit123@gmail.com E-mail:	AEC

			arpitajohri@rediffmail.com	
152	Dr. R.K. Deolia (Indian)	Department of Applied Sc. Anand Engineering College, Agra	deolia_rajesh@yahoo.com 09319429089	AEC
153	Mohd. Salim Ahamad (Indian)	Department of Applied Sc., Anand Engg. College, Agra	mohdsalim10@yahoo.com	AEC
154	Ms. Vandana Vikas Thakare (Indian)	Anand Engineering College, Keetham, Agra, India, 9411961534	vandanavt_19@rediffmail.com	AEC
155	Mr. Anshu Kumar (Indian)	HIT GR Noida	anshukumar.sgi@gmail.com, 09711826692	HIT
156	Mr. S. Mukharjee (Indian)	HIT GR Noida	mukharjee80@yahoo.co.in, 09910973215	HIT
157	Dr. Khursheed Alam (Indian)	HIT GR Noida	khursheed_alam55@yahoo.com	HIT
158	Ms. Priti Gupta (Indian)	HCST Agra	joinpriti@gmail.com	HCST
159	Ms.Kirti Mathuria (Indian)	EE Department HCST Agra		HCST
160	Mr. Neeraj Menhas (Indian)	HCST, Farah Mathura		HCST
161	Ms. RSS Subramanan (Indian)	HOD Electrical & Electronics Department, HCST Farah Mathura		HCST
162	Mrs. Ridhi Bindra (Indian)	IT Department		BMAS
163	Mr. Brajesh Sharma (Indian)	IT Department		BMAS
164	Mr. Pratap Sakhare (Indian)	CS Department		BMAS
165	Ms. Tulika Srivastava (Indian)	CS Department		BMAS
166	Mr. Vinay Tomar (Indian)	EC Department		BMAS
167	Mr. Ashish Chandiok (Indian)	EC Department		BMAS
168	Mr. Tarun Rathi (Indian)	EC Department		BMAS
169	Ms. Charanpriya (Indian)	EN Department		BMAS
170	Mr. Mudit Saxena (Indian)	EN Department		BMAS
171	Mr. Ajay Verma (Indian)	EN Department		BMAS
172	Mr. Akash Chaudhary (Indian)	EN Department		BMAS
173	Mr. V. Srikanth (Indian)	EN Department		BMAS
174	Mrs. Neema Verma (Indian)	EI Department		BMAS
175	Dr. N.P. Singh (Indian)	Applied Science		BMAS
176	Mrs. Arti Singh (Indian)	Applied Science		BMAS
177	Mr. Vijay V Singh (Indian)	Applied Science		BMAS
178	Mr. Ashutosh Srivastava (Indian)	Applied Science		BMAS
179	Mr. N.K. Singh (Indian)	Applied Science		BMAS
180	Prof. S.K. S. Rathore (Indian)	Applied Science		BMAS
181	Mr. Awadhesh Pandey (Indian)	Applied Science		BMAS

182	Dr. A.K. Bhattacharya (Indian)	Applied Science		BMAS
183	Mr. S.K. Singh (Indian)	Applied Science		BMAS
184	Dr. C.K. Singh (Indian)	MBA Department		BMAS
185	Mr. Neeraj Gogia (Indian)	MBA Department		BMAS
186	Mrs. Gunjan Bhatnagar (Indian)	MBA Department		BMAS
187	Mrs. Raju Ganesh Sundar (Indian)	T & P Department		BMAS
188	Mr. Vineet Arora (Indian)	ME Department		BMAS
189	Mr Mohammad Kamaruddin (Indian)	Electronics & Communication Department,	qamruddins@gmail.com	BMAS

AUTHOR INDEX

A

Achuthan, P., 105
Ali, M., 177
Aoyama, K., 80
Arora, S., 169
Asghar, K., 411
Aslan, Z., 229

B

Barman, M., 434
Bhattacharya, P., 3
Bhunia, C. T., 434
Bokhari, A. H., 411
Brokate, M., 332

C

Chandiok, A., 284
Chowdhry, B. S., 113

D

Dayo, K., 113
Dhaliwal, S. S., 169
Dikshit, H. P., 18

F

Feichtinger, H. G., 189
Fiol, C., 303

G

Govindarajan, P., 3
Gupta, A., 41

H

Hasan, A., 284

I

Irfan, S. S., 365
Ishwar, B., 456

J

Jeswani, J. K., 113

K

Kandar, S., 434
Karuppath, N., 105
Khanday, M. A., 375
Khodabakhshi, A. H., 41
Koga, T., 80
Kukreja, V. K., 169
Kukreja, V., 386
Kumar, D., 390
Kumar, J., 272
Kumar, S., 456
Kumar, V., 390

L

Lozi, R., 303

M

Mañuch, J., 41
Manchanda, P., 253, 272
Meenakshi, P., 253
Mehra, M., 241
Melek, M., 229
Messaoudi, S. A., 123
Mittal, A., 386
Mittelmann, H., 65
Mudgai, B. S., 400
Mustafa, M. I., 123

529